Vince

2390 A

W Broadway

734-5250

FRI   Ceasar's

KEG

Harnby & Dunsmuir

ELECTRONIC CIRCUITS

Digital and Analog

ELECTRONIC CIRCUITS

Digital and Analog

CHARLES A. HOLT

Virginia Polytechnic Institute and State University

JOHN WILEY & SONS   New York   Santa Barbara   London   Sydney   Toronto

Library of Congress Cataloging in Publication Data:

Holt, Charles Asbury, 1921—
  Electronic circuits.

  Includes index.
  1. Electronic circuit design.   2. Integrated circuits.   3. Digital electronics.   I. Title.
TK7867.H64   621.3815   77-11654
ISBN 0-471-02313-2

Printed in the United States of America

10 9 8 7 6 5 4 3 2 1

To my wife JOSEPHINE
and my children
CHARLES, JOEL, WINIFRED, and SARAH

Preface

This book is an introduction to electronics. Its objective is to present in a clear, organized, and logical manner the fundamentals essential to electronic circuit design and analysis, both analog and digital. Although *Electronic Circuits* is planned for use in a junior-level course in electronics, it will also be useful to practicing engineers who wish to keep abreast of developments in this field. The reader is expected to have a knowledge of elementary calculus and electric network theory, as well as enthusiasm for learning about an exciting field of engineering.

It is interesting to note that in a span of only a few years several logic families have been introduced, become very popular, and then faded. Examples are diode-transistor logic, resistor-transistor logic, and standard transistor-transistor logic. Today new logic systems are being developed, and some have considerable potential for increasing substantially the capabilities of electronic systems. In view of this it is imperative that young electronics engineers be educated so that they acquire a firm grasp of fundamentals that apply not only to current technologies but also to technologies that surely will emerge.

The heart of semiconductor electronics is the bipolar junction transistor. Bipolar transistors employed in logic gates operate not only in the active region normally encountered in linear circuitry but also in the saturation, inverse, and cutoff regions. The engineer who is to design circuits intelligently with these gates must understand their operation and their characteristics. This requires a knowledge of the physical mechanisms occurring within the transistor in each of the four operating modes. Engineers must be equally familiar with the field-effect transistor, which is also of immense importance. Accordingly, the four chapters that constitute Part 1 of the book examine these devices carefully, along with films, hybrids, and integrated circuit technology.

The seven chapters of Part 2 treat digital circuitry. This material is placed ahead of that on analog circuits because electronics today is predominantly digital and becoming even more so as time progresses. Analog circuits are presented in Part 3, with operational amplifiers introduced early. OP AMP circuits and applications are emphasized throughout the analog section.

A wide range of topics is presented in this book, and more material is included than would normally be covered in a one-year introductory course. However, the book is organized so that it can be used in courses confined to analog circuits, or digital circuits, or portions of both. In order to provide flexibility, special care has been exercised in the preparation of Part 3 so that it can directly follow Part 1.

The book is oriented toward devices and technologies that are important today, with special emphasis given to circuit design. MOSFET logic gates are carefully examined, including PMOS, NMOS, CMOS, and dynamic MOS networks. The new logic I^2L is analyzed in some detail. Throughout the book the orientation is strongly toward the integrated circuit. Devices and circuits are investigated by means of numerical analyses, and numbers are selected to correspond closely with practical integrated-circuit values.

In 1977 a discernable trend is developing in industry to switch to MOSFET technology at the expense of bipolar technology. MOSFET circuits are becoming popular and common. Three chapters (3, 8, and 17) are wholly devoted to field-effect electronics, and numerous applications of FET digital and analog networks are presented in these and other chapters. It is well recognized that the bipolar junction transistor is superior to FET devices in gain and speed. It remains dominant in analog circuitry, and bipolar digital technology is making a considerable resurgence as newer types of integrated circuits become available. Thus extensive coverage of both bipolar and field-effect devices and circuits is provided, including networks containing both types of active devices.

The end-of-chapter problems are an integral part of the book. Each problem has been carefully prepared either to illustrate an important principle or to present new and meaningful material. The problems are arranged by section number and are closely related to the associated section. Students are strongly encouraged to work as many of them as possible, and they should endeavor to absorb the full significance of the results.

Although preparation of this book has required many hours of diligent work, it has been an enjoyable experience. My greatest hope is that the book will contribute substantially to the progress of engineering education.

Blacksburg, Va.   Charles A. Holt
1977

Suggestions for Course Organization

Electronic Circuits has been designed for flexibility. The contents of Part 2 on digital circuits are not in any way prerequisite to those of Part 3. Thus there are various ways in which the material may be organized in an introductory course. Individual instructors will undoubtedly develop their own course organization, after considering prerequisite material. The comments in this section may be useful as guidelines.

All introductory courses should begin with Chapters 1 and 2, which examine the bipolar junction transistor. However, this material can be covered fairly quickly. When developments based on the concepts of Chapters 1 and 2 appear in later chapters, the student should be encouraged to review these concepts. The material of Chapter 3 on Field Effect Transistors need not be covered before Chapter 8 of Part 2 and Chapter 17 of Part 3. Chapter 4 is a one-lesson reading assignment on integrated-circuit technology.

For a general course of broad scope *the contents of Parts 2 and 3 can be interspersed as desired.* One suggestion is to study combinational logic, transistor switching, and BJT logic gates in Chapters 5, 6, and 7 and then advance to Chapters 12 through 16 on BJT models, basic amplifier configurations, OP AMPs, bias circuits, and audio power amplifiers. After this, the program might return to the digital section. A first course designed to emphasize digital circuits should follow the material in the order presented.

In some cases it may be desirable to lead the student rather quickly into the design and analysis of analog circuits. This can be accomplished by proceeding from Chapter 2 directly to Chapter 12, which treats incremental models. The first two sections of Chapter 12 can be covered quickly. After studying the basic CE, CC, and CB configurations in Chapter 13, the student encounters the operational amplifier in Chapter 14, along with numerous important and interesting applications. Upon completion of Chapter 14 it might be well to return to the Combinational Logic Design of Chapter 5. Many students are quite interested in OP AMPs and also digital networks, and the approach suggested here is attractive because of its appeal to these students.

It is believed the book may prove beneficial in certain specialized courses. For example, the contents of Chapters 18 through 24 are appropriate for an advanced course emphasizing amplifier design that uses feedback techniques, along with various applications of linear integrated circuits. A course on field-effect electronics can be organized around Chapters 3, 8, and 17, with additional topics selected from Chapters 9, 10, 11, and 18.

C. A. H.

Acknowledgments

The author wishes to express his gratitude to the staff, faculty, and secretaries of the Department of Electrical Engineering at Virginia Polytechnic Institute and State University for their assistance and encouragement of this project. Among those who gave suggestions and technical advice are Professors W. A. Blackwell, F. G. Gray, H. L. Krauss, E. A. Manus, and H. R. Skutt. Some significant contributions came from a number of graduate students who read many chapters of the manuscript and offered numerous suggestions; I am especially grateful to R. I. Bedia and G. F. Henry of the General Electric Company, T. F. Morris of NASA, S. L. Witzel of Western Electric, and F. W. Colliver, C. C. Hupfield, F. W. Phillips, and S. J. Showalter. Several helpful computer programs were developed and run by Wing Moy.

Particularly important to the successful completion of the project was the strong support given me by editor and friend Gene A. Davenport. Also, critical comments from anomymous reviewers led to substantial improvements during the final revision, and additional improvements resulted from the very competent copyediting by Pamela A. Goett.

Special thanks to go Sandy Crigger and Mary Brent, who typed the greater part of the manuscript with unusual skill, diligence, and dedication. Finally, I am indebted to my wife and children, who not only tolerated my long hours of manuscript work but also encouraged and assisted me in every possible way.

C. A. H.

Contents

Chapter Sixteen
Audio Power Amplifiers
492

Chapter Seventeen
FET Linear Circuits
541

Chapter Eighteen
Frequency Response
579

Chapter Nineteen
Wideband Multistage Amplifiers
619

Chapter Twenty
Feedback Amplifiers
642

Chapter Twenty One
Frequency and Transient Response of Feedback Amplifiers
684

Chapter Twenty Two
Oscillators and Tuned Amplifiers

737

Chapter Twenty Three
Applications of Integrated Circuits

777

Chapter Twenty Four
Voltage Regulators

811

Appendix A
Computer Analysis of Inverter of Section 6-3

845

Appendix B
Two-Port Matrix Parameters

847

Index

849

PART ONE
TRANSISTOR FUNDAMENTALS AND INTEGRATED CIRCUIT TECHNOLOGY

Chapter One
The PN Junction

This book develops, in a logical manner, the basic theories and principles necessary for the intelligent design and proper utilization of electronic circuits. This necessitates a brief study of *device physics*, which treats the internal conduction processes of the various devices that are important to us, leading to the development of precise relationships between the terminal voltages and currents. *Solid-state physics*, which explains the conduction processes from the viewpoint of the band theory of solids, based on quantum mechanics, is of the greatest importance to many electronics engineers, particularly those developing new devices. However, it is perhaps best studied *after* an introductory course on electronic circuits.

In the development of device physics it is necessary to present without proof a few principles and laws from the realm of solid-state physics. These will be referred to as *postulates*, and they serve as the foundation that is required for the study of transistors.

Semiconductor physics is applied here to the PN junction, and in the following two chapters it is applied to both the bipolar junction transistor (BJT) and the field-effect transistor (FET). The analysis and design of electronics circuits is the subject of most of the remainder of the book.

Units used are those of the *Système International (SI)*. The basic SI units are the *meter, kilogram, second, ampere*, and *kelvin*.

1-1. SEMICONDUCTOR CHARGE CARRIERS

A crystal is a solid whose atoms are arranged in a symmetrical, periodic array. Metals, semiconductors, and many insulators are crystalline. At normal temperatures semiconductors are characterized by conductivities that are intermediate between those of metals and those of good insulators. Of special interest to us are the semiconductors silicon and germanium.

Postulate 1. A semiconductor has two types of charge carriers—the hole with charge $+q$, *and the free electron with charge* $-q$, *with* $q = 1.6 \times 10^{-19}$ C. *The hole density in holes per cubic meter is designated by the symbol* p, *for positive. The free-electron density is designated by the symbol* n, *for negative. Holes and free electrons move through a crystal in a random manner at high speeds, of the order of* 10^5 m/s, *encountering frequent collisions with the atoms. An applied electric field exerts forces on these charge carriers, causing them to acquire small drift velocities, resulting in hole and electron* drift currents *in the direction of the applied field.*

Postulate 2. A semiconductor may have two types of immobile ionized impurity atoms— the donor ion with charge $+q$, *and the acceptor ion with charge* $-q$. *The donor-ion density, in donors per cubic meter, is designated by the symbol* N_d, *and the acceptor-ion density is* N_a. *A semiconductor with donor and acceptor atoms is said to be* doped *with impurities; these* dopants *are rigidly fixed in the crystal structure, unable to participate in the conduction process. At a temperature of zero kelvin* $(-273°C)$ *none of the impurities are ionized, and* $N_d = N_a = 0$. *Above about 50 K nearly all are ionized, and* N_d *and* N_a *become constants independent of temperature. Charged particles other than holes, free electrons, donors, and acceptors are effectively neutralized and can be ignored insofar as our study of device physics is concerned.*

A semiconductor not subjected to external disturbing influences, such as an applied electric field or high-energy incident radiation, is said to be in *equilibrium*. The equilibrium hole and free-electron densities are denoted by the symbols p_o and n_o, respectively. When a voltage is applied to the terminals of a semiconductor device, the carrier densities p and n are, of course, no longer necessarily equal to their equilibrium values.

Postulate 3. Regardless of the number and types of dopants, the product of the equilibrium carrier densities is

$$p_0 n_0 = n_i^2 \tag{1-1}$$

with n_i^2 in silicon representing the expression

$$n_i^2 = 15 \times 10^{44} T^3 e^{-14000/T} \text{ m}^{-6} \tag{1-2}$$

T is the absolute temperature in kelvin units with one unit corresponding to 1°C.

An *intrinsic semiconductor* is one in which p_o and n_o are equal, or at least nearly so. For this case it follows from (1-1) that

$$p_o = n_o = n_i \tag{1-3}$$

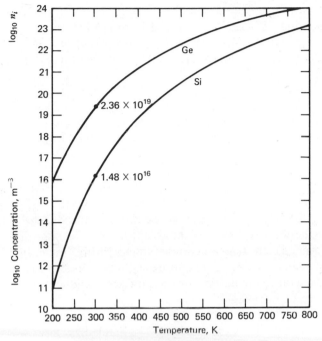

Figure 1-1   Temperature dependence of n_i.

with n_i called the *intrinsic carrier concentration*. At a room temperature of 300 K the intrinsic carrier concentration in silicon is 148×10^{14} m^{-3}. In the vicinity of room temperature the quantity T^3 in (1-2) varies slowly in comparison with the exponential term; hence the temperature dependence of both n_i and $n_i{}^2$ is approximately exponential. The curves of Fig. 1-1 show the variation of n_i with temperature for both silicon and germanium. Equation 1-2 was used to determine the curves, with the constants 15 and 14,000 changed to 3.1 and 9100, respectively, for germanium.

In a uniformly doped semiconductor in equilibrium the symmetrical arrangement of the uniformly distributed ions, combined with the fact that the net electric charge in the crystal is zero, requires that the positive charge density at each point be precisely equal to the negative charge density. Reference is to a "classical point," defined as a region sufficiently large to contain significant numbers of the charged particles but with dimensions very small compared with important physical dimensions of a practical problem. Accordingly,

$$p_o + N_d = n_o + N_a \tag{1-4}$$

This is the *equation of electric neutrality*. For specified dopant densities N_a and N_d the equilibrium carrier densities in silicon can be found at any specified temperature from (1-1), (1-2), and (1-4).

Suppose a semiconductor is doped so that the acceptor ions are more numerous than the donor ions. Then $N_a > N_d$, and from (1-4) we note that $p_o > n_o$. Such a semiconductor is said to be *P-type*. The holes are the *majority carriers* and the electrons are the *minority carriers*. Let us define the *effective density* N_A of the acceptor ions as

$$N_A = N_a - N_d \qquad (1-5)$$

P-type semiconductors are commonly doped so that $N_A \gg n_i$ except at very high temperatures. Thus, the equilibrium carrier densities are determined from (1-1) and (1-4) to be approximately

$$p_o \approx N_A \quad \text{and} \quad n_o \approx \frac{n_i^2}{N_A} \qquad (1-6)$$

Clearly, in the operating temperature range p_o is a constant independent of temperature and equal to the effective density of the acceptor ions. The minority-carrier density is, however, a sensitive function of temperature, increasing rapidly as T increases. At sufficiently high temperatures $n_i \gg N_A$, and from (1-1) and (1-4) it is easy to deduce that $p_o \approx n_i$. For this case the semiconductor is no longer P-type, but has become intrinsic.

An *N-type* semiconductor is one with more donors than acceptors; that is, $N_d > N_a$. Let the *effective density* N_D of the donor ions be defined as

$$N_D = N_d - N_a \qquad (1-7)$$

In N-type semiconductors used in diodes and transistors the doping is such that $N_D \gg n_i$, except at very high temperatures. Therefore, in the operating temperature range, it follows from (1-1) and (1-4) that

$$n_o \approx N_D \quad \text{and} \quad p_o \approx \frac{n_i^2}{N_D} \qquad (1-8)$$

The free electrons are the majority carriers, which have a density equal to N_D at all operating temperatures. Both P-type and N-type semiconductors are said to be *extrinsic*, because their electrical properties are determined mainly by their impurities.

In intrinsic semiconductors the free-electron and hole densities are equal, and both increase rapidly with temperature. *However, at normal temperatures in reasonably doped semiconductors the majority-carrier density equals the constant effective density of the impurity ions, but the minority-carrier density is a sensitive function of temperature.* At sufficiently high temperatures all semiconductors become intrinsic. If we define the upper limit of the extrinsic temperature range as that temperature at which the minority-carrier density is 10% of the majority-carrier density, it can be shown (see P1-3) that this limit is the temperature at which n_i equals $0.35N_A$ or $0.35N_D$. Semiconductor devices with dopants do not operate properly above this range.

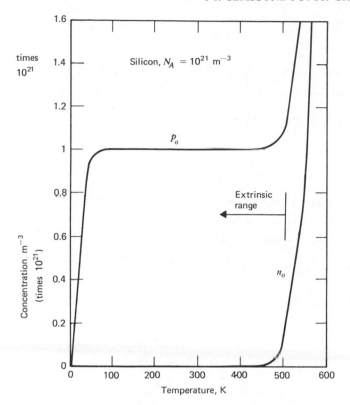

Figure 1-2   Temperature dependence of carrier densities.

For a specified impurity concentration the curves of Fig. 1-1 can be used to determine the upper limit of extrinsic operation. For example, if a P-type semiconductor has $N_A = 10^{21}$ m^{-3}, the limit is the temperature at which n_i is 0.35×10^{21}, or $\log_{10} n_i = 20.54$. From Fig. 1-1 this temperature is found to be 357 K (84°C) for germanium and 499 K (226°C) for silicon. The higher temperature limit for silicon is an advantage. The curves of Fig. 1-2 show the temperature dependence of the carrier densities of a P-type silicon semiconductor with $N_A = 10^{21}$; because of the scale that is used, the rapid variation of n_o with T is obscured in the extrinsic range.

DISCUSSION OF THE POSTULATES

The postulates that have been presented can be explained in a rigorous manner only by an in-depth study of the band theory of solids based on quantum mechanics. However, simple explanations can be obtained from a simple, but crude, crystal model. Figure 1-3 shows a two-dimensional model of a pure

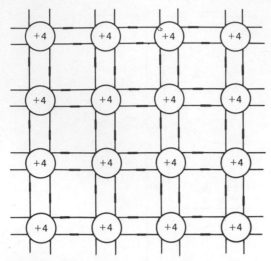

Figure 1-3 Schematic two-dimensional model of a semiconductor crystal.

silicon (or germanium) crystal. Each atom is surrounded by four nearest neighbors, and the four valence electrons per atom are shared equally with the nearest neighbors. Consequently, each *covalent bond* between an atom and one of its neighbors contains two electrons. A circle represents an ion with charge $4q$. At zero kelvin all valence electrons are in the covalent bonds, and these electrons are not free to respond to the force of an applied electric field.

As the temperature is raised above absolute zero, some electrons in the covalent bonds acquire sufficient thermal energy from the vibrating atoms to break loose from the bonds, becoming free electrons. The energy required to break a bond is about 1.1 eV in silicon and about 0.7 eV in germanium (1 eV = 1.6×10^{-19} J); these different energies account for the difference in the intrinsic carrier concentrations of these materials. At room temperature the number of broken bonds in silicon is only about one out of every twenty trillion bonds, but even so, the effect on the electrical properties of the crystal is considerable.

An electron that breaks loose from its covalent bond becomes a free electron. The vacancy left behind has a net positive charge q, because the electron left a region that was electrically neutral; this vacancy is called a hole. Once a hole is created, a valence electron from another covalent bond, without acquiring enough energy to become a free electron, readily moves into the vacancy, and the hole has moved to a new location. In this manner, utilizing the "bound" electrons in the covalent bonds, holes move through the crystal structure. In fact, both holes and free electrons travel rapidly and aimlessly, colliding frequently with the atoms. When an electric field is applied, they acquire drift velocities that are very much smaller than the thermal velocities. Holes and free electrons behave quite similarly, but the hole is positive and the electron is negative.

The thermal breaking of the covalent bonds creates holes and free electrons in equal numbers. The rate at which the bonds are broken increases rapidly with temperature. The reverse process also occurs; that is, *recombination of a hole and a free electron takes place when a free electron moves into a vacancy in a covalent bond*. When a hole and a free electron recombine, both simply disappear. At a constant temperature the hole and free-electron densities are constant, and this requires equal thermal generation and recombination rates.

An N-type semiconductor has donor impurities. These are elements with five valence electrons per atom, such as phosphorus, arsenic, and antimony. Each donor fits into the crystal structure, replacing a silicon atom, but there is one electron remaining after the covalent bonds are filled. This negative electron is attracted to the positive ion, but only loosely, and at normal temperatures, it acquires enough thermal energy to break away and become a free electron. The energy required is only about 0.01 eV. The donor atom left behind is now ionized. At temperatures above about 50 K nearly all the donors are ionized, *with each donor contributing one free electron*. The contribution of free electrons from the thermal breaking of the covalent bonds is usually negligible in comparison.

A P-type semiconductor has acceptor impurities, these being elements with three valence electrons per atom, such as boron, indium, and aluminum. An acceptor atom in the crystal structure has one valence electron less than required to complete the covalent bonds. The resulting vacancy is bound to the atom at 0 K. An energy of about 0.01 eV is required to transport an electron from a nearby covalent bond into this vacancy, and this energy is not available at 0 K. At operating temperatures, however, the energy is plentiful, electrons from covalent bonds fill the vacancies of the acceptors, and the acceptors are ionized. *This process generates one hole per acceptor ion*. Holes are also generated, of course, by the thermal breaking of the covalent bonds, but for P-type semiconductors with normal doping and at normal temperatures the holes contributed in this manner are insignificant.

If both donors and acceptors are present, donors give their extra electrons to an equal number of acceptors. This process creates neither holes nor free electrons, and the equilibrium carrier densities depend on the effective density N_A or N_D.

The equation $p_o n_o = n_i{}^2$ implies that the pn product is independent of the number and types of impurity atoms present. Suppose acceptors are introduced into pure silicon during the manufacturing process. These acceptors contribute holes, causing p_o to increase. More holes make chance encounters between holes and free electrons more likely. Hence recombination proceeds more rapidly, and a smaller n_o results. Although both p_o and n_o depend on the impurities, their product does not.

A semiconductor has about 10^{29} atoms per cubic meter. Dopant densities are at least a few orders of magnitude smaller. A typical impurity density of 10^{21} atoms per cubic meter corresponds to one donor or acceptor per 100 million atoms of the crystal. Such minute quantities of impurities have enormous effect on the electrical properties of the material.

1-2. EXCESS CARRIERS

When nonequilibrium conditions exist, the hole density p and the free-electron density n may be greater than, equal to, or less than their respective equilibrium values p_o and n_o. For example, holes can be introduced into or withdrawn from a semiconductor by a voltage applied to a suitable metal or semiconductor contact. The photons of incident light can break covalent bonds, producing pairs of holes and free electrons. The *excess* hole density $p - p_o$ and the *excess* electron density $n - n_o$ are normally not exactly the same at each point of a semiconductor; their difference represents a separation of charge, which results in an electric field with associated drift current. However, only a *very slight difference* is required to produce an appreciable drift current that tends to equalize the excess densities. From a somewhat different viewpoint, the electrostatic attractive force between unlike charges prevents any appreciable charge separation within a semiconductor.

Postulate 4. In a uniformly doped semiconductor the excess hole and free-electron densities are everywhere approximately equal; that is,

$$p - p_0 \approx n - n_0 \tag{1-9}$$

In practical semiconductors the difference between the excess hole density and the excess electron density at a point is always well less than 1 % of either excess density at that point. However, the very small differences produce an electric field that may be important. The postulate is reasonable even for nonuniform doping, provided the doping gradient is not too large, but it does not apply to the immediate region of a junction across which the dopant changes rather abruptly.

The addition of (1-4) and (1-9) yields

$$p + N_d \approx n + N_a \tag{1-10}$$

This *quasi-neutrality equation* is an excellent approximation, and of course the neutrality equation (1-4) is exact.

Suppose p and n are functions of distance x. Differentiation of (1-10) gives

$$\frac{dp}{dx} \approx \frac{dn}{dx} \tag{1-11}$$

In uniformly doped semiconductors the hole and free-electron gradients are everywhere nearly equal. Curves showing the variation of p and n with distance x must have the same slope at each value of x, as shown in Fig. 1-4.

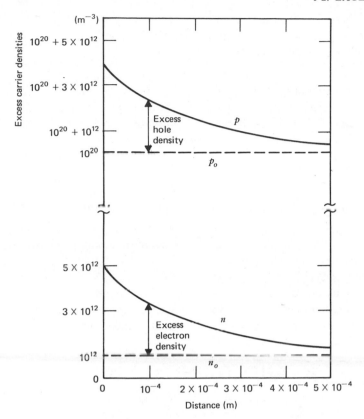

Figure 1-4   Positive excess carrier densities in a P-type semiconductor.

The processes of creating excess carriers, such as the application of a voltage to a PN junction or exposure of the semiconductor to incident light, are referred to as *injection*. In a reasonably doped P-type semiconductor (1-9) becomes $p \approx p_o + n$, because n_o is negligible compared with p_o. Suppose injection is so great that the minority-carrier density n becomes appreciable to or greater than the equilibrium hole density p_o. For this case the majority and minority carrier densities are the same order of magnitude, and injection is said to be *high-level*. On the other hand, if injection is small, so that n is much less than p_o, then $p \approx p_o$ and $p \gg n$. This is *low-level injection*.

To be more precise, let us agree that low-level injection exists whenever the minority-carrier density is everywhere less than 5% of the majority-carrier density, or

$$\frac{\text{minority-carrier density}}{\text{majority-carrier density}} < 0.05 \qquad (1\text{-}12)$$

Note that injection in the semiconductor with carrier densities plotted in Fig. 1-4 is definitely low-level. Henceforth, our attention will be confined mainly to uniformly doped semiconductors in low-level injection.

Example

Suppose a silicon semiconductor has $n_i = 10^{16}$ and $N_D = 10^{20}$ m^{-3}. Injection produces an excess-carrier density at a point equal to 10^{18} m^{-3}. At the point determine the ratio p/n and the percentage increases in p and n resulting from injection.

Solution

The semiconductor is N-type, with $n_o = 10^{20}$ and $p_o = 10^{12}$ m^{-3}. Therefore, $n = 10^{20} + 10^{18} = 101 \times 10^{18}$, and $p = 10^{12} + 10^{18} \approx 10^{18}$. The ratio $p/n = 0.01$. Clearly, injection is low-level.

The percentage increase in the minority-carrier density is $(p - p_o)/p_o \times 100\%$, or $10^8\%$. However, the percentage increase in the majority-carrier density is $(n - n_o)/n_o \times 100\%$, which is only 1%.

It should be noted that the injection of equal numbers of holes and electrons produced an enormous percent increase in the minority-carrier density but a rather negligible percent increase in the majority-carrier density. Of course if injection is low-level, the majority-carrier density cannot have increased more than 5%, using the definition given by (1-12).

1-3. DIFFUSION AND DRIFT CURRENTS

Diffusion current is the flow of electric charge caused by the influence of a concentration gradient, whereas drift current is the flow of charge caused by the driving force of an electric field. Whenever carrier distributions are non-uniform, there are both hole and electron diffusion currents. Drift currents exist when there is an electric field, regardless of whether the field results from an applied voltage or from a separation of charge within the semiconductor. Frequently both diffusion and drift currents are present in semiconductors, and the total current is their sum.

DIFFUSION CURRENTS

Using a familiar example to illustrate the diffusion process, let us consider the effect of releasing a small quantity of some pungent gas, such as sulfur dioxide, in a corner of a room. We all know that someone on the other side of the room would soon smell the gas, even if there are no circulating air currents. The random thermal motion of the gas molecules causes the sulfur dioxide gas to spread, or *diffuse*, away from the region of concentration. This net flow across a specified surface constitutes a diffusion current of sulfur dioxide molecules. After a period of time the gas density becomes uniform, and the current no longer exists.

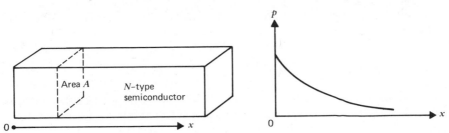

Figure 1-5   N-type semiconductor bar with nonuniform hole density.

Suppose the hole density p in an N-type semiconductor bar of cross-sectional area A is a function of distance x along the length of the bar, as shown in Fig. 1-5. Let us assume that there is no applied voltage and that the far more numerous free electrons neutralize the holes so completely that there are no appreciable electric forces on the holes. As a consequence, the hole drift current is negligible.

The holes move in the crystal structure at high speeds, colliding frequently with the vibrating atoms. After each collision a hole has an equal likelihood of going in any direction. Although the motion is completely random, more holes move in the positive x direction than in the opposite direction, because the hole density decreases as x increases. This diffusion process is similar to that of the sulfur dioxide molecules in the example considered. The holes diffuse to the right, and the hole diffusion current in the x direction is positive. As this current is proportional to the hole charge q, the cross-sectional area A, and the concentration gradient dp/dx, the hole diffusion current in the x direction can be expressed in the form

$$(I_h)_{\text{diffusion}} = -qAD_h \frac{dp}{dx} \qquad (1\text{-}13)$$

with the constant D_h of proportionality denoting the *hole diffusion constant* in m^2/s. Since dp/dx is negative, (1-13) correctly gives a positive hole current.

The free electrons in the semiconductor bar, having the same spatial distribution as the holes, also diffuse in the positive x direction. However, because the electronic charge is negative, the electron diffusion current in the positive x direction is negative. This current is

$$(I_e)_{\text{diffusion}} = qAD_e \frac{dn}{dx} \qquad (1\text{-}14)$$

with D_e denoting the *electron diffusion constant*. Note that dn/dx is negative, giving a negative I_e. Equations 1-13 and 1-14 give the diffusion currents in the positive x direction, regardless of the manner in which p and n vary with x.

In uniformly doped semiconductors $dp/dx \approx dn/dx$ by (1-11), and the ratio of the diffusion currents is

$$\left(\frac{I_h}{I_e}\right)_{\text{diffusion}} = \frac{-D_h}{D_e} \tag{1-15}$$

The hole and electron diffusion currents in the same direction always have opposite signs. Alternatively, the positive hole-diffusion current and the positive electron-diffusion current are always in opposite directions. The ratio of their magnitudes equals the ratio of their diffusion constants.

DRIFT CURRENTS

Now suppose a voltage is impressed across the ends of the semiconductor bar of Fig. 1-5. The electric field E_x imparts a drift velocity to the charge carriers, with the holes drifting in one direction and the electrons drifting in the opposite direction. Because the electrons carry negative charge, both holes and electrons contribute positively to the total drift current. This current in amperes is the product of the cross-sectional area A, the conductivity σ (mhos/m), and the field E_x (V/m), or $I = A\sigma E_x$.

The conductivity σ is

$$\sigma = qp\mu_h + qn\mu_e \tag{1-16}$$

The products qp and qn are the magnitudes of the charge densities, and the hole and electron *mobilities* μ_h and μ_e are the magnitudes of the respective drift velocities divided by the electric field. Mobility is independent of the electric field and has the unit m^2/V-s. The hole and electron drift currents in the x direction are:

$$(I_h)_{\text{drift}} = qAp\mu_h E_x \qquad (I_e)_{\text{drift}} = qAn\mu_e E_x \tag{1-17}$$

The net hole and electron currents in the positive x direction are found by adding the drift and diffusion currents, giving

$$I_h = qA\mu_h pE_x - qAD_h \frac{dp}{dx} \tag{1-18}$$

$$I_e = qA\mu_e nE_x + qAD_e \frac{dn}{dx} \tag{1-19}$$

The total current I is, of course, the sum of I_h and I_e.

Postulate 5. The mobilities and the diffusion constants in a given material are related as follows:

$$\frac{\mu_h}{D_h} = \frac{\mu_e}{D_e} = \frac{q}{kT} \tag{1-20}$$

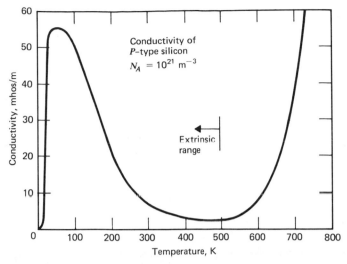

Figure 1-6   Approximate temperature dependence of the conductivity of P-type silicon with $N_A = 10^{21}$ m^{-3}.

50 K to 500 K the conductivity drops because of the decreasing mobilities. The sharp rise above 500 K is a consequence of the semiconductor changing from extrinsic to intrinsic operation, as shown in Fig. 1-2.

In a uniformly doped P-type semiconductor in low-level injection the electron drift current is much smaller than the hole drift current and can therefore be neglected. This is evident from (1-17), recognizing that the mobilities are the same order of magnitude whereas $n \ll p$. Frequently, operation is such that the much larger hole current is canceled, or nearly so, by the hole diffusion current. Even so, a little reflection shows that the electron drift current is still negligible. It is small compared with the hole drift current. Therefore, it is small compared with the equal-and-opposite hole diffusion current, and this component is the same order of magnitude as the electron diffusion current by (1-15). Consequently, the electron diffusion current dominates. Similar reasoning applied to N-type semiconductors shows that the hole drift current is negligible. As stated previously, our future investigations will treat uniformly doped semiconductors in low-level injection. Accordingly, *we shall henceforth completely ignore minority-carrier drift currents.*

1-4. MINORITY-CARRIER DIFFUSION EQUATION

Let us now examine further the processes of generation and recombination of charge carriers in semiconductors. If a uniformly doped P-type semiconductor

This is the *Einstein relation*. Its existence is due to the fact that both drift and diffusion processes that are limited by collisions between the carriers and the vibrating atoms. constant k is Boltzmann's constant (1.38×10^{-23} J/K). At a room temperature of 290

$$\frac{q}{kT} = 40 \text{ V}^{-1} \tag{1-}$$

At this temperature the mobility of each carrier is 40 times its diffusion constant.

In pure silicon at 300 K the hole mobility is 0.048 m²/V-s and the fre electron mobility is 0.135 m²/V-s. In accordance with the Einstein relation tl hole and electron diffusion constants are 0.00124 and 0.00349 m²/s, respectivel In doped silicon the mobilities and diffusion constants are lower, and the values depend on the type and density of the impurity. They also vary wit temperature. As the temperature rises, the atoms vibrate more vigorously. Thi results in more frequent collisions between the charge carriers and the atom! For a given electric field the drift velocities decrease. Hence the mobilitie decrease as the temperature increases. At very low temperatures the mobilitie are quite large, and in a pure semiconductor the mobilities approach infinit) as the temperature approaches absolute zero. Some books present curves showing experimentally determined mobilities of silicon and germanium at room temperature as functions of the dopant densities, along with approximate relations for adjusting the values to other temperatures.[1] The diffusion constants are found from the Einstein relation.

The mobilities of the carriers in germanium are several times greater than those in silicon. This advantage allows faster switching speed and higher frequency operation. In pure germanium at 300 K the hole and electron mobilities are 0.19 and 0.39 m²/V-s, respectively. However, because of simplified processing and lower sensitivity to typical operating temperatures, integrated circuits and most discrete transistors are made of silicon.

The conductivity of an intrinsic semiconductor, called the *intrinsic conductivity* σ_i, is

$$\sigma_i = qn_i(\mu_e + \mu_h) \tag{1-22}$$

This conductivity increases rapidly as the temperature rises, because the temperature dependence of n_i is considerable. A resistor made of a piece of intrinsic semiconductor with two leads attached is called a *thermistor* and is normally used for temperature measurement or compensation.

The conductivities of well-doped P-type and N-type semiconductors are $qN_A\mu_h$ and $qN_D\mu_e$, respectively. A curve showing the variation of conductivity with temperature is shown in Fig. 1-6. At 0 K the absence of charge carriers makes the semiconductor a perfect insulator. As the temperature is raised to 50 K, the ionization of the impurities causes the conductivity to rise. From

---

[1] A superscript numeral refers to the corresponding end-of-chapter reference.

is illuminated with constant ionizing radiation, covalent bonds are broken at a steady rate, producing additional holes and free electrons. This generation process, which is in addition to thermal generation, continues as long as there is incident radiation. Assuming injection is low level, the hole density is not appreciably changed, but the electron density may increase by several orders of magnitude. The excess free electrons that appear as a consequence of this illumination make suitable chance encounters between holes and electrons more likely, resulting in an increased recombination rate. Under steady conditions the recombination rate is precisely equal to the sum of the generation rates caused by both the radiation and the thermal process.

Let us suppose that the illumination is such that the excess electrons are uniformly distributed throughout the material, and that at time zero the radiation is removed. Immediately the generation rate is reduced to the thermal generation rate. Because the recombination rate is now greater than this, the carrier densities decrease with time, and as they decrease, the recombination rate also decreases. Ultimately the carrier densities reach their equilibrium values.

At time zero, just after the radiation is removed, the rate of decrease is a maximum. As the excess free electrons disappear, the difference between the recombination and generation rates becomes smaller, and the rate of decrease is reduced. It appears plausible that the rate is approximately proportional to the *excess free-electron density* $n - n_o$. This is indeed a valid assumption, which leads to an excess density that decreases exponentially with time. Mathematically

$$\text{the } net \text{ recombination rate} \approx \frac{n - n_o}{\tau} \qquad (1\text{-}23)$$

with the constant τ denoting the *excess-carrier lifetime*. If the excess-carrier density is positive at a point, the net recombination rate is also positive. Under equilibrium conditions the net recombination rate is zero. For a negative excess-carrier density the net recombination rate is negative; in this case holes and free electrons are thermally generated more rapidly than they are annihilated by recombination. The excess-carrier lifetime depends on the temperature and the properties of a particular specimen. It is usually less than 500 microseconds ($1 \ \mu s = 10^{-6} \ s$) and greater than a few nanoseconds ($1 \ ns = 10^{-9} \ s$).

It is possible to withdraw essentially all the free electrons from a P-type semiconductor. One way is simply to reverse bias both junctions of an NPN transistor. In such a case there can be no recombination, and from (1-23) the generation rate is found to be n_o/τ. Because this represents the rate at which the covalent bonds are broken thermally, it is unrelated to the actual free-electron density. Accordingly, the individual terms on the right side of (1-23) have the following interpretation:

$$\text{recombination rate} = \frac{n}{\tau} \qquad \text{generation rate} = \frac{n_o}{\tau} \qquad (1\text{-}24)$$

The preceding discussion applies, of course, to N-type semiconductors as well as to P-type except that n and n_o in (1-24) must be replaced with p and p_o, respectively. Usually (1-23) would also be expressed in terms of the excess hole density even though the excess free-electron and hole densities are equal by (1-9). The results are presented here as a postulate.

Postulate 6. In uniformly doped semiconductors in which carriers are generated by the thermal process only, the recombination rate at a point equals the minority carrier density divided by the lifetime τ, and the generation rate is the equilibrium minority-carrier density divided by τ. It follows that the net recombination rate is the excess carrier density divided by τ.

THE DIFFUSION EQUATION

The analytical determination of the currents of semiconductor devices, when voltages are applied to the terminals, requires knowledge of the spatial variation of the charge carriers. Essential information is obtainable from the transistor diffusion equation, and the one-dimensional form of this equation will now be derived.

Shown in Fig. 1-7 is a small specimen of an N-type semiconductor that is uniformly doped and in low-level injection. The manner in which the excess carriers were injected is not important; if we wish, we can assume they were injected at one end of the specimen by a pulse of incident radiation. We shall assume that p_n and n_n, with the subscript n used to denote an N-type region, are functions only of distance x and time t.

The hole charge per second that passes through the cross-sectional area A at the coordinate x is the hole current $I_h(x)$, and the rate at which holes pass through A is $(1/q)I_h(x)$ holes per second. The rate at which holes pass through the cross-section at $x + \Delta x$ is $(1/q)I_h(x + \Delta x)$. Consequently, the net rate at which holes flow out of the small parallelepiped between x and $x + \Delta x$ is

$$\left(\frac{1}{q}\right)[I_h(x + \Delta x) - I_h(x)]$$

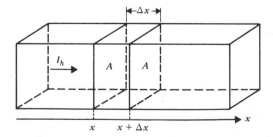

Figure 1-7   Small specimen of N-type semiconductor.

Dividing this by the volume $A \, \Delta x$ and letting Δx approach zero give the outward hole flow per unit volume to be

$$\left(\frac{1}{qA}\right) \frac{\partial I_h}{\partial x} \text{ holes per second per cubic meter}$$

The partial derivative is used, because the carrier densities and I_h are functions of the two variables x and t. Because I_h is the minority-carrier current and injection is low-level, the hole current is $-qAD_h \partial p_n/\partial x$. Substituting this for I_h gives:

$$\text{outward hole flow} = -D_h \frac{\partial^2 p_n}{\partial x^2} \text{ holes per second per cubic meter} \quad (1\text{-}25)$$

If the outward hole flow per unit volume, given by (1-25), is added to the net hole recombination rate, the result is $-\partial p_n/\partial t$, which denotes the total rate at which holes disappear from the differential volume $A \, dx$, in holes per second per m^3. Hence,

$$\frac{-\partial p_n}{\partial t} = -D_h \frac{\partial^2 p_n}{\partial x^2} + \frac{p_n - p_{no}}{\tau} \quad (1\text{-}26)$$

A *diffusion equation* is a partial differential equation with space derivatives and a first partial time derivative. Accordingly, (1-26) is the *minority-carrier diffusion equation*.* A similar equation applies to the free-electron density n_p in a P-type semiconductor, using the electron diffusion constant D_e and the P-region excess-carrier lifetime.

Our applications of the diffusion equation will be only to situations in which p_n is independent of time. For this case the time derivative in (1-26) is zero, and the general solution of the ordinary differential equation that results is

$$p_n = C_1 e^{-x/L_h} + C_2 e^{x/L_h} + p_{no} \quad (1\text{-}27)$$

with the *hole diffusion length* $L_h = (D_h \tau)^{1/2}$ and with C_1 and C_2 denoting arbitrary constants. The corresponding solution for the free-electron density in a P-type semiconductor is

$$n_p = C_3 e^{-x/L_e} + C_4 e^{x/L_e} + n_{po} \quad (1\text{-}28)$$

with the *electron diffusion length* $L_e = (D_e \tau)^{1/2}$.

Equations 1-27 and 1-28 apply to the minority carriers of uniformly doped semiconductors in low-level injection. Henceforth, whenever we need to find a carrier density as a function of distance x, we shall begin with (1-27) or (1-28). Suitable *boundary conditions*, which depend on the particular situation, must

---

*The diffusion equation can be expressed in three dimensions by replacing the second partial x derivative in (1–26) with $\nabla^2 p_n$, with ∇^2 denoting the Laplacian operator *div grad*.

be utilized to determine the arbitrary constants. The majority-carrier densities are readily found from the minority-carrier densities and our knowledge that the excess carrier densities are approximately equal in accordance with (1-9).

1-5. THE PN JUNCTION

A PN junction is present when a single crystal of a semiconductor contains both a P-type and an N-type region. The metallurgical junction is the surface where the effective doping density changes from N_A to N_D. Most semiconductor devices have one or more PN junctions, and a knowledge of junction theory is essential for their careful study. Our purpose here is to learn the important attributes and properties of the junction and to develop a strong, intuitive understanding of the conduction mechanism.

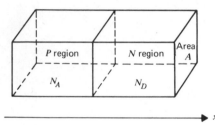

Figure 1-8   PN junction.

For simplicity we shall consider a junction with the geometry of Fig. 1-8. The cross-sectional area A is constant, and each region is many diffusion lengths long. The analysis will be one-dimensional with the charge-carrier densities assumed to be functions only of distance x. Both regions are uniformly doped. The effective doping densities are much greater than the intrinsic carrier concentration, and the transition from N_A to N_D occurs abruptly at the metallurgical junction.

THE QUALITATIVE PROCESS

A PN junction cannot be formed by pressing together a P-type and an N-type semiconductor, because such a boundary would have far too many imperfections. Nevertheless, let us imagine that a junction is made in this manner without imperfections. At once, many of the P-region holes that are located immediately adjacent to the junction wander across the junction, simply as a result of their random motion. As these holes enter the N region and become minority carriers, most of them are quickly annihilated by recombination with free electrons.

Similarly, many of the N-region free electrons immediately adjacent to the junction cross into the P region; almost at the point of entry they are destroyed by recombination with the far more numerous holes. The rather obvious result of this process is that, in extremely thin regions adjacent to the metallurgical junction, the majority-carrier densities almost instantly become quite small compared with the dopant densities. Of course the minority-carrier densities increase in order to keep the *pn* product constant.

At this point in the transient process practically none of the immobile impurity ions located immediately adjacent to the junction are neutralized, because most of the majority carriers, which normally neutralize these ions, have disappeared. Consequently, present are thin layers of equal and opposite charge, one layer on each side of the junction. This *dipole layer* produces an electric field that is confined to the region of the dipole layer and is directed from the positive donor ions of the N region toward the negative acceptor ions of the P region. The direction is such that the flow of majority carriers is impeded. Yet the flow continues but at a reduced rate, and the sharp decrease in each majority-carrier density occurs at slightly greater and greater distances in the P and N regions. Thus the dipole layer of unneutralized impurity ions expands. The increasing separation of charge causes the field within the dipole layer to increase, and the flow of majority carriers across the junction continues to decrease.

Let us digress for a moment to consider the flow of the minority carriers across the junction. Initially, before the electric field has grown appreciably, this flow is insignificant in comparison with the majority-carrier flow, because the majority carriers are much more numerous. Furthermore, the flow is unaffected by the presence of the dipole layer and its electric field. This is a consequence of the fact that the field is confined to the dipole layer and is directed so as to sweep minority carriers across the junction. Therefore, *the rate at which they cross the junction is precisely equal to the rate at which they simply happen by chance to wander into the dipole layer*, and these chance happenings depend only on the random motion of the minority carriers in the neutral regions.

In equilibrium the net hole current across the junction must be zero, and the electron current must also be zero. Accordingly, the expansion of the dipole layer continues until its associated electric field builds up to a value that impedes the flow of holes from the P to the N region just the right amount. This is the amount needed to make this flow exactly equal to the flow of minority-carrier holes from the N to the P region. When this condition has been reached, the junction is in equilibrium. Similar comments apply, of course, to the movement of free electrons across the junction.

Now suppose a battery is connected to the external terminals of the PN junction. The applied voltage causes the field in the vicinity of the junction to increase or decrease, depending on the polarity of the voltage. The dipole layer either expands or contracts in order to adjust to the new field, and a change occurs in the rate at which majority carriers cross the junction, Because the flow of minority carriers across the junction is not influenced by the field, an imbalance exists and a current results. This situation is analyzed in more detail later in this chapter.

In Fig. 1-9 is shown a PN junction with an applied terminal voltage V and a current I defined. If V is positive, I is positive, and this is the *forward-bias* condition. In electronics the word *bias* is used with reference to the dc operating conditions of an electronic device. The forward-bias condition is easily remembered, for the voltage-source connections are positive-to-P and negative-to-N. A negative V, which gives a negative I, is the *reverse-bias* condition.

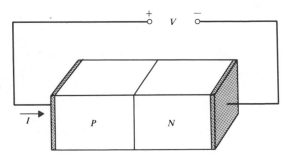

Figure 1-9 V and I of a PN junction.

Equilibrium exists when V is zero, and the voltage across the junction is called the *built-in junction voltage* ψ_0. It is sometimes called the *contact potential*, which is the term commonly used for the junction voltage that is present whenever two dissimilar metallic conductors are in contact. There is no current in the shorted electric circuit, because the contact potentials at the metal-semiconductor contacts precisely cancel the effect of the built-in junction voltage.

When the junction is forward biased, the positive voltage V superimposes an applied electric field on the built-in electric field of the junction. The applied field is positive in the direction from the P region to the N region, but the built-in field is positive in the opposite direction. Hence the field in the junction is reduced and so is the junction voltage. When V is negative, the applied field reinforces the built-in field, and the junction voltage is increased. If we assume negligible ohmic IR drops in the closed circuit, then for both positive and negative V the total junction voltage ψ is

$$\psi = \psi_o - V \qquad (1\text{-}29)$$

In most practical situations the *IR* drops are small, and the voltage applied directly to the junction is approximately the terminal voltage. At the metallurgical junction the electric field is always directed from the positive donor ions of the N region toward the negative acceptor ions of the P region, regardless of the magnitude and polarity of the terminal voltage. For an applied voltage greater than ψ_o the current and the *IR* drops become large, and *V* in (1-29) must be interpreted as the applied voltage less the ohmic drops. *In* (1-29) *the voltage V can never be greater than ψ_o.*

THE SPACE-CHARGE LAYER

We now proceed with an approximate analysis of the space-charge layer. For this purpose it is appropriate to present a postulate based on the preceding qualitative discussion.

Postulate 7. On each side of the metallurgical junction is a thin layer of unneutralized impurity ions, and the charges on the two sides are equal and opposite. Within the space-charge layer the majority-carrier densities are considerably smaller than the effective dopant densities. Also, the electric field is very intense compared with the field of the so-called neutral regions outside the layer. In the neutral regions the effects of the change in the dopants are negligible, and quasi-neutrality prevails.

The space-charge layer is so named because the unneutralized acceptor ions on the P side constitute a thin layer of negative charge, and the unneutralized donors on the N side constitute a thin layer of positive charge. It is sometimes referred to as the *depletion region*, and the assumption that the carrier densities are small compared with the dopant densities is known as the *depletion approximation*. This approximation is usually quite reasonable for reverse-biased junctions. It is valid also for junctions in equilibrium, provided the two regions have dopant densities that differ by no more than about 3 to 1. Yet it is rather poor for forward-biased junctions, and invalid for forward-biased junctions that are *asymmetrically doped*, that is, having one side doped much more heavily than the other. Since quasi-neutrality is postulated for the neutral regions, all developed equations apply. However, within the space-charge layer where conditions are far from neutral, those equations based on uniform doping are inapplicable.

Figure 1-10 shows a side view of the junction of Fig. 1-9. The length of the space-charge layer is greatly exaggerated relative to the lengths of the P and N regions, in order to show clearly the unneutralized acceptor and donor ions and the dimensions l_p and l_n. The electric field *E* is properly directed from the positive donor ions to the negative acceptor ions, regardless of the applied voltage, and *E* is the negative of E_x. The voltage rise from the neutral P region to the neutral N region is positive.

The electric field in the space-charge layer can be determined by applying Gauss's law, a fundamental equation of electromagnetic field theory. Because

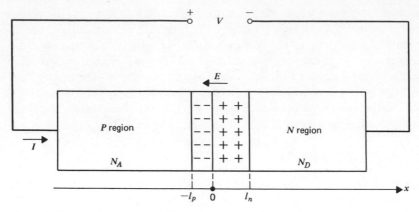

Figure 1-10   Junction side view with enlarged space-charge layer.

E is the negative of E_x, the one-dimensional form of the equation applicable to the layer is

$$\frac{dE}{dx} = \frac{-\rho}{\varepsilon} \tag{1-30}$$

with ρ denoting the charge density at a point and with ε denoting the permittivity of the semiconductor. By Postulate 7 the majority-carrier densities in the layer are small compared with the dopant densities. Therefore, the charge density ρ in (1-30) is the charge density of the impurity ions, as shown in Fig. 1-11.

It is evident from Postulate 7 that the electric field in each neutral region is negligible in comparison with that of the space-charge layer. Accordingly, the constant of integration that results when (1-30) is integrated is selected to make the field zero at each end of the layer.

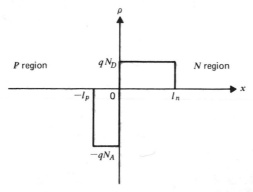

Figure 1-11   Charge density ρ versus distance.

For values of x from $-l_p$ to zero the charge density is $-qN_A$. Substituting into (1-30), integrating, and selecting the constant so that E is zero at $x = -l_p$, we obtain

$$E = \frac{qN_A}{\varepsilon}(l_p + x) \qquad -l_p \leq x \leq 0 \tag{1-31}$$

For values of x from zero to l_n the charge density is qN_D. Substituting into (1-30), integrating, and selecting the constant so that E is zero at $x = l_n$, we find that

$$E = \frac{qN_D}{\varepsilon}(l_n - x) \qquad 0 \leq x \leq l_n \tag{1-32}$$

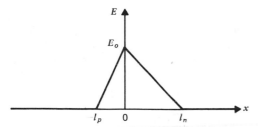

Figure 1-12   Electric field E in space-charge layer.

A plot of E versus x is shown in Fig. 1-12. Note that *the maximum field occurs precisely at the metallurgical junction* at $x = 0$. From (1-31) and (1-32) the maximum field E_o is found to be

$$E_o = \frac{qN_A l_p}{\varepsilon} = \frac{qN_D l_n}{\varepsilon} \tag{1-33}$$

We note from Fig. 1-11 that the magnitude $qN_A l_p A$ of the negative charge of the acceptor ions equals the positive charge $qN_D l_n A$ of the donor ions. Therefore, $N_A l_p = N_D l_n$, or

$$\frac{l_p}{l_n} = \frac{N_D}{N_A} \tag{1-34}$$

This result also follows directly from (1-33). It states that the *ratio of the lengths* l_p *and* l_n *of the space-charge layer equals the inverse of the ratio of the dopant densities of the regions.* The layer penetrates the greater distance into the more lightly doped region.

The total junction voltage $(\psi_o - V)$ is the integration with respect to x of the field E across the space-charge layer. Adding the integral of E from $-l_p$ to 0, using (1-31), to the integral from 0 to l_n, using (1-32), we find that

$$\psi_o - V = \frac{q}{2\varepsilon}(N_A l_p{}^2 + N_D l_n{}^2) \qquad (1\text{-}35)$$

Equations 1-34 and 1-35 are easily solved for l_p and l_n, and the sum of these is the total length l of the layer. The result can be put in the form:

$$l = \left[\left(\frac{2\varepsilon}{q}\right)\left(\frac{1}{N_A} + \frac{1}{N_D}\right)\right]^{1/2}(\psi_o - V)^{1/2} \qquad (1\text{-}36)$$

From (1-33), (1-34), and (1-36) it is easy to deduce that

$$E_o = \left[\frac{2qN_A N_D}{\varepsilon(N_A + N_D)}\right]^{1/2}(\psi_o - V)^{1/2} \qquad (1\text{-}37)$$

In (1-36) and (1-37) the quantities in brackets are constants for a given semiconductor junction. The length l and the maximum field E_o are directly proportional to the square root of the junction voltage $(\psi_o - V)$ for the abrupt junction being investigated.

Example

In SI units a silicon PN junction has $N_A = 10^{23}$, $N_D = 10^{22}$, $\psi_o = 0.73$, and permittivity $\varepsilon = 10^{-10}$. Determine l_p, l_n, and E_o for a forward bias of 0.6 V and also for a reverse bias of 10 V.

Solution

The length l is $0.371(0.73 - V)^{1/2}$ μm from (1-36). Because the ratio of l_p to l_n is 0.1 by (1-34) and their sum is l, it follows that l_p is $l/11$ and l_n is $10l_p$. For a forward bias of 0.6 V, l_p is found to be 0.012 μm and l_n is 0.12 μm. With $V = -10$ these lengths become 0.11 and 1.1 μm, respectively. *Note that the length of the layer increases considerably when the bias is changed from forward to reverse. Also, most of the depletion region resides in the N-type semiconductor, which has the lower dopant density.*

The maximum field intensity E_o, which occurs at the metallurgical junction, is $5.39(0.73 - V)^{1/2}$ MV/m. For $V = 0.6$ the field is 1.9 MV/m and for $V = -10$ it is 17.7 MV/m. *Note the increase in E_o when the bias is changed from forward to reverse and also note that the field is extremely large.* In fact, as indicated in Sec. 1-9 the junction is near breakdown at the higher field intensity.

1-6. JUNCTION VOLTAGE AND BOUNDARY RELATIONS

In equilibrium the carrier densities p_o and n_o vary with distance within the space-charge layer, and an intense electric field is present. Although there is no net hole current or electron current, there are rather large diffusion and drift components. Let us consider the hole currents. The diffusion current $-qAD_h\, dp_o/dx$ in the positive x direction must equal the drift current $qAp_o\mu_h E$ in the negative x direction. The positive field E is $-E_x$. Thus

$$D_h \frac{dp_o}{dx} = -p_o\mu_h E \qquad (1\text{-}38)$$

By rearranging terms and utilizing the Einstein relation, this becomes

$$\frac{dp_o}{p_o} = \frac{-qE}{kT}\, dx \qquad (1\text{-}39)$$

Let us integrate (1-39) through the space-charge layer from the neutral P region to the neutral N region. The p_o limits are p_{po} and p_{no}, and integration of the left side of (1-39) gives the expression $\ln(p_{po}/p_{no})$. Because p_{po} equals N_A and p_{no} equals n_i^2/N_D, this becomes $\ln(N_A N_D/n_i^2)$. The integral of $-E\,dx$ through the layer is the junction voltage ψ_o, and the right side of (1-39) becomes $q\psi_o/kT$. Equating these expressions and solving for ψ_o, we find that

$$\psi_o = \frac{kT}{q} \ln\!\left(\frac{N_A N_D}{n_i^2}\right) \qquad (1\text{-}40)$$

At a given temperature the built-in voltage of a junction with specified dopant densities can be determined from (1-40). Because of the rapid increase in n_i with temperature, the built-in junction voltage ψ_o drops as T rises, typically about 2 mV per degree Celsius.

CARRIER DENSITIES AND CURRENTS OF THE SPACE-CHARGE LAYER

Equation 1-39 can be used to find p_o as a function of x within the space-charge layer. To do this we replace E with the proper function of x, using (1-31) and (1-32), and integrate from $-l_p$ to x. We will not proceed with this mathematical exercise, but the result for a silicon junction at 300 K with $N_A = 2 \times 10^{22}$ and $N_D = 10^{22}$ is shown in Fig. 1-13. The equilibrium free-electron density n_o was found from n_i^2/p_o.

The space-charge layer of a junction is often said to be depleted of charge carriers. This is not intended to signify that all carriers are absent. It simply implies that the carrier densities are small compared with the immobile ion density. In the neutral regions these ions are neutralized by the carriers, but such is not the case within the layer. As indicated by the curves of Fig. 1-13 there are numerous carriers present. Because the fixed ions produce an intense electric field, a large drift current is present. There is also a large diffusion current in the opposite direction as a consequence of the enormous carrier gradient. To illustrate, in Fig. 1-13 the hole density changes by more than 10^{22} in a distance

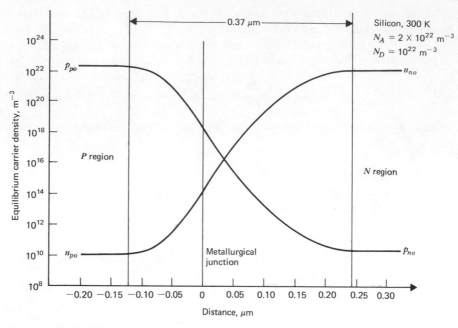

Figure 1-13 Equilibrium carrier densities in space-charge layer.

of only 0.37 μm. Of course the drift and diffusion currents exactly cancel under equilibrium conditions and they almost cancel when the junction carries an appreciable current.

For example, suppose the net current of a forward-biased junction is 1 mA. At the metallurgical junction the hole diffusion current might be 1001 mA with an opposite drift current of 1000 mA. The difference gives the actual current. In general, the very thin space-charge layer has an intense electric field caused by unneutralized donor and acceptor ions. The carrier gradients are exceedingly large. Appreciable equal-and-opposite drift and diffusion currents are present under equilibrium conditions. The effect of an applied voltage is to produce a small imbalance in these components, and this imbalance is the net current.

A study of Fig. 1-13 shows that the ends of the space-charge layer are not appreciably depleted of charge carriers. For a forward bias the curves for p and n lie above those for p_o and n_o, respectively, and the end-regions that are not appreciably depleted are widened somewhat. If the doping of the P-region is greatly increased, say to 10^{24}, the major changes in Fig. 1-13 would be an increase in p_{po} and a shift in the location of the metallurgical junction to the left. Clearly, a substantial percentage of the acceptor atoms in the layer are now neutralized. In addition, the hole density at the metallurgical junction is now greater than N_D, and the depletion approximation is no longer reasonable.

BOUNDARY CONDITIONS

By replacing N_A in (1-40) with p_{po} and N_D/n_i^2 with $1/p_{no}$, we find that p_{no} can be expressed as

$$p_{no} = p_{po} e^{-q\psi_o/kT} \qquad (1-41)$$

The left side of (1-41) is the hole density at the edge of the space-charge layer in the N region. If the junction voltage ψ_o is changed to $\psi_o - V$ by the addition of an applied voltage V, the hole density at the edge of the layer in the N region also changes. We might reasonably expect (1-41) to apply to the nonequilibrium case provided ψ_o is replaced with $\psi_o - V$. Of course, p_{no} at the edge is no longer the equilibrium value, and thus a different symbol must be selected. The symbol that will be used to denote the hole density in the neutral N region *at the edge of the layer* is $p_n(0)$. Then (1-41) with $\psi_o - V$ substituted for ψ_o becomes

$$p_n(0) = p_{po} e^{-q(\psi_o - V)/kT} \qquad (1-42)$$

With the aid of (1-41) this can be expressed in the form

$$p_n(0) = p_{no} e^{qV/kT} \qquad (1-43)$$

Similarly, in the P region at the edge of the layer we find that

$$n_p(0) = n_{po} e^{qV/kT} \qquad (1-44)$$

Although (1-43) and (1-44) have not been rigorously deduced, they are indeed adequate approximations for PN junctions with applied voltages that are restricted so as to maintain low-level injection. Accordingly, we shall assume the following postulate.

Postulate 8. At each of the edges of the space-charge layer of a PN junction the minority-carrier density $p_n(0)$ or $n_p(0)$ equals the product of the equilibrium minority-carrier density and the exponential of qV/kT, provided V is such that injection is low-level.

Equations 1-43 and 1-44 are the boundary conditions relating the minority-carrier densities at the edges of a space-charge layer to the applied voltage V. They are used in the next section to derive the diode equation and in the next chapter to develop transistor theory.

Example

At 300 K a silicon PN junction has $N_A = 10^{24}$ and $N_D = 10^{22}$ m^{-3}. (a) Calculate the built-in junction voltage. (b) Determine the maximum voltage that can be applied with low-level injection maintained everywhere.

Solution

(a) At 300 K, n_i is 1.48×10^{16} from Fig. 1-1. With $q/kT = 40$ at 290 K by (1-21), it is $40 \times (290/300)$, or 38.7, at 300 K. From (1-40) the built-in junction voltage is found to be 0.81 V.

(b) Because the N region has the smaller dopant density, p_{no} is greater than n_{po} and examination of (1-43) and (1-44) shows that injection into the N region is greater than that into the P region. Thus we need to consider only the value of $p_n(0)$, which must be less than $0.05n_{no}$ by (1-12). Therefore, at the maximum voltage $p_n(0)$ equals $0.05N_D$, or 5×10^{20} m^{-3}. Equating this to $p_{no} \exp qV/kT$ with $p_{no} = n_i^2/N_D$, or 2.19×10^{10}, the maximum voltage is determined to be 0.62 V. Above this value injection is high-level according to (1-12).

1-7. THE DIODE EQUATION

A PN junction is illustrated in Fig. 1-14. Injection is low-level and the IR drops in the neutral regions and at the metal-semiconductor contacts are assumed to be negligible. Therefore, the terminal voltage V is the same as the applied junction voltage. Both neutral regions are many diffusion lengths long. The point $x = 0$ is located at the space-charge layer, which is so thin that it is represented in the figure simply as a plane. The length l of this layer is perhaps less than 1% of a diffusion length, whereas the neutral regions are many diffusion lengths long, say five or more.

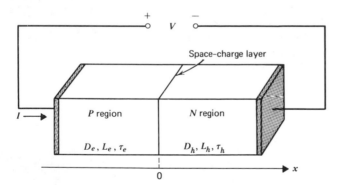

Figure 1-14 PN junction.

Because injection is low-level, the diffusion equation applies, and the hole density p_n in the N region is

$$p_n = C_1 e^{-x/L_h} + C_2 e^{x/L_h} + p_{no} \tag{1-45}$$

The constant C_2 in (1-45) must be zero. Otherwise, the equation gives the absurdity of a hole density p_n increasing exponentially with distance x at large distances (many diffusion lengths) from the junction. The boundary condition of (1-43) requires p_n to be p_{no} times $\exp(qV/kT)$ at $x = 0$, understood here to be in the N region at the edge of the layer. From this condition C_1 is easily found, and (1-45) becomes

$$p_n = p_{no}(e^{qV/kT} - 1)e^{-x/L_h} + p_{no} \qquad (1\text{-}46)$$

The hole density is an exponential function of x, approaching the equilibrium value at large distances from the junction.

A similar procedure can be used to find the free-electron density n_p in the P region. The result is

$$n_p = n_{po}(e^{qV/kT} - 1)e^{x/L_e} + n_{po} \qquad (1\text{-}47)$$

The electron density n_p is $n_{po} \exp(qV/kT)$ at the edge of the layer $(x = 0)$ in the P region, as required by the boundary condition of (1-44), and n_p approaches n_{po} for large values of negative x.

In the N region the hole current is $-qAD_h \, dp_n/dx$, because the minority-carrier drift current is negligible when injection is low-level. Using (1-46), the hole current in the positive x direction is found to be

$$I_h = \frac{qAD_h p_{no}}{L_h}(e^{qV/kT} - 1)e^{-x/L_h} \qquad (1\text{-}48)$$

In the P region the electron current is $qAD_e \, dn_p/dx$. Using (1-47), this current in the positive x direction is found to be

$$I_e = \frac{qAD_e n_{po}}{L_e}(e^{qV/kT} - 1)e^{x/L_e} \qquad (1\text{-}49)$$

For an applied voltage V, (1-48) and (1-49) give the *minority-carrier* currents as functions of distance. The total current I as a function of V is desired. This current is, of course, independent of x.

For a forward bias the hole density at each point of the space-charge layer is greater than its equilibrium value, and the net recombination rate is positive. Consequently, some of the holes that flow into the space-charge layer from the P region are lost by recombination. However, because the layer is so thin, the net recombination of holes has negligible effect on the hole current, and a similar statement applies to the electron flow. Therefore, the hole and electron currents in the layer are independent of distance x.

For a reverse bias the hole and electron densities at each point of the layer are less than their equilibrium values, and the *net generation* rate is positive. Thermally generated holes and free electrons are swept out of the layer, in opposite directions, by the intense field. Both contribute positively to the reverse current. This generation current is typically of the order of a nanoampere $(1 \text{ nA} = 10^{-9} \text{ A})$, and may or may not be significant. For convenience we assume the following postulate.

Postulate 9. Recombination and thermal generation of charge carriers in the space-charge layer of a PN junction have negligible effect on the hole and electron currents.

A direct deduction from Postulate 9 is that the electron and hole currents in a space-charge layer are independent of distance, assuming that the only generation of carriers in the layer is solely thermal. Accordingly, the hole current at the metallurgical junction is given by (1-48) with $x = 0$, and the electron current is given by (1-49) with $x = 0$. The addition of I_h and I_e at the metallurgical junction gives the total current I, and the result is

$$I = I_s(e^{qV/kT} - 1) \tag{1-50}$$

with

$$I_s = qA\left(\frac{D_h p_{no}}{L_h} + \frac{D_e n_{po}}{L_e}\right) \tag{1-51}$$

Equation 1-50 is the *diode equation*, and I_s is the *saturation current* of the *PN* junction, which is the current when $\exp(qV/kT)$ is very small compared with unity. The diffusion constants and diffusion lengths in (1-51) are those of the *minority* carriers.

It should be carefully noted that *the total junction current was found by determining the minority-carrier currents and adding these at the edges of the space-charge layer.* The minority-carrier currents, which are diffusion currents when injection is low-level, were easily obtained by utilizing the general solution of the diffusion equation and the boundary conditions. This procedure will be employed again in the next chapter, to find the currents of a bipolar junction transistor.

Example

Given a silicon PN junction with $N_A = 9 \times 10^{22}$ and $N_D = 2 \times 10^{22}$ SI units. In the P region the hole and electron mobilities are 0.035 and 0.05 m²/V-s, respectively. The N-region hole and electron mobilities are 0.03 and 0.09, respectively. Assume an excess-carrier lifetime of 1 μs in both regions. The cross-sectional area A is 10^{-6} m². At 300 K find the current I as a function of V. The diode is shown in Fig. 1-14.

Solution

From the Einstein relation the minority-carrier diffusion constants are found to be $D_h = 0.000776$ and $D_e = 0.00129$ m²/s. From these and the lifetime the minority-carrier diffusion lengths are calculated to be $L_h = 28$ μm and $L_e = 36$ μm. At 300 K, n_i is 1.48×10^{16}. The equilibrium minority-carrier densities are easily determined to be $p_{no} = 1.10 \times 10^{10}$ and $n_{po} = 2.43 \times 10^9$. Substitution into (1-51) gives a saturation current of 0.0627 pA. The diode equation becomes

$$I = 6.27 \times 10^{-14}(e^{38.7V} - 1)\text{A} \tag{1-52}$$

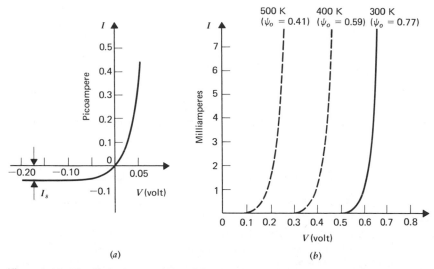

Figure 1-15   The diode characteristic. (a) Small voltages. (b) The forward characteristic.

Figure 1-15 shows two plots of this equation, each to a different scale. The dashed curves of Fig. 1-15b represent the characteristic at elevated temperatures.

For reverse-bias voltages greater in magnitude than about 0.1 V the current saturates at the saturation current I_s. This is clearly indicated in Fig. 1-15a. Because the minority-carrier densities in the expression for I_s are each proportional to $n_i{}^2$, the saturation current is proportional to $n_i{}^2$. For this reason I_s *in germanium devices at room temperature is typically about a million times greater than* I_s *in silicon devices, and the saturation currents of both are very temperature sensitive.*

At room temperature the current of a silicon junction is substantial only for forward-bias voltages greater than about 0.6 V, as indicated by Fig. 1-15b. Because of the much larger saturation current, the break voltage of a germanium device is considerably smaller, being usually about 0.2 V at room temperature. As indicated by Fig. 1-15b, the characteristic curve shifts to the left approximately 2 mV per degree Celsius as T increases. The built-in junction voltage ψ_o decreases, and the values are given on the curves. A rise in temperature causes an increase in I_s and a decrease in the exponential term of the diode equation, but the change in I_s is dominant.

The PN junction conducts much more readily in the forward direction (I positive) than in the reverse direction (I negative). A junction with two external leads can be used as a valve that allows appreciable current in only one direction. Such a device is called a *diode*, commonly represented by the

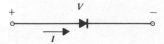

Figure 1-16   Diode symbol.

symbol of Fig. 1-16. Diodes have many uses, one of the most important being the conversion of alternating voltage into direct voltage, a process known as *rectification*.

1-8. THE FORWARD-BIASED DIODE

To illustrate the processes that occur in a forward-biased PN junction, the diode of the example of Sec. 1-7 will be analyzed with a voltage V equal to 0.65 V corresponding to a forward current of 5.27 mA. The minority-carrier densities as functions of x are easily found from (1-46) and (1-47). The majority-carrier densities are then found from the requirement that the excess-carrier densities must be equal, in accordance with (1-9). These carrier densities are plotted in Fig. 1-17. In order to use a linear scale it was necessary to break the ordinate at two points.

From Fig. 1-17 we note that injection is low-level, because the minority-carrier density is everywhere less than 5% of the majority-carrier density. A slight increase in the applied voltage would, however, produce high-level injection, making the use of the diffusion equation invalid. The length l of the space-charge layer is approximately 0.1 μm and shows in Fig. 1-17 only as the width of the line that constitutes the ordinate. The injection of holes into the N region from the P region is considerably greater than the electron injection into the P region from the N region. This is because the P region is more heavily doped (see P1–18).

The injection of holes into the N region and electrons into the P region is a consequence of the reduced electric field in the space-charge layer. The field with a forward bias is less intense than the built-in field. The reduced field has no effect on the minority-carrier flow across the junction, but makes passage of majority carriers considerably easier. In equilibrium ($V = 0$) only the most energetic of the majority carriers are able to traverse the layer against the retarding force of the built-in field. The forward bias reduces the junction voltage ($\psi_o - V$), allowing many majority carriers with lesser kinetic energies to surmount the lowered barrier of the junction voltage. Those crossing the barrier become minority carriers that diffuse into the neutral regions. As they diffuse away from the space-charge layer, the excess carriers are lost by recombination. Because the minority-carrier densities at the edges of the layer are exponential functions of V, slight changes in the forward-bias voltage produce large changes in the densities.

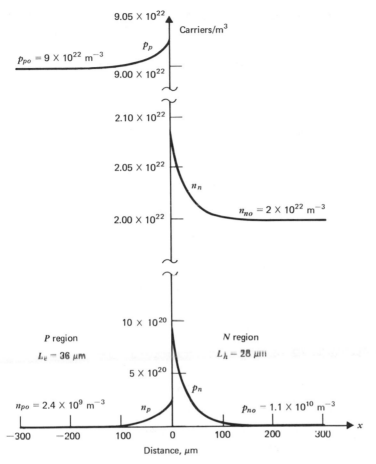

Figure 1-17   Carrier densities of forward-biased diode in m^{-3}.

The minority-carrier currents are positive in the x direction and each is proportional to the magnitude of the slope of the curve of the minority-carrier density versus x. Hence, from Fig. 1-17 it is rather evident that the minority-carrier currents become insignificant at distances greater than three or four diffusion lengths from the junction. The majority-carrier currents consist of both diffusion and drift components.

From our knowledge that the minority-carrier drift currents are negligible, we can find the hole current in the N region from the expression for p_n. The hole current, multiplied by $-D_e/D_h$, gives the electron diffusion current, and the total current is obtained from the diode equation. As this is the sum of the hole current, the electron diffusion current, and the electron drift current, the latter component is easily determined (see P1-22). A similar procedure can be used to find the components of the currents in the P region (see P1-24). The results are shown in Fig. 1-18. The majority-current drift and diffusion components are plotted as dashed lines, and the distance scale is identical with that of Fig. 1-17. It should be noted that *the current in each region, at distances greater than three or four diffusion lengths away from the junction, is almost entirely a majority-carrier drift current.*

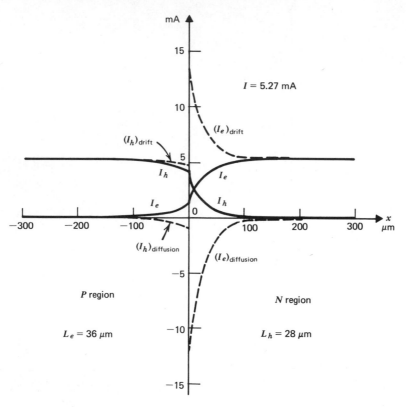

Figure 1-18   Currents of forward-biased diode.

For a given applied voltage V the forward current of an actual diode is less than that predicted by the diode equation. This is due to the IR drops in the neutral regions and at the metal-semiconductor contacts. Because of the ohmic voltage drops, the voltage applied to the junction is less than the terminal voltage, as indicated in Fig. 1-19. At the higher currents injection becomes high level. Although this tends to increase the current, the IR drops that tend to decrease the current are usually more significant.

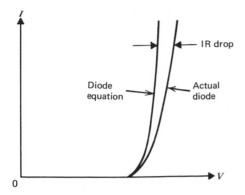

Figure 1-19   Actual diode characteristic compared with plot of diode equation.

1-9. THE REVERSE-BIASED DIODE

Diodes are often biased with a negative applied voltage V. Assuming the reverse bias is appreciable, with the magnitude of V greater than 0.1 V, the exponential of qV/kT is very much smaller than unity. Consequently, in comparison with the equilibrium values the minority-carrier densities at the edges of the space-charge layers are essentially zero in accordance with the boundary conditions of (1-43) and (1-44). As distance from the layer increases, the carrier densities approach their equilibrium values at rates depending on the diffusion lengths. Plots can be made from (1-46) and (1-47). Assuming the reverse bias is appreciable, the sketches are independent of V, with the minority-carrier densities rising from zero at the junction to their equilibrium values a few diffusion lengths away.

We can also determine and sketch the current components as functions of distance, and these are also independent of V. The procedure is identical to that of the preceding section, with $I = -I_s$. For the diode examined there the saturation current I_s is 0.06 pA. Thus the drift and diffusion components are only fractions of a picoampere. At room temperature such currents are normally several orders of magnitude less than the actual diode current of perhaps a nanoampere or more. Thus the theoretical results are invalid for reasons now to be examined.

THERMAL GENERATION CURRENT

In the development of the diode equation we assumed that recombination and thermal generation of charge carriers in the space-charge layer have negligible effects. This was stated as Postulate 9 of Sec. 1-7. When the junction is reverse biased, the carrier densities in the layer are less than the equilibrium values and there is a positive net thermal generation rate producing free electrons and holes. These carriers are pulled in opposite directions by the intense field, contributing to the current. In fact, in silicon at room temperature this thermal generation current is dominant.

As long as the thermal generation current is appreciable to or greater than I_s, the diode will not saturate as predicted by the ideal diode equation. This current rises with increased reverse bias because the volume of the space-charge layer becomes greater. It can be shown that the thermal generation rate in the layer is approximately n_i/τ, with τ denoting the effective lifetime in the layer. The value of I_s is proportional to n_i^2. Consequently, at low temperatures the thermal generation current dominates, and at high temperatures it is negligible.

For silicon junctions the greater part of the reverse current is the thermal generation current up to temperatures substantially above room temperature. In Fig. 1-20 is shown the reverse characteristic of a silicon diode. As indicated,

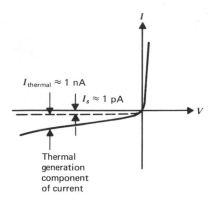

Figure 1-20 Reverse characteristic of silicon diode at 300 K.

a diode with a saturation current of a picoampere or less is likely to have a thermal generation current of about a nanoampere, which is a thousand times I_s. In germanium devices, which have saturation currents of the order of microamperes, the thermal generation current is important only at temperatures well below room temperature. Sometimes of considerable significance is the surface leakage current across the reverse-biased junction. This current is a function of the voltage, and it is minimized by proper surface treatment. Still, in many cases it is larger than the thermal generation current. Of course when a diode is forward biased, the diode equation applies *with* I_s *denoting the theoretical saturation current*; the effects of surface leakage and recombination and generation within the space-charge layer are usually negligible provided the current is not exceedingly small.

BREAKDOWN VOLTAGE

There is yet another effect on the reverse current. If the voltage exceeds the *reverse breakdown voltage*, a large current can flow. The reverse breakdown characteristic is shown in Fig. 1-21.

Two distinct mechanisms can cause breakdown. One of these is called *avalanche breakdown.* A high junction voltage produces an intense electric field in the space-charge layer, and the minority carriers that enter the layer from the neutral regions are strongly accelerated between collisions. During a collision a carrier may have sufficient kinetic energy to dislodge an electron from a covalent bond, thus generating a free electron and a hole. The intense field sweeps the hole in one direction and the electron in the other. These secondary carriers can, before leaving the layer, produce additional carriers in the same manner. For a sufficiently intense field, in the range of 2×10^7 V/m, the reverse current is multiplied by this method of carrier generation within the layer, and the multiplication can become quite large.

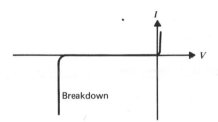

Figure 1-21   Diode characteristic including breakdown.

The second mechanism is called *Zener breakdown.* For extremely intense fields, over 5×10^7 V/m, some electrons in the covalent bonds may simply be pulled out of their bonds by the force of the field. Hence there is carrier generation in the layer as before, but this process does not involve collisions. Neither of the two breakdown mechanisms is harmful to a junction. When taken out of breakdown, a diode behaves normally.

Most PN junctions have breakdown caused by the avalanche-multiplication process. The breakdown voltage is very stable and makes a useful reference voltage in electronic circuitry. Special devices for this purpose are often called *Zener diodes,* a misnomer when breakdown results from the avalanche effect. They are made of silicon, which has a sharper breakdown than germanium. The avalanche voltage is usually from 8 to 1000 or more volts. Reducing the dopant density in the more lightly doped neutral region increases the breakdown voltage (see P1-26).

Zener breakdown occurs when both neutral regions are heavily doped, which gives a very thin space-charge layer. The layer is so thin that the carriers passing through it do not have enough collisions to produce significant numbers of secondary carriers. Because the built-in field is so intense, breakdown occurs at about 5 V or less. For breakdown between 5 and 8 V both avalanche and Zener effects may be significant.

Zener diodes are used in many networks, both discrete and integrated, to provide a fixed voltage drop along with a low-resistance path for alternating currents. The input and output terminals of many integrated circuits, especially those with field-effect devices, are often protected from excessive voltage by means of Zener diodes. They are commonly employed in voltage regulators to establish stable reference voltages. These and other applications, along with further discussions of important characteristics of these devices, are presented in later chapters.

1-10. JUNCTION CAPACITANCE

SPACE-CHARGE CAPACITANCE

There are charges stored in the space-charge layer and in the neutral regions of a PN junction. Furthermore, these charge stores are functions of the applied voltage V. If V changes with time, the terminal current will have components that supply the *changes* in these stored charges, in addition to the components considered in our development of the diode equation. We might view a PN junction as a capacitor that has considerable leakage current in the forward direction. In this section the stored charges will be expressed in terms of the voltage V, and the junction capacitance will be determined.

Let us first direct our attention to the charge Q of the space-charge layer, with Q defined as the *negative* charge of the unneutralized acceptor ions, as shown in Fig. 1-22a. This definition is convenient and it conforms to the usual

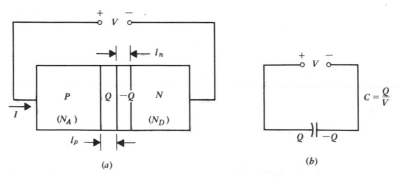

Figure 1-22 (a) Space-charge-layer charge Q (negative). (b) Capacitor charge Q.

convention with regard to the charge and voltage of a capacitor, as indicated in Fig. 1-22b. Note that an increase in V gives a *positive* increase in Q, with the junction charge Q becoming less negative because of the movement of majority carriers into the extremities of the layer so as to reduce its length l.

The charge Q equals $- qN_A Al_p$, as indicated by Fig. 1-11 of Sec. 1-5. Let Q_o denote the charge when $V = 0$. Then $Q - Q_o$ denotes the increase in the stored charge when the voltage increases from 0 to V. With l_p expressed in terms of V with the aid of (1-34) and (1-35), this increase is easily found to be

$$Q - Q_o = a(\sqrt{\psi_o} - \sqrt{\psi_o - V}) \qquad (1\text{-}53)$$

with

$$a = A\left(\frac{2q\varepsilon N_A N_D}{N_A + N_D}\right)^{1/2} \qquad (1\text{-}54)$$

A plot of $Q - Q_o$ versus V is shown in Fig. 1-23 for the diode of the example of Sec. 1-7. The *space-charge-layer capacitance* is the ratio of the charge $Q - Q_o$ supplied by the source to the applied voltage V, or $(Q - Q_o)/V$. This ratio is positive for all values of V. Because the plot of $Q - Q_o$ versus V is not a straight line, as shown by the curve of Fig. 1-23, this capacitance is said to be nonlinear.

If the junction voltage of Fig. 1-22 is changing with time t, both the current I and the stored charge Q change with time. Let $i_c(t)$ denote the component of the junction current I that increases the stored charge $Q(t)$ of the layer. Then

$$i_c(t) = \frac{dQ}{dt} = \left(\frac{dQ}{dV}\right)\frac{dV}{dt} \qquad (1\text{-}55)$$

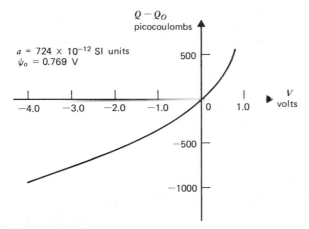

Figure 1-23   Charge-supplied-to-layer versus V.

The ratio dQ/dV, which is the slope of the curve of Fig. 1-23 at a specified voltage, is called the *incremental space-charge capacitance* C_t. It is also referred to as the *depletion-region capacitance*, or the *transition capacitance*. Thus

$$C_t = \frac{dQ}{dV} \qquad \text{and} \qquad i_c(t) = C_t \frac{dV}{dt} \qquad (1\text{-}56)$$

The current i_c is identified on the equivalent circuit of Fig. 1-24. In this diagram the actual diode is represented as an ideal device described by the diode equation in parallel with two capacitances. One of these is C_t and the other one is C_d, which accounts for charge stores in the neutral regions.

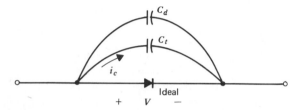

Figure 1-24   Diode and junction capacitances.

The incremental capacitance is easily determined as a function of V by differentiating (1-53) with respect to V. With the aid of (1-36) and (1-54) the result can be expressed in terms of the length l of the layer. We find that

$$C_t = 0.5a(\psi_o - V)^{-1/2} = \frac{\varepsilon A}{l} \qquad (1\text{-}57)$$

The expression $\varepsilon A/l$ is identical to that for the capacitance of two closely spaced parallel plates separated a distance l by a dielectric of permittivity ε. A small change in V produces a small change in Q. Because the change in Q occurs at the extremities of the layer, a distance l apart, the expression $\varepsilon A/l$ is certainly reasonable.

The space-charge capacitance of an abrupt junction is inversely proportional to the square root of the total junction voltage. It increases as V increases, being larger for a forward-biased junction than for a reverse-biased junction. This is quite evident from inspection of the curve of Fig. 1-23, as C_t is the slope of this curve. For the diode used to obtain the data of the curve, C_t is 1050 pF for a forward bias of 0.65 V and 110 pF for a reverse bias of 10 V. *Varactors* are PN junctions designed for use as variable capacitive reactors. The junction is reverse biased, and therefore the leakage current is small. Adjustment of the reverse-bias voltage provides control of the capacitance.

DIFFUSION CAPACITANCE

Charges are also stored in the neutral regions. The curves of Fig. 1-17 of Sec. 1-8 show the carrier densities as functions of distance for a forward-biased junction. In each region the excess holes and free electrons represent equal-and-opposite charges, and the stored charge is proportional to the exponential of qV/kT. The charge-to-voltage ratio is referred to as a *diffusion capacitance*. When a forward bias is applied suddenly to a PN junction by means of a switch, a large current flows momentarily, and this current supplies the charge stores of the neutral regions as well as the charge store of the space-charge layer.

Of particular importance is the *incremental diffusion capacitance dQ/dV* at a specified bias point, with Q denoting the charge store in the neutral regions. This is shown in Fig. 1-24 as C_d. The current through C_d is that component of the diode current that supplies changes in the charge stores of the neutral regions. It is proportional to the exponential of qV/kT. Diffusion capacitances are negligible for reverse-biased junctions. Unlike the positive and negative charges of common capacitors, which have a dielectric separating the charges, the holes and electrons associated with diffusion capacitance intermingle within the same region.

REFERENCES

1. R. B. Adler, A. C. Smith, and R. G. Longini, *Introduction to Semiconductor Physics*, Wiley, New York, 1964.
2. M. F. Uman, *Introduction to the Physics of Electronics*, Prentice Hall, Englewood Cliffs, N.J., 1974.
3. A. G. Grove, *Physics and Technology of Semiconductor Devices*, Wiley, New York, 1967.
4. L. P. Hunter, *Introduction to Semiconductor Phenomena and Devices*, Addison-Wesley, Reading, Mass., 1966.

PROBLEMS

Section 1-1

1-1. At 450 K find p_o and n_o in a silicon semiconductor with $N_a = 5 \times 10^{23}$ and $N_d = 3 \times 10^{23}$ m^{-3}.

1-2. The expression $(1/n_i^2)(dn_i^2/dT)$ denotes the fractional increase in n_i^2 per degree Celsius rise in temperature. Using (1-2) deduce that this is $T^{-2}(B + 3T)$, with $B = 14,000$ for silicon and 9100 for germanium. Show that the percent increases in n_i^2 per degree Celsius for silicon and germanium at 300 K are 16.6% and 11.1%, respectively.

1-3. Using (1-1), (1-4), and (1-5), deduce for a P-type semiconductor with $p_o = a^2 n_o$ that $n_i = aN_A/(a^2 - 1)$. Then show that the upper limit of the extrinsic temperature range, defined as that temperature at which $n_o = 0.1\, p_o$, is the temperature at which $n_i = 0.35 N_A$. Also, for both silicon and germanium find in degrees Celsius the upper temperature limit of the extrinsic temperature range, assuming 4×10^{22} donors/m^3 and 3×10^{22} acceptors/m^3. Use Fig. 1-1. (*Ans*: 312, 152).

1-4. A silicon semiconductor with $N_A = 10^{22}$ m^{-3} is in the intrinsic temperature range with $p_o = 1.1 n_o$. Calculate the temperature T in degrees Celsius. Refer to Pl-3 and use Fig. 1-1.

Section 1-2

1-5. A silicon semiconductor with $N_A = 10^{21}$ has $n_i = 10^{16}$ SI units. Carrier injection gives an excess free-electron density in the region of positive x that is $10^{23} \exp(-20000x)$. Find the hole density p as a function of x. Calculate the ratio p/n at $x = 0$ and state whether injection is low-level or high-level.

1-6. The electron density $n = n_o[1 - \exp(-10^4 x)]$ for values of x from 0 to 0.0005 in the P region of a silicon device. If $N_A = 10^{21}$ and $n_i = 10^{16}$, calculate the excess-carrier density at $x = 0$, and plot p and n versus x in a manner similar to that of Fig. 1-4. Units are SI. (Ans: -10^{11}).

Section 1-3

1-7. (a) A thermistor made of pure silicon has a uniform cross-sectional area A of 10^{-4} and a length l of 0.003. At 300 K, $\mu_h = 0.048$ and $\mu_e = 0.135$, and the mobility ratio $\mu(T)/\mu(300\text{ K})$ equals the ratio $(300/T)^{2.6}$. Calculate the resistance $(l/\sigma A)$ at 300 K and also at 350 K.
(b) Repeat (a) for a P-type semiconductor with $N_A = 10^{21}$, assuming the mobilities are the same as for pure silicon. Tabulate results of (a) and (b). Units are SI.

1-8. (a) An N-type semiconductor with $N_D = 10^{21}$ has a cross-sectional area $A = 10^{-7}$ and carries a current of 1 mA. Calculate the drift velocity $(\mu_e E_x)$ of the free electrons and find the time required for the electrons to drift through a distance of 1 cm.
(b) Copper has a free-electron density $n \approx 10^{29}$. If a copper conductor also has $A = 10^{-7}$ and $I = 1$ mA, find the time in hours required for an electron to drift through a distance of 1 cm. Unstated units are SI.

1-9. Suppose the steady ionizing radiation incident on the end of the silicon bar of Fig. 1-25 injects excess carriers such that

$$n = 10^{21} e^{-50000x} + 10^9 \quad \text{m}^{-3}$$

(a) As functions of x calculate the electron current I_e and the hole diffusion current, in the positive x direction. Give currents in mA.
(b) The total current I is zero, because there is no closed electric circuit. Deduce that the hole drift current is $16e^{-50000x}$ mA.

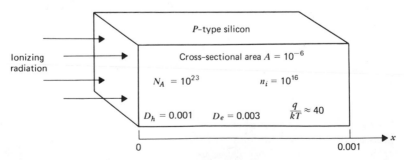

Figure 1-25   Optical injection into silicon bar (SI units).

1-10. This is a continuation of Pl-9. (a) From the hole drift current calculate the electric field E_x as a function of x.
(b) From Gauss's law in electromagnetic theory the net charge density ρ is $\varepsilon\,dE_x/dx$, with ε denoting the permittivity. For silicon $\varepsilon \approx 10^{-10}$ F/m. Calculate ρ as a function of x.
(c) Calculate the ratio ρ/(excess hole charge density). Note that the net charge density ρ, which produces the electric field and its associated hole drift current, is very small compared with the excess charge density, and quasi-neutrality applies. (*Ans*: -7.81×10^{-7}).

Section 1-4
1-11. In Fig. 1-25 suppose the radiation breaks covalent bonds at the end of the bar, producing an excess-carrier density of 10^{20} at $x = 0$. The excess-carrier lifetime is $1/30$ μs. Because the other end of the bar is many diffusion lengths away from the disturbance, the carrier densities approach their equilibrium values for large x. As functions of x find the free-electron density n, the electron current I_e, and the hole drift and diffusion currents in the positive x direction. [*Ans*: hole drift $I = 3.2\exp(-10^5 x)$ mA].
1-12. A uniformly doped semiconductor has $N_D - 10^{??}$, $n_i = 10^{16}$, $D_h = 0.001$ and $\tau = 10$ μs. Assume the hole densities at $x = 0$ and at $x = 5 \times 10^{-5}$ are known to be 10^{20} and 10^{10}, respectively. Find p and n as functions of x. Units are SI.

Section 1-5
1-13. From Gauss's law derive the electric field of (1-31) and (1-32). Sketch and dimension the field for a silicon junction ($\varepsilon - 10^{-10}$) having $N_A - 2 \times 10^{14}$, $N_D = 10^{22}$, $l = 1.5 \times 10^{-7}$, and $\psi_o = 0.7$. The junction is forward biased. Also, calculate the applied voltage V. Units are SI.
1-14. For the junction of Pl-13 what voltage V will give a total length l of the layer that is increased tenfold to 15×10^{-7}? For this voltage what is the maximum field E_o? (*Ans*: 16 MV/m).
1-15. Calculate the total length l of the space-charge layer and the maximum field E_o of a silicon junction ($\varepsilon = 10^{-10}$) under equilibrium conditions for each of the following cases:
(a) lightly doped, $N_A = N_D = 10^{19}$; $\psi_o = 0.35$
(b) heavily doped, $N_A = N_D = 10^{23}$; $\psi_o = 0.81$
Compare results. Units are SI.
1-16. A *graded* PN junction is one in which the change in impurity concentration from P-type to N-type is gradual rather than abrupt. Let us assume that this change is linear and that the depletion approximation applies. For such conditions the requirement for equal-and-opposite charge stores in the layer dictates that the layer must extend equal distances into each region. Assume that the charge density ρ resulting from the unneutralized ions is $\rho = ax$ from $x = -\frac{1}{2}l$ in the P region to $x = \frac{1}{2}l$ in the N region.
(a) Beginning with Gauss's law and following the procedure of Sec. 1-5, determine the maximum field E_o and the length l of the layer in terms of the positive constant a, the permittivity ε, and the junction voltage $\psi_o - V$.
(b) Calculate E_o and l for a silicon junction ($\varepsilon = 10^{-10}$) with $a = 12 \times 10^8$, $\psi_o = 0.7$, and $V = -26.3$. Units are SI.

Section 1-6

1-17. Show that $d\psi_o/dT = \psi_o/T - (kT/qn_i{}^2)d(n_i{}^2)/dT$. Use the results of P1-2 and calculate $d\psi_o/dT$ in mV/K for a silicon junction at 300 K with $\psi_o = 0.77$ V.

1-18. Show that the excess-carrier densities on the two sides of a space-charge layer are related by

$$n_p(0) - n_{po} = \frac{n_{po}}{p_{no}} [p_n(0) - p_{no}]$$

From this result deduce that the injection of excess carriers is always greater into the more lightly doped region.

1-19. The neutral N region of Fig. 1-26 extends from $x = 0$ to $x = 10^{-5}$. The left PN junction is forward biased, with $V_1 = 0.6$ V. The right junction is reverse biased, with $V_2 = -10$ V. Assume $q/kT = 40$, $A = 10^{-7}$, $p_{no} = 10^{10}$, $D_h = 0.001$, and $L_h = 10^{-4}$.
(a) From the boundary conditions at the junctions evaluate C_1 and C_2 in (1-27) and express p_n as a function of x. Note that p_{no} in (1-27) is negligible.
(b) Calculate the hole current in the positive x direction at $x = 0$ and also at $x = 10^{-5}$. Units are SI. (*Ans*: 0.425, 0.423 mA).

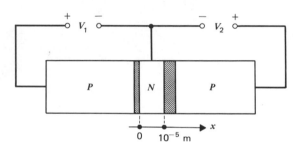

Figure 1-26   Two-junction device.

Section 1-7

1-20, Using (1-2), eliminate $n_i{}^2$ in (1-40) for ψ_o. Then show that

$$\frac{d\psi_o}{dT} = \frac{k}{q}(\ln N_A N_D - 3 \ln T - 107)$$

For the silicon diode of the example of Sec. 1-7 calculate $d\psi_o/dT$ at 300 K, 400 K, and 500 K. Use the results to verify the changes in ψ_o indicated on Fig. 1-15b. ($k/q = 8.6 \times 10^{-5}$ SI units.)

1-21. At 290 K ($q/kT = 40$) the saturation currents of a germanium and a silicon diode are 1 μA and 0.5 pA, respectively. If the two diodes are connected in series and supplied with a forward current of 1 mA, what are their respective junction voltages?

Section 1-8

1-22. For the diode of the example of Sec. 1-7 the current I is found from (1-52) to be 5.27 mA when V is 0.65 V.
(a) In SI units with the x axis defined in Fig. 1-17, show that

$$p_n = 9.25 \times 10^{20} \exp(-35700x) + 11 \times 10^9$$

(b) Find n_n as a function of x.

(c) As functions of x in the N region, determine the following currents in mA in the positive x direction: hole current, electron diffusion current, electron current, electron drift current [Ans: $5.27 + 8.20 \exp(-35700x)$].

1-23. This is a continuation of P1-22. From the electron drift current calculate the electric field E_x as a function of x in the neutral N region and find the maximum value of this field. Assume $n_n = n_{no}$.

1-24. For the diode of the example of Sec. 1-7 the current I is found from (1-52) to be 5.27 mA when $V = 0.65$.

(a) In SI units with the x axis defined in Fig. 1-17, show that

$$n_p = 2.04 \times 10^{20} \exp(27800x) + 2.43 \times 10^9$$

(b) Find p_p as a function of x.

(c) As functions of x in the P region, determine the following currents in the positive x direction: electron current, hole diffusion current, hole current, hole drift current. (Ans: $5.27 - 0.35 \exp 27800x$ mA).

Section 1-9

1-25. A Zener diode has $N_A = 10^{23}$ and $N_D = 10^{21}$. Assume that avalanche breakdown occurs when the field reaches 2×10^7 V/m. Calculate the breakdown voltage. Units are SI and ε is 10^{-10}. (Ans: 126).

1-26. A Zener diode is to be designed to breakdown at 1000 V. Assume that avalanche breakdown occurs at a field intensity of 2×10^7 V/m. If $N_A = 10^{23}$ m^{-3}, what dopant density should be used for N_D? ($\varepsilon = 10^{-10}$).

1-27. A silicon diode ($\varepsilon = 10^{-10}$) has $N_A = N_D = 10^{24}$ m^{-3}. Calculate the equilibrium values of E_o and l and state whether breakdown would most probably be the avalanche or Zener mechanism. Justify your selection. $T = 300$ K.

1-28. For the diode of the example of Sec. 1-7 with a reverse bias of 1 V, in the N region determine

(a) p_n and n_n as functions of x, with $x = 0$ at the junction.

(b) the hole and electron currents in picoamperes at $x = 0$, in the positive x direction. Assume thermal-generation and surface-leakage currents are negligible.

(c) the electron drift and diffusion currents in picoamperes at $x = 0$, in the positive x direction (same assumption). (Ans: 0.146, -0.160).

Section 1-10

1-29. A diode has $\psi_o = 0.70$ and $a = 10^{-10}$ SI units with a defined by (1-54). For bias voltages of 0.54 and -24.3 calculate the incremental space-charge capacitance in picofarads. If the applied voltage is $-24.3 + 0.1 \cos 10^6 t$, find the time-changing component of the diode current due to the junction capacitance. This is the current i_c in Fig. 1-24. (Ans: $-\sin 10^6 t \ \mu$A).

1-30. For the graded junction of P1-16 the length l of the layer is $(12\varepsilon/a)^{1/3}(\psi_o - V)^{1/3}$. Using the data of P1-16(b) calculate the incremental space-charge capacitance. The cross-sectional area A is 10^{-7} m^2.

Chapter Two
The Junction Transistor

The junction transistor consists of a single crystal of silicon (or possibly germanium) with two PN junctions that are separated by a very thin neutral section. The three neutral regions may be arranged PNP or NPN. This chapter treats the NPN device, but the analysis of PNP transistors is analogous, and the results are stated for both cases.

First we investigate mathematically the behavior of a transistor, finding the terminal currents as functions of the applied voltages, following a procedure quite similar to that used in the preceding chapter. We then use the results to aid our understanding of the physical processes that occur within the device. The dc equations derived are used in later chapters to examine the behavior of digital logic gates and to develop the important hybrid-π linear model of a transistor.

To simplify the analysis we employ a device with the geometry of Fig. 2-1. The N region on the left is the *emitter*, the thin P region is the *base*, and the N region on the right is the *collector*. The usual linear-amplifier bias arrangement is illustrated, consisting of a forward bias on the emitter-base junction, or *emitter junction*, and a reverse bias on the collector-base junction, or *collector junction*.

The forward bias produces electron injection from the emitter into the base, where the injected electrons diffuse away from the emitter junction. Some are lost by recombination with holes, similar to the process that occurs in the P region of a diode. Most of the injected electrons, however, diffuse into the space-charge layer of the reverse-biased junction, and the intense field sweeps them into the collector.

Hole injection from the base into the emitter is rather negligible because of the relatively light dopant density of the base. Consequently, most of the terminal base current is simply the flow of holes into the base to replace those lost by recombination with electrons, and this current is small compared with the electron current of the collector. The current of the base terminal can be used to control the considerably larger collector current. The base terminal may be thought of as the "control handle" of a valve, and either the base current or the emitter-base voltage may be regarded as the control variable.

The main current in each region is an electron current. Because this current is controlled by hole flow into the base, transistor action depends on both types of charge carriers. Consequently, the transistor is called a *bipolar* junction transistor (BJT). An example of a *unipolar* device is a field-effect transistor, which has a current of either holes or free electrons, but not both.

We utilize simplifying restrictions and assumptions similar to those used in our analysis of the PN junction in the preceding chapter. The cross-sectional area A of the transistor of Fig. 2-1 is constant, and the carrier densities are functions only of x. Each region is uniformly doped. The junctions are abrupt, and the voltages applied to the junctions are approximately equal to the terminal voltages. The analysis is restricted to low-level injection. In the extremely thin

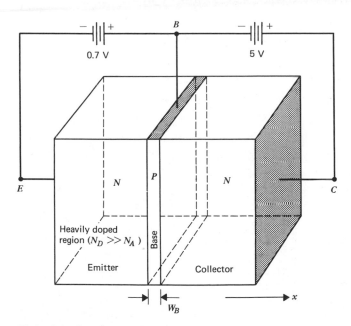

Figure 2-1   Junction transistor with normal bias.

space-charge layers the net generation of charge carriers is assumed to be inconsequential, including both thermal generation and avalanche multiplication. The emitter and collector regions are many diffusion lengths long.

There are two restrictions in addition to those used in the diode analysis. One of these is the assumption that the emitter is much more heavily doped than the base. The other is the assumption that the distance W_B between the metallurgical junctions is small compared with the base-region minority-carrier diffusion length. In order to be precise let us specify that

$$n_{eo} > 100p_{bo} \qquad (2\text{-}1)$$

$$W_B < 0.3L_b \qquad (2\text{-}2)$$

with the subscripts e and b denoting the emitter and base regions, respectively, and the second subscript o used to indicate that the carrier densities are the equilibrium values. This notation will be used consistently, along with the subscript c to denote the collector region. Unless specifically stated otherwise, diffusion lengths and diffusion constants are those of the minority carriers; for example, D_e is the *hole* diffusion constant in the N-type emitter and D_c is that of the collector.

A one-dimensional idealized transistor has been defined. We now proceed to find the carrier-density distributions and the minority-carrier currents in each neutral region of this *ideal transistor*. Then we express the terminal currents as functions of the terminal voltages, and carefully analyze the results.

2-1. THE MINORITY-CARRIER CURRENTS

In Fig. 2-2 is shown a view of the transistor with the terminal voltages and currents defined. The relative dimensions of the neutral base and the space-charge layers are greatly exaggerated in the x direction. V_{BE} is the voltage drop

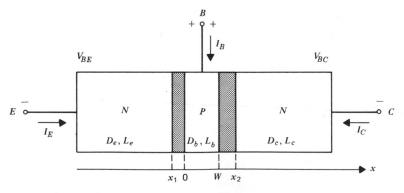

Figure 2-2　Terminal currents and voltages of transistor.

from the base terminal to the emitter, and a positive V_{BE} is a forward bias. V_{BC} is the voltage drop from the base to the collector, and a positive V_{BC} is also a forward bias. The reference directions of the currents are into the device, in accordance with convention. The distance W_B between the metallurgical junctions is not shown, but W_B is a constant. On the contrary, the base width W is a function of V_{BE}, V_{BC}, and temperature T, because the penetrations of the space-charge layers into the P region depend on these variables. Of course, W is less than W_B.

The procedure for finding the hole currents in the N-type emitter and collector regions is identical to that employed in the analysis of the N region of the diode. In each region we begin with the hole distribution $p(x)$ given by the general solution (1-27) of the minority-carrier diffusion equation. Evaluation of the constants is accomplished by applying the boundary condition of (1-43), which applies at the edge of the space-charge layer, and by using our knowledge that $p(x)$ approaches its equilibrium value at large distances from the layer. From the expressions for $p(x)$ the hole diffusion currents are determined (see P2-1), and of course, the hole drift currents are negligible. The results are

$$\text{emitter } I_h(x) = -\frac{qAD_e p_{eo}}{L_e}(e^{qV_{BE}/kT} - 1)e^{(x-x_1)/L_e} \tag{2-3}$$

$$\text{collector } I_h(x) = \frac{qAD_c p_{co}}{L_c}(e^{qV_{BC}/kT} - 1)e^{-(x-x_2)/L_c} \tag{2-4}$$

These equations are similar to the expression (1-48) for the hole current in the N region of a diode.

The analysis of the base differs from that of the other regions only because of the different boundary conditions. The electron density in the base is

$$n_b(x) = C_1 e^{-x/L_b} + C_2 e^{x/L_b} + n_{bo} \tag{2-5}$$

To determine C_1 and C_2 in (2-5) we employ the boundary conditions at $x = 0$ and at $x = W$. These are

$$n_b(0) = n_{bo}e^{qV_{BE}/kT} \qquad \text{and} \qquad n_b(W) = n_{bo}e^{qV_{BC}/kT}$$

It is convenient to use hyperbolic functions. The hyperbolic sine and cosine functions are defined by the following expressions:

$$\sinh z = \frac{1}{2}(e^z - e^{-z}) = z + \frac{z^3}{3!} + \frac{z^5}{5!} + \cdots \tag{2-6}$$

$$\cosh z = \frac{1}{2}(e^z + e^{-z}) = 1 + \frac{z^2}{2!} + \frac{z^4}{4!} + \cdots \tag{2-7}$$

From these we note that $\sinh 0 = 0$, $\cosh 0 = 1$, the derivative of $\sinh z$ gives $\cosh z$, and the derivative of $\cosh z$ gives $\sinh z$. The functions $\tanh z$, $\coth z$, sech z, and csch z are defined in terms of the hyperbolic sine and cosine functions in a manner similar to the corresponding trigonometric functions. Finding C_1 and C_2 from the boundary conditions and replacing them in (2-5), we can with a little manipulation (see P2-2) express the result in the form:

$$n_b(x) = \frac{n_{bo}}{\sinh W/L_b}\left[\left(\sinh \frac{W-x}{L_b}\right)(e^{qV_{BE}/kT} - 1)\right.$$

$$\left. + \left(\sinh \frac{x}{L_b}\right)(e^{qV_{BC}/kT} - 1)\right] + n_{bo} \tag{2-8}$$

Inspection readily verifies that the boundary conditions are satisfied. The electron current is $qAD_b \, dn_b/dx$, and using (2-8) we obtain:

$$I_e(x) = \frac{-qAD_b n_{bo}}{L_b \sinh W/L_b}\left[\left(\cosh \frac{W-x}{L_b}\right)(e^{qV_{BE}/kT} - 1)\right.$$

$$\left. - \left(\cosh \frac{x}{L_b}\right)(e^{qV_{BC}/kT} - 1)\right] \tag{2-9}$$

The terminal currents of the transistor can be found from (2-9) and the expressions of (2-3) and (2-4) for the hole currents of the emitter and collector.

2-2. THE TERMINAL CURRENTS

In the emitter at the edge of the space-charge layer the hole current is given by (2-3) with $x = x_1$, and the electron current in the base at $x = 0$ is obtained from (2-9). The sum of these is the current across the emitter-base metallurgical junction, and this is I_E. In the collector at the edge of the space-charge layer the hole current is given by (2-4) with $x = x_2$, and the electron current in the base at $x = W$ is determined from (2-9). The sum of these is the current across the collector-base metallurgical junction, which is $-I_C$. The results can be expressed (see P2-5) in the forms

$$I_E \doteq -I_{ES}(e^{qV_{BE}/kT} - 1) + \alpha_R I_{CS}(e^{qV_{BC}/kT} - 1) \tag{2-10}$$

$$I_C = \alpha_F I_{ES}(e^{qV_{BE}/kT} - 1) - I_{CS}(e^{qV_{BC}/kT} - 1) \tag{2-11}$$

with I_{ES}, I_{CS}, α_F, and α_R defined by the relations

$$I_{ES} = \frac{qAD_b n_{bo}}{L_b} \coth \frac{W}{L_b} + \frac{qAD_e p_{eo}}{L_e} \tag{2-12}$$

$$I_{CS} = \frac{qAD_b n_{bo}}{L_b} \coth \frac{W}{L_b} + \frac{qAD_c p_{co}}{L_c} \tag{2-13}$$

$$\alpha_F I_{ES} = \alpha_R I_{CS} = \frac{qAD_b n_{bo}}{L_b \sinh W/L_b} \tag{2-14}$$

Equations 2-10 and 2-11 are the *Ebers-Moll equations*.[4] They relate the terminal currents and voltages. The quantities I_{ES}, I_{CS}, α_F, and α_R are functions of the base width W, and W depends, at least to a small extent, on the voltages V_{BE} and V_{BC}.

With a little manipulation the Ebers-Moll equations can be expressed as

$$I_E = -\alpha_R I_C - I_{EO}(e^{qV_{BE}/kT} - 1) \tag{2-15}$$

$$I_C = -\alpha_F I_E - I_{CO}(e^{qV_{BC}/kT} - 1) \tag{2-16}$$

with the saturation currents I_{EO} and I_{CO} defined by

$$I_{EO} = (1 - \alpha_F \alpha_R)I_{ES} \tag{2-17}$$

$$I_{CO} = (1 - \alpha_F \alpha_R)I_{CS} \tag{2-18}$$

Equation 2-15 is derived by multiplying (2-11) by α_R and adding the result to (2-10), and (2-16) is found by multiplying (2-10) by α_F and adding to (2-11). A circuit model based on this set of equations is shown in Fig. 2-3. The terms I_{EO} and I_{CO} are the saturation currents of the diodes in this model.

From (2-14), (2-17), and (2-18) it is clear that

$$\alpha_F I_{ES} = \alpha_R I_{CS} \tag{2-19}$$

$$\alpha_F I_{EO} = \alpha_R I_{CO} \tag{2-20}$$

These are the *reciprocity relations* and they apply to actual transistors. The static transistor models contain four parameters—α_F, α_R, I_{ES}, and I_{CS} (or I_{EO} and I_{CO}). Only three are independent, however. The parameters α_F and α_R are often called the forward and reverse short-circuit dc current gains, respectively (see P2-7). We refer to I_{ES} and I_{CS} as the respective emitter-junction and collector-junction short-circuit saturation currents, and I_{EO} and I_{CO} are the open-circuit saturation currents (see P2-8). These names are not especially

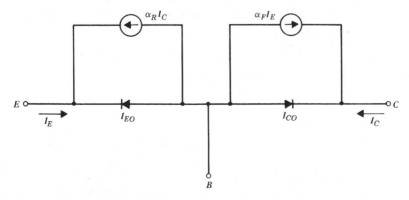

Figure 2-3   Ebers–Moll model for an NPN transistor.

appropriate because the parameters are functions of the base width W, which depends on the voltages. Measurements of the parameters should be made with the proper voltages applied, not with a terminal open or shorted.

The base width $W = W_B - l_{be} - l_{bc}$, with l_{be} denoting the length of the space-charge layer that is in the P-type base at the emitter junction and l_{bc} denoting the corresponding length at the collector junction. The lengths are given indirectly by (1-34) and (1-35) of Sec. 1-5. Because the emitter is much more heavily doped than the base, W can be expressed in terms of V_{BE}, V_{BC}, and the built-in junction voltages ψ_{oe} and ψ_{oc} as follows:

$$W = W_B - \sqrt{\frac{2\varepsilon}{qp_{bo}}} \left[\sqrt{\psi_{oe} - V_{BE}} + \sqrt{\frac{n_{co}}{n_{co} + p_{bo}}} (\psi_{oc} - V_{BC}) \right] \quad (2\text{-}21)$$

This shows the relationship between W and the applied voltages.

An explicit expression for α_F is obtained by dividing the right side of (2-14) by the expression for I_{ES} given by (2-12), with the result that

$$\alpha_F = \left[\cosh \frac{W}{L_b} + \frac{D_e p_{eo} L_b}{D_b n_{bo} L_e} \sinh \frac{W}{L_b} \right]^{-1} \quad (2\text{-}22)$$

By (2-2) the ratio $W/L_b < 0.3$. Therefore, the hyperbolic sine of W/L_b is approximately equal to its small argument by (2-6), and the hyperbolic cosine is slightly greater than unity, being approximately $1 + W^2/2L_b^2$ by (2-7). Because n_i^2 equals both $p_{eo}n_{eo}$ and $p_{bo}n_{bo}$, the ratio p_{eo}/n_{bo} in (2-22) can be replaced with p_{bo}/n_{eo}. Making these substitutions, we obtain

$$\alpha_F \approx \left[1 + \frac{W^2}{2L_b^2} + \frac{D_e p_{bo} W}{D_b n_{eo} L_e} \right]^{-1} \quad (2\text{-}23)$$

In (2-23) the quantity $W^2/2L_b^2$ must be less than 0.045 in accordance with (2-2) and our knowledge that W is less than W_B. Still smaller is the term with the ratio p_{bo}/n_{eo}, because this ratio is less than 0.01 by (2-1). We know that the reciprocal of $(1 + a)$ is approximately equal to $(1 - a)$ provided a is small compared with unity. It follows that (2-23) can be expressed as

$$\alpha_F \approx 1 - \frac{W^2}{2L_b^2} - \frac{D_e p_{bo} W}{D_b n_{eo} L_e} \quad (2\text{-}24)$$

Typically, α_F is between 0.95 and unity. The right side of (2-23) represents α_R provided D_e, n_{eo}, and L_e are replaced with D_c, n_{co}, and L_c, respectively. Because the collector is not always heavily doped compared with the base and also because of the nonuniform cross section of practical transistors, α_R may be considerably less than unity.

2-3. THE BASE CURRENT I_B

The base terminal of an NPN transistor is connected to a P-type region. Although current in the base wire consists of electron flow, within the semiconductor the current I_B of the terminal is a flow of holes into or out of the P region. For the dc case that we are considering, this flow must be such as to maintain constant the hole density at each point of the base. Holes flow into or out of the base not only through the base terminal but also through the two space-charge layers. They also enter or leave at each point of the base by the processes of thermal generation and recombination. For most bias arrangements the base current I_B of an NPN transistor is positive. By adding the Ebers-Moll equations (2-10) and (2-11) and eliminating the saturation currents and the alphas with the aid of (2-12), (2-13), and (2-14), we obtain the sum of $-I_E$ and $-I_C$, which is I_B. The result is

$$I_B = I_{BB} + I_{BE} + I_{BC} \qquad (2\text{-}25)$$

with

$$I_{BB} = \frac{qAD_b n_{bo}(\cosh W/L_b - 1)}{L_b \sinh W/L_b}(e^{qV_{BE}/kT} + e^{qV_{BC}/kT} - 2) \qquad (2\text{-}26)$$

$$I_{BE} = \frac{qAD_e p_{eo}}{L_e}(e^{qV_{BE}/kT} - 1) \qquad (2\text{-}27)$$

$$I_{BC} = \frac{qAD_c p_{co}}{L_c}(e^{qV_{BC}/kT} - 1) \qquad (2\text{-}28)$$

Let us now refer to (2-9), which gives the electron current $I_e(x)$ in the base. We note that subtraction of $I_e(0)$ from $I_e(W)$ yields (2-26) for I_{BB}. Therefore,

$$I_{BB} = I_e(W) - I_e(0) \qquad (2\text{-}29)$$

The electron current at $x = W$ is the net rate at which negative charge enters the base from the flow of electrons across the collector junction, whereas $I_e(0)$ is the net rate at which negative charge leaves the base due to electron flow across the emitter junction. Any difference is the result of recombination and generation of electron charge within the base. Holes and free electrons are destroyed only by recombination with one another, and they are thermally generated in pairs. *Thus we interpret the component* I_{BB} *of the current* I_B *as the hole current that results from recombination and generation within the base region.* A positive net recombination rate signifies that holes are annihilated faster than they are generated. Hence, holes must flow into the base to replace those destroyed, and I_{BB} is positive. On the other hand, a positive net generation rate must be accompanied by a flow of holes out of the base, giving a negative I_{BB}. Whether I_{BB} is positive, negative, or zero depends on whether the average excess-carrier

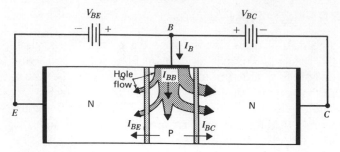

Figure 2-4 Hole flow and components of I_B, both junctions forward biased.

density in the base is positive, negative, or zero. Because $W/L_b < 0.3$, the component I_{BB} given by (2-26) can be put in the form:

$$I_{BB} = \frac{qAWn_{bo}}{2\tau_b}(e^{qV_{BE}/kT} + e^{qV_{BC}/kT} - 2) \qquad (2\text{-}30)$$

The component I_{BE} of (2-27) is exactly equal to the emitter-junction hole current out of the base, as given by the negative of (2-3) with $x = x_1$. Likewise, I_{BC} *of (2-28) is the collector-junction hole current out of the base*, as given by (2-4) at $x = x_2$. All three components of I_B are positive when both junctions are forward biased. This condition is illustrated in Fig. 2-4, which shows the positive currents with conventional arrows and also the hole flow. Note that the hole flow into the base region is supplied through the base terminal. Also, as indicated on the sketch the component I_{BC} is the largest, being over 90% of the total base current in many practical transistors having both junctions forward biased equally.

When both junctions are reverse biased, each of the components is negative and small. The hole flow of Fig. 2-4 is reversed.

With the normal bias arrangement (V_{BE} positive, V_{BC} negative), the flow is that of Fig. 2-5. Because the collector junction is reverse biased, I_{BC} is negligible. Note that the current I_B into the base is positive. The base current I_B is a drift

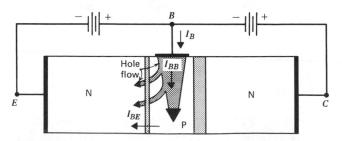

Figure 2-5 Hole flow and components of I_B, normal bias.

current and it produces an ohmic voltage drop which we have chosen to neglect for the present.

THE ELECTRON DENSITY

The behavior of an NPN transistor with specified bias voltages or currents is perhaps most easily ascertained by examination of the free-electron density in the base. This is given as a function of x by (2-8). Because the arguments of the hyperbolic sines in (2-8) are each less than 0.3 by (2-2), these hyperbolic functions can be replaced by their arguments with but slight error. *The result is the equation of a straight line having values at the ends of the base that satisfy the boundary conditions.*

The degree of accuracy that can be expected from the straight-line approximation is high. For example, selecting the maximum value of $0.3L_b$ for W, which gives the greatest error, the exact equation (2-8) for n_b and the straight-line approximation having the same end points differ at the midpoint of the base by only 1.1% provided at least one junction is forward biased (see P2-12). This difference is even less for the smaller base widths usually encountered. The true curve of n_b versus distance approximates a straight line so closely that it is not possible to differentiate between the two with the scales used in this book. As a consequence, a straight line drawn between the proper end points will be considered the true curve. Still, to show how the slope changes with distance x, if this information is important, a dashed curve will be drawn between the proper end points, with the curvature exaggerated. Several such curves are presented in the remainder of this chapter.

2-4. STATIC CHARACTERISTIC CURVES

A useful description of the dc behavior of a particular transistor requires the presentation of families of measured static characteristics. A set of curves may relate any two of the terminal voltages and one of the terminal currents. Such a set is called a common-emitter (CE), common-base (CB), or common-collector (CC) family, depending on which of the three terminals is common to the two voltages. An alternate choice for a family is one that relates one voltage and two currents; for this case the designation CE, CD, or CC is employed on the basis of the terminal current that is not included. For the CE family the base is referred to as the input and the collector the output; for the CB family the emitter is the input and the collector the output; and for the CC family the base is the input and the emitter is the output. The term *input characteristics* is often used when two of the three variables are input quantities, and the term *output characteristics* implies that two of the variables are output quantities. For example, curves relating V_{CE}, V_{BE}, and I_B are CE input characteristics, and curves relating V_{CE}, I_B, and I_E are CC output characteristics.

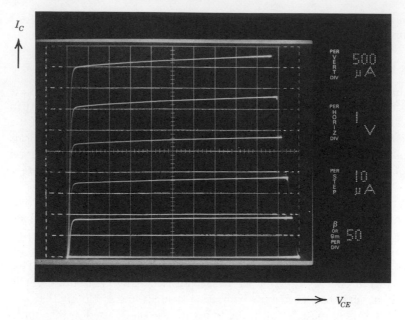

Figure 2-6   CE output characteristics of an NPN silicon transistor.

Figure 2-6 is an oscilloscope photograph of a family of CE output character-istics for an NPN silicon transistor. The range of the collector current I_C is 0 to 5 mA and that of the collector-emitter voltage V_{CE} is 0 to 10 V. Each curve is a trace of I_C versus V_{CE} for a constant base current I_B. The six curves have base currents of 0 to 50 μA in 10 μA steps. Because the characteristics give no information about the other terminal voltages, a second family is necessary for a complete description of the dc behavior of the transistor.

The characteristic curve along the horizontal axis of the cathode-ray-oscillo-scope (CRO) grid has zero base current. With no hole flow into the base to support recombination, the excess hole store is zero. Hence the collector current is approximately zero for all values of V_{CE}.

Let us examine the third curve above the bottom one, which has a constant base current of 30 μA. At the extreme left the slope of the characteristic is very large. At the breakpoint the collector current is 2.7 mA and V_{CE} is 0.2 V. To the right of this point the slope is quite small.

At the point at which V_{CE} is zero, the junction voltages V_{BE} and V_{BC} are exactly equal, because V_{CE} equals the difference $V_{BE} - V_{BC}$. The junctions are forward biased to the extent required to give the 30 μA base current. In this transistor the dopant density N_D of the collector is approximately the same as N_A of the base. Also, the collector-junction area is several times greater than that of the emitter junction. Consequently, the largest component of I_B is the hole current I_{BC} across the collector junction. The forward bias gives appreciable hole injection into the collector. The hole current I_{BE} across the forward-biased emitter junction is fairly insignificant because of the heavy doping of the emitter region and the small junction area. With assumed junction voltages of 0.60 V, I_{BC} is perhaps 90% of I_B, I_{BB} is 9%, and I_{BE} is 1%. Because of the equal junction voltages, the free-electron density in the base is constant, independent of distance x, and there is no gradient of n_b. Thus the collector current appears as zero on the scale used.

Now suppose V_{CE} is raised to the breakpoint value of 0.2 V. With a collector current of 2.7 mA approximately 0.05 V is dropped across the ohmic resistance of the collector region, and the additional 0.15 V is the difference between the junction voltages. The forward-bias voltages V_{BE} and V_{BC} are perhaps 0.68 and 0.53, respectively. The value of V_{BC} has dropped 0.07 V, and because I_{BC} is proportional to $\exp 40V_{BC}$, this current is now only about 5% of I_B. The average value of the excess-carrier density in the base is much larger than before, due to the increased V_{BE}. Hence the recombination rate is greater, and the recombination current I_{BB} is almost 95% of I_B. The 2.7-mA collector current is a consequence of the large free-electron gradient in the base. Even though V_{BE} and V_{BC} differ by only 0.15 V, the free-electron density at the emitter end of the base is 400 times greater than that at the collector end. Throughout the nearly vertical portion of the characteristic at the extreme left, both junctions are forward biased with voltages differing by 0.2 V or less. This portion of the characteristic is the *saturation region*, and the transistor is said to be *saturated*.

Beyond the breakpoint, conditions change very little. As V_{CE} is raised from 0.2 to 0.7 V, the emitter-junction voltage stays constant at about 0.7 V and the collector-junction voltage drops from about 0.5 V to zero. For larger values of V_{CE} the collector junction is reverse biased. For example, with $V_{CE} = 6$ the collector-junction voltage V_{BC} is -5.3. The change in V_{BC} has very little effect on the free-electron gradient in the base. In comparison with the value of n_b at the emitter edge of the base, the value at the collector edge is essentially zero for all collector-emitter voltages greater than 0.2 V. The region to the right of the breakpoint is the *normal*, or *active*, region of operation. Transistors in linear circuits are biased in the active mode.

The 30-μA characteristic in the active mode has a small positive slope. In this region almost all of I_B is the recombination current I_{BB}, and this is constant. It follows that the average excess-carrier density in the base is constant. Shown in Fig. 2-7 are active-mode plots of n_b versus x for 0.2 and 6 V. Because n_{bo} is negligible, the excess-carrier density equals n_b, and the areas under the two curves are equal. However, the 6 V curve has the steeper slope. At the higher voltage the space-charge layer of the collector junction is wider, which gives a reduced base width W. The smaller W requires a slightly greater value of p_b at $x = 0$.

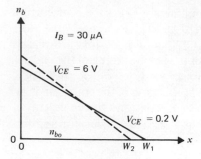

Figure 2-7 Active-mode plots of n_b versus x with I_B constant.

The decrease in W as the reverse bias on the collector junction is increased is referred to as *base-width modulation*, and this effect accounts for the positive slope of the characteristic. Base-width modulation is undesirable. It is reduced in transistors with lightly doped collector regions, because such transistors have most of the space-charge layer within the collector. The active-region collector-characteristic curves of transistors with lightly doped collectors are nearly horizontal.

An increase in I_B results in an increase in I_C. For example, with $V_{CE} = 5$ the collector current of the transistor of Fig. 2-6 rises from 2.8 to 3.7 mA when I_B is changed from 30 to 40 μA. The larger I_{BB} requires a greater excess-carrier density in the base as indicated in Fig. 2-8, and the increased free-electron gradient gives a larger I_C. In fact, if the base current is doubled, n_b at each x coordinate is also doubled and so is the collector current. This explains the spacing between the curves of Fig. 2-6.

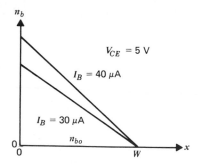

Figure 2-8 Active-mode plots of n_b versus x with constant V_{CE}.

TEMPERATURE EFFECTS

Characteristic curves showing a terminal current as a function of V_{BE} with the collector voltage held constant, such as those of Figs. 2-9c and e, have an exponential shape similar to the diode characteristic. Such curves are temperature sensitive, primarily because the saturation currents that appear in the Ebers-Moll equations are proportional to n_i^2. As in the case of the diode characteristic, these curves shift to the left about 2 mV per kelvin unit increase in temperature, for both silicon and germanium transistors.

Let us consider the effect of temperature on the CE output characteristics. For sufficiently large values of V_{CE} (one or more volts) the collector junction is reverse biased, and the Ebers-Moll equation (2-16) becomes

$$I_C = -\alpha_F I_E + I_{CO} \qquad (2\text{-}31)$$

By replacing I_E with $-(I_B + I_C)$ and solving for I_C, we find that

$$I_C = \beta_F I_B + (\beta_F + 1)I_{CO} \qquad (2\text{-}32)$$

with

$$\beta_F = \frac{\alpha_F}{1 - \alpha_F} \qquad (2\text{-}33)$$

In silicon transistors at room temperature the current I_{CO} as defined by (2-18) and (2-13) is typically a small fraction (0.100 to 0.001) of a picoampere for a device rated in milliamperes, and because it is proportional to n_i^2, I_{CO} increases rapidly with temperature. However, *when the collector junction is reverse biased*, I_{CO} in (2-32) properly includes a thermal generation component resulting from carrier generation within the space-charge layer, as well as a surface leakage component. These components of the saturation current of a reverse-biased junction were discussed in Sec. 1-9 for the diode. They were not considered in the analysis of the idealized transistor presented in Secs. 2-1 and 2-2. Accordingly, I_{CO} is actually several orders of magnitude greater than the theoretical value, provided the collector junction is reverse biased, and it does not vary with temperature as rapidly as does n_i^2. For most silicon devices I_{CO} is negligible below 125°C as long as the collector current I_C is above the microampere range and the power level is low. In such cases I_C is $\beta_F I_B$, and the temperature effects on the CE output characteristics depend on β_F (see P2-16).

To a first approximation, α_F is $(1 - W^2/2L_b^2)$ and β_F is $2L_b^2/W^2$. The base width W is a function of the built-in junction voltages, which depend on T, but these are fairly negligible effects. The electron diffusion length L_b in the P-type base is the square root of the product $D_b\tau_b$. We know that D_b is a function of temperature, and even more important is the temperature dependence of the lifetime τ_b. There is no theoretical expression that accurately predicts the manner in which τ_b varies, but manufacturers often give curves on transistor data sheets

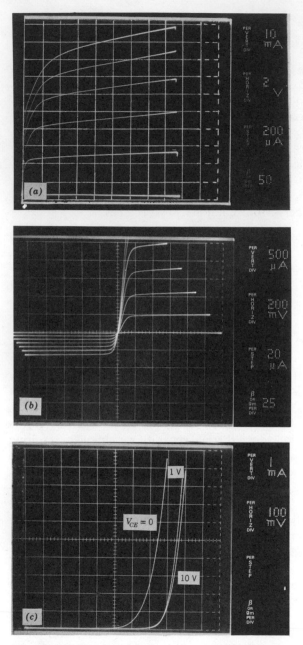

Figure 2-9   Static characteristics of silicon NPN transistors.
(a) CE output characteristics, I_C versus V_{CE} (I_B constant).
(b) Inverse and active regions. (c) CE input characteristics, I_B versus V_{BE} (V_{CE} constant).

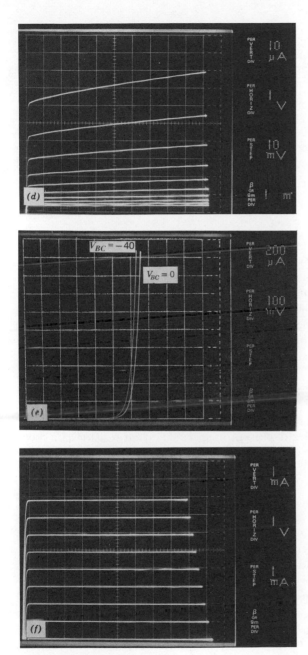

Figure 2-9 (d) CE output characteristics, I_C versus V_{CE} (V_{BE} constant). (e) CB input characteristics, I_E versus V_{BE} (V_{BC} constant). (f) CB output characteristics, I_C versus $-V_{BC}$ (I_E constant).

showing the variation of β_F with temperature. In silicon transistors the lifetime usually increases as the temperature rises above room temperature, causing β_F to increase. *This results in a greater spacing between successive* CE *output curves.* As the temperature of a silicon transistor is increased, the zero base-current characteristic does not change for low and moderate values of T, because I_{CO} in (2-32) is negligible. The only noticeable difference in the family is the increased spacing between the curves at the higher temperatures. In germanium transistors, however, the entire family of curves is shifted upward because of the increase in I_{CO}. New curves for negative I_B appear between the zero base-current curve and the abscissa.

OTHER FAMILIES

There are many different types of junction transistors with different characteristics. The various families of static characteristics of Fig. 2-9 are fairly representative of an NPN silicon transistor at room temperature.

The transistor with the curves of Fig. 2-9a is intended for operation at low currents. At high currents the ohmic drop in the collector region is appreciable, which accounts for the gradual transition between saturation and active-mode operation. The reduced spacing between the curves at higher currents indicates lower values of α_F and β_F. This effect is a consequence of high-level injection, which causes the hole density to rise substantially above its equilibrium value. The result is increased hole injection into the emitter. For a constant base current the larger I_{BE} subtracts from the recombination current I_{BB}, lowering α_F. The maximum rated collector current, as specified on a transistor data sheet, is usually determined by the current that gives an α_F unacceptably low. Further discussion of the variation of α_F and β_F with I_C is presented in Sec. 15-2.

Figure 2-9b clearly shows the saturation region for both positive and negative V_{CE}. The slopes of the curves in saturation are affected appreciably by the ohmic resistance of the collector, typically tens of ohms. To the left of the breakpoints in the third quadrant is the *inverse* region. Here the transistor operates with the emitter and collector interchanged in effect. Electrons are injected from the collector into the base, and they are collected by the emitter. The reduced spacing indicates a low α_R.

The CE input characteristics of Fig. 2-9c have the expected exponential form of the diode equation. At the left is the curve for $V_{CE} = 0$. The transistor is saturated with the junctions forward biased the same amount. For a fixed value of V_{BE} both I_{BB} and I_{BC} are greater in saturation than in active-mode operation. This explains why the curve is to the left of the 1 V and 10 V active-mode curves. The slight separation between the 1 V and 10 V curves is a consequence of base-width modulation (see P2-18).

The CE output characteristics of Fig. 2-9d have V_{BE} as the constant quantity. For a fixed V_{CE} the collector current is an exponential function of V_{BE}, with very little current when this voltage is less than 0.5 V. The positive slope in the active region is caused by base-width modulation.

In Fig. 2-9e and f are CB input and output characteristics. The input characteristics with V_{BC} constant have an exponential shape with base-width modulation accounting for the small horizontal separation between the two active-mode curves. In Fig. 2-9f the emitter current is the constant parameter, and the active region is to the right of the I_C axis. The collector current is nearly equal to the magnitude of I_E in accordance with (2-31). The saturation region is in the second quadrant, with V_{BC} positive.

These and other families can be plotted point-by-point from the two Ebers-Moll equations provided the basic transistor parameters are known. The equations are somewhat complicated if the effects of base-width modulation are to be included, because the base width W is a function of the junction voltages. The difficulty is the fact that the lifetimes, the mobilities, and other parameters are not easily measured or determined from data sheet information. Furthermore, the equations do not include certain second-order effects, such as high-level injection, ohmic resistances, carrier generation and recombination within the space-charge layers, and others. Under certain conditions some of these can become first-order effects. An electronic curve tracer that visually displays a family of curves which can be photographed is clearly preferred.

2-5. REGIONS OF OPERATION

CUTOFF

In addition to the saturation, active, and inverse regions of operation, there is *cutoff*. In cutoff the bias voltages are so small that the terminal currents are negligible, and the transistor is said to be *off*. At room temperature this is usually the case if V_{BE} and V_{BC} are each less than about 0.55 V. Certainly a transistor with both junctions reverse biased is in the cutoff mode provided there are no appreciable avalanche-breakdown effects.

Suppose an NPN transistor is *off* with both junctions reverse biased by at least 0.1 V. With the aid of the reciprocity relation of (2-19) the Ebers-Moll equations (2-10) and (2-11) can be expressed in the form

$$I_E = I_{ES}(1 - \alpha_F) \qquad I_C = I_{CS}(1 - \alpha_R) \qquad (2\text{-}34)$$

At most operating temperatures these currents are so small that *the three terminals of an off transistor are simply open circuits.*

The free-electron density n_b in the base of a transistor with reverse-biased junctions is sketched in Fig. 2-10. With n_b everywhere much less than n_{bo}, recombination is insignificant. Thus the thermally generated electrons in the base must flow into the emitter and collector regions at a rate equal to the generation rate. This justifies the shape of the dashed curve, recalling that

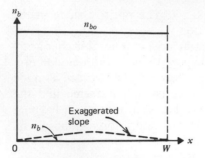

Figure 2-10   Free-electron density in
base with both junctions reverse biased.

carriers diffuse away from a concentration. Of course, the shape can also be
determined from (2-8). By (1-24) the rate at which the covalent bonds are broken
is n_{bo}/τ_b. With negligible recombination it follows that the product of this and q
and the volume AW of the base represents the hole current out of the base
terminal caused by thermal generation in the base, and $-I_{BB}$ is $qAWn_{bo}/\tau_b$. Also
contributing to $-I_B$ is hole flow into the base from the emitter and collector
regions from carrier generation within the space-charge layers.

Now let us consider the transistor with the collector junction reverse biased
and the emitter open, as shown in Fig. 2-11a. Note the symbol that represents an
NPN transistor. The arrow on the emitter points in the direction of forward
current flow across the PN junction. The free-electron distribution is illustrated
in Fig. 2-11b, and the sketch indicates that the emitter junction is automatically
reverse biased by perhaps 0.1 V. If we ignore the very small hole current of the
junction, then the electron current must be zero. Thus the curve of n_b has zero
slope at $x = 0$. With the base nearly depleted of charge carriers there is no
appreciable recombination. The current I_{CO} consists of the sum of the electron

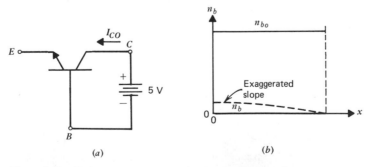

Figure 2-11   Emitter open, collector junction reverse biased.

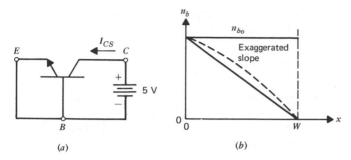

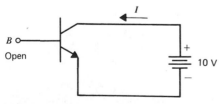

Figure 2-12   Emitter shorted to base, collector junction reverse biased.

generation current of the base, the collector-junction hole current, the surface leakage current, and the current created by carrier generation within the space-charge layer. This latter component is dominant in silicon transistors at room temperature.

Next, suppose the emitter-base terminals are shorted with the collector junction reverse biased, as shown in Fig. 2-12a. The electron current at the collector junction is caused by the diffusion of electrons through the base from emitter to collector, as indicated by the sketch of Fig. 2-12b. From the slope n_{bo}/W, the electron current in the negative x direction is found to be $q A D_b n_{bo}/W$. The shape of the dashed curve indicates that the electron current increases as x increases, because electrons are added to the flow by the positive net generation rate. The current I_{CS} is the sum of this electron current, the hole current across the junction, the surface leakage current, and that contributed by carrier generation within the space-charge layer. Comparison of Figs. 2-11 and 2-12 reveals a clear distinction between the open-circuit and short-circuit saturation currents I_{CO} and I_{CS}. A similar distinction may be made for I_{EO} and I_{ES}.

In order to obtain the collector output characteristic for $I_B = 0$ the bias circuit of Fig. 2-13 would be used with an adjustable voltage supply. The applied voltage provides a small forward bias to the emitter junction and a large reverse bias to the collector junction. Using the Ebers-Moll equations (2-10) and (2-11) with the exponentials of qV_{BC}/kT eliminated, setting the sum

Figure 2-13   Transistor with open base.

of I_E and I_C equal to zero, and solving for exp qV_{BE}/kT, we obtain with the aid of the reciprocity relation (2-19) the expression

$$e^{qV_{BE}/kT} = 2 + \frac{I_{CS}/I_{ES} - 1}{1 - \alpha_F} \tag{2-35}$$

In practical transistors I_{CS} is greater than I_{ES} because of the lower dopant density in the collector compared with that of the emitter and also because of the larger collector-junction area. Assuming I_{CS} is $5I_{ES}$, α_F is 0.98, and q/kT is 40, from (2-35) the junction voltage V_{BE} is found to be 0.13, and V_{BC} is -9.87. The forward bias of the emitter junction is too low to give a significant collector current. *Thus the silicon transistor is effectively in cutoff when the base terminal is open* (see P2-22).

SATURATION

In the saturation mode both junctions are forward biased. Plots of n_b versus x for three saturation-mode bias conditions are shown in Fig. 2-14. The terms *forward* and *reverse* are used to denote diffusion in the positive and negative x directions, respectively, as illustrated. The shape of the dashed curve of Fig. 2-14*a* is obvious when it is realized that some of the electrons diffusing across the base are lost by recombination, causing the electron current to decrease as x increases. The current is, of course, proportional to the slope. The dashed curve of Fig. 2-14*b* is similarly explained. In Fig. 2-14*c* the shape of the dashed curve is due to the flow of electrons into the base from both the emitter and collector, replacing those lost by recombination.

Let us again consider a characteristic curve of I_C versus V_{CE} with I_B fixed at 30 μA. Above 0.2 V the collector current is fairly constant at perhaps 3 mA, and the hole current I_{BC} across the collector junction is negligible. At 290 K the respective free electron densities $n_b(0)$ and $n_b(W)$ at the emitter and collector

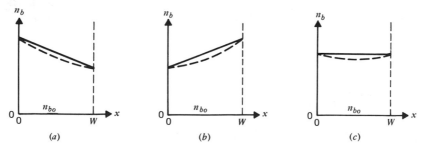

Figure 2-14   Free-electron density plots, both junctions forward biased (dashed curves have exaggerated slopes). (*a*) Forward saturation mode. (*b*) Reverse saturation mode. (*c*) $V_{BE} = V_{BC}$.

ends of the neutral base region have a ratio $(\exp 40V_{BE})/(\exp 40V_{BC})$, which equals $\exp 40V_{CE}$. When V_{CE} is 0.2, making V_{BE} greater than V_{BC} by this amount, the ratio is 3000. Thus operation is still essentially that of the active mode. However, as V_{CE} is reduced from 200 mV, the collector current I_C drops, and substantial changes occur in the charge stores. This is the saturation mode. The forward bias of the emitter junction drops while that of the collector junction rises. *The combined effects are to reduce the electron flow through the base and to increase the hole flow across the collector junction.* When these flows become equal at a value of V_{CE} of perhaps 40 mV, the collector current is zero. The holes and electrons injected into the collector replace those lost by recombination, and there is appreciable excess charge stored in the neutral collector region.

The preceding discussion shows that *a transistor with one junction forward biased and either the collector or emitter open is in the saturation mode.* The junction voltages differ by only a few tens of millivolts or less. With the collector or the emitter open, the greater part of I_B is I_{BC} (see P2-19). For a constant I_R the recombination current I_{BB} in saturation is much less than its active-mode value, and of course, this implies that the charge store in the base is lower. In heavy saturation most of the excess charge store is in the neutral collector region near the space-charge layer.

ACTIVE MODE

In the normal, or active, region the emitter junction is appreciably forward biased, and the collector-junction voltage V_{BC} is small or negative. This region is especially important, because operating conditions are confined to the active mode when a transistor is used to amplify signals.

Assuming I_{CO} is negligible as it normally is, we find from (2-32) that the active-mode currents are related by

$$I_C = \beta_F I_B \qquad \text{with} \qquad \beta_F = \frac{\alpha_F}{1 - \alpha_F} \qquad (2\text{-}36)$$

As β_F is the ratio I_C/I_B of the dc currents, it is called the *CE dc current gain.* From (2-31) it follows that

$$I_C = -\alpha_F I_E \qquad \text{with} \qquad \alpha_F = \frac{\beta_F}{\beta_F + 1} \qquad (2\text{-}37)$$

As α_F is the ratio $-I_C/I_E$, it is the *CB dc current gain.* For typical values of α_F between 0.950 and 0.997 the corresponding values of β_F are 19 to 332. Because large values of β_F are desirable, α_F should be nearly unity. Some integrated-circuit devices known as *supergain* transistors have betas of 5000 or more, but most transistors have betas between 40 and 100.

The active-mode base current I_B consists of the flow of holes into the P-type base to replace those lost by recombination and those injected into the emitter. It is the sum of I_{BB} and I_{BE}, which are found from (2-30) and (2-27) to be

$$I_{BB} = \left(\frac{qAWn_{bo}}{2\tau_b}\right)e^{qV_{BE}/kT} \tag{2-38}$$

$$I_{BE} = \left(\frac{qAD_e p_{eo}}{L_e}\right)e^{qV_{BE}/kT} \tag{2-39}$$

For the usual case of low-level injection, I_{BE} is very small compared with I_{BB}, and I_B is approximately I_{BB}.

Figure 2-15 shows an NPN transistor biased in the normal region, along with the corresponding free-electron distribution in the base. From this figure the collector current, which is approximately equal to the electron diffusion current of the base, is found to be

$$I_C = \left(\frac{qAD_b n_{bo}}{W}\right)e^{qV_{BE}/kT} \tag{2-40}$$

It is easily shown from (2-38) through (2-40) that α_F as given by (2-24) can be expressed in the form

$$\alpha_F = 1 - \frac{I_{BB}}{I_C} - \frac{I_{BE}}{I_C} \tag{2-41}$$

The collector current is slightly less than the emitter current because of I_{BB} and I_{BE}. These are kept small compared with I_C by designing transistors with W small compared with the diffusion length L_b and by doping heavily the emitter region so as to minimize I_{BE}. The ratio I_{BB}/I_C is $W^2/2L_b^2$ and is called the *base defect*. The *emitter defect* I_{BE}/I_C is $D_e p_{bo} W/(D_b n_{eo} L_e)$.

The hole distribution in the base has the same slope as the free-electron distribution of Fig. 2-15b. However, because the hole current I_{BC} across the

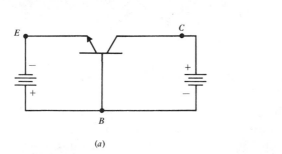

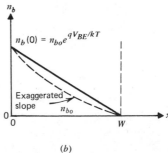

(a)   (b)

Figure 2-15   NPN transistor biased in the normal region. (*a*) Circuit. (*b*) Free-electron density in base.

collector junction is negligible, it follows that *the x-directed hole drift and diffusion components cancel one another.* For this reason we have concentrated on the base minority-carrier current. The currents in the emitter and collector regions are almost entirely electron drift currents, and the x-directed current through the base is an electron diffusion current (see P2-21).

INVERSE MODE

The inverse region is similar to the active region, but the roles of emitter and collector are interchanged. Most transistors have a larger cross-sectional area at the collector junction than at the emitter junction in order to collect more efficiently the electrons injected into the base from the emitter. In the inverse region many of the electrons injected from the collector into the base diffuse to the surface where they are lost by recombination. The injection efficiency in the inverse mode is usually low because the collector is not as heavily doped as the emitter. Thus α_R is less than α_F. This is indicated by the reduced spacing of the characteristic curves in the third quadrant of Fig. 2-9b. The inverse mode is especially important in certain types of logic networks.

2-6. BREAKDOWN VOLTAGES

PUNCH-THROUGH

We have learned that base-width modulation causes the collector characteristic curves in the active region to rise as V_{CE} is increased. The base width W decreases, and the free-electron density $n_b(0)$ at the emitter edge of the base must rise so that I_B remains constant. Eventually, W approaches zero, $n_b(0)$ becomes very large, and the electron diffusion current in the base theoretically approaches infinity. This is *punch-through.* The punch-through voltage V_P is the value of $|V_{BC}|$ that reduces the base width W to zero (see P2-23).

At punch-through the two space-charge layers meet. Their electric fields are in opposite directions. A slight overlap reduces the barrier of the emitter junction. Electron injection from the emitter through the overlapping space-charge layers and into the base is very large. In effect, the transistor has become a short-circuit. The applied collector-junction voltage must never exceed V_P, or otherwise the heat generated by the large current will destroy the transistor.

As punch-through is approached, the excess-carrier store in the base is lowered, reducing I_{BB}. However, I_{BE} rises as a consequence of high-level injection. The sum of the two components equals the constant current I_B.

AVALANCHE MULTIPLICATION

The process of avalanche multiplication, which was discussed in Sec. 1-9, occurs at the collector or emitter junction when the respective reverse-bias voltage is sufficiently large. Let us consider the active region. A primary carrier crossing the space-charge layer of the collector junction may, during a collision with an atom, dislodge an electron from a covalent bond. This generates a hole and a free electron. These secondary carriers may produce additional carriers in the same manner. The generated electrons are swept by the intense field into the collector, and the holes flow into the base. This process increases the collector current.

In the active mode I_C is $-\alpha_F I_E$ by (2-37), assuming I_{CO} is negligible. When avalanche multiplication is significant, the expression for I_C must be multiplied by the avalanche multiplication factor M, giving

$$I_C = -M\alpha_F I_E \qquad (2\text{-}42)$$

Substituting $-I_C - I_B$ for I_E and solving for I_C, we obtain

$$I_C = \frac{M\alpha_F I_B}{1 - M\alpha_F} \qquad (2\text{-}43)$$

The factor M has the approximate form

$$M = \frac{1}{1 - (V_{CB}/V_a)^n} \qquad (2\text{-}44)$$

with V_a denoting the *avalanche breakdown voltage* and with n denoting a constant which is usually between 1.5 and 6.5. The avalanche voltage V_a is the value of V_{CB} that makes M infinite. Both V_a and V_{CB} are positive, and the collector junction is reverse biased.

Typical CB output characteristic curves of an NPN transistor with a breakdown voltage of 100 are shown in Fig. 2-16a. The curves are plots of I_C versus

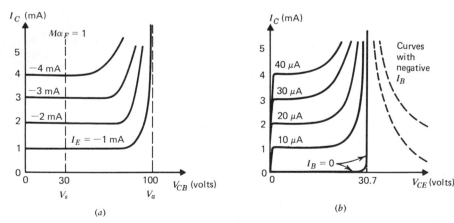

(a)

(b)

Figure 2-16   NPN characteristics showing avalanche breakdown. (a) CB output characteristics. (b) CE output characteristics.

V_{CB} with I_E constant. All curves approach infinity as V_{CB} approaches 100 V. On a data sheet the breakdown voltage is referred to as $V_{(BR)CBO}$, with the subscript C referring to the collector, the next subscript B identifying the reference terminal, and the last subscript O signifying that the third terminal is open ($I_E = 0$). This breakdown voltage might be either the punch-through voltage V_P or the avalanche voltage V_a, depending on the design parameters of the particular transistor. In either case the CB output characteristics are similar to those of Fig. 2-16a.

Let us assume that the punch-through voltage is substantially greater than the avalanche breakdown voltage. For this case the curves of Fig. 2-16a are described mathematically by (2-42). When V_{CB} is 100, the multiplication factor M is theoretically infinite. Avalanche breakdown in the space-charge layer of the collector junction gives a collector current that approaches infinity. The base current approaches minus infinity. It consists of hole flow from the space-charge layer into the base and out of the base terminal. The emitter current is maintained constant. Even the characteristic for $I_E = 0$ rises to infinity as V_{CB} approaches V_a. For this case we must add the term MI_{CO} to the right side of (2-42). As indicated on the characteristics of Fig. 2-16a, the effective alpha $M\alpha_F$ is unity at a voltage of 30. This is the *sustaining voltage* V_s.

The sustaining voltage V_s is the value of V_{CB} that makes $M\alpha_F$ unity. When V_{CB} is less than V_s, the effective alpha is less than unity, the collector current is less than the emitter current, and I_B is positive. This is the usual condition. With V_{CB} greater than V_s the effective alpha is greater than unity, the collector current is greater than the emitter current, and I_B is negative. Normally this condition is avoided, and the reason becomes evident when we examine the CE characteristics of Fig. 2-16b.

The CE characteristics are described mathematically by (2-43). Both sets of characteristics apply to the same transistor. However, the CE curves enter breakdown at a value of V_{CE} equal to the sum of V_{BE} and the sustaining voltage. This assumes that breakdown is due to avalanche multiplication, not punch-through. It is evident from (2-43) that I_C approaches infinity as the effective alpha $M\alpha_F$ approaches unity with I_B maintained constant. Physically, the effect is easily explained. As the sustaining voltage is approached with I_B held constant, the hole flow into the base from avalanche multiplication approaches the precise rate needed to replace those lost by recombination for any selected value of I_C. However, because the base current is fixed, there is additional hole flow into the base, which is in excess of that required. Hence the charge store rises, increasing the free-electron gradient and the current. The effect is not observed in the CB family. Here the base current changes so as to keep the charge store and the emitter current constant in the vicinity of V_s.

The breakdown voltage of the CE characteristics of Fig. 2-16b is specified on data sheets as $V_{(BR)CEO}$. This is either the approximate sustaining voltage or the punch-through voltage, whichever is lower. Usually V_s is from 10 to 40%

of the avalanche breakdown voltage V_a. Operation of a transistor at voltages greater than V_s and V_a will not damage the device if the dissipated power and the collector current are kept below allowed values. However, the applied voltage should never exceed the punch-through value. In linear amplifiers transistors are operated at voltages well below the sustaining voltage in order to avoid the considerable nonlinearity displayed in the CE output characteristics in the vicinity of V_s.

2-7. JUNCTION CAPACITANCE

Charges stored in the space-charge layers and neutral regions of transistors are nonlinear functions of the junction voltages. They strongly influence the switching speed of a digital logic network and the frequency response of an amplifier. In Chapter 6 a charge-control model is developed and used to determine the properties of a transistor that is switched *on* and *off* between saturation and cutoff. Hence we shall confine our attention here to incremental capacitances employed in small-signal linear models of transistors biased in the active mode.

Let us first consider the charge stores of the space-charge layers. For an abrupt junction with uniform doping in the P and N regions the increase in the stored charge when the junction voltage V is applied is given by (1-53) of Sec. 1-10. However, most integrated-circuit transistors have junctions that more nearly approximate the graded junction of P1-16, and expressions for the charge stores of this type of junction are used here (see P2-26). Let q_{SE} denote the charge store of the emitter-junction space-charge layer due to the applied voltage V_{BE} and let q_{SC} denote the corresponding charge of the collector-junction space-charge layer. Then

$$q_{SE} = a_1[\psi_{oe}^{2/3} - (\psi_{oe} - v_{BE})^{2/3}] \qquad (2\text{-}45)$$

$$q_{SC} = a_2[\psi_{oc}^{2/3} - (\psi_{oc} - v_{BC})^{2/3}] \qquad (2\text{-}46)$$

with a_1 and a_2 denoting constants and with ψ_{oe} and ψ_{oc} denoting the respective built-in junction voltages.

At the emitter junction the incremental space-charge capacitance C_{te} is dq_{SE}/dv_{BE} evaluated at the bias voltage. It follows from this and (2-45) that

$$C_{te} \propto (\psi_{oe} - V_{BE})^{-1/3} \qquad (2\text{-}47)$$

Similarly,

$$C_{tc} \propto (\psi_{oc} - V_{BC})^{-1/3} \qquad (2\text{-}48)$$

For abrupt junctions the exponents are -0.5.

In the active mode the only appreciable excess-charge store in the neutral regions is that of the base. There is no appreciable charge store in the emitter because the large dopant density nearly eliminates injection into this region.

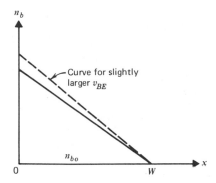

Figure 2-17 Active-mode free-electron density in base.

Also, the excess charge of the collector is negligible because of the reverse bias. The free-electron density in the base is plotted in Fig. 2-17. At $x = 0$ the electron density is $n_{bo} \exp qv_{BE}/kT$, and the stored excess charge q_F is

$$q_F = 0.5qAWn_{bo}e^{qv_{BE}/kT} \qquad (2\text{-}49)$$

The excess free-electron and hole charges are equal and opposite. With the collector voltage constant, an incremental increase in v_{BE} gives a small increase in the free-electron density at each point of the base. This is indicated by the dashed line of Fig. 2-17. There is, of course, an identical change in the hole density. The ratio of the incremental change in q_F to the corresponding change in v_{BE} is a measure of the junction diffusion capacitance. This ratio is found by differentiating (2-49) with respect to v_{BE}. With the aid of (2-40) we find that dq_F/dv_{BE} is $(0.5W^2/D_b)(qI_C/kT)$. A more rigorous analysis based on the ac diffusion equation gives the same result except that the factor 0.5 is $\frac{1}{3}$. Thus the emitter-junction incremental *diffusion capacitance* C_d is

$$C_d = \left(\frac{W^2}{3D_b}\right)\left(\frac{qI_C}{kT}\right) \qquad (2\text{-}50)$$

Of particular importance is the observation that C_d is proportional to the bias current I_C.

The total incremental capacitance C_π of the emitter junction is the sum of the space-charge capacitance C_{te} and the diffusion capacitance C_d. Although both are important, the diffusion capacitance usually is dominant, perhaps being as much as 90% of C_π. This capacitance is typically tens of picofarads for transistors used in small-signal circuits at low current levels.

The total incremental capacitance C_μ of the collector junction is simply the space-charge capacitance C_{tc}, except for a diffusion capacitance that is usually negligible. Because the collector junction is reverse biased, C_μ is normally less

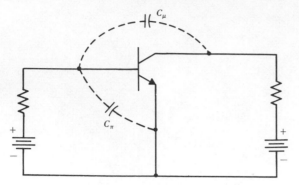

Figure 2-18   Active-mode incremental capacitances.

than the space-charge capacitance of the forward-biased emitter junction even though the collector area is greater. Typically, C_μ has values in the range of 2 to 5 pF for low-current transistors. This capacitance is especially objectional in common-emitter amplifiers. It provides a feedback path from the output collector to the input base, thereby reducing the bandwidth. Its effect may be considerably greater than that of the much larger C_π. The collector junction is often reverse biased with a rather large voltage to minimize C_μ. The fact that this is possible is evident from (2-48). The active-mode incremental capacitances are shown in Fig. 2-18.

The diffusion capacitance of the collector junction is due to base-width modulation. A change in V_{BC} changes the base width W, thereby producing a very small increment in the charge store of the neutral base. However, base-width modulation is typically small enough to make this component of C_μ rather insignificant.

2-8. PNP TRANSISTORS

In a PNP transistor the x directed currents in the emitter, base, and collector are primarily hole currents, and the transverse base current is one of electrons. In the active mode holes are injected from the emitter into the base, moving by diffusion through this region to the collector. Some are lost in the base by recombination with electrons, and the negative base current I_B supplies electrons to replace those lost by recombination and by injection into the emitter. Here I_E is positive and I_C is negative. The Ebers-Moll equations for a PNP transistor are

$$I_E = I_{ES}(e^{qV_{EB}/kT} - 1) - \alpha_R I_{CS}(e^{qV_{CB}/kT} - 1) \qquad (2\text{-}51)$$

$$I_C = -\alpha_F I_{ES}(e^{qV_{EB}/kT} - 1) + I_{CS}(e^{qV_{CB}/kT} - 1) \qquad (2\text{-}52)$$

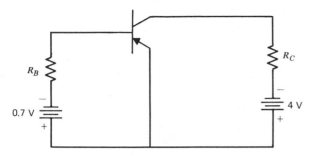

Figure 2-19 PNP transistor biased in the active mode.

The alternate forms are

$$I_E = -\alpha_R I_C + I_{EO}(e^{qV_{EB}/kT} - 1) \qquad (2\text{-}53)$$

$$I_C = -\alpha_F I_E + I_{CO}(e^{qV_{CB}/kT} - 1) \qquad (2\text{-}54)$$

The equations for α_F, α_R, and the saturation currents that were presented for the NPN transistor are applicable to PNP devices simply by interchanging n and p. The Ebers-Moll circuit model of Fig. 2-3 applies to the PNP transistor provided each diode is reversed. No changes in the reference currents should be made. A PNP transistor biased in the normal region is shown in Fig. 2-19. The arrow on the emitter points in the direction of easy current flow across the PN junction. Incremental capacitances C_π and C_μ, which are used in small-signal circuit models, are similar to those of the NPN device.

REFERENCES

1. P. E. Gray and C. L. Searle, *Electronic Principles*, Wiley, New York, 1969.
2. J. Lindmayer and C. Y. Wrigley, *Fundamentals of Semiconductor Devices*, D. Van Nostrand, Princeton, N.J., 1965.
3. C. L. Searle, A. R. Boothroyd, E. J. Angelo, Jr., P. E. Gray, and D. O. Pederson, *Elementary Circuit Properties of Transistors*, Wiley, New York, 1964.
4. J. J. Ebers and J. L. Moll, "Large-Signal Behavior of Junction Transistors," *Proc. IRE*, **42**, 1761–1772, Dec. 1954.

PROBLEMS

Section 2-1

2-1. With reference to Fig. 2-2 derive (2-3) and (2-4), which give the hole currents in the emitter and collector regions. Begin with the general solution (1-27) of the minority-carrier diffusion equation.

2-2. Apply the boundary conditions to (2-5) to find C_1 and C_2 and use the result to show that $n_b(x)$ can be expressed in the form of (2-8).

2-3. An NPN silicon transistor at 290 K ($q/kT = 40$) has $D_b = 0.001$, $L_b = 10^{-5}$, $A = 10^{-7}$, $W = 10^{-6}$, and $n_{bo} = 10^{11}$. The bias voltages are $V_{BE} = 0.4605$ and $V_{BC} = -5$. (a) Assuming n_b is a linear function of x, find the values of n_b at the end points of the base and sketch n_b versus x. Using the slope of the curve, calculate the approximate electron current in mA. (b) As $W \ll L_b$, show from (2-9) that

$$I_e \approx -\left(\frac{qAD_b n_{bo}}{W}\right)(e^{qV_{BE}/kT} - e^{qV_{BC}/kT})$$

Calculate I_e from this equation. Units are SI.

2-4. At 290 K an NPN transistor has $V_{BE} = 0.01733$ V and $V_{BC} = -5$ V. From (2-9), find the ratio x/W at which the electron current in the base is a minimum. Calculate this minimum electron current in picoamperes using the data of P2-3.

Section 2-2

2-5. Derive the Ebers–Moll (2-10) and (2-11) from (2-3), (2-4), and (2-9).

2-6. An NPN silicon transistor ($\varepsilon = 10^{-10}$) has dopant densities of 10^{25}, 2×10^{22}, and 10^{22} in the emitter, base, and collector, respectively, and the distance between the metallurgical junctions is 10^{-6}. The built-in junction voltage ψ_{oe} is 0.90, and ψ_{oc} is 0.72. If V_{BE} is 0.65, what is the *punch-through voltage*, defined as the value of $|V_{BC}|$ that reduces the base width to zero? Units are SI.

2-7. In the Ebers–Moll model of Fig. 2-3 connect a forward-bias dc source between terminals E and B with terminals C and B shorted. Then show that the common-base forward current gain $-I_C/I_E$ is α_F. In a similar manner show that the common-base reverse short-circuit current gain $-I_E/I_C$ is α_R.

2-8. (a) In the Ebers–Moll model of Fig. 2-3 apply a reverse bias to terminals E and B with terminals C and B shorted, and show with the aid of (2-17) that $I_E = I_{ES}$. (b) Apply a reverse bias to terminals C and B with the emitter E open, and find I_C.

2-9. The emitter junction of an NPN transistor is forward biased and the collector junction is reverse biased. Employing the Ebers–Moll model of Fig. 2-3, show that I_C is the sum of $\beta_F I_B$ and $(\beta_F + 1)I_{CO}$ with β_F equal to $\alpha_F/(1 - \alpha_F)$. Assuming I_{CO} is negligible, use this result to explain why I_C increases as V_{BC} is made more negative, with I_B maintained constant.

2-10. Replace the transistor of Fig. 2-20 with the Ebers–Moll model of Fig. 2-3. Noting that I_{CO} is negligible compared with I_C, solve for V_{BE} at 290 K assuming α_F is 0.96, α_R is 0.40, and I_{EO} is 10^{-15} A.

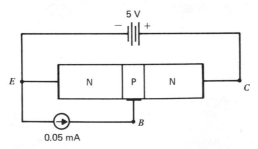

Figure 2-20   For P2-10.

Section 2-3

2-11. An NPN transistor with normal bias voltages has $I_e(0) = -1.010$ mA, $I_e(W) = -1.000$ mA, and a hole current of -0.002 mA across the emitter junction. (a) Calculate the net rate, in holes per second, at which holes are destroyed in the neutral base between $x = 0$ and $x = W$. (b) Determine I_{BB}, I_{BE}, and I_{BC}. (c) Find I_E, I_B, and I_C in mA.

2-12. (a) Replace the hyperbolic sines in (2-8) with their arguments and show that the result is an equation of a straight line having values at the ends of the base that satisfy the boundary conditions. (b) With $W = 0.3L_b$, find n_b at the midpoint of the base using (2-8) and also using the result of (a). Show that the two values differ by only 1.1% provided at least one junction is forward biased.

2-13. An NPN transistor has its base open ($I_B = 0$) and $V_{CE} = 5$ V. In (2-30), (2-27), and (2-28), assume that $qAWn_{bo}/2\tau_b$ is 0.10 pA, $qAD_e p_{eo}/L_e$ is 0.01 pA, and $qAD_c p_{co}/L_c$ is 0.03 pA. With $q/kT = 40$, find the junction voltages V_{BE} and V_{BC}, and the currents I_{BB}, I_{BE}, and I_{BC} in pA. Assume junction leakage currents and space-charge-layer generation currents are negligible.

2-14. From the expression for $n_b(x)$, given by (2-8), deduce that $n_b(0)$ and $n_b(W)$ are very small compared with n_{bo} when both junctions are appreciably reverse biased. Also, using the first two terms of the infinite series for the hyperbolic sines, show that $n_b(W/2)$ is $n_{bo}W^2/8L_b^2$ for the same bias condition, provided $W \ll L_b$.

Section 2-4

2-15. Find the approximate values of β_F and α_F for the transistor of Fig. 2-6 at 5 V with a quiescent current between 3 and 4 mA. Refer to (2-32) and (2-33).

2-16. An NPN silicon transistor at 300 K has $I_B = 10\ \mu A$, $I_{CO} = 1$ nA, and $\beta_F = 100$. (a) With I_B maintained constant, at what temperature in °C does the term $(\beta_F + 1)I_{CO}$ in (2-32) become 10% of I_C, assuming β_F is 200 at the higher temperature? (b) Repeat with I_B maintained constant. Assume I_{CO} doubles for each 10°C rise in temperature; i.e., $I_{CO2} = 2^{0.1(T_2 - T_1)}I_{CO1}$.

2-17. Although the CE output characteristics, such as those of Fig. 2-9a, appear to pass through the origin, this is not the case. When I_C is zero, V_{CE} is slightly positive, and this voltage is called the *offset voltage*. Using the Ebers–Moll equations calculate the offset voltage in mV at 290 K for an NPN transistor with $I_B = 50\ \mu A$ and having $I_{ES} = 0.5$ pA, $I_{CS} = 1$ pA, and $\alpha_F \approx 1$. (*Ans*: 17.3)

2-18. (a) For an NPN transistor in the active mode show that

$$I_B = qA\left(\frac{0.5Wn_{bo}}{\tau_b} + \frac{D_e p_{eo}}{L_e}\right)e^{qV_{BE}/kT}$$

Use this result to explain why the 10 V curve of Fig. 2-9c is below the 1 V curve. Also, give a physical explanation. (b) Noting the effect of base-width modulation on the emitter current, explain why the zero-volt curve of Fig. 2-9e is below the 40 V curve.

Section 2-5

2-19. An integrated-circuit NPN transistor has respective saturation currents I_{ES} and I_{CS} of 5.1×10^{-16} and 25×10^{-16} A, and the respective values of α_F and α_R are 0.98 and 0.20. At 290 K the junction hole currents are given by (2-27) and (2-28) with $qAD_e p_{eo}/L_e$ equal to 10^{-18} A and $qAD_c p_{co}/L_c$ equal to 2×10^{-15} A. The transistor is driven with a base current of 50 μA with the collector open-circuited. (a) Calculate the junction voltages V_{BE} and V_{BC}, and determine V_{CE} assuming internal IR drops are negligible. (b) Find the ratio $n_b(0)/n_b(W)$. (c) Calculate the hole currents I_{BB}, I_{BE}, and I_{BC} in μA. (d) Determine the electron currents in μA across each junction (*Ans*: 49.9, 48.8).

2-20. An NPN transistor with $\alpha_F = 0.96$ and $q/kT = 40$ has an open emitter ($I_E = 0$). Using the Ebers–Moll equation (2-10) and the reciprocity relation, calculate V_{BE} when V_{BC} is 0.65 V and also when it is -5 V.

2-21. An NPN transistor biased in the normal region has an emitter current of -3 mA. Assume the base and emitter defects of (2-41) are negligible, and the electron mobility is everywhere three times the hole mobility. In the positive x direction determine in the emitter, base, and collector regions, respectively, the following currents in mA: (a) electron drift, (b) electron diffusion, (c) hole drift, (d) hole diffusion. Tabulate results.

2-22. The base of an NPN transistor at 290 K is open and V_{CE} is 10 V, as shown in Fig. 2-13. Assume α_F is 0.98 and α_R is 0.49. Calculate V_{BE}, and from the Ebers–Moll equation for I_C show that I_C is $52I_{ES}$.

Section 2-6

2-23. The dopant densities of the emitter, base, and collector of an NPN silicon transistor ($\varepsilon = 10^{-10}$) are 10^{25}, 10^{23}, and 10^{23}, respectively. At 290 K with $V_{BE} = 0.7$, calculate the punch-through voltage assuming the distance between the metallurgical junctions is (a) 10^{-6} and (b) 0.5×10^{-6}. Units are SI. [*Ans*: (b) 31].

2-24. Deduce that $V_s = V_a(1 - \alpha_F)^{1/n}$. With $\alpha_F = 0.98$ calculate the sustaining voltage of a transistor having an avalanche breakdown of 80 V. Assume $n = 3$ in (2-44).

2-25. (a) Add MI_{CO} to the right side of (2-42) and determine the expression for I_C corresponding to (2-43). (b) Assume I_{CO} is 1 nA, α_F is 0.98, V_a is 80, and n in (2-44) is 4. With $I_B = 0$ calculate I_C in mA when V_{CB} is exactly 0.99999 times the sustaining voltage V_s. Refer to P2-24.

Section 2-7

2-26. In a graded PN junction the charge density ρ caused by unneutralized ions within the space-charge layer is $\rho = ax$ from $x = -0.5l$ in the P region to $x = 0.5l$ in the N region, with a denoting a constant. From this and the expression of P1-30 for l, deduce (2-46) which gives the charge q_{SC} supplied to the space-charge layer when v_{BC} is applied. Also, find a_2 in (2-46) in terms of a and ε.

2-27. A transistor in the active mode has $C_\mu = 5$ pF when V_{CE} is 0.7 V. The forward bias on the emitter junction is 0.7 and the built-in voltage of the collector junction is 0.9. V_{CE} is increased to 10 V with V_{BE} kept constant at 0.7 V. (a) Determine C_μ in pF, assuming graded junctions. (b) Discuss the effect on C_π, assuming V_{BE} is adjusted slightly so as to keep the collector current constant.

Section 2-8

2-28. A PNP transistor biased in the normal region has an emitter current of 1 mA. Assume the base and emitter defects of (2-41) are negligible, and the electron mobility is everywhere three times the hole mobility. In the positive x direction determine in the emitter, base, and collector regions, respectively, the following currents in mA: (a) hole drift, (b) hole diffusion, (c) electron drift, (d) electron diffusion (*Ans*: 0, -3, 0). Tabulate results.

Chapter Three
Field Effect
Transistors

The field effect transistor, or FET, finds its greatest use in integrated circuits, especially in the digital area. The circuitry of a single chip often contains several thousand FETs, which are used not only as active devices but also as resistors and capacitors. In comparison with most bipolar ICs, FET integrated circuits can be produced with greater complexity at lower cost, although they suffer the disadvantage of slower switching speed. However, in industrial controls, automatic instrumentation, optoelectronics, telephone switching, electronic calculators, and many additional applications outside the central processors of large computers, the speed of the system is determined by factors other than the speed of the electronics, and the cost advantage is of first importance. Furthermore, improvements in the technology are leading to high-performance FET circuits that operate at high speeds.

In linear electronics the FET has special properties of particular significance. One of these is the extremely high input resistance of the insulated-gate field-effect transistor (IGFET), more commonly referred to as a MOSFET. The junction field-effect transistor (JFET) also has high input resistance, and low noise as well. These and other properties are exploited in various applications in linear circuits, many of which will be examined in later chapters. The difference in the cost is usually not a significant factor in the selection of the type of a discrete device to be used in a circuit. For most purposes the BJT has superior performance. The *transconductance* of a device is the incremental change in its output current divided by the corresponding change in the input voltage. The *transition frequency* ω_T, which is the ratio of the transconductance to the sum of the device incremental capacitances, is an excellent figure of merit, and this figure of merit is an order of magnitude higher for the BJT than for the FET. The transistor capacitances have a dominant effect on the bandwidth. For the same bandwidth the BJT has a larger transconductance and hence more voltage gain than the FET, which is a major consideration in amplifiers. In addition, the low saturation voltage of the BJT combined with its higher gain makes it generally more suitable for use in power amplifiers. However, power FETs are not as susceptible to thermal runaway and can provide less distortion, both of which are desirable in audio amplifiers. A thorough understanding of the unique characteristics of field-effect transistors is essential for proper utilization of these devices in electronic circuits. We begin our study with the most important member of the family—the enhancement-mode MOSFET.

3-1. THE ENHANCEMENT-MODE MOSFET—THEORY

A simplified structure of a p-channel MOSFET is illustrated in Fig. 3-1. The metallic *gate* is electrically isolated from the silicon by a very thin insulator. This insulator is usually an oxide, and the most common choice is silicon dioxide SiO_2. The letters MOS represent *metal-oxide-semiconductor*, and the device is often referred to as a MOS transistor, or simply *MOST.*

Each of the two metallic layers of Fig. 3-1 located on opposite sides of the gate is in electrical contact with separate P regions, which constitute the *source S* and the *drain D*. These are shown in the cross-sectional view of Fig. 3-2, along with the metallic layer on the bottom that provides an electrical connection to the substrate U. The voltage applied to the substrate terminal is almost always such

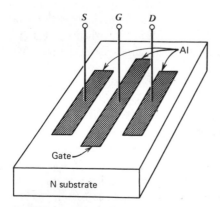

Figure 3-1   MOSFET structure, p-channel (detail shown in Fig. 3-2).

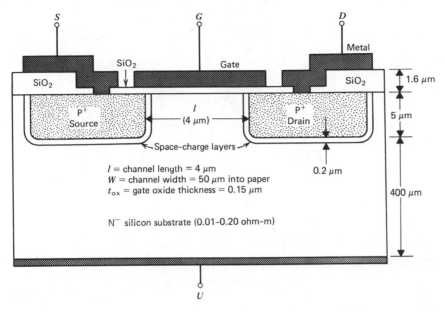

Figure 3-2   Enhancement-mode p-channel MOSFET (all voltages zero).

as to give either a reverse bias or zero bias to each junction. Normally this is accomplished by joining together the substrate and source terminals, and this connection is sometimes made within the device during the manufacturing process. In discrete MOSFETs the substrate is likely to be connected to the metal case.

The sketch of Fig. 3-2 is not to scale. The dimensions are given in micrometers, or *microns*, for a typical MOSFET in a digital integrated circuit. As indicated, the *channel length l* is the distance between the doped source and drain regions, and *l* may be substantially greater than or less than the stated 4 μm. The *channel width W* is the width of each P region along the surface in the direction normal to the length *l*. It may be more or less than the 50 μm shown on the figure. In a power device designed to handle a large current the width would be much greater. The gate-oxide thickness t_{ox} is generally between 0.08 and 0.20 μm, a thickness only about 400 to 1000 atomic diameters. Both P regions are heavily doped, and the N substrate is lightly doped, having a resistivity of from 0.01 to 0.20 ohm-m. Consequently, the space-charge layers of the PN junctions lie wholly within the substrate.

Let us suppose that the substrate and gate terminals are connected to the source, along with a battery between *D* and *S* that makes $V_{DS} = -6$ V. The negative drain voltage tends to drive an internal current from the source through the silicon substrate to the drain. However, the current is practically zero,

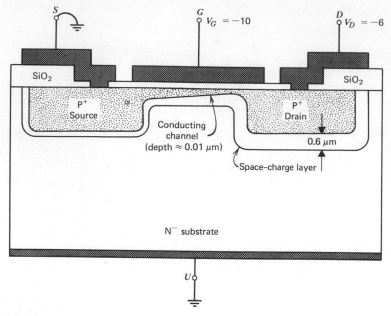

Figure 3-3   p-channel MOSFET with applied voltages.

because the voltage across the source-substrate junction is zero and the other junction is reverse biased. The two junctions are arranged back-to-back in the source-drain circuit. This is somewhat similar to a PNP junction transistor with no base terminal.

Now suppose V_{GS} is changed from zero to -10 V with V_{DS} remaining at -6 V. This is illustrated in Fig. 3-3. The negative gate attracts positive holes that are present in the N substrate and in the source and drain regions and repels free electrons. In a thin surface region the equilibrium hole density in the N substrate becomes greater than the equilibrium free-electron density, with the result that this surface region has changed from N-type to P-type. As shown in the figure, there is a continuous P region from the source to the drain, and V_{DS} causes holes to flow from the source through the channel to the drain. This is a majority-carrier current which flows by the drift process. It crosses no space-charge layers, and there are no diffusion currents. Because their operation depends on only a single type of charge carrier, FETs are *unipolar transistors*. In contrast, the bipolar junction transistor requires both hole and electron currents.

The metal gate, the oxide insulator, and the channel form a capacitor. Within the insulator of this capacitor is an electric field associated with the gate voltage. The field controls the channel conductance and consequently the current. As we shall see, this *field effect* can be used to amplify signals. With zero gate voltage and V_{DS} negative, there is no conduction, but when V_{GS} is sufficiently negative, a channel is formed and current flows. The more negative we make V_{GS} the greater the current. Thus the gate voltage *enhances* the conductance. An FET that conducts appreciably only when a nonzero voltage is applied to the gate is called an *enhancement-mode field-effect transistor*.

It is informative to compare the basic operation of a p-channel device with that of a PNP junction transistor. In the p-channel MOSFET the holes flow from the source through the channel to the drain, with the flow controlled by the gate voltage. In the PNP transistor holes flow from the emitter through the base to the collector, with the flow controlled by the base current. Accordingly, there is a functional correspondence between the source and the emitter, the gate and the base, and the drain and the collector.

Let us examine in more detail the formation of the channel. Regardless of the type of substrate, the silicon dioxide that coats the surface of the MOSFET nearly always contains positive immobile ions, which enter the oxide during the manufacturing process. These positive ions attract the free electrons of the silicon and repel the holes. This is shown in Fig. 3-4 for a substrate with a donor density

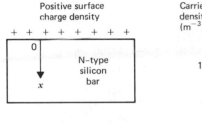
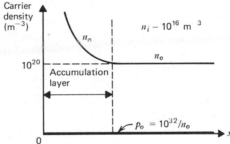

Figure 3-4   Accumulation layer.

of 10^{20} m^{-3}, with n_i assumed to be 10^{16}. Note the rise in the equilibrium free-electron density n_o near the surface. In this region the hole density p_o is lower in accordance with the relation $p_o n_o = n_i^2$. Because the curve of p_o-versus-x is practically on the axis, the reduction in p_o is not evident. The increased majority carrier density near the surface is called an *accumulation layer*. With a gate-source voltage of zero the accumulation layer is present, and there is no channel.

For V_{GS} negative, the negative charge deposited on the metal gate tends to cancel the effects of the fixed positive ions within the silicon dioxide. The electric field that penetrates into the surface region of the substrate depends on the *net* surface charge. For a small negative V_{GS} of perhaps -1 V, the net charge density is negative, though small, as shown in Fig. 3-5. Now the free electrons of the

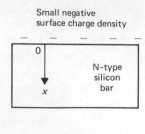
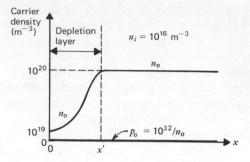

Figure 3-5   Depletion layer.

silicon are repelled and holes are attracted. However, n_o is still greater than p_o. As shown in the figure, at the surface at $x = 0$ the equilibrium free-electron density has been reduced to 10% of the donor density, and of course, p_o has increased to 10^{13}. In the region from 0 to x' the semiconductor is still N-type, but the region is somewhat depleted of charge carriers and a *depletion layer* exists. A space charge is present due to the unneutralized donors.

In order to form a channel the magnitude of V_{GS} must be increased sufficiently to form an *inversion layer*, as shown in Fig. 3-6. Here the negative surface charge density is large, producing a strong field that makes p_o greater than n_o within a thin surface layer. The curves of p_o and n_o cross at x_1, where each carrier density equals n_i. From 0 to x_1 the silicon has become P-type, because p_o is greater than n_o, and this is the conducting channel. From x_1 to x_2 there is a depletion, or space-charge, layer. This layer is shown in the cross section of Fig. 3-3, where we observe that *a single continuous depletion region isolates the source, channel, and drain from the bulk substrate*. The increased depth of the layer at the drain is a consequence of the 6 V of reverse bias, whereas the source and substrate terminals are connected together.

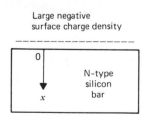
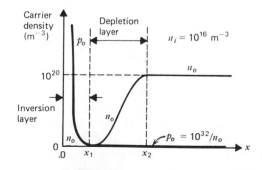

Figure 3-6   Inversion layer.

3-2. THE ENHANCEMENT-MODE MOSFET—CHARACTERISTICS

P CHANNEL

A family of drain characteristics is presented in Fig. 3-7. There is no drain current for very small values of v_{GS} because there is no inversion layer at the source end. The minimum gate voltage required for formation of the channel is known as the *threshold voltage* V_T.

Theoretical expressions for the static drain current i_D can be derived. It is necessary to make a number of simplifying approximations, and the results depend on these. For most p-channel enhancement-mode MOSFETs the following equations are reasonably satisfactory, with the understanding that v_{GS} and v_{DS} are restricted to *negative* values:

$$i_D = 0 \quad \text{for} \quad |v_{GS}| < |V_T| \tag{3-1}$$

$$i_D = k[2(v_{GS} - V_T)v_{DS} - v_{DS}^2] \quad \text{for} \quad |v_{DS}| \leq |v_{GS} - V_T| \tag{3-2}$$

$$i_D = k(v_{GS} - V_T)^2 \quad \text{for} \quad |v_{DS}| > |v_{GS} - V_T| \tag{3-3}$$

with

$$k = -\left(\frac{\mu_h \varepsilon_{ox}}{2t_{ox}}\right)\left(\frac{W}{l}\right) \tag{3-4}$$

The constant k depends on the hole mobility μ_h in the channel, the permittivity ε_{ox} and the thickness t_{ox} of the gate insulator, and the ratio of the channel width

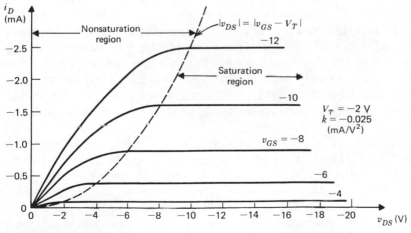

Figure 3-7   Drain characteristics, p-channel enhancement-mode MOSFET.

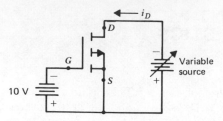

Figure 3-8   Circuit with a p-channel enhancement-mode MOSFET.

W to the channel length l. In an integrated circuit (IC) the constant k is typically about $-0.002\ W/l$ mA/V², with the circuit designer selecting the desired W/l ratio for each transistor of the IC. The curves of Fig. 3-7 were plotted from the preceding equations with $V_T = -2$ V and $k = -0.025$ mA/V².

Let us examine the characteristic curve of Fig. 3-7 having $v_{GS} = -10$ V. The data for the curve could be obtained from the circuit of Fig. 3-8. Note the symbol for the enhancement-mode device. The broken line between the source and the drain symbolizes the lack of a conducting path with no gate voltage. The gate is properly represented as being isolated from the source and drain, and the location of the gate terminal identifies the source when G, S, and D are omitted. The arrow on the substrate terminal, always drawn on device symbols in the direction of easy current flow, informs us that the substrate is N-type, and hence the MOSFET is a p-channel device. An n-channel enhancement-mode MOSFET has the same symbol, except the arrow is reversed. Oftentimes the substrate connection is not shown. In Fig. 3-8, v_{GS} is -10 and v_{DS} is variable.

Suppose v_{DS} is varied from 0 to -18 V. With v_{DS} zero there is no current through the channel. Both v_{GS} and v_{GD} are -10 and the height of the inversion layer is uniform. The point on the characteristic is precisely at the origin. The fact that FET characteristics pass precisely through the origin is sometimes important. As shown in P2-17, this is not the case for the BJT. Let us change v_{DS} from 0 to -6 V. The drain current rises, but the curve is not a straight line because the channel conductance decreases as the voltage is raised. At $v_{DS} = -6$ V, v_{GS} is still -10 but v_{GD} is only -4, and the inversion layer at the drain end is not so strong. This situation is illustrated in Fig. 3-3. At -8 V, v_{GD} becomes -2, which is the threshold voltage V_T.

For larger values of $|v_{DS}|$ the current is saturated at -1.6 mA. There is now no inversion layer at the drain end of the channel, and the channel is said to be pinched off. The pinched off portion extends toward the source, as illustrated in Fig. 3-9 for $v_{DS} = -18$. In this region there is an extremely large current density with a very small cross section, and the flow is through a space-charge layer. At the left side of the pinch-off region the electric potential is -8 V, which gives the threshold voltage across the oxide, and the potential is -18 at the drain end.

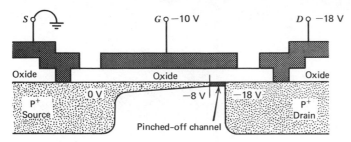

Figure 3-9   MOSFET with pinched-off channel.

Thus there is a 10 V drop across this very short length. If v_{DS} is changed to -20, the length of the pinch-off region increases a small amount and the drop across it rises to 12 V.

In the actual channel there is very little change. At each point the channel height is nearly independent of v_{DS}. The effective length from the source to the pinched off region is slightly reduced as $|v_{DS}|$ is increased, giving a small reduction in the resistance. Because the voltage across this resistance remains constant at 8 V, there is a small increase in the current as $|v_{DS}|$ increases. This is not shown in the characteristics of Fig. 3-7, and it is not indicated in (3-3). However, in the *saturation region*, defined as that region of operation in which the channel is pinched off at the drain end, which is the flat region to the right of the knees of the curves of Fig. 3-7, each characteristic curve does indeed have a positive slope, though small. This slope is greater for devices with very short channels because of the increased effect of pinch-off on the effective channel length. The saturation region is also referred to as the *pinch-off region*, and the threshold voltage V_T is often called the *pinch-off* voltage V_P, particularly in depletion-type FETs which are examined in the next section.

Shown in Fig. 3-10 is the saturation-region static transfer characteristic i_D-versus-v_{GS} for the MOSFET of Fig. 3-7, along with a plot of $\sqrt{|i_D|}$ versus v_{GS}. By (3-3) i_D is proportional to $(v_{GS} - V_T)^2$. This *square law* is especially important in certain applications, such as RF amplifiers and mixers. With v_{DS} constant, the drain current increases in magnitude as $|v_{GS}|$ increases, due to the stronger inversion layer giving a greater channel depth. From experimental data relating i_D and v_{GS} it is easy to determine both k and V_T. A plot of $\sqrt{|i_D|}$ versus v_{GS} is a straight line as indicated in the figure. The constant k is found from the slope, and V_T is found from the intercept with the voltage axis. Actually, the transition between the cutoff and saturation regions is gradual. However, V_T is precisely defined as the intercept of the straight-line portion extended to the axis even though the true curve deviates from the straight line in this region.

Both k and V_T decrease in magnitude as the temperature T in kelvin units increases. All the parameters in (3-4) for k are constants except the mobility.

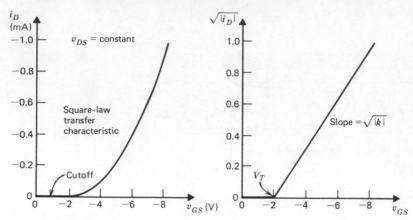

Figure 3-10   Saturation-region transfer characteristic.

The mobility is approximately proportional to $T^{-3/2}$ and so is k. This relation can be used to calculate k at a temperature different from the one at which the measurement was made. An increase in T causes the hole density in the N substrate to rise. The increased number of holes makes easier the formation of an inversion layer, which is one reason why V_T decreases in magnitude. Another reason is the effect of temperature on the trapping of carriers due to imperfections in the surface region. Because FETs do not depend on minority-carrier currents and lifetimes, they are not nearly so temperature sensitive as BJTs. In Fig. 3-10 the abscissa intercept shifts to the left and the slope decreases as the temperature rises.

CAPACITANCES

In digital circuits a disadvantage of the p-channel MOSFET made by the standard process is the relatively high threshold voltage. Whereas it takes only about 0.6 V applied to the base of a BJT to cause it to turn *on*, or conduct, the MOSFET requires from 2 to 4 V. The metal gate, the thin insulator, and the substrate form a capacitor, and the charge on the gate equals the product of the capacitance C and the gate voltage. Increasing the capacitance C per unit area gives more charge per applied volt, and the more effective gate has a reduced threshold voltage.

An obvious way to increase C is to utilize a thinner oxide layer. However, there is a limit here. Too thin an oxide results in pinholes as well as greater susceptibility to rupture by static charges. An alternate way to increase C is to use an insulator with a very high dielectric constant. A layer of silicon nitride (Si_3N_4) is satisfactory for this purpose except for difficulties that arise from carrier trapping at the interface of the substrate and the nitride. These are resolved by using a 0.04 μm layer of silicon nitride over a 0.06 μm layer of silicon dioxide next to the substrate. *Ion implantation* can be employed to achieve lowered threshold voltages. This consists of exposing the oxide-coated surface to incident high-energy boron ions. These negative ions pass through the thin gate oxide into the surface region of the substrate, where they serve to neutralize at least partially the positive ions always present in the oxide layer. Thermal methods are also used to introduce negative ions into either the oxide insulator or the surface region of the substrate. With these processes threshold voltages below 1 V are possible. Ion implantation also significantly reduces the required surface area of a device, thereby reducing capacitance, improving high-frequency performance, and allowing greater circuit densities. It is discussed further in Sec. 4-4.

Because the capacitor formed by the gate, the oxide, and the conducting channel is similar in geometry to a parallel-plate capacitor, the capacitance C_{ox} per unit area is the ratio ε_{ox}/t_{ox}. For silicon dioxide the dielectric constant is 4, and ε_{ox} is $4\varepsilon_0$, with $\varepsilon_0 = 8.854 \times 10^{-12}$ F/m. Typically, C_{ox} is about 3×10^{-4} pF per square micrometer. In small-signal amplifiers the transistors are normally biased in the saturation region, and the incremental capacitances C_{gs} and C_{gd} are used in the models. When v_{GS} is varied with v_{GD} constant, the channel height changes as does the stored charge. It is not difficult to show that C_{gs} approximately equals the sum of $\frac{2}{3}C_{ox}$ and the capacitance between the metal gate and the source, including both the doped P region and its metal terminal. However, when v_{GD} is varied with v_{GS} constant, conditions in the channel do not change appreciably, which is indicated in Fig. 3-9. Hence, C_{ox} does not contribute to C_{gd}, and C_{gd} is simply the capacitance between the metal gate and the drain. This is not the case in the nonsaturation region, where C_{gd} is considerably larger because of the effect of the charges stored in the channel. There are also capacitances associated with the space-charge layers.

Data sheets specify typical and maximum values for C_{iss} and C_{rss}. According to conventional subscript notation, the first of these is the common-source (CS) short-circuit incremental input capacitance, which is the sum of C_{gs} and C_{gd}. The second is the CS incremental reverse transfer capacitance, which is C_{gd}. Sometimes curves are presented showing variations with bias conditions, such as the one of Fig. 3-11. For positive v_{GS}, the capacitance C_{ox} is that of the gate, the oxide, and the accumulation layer. In the vicinity of the dip there is neither an accumulation nor an inversion layer, and C_{ox} is in series with the depletion-layer capacitance. To the right of the dip the inversion layer is present.

MOS integrated circuits use MOS capacitors. One type is simply a MOSFET with the source and drain terminals connected together. The capacitance between the gate and source is similar to that of Fig. 3-11. Another type, which is often employed when one side of C is connected to ground, is the isolated capacitor that utilizes the capacitance between the gate and the substrate, with the source and drain dopants omitted. A third type often used in linear bipolar integrated circuits is made by extending the source doping so as to include the entire region

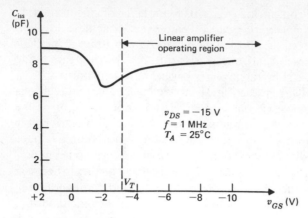

Figure 3-11 C_{iss} versus v_{GS}.

normally occupied by the channel, with the drain omitted. This is essentially a parallel-plate linear capacitor, consisting of the metal gate, the oxide, and a doped P or N region. Such a MOS capacitor is shown in Fig. 4-5 of Sec. 4-2.

THE NONSATURATION REGION

For values of $|v_{DS}|$ less than $|v_{GS} - V_T|$ the channel is not pinched off anywhere, and operation is in the *nonsaturation*, or *triode*, or *linear*, region. The channel conductance G is the ratio i_D/v_{DS}. From (3-2) we obtain

$$G = 2k(v_{GS} - V_T - \tfrac{1}{2}v_{DS}) \tag{3-5}$$

With $v_{DS} = 0$, the conductance is designated G_o, which is

$$G_o = 2k(v_{GS} - V_T) \tag{3-6}$$

For v_{DS} sufficiently small and with v_{GS} constant, the channel has the properties of a *linear resistor* with an ohmic value dependent on v_{GS}. Thus the MOSFET can be used as a voltage-controlled variable resistor (see P3-4).

N CHANNEL

The discussion has focused thus far on the p-channel device. The characteristics of n-channel enhancement-mode MOSFETs are similar. The substrate is P-type, the source and drain are N-type, and the normal current is a flow of electrons from the source, through the channel, to the drain. This flow is similar to that of an NPN transistor. The drain current i_D and the voltages v_{DS} and v_{GS}

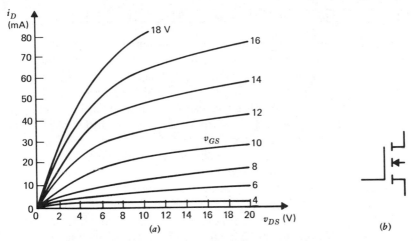

Figure 3-12   n-channel enhancement-mode MOSFET. (*a*) Drain characteristics. (*b*) Symbol.

are positive, as shown in Fig. 3-12. All equations of this section are applicable except that the constant k of (3-4) becomes

$$k = \left(\frac{\mu_e \ell_{ox}}{2t_{ox}}\right)\left(\frac{W}{l}\right) \tag{3-7}$$

The n-channel device has the advantage of higher frequency response in amplifiers and greater switching speed in digital networks. This is due to the higher mobility of the electrons, which also allows physically smaller channels with correspondingly lower capacitance. The manufacture of n-channel enhancement-mode integrated circuits is more complicated, because the positive charge of the silicon-dioxide layer tends to create an inversion layer with zero gate voltage. Special processing is required to raise the threshold voltage to a positive value. Ion implantation can accomplish this. Another common technique is the use of gates made of doped silicon, rather than aluminum. Silicon gates (SG) raise the threshold voltage substantially, perhaps a volt or more. They are also occasionally used with PMOS, making V_T *less negative* as desired. Further discussion is presented in Sec. 8-4.

The earliest and cheapest MOS integrated circuits were made exclusively of p-channel enhancement-mode devices, but the trend is to employ n-channel devices in order to achieve high performance. Complementary MOSFET (CMOS) circuits consisting of matched p-channel and n-channel devices are popular. These have advantages of high speed and very low power dissipation. CMOS circuits are examined in Chapter 8.

3-3. THE DEPLETION-MODE MOSFET

A *depletion-mode* field-effect transistor is one in which an applied voltage v_{DS} produces appreciable drain current with zero gate-source voltage. The channel is usually a very thin doped region, but it may be simply an inversion layer produced by ions in the SiO_2 layer. Because the positive ions that are normally present in the oxide attract electrons and repel holes, n-channel depletion devices with P-type substrates are easier to make, and consequently, most depletion-mode MOSFETs are n-channel.

The geometry is illustrated in Fig. 3-13. The n-channel is doped, and there are depletion regions both above and below the channel. The value of V_{GS} is -3 and V_{GD} is -7. These negative voltages repel electrons, producing a depletion region between the silicon dioxide layer and the channel. If the gate is shorted to the source with $V_{DS} = 4$, electrons flow from the source through the doped channel to the drain, and the drain current is positive. A negative voltage now applied to the gate increases the portion of the channel that is depleted of charge carriers, and the current drops. Hence the name *depletion-mode*.

It should be noted that the metal gate does not overlap the doped drain region. In the enhancement-mode MOSFET the gate must overlap both the source and the drain in order to obtain an inversion-layer channel connecting the two. This is, of course, unnecessary in the depletion-mode MOSFET, and the device is constructed with a reduced gate-drain capacitance, *which makes it more suitable for high-frequency amplification.*

A typical family of drain characteristics is presented in Fig. 3-14. With v_{DS} positive and v_{GS} less than -5, which is the *pinch-off* voltage V_P (or threshold voltage), the channel is completely pinched off throughout the region from source to drain, and there is no current. As v_{GS} is increased from -5 to 0, the drain current rises. This is the depletion-mode region of operation. For positive values of v_{GS} the current is even greater, because the positive gate attracts more electrons

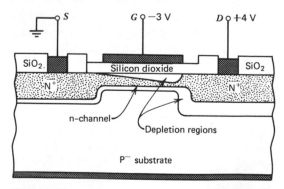

Figure 3-13   n-channel depletion-mode MOSFET.

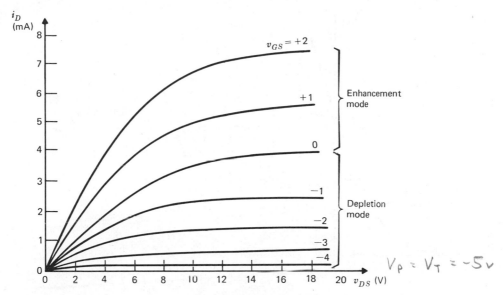

Figure 3-14   Drain characteristics, n-channel depletion-mode MOSFET.

into the n-channel, enhancing its conductance. This is the enhancement-mode region of operation. Thus a depletion-mode MOSFET can operate in the enhancement-mode, though this is not common.

Let us consider the -2 V characteristic of the figure. Increasing v_{DS} from zero causes the drain current to rise. We might expect the channel to be pinched off and the current to be saturated when v_{DS} is 3 V, which makes $v_{GD} = V_P$. However, the offset gate leaves a portion of the channel between the point of pinch-off and the drain, and a voltage exists across this length. Hence saturation occurs at about 5 V. The saturation and nonsaturation regions of the family of curves are defined similarly to those of the enhancement-mode MOSFET.

THE SQUARE LAW

The numbered equations of the preceding section apply to all MOSFETs. However, it is convenient to express (3-3) in the form:

$$i_D = I_{DSS}\left(1 - \frac{v_{GS}}{V_P}\right)^2 \tag{3-8}$$

This is the saturation-region drain current of a *depletion-mode* MOSFET. The term I_{DSS} is the product kV_P^2, and V_P corresponds to V_T. For an n-channel device I_{DSS} is positive, V_P is negative, and (3-8) is valid for v_{DS} greater than $(v_{GS} - V_P)$.

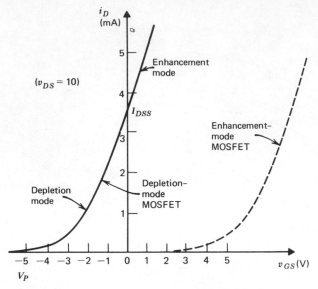

Figure 3-15   Saturation-region static transfer characteristic, n-channel.

Because I_{DSS} denotes i_D with $v_{GS} = 0$, (3-8) applies to an enhancement-mode device only if I_{DSS} is replaced with kV_P^2.

The static transfer characteristic of Fig. 3-15 shows the depletion and enhancement modes of operation, the square-law curve, and both I_{DSS} and V_P. For comparison, an enhancement-mode characteristic is given as a dashed line. Its I_{DSS} is the very small cutoff-mode current, perhaps a nanoampere, with $v_{GS} = 0$.

BREAKDOWN

The drain characteristics of Fig. 3-14 do not show the breakdown region. With v_{GS} constant, if we increase v_{DS} until breakdown is approached, each characteristic will turn upward rather sharply. The large increase in i_D is caused by one of two possible mechanisms. *Avalanche* breakdown occurs at the reverse-biased PN junction between the substrate and the drain. This is similar to that discussed for the diode in Sec. 1-9. The other mechanism, which usually dominates for channel lengths less than about 10 μm, is *punch-through*, which was discussed in Sec. 2-6 with regard to the BJT. The reverse bias at the drain PN junction produces a depletion layer with a length that increases as v_{DS} increases. If this layer reaches the source, the barrier at the source junction is broken down and an effective short-circuit appears between the source and drain. Both types of breakdown are influenced by the geometry of the metal gate and its thin oxide with the result that the so-called *gate-modulated* breakdown voltage is substantially lower than normally would be expected. Similar effects occur in enhancement-mode MOSFETs.

Some form of gate protection for the MOS transistor is desirable. The breakdown field in silicon dioxide is about 10^9 V/m. For an oxide thickness of 0.1 μm the breakdown voltage would be 100 V. Allowing a wide safety margin in consideration of the fact that the oxide layer is not perfectly uniform, manufacturers specify a maximum gate voltage, from the gate to any other terminal, between about 25 and 80 V. Special precautions must be taken to avoid exposure of the gate to stray electrostatic charges. While handling MOS devices, their leads should be kept short-circuited. Protective devices incorporated into the circuit give partial protection. In the n-channel depletion-mode MOSFET of Fig. 3-16 a back-to-back diode arrangement is used for this purpose. As long as the gate voltage is not excessive, the diode arrangement is an open-circuit, but the breakdown voltage of each diode is below the maximum allowable. Unfortunately, this protection increases the gate capacitance which reduces switching speeds and high-frequency response. Frequently the diodes are integrated into the circuit during the manufacturing process. Note that the symbol for a depletion-mode MOSFET has a continuous line between the source and drain, symbolizing a conducting path with the gate shorted to the source.

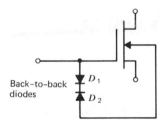

Back-to-back diodes D_1 D_2

Figure 3-16 Depletion-mode n-channel MOSFET.

DISCUSSION

There are two reasons for favoring enhancement-mode MOSFETs over depletion-mode devices in integrated circuits. The first of these is the fact that the gate and drain terminals of an enhancement-mode MOSFET in the CS configuration can have the same dc voltage. This is a great advantage. Cascaded stages can be connected directly together without interstage circuitry to change the dc level. Simpler circuit layouts are possible, and dc logic is practical. With BJTs and depletion-mode FETs, including both insulated-gate and junction types, such interstage circuitry is unavoidable. However, dc logic interconnection can be retained and other advantages gained by using depletion-mode

MOSFETs as "load" devices. This is often done, employing ion implantation to form simultaneously both enhancement and depletion MOSFETs within one substrate. The second reason is the reduced number of processing steps resulting from the absence of a diffused channel. This lowers the cost and increases the production yield.

There are a number of applications that require amplifiers with very high input resistance and also low noise. Examples are the input amplifier stage of a radio receiver, audio amplifiers, and amplifiers for electrometers and radiation detectors. The MOSFET excels with respect to input resistance, but its noise characteristics are not ideal. At frequencies from nearly zero to 20 kHz or more, 1/f noise is dominant and quite objectional. The noise is created by fluctuations in the number of charge carriers that occupy surface traps present at the interface between the oxide and the substrate. At high frequencies thermal noise in the channel resistance is dominant, though reasonably low. In comparison with the BJT, with respect to noise the MOSFET is generally superior above a few MHz when the source resistance is larger than about 50 kΩ.

Matched pairs of MOS transistors on a single monolithic chip are commercially available, as well as IC chips with many MOSFETs. Provision is made for external connections to the gates and various other terminals to allow flexibility in circuit design. Then there is the n-channel *dual-gate* depletion-type MOSFET, consisting in effect of two devices with their channels in series. The same diffused N region serves as the drain of one and the source of the other. The arrangement can be used as composite MOS transistor having a very small C_{gd} and a large incremental transconductance. Because applications are primarily in radio-frequency circuits, further discussion is deferred to Sec. 22-7.

3-4. THE JFET

The basic principle of operation of a p-channel junction field-effect transistor will be discussed briefly with reference to Figs. 3-17, 3-18, and 3-19. In the top

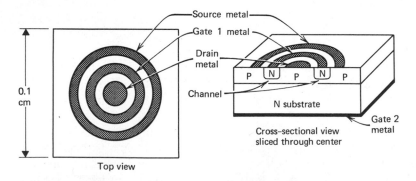

Figure 3-17 p-channel JFET with circular geometry.

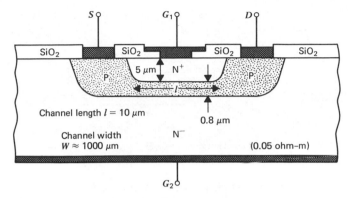

Figure 3-18   Channel cross-section.

view of Fig. 3-17 a circular geometry is shown. Other configurations are also widely used. For example, many JFETs have a winding gate metallization, which zig-zags back and forth while completely encircling either the source or the drain. In each case the geometry is such that holes flowing from the source to the drain must pass through a p-channel that is sandwiched between two N regions, with the junctions reverse biased. The p-channel connecting the P-type source and drain regions is shown in the cross-sectional view of Fig. 3-17. Gate G_2 is usually connected to the source S.

GEOMETRY

In Fig. 3-18 some typical dimensions in micrometers are indicated on the sketch of the channel cross-section. The channel length l varies considerably from one device to another. The width W, given as 1000 μm, is normal to the view. For the circular geometry of Fig. 3-17, W is the circumference of the circular gate metallization. Because the drain current is proportional to the channel width, a power device would be designed so that W is large. The channel height h is 0.8 μm, or 8×10^{-7} m. We learned in Sec. 1-6 that the thickness of the space-charge layer of a PN junction in equilibrium is typically about 3×10^{-7} m, more or less depending on the dopant densities. The space-charge layers at both the top and bottom of the channel are shown in Fig. 3-19.

The left side of Fig. 3-19 applies for V_{GS} and V_{DS} both zero. At the top gate-channel junction essentially the entire depletion layer is within the more lightly doped P-type silicon channel. At the substrate-channel junction a significant portion of the depletion layer, perhaps 20%, is within the P-type silicon, because the dopant densities on opposite sides of the junction are about the same order of magnitude, although the substrate is the more lightly doped region. Together, the two depletion layers reduce the height of the conducting channel to about

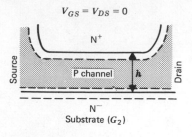
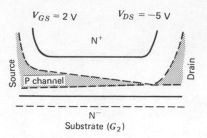

Figure 3-19   Cross-sections of channel and space-charge layers.

three-quarters of h. The right side of the figure shows the channel in the pinch-off condition. Across the gate-source end of the junction there is a reverse bias of 2 V, and across the drain-source end is a reverse bias of 7 V ($V_{GD} = 7$). Both depletion layers are wider. The height of the conducting channel is now a function of distance, with this height being zero at the drain end. As V_{DS} is increased further in magnitude, the additional voltage appears across the slightly increased length of the pinched-off channel that lies in the drain region. Within the length l of the channel there is very little change in either the dimensions or the voltage, and the current is saturated at a nearly constant value.

CHARACTERISTICS AND EQUATIONS

The drain characteristics of an n-channel JFET are shown in Fig. 3-20. These are similar to those of the depletion-mode MOSFET. For v_{GS} more negative than -5 there is no current. This is the cutoff region. In the nonsaturation region where v_{DS} is less than $v_{GS} - V_P$, the drain current is

$$i_D = \frac{I_{DSS}}{V_P^2} \left[2(v_{GS} - V_P)v_{DS} - v_{DS}^2 \right] \tag{3-9}$$

This is the MOSFET equation (3-2) with k replaced with I_{DSS}/V_P^2. Its justification here is based on the fact that it yields results that are usually in reasonably good agreement with the measured behavior of most real JFETs. In fact, it gives better results than various theoretical equations, although in some cases the coefficient outside the brackets is substantially too large.

In the saturation region where the curves are flat the square law (3-8) applies. The characteristics have the illustrated small positive slope, corresponding to a rather large incremental output resistance r_{ds}, perhaps 50 kΩ. The characteristics show that breakdown occurs at $V_{DG} = 25$, and the value of v_{DS} at breakdown depends on v_{GS} as indicated. The characteristics of MOS transistors in breakdown are somewhat similar. Occasionally a supply voltage V_{DD} that is greater than the breakdown value is used. In such cases care must be exercised in the

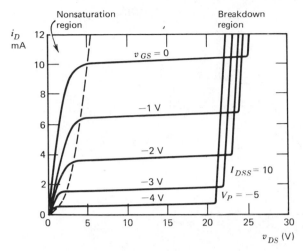

Figure 3-20   CS drain characteristics, n-channel JFET.

design of the bias and signal circuitry to ensure that operation is wholly confined to the saturation region.

A useful approximate expression for V_P is easily deduced by assuming uniformly doped regions. With substrate doping much less than that of the channel, pinch-off occurs when the depletion layer of the G_1-channel junction has a thickness equal to the channel height h. Refer to Fig. 3-19. From (1-36) of Sec 1-5, with l replaced with h, N_D infinite, and V replaced with $-|V_P|$, we obtain

$$|V_P| = \frac{qNh^2}{2\varepsilon} - \psi_o \qquad (3\text{-}10)$$

with N denoting the dopant density of the channel. Of course, ψ_o is the built-in junction voltage of the G_1-channel junction, given by (1-40) of Sec. 1-6. The expression is only approximate, because the doping is nonuniform and the width of the depletion layer of the substrate junction was not considered.

A theoretical expression for I_{DSS} is

$$|I_{DSS}| = \frac{2}{3}\left(\frac{\sigma h W}{l}|V_P'|\right) \qquad (3\text{-}11)$$

As we might expect, I_{DSS} is proportional to the channel conductivity σ, the channel height h, and the width W, and it is inversely proportional to the length l. Similarly identified field-effect transistors from the same manufactured batch have wide variations in I_{DSS} and V_P. The proportionality between I_{DSS} and V_P indicates that a JFET with a large I_{DSS} also has a large V_P, and vice versa. This is an important consideration in bias-circuit design.

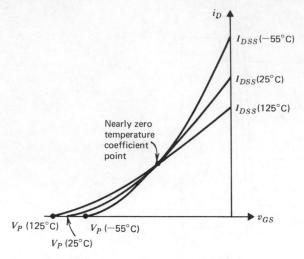

Figure 3-21   Temperature effect on transfer curves.

In the saturation region the transfer characteristic depends on temperature as indicated in Fig. 3-21. From (3-10) we note that $d|V_P|/dT$ equals $-d\psi_o/dT$. We have learned that ψ_o decreases about $2\ \text{mV}/°\text{C}$ as T rises. Accordingly, the pinch-off voltage of a JFET increases as T increases, at a rate of about $2\ \text{mV}/°\text{C}$. I_{DSS} is proportional to the product of the carrier mobility and V_P, as evident from (3-11). These change in opposite directions as T rises. In most cases the effect of the decrease in mobility is greater than that caused by the increase in V_P, provided the pinch-off voltage is greater than about 0.6 V, and I_{DSS} drops as T rises. In dc amplifiers it is very important to maintain a quiescent current that does not drift appreciably with temperature, because a slight change in I_D will give a dc output unrelated to the dc input signal. Excellent temperature stability results from selecting a JFET with a reasonably large V_P and choosing the drain current at the crossover of Fig. 3-21, with the proper value obtained from the data sheet or by laboratory experiment.

INCREMENTAL PARAMETERS

The slope $\partial i_D/\partial v_{DS}$ of a drain characteristic at a particular bias point is the incremental *output conductance* g_{ds}, and its reciprocal is the *output resistance* r_{ds}. It is clear from Fig. 3-20 that the incremental output resistance is quite large in the saturation region. Ideally, it should be infinite, but actually it is large and almost inversely proportional to I_D.

The incremental *transconductance* g_{fs} is the slope $\partial i_D/\partial v_{GS}$ of the transfer characteristic at the bias point. Because a large g_{fs} indicates that the gate voltage exercises strong control over the drain current, large values are desired. The curves of Fig. 3-21, which apply to the saturation region, show that g_{fs} decreases as temperature rises. It is easy to deduce from the square law of (3-8) that the transconductance in the saturation region is

$$g_{fs} = \frac{2}{|V_P|} \sqrt{I_D I_{DSS}} \tag{3-12}$$

The transconductance is proportional to the square root of I_D. With $v_{GS} = 0$ the drain current is I_{DSS}. At this point the transconductance, designated g_{fso}, is a maximum for operation in the depletion mode. From (3-11) and (3-12) it is evident that

$$g_{fso} = \left| \frac{2I_{DSS}}{V_P} \right| = \frac{4}{3} \left(\frac{\sigma hW}{l} \right) \tag{3-13}$$

Because both g_{fso} and I_{DSS} are relatively easy to measure, compared with V_P, the pinch-off voltage is often determined from these measured quantities with the aid of (3-13). In accordance with this equation, a straight line tangent to the transfer curve at I_{DSS} intercepts the v_{GS} axis at $\frac{1}{2}V_P$. Data sheets refer to g_{fs} as $|y_{fs}|$ at 1 kHz, and normally specify minimum, typical, and maximum values of g_{fso}. Sometimes included are curves of g_{fs} versus v_{GS} at various temperatures. The y parameters are discussed in Chapter 12.

The reciprocal of the incremental output resistance r_{ds} is g_{ds}, referred to on data sheets as g_{os}, or $|y_{os}|$ at 1 kHz. The maximum value of g_{ds} is specified, along with a typical value. Of the other incremental parameters specified, C_{iss} and C_{rss} are particularly important. As noted in Sec. 3-2, the incremental capacitances C_{gs} and C_{gd} can be determined from the relations

$$C_{iss} = C_{gs} + C_{gd} \qquad C_{rss} = C_{gd} \tag{3-14}$$

The value of C_{gs} is determined by the charge Q stored in the space-charge layer of the gate-channel junction. With v_{DS} constant, a change dv_{GS} changes the thickness of the space-charge layer through the channel length, and the ratio dQ/dv_{GS} is the major component of C_{gs}. There is also a small parasitic capacitance directly between the source and gate metallizations. Typical values are from 3 to 20 pF. The value of C_{gs} is approximately proportional to the product σlW and inversely proportional to the channel height h.

In the saturation region C_{gd} is smaller than C_{gs} because a change in v_{DS} has practically no effect on the charge within the actual channel. However, the space-charge layer extends beyond the channel into the drain region, and the charge stored here is dependent on v_{DS}. There is also parasitic capacitance between the gate and drain metallizations. Because it provides signal feedback from the drain output to the gate input of a CS amplifier, C_{gd} is especially objectional. In the geometry of Fig. 3-17 the smallest metallic area was chosen for the drain terminal. Not only does this maximize g_{fs} by minimizing the ohmic resistance in series with the source terminal but it also gives a minimum value for C_{gd}, with typical values from 0.5 to 6 pF. Both C_{gd} and C_{gs} decrease as v_{GS} increases in magnitude, and both are slightly temperature dependent. These are basic characteristics of a space-charge-layer capacitance, as indicated by (1-57) of Sec. 1-10.

Operation in the enhancement mode with v_{GS} greater than 0.5 V is not feasible. This would appreciably forward bias the PN junction, producing a direct gate current and causing the incremental input resistance between the gate and the source terminals to drop to a low value. In the depletion mode the junctions are reverse biased. Although I_G is typically less than a nanoampere at 25°C, at elevated temperatures it may be significant. Data sheets give typical and maximum values of I_{GSS} at room temperature, along with a curve showing the variation with T. With each 10°C rise in temperature I_{GSS} approximately doubles.

NOISE

In Sec. 15-2 we shall briefly examine 1/f noise, thermal noise, and shot noise. All three types are generated within a JFET as in all transistors. The sporadic nature of the thermal generation and recombination processes, especially within the space-charge layers, produces 1/f noise that dominates at low frequencies. Unlike the MOS transistors, this noise in JFETs is usually negligible above a few hundred hertz. Surface traps in MOSFETs contribute substantially to 1/f noise, and such surface traps are not present in the channels of JFETs. There is thermal noise caused by the resistance of the conducting channel. Shot noise results from the gate current passing through the space-charge layer. This is considerably less than the shot noise of a BJT because the gate current is so small. When the source resistance is large, the JFET has a low noise figure in comparison with the BJT, and this is certainly one of its outstanding characteristics. The attributes of low noise and high input impedance make the JFET desirable as the input stage for many audio amplifiers. Both JFETs and MOSFETs are used in narrow-band RF amplifiers when low noise is especially desirable, as in the first stage of a multistage amplifier. However, the depletion-mode MOSFET is becoming dominant for this application because it has significantly lower capacitances.

DISCUSSION

The designer of a JFET usually seeks large values of I_{DSS}, g_{fso}, and r_{ds}, along with small values of V_P, C_{gs}, and C_{gd}. A high breakdown voltage is desirable. By (3-11) and (3-13) both I_{DSS} and g_{fso} are proportional to the channel height-to-length ratio h/l, and the incremental capacitances are inversely proportional to this ratio. Thus h/l should be as large as possible. However, the minimum l is limited by the manufacturing process, and very short channels have lower values of r_{ds}. Unfortunately, a large h gives a large V_P according to (3-10). Small pinch-off voltages are advantageous because they permit operation in the saturation region with low values of v_{DS}. Increasing the dopant density of the channel increases both I_{DSS} and g_{fso} by (3-11) and (3-13) but adversely affects V_P, the capacitances, and the breakdown voltage. I_{DSS}, g_{fso}, and the capacitances are proportional to the channel width W. Clearly the design is a compromise that depends on the intended use of the device.

Many junction field-effect transistors are fabricated by diffusing the desired dopants into a planar epitaxial layer grown on the surface of a silicon bar, or substrate. The method is discussed briefly in Sec. 4-1. For an n-channel device the substrate is P-type and the epitaxial layer is N-type. A carefully controlled P-type diffusion into a portion of the layer forms the gate, with a narrow n-channel remaining between the gate and the substrate. This process provides better control over the geometry and conductivity of the channel than does the planar diffused technique without an epitaxial layer. Although the source and gate terminals are in contact with the N-type layer, a small N^+ diffusion is carried out in the regions below the aluminum contacts to provide a good ohmic contact and to avoid any possibility of formation of P-type silicon by migration of acceptor aluminum atoms into the epitaxial layer. In addition, a deep P-type isolation diffusion is made at the two ends of the bar, with this diffusion extending into the P substrate.

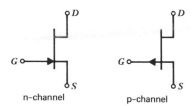

n-channel   p-channel

Figure 3-22   JFET symbols.

The symbols for JFETs are shown in Fig. 3-22. The location of the gate terminal identifies the source and the arrow indicates the direction of forward current flow.

CURRENT-REGULATOR FET DIODES

Among the many applications of JFETs are their use to provide a constant current to a variable load. This is accomplished by shorting the gate to the source and maintaining v_{DS} greater than the pinch-off voltage. Although the designer can employ any JFET or depletion-mode MOSFET in this manner, commercial devices known as *current-regulator diodes* have on-chip metalization that short-circuits the gate to the source. Such current regulators are often used to establish the proper direct current to a discrete transistor of a linear amplifier, and they are

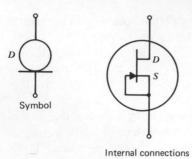

Symbol

Internal connections

Figure 3-23   Current-regulator diode.

sometimes fabricated as part of a monolithic integrated circuit for the same ·purpose. There are many other applications, including generation of various types of waveforms (see P3-15). The symbol is shown in Fig. 3-23.

REFERENCES

1. W. Gosling, W. G. Townsend, and J. Watson, *Field-Effect Electronics*, Wiley-Interscience, New York, 1971.
2. R. P. Nanavati, *Semiconductor Devices*, Intext Educational Publishers, New York, 1975.
3. D. J. Hamilton and W. G. Howard, *Basic Integrated Circuit Engineering*, McGraw-Hill, New York, 1975.

PROBLEMS

Section 3-2

3-1. In Fig. 3-7 determine the value of v_{GD} at points on the dashed curve that separates the saturation and nonsaturation regions. Also, directly from the characteristics find the incremental transconductance g_{fs} in mmhos at $v_{DS} = -10$ and $v_{GS} = -8$, with g_{fs} denoting the slope of the transfer characteristic i_D-versus-v_{GS} with v_{DS} constant.

3-2. A p-channel enhancement-mode MOSFET has v_{DS} constant at -10 V. For $|v_{GS}| = 4.55, 4.35, 4.15, 3.95, 3.80,$ and 3.60 the respective experimental values of $|i_D|$ are 16, 9, 4, 1, 0.25, and 0.04 μA. Plot $\sqrt{|i_D|}$ versus v_{GS}. Determine k in μA/V^2, V_T, and i_D in μA at $v_{GS} = V_T$.

3-3. For $v_{GS} = -6 + V_m \cos \omega t$ in an amplifier using the MOSFET in Fig. 3-7, determine the maximum value of the amplitude V_m of the incremental signal for an output harmonic distortion not greater than 5%. The output voltage is proportional to i_D, and operation is in the saturation region. Refer to (16-19). (*Ans*: 0.8).

3-4. Shown in Fig. 3-24 is a voltage-divider attenuator using a MOSFET having $k = -0.05$ mA/V^2 and $V_T = -3$ V. The input phasor voltage V_i is 10 mV. Find V_0 in mV for V_{dc} equal to 3.5 V and also for 23 V. Assume the device geometry is symmetrical with respect to the source and drain.

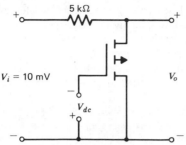

Figure 3-24 Attenuator.

3-5. The enhancement-mode MOSFET of Fig. 3-25 has $k = -0.05$ mA/V^2 and V_T is -4. (a) Deduce that the transistor is in the saturation mode for all values of R. (b) Determine V when R is infinite, 100 kΩ, 20 kΩ, and 2.5 kΩ. ($Ans: -6, -5, -4, -2$).

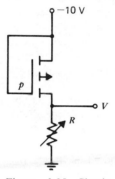

Figure 3-25 Circuit for P3-5 and P3-8.

Section 3-3

3-6. In Fig. 3-26 is a MOSFET used as a current source that supplies a current to the variable load R. I_{DSS} is 2 mA and V_P is -4. (a) For what range of resistance values in kilohms is the current through R constant at 2 mA? (b) Calculate I in mA for R equal to 3, 5, and 10 kΩ.

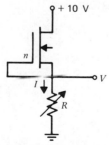

Figure 3-26 Circuit for P3-6, P3-7, and P3-8.

3-7. The MOSFET of Fig. 3-26 has $I_{DSS} = 1$ mA and $V_P = -2$. (a) For what range of resistance values in kilohms does the MOSFET supply a constant current of 1 mA to the load R? (b) Calculate I in mA for R equal to 0, 8, and 25 kΩ (*Ans*: 1, 1, 0.38).

3-8. The MOSFETs of Figs. 3-25 and 3-26 have identical threshold voltages of -4. Assume R is infinite in each circuit. (a) In Fig. 3-25 determine V. Also, what is V if the gate is disconnected from the drain and connected to -20 V? (b) In Fig. 3-26 determine V.

3-9. From (3-8) find the slope of a transfer characteristic, such as that of Fig. 3-15, at the point where it crosses the i_D axis. Then deduce that a straight line through this point with the slope of the characteristic intercepts the v_{GS} axis at $\frac{1}{2}V_P$.

Section 3-4

3-10. Using the drain characteristics of Fig. 3-14, calculate the incremental transconductance g_{fs} in mmhos and the incremental r_{ds} in kΩ at the point $v_{DS} = 8$, $v_{GS} = -1$.

3-11. For the channel cross-section of Fig. 3-18 with the given dimensions, assume uniform dopant densities of 10^{25} and 10^{22} m^{-3} in the N$^+$ gate and channel regions, respectively. Also assume the substrate doping is very light, $q/kT = 40$, $n_i = 10^{16}$, $\varepsilon = 10^{-10}$, and $\mu_h = 0.04$ SI units in the channel. Calculate V_P and I_{DSS} in mA. (*Ans*: 4.26, -14.5).

3-12. (a) Let G_o denote the channel conductance of a JFET with $|v_{DS}|$ very small compared with $|v_{GS} - V_P|$. Deduce that G_o equals $G_{oo}(1 - v_{GS}/V_P)$, and show from (3-9) that the conductance G_{oo} with v_{GS} zero equals g_{fso}. (Actually, G_{oo} may be a good bit smaller than g_{fso}). (b) Calculate the source-drain resistance in kilohms for $I_{DSS} = 2$ mA, $V_P = -4$, $v_{DS} < 0.1$, and v_{GS} equal to -1 and also -2.

3-13. For the drain characteristics of Fig. 3-20 estimate from the curves the numerical values of G_{oo} (see P3-12) and both g_{fs} and r_{ds} at $v_{GS} = -1$ and $v_{DS} = 10$. Then calculate G_{oo} and g_{fs} using the indicated values of I_{DSS} and V_P. Compare results.

3-14. From (3-11) deduce that

$$\frac{1}{I_{DSS}}\frac{dI_{DSS}}{dT} = \frac{1}{V_P}\frac{dV_p}{dT} + \frac{1}{\mu}\frac{d\mu}{dT}$$

with μ denoting the carrier mobility in the channel. Assume ψ_o in (3-10) decreases 2 mV/°C rise in temperature and assume μ is proportional to $T^{-1.5}$. At 300 K, calculate the value of V_P that makes I_{DSS} independent of temperature.

3-15. In the circuit of Fig. 3-27 the 1 kHz square-wave input varies between ±10 V and each

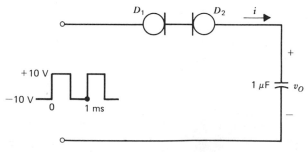

Figure 3-27   Circuit for P3-15.

n-channel current-regulator diode has $I_{DSS} = 5$ mA and a pinch-off voltage of 4 V. When one device conducts in the normal direction for current regulation, the other is in the enhancement mode with its source and drain interchanged. Sketch and dimension the output voltage v_O throughout two periods. The average value of v_O is zero.

Chapter Four
Integrated Circuit Technology

The technologies for fabricating microelectronic networks impose various restraints on circuit design, and network performance depends on the processing procedure. Thus a knowledge of the methods are of considerable importance to the electronics engineer. We shall briefly examine the fabrication of monolithic, thick film, and thin film integrated circuits and consider their relative advantages and disadvantages with regard to circuit performance. Additional techniques are described in later chapters. It is convenient to begin with a discussion of the manufacture of a discrete transistor.

4-1. TRANSISTOR FABRICATION

Most transistors are made of silicon by a set of techniques known as *planar diffusion technology*. A common method of fabricating a planar-diffused NPN transistor is briefly described here.

Within a glass or quartz container, known as a *reactor*, is placed a single-crystal N^+-type silicon wafer, called the *substrate*, along with a number of other identical wafers. The symbol N^+ signifies that the substrate is heavily doped with donors. Mounted around the reactor are radio-frequency induction coils that heat the wafer to a temperature above 1000°C. A stream of hydrogen gas containing silicon tetrachloride ($SiCl_4$) and donor-type dopants is passed through the reactor. The gas deposits a mixture of silicon and donor atoms on the surface of the wafer, and a moderately doped N-type silicon layer is grown on the heavily

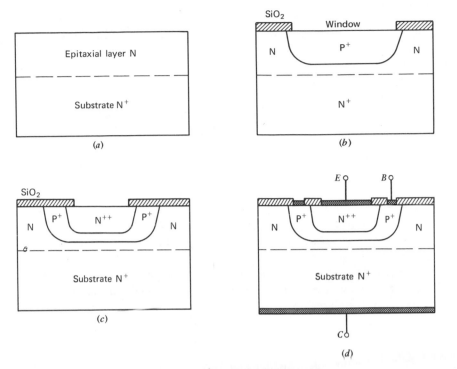

Figure 4-1   Development of NPN diffused-planar transistor. (a) Substrate and layer. (b) PN Junction. (c) NPN regions. (d) NPN transistor with leads.

doped substrate. A single crystal is formed by the substrate and the layer, which is referred to as an *epitaxial layer* (epitaxial is Greek, meaning "arranged upon"). The substrate with its epitaxial layer is shown in Fig. 4-1a.

Next, the wafer is placed in an oxidation furnace and heated to about 1000°C. The epitaxial layer is exposed to a gas containing an oxidizing medium such as oxygen or water vapor, and a film of silicon dioxide (SiO_2) forms on the surface of the layer. The use of a thin silicon-dioxide layer is basic to the planar technology. After the oxidation process, a small window is cut through the oxide layer by dissolving a portion of the oxide with hydrofluoric acid. The sample is then put into a diffusion furnace, through which flows a gas containing acceptors. At about 1000°C the acceptors diffuse into the epitaxial layer through the surface exposed by the window, with diffusion over the rest of the surface blocked by the silicon-dioxide layer. After an appropriate period of time the acceptors are removed from the gas and the solid-state diffusion of the acceptors into the layer continues until the heat is removed. There is now a PN junction, as shown in Fig. 4-1b, with the P-region doped heavily in comparison with the epitaxial layer, as indicated by the symbol P^+.

The surface is again oxidized, and a second window, smaller than the first, is cut through the oxide layer. The sample is placed once more in a diffusion furnace, but this time the gas contains donors that diffuse into the P^+ region, developing an even more heavily doped N^{++} region as shown in Fig. 4-1c. The oxide layer is regrown, suitable windows are opened, and metallic contacts are deposited. Also, a metallic film is deposited on the substrate, and leads are attached to all three regions, as shown in Fig. 4-1d. Note that the junctions and their space-charge layers are under the oxide layer, which reduces surface leakage currents.

Figure 4-2a provides a cross-sectional view of a silicon epitaxial planar-diffused NPN transistor with specified dopant densities and dimensions, and Fig. 4-2b displays a typical top view of the device. Actually, the diffusion process that has been described produces nonuniform doping in the base, with the base more heavily doped in the region near the emitter. Base-width modulation is minimized by the lightly doped epitaxial layer, because the collector-junction space-charge layer penetrates farther into the layer than into the base. Also, the light doping of the layer gives a high avalanche-breakdown voltage, as evidenced by examination of (1-37), and it gives an incremental capacitance for the collector junction that is small, in accordance with (1-54) and (1-57) of Sec. 1-10. Even though the layer has a large resistivity, its resistance is small because it is so thin, and the substrate also has small resistance because it is heavily doped.

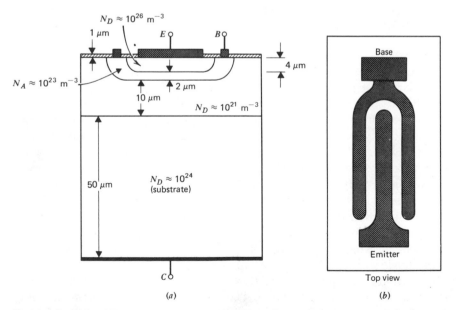

Figure 4-2   Planar NPN transistor. (a) Cross-section (collector-junction area $\approx 10^{-7}$ m^2; emitter-junction area $\approx 10^{-8}$ m^2). (b) Top view.

Transistors are sometimes made without the epitaxial layer. When this is done, a lightly doped collector is usually employed in order to minimize base-width modulation, to provide a sufficiently high breakdown voltage, and to reduce the collector-junction capacitance. The larger collector resistance due to the light dopant density is undesirable.

The extensive use of silicon in preference to germanium is in large part attributed to the natural oxide that allows easy fabrication of planar-diffused devices. Silicon dioxide has a very high melting point, is noncrystalline, and can be heated red hot and plunged into water without cracking. It constitutes from 50 to 96% of all glasses. Silicon dioxide adheres to the surface of silicon, masks diffusion, reduces surface leakage currents, and provides surface protection from environmental effects. There are additional reasons for using silicon. Diodes and transistors made of silicon are useful at higher temperatures than similar germanium devices. Also, they have lower saturation currents, higher breakdown voltages, and sharper avalanche breakdown. However, the smaller mobilities of the charge carriers are disadvantages.

4-2. MICROELECTRONICS

Complex electronic circuits made of discrete components have many interconnections, and each is a potential source of trouble. Failure of electronic equipment is usually due to a poor connection. In an integrated circuit the connections between the components are made internally during the fabrication process, and the number of external connections is thereby minimized. This is certainly a major accomplishment of the integrated circuit. Along with the *greatly increased reliability* are the additional advantages of *small size* and *low cost*, which are also of immense importance.

The technology associated with electronic components and systems made from extremely small parts is known as *microelectronics*. It includes minicomponents, such as the multilayer ceramic capacitor chip and the miniaturized transistor chip. It also includes multichip circuits consisting of two or more semiconductor wafers having single elements or simple circuits, with the wafers interconnected and encapsulated in a single package. However, its major area is that of the *integrated circuit*. This is defined as an assembly of elements made in a single structure that cannot be divided without destroying its usefulness.

There are two basic subdivisions of integrated circuits. One of these is the *film IC*. Film circuits are fabricated by depositing components and metallic interconnections, which have the form of a film, directly on the surface of an insulator substrate. The active devices are usually assembled into the circuit as discrete elements on separate wafers. There are two kinds of film circuits, and these are referred to as *thick film* and *thin film*. Film techniques are discussed the next section. The other basic type is the *monolithic IC* in which all elements, both active and passive, are formed at the surface of a single crystal by planar diffusion techniques. The individual elements are interconnected by metallic strips that are deposited on the oxide coating and exposed semiconductor surface.

A *hybrid integrated circuit* is one that is fabricated by a combination of film and monolithic techniques or one consisting of either a film or monolithic circuit combined with discrete components and packaged as a unit. An example is the use of the planar diffusion process to form the active components in a silicon wafer, with the passive components and interconnection wiring pattern deposited on the surface of the oxide that covers the wafer. Also, hybrid film circuits using discrete diodes and transistors, and perhaps several discrete resistors and capacitors, are quite common.

MONOLITHIC IC TECHNOLOGY

Monolithic integrated circuits are made by the diffused-planar process described in Sec. 4-1. Transistors, diodes, resistors, and capacitors are formed in a single crystal by an appropriate sequence of diffusions that take place through the top surface. Element contact regions are interconnected by a metallic-film wiring pattern deposited on the oxide layer, and this metallization pattern also furnishes the top electrodes of capacitors.

Let us consider a typical fabrication process. A 0.03 cm slice of P-type silicon, which was sawed from a cylindrical crystal of 5 cm diameter, is polished and etched until the surface is smooth and shiny and its thickness is reduced to 0.01 cm. An N-type epitaxial layer with a thickness of 0.001 cm is grown on the top surface, and this is coated with a 0.0001 cm layer of silicon dioxide. A drop of photoresist is placed on the wafer, which is spun so that centrifugal force gives a uniform coating. One thousand identical circuits are to be made from the wafer. A mask with 1000 tiny individual patterns is placed over it and exposed to light. The light causes the areas under the openings of the mask to become insoluble to the developer used to wash away the unexposed photoresist. Hydrofluoric acid then removes the silicon dioxide of the areas not protected by photoresist. The wafer is placed in a diffusion furnace with gas containing boron atoms, and P-type regions form in the silicon immediately below the windows, with the depth of each extending beyond the N-type epilayer into the P-type substrate.

The pattern of the first diffusion is such that a large number of square N-type sections remain in the epilayer, with each one surrounded by a P-type semiconductor, as shown in Fig. 4-3. Between any two N regions are both an NP and a PN junction, and these are equivalent to a pair of back-to-back diodes. Thus the N regions, which are formed as squares in this example, *are electrically isolated from one another*. They are isolated from the substrate as well, provided it is connected to the most negative voltage of the circuit.

Upon completion of the isolation diffusion the oxide coating is regrown over the surface, closing the windows. The entire diffusion process is repeated with a

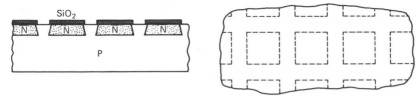

Figure 4-3   Cross-section and top view of isolation islands.

mask now designed to give P-type base regions and resistors. Each active and passive component of the IC will be within an isolation island. Following the base and resistor diffusion is the N-type emitter diffusion. The bottom electrodes of capacitors, each of which consists of an N^+ region within an isolation island, are formed simultaneously with the emitters. In Fig. 4-4 are shown cross-sectional and top views of a capacitor, a resistor, and a transistor (see P4-2).

After the emitter diffusion and reoxidation, the photoresist method is used to etch windows for electrical contacts to the base, emitter, and collector areas, to the resistor terminals, and to the bottom electrodes of the capacitors. In a vacuum chamber aluminum vapor is condensed on the surface, making electrical contact with the silicon in the areas of the windows. This metal layer also forms the top electrodes of the capacitors and supplies pads for electrical connections to the IC. Using a metallization mask, the aluminum is removed from unwanted areas, and the metallic interconnection pattern is complete. The oxide below the

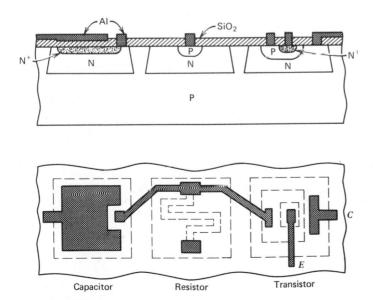

Figure 4-4   Cross-section and top view of capacitor, resistor, and transistor.

metal serves as an insulator, provides dielectrics for the capacitors, and protects the surface. The fabrication process is completed by testing the individual circuits, cutting the wafer into individual chips, mounting each chip on a metallized ceramic substrate of a selected package, connecting the pads of the chip to the leads of the package using very fine wire, and sealing the package. This same batch process is used to make discrete transistors.

MONOLITHIC COMPONENTS

The resistors, such as the one shown in Fig. 4-4, may typically have a width of 0.002 cm with 100 ohms in a segment of length equal to the width. Thus a 5 kΩ resistor has a length 50 times the width. For small values the P channels of the resistors are wider and shorter, and for values of only a few ohms the high-concentration emitter diffusion is used to form N-type resistors. It is feasible to include several resistors in the same isolation region. Normally, monolithic integrated circuits are designed to avoid resistances less than an ohm and greater than 20 kΩ. The surface area of a 1 kΩ resistor is approximately that required for a transistor. A tolerance of $\pm 10\%$ is about as good as can be expected for a diffused resistor. However, a ratio of two resistors can be fixed with an accuracy of nearly 2%. Consequently, integrated circuits are often designed so that their performance depends more on resistance ratios than on actual resistance values. Thin-film resistors are sometimes used with monolithic ICs, with the resistors either deposited on the oxide layer or located on a separate substrate. These have the advantages of reduced stray capacitance and leakage current, a wider range of more precise values, lower noise, and smaller temperature coefficients. A disadvantage is the additional processing required.

The capacitors considered in the fabrication sequence are called *MOS capacitors*. They are limited to a few hundred picofarads because larger values require too much surface area. A second type, the *junction capacitor*, is also used. It consists of a reverse-biased PN junction, and the capacitance depends on the bias voltage with values restricted to about 100 pF. The cross-sections of both types are shown in Fig. 4-5. As with resistances, the ratio of two capacitances is

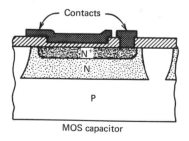

MOS capacitor

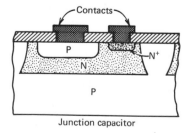

Junction capacitor

Figure 4-5   IC capacitors.

fixed more accurately than the magnitude of either. Thin-film capacitors can also be used. This type consists of two metal electrodes separated by a dielectric, with the bottom electrode deposited on the top side of the silicon dioxide layer of the IC. Advantages are low leakage current, reduced stray capacitance, a wider range of more precise values, and efficient use of chip area, but additional processing steps are required. Monolithic integrated circuits are designed to minimize the number of capacitors, and when a capacitance in excess of a few hundred picofarads is needed, provision is made for external connection of a discrete component.

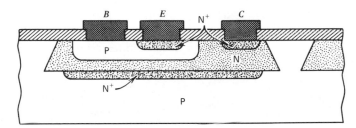

Figure 4-6   IC transistor.

Figure 4-6 shows the cross-section of an NPN integrated-circuit transistor. The geometry is similar to that of the discrete planar transistor. However, the collector contact is at the top surface rather than at the bottom of the substrate, which results in an increase in the ohmic resistance in series with the collector. To minimize this resistance, a heavily doped N^+ buried layer is often used, providing a low-resistance path for the collector current. It is formed by diffusion into the substrate before the epitaxial layer is grown. The N^+ region just below the collector metallization provides a suitable contact and prevents the possible formation of a P region due to migration of aluminum ions into the silicon. The N-type collector is also the isolation island. Two or more transistors with common collectors can be formed in the same isolated region simply by diffusing the bases in different areas of the region. Because this conserves surface area, *integrated circuits are designed with common-collector transistors whenever possible.*

The use of NP and PN junctions for isolation of elements is called *diode* isolation. There are other methods. One of these, known as *oxide* isolation, provides single-crystal N regions that are isolated from a polycrystalline substrate by a layer of silicon oxide that surrounds each one. A third method, called *beam-lead* isolation, forms the elements in a silicon wafer with an unusually heavy metallization for interconnection. Then an etchant is used on the back side to remove the silicon between the elements, leaving only the isolated components supported by the metallic connections. The spaces between the elements can be filled with plastic.

JFETs and MOSFETs are also fabricated for integrated circuits by the planar diffusion process. Some ICs contain both field-effect and bipolar transistors, and the FETs are made simultaneously with the other elements. Many circuits are composed entirely of MOSFETs, which can be used as resistors and capacitors as well as for active devices. A MOS resistor consists of the channel between the source and the drain, and the resistance is a function of the bias voltages. The MOS capacitor has been discussed and is shown in Fig. 4-5. Fortunately, MOSFETs are self-isolating because of an oxide layer below the gate, a PN junction associated with the channel, and back-to-back diodes of PN junctions between the source and drain. Because isolation islands are not required and because a MOS resistor uses considerably less surface area than does an equivalent diffused resistor, the element density of a MOS integrated circuit is normally much higher than that of a BJT circuit, with reduced cost per element. An exception is I^2L logic circuitry, discussed in Sec. 7-2.

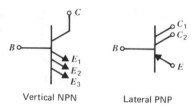

Vertical NPN   Lateral PNP

Figure 4-7   Transistors with multiple emitters and collectors.

IC transistors are sometimes made with two or more emitters. The three-emitter transistor of Fig. 4-7 is fabricated by diffusing into the P-type base three separate N regions, each with their own electrical contact. The sum of the emitter currents equals the sum of I_C and I_B, and each emitter current is separately controlled (see P4-3). Also useful are transistors with multiple collectors such as the one shown in Fig. 4-7. This transistor has single emitter and base diffusions, but there are two separate collector regions.

PNP TRANSISTORS

Although IC active devices are normally NPN, it is often desirable to have one or more PNP transistors on the same chip. Frequently, these are *lateral transistors*, with a configuration as illustrated by the side view of Fig. 4-8. The current of such a device flows laterally, or parallel to the surface, from emitter to collector. Many of the charge carriers injected from the emitter are lost by recombination

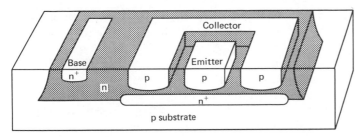

Figure 4-8   PNP lateral transistor.

in the base. This is attributable partly to the geometry and partly to the larger base widths, typically 3 μm compared with 1 μm for NPN devices. As a consequence, the betas of lateral transistors are low, having values from 1 to 10, and the cutoff frequencies are also low. Lateral transistors with multiple collectors are made by dividing the collector into segments. In active-mode operation current division between the collectors depends on the geometry (see P4-4).

PNP devices with higher betas are frequently required in integrated circuits. For example, many operational amplifiers employ in the output stage an NPN and a PNP transistor in a *complementary emitter-follower* configuration, and these devices should have matched characteristics although considerable mismatching can be tolerated. In such cases it is customary to fabricate a vertical PNP transistor, or *substrate transistor*.

The cross-section of a substrate device is shown in Fig. 4-9. The emitter p-diffusion is performed along with the NPN base diffusion. The active base lies in the epitaxial layer between the emitter P region and the P substrate, and this substrate serves as the collector. For an epitaxial layer of 8 μm depth, the base width is about 6 μm, which compares unfavorably with the typical base width of 0.5 μm for NPN devices. However, the excess-carrier lifetime in the epitaxial layer is large, and values of β from 20 to 50 are attainable. Although low, these values give acceptable performance.

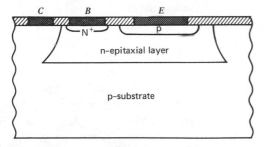

Figure 4-9   Cross section of substrate PNP transistor.

In addition to the low beta, a substrate transistor has other disadvantages. Because the collector is the substrate, it must usually be connected to the negative voltage supply of an integrated circuit. As we shall see in Chapter 13, this restricts the device to operation as an emitter follower. Another disadvantage is the larger ohmic base resistance compared with that of NPN transistors, due to the high resistivity of the epitaxial layer. This large base resistance gives appreciable *current crowding* (as discussed in Sec. 12-3) at much lower current levels, thereby causing a decrease in beta. Also, in conjunction with a larger collector-junction capacitance, the high-frequency performance is degraded. Finally, the emitter efficiency is moderate at best, because of the selection of the dopant density of the p-diffusion to optimize performance of the NPN devices of the IC.

Standard monolithic integrated circuits require four diffusions and seven masking steps. The diffusions are those for the buried layers, the isolation islands, the base regions, and the emitters. The masks provide openings for the four diffusions, the capacitor locations, the ohmic contacts, and the metallization patterns. Resistors, lateral transistors, and substrate transistors can be included. The monolithic circuits described here consist of isolated "discrete" components that are wired together. They differ from discrete circuits in that the components are made on a single chip, with surface metallization accomplishing the interconnection between components. As we shall see in Chapter 7, I^2L circuitry is truly integrated.

4-3. FILM INTEGRATED CIRCUITS

THICK-FILM TECHNOLOGY

The thick-film process is relatively simple and low in cost. The insulating substrate is usually an aluminum-oxide ceramic containing small amounts of other oxides. Typical substrate dimensions are $1 \times 1 \times 0.1$ cm. Screen printing is used to deposit the passive elements on the substrate. The screen is woven from stainless steel wire with about 100 meshes per cm and it is coated with a photo-sensitive emulsion. Photolithographic techniques similar to those used to make masks in the planar diffusion process are employed to form patterns on the screen, with the patterns being areas of open mesh. The screen is mounted in a holder that places it in a position which is slightly above and parallel to the substrate. A squeegee supplies conductive ink, or paste, to the screen. The rubber blade of the squeegee slides over the screen with sufficient pressure to depress only the portion of the screen that is directly under the blade so that it contacts the substrate, and ink is pushed through the mesh openings of the pattern. The squeegee moves on, leaving ink that adheres to the substrate. The thick film of ink is then dried in an oven at a temperature not over 125°C. The screening process can be repeated to add additional elements to the circuit. Different inks are used for conductors, resistors, and capacitor dielectrics.

A typical fabrication process might consist of printing in sequence the bottom electrodes of the capacitors, the dielectrics, the top electrodes and the conduction pattern, the bonding pads, resistor pattern 1, and resistor pattern 2. At various intervals in the process the prints are *fired* by placing the substrate in a high-temperature furnace. This firing develops desirable physical and electrical properties in the film, which has a thickness of about 0.001 cm. Printing and firing are followed by trimming techniques that adjust the resistors and capacitors to fairly precise values. A resistor that is to be trimmed is purposely made too low in value, and trimming is accomplished by removing portions of resistive ink until the proper value is reached. A capacitor to be trimmed is made too large, and trimming consists of removing part of an electrode. After trimming, a protective coating is printed over the entire surface and fired. Then appropriate active and passive components are attached, and the IC is packaged and tested.

THIN-FILM TECHNOLOGY

Thin-film ICs are fabricated in a vacuum. This is a disadvantage because high-vacuum equipment is expensive. Two basic processes are used to deposit on the substrate the films used for conductors, resistors, and dielectrics. One of these is *vacuum evaporation*, in which the material to be deposited is heated in a vacuum until it begins to evaporate; the atoms leave, traveling in straight lines, and those that strike the nearby substrate form a thin film. The other process, referred to as *sputtering*, dislodges the particles of the material by bombarding it with energetic gas ions, and some of the dislodged particles strike the nearby substrate, again forming a thin film. The pattern on the substrate can be defined by placing a mask over the substrate before the film is deposited, with the mask allowing the formation of the film only through the open areas. A second technique is to permit the film to cover the entire surface, with photoresist and etching processes employed to form the pattern. The deposited film is quite thin, typically about 10^{-5} cm.

Substrates of glass or glazed aluminum oxide are often used. In fact, the thin film can be deposited on the oxide layer of a silicon substrate, with the active devices formed within the silicon. An advantage of this is that the resistors of the thin film are superior in quality to those that might be made within the silicon. However, because silicon substrates are expensive, passive substrates are normally preferred, with active devices added as discrete chips. This is especially true for complex circuits requiring large areas. Gold, tantalum, and aluminum films are satisfactory for conductors, and tantalum and other materials are used for resistors. The oxides of silicon and tantalum are among the materials suitable for the dielectrics of capacitors.

A typical processing sequence consists of first coating the alumina substrate with gold film by vacuum evaporation. A photoresist technique etches away all areas except those required for conductors, capacitor electrodes, resistor termination pads, and semiconductor chip mounting pads. Next, the resistor film is formed by sputtering tantalum through an appropriate mask. This is followed by depositing the capacitor dielectrics by sputtering silica through a second mask containing the capacitor pattern, and then the top electrodes are formed by gold evaporation, again using a suitable mask. If necessary, components can be trimmed. Discrete components are added, and the device is packaged and tested. A substantial number of thin-film circuits can be fabricated simultaneously on a substrate several inches square, and the individual circuits are cut apart after the last vacuum operation.

CAPABILITY COMPARISONS

From the discussions of the fabrication procedures for monolithic, thin-film, and thick-film ICs many of the advantages and disadvantages should be apparent. In general, monolithic circuits are superior with regard to low cost in high-volume production, extremely small size, and suitability for digital networks. Consequently, monolithic ICs find much greater use than the other types, and their use is growing rapidly. However, both thin-film and thick-film hybrid circuits are indeed important and find extensive use in many specialized applications.

Thin-film hybrid circuits are less expensive than monolithic for small production runs. In comparison with diffused resistors the thin-film resistors are of high quality, having good tolerance capability, excellent temperature stability, low stray capacitance, and minimum noise, and they can be fabricated with a wide range of values. Thin-film hybrids are excellent for operation at high voltages and power. Their precise geometry and low-loss substrates are ideal for the microwave range, from ultra high frequencies to many GHz, and they find widespread application. Because the hybrid circuits have considerable flexibility and precision, they are especially suitable for high-performance linear circuits.

Thick-film hybrids are by far the least expensive for small production runs. They provide the widest range of resistance values, and the quality of the passive components is nearly as good as that of thin film. They are excellent for operation at high voltages and power and useful at frequencies up to several GHz. Many demanding analog circuits utilize thick film.

4-4. ION IMPLANTATION

Although integrated circuits of high quality are made by the planar diffusion process, the method has certain deficiencies. Let us refer to Fig. 4-10 which shows a portion of a typical integrated-circuit transistor drawn approximately to scale.

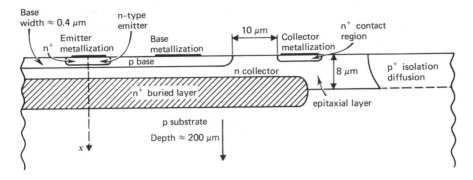

Figure 4-10   A portion of a cross-section of an IC transistor drawn to scale, with the oxide omitted.

The portion omitted at the left side of the x axis is identical to that of the right side.

The N-type epitaxial layer grown on the P-type substrate has a uniform dopant density, usually about 10^{21} donors/m^3. The first diffusion step after the buried-layer and isolation diffusions forms the P-type base region. During the process, acceptors enter the silicon and diffuse downward from the surface. These must be sufficiently dense to neutralize the donor ions of the base region with enough additional ions to provide the desired effective value N_A. To form the emitter region a second diffusion is required, and the donors must be even denser than N_A. Even though donors and acceptors neutralize one another, the large number of impurities present have degrading effects on both the mobilities and the minority-carrier lifetimes. The impurity profiles are sketched in Fig. 4-11. The

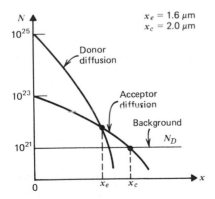

Figure 4-11   Impurity profiles of transistor of Fig. 4-10.

x axis corresponds to that of Fig. 4-10, with x_e denoting the distance from the surface to the emitter junction and x_c denoting the distance to the collector junction. Note the logarithmic scale of the ordinate.

A second problem associated with diffusion is the difficulty of controlling carefully the *sheet resistance* R_s. This is the resistance of a doped region located at the surface with a surface geometry being that of a square. Because the length and the width are equal, R_s does not depend on the size of the square. It does depend on the dopant density and the depth of diffusion. The resistance of a diffused resistor of length L and width W is, of course, the product of R_s and L/W. It is frequently desirable to obtain a high sheet resistance in a P region commonly used to fabricate a resistor, and this is particularly difficult.

Ion implantation is a technique that injects ions into silicon in a manner that overcomes the difficulties mentioned. It allows the fabrication of P-type layers at the surface having shallow depths and low dopant concentrations. A well-controlled sheet resistance of 10^4 ohms per square is easily obtained (see P4-5). Impurity profiles can be controlled so as to give optimum values for the semiconductor parameters, and dimensions of the doped regions are fixed with greater precision. As a consequence, the important parameter β_F is held within close limits. The technique allows the fabrication of microwave transistors, and digital networks made with ion implantation have better performance and higher speed. Also, it is possible to control precisely the threshold voltages of MOSFET transistors. However, ion implantation increases processing costs.

Basically, the technique consists of injecting ions into the silicon by exposing the surface to high-energy ions, such as boron for P-type diffusion and phosphorus for N-type. Ions emitted from a source are accelerated by an electric field. The ion beam is directed to the proper location on the silicon wafer by conventional "windows," where the ions enter the silicon. They are implanted at a depth that depends on the accelerating electric field, and the dopant density depends on the beam current. By controlling the field and the current, it is possible to obtain almost any desired impurity profile. The ions can be injected into the silicon through the normal oxide layer. A mask pattern can be obtained by using either an unusually thick oxide layer or a metallization, with windows established by photolithography. Improvements in ion-implantation technology are leading to high-quality integrated circuits.

REFERENCES

1. R. P. Nanavati, *Semiconductor Devices*, Intext Educational Publishers, New York, 1975.
2. D. J. Hamilton and W. G. Howard, *Basic Integrated Circuit Engineering*, McGraw-Hill, New York, 1975.
3. J. Lindmayer and C. Y. Wrigley, *Fundamentals of Semiconductor Devices*, Van Nostrand, Princeton, N.J., 1965.

4. D. W. Hammer and J. V. Biggers, *Thick-Flim Hybrid Microcircuit Technology*, Wiley-Interscience, New York, 1972.
5. R. G. Hibberd, *Integrated Circuits*, McGraw-Hill, New York, 1969.

PROBLEMS

4-1. Identify each mask and the type and purpose of each diffusion required in the manufacture of an IC from a slice of P-type silicon. The circuit includes buried layers, an epitaxial layer, NPN and PNP transistors, diffused resistors, and MOS capacitors.

4-2. Sketch the schematic diagram of the integrated circuit of Fig. 4-4.

4-3. An NPN transistor with three identical emitters has $V_{BE1} = 0.7$, $V_{BE2} = 0.6$, and $V_{BE3} = 0.5$ V. If the collector current is 10 mA, calculate the emitter currents in mA. Assume negligible base current, and q/kT is 40. (*Ans*: $I_{F3} = -0.00329$).

4-4. The emitter of a PNP lateral transistor, such as that of Fig. 4-8, is driven with a current of 15 mA from a current source. The collector is divided into two separate P-type regions with a geometry such that I_{C1} is twice I_{C2}. For $\beta_F = 4$, with beta defined as the ratio of the sum of the collector currents to the base current, calculate I_{C1} and I_{C2} in mA.

4-5. An IC resistor, such as that of Fig. 4-4, is required with a value of 40 kΩ. If the width is 0.002 cm, calculate the length in cm assuming the sheet resistance R_s is the typical value of 100 ohms per square for a diffused resistor. Repeat with ion implantation employed, giving a sheet resistance of 10,000 ohms.

PART TWO
Digital Circuits

Chapter Five
Combinational
Logic Design

In contrast to analog circuits, which utilize continuous variables, in digital circuits the electrical variables are usually restricted to *two discrete values*, with reasonable tolerances for component variations and noise. The two values are arbitrarily designated as *state* 0 and *state* 1. They are frequently, but not always, controlled by a transistor driven out of its normal region of operation so that it acts approximately either as a short-circuit (ON) or an open-circuit (OFF). Complicated systems can be built by properly arranging large numbers of a few elementary circuits, called *logic gates*, which implement desired logical operations of the 0 and 1 states.

Combinational logic systems consist of gates with outputs that depend on the inputs present *at the time*. In contrast, the outputs of the gates of *sequential* systems depend not only on the inputs present but also on the past history, or *sequence*, of the inputs. Sequential systems have memory capability and are the subject of Chapter 9. We confine our attention here to combinational networks.

A pulse generator, called a *clock*, is often used to control the timing of a digital system. In this case the output of a gate is allowed to change states only when a pulse is present, and the speed of the system is fixed by the clock frequency. Because the pulse generator synchronizes the timing, such systems are referred to as *synchronous*. Networks not clock-controlled are *asynchronous*. Although the sequential networks of a digital system are usually synchronous, the combinational networks are oftentimes asynchronous.

In this chapter we shall examine the basic types of gates and their interconnection to accomplish a desired logic function. The detailed circuitry of the gates of the various families of logic systems, such as TTL and CMOS, is investigated in the following three chapters.

5-1. BASIC LOGIC GATES

THE BINARY SYSTEM

A knowledge of the binary system is essential to an understanding of the operation of digital networks. The familiar decimal system with base 10 has ten digits 0 to 9. The first ten numbers go from 0 to 9. The next ten consist of a 1 in the second column with the same digits 0 to 9 in the first, and so on. The number 4507 represents the sum of 4×10^3, 5×10^2, 0×10^1, and 7×10^0.

The binary system has the base 2 with only two digits 0 and 1. The first two numbers are 0 and 1. The next two consist of a 1 in the second column with the same digits 0 and 1 in the first. The binary number 1101 is converted to decimal form as follows:

$$1101 = 1 \times 2^3 + 1 \times 2^2 + 0 \times 2^1 + 1 \times 2^0 \qquad (5\text{-}1)$$

Note that the powers of 2 begin at zero for the binary digit at the extreme right. A zero digit contributes nothing. Thus the binary number 10001001 represents the decimal number $2^7 + 2^3 + 1$, or 137. A binary digit 0 or 1 is called a *bit*, and a group of bits representing a number, instruction, or other quantity of significance is a *word*. A word of eight bits is often called a *byte*.

The inverse procedure can be used to change a decimal number to binary form. For example, 137 is greater than 2^7 and less than 2^8. Therefore, an eight-digit binary number is required with the leftmost bit a one. As 2^7 is 128, additional ones must be included at the proper locations to give the difference between 137 and 128. There is a simpler method illustrated in Fig. 5-1. The decimal number 137 is placed at the right and divided by 2, giving 68 with a remainder of 1. These are entered in the adjacent column as shown. Then 68 is divided by 2 giving 34 with a remainder of 0, and these are entered in the next column. The procedure continues until the entry in the top row is 0. The bits of the bottom row represent the binary number.

Divide by 2	0	1	2	4	8	17	34	68	137	decimal
Remainder	1	0	0	0	1	0	0	1		binary

Figure 5-1 Decimal-to-binary conversion.

VOLTAGE MARGIN

Logical states 0 and 1 are represented in most systems by voltage levels. With *positive* logic the high-level voltage V_H is designated state 1 and the low-level voltage V_L is state 0, as illustrated in Fig. 5-2a. This is reversed in *negative-logic* systems, with the high state being a logical 0 and the low state a logical 1. For example, suppose V_H is 0 V and V_L is -12 V. Then for positive logic an output of 0 V denotes state 1 and an output of -12 V is state 0, whereas in a negative-logic system the -12 V is state 1 and 0 V is state 0. The designer may select either positive or negative logic, and neither has an inherent advantage. Unless specifically stated otherwise, we shall assume positive logic. In Fig. 5-2b is a practical representation in which each state consists of a voltage range separated by a voltage margin. This margin must be sufficiently large to ensure that the gate inputs within the allowed ranges produce the proper outputs, also within the ranges, in spite of component variations, power-supply fluctuations, noise, and other factors.

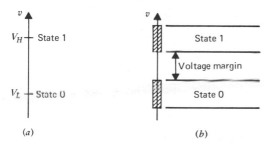

Figure 5-2   States 0 and 1 for positive logic. (*a*) Discrete states. (*b*) Practical representation.

Various sources of noise may affect the selection of the voltage margin. External noise generated by electric motors, arcing relay contacts, circuit breakers, radio-frequency generators, and other equipment can enter the logic system by radiation or through coupling with the power line. Improper ground connections result in the generation of undesirable voltages. Then there is *crosstalk,* which is internal noise caused by capacitive and inductive coupling between signal lines. A pulse along a sending line can be coupled into a receiving line, perhaps causing a logic state to be in error. Coaxial lines, twisted-pair lines, and straight wires are used for signals. These are often long enough so that traveling waves are present, and reflections due to mismatched impedances produce overshoot and ringing in a transmitted pulse. Reasonable efforts must be made, of course, to minimize noise, such as shielding, decoupling, proper grounding, and good circuit design.

THE NOT GATE

The NOT gate has one input and one output, and the output state Y is opposite that of the input A. Thus if A is 0, the output Y is 1, and if A is 1, then Y is 0. A bar over a symbol representing a logical state indicates the opposite state. Thus $Y = \bar{A}$, with $\bar{A}$ said to be NOT A, or the *complement* of A. It is often read "A bar." The operation is referred to as *inversion*, or *negation*, and a NOT gate is also called a digital *inverter*.

A logic expression can be represented by a *truth table*, which gives the logic state of the output for every possible combination of the input variables. Because the inverter has a single input variable with two possible states, the table has only two rows. In Fig. 5-3 are shown the symbol for the NOT gate,

A	Y
0	1
1	0

Figure 5-3   NOT-gate symbol, truth table, and alternate symbol.

the truth table representing the expression $Y = \bar{A}$, and an alternate symbol. *The small circle on the symbols denotes inversion.* It is used extensively at input terminals, output terminals, or both, whenever one wishes to indicate that the variable is the complement of that which would exist in the absence of the circle. With the circle omitted on the symbols of Fig. 5-3, the output is simply the input A. With the circle at the output, the output is complemented to give $\bar{A}$, and when it is at the input, the input is complemented, again giving $\bar{A}$ at the output. A circle at the input and also at the output indicates inversions at both input and output, and for this case Y is A. Double inversion of a variable gives the variable; that is, $\bar{\bar{A}} = A$.

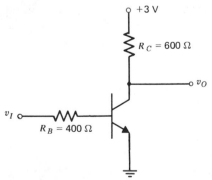

Figure 5-4   A digital inverter.

A possible circuit for a digital inverter is shown in Fig. 5-4. The value of V_H is 3 V representing a logical 1 and V_L is 0.2 V representing a logical 0. When v_I is in the 1-state at 3 V, the transistor is in saturation, which places v_O in the zero state at about 0.2 V. However, when v_I is a logical 0 at 0.2 V, the transistor is *off*. For this case there is no current through R_C, and v_O is a logical 1 at 3 V.

THE AND GATE

This gate has several inputs but only one output. The output Y is a logical 1 if and only if all inputs are 1. Thus any input of 0 makes $Y = 0$ regardless of the states of the other inputs. For an AND gate with three inputs A, B, and C the output Y is expressed as ABC, often read as "Y is equivalent to A and B and C." The AND operation performs *Boolean multiplication*. Shown in Fig. 5-5 are the symbol and truth table for a three-input AND gate. There are 2^3 possible combinations of the input variables. Consequently, the truth table has eight rows of data. For n inputs there are 2^n rows.

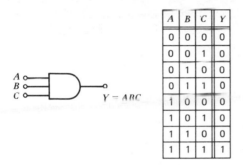

A	B	C	Y
0	0	0	0
0	0	1	0
0	1	0	0
0	1	1	0
1	0	0	0
1	0	1	0
1	1	0	0
1	1	1	1

$Y = ABC$

Figure 5-5   AND gate, symbol and truth table.

Boolean algebraic equations are those relating logical expressions. Several Boolean identities involving the AND and NOT operations are as follows:

$$ABC = (AB)C \qquad AA = A \qquad A1 = A$$
$$AB = BA \qquad A\bar{A} = 0 \qquad A0 = 0 \qquad (5\text{-}2)$$

It is evident from the truth table of Fig. 5-5 that the *Boolean product* ABC *follows the rules of ordinary multiplication*. The product is zero if any of the variables is zero and it is 1 if all variables are 1. From this the relations of (5-2) are obvious, recognizing that each variable is either 0 or 1.

THE NAND GATE

Connection of a NOT gate to the output of an AND gate gives inversion, and the combination is a NOT–AND, or NAND, gate. The symbol for a two-input NAND gate and the truth table for the logic expression $Y = \overline{AB}$ are shown in

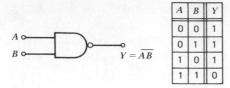

A	B	Y
0	0	1
0	1	1
1	0	1
1	1	0

$Y = \overline{AB}$

Figure 5-6 NAND gate, symbol and truth table.

Fig. 5-6. The small circle at the output of the AND symbol signifies, of course, that AB is inverted. Thus the output $\overline{AB}$ is the complement of the product AB, and this is the NAND operation. A NAND gate with all input terminals connected to form a single input is an inverter.

The transistor inverter of Fig. 5-4 can be made into a NAND gate by adding one or more transistors in series with the one present. In Fig. 5-7 is a circuit for a three-input NAND gate. Clearly, all inputs must be high-level in order to give a low-level output. NAND gates are frequently easier to make than AND gates.

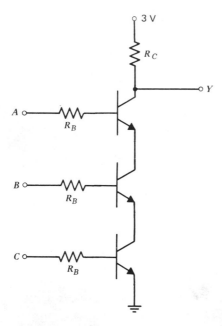

Figure 5-7 Three-input NAND circuit.

THE OR GATE

Boolean addition is accomplished by the OR gate which has an output state of 1 provided one or more inputs are 1. Only when all inputs are logical zeros is the output zero. With three inputs the output Y is expressed as $A + B + C$, often read as " Y is equivalent to A or B or C." In Fig. 5-8 are the symbol and truth table of a two-input OR gate. Boolean addition is similar to ordinary addition except that $1 + 1$ equals 1.

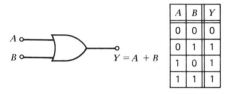

A	B	Y
0	0	0
0	1	1
1	0	1
1	1	1

$Y = A + B$

Figure 5-8   OR gate, symbol and truth table.

The following Boolean identities involving the OR and AND operations are useful in designing logic systems:

$$A + B = B + A \qquad A + B + C = (A + B) + C$$
$$A + A = A \qquad\qquad A + AB = A$$
$$A + 1 = 1 \qquad\qquad A + BC = (A + B)(A + C)$$
$$A + 0 = A \qquad\qquad A(B + C) = AB + AC$$

$$(5\text{-}3)$$

The expression $A + AB$ equals $A(1 + B)$, and $1 + B = 1$. By similar reasoning the sum of AB and $ABCD$ is AB. These and the relations of (5-3) are easily verified by application of the rules of Boolean multiplication and addition. An alternate procedure, which is rigorous and direct, is to form a truth table with columns giving expressions that are supposed to be equal. For example, consider the table of Fig. 5-9. The two columns on the left contain all possible combinations of A and B. The column AB is found from the AND operation,

A	B	AB	A + AB
0	0	0	0
0	1	0	0
1	0	0	1
1	1	1	1

Figure   5-9   Truth table showing that $A = A + AB$.

and the one on the right is determined from columns A and AB, using Boolean addition. Columns A and $A + AB$ are identical, and this logic expression of (5-3) is verified.

THE NOR GATE

Inversion at the output of an OR gate, which can be implemented with an inverter, gives a NOT–OR, or NOR, gate. For two inputs A and B the output is the complement of $A + B$, or $\overline{A + B}$. The symbol is that of an OR gate with a circle at the output to indicate inversion. This is shown in Fig. 5-10 along with the truth table. The output is 1 if and only if both inputs are zeros.

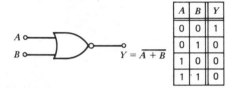

A	B	Y
0	0	1
0	1	0
1	0	0
1	1	0

$Y = \overline{A + B}$

Figure 5-10   NOR gate, symbol and truth table.

The inverter circuit of Fig. 5-4 is converted into a three-input NOR gate by adding two additional transistors, with the three transistors arranged with their collector-emitter terminals in parallel. The network is shown in Fig. 5-11. Clearly, if any one of the transistors is saturated, the output is low-level, and the NOR operation is accomplished. With the inputs connected to form a common terminal, the circuit is that of an inverter. Thus an integrated-circuit NOR gate can be used as an inverter. By utilizing an inverter at the output, a NOR gate is converted into an OR gate.

For most of the various logic families the gates that have been described are available in integrated-circuit (IC) packages. For example, a Dual Four-Input

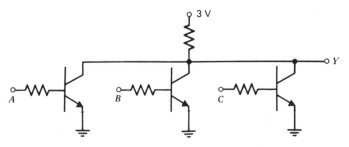

Figure 5-11   Three-input NOR circuit.

NOR IC contains two separate NOR circuits, each with four inputs and one output connected to pins for external connections. A Quad NAND package has four separate NAND gates. For lowering the voltage level of a COS/MOS logic system to that required for interfacing with a TTL system, there is a COS/MOS Hex Buffer/Converter. This consists of either six inverters for inversion buffering with a voltage-level change or six double inverters that serve the same purpose without inversion.

Complex logical functions are accomplished by various arrangements of the basic NOT, AND, NAND, OR, and NOR gates. In fact, a complete system can be formed using NAND gates exclusively, or perhaps NOR gates. As we shall see, the type gate or gates depend on the logic family employed and the choice of positive or negative logic. Integrated circuits with interconnected gates are common, and a number of these are examined in later chapters. Fewer than 13 gates on a monolithic chip, such as those ICs mentioned earlier, is considered as *small-scale integration* (SSI). Many chips have from 13 to 100 gates, which constitutes *medium-scale integration* (MSI), and those IC packages with gates in excess of 100 are regarded as having *large-scale integration* (LSI).

A circuit designed to perform the AND operation with positive logic gives the OR operation in a negative-logic system. Likewise, a positive-logic OR gate is a negative-logic AND gate. Similar statements apply to NAND and NOR circuits. Clearly, the effect of changing the logic designation is to switch the 0 and 1 states. As an illustration, interchanging the zeros and ones of the NOR truth table of Fig. 5-10 gives the NAND table of Fig. 5-6. The positive-logic NAND circuit of Fig. 5-7 is a NOR gate in a negative-logic system, and the NOR circuit of Fig. 5-11 is a NAND gate in a negative-logic system.

5-2. DEMORGAN'S LAWS

By means of a truth table it is easy to show that

$$\overline{A + B} = \bar{A}\bar{B} \tag{5-4}$$

Because the identity applies to any variables A and B, we can replace B with the variable $(B + C)$, giving

$$\overline{A + (B + C)} = \bar{A}\,\overline{B + C} \tag{5-5}$$

By application of (5-4) to the term $\overline{B + C}$, we obtain

$$\overline{A + B + C} = \bar{A}\bar{B}\bar{C} \tag{5-6}$$

Clearly, this procedure can be extended again and again as often as desired to include any number of variables.

Suppose A and B in (5-4) are replaced with $\bar{A}$ and $\bar{B}$. The relation becomes

$$\overline{\bar{A} + \bar{B}} = \bar{\bar{A}}\bar{\bar{B}} = AB \tag{5-7}$$

Inversion of the equation and interchanging the two sides of the equality, we find that

$$\overline{AB} = \bar{A} + \bar{B} \tag{5-8}$$

Now replace B with BC, which gives

$$\overline{A(BC)} = \bar{A} + \overline{BC} \tag{5-9}$$

Upon application of (5-8) to $\overline{BC}$, we obtain

$$\overline{ABC} = \bar{A} + \bar{B} + \bar{C} \tag{5-10}$$

As before, this procedure can be extended to include additional variables.

Equations 5-6 and 5-10 in terms of an unspecified number of variables become

$$\overline{A + B + C + \cdots} = \bar{A}\bar{B}\bar{C}\cdots \tag{5-11}$$

$$\overline{ABC\cdots} = \bar{A} + \bar{B} + \bar{C} + \cdots \tag{5-12}$$

These are *DeMorgan's laws*. They are extremely useful in simplifying logic expressions and logic diagrams.

SUM-OF-PRODUCTS FORM

A logic equation can always be expressed in the form of a sum of products, with the function Y written as the sum of terms, each of which consists of a product of logic variables. An example is the relation

$$Y = A + \bar{A}B\bar{C} + \bar{A}\bar{B} \tag{5-13}$$

Each of the product terms is referred to as a *minterm*. This particular function can be implemented by either logic configuration of Fig. 5-12. As we shall see in Example 1 of the next section, (5-13) can be simplified considerably.

In Fig. 5-12a the respective outputs of AND gates 1, 2, and 3 are A, $\bar{A}B\bar{C}$, and $\bar{A}\bar{B}$. Of course the first AND gate can be eliminated. The three-input OR gate gives the desired function Y. In Fig. 5-12b the respective outputs of NAND gates 1, 2, and 3 are $\bar{A}$, $\overline{\bar{A}B\bar{C}}$, and $\overline{\bar{A}\bar{B}}$. The first NAND gate is simply an inverter. It can be eliminated because $\bar{A}$ is assumed available as an input variable. The output of NAND gate 4 is

$$Y = \overline{(\bar{A})(\overline{\bar{A}B\bar{C}})(\overline{\bar{A}\bar{B}})} = A + \bar{A}B\bar{C} + \bar{A}\bar{B} \tag{5-14}$$

The right side of (5-14) is deduced from DeMorgan's law (5-12), with A, B, and C in (5-12) replaced with $\bar{A}$, $\overline{\bar{A}B\bar{C}}$, and $\overline{\bar{A}\bar{B}}$.

The configurations of Fig. 5-12 are referred to as *two-level* logic systems, because each input passes through two logic operations before reaching the output. When a gate switches state, the switching is not instantaneous and there is a small gate propagation delay. These time delays limit the speed of the system.

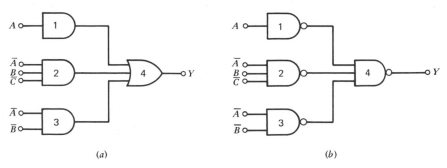

(a)   (b)

Figure 5-12   Equivalent logic configurations that implement (5-13). (a) AND–OR network. (b) NAND–NAND network.

In a two-level configuration each input has a maximum of two such delays. Thus it is usually best to design a network that implements a specified logic expression so that there are no more than two propagation delays. It is evident from (5-13) and Fig. 5-12 that *the sum-of-products form of a logic expression can always be implemented by a two-level system*, provided both the variables and their complements are available as inputs. If (5-13) had additional product terms on the right side, the diagram of Fig. 5-12 would have additional input gates, but the logic would remain two-level.

The equivalence of the logic configurations of Fig. 5-12 should be carefully noted. In general, any two-level AND–OR network such as that of Fig. 5-12a can be replaced with a two-level NAND network, and *the output Y is the sum of the products of the inputs of the first-level gates.*

Conversion of a logic expression into a sum-of-products form can be accomplished by means of Boolean identities and DeMorgan's laws. The technique is illustrated by the example that follows.

Example
Find a sum-of-products form for

$$Y = \overline{\overline{AB + C} + A(\bar{B} + \bar{C})} \qquad (5\text{-}15)$$

Solution
It is usually best to remove the top bar first. This is accomplished by applying DeMorgan's law (5-11), and we obtain

$$Y = (AB + C)\overline{A(\bar{B} + \bar{C})} \qquad (5\text{-}16)$$

The top bar of this expression is eliminated by using DeMorgan's law (5-12), giving

$$Y = (AB + C)(\bar{A} + \overline{\bar{B} + \bar{C}}) \qquad (5\text{-}17)$$

Next, we apply (5-11), and Y becomes

$$Y = (AB + C)(\bar{A} + BC) \tag{5-18}$$

By an identity of (5-3) we note that this can be multiplied term-by-term, and we find that

$$Y = A\bar{A}B + ABBC + \bar{A}C + BCC \tag{5-19}$$

The result is a sum-of-products form. However, using identities of (5-2) and (5-3), we deduce from (5-19) that

$$Y = \bar{A}C + BC \tag{5-20}$$

A sum-of-products form of a logic expression is easily found from a truth table. Suppose we wish to find the logic expression represented by the truth table of Fig. 5-13. Let us consider each row having $Y = 1$. The second data row has $A = 0$, $B = 0$, and $C = 1$. The minterm $\bar{A}\bar{B}C$ gives the proper value of 1 for Y. For all other combinations of A, B, and C the expression is zero. The corresponding minterm for the fourth data row is $\bar{A}BC$, which gives $Y = 1$ for the values of A, B, and C of this row. The bottom-row expression is ABC. It follows that

$$Y = \bar{A}\bar{B}C + \bar{A}BC + ABC \tag{5-21}$$

This can be simplified. In fact, it is equivalent to the expression of (5-20) of the previous example (see P5-4).

The procedure is general. For each row of a truth table having $Y = 1$ we form a minterm consisting of the variables, using the complement of a variable that is zero. The sum of the minterms is Y. A direct method for converting any logic expression into a sum-of-products form is now evident. We simply develop the truth table and use the table to find a sum-of-products form. If possible, the expression should be simplified before designing the logic network.

A	B	C	Y
0	0	0	0
0	0	1	1
0	1	0	0
0	1	1	1
1	0	0	0
1	0	1	0
1	1	0	0
1	1	1	1

Figure 5-13
Truth table.

PRODUCT-OF-SUMS FORM

We have found that a logic equation can be implemented with a two-level arrangement of NAND gates. It is also possible to accomplish this with NOR gates. The procedure is to express the function Y as a product of terms, each of which consists of a sum of one or more input variables. To illustrate, we shall implement with NOR gates the equation

$$Y = AB + AC + \bar{B}C \qquad (5\text{-}22)$$

The first step is to find the *complement* $\bar{Y}$ as a sum of products. Using DeMorgan's law (5-11), the complement can be put in the form

$$\bar{Y} = \overline{AB + AC + \bar{B}C} = \overline{AB}\,\overline{AC}\,\overline{\bar{B}C} \qquad (5\text{-}23)$$

The top bars are eliminated by application of (5-12), giving

$$\bar{Y} = (\bar{A} + \bar{B})(\bar{A} + \bar{C})(B + \bar{C}) \qquad (5\text{-}24)$$

Term-by-term multiplication and utilization of Boolean identities give the result:

$$\bar{Y} = \bar{A}B + \bar{A}\bar{C} + B\bar{C} \qquad (5\text{-}25)$$

Now that $\bar{Y}$ has been determined as a sum of products, the desired form for Y is easily obtained by applying DeMorgan's laws to (5-25). We find that

$$Y = (A + \bar{B})(A + C)(B + C) \qquad (5\text{-}26)$$

Neither (5-25) nor (5-26) is in its simplest form. This is shown in Example 2 of the next section.

The logic configurations of Fig. 5-14 implement (5-26). The network of Fig. 5-14a consists of three OR gates in the first logic level, and these are followed by an AND gate. It is apparent that this OR–AND arrangement satisfies (5-26). The configuration of Fig. 5-14b utilizes NOR gates exclusively. Because it also implements (5-26), it is equivalent to the OR–AND network (see P5-11).

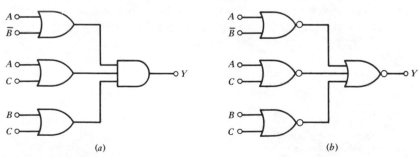

(a)   (b)

Figure 5-14 Equivalent logic configurations that implement (5-26). (a) OR–AND network. (b) NOR–NOR network.

The procedure is general. First, the complement $\overline{Y}$ is expressed as a sum of products, such as that of (5-25). A truth table relating $\overline{Y}$ to the inputs can be used for this purpose (see P5-12). Next, DeMorgan's laws are applied to $\overline{Y}$ to obtain Y in the form of a product of sums, such as that of (5-26). Each product term corresponds to an input NOR gate of a two-level configuration similar to that of Fig 5-14b.

5-3. KARNAUGH MAPS

Although we have learned ways to design two-level combinational networks that will implement a desired logic function, the techniques do not always give the best solution. There are usually a number of equivalent logic expressions. The simplest circuit is obtained from the *minimal* expression, defined as the one with the smallest number of terms. When NAND logic is used, we want to minimize the sum-of-products expression for Y, and when NOR logic is selected, the sum-of-products expression for $\overline{Y}$ is minimized. Often the minimal equation can be found by application of Boolean identities to a specified logic relation. However, this is not always easy and may require considerable intuition. Fortunately there is a direct method known as *Karnaugh mapping*.

THREE-VARIABLE MAPS

A Karnaugh map is essentially a truth table arranged in a form suitable for determining by inspection the minimal sum-of-products expression of a logic function. An example is presented in Fig. 5-15. The expression given in the figure is that of (5-21) of the preceding section.

The expression for Y has three variables. Two of these, A and B, are represented at the top of the map. Their respective values are 00, 01, 11, and 10 as indicated at the top boundary. The variable C has its logic states 0 and 1 along the side boundary. The values must be arranged exactly as shown, because this arrangement gives only a single variable change when passing from any square to an adjacent one. For example, consider the third square of the top row with ABC represented by 110. To the immediate left the square has ABC represented by

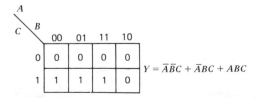

Figure 5-15   Logic equation and Karnaugh map.

010, and only A has changed. The square to the right has ABC represented by 100, and only B has changed. The adjacent square at the bottom has ABC represented by 111, and C has changed. *Squares at the extreme left and right are considered adjacent.* The entries in the blocks are the values of Y. When $A = 0$, $B = 0$, and $C = 1$, the function Y is 1. Thus a logical 1 is entered in the lower left corner of the map. The ones and zeros of the other squares are determined similarly. *The zeros are often omitted*, with a blank space understood to represent a zero.

The three-variable Karnaugh map has eight squares corresponding to the eight possible combinations of the three variables. Each square is equivalent to one row of a truth table. We found earlier that we can express Y as a sum of products from inspection of a truth table. This can also be done from a Karnaugh map. The logical 1 in the lower left corner is in the block with ABC represented by 001. As in the case of the truth table, this leads to the product term $\overline{A}\overline{B}C$. The 1 in the adjacent block leads to $\overline{A}BC$, and the third 1 leads to ABC. The sum of these represents Y. *Each square leads to a product term of three variables.*

Consider the logical 1's in the two bottom squares at the left side. These correspond to the sum of $\overline{A}\overline{B}C$ and $\overline{A}BC$. This sum equals $\overline{A}C(\overline{B} + B)$, or $\overline{A}C$. In the expression for Y the product term $\overline{A}C$ can be substituted for the sum of the two product terms. Furthermore, this deduction is obvious from the map. Because the two squares contain 1's with B equal to both 0 and 1, it is evident that the terms represented by these adjacent squares are independent of B. As A is 0 and C is 1, the two adjacent squares represent the expression $\overline{A}C$. This observation can be generalized as follows. *Any two adjacent squares represent a single product term of two variables, with the deleted variable being the one that changes from one square to the next.*

It follows that the logical 1's of the center two squares of the bottom row correspond to the product BC, which represents the expression $\overline{A}BC + ABC$. Accordingly, $Y = \overline{A}C + BC$, and this is the minimal expression. The logical-1 square between the other two was used twice. This is equivalent to using the product term $\overline{A}BC$ in the expression for Y twice, once in conjunction with $\overline{A}\overline{B}C$ to obtain $\overline{A}C$ and again in conjunction with ABC to obtain BC. Clearly this is permissible, because adding an additional product term $\overline{A}BC$ to Y does not change the logic, recalling the identity $A + A = A$. *Thus logical expressions developed from a Karnaugh map can utilize overlapping squares.*

We have found that a logical 1 square in a three-variable map represents a product term having three variables. Two adjacent squares correspond to a term with two variables, and squares at the left side are considered adjacent to those at the right side. In Fig. 5-16 are shown Karnaugh maps with logical 1's in four adjacent squares. The zeros are omitted. In Fig. 5-16a, the expression is $Y = C$, independent of A and B. In Fig. 5-16b, Y is independent of A and C and its value is the complement of B. In Fig. 5-16c, Y is B. Four adjacent squares in a three-variable map corresponds to a "product" term of a single variable.

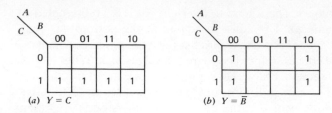

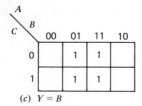

(c) Y = B

Figure 5-16   Karnaugh maps. (a) $Y = C$. (b) $Y = \bar{B}$. (c) $Y = B$.

In the event all eight squares are filled with 1's, $Y = 1$ and Y is independent of A, B, and C. Eight zeros correspond to $Y = 0$.

In developing the minimal expression from a three-variable Karnaugh map one should combine adjacent squares with logical 1's into groups of two, or if possible, into groups of four or perhaps eight. The larger the groups, the simpler is the logic expression. Overlapping squares are permissible and squares at the left and right are considered adjacent. Two examples are presented here.

Example 1

For $Y = A + \bar{A}B\bar{C} + \bar{A}\bar{B}$, which is (5-13) of the previous section, determine the minimal expression.

Solution

The Karnaugh map and the minimal expression are those of Fig. 5-17. The four 1's of the right half of the map represent A, those of the extreme left and

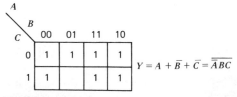

$Y = A + \bar{B} + \bar{C} = \overline{\bar{A}BC}$

Figure 5-17   Karnaugh map and minimal expression.

right columns represent $\bar{B}$, and those of the top row represent $\bar{C}$. Accordingly, Y is the sum of these, and this sum is equivalent to $\overline{\bar{A}BC}$. Thus the network of Fig. 5-12b, designed in the previous section for implementation of this logic, can be simplified to a one-level system consisting of a single NAND gate with three inputs.

Example 2

For $\bar{Y} = \bar{A}B + \bar{A}\bar{C} + \bar{B}\bar{C}$, which is (5-25) of the previous section, find the minimal expression for $\bar{Y}$ and determine the simplest NOR–NOR network for implementing the logic for Y.

Solution

The Karnaugh map and the minimal expression for $\bar{Y}$ are those of Fig. 5-18. The proper way to group the ones is shown. From the minimal expression for $\bar{Y}$ given in the figure it follows that Y is equivalent to $(A + \bar{B})(B + C)$. In the previous section it was shown that the networks of Fig. 5-14 implement this logic. From the results of this example it is evident that the gate with inputs A and C can be eliminated in each of the networks without changing the output Y. The advantage of using the minimal expression for network design is considerable.

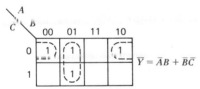

Figure 5-18 Karnaugh map and minimal expression.

FOUR-VARIABLE MAPS

Mapping of logic expressions with four variables is similar. As before, A and B are arranged along the top boundary and C and D are arranged along the left side with values 00, 01, 11, and 10 from top to bottom. There is a variable change of one from the left square of a row to the right square, and there is a change of one from the bottom square of a column to the top square. Consequently, the left and right columns are considered to be adjacent, as before, and so are the bottom and top rows.

Each logical 1 square corresponds to a product term of four variables, two adjacent squares correspond to a term of three variables, four adjacent squares represent a term of two variables, and eight squares represent a single variable. If all 16 blocks contain 1's, the expression is $Y = 1$. Let us consider three examples.

Example 3

Find the minimal sum-of-products form for

$$Y = \bar{A}B(\bar{C} + D) + \bar{A} + B(CD) + AD(B\bar{C} + \bar{B}C) + \bar{A}BC\bar{D}$$

Solution

The Karnaugh map and the minimal expression are given in Fig. 5-19. The 1's in the top and bottom rows correspond to the minterm $\bar{A}B\bar{D}$. Those of the left and right columns give $\bar{B}CD$, and the remaining logical 1 represents $AB\bar{C}D$. The logic can be implemented by a NAND–NAND network. The first logic level would consist of three NAND gates with 3, 3, and 4 inputs.

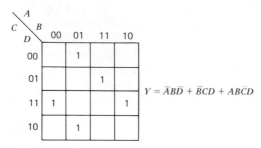

$$Y = \bar{A}B\bar{D} + \bar{B}CD + ABCD$$

Figure 5-19 Karnaugh map and minimal expression.

Example 4

$Y = AB\bar{C} + B\bar{C}\bar{D} + \bar{B}C$. Design a two-level NOR network to implement this logic.

Solution

The Karnaugh map for the complement $\bar{Y}$ and a minimal sum-of-products form are shown in Fig. 5-20. The logical 1 in the square of the second row and second column can be grouped as indicated, giving the product term $\bar{A}BD$, or it

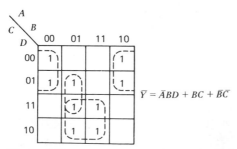

$$\bar{Y} = \bar{A}BD + BC + \bar{B}\bar{C}$$

Figure 5-20 Karnaugh map and minimal expression.

can be grouped with the one to its left, which gives $\bar{A}\bar{C}D$. Either is acceptable. The bottom group of four squares yields the product term BC, and the remaining group gives $\bar{B}\bar{C}$.

By applying DeMorgan's laws to the expression for $\bar{Y}$, we find the sum-of-products form for Y to be

$$Y = (A + \bar{B} + \bar{D})(\bar{B} + \bar{C})(B + C)$$

The first NOR level of the network will have three gates with 3, 2, and 2 inputs.

Example 5
$Y = D(A + \bar{A}\bar{B} + \bar{A}B\bar{C})$. Design a network to implement this function using NOR gates.

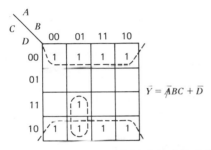

$\bar{Y} = \bar{A}BC + \bar{D}$

Figure 5-21 Karnaugh map and minimal expression.

Solution
The Karnaugh map for the complement $\bar{Y}$ and the minimal sum-of-products form are shown in Fig. 5-21. The group of two 1's corresponds to $\bar{A}BC$ and the eight adjacent squares with logical 1's give $\bar{D}$. Using DeMorgan's laws we find that Y is $D(A + \bar{B} + \bar{C})$. This can be implemented with a two-level logic network made with two NOR gates. If the complements of B, C, and D are unavailable, these inputs must be inverted. The network is shown in Fig. 5-22.

We have examined Karnaugh maps for logic expressions with three and four variables. The technique can be extended to the simplification of logic functions with five and six variables, but other minimization methods are usually preferred

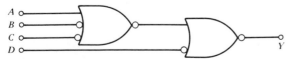

Figure 5-22 NOR–NOR network giving
$Y = D(A + \bar{B} + \bar{C})$.

for functions with more than four variables. Karnaugh mapping can be used to simplify a function of only two variables, with a 2×2 arrangement of squares (see P5-18). Seldom is this advantageous.

5-4. OTHER COMBINATIONAL NETWORKS

From the concepts that have been presented, it is a routine process to analyze a given logic configuration to determine if it can be simplified. To illustrate the technique, the network of Fig. 5-23 is examined. It is assumed that only the true variables are available as inputs. We observe that the system is three-level, not counting the inverters, and a two-level network is possible. A good procedure is to find the logic expression for Y. Then Karnaugh mapping can be used to determine the simplest network.

The inputs to the top gate are A and B, but because the circle at input B denotes inversion, the inputs to the NOR gate are A and $\bar{B}$. The output E is $\bar{A}B$, after application of DeMorgan's law. The respective outputs F and G are similarly found to be $\bar{A}D$ and $\bar{B}D$. Using DeMorgan's law twice, the output J of the second-level NOR gate is determined to be $(A + \bar{B})(A + \bar{D})(B + \bar{D})$. This product of sums is as expected from the NOR–NOR configuration.

The NAND gate at the first level has inversion at both inputs, and such a gate performs the OR operation, with H equivalent to $B + D$. Of course, I is $\bar{C}$. With the aid of DeMorgan's laws, the output K of the second-level NAND gate is found to be $\bar{B}\bar{D} + C$. With inversion at both inputs the output NOR gate performs the AND operation, with $Y = JK$, or

$$Y = (A + \bar{B})(A + \bar{D})(B + \bar{D})(\bar{B}\bar{D} + C) \tag{5-27}$$

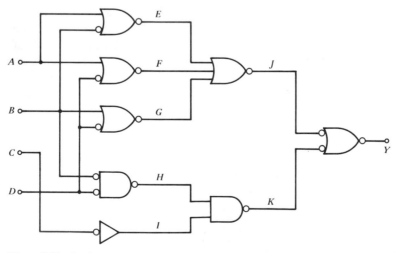

Figure 5-23   Logic network.

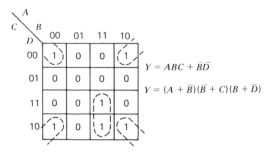

Figure 5.24   Karnaugh map for network of Fig. 5-23.

The Karnaugh map is shown in Fig. 5-24, with the entries determined from (5-27). Because the four corner squares are adjacent, the minimal sum-of-products form for Y consists of only two terms. A similar form is obtained for $\overline{Y}$ from the zeros of the map, which are the 1's for $\overline{Y}$. Each zero can be clustered within a group of four. The product-of-sums for Y is found from $\overline{Y}$ and De-Morgan's laws. The results are

$$Y = ABC + \overline{B}\overline{D} = (A + \overline{B})(\overline{B} + C)(B + \overline{D}) \qquad (5\text{-}28)$$

Each expression for Y requires inversion of B and D. The two-level NAND configuration requires fewer gates than the corresponding NOR circuit. The simplified network with NAND gates is shown in Fig. 5-25.

THE EXCLUSIVE-OR GATE

This widely used gate is defined by the Boolean expression

$$Y = A \oplus B = A\overline{B} + \overline{A}B = (A + B)(\overline{A} + \overline{B}) \qquad (5\text{-}29)$$

The circled plus sign is conventional for the Exclusive-OR operation. From the sum-of-products form the operation can be implemented with a two-level NAND network, and the product of sums allows implementation with a NOR

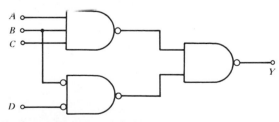

Figure 5-25   Simplified network.

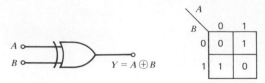

Figure 5-26 Exclusive-OR symbol and Karnaugh map.

configuration. There are also other equivalent networks (see P5-20). The symbol and Karnaugh map are shown in Fig. 5-26.

The Exclusive-OR gate has two inputs, and its output is a logical 1 if and only if the inputs are unequal. It can be used as an *inequality comparator* because Y is 1 only when inputs A and B are unequal. The gate is often found within the arithmetic section of a digital system. Available are medium-scale-integrated (MSI) circuits with a quad of Exclusive-OR gates. Each of the four gates has two input terminals and one output terminal connected to package pins, and the other two pins of the 14-pin package are for the supply voltage.

THE EXCLUSIVE-NOR GATE

Inversion of the output of an Exclusive-OR gate gives the Exclusive-NOR gate. The output Y can be expressed in the form

$$Y = \overline{A \oplus B} = AB + \bar{A}\bar{B} = (A + \bar{B})(\bar{A} + B) \tag{5-30}$$

The function can be implemented, using the logic of (5-30), by means of two-level NAND or NOR networks, and there are other equivalent configurations (see P5-21). The symbol and Karnaugh map are shown in Fig. 5-27.

As indicated by the Karnaugh map, the output of the gate is a logical 1 if and only if the two inputs are equal. Thus the gate can be used as an equality detector. It is commonly referred to as a *digital comparator*, or *coincidence circuit*. MSI digital comparators are available which compare the corresponding bits of two words. One output indicates equality of the words, and if the words are unequal, two other outputs indicate which of the words is the greater.

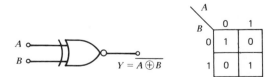

Figure 5-27 Exclusive-NOR symbol and Karnaugh map.

THE BINARY ADDER

Combinational networks are used in digital systems to add binary numbers. Shown in Fig. 5-28 is the addition of the decimal numbers 23 and 53 and their binary equivalents. The least significant bit (LSB) of each of the binaries is a one, and the sum of 1 and 1 is the binary 10. Thus the LSB of the sum is 0, and the 1 is a carry to the second column from the right. The bits of this column are 1 and 0, which added to the carry again gives 10. A one is carried to the third column, and the bits here along with the carry add to 11. As before, a 1 is carried over. As the sum of the bits and the carry of the fourth column is 1, the carry to the fifth column is a zero. The process continues to completion.

Decimal	Binary							
23	0	0	1	0	1	1	1
53	0	1	1	0	1	0	1
76	1	0	0	1	1	0	0

Figure 5-28   Decimal and binary addition.

A binary one-bit full adder is a network that adds the corresponding two bits A and B of one column to the input carry C from the adjacent column, giving the one bit S of the sum and the output carry K. The truth table relating the outputs S and K to the inputs A, B, and C is shown in Fig. 5-29, along with Karnaugh maps for S and K obtained from the truth table.

From the maps the minimal sum-of-products forms for S and K are found to be

$$S = ABC + \bar{A}\bar{B}C + \bar{A}B\bar{C} + A\bar{B}\bar{C} \tag{5-31}$$

$$K = AB + AC + BC \tag{5-32}$$

Inputs			Outputs	
A	B	C	S	K
0	0	0	0	0
0	0	1	1	0
0	1	0	1	0
0	1	1	0	1
1	0	0	1	0
1	0	1	0	1
1	1	0	0	1
1	1	1	1	1

Map for S

C \ AB	00	01	11	10
0	0	1	0	1
1	1	0	1	0

Map for K

C \ AB	00	01	11	10
0	0	0	1	0
1	0	1	1	1

Figure 5-29   Truth table and Karnaugh maps for adder.

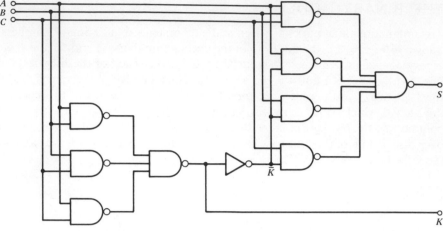

Figure 5-30   One-bit full adder.

Each of these can be implemented by a two-level NAND network along with the necessary inverters. However, a better solution is obtained by noting that

$$S = ABC + (A + B + C)\overline{K} \tag{5-33}$$

This expression is easily verified, using (5-31) and (5-32).

The one-bit full adder of Fig. 5-30 implements (5-32) and (5-33). The output carry K is generated by a two-level NAND network based on (5-32). The inverter gives the complement of K required for implementation of (5-33). By expressing (5-33) in the sum-of-products form we find that its implementation requires four NAND gates at the first level with inputs ABC, $A\overline{K}$, $B\overline{K}$, and $C\overline{K}$. The NAND configuration is not the only practical one. Many adders are made of AND, NOR, Exclusive-OR, and other gates. A common MSI integrated circuit contains four one-bit full adders on a single chip.

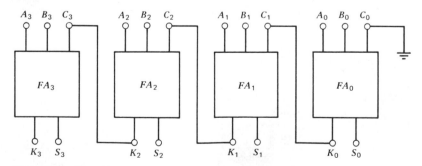

Figure 5-31   Four-bit parallel binary adder.

Shown in Fig. 5-31 is a four-bit parallel binary adder consisting of four one-bit full adders. The least significant bits are those of the full adder FA_0. Its input carry C_0 is grounded. The output word $K_3S_3S_2S_1S_0$ represents the sum of the input binary words $A_3A_2A_1A_0$ and $B_3B_2B_1B_0$.

A single full adder can add two binary words provided the respective bits are supplied in *serial* form. In this case the inputs A and B are synchronous pulse trains on separate lines. The output carry bit at K is fed back to the input carry terminal C through a time delay network, which introduces a delay equal to the interval between successive bits. The output bit train at S is fed into a *register* that stores the bits of the sum. A serial adder is slower than the parallel adder of Fig. 5-31, but it is less expensive.

REFERENCES

1. H. J. Nagle, Jr., B. D. Carroll, and J. D. Irwin, *An Introduction to Computer Logic*, Prentice-Hall, Englewood Cliffs, N.J., 1975.
2. F. J. Hill and G. R. Peterson, *Introduction to Switching Theory and Logical Design*, Wiley, New York, 1968.
3. J. B. Peatman, *The Design of Digital Systems*, McGraw-Hill, New York, 1972.
4. W. G. Oldham and S. E. Schwarz, *An Introduction to Electronics*, Holt, Rinehart and Winston, New York, 1972.

PROBLEMS

Section 5-1

5-1. (a) Find the decimal numbers corresponding to the following binary numbers: 10, 11010011, and 101010101. (b) Find the binary numbers corresponding to 25, 64, and 255. [*Ans*: (a) 2, 211, 341].

5-2. Prepare a table giving the binary numbers corresponding to decimal numbers 0 through 20.

5-3. Use a truth table to verify that $A + B + C = \overline{\overline{A}\,\overline{B}\,\overline{C}}$.

5-4. (a) Noting that $A + \overline{A} = 1$ and using the relation $A + A = A$, deduce that $\overline{A}BC + ABC + A\overline{B}C$ equals $\overline{A}C + BC$. Do not use a truth table. (b) Using the other identities of (5-3), verify that $A + BC$ equals $(A + B)(A + C)$.

5-5. In the NOT circuit of Fig. 5-4 change the supply from 3 to 7 V and change R_L to 1 kΩ and R_B to 10 kΩ. Assume the idealized transistor has $\beta_F = 50$ and a voltage V_{BE} that is 0.6 V in the active mode. In saturation, both junction voltages are 0.6 V and in cutoff the currents I_B and I_C are zero. (a) Find v_O as a function of v_I for active-mode operation and calculate the active-mode range of values for v_I. (b) Sketch and dimension the voltage transfer characteristic for values of v_I between -3 and $+3$ V. (c) Calculate v_O for $v_I = 0$ (state 0) and also for $v_I = 7$ (state 1).

5-6. Add to a two-input NAND gate small circles at both inputs, denoting inversion, and find the output Y in terms of the inputs A and B. By means of a truth table deduce that the circuit is an AND gate.

5-7. A NAND gate with inputs A and B and a second NAND gate with inputs A and C have their outputs feeding into a third NAND gate. Draw the configuration and deduce by means of a truth table that the output Y is $AB + AC$.

Section 5-2

5-8. Draw an OR gate and an AND gate, each with inputs A, B, and C. Add inversion (negation) circles at each of the inputs. Deduce that the modified OR and AND gates are NAND and NOR gates, respectively.

5-9. Deduce that a three-input AND gate with inversion at *all* terminals becomes an OR gate. Also, show that a three-input OR gate with inversion at *all* terminals is an AND gate.

5-10. Given $Y = \bar{A}(\bar{B} + \bar{C}) + \overline{\bar{A} + \bar{B} + \bar{C}}$. Find Y as the sum of three minterms and sketch the two-level NAND configuration that implements the logic. The true variables and their complements are available as inputs.

5-11. Deduce the logic equation for Y directly from the network of Fig. 5-14b. With the aid of DeMorgan's laws, show that Y can be put in the form of (5-26).

5-12. Given $Y = \bar{A}\bar{B}\bar{C} + \bar{A}BC + A\bar{B}\bar{C} + A\bar{B}C + ABC$. (a) From a truth table find the complement $\bar{Y}$ as a sum of products. (b) Next, determine Y as a product of sums. (c) Sketch a two-level NOR network that implements the logic.

Section 5-3

5-13. $Y = \overline{AB + C} + A(\bar{B} + \bar{C})$. (a) Develop the Karnaugh map directly from the expression for Y and deduce the minimal sum-of-products form. Compare the result with that found in the example of Sec. 5-2. Also, sketch a two-level NAND network implementing the logic. (b) From the Karnaugh map of (a), determine the map for $\bar{Y}$ and use the result to find the simplest two-level NOR network implementing the logic for Y. Sketch the network. True variables and complements are available as inputs.

5-14. In the truth table of Fig. 5-13 change the entries in the Y column to 11101101 from top to bottom. (a) Design and sketch the simplest two-level NAND network that implements the logic. (b) Repeat for a NOR network. True variables and their complements are available as inputs.

5-15. In the truth table of Fig. 5-13 change the entries in the Y column to 11110100 from top to bottom. (a) Design and sketch the simplest two-level NAND network that implements the logic. (b) Repeat for a NOR network. True variables and their complements are available as inputs.

5-16. $Y = AB + \bar{A}C\bar{D} + \bar{A}B\bar{C} + BCD$. Design a two-level NOR–NOR network to implement the logic, using a minimal number of gates and terminals. True variables and their complements are available as inputs.

5-17. A Karnaugh map for Y, which is a function of A, B, C, and D, has 1's in all squares except the third square from the left in the third row from the top. Show that each logical 1 can be associated with a group of eight adjacent squares and deduce that the logic can be implemented with a single NAND gate and also with a single OR gate if complements of the inputs are available. Sketch the gates.

5-18. $Y = \overline{\bar{A}\bar{B}AB} + (\bar{A} + B)(AB + \overline{AB})$. (a) Using a Karnaugh map, simplify the expression for Y. Also, directly from the map for Y deduce the simplified expression for $\bar{Y}$. (b) An IC package containing three NAND gates, each with two inputs, is used to implement the logic. Show the connections. No input terminal should be left open. Repeat for an IC package containing two NOR gates, each with two inputs. The complements of A and B are not available.

Section 5-4

5-19. In Fig. 5-23 of Sec. 5-4 change all NOR gates to NAND gates and all NAND gates to NOR gates, but do not change any inversions. From the Karnaugh map for Y, find and sketch two-level NAND–NAND and also NOR–NOR networks. Each should contain the fewest number of gates with the fewest number of inputs.

5-20. Connect the A and B inputs of a two-input NOR gate with output Y_1 to the respective A and B inputs of a two-input AND gate with output Y_2. Feed Y_1 and Y_2 into another two-input NOR gate of output Y. With the aid of DeMorgan's laws deduce that Y is the Exclusive-OR function of (5-29).

5-21. (a) Connect the inputs A and B of a NAND gate through inverters to the two inputs of a second NAND gate. Both outputs feed into another two-input NAND gate of output Y. Using DeMorgan's laws, deduce that Y is the Exclusive-NOR function of (5-30). (b) Repeat (a) with the first two NAND gates replaced with OR gates.

5-22. For the one-bit full adder described by the truth table of Fig. 5-29, determine S and K in the product-of-sums form. Directly from the results, design a one-bit full adder using NOR gates and inverters exclusively. The inputs are the true variables A, B, and C.

Chapter Six
Logic-Gate Switching

There are several major families of logic systems. Some use bipolar junction transistors, such as TTL, and others employ field-effect devices, such as COS/MOS systems. Switching parameters largely determine the family one should use for a particular application. The high-level and low-level voltages of the logical 1 and 0 states differ from one family to another. The difference between these levels is an indication of the noise voltage that can be tolerated, and this is a major factor influencing the choice of a system, especially in noisy industrial environments. In many situations the switching speed is of vital importance. For example, the arithmetic section of a large computer system must operate at the highest possible speed. These and other switching characteristics are investigated here. An elementary BJT inverter gate is employed to illustrate the basic concepts. The gate circuits and properties of the more popular logic families are considered in following chapters.

An important characteristic of a logic family is the static transfer curve relating the output voltage v_O to the input voltage v_I of an inverter. An idealized curve is represented by the solid line of Fig. 6-1, with the dashed line being more realistic. As indicated, the high-level voltage V_H is 8 and the low-level voltage V_L is 2. Switching between the states occurs at the midpoint of 5 V, which provides a safety margin of 3 V for each state. Suppose switching occurred at 7 V instead of 5. In this case the voltage margin is only 1 V when the input is at the high-level value of 8 V. A 1 V noise signal could cause the output to switch erroneously. Thus it is evident that switching should ideally occur at the midpoint between the

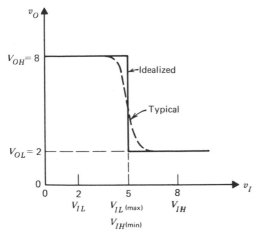

Figure 6-1   Idealized static transfer characteristic of an inverter.

high and low values. It is also apparent that the safety margin is reduced when the characteristic is that of the dashed line. This curve has a finite transition region, and within this region there may be ambiguity in the output state.

From our study of the NOT gate we know that, when the input is low-level, the output should be high-level. Therefore, the 5 V transition point of the ideal curve represents the maximum allowed value of the low-level input voltage V_{IL}. For values of v_I less than $V_{IL(max)}$ the output is properly in the high state. Because v_I is typically the output of an identical preceding stage, it follows that the low-level output voltage V_{OL} must be less than $V_{IL(max)}$, and the difference must be greater than any noise voltage present.

When the input is high-level, the output of the inverter should be low-level. Consequently, the 5 V transition point also represents the minimum allowed value of the high-level input voltage V_{IH}. For v_I greater than $V_{IH(min)}$, the output is properly in the low state. Recognizing again that v_I is typically the output of an identical preceding stage, it is clear that the high-level output voltage V_{OH} must be greater than $V_{IH(min)}$ by at least the amount of any possible noise voltage.

Practical inverters do not switch at precisely the midpoint between the low and high states. The transition region is finite and the voltage of a state can be specified only within a certain range. Furthermore, differences between the low-level and high-level voltages are sometimes small, and noise is often a major problem. For each logic family introduced in this and later chapters, an inverter characteristic is examined. As shown by the discussion here, such a curve reveals much information.

6-1. AN ELEMENTARY LOGIC GATE

FAN-IN AND FAN-OUT

In order to introduce certain basic and important properties of logic gates the elementary NOR gate of Fig. 6-2 will be examined. It has a *fan-in* of 3, which means there are three inputs A, B, and C. Furthermore, because it drives five identical gates it is said to have a *fan-out* of 5.

Suppose one or more inputs of the gate of Fig. 6-2 are at the high level of 1 V or more. This causes at least one of the transistors to be saturated, providing an output voltage v_O at the saturation value of v_{CE}, or approximately 0.2 V. However, if all inputs are low-level at 0.2 V, the three transistors are in cutoff, giving a logical 1 state at the output. This high level causes the transistors of the five load gates, which are assumed to be identical to the one shown, to be saturated. The resulting load current makes v_O less than V_{CC}. For the values shown, with each v_{BE} of a saturated load transistor assumed to be 0.8 V, the output voltage is determined to be 1.06 V (see P6-1). Clearly, the high-level voltage is lowered as the fan-out is increased, and this limits the allowed fan-out. Fan-outs up to about 5 are reasonable for the illustrated gate.

PULL-UP RESISTOR

A logical 1 at the input of a transistor turns it *on*, causing the transistor to conduct. The voltage across R_C is 2.8 V, assuming v_{CE} is 0.2 V, and the current is limited to 4.7 mA. When the gate switches *off*, the current through R_C drops and v_O rises. Consequently, R_C is sometimes referred to as a *pull-up* resistor. Pull-up resistors are always connected to *positive* voltages, whereas *pull-down* resistors, which we encounter later, are always connected to *negative* supplies.

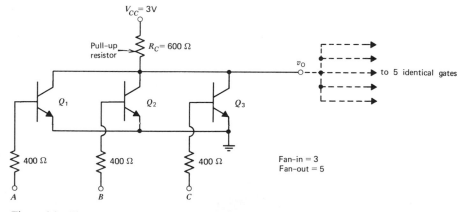

Figure 6-2 Elementary NOR gate (positive logic).

CURRENT HOGGING

The 400 ohm resistors in series with the inputs assist in dividing the gate input currents evenly between those gates in saturation. Without them, mismatched input devices might cause one gate to carry most of the current, and this so-called *current hogging* seriously degrades the transient response. Consequently, the resistances should not be too small. Because they reduce switching speed, they must not be larger than necessary.

WIRE-AND LOGIC

The logic system with the basic gate of Fig. 6-2 can utilize *wire-AND* logic. This is also called *implied-AND* logic and is illustrated in Fig. 6-3. The outputs of the two gates are simply tied together, and the connection makes $Y = Y_1 Y_2$. Only when both transistors are *off* is the output high. Unfortunately, the connection places the two 600-ohm pull-up resistors in parallel, giving an effective R_C of 300 ohms. Application of wire-AND logic to more than two gates further reduces R_C. Certainly each transistor must be able to sink the greater current without appreciable rise in the saturation value of v_{CE}. The difficulty can be avoided by using *open-collector* gates. These are integrated circuits with the collector resistors omitted. The proper resistance *must be added externally.*

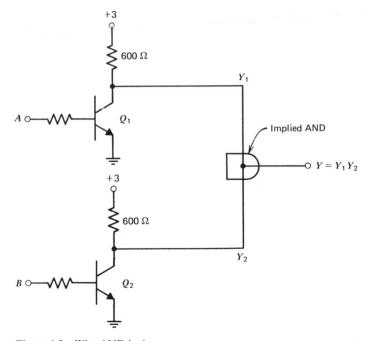

Figure 6-3 Wire-AND logic.

The collectors of several open-circuit gates can be connected together in conjunction with a single pull-up resistor giving wired-AND logic, and the designer selects the appropriate value of the resistor (see P6-2).

NOISE MARGINS

The maximum and minimum static transfer characteristics of the gate of Fig 6-2, with the inputs connected so as to form an inverter, are shown in Fig. 6-4. A typical curve lies between the extremes. With the input at a low level of 0.2 V the transistors are in cutoff, giving a high-level output of about 1 V. In the transition region between the two breakpoints, which occur at inputs of 0.7 V and 0.8 V approximately, the transistors are in the active mode, and the load gates change from *on* to *off* when v_O drops below 0.7 V. With v_I in the high-level region the transistors are saturated.

The noise immunity of a gate is commonly specified in terms of *noise margins*. At the bottom of Fig. 6-5 are shown two inverting gates with a noise source between them. Indicated above gate 1 are the low-level and high-level output ranges of v_{O1}, as *guaranteed by the manufacturer*, and above gate 2 are the *permissible* low-level and high-level input ranges of v_{I2} that give the proper guaranteed outputs of this gate. Each gate has the extreme characteristics of Fig. 6-4.

Suppose the output v_{O1} of gate 1 is a logical 1, having a high-level output with a value greater than $V_{OH(min)}$. To assure that the input v_{I2} of gate 2 is within the permissible high-level range, it is evident that $V_{OH(min)}$ must be greater than the minimum value $V_{IH(min)}$ of the input high-level range by the amount of any noise voltage V_N. The voltages $V_{OH(min)}$ and $V_{IH(min)}$ are precisely defined by the

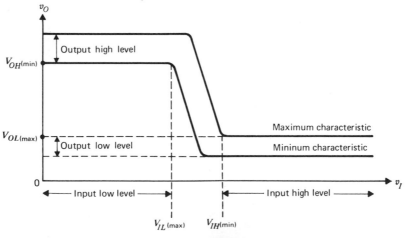

Figure 6-4   Gate extreme transfer characteristics.

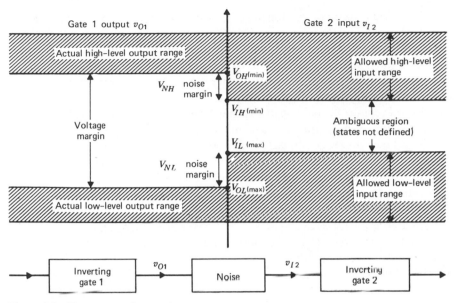

Figure 6-5   Worst-case noise margins.

appropriate breakpoints of Fig. 6-4 and they are indicated on Fig. 6-5. Thus we require that

$$V_{OH(\text{min})} > V_{IH(\text{min})} + V_N \qquad (6\text{-}1)$$

Now suppose the output v_{O1} of gate 1 is a logical 0 with a low-level output less than $V_{OL(\text{max})}$. The input v_{I2} of gate 2 is certainly within the permissible low-level range provided the sum of $V_{OL(\text{max})}$ and the noise voltage V_N is less than the maximum value $V_{IL(\text{max})}$ of the input low-level range. Both $V_{OL(\text{max})}$ and $V_{IL(\text{max})}$ are defined by the appropriate breakpoints of Fig. 6-4 and are indicated on Fig. 6-5. The requirement is that

$$V_{OL(\text{max})} + V_N < V_{IL(\text{max})} \qquad (6\text{-}2)$$

The maximum V_N satisfying the inequality of (6-1) is the *dc high-level noise margin* V_{NH}, and the maximum V_N satisfying (6-2) is the *dc low-level noise margin* V_{NL}. These noise margins are shown on Fig. 6-5. Mathematically,

$$V_{NH} = V_{OH(\text{min})} - V_{IH(\text{min})} \qquad (6\text{-}3)$$

$$V_{NL} = V_{IL(\text{max})} - V_{OL(\text{max})} \qquad (6\text{-}4)$$

Although these are dc margins, the high speed of IC logic systems justifies treating most noise signals as dc insofar as noise margins are concerned.

In some logic systems the transition between transistor modes of operation is more gradual than that shown in Fig. 6-4. For such cases the breakpoints used to determine the voltages of (6-3) and (6-4) are taken to be the *unity-gain* points, defined as the points of the transfer characteristic at which the magnitude of the differential gain dv_O/dv_I of the gate is unity. At each unity-gain point the slope of the curve is -1 (see P6-3).

Noise margins are usually given on data sheets in volts. Assumed are *worst-case* conditions, which are the most unfavorable connections of the input terminals and the poorest combination of circuit and device parameters, along with a maximum fan-out. Although such specifications are highly conservative, worst-case design assures proper performance. Actual noise margins are commonly greater than those specified by a considerable amount.

For the gate of Fig. 6-2 values of $V_{OH(min)}$ and $V_{IH(min)}$ are about 1.0 and 0.85 V, giving a high-level noise margin of 0.15 V. Thus the output of a gate in the logical 1 state can tolerate only 150 mV of noise on the line connecting the output to the next gate with a certainty that the driven gate will not err. The low-level noise margin is substantially greater. For $V_{IL(max)} = 0.7$ and $V_{OL(max)} = 0.3$, the low-level noise margin is 0.4 V.

The gate which has been examined here is the basic gate of *resistor-transistor logic (RTL)*. As implied by the name, it consists of resistors and transistors only. Although widely used in the past and simple in structure and theory, it is inferior to gates used in new designs today.

6-2. DC ANALYSIS

It is easy to deduce the transfer characteristic of a logic gate from the Ebers-Moll equations. The gate to be analyzed is that of Fig. 6-6, which has a fan-in of 3 and a fan-out m of 5. Each of the five load gates has a fan-in n of 2. The two transistors of each load gate are assumed to have the same input voltage. Accordingly, the current through R_L of the first load gate divides equally between Q_4 and Q_5, and the situation is similar for the other load gates. Both Q_2 and Q_3 are *off*, and the objective here is to plot a curve of v_O versus the input v_I of transistor Q_1. To simplify the equations we shall assume the transistors are identical. The respective saturation currents I_{ES} and I_{CS} are 2×10^{-15} and 10^{-14} A, α_F is 0.98, α_R is 0.20, and q/kT is 40. The small saturation currents and the small reverse current gain α_R are typical of IC transistors. For simplicity the ohmic resistance of the collector, typically 10 ohms, and the leakage current across the reverse-biased PN junction between the collector and the substrate are assumed to be negligible. Their effects on the results are second-order.

For an NPN transistor the Ebers–Moll equations are

$$i_E = -I_{ES}(e^{qv_{BE}/kT} - 1) + \alpha_R I_{CS}(e^{qv_{BC}/kT} - 1) \qquad (6\text{-}5)$$

$$i_C = \alpha_F I_{ES}(e^{qv_{BE}/kT} - 1) - I_{CS}(e^{qv_{BC}/kT} - 1) \qquad (6\text{-}6)$$

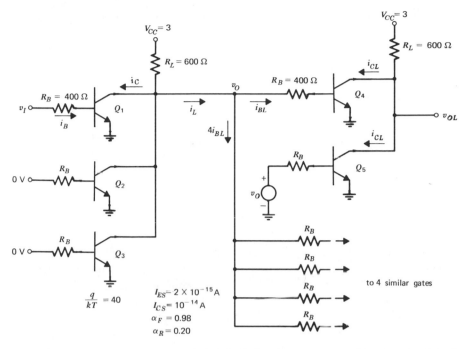

Figure 6-6   Logic gate with typical load. Gate fan-in is 3, fan-out m is 5, and fan-in n of each load gate is 2.

These are (2-10) and (2-11) of Sec. 2-2. Replacing v_{BC} in (6-6) with $v_{BE} - v_{CE}$ and solving for v_{BE}, we obtain

$$v_{BE} = \frac{kT}{q} \ln\left[\frac{i_C + \alpha_F I_{ES} - I_{CS}}{\alpha_F I_{ES} - I_{CS}\exp(-qv_{CE}/kT)}\right] \qquad (6\text{-}7)$$

From (6-5) and (6-6) the base current i_B is found to be

$$i_B = I_{ES}(1 - \alpha_F)(e^{qv_{BE}/kT} - 1) + I_{CS}(1 - \alpha_R)[e^{q(v_{BE} - v_{CE})/kT} - 1] \qquad (6\text{-}8)$$

Data for the transfer characteristic are calculated by initially assuming a numerical value for the output voltage v_{OL} of a load gate. The collector current i_{CL} of the load gate is $(V_{CC} - v_{OL})/nR_L$, or $(3 - v_{OL})/1200$. Thus from (6-7) the base-emitter voltage v_{BEL} of the gate is

$$v_{BEL} = \frac{1}{40} \ln\left[\frac{(3 - v_{OL})/1200 - 8.04 \times 10^{-15}}{1.96 \times 10^{-15} - 10^{-14}\exp(-40v_{OL})}\right] \qquad (6\text{-}9)$$

Next we find i_{BL} of the load gate using (6-8), which becomes

$$i_{BL} = 4 \times 10^{-17}e^{40v_{BEL}} + 8 \times 10^{-15}e^{40(v_{BEL} - v_{OL})} - 8.04 \times 10^{-15} \qquad (6\text{-}10)$$

From the calculated values of v_{BEL} and i_{BL} the output voltage v_O and the load current i_L are found from the relations:

$$v_O = 400i_{BL} + v_{BEL} \qquad i_L = 5i_{BL} \qquad (6\text{-}11)$$

These are evident from the network of Fig. 6-6.

The collector current i_C of Q_1 is $(3 - v_O)/600$ minus i_L, and (6-7) gives the base-emitter voltage v_{BE} of this transistor to be

$$v_{BE} = \frac{1}{40} \ln\left[\frac{(3 - v_O)/600 - i_L - 8.04 \times 10^{-15}}{1.96 \times 10^{-15} - 10^{-14} \exp(-40v_O)}\right] \qquad (6\text{-}12)$$

The input v_I is $v_{BE} + 400i_B$, and with the aid of (6-8) we deduce that v_I is

$$v_I = v_{BE} + 400[4 \times 10^{-17}e^{40v_{BE}} + 8 \times 10^{-15}e^{40(v_{BE}-v_O)} - 8.04 \times 10^{-15}] \qquad (6\text{-}13)$$

The calculated values of v_O from (6-11) and v_I from (6-13) give one point on the transfer characteristic. For example, suppose we select a value of 2 for v_{OL}. Then v_{BEL} is 0.669 from (6-9), i_{BL} is 0.000017 from (6-10), v_O is 0.676 and i_L is 0.000085 from (6-11), v_{BE} is 0.707 from (6-12), and v_I is 0.738 from (6-13). Thus for $v_I = 0.738$ the output v_O is 0.676. Additional points of the transfer characteristic are found by choosing additional values for v_{OL} and repeating the procedure (P6-6). Calculations reveal that v_{OL} is restricted to values between 0.07658 and

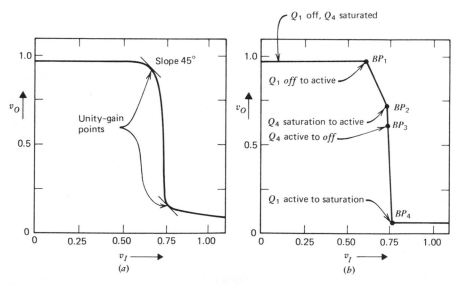

Figure 6-7 Calculated transfer characteristic. (*a*) Actual characteristic. (*b*) Piecewise linear characteristic.

$3(1 - 10^{-11})$, with values outside this range resulting in logarithms of negative numbers in the calculation procedure.

Although the method based on the Ebers–Moll equations is reasonably accurate, simple, and direct, it does not clearly reveal the effect of the circuit parameters on the transfer characteristics. A less accurate approach, referred to as *piecewise-linear*, gives more meaningful results. Shown in Fig. 6-7 is the calculated transfer characteristic of the RTL gate of Fig. 6-6, along with the piecewise-linear characteristic which we shall now investigate.

PIECEWISE LINEAR ANALYSIS

As indicated in Fig. 6-7b there are four distinct breakpoints in the transfer characteristic. The operating modes of the transistors relative to the breakpoints are given in Table 6-1. When v_I is low, the input transistor Q_1 is *off* and the load transistor Q_4 is saturated. Of course, in accordance with the specifications that have been made, all the load transistors have the same operating region as Q_4. When v_I is increased to about 0.6 V at BP_1, transistor Q_1 changes from cutoff to active-mode operation. At BP_2 the increased collector current of Q_1 gives an output voltage so low that the base current of Q_4 is no longer adequate to produce saturation, and Q_4 moves into the active region. A further drop in v_O causes Q_4 to be cut off at BP_3, and at BP_4 the input voltage v_I is large enough to saturate Q_1.

The piecewise-linear procedure consists of determining the values of v_I and v_O at the four breakpoints. Then the transfer characteristic is approximated by straight-line segments drawn from the breakpoints as illustrated in Fig. 6-7b. We note from Table 6-1 that the input transistor is in the active mode at each breakpoint. Therefore, v_{BE} is kT/q times $\ln(i_C/\alpha_F I_{ES})$ from the active-mode Ebers–Moll equation. From this and a loop equation around the base-emitter circuit of Q_1, with i_B replaced with i_C/β_F, we find that

$$v_I = \frac{i_C R_B}{\beta_F} + \frac{kT}{q} \ln\left(\frac{i_C}{\alpha_F I_{ES}}\right) \tag{6-14}$$

We now determine i_C and v_O at each breakpoint, using (6-14) to find v_I.

Table 6-1 Transistor Operating Modes

Breakpoint Region	Gate Transistor Q_1	Load Transistors
v_I low	off	saturation
BP_1 to BP_2	active	saturation
BP_2 to BP_3	active	active
BP_3 to BP_4	active	off
v_I high	saturation	off

BREAKPOINT 1

With reference to Fig. 6-6 with transistor Q_1 *off* so that i_C is zero and with transistor Q_4 saturated, application of Kirchhoff's voltage law from the V_{CC} supply of the input gate through R_L, R_B, and the base-emitter of Q_4 gives

$$V_{CC} = R_L i_L + \frac{R_B i_L}{m} + V_{BEL}'$$

with V_{BEL}' denoting the saturation value of the base-emitter voltage of Q_4. The fan-out m is 5. We solve this for i_L and find v_O from $V_{CC} - R_L i_L$. The result is

$$v_O = \frac{V_{CC} R_B + m V_{BEL}' R_L}{m R_L + R_B} \qquad (6\text{-}15)$$

When the collector current i_C of Q_1 increases to about 1% of V_{CC}/R_L, the transistor is approaching the active region. Thus a reasonable choice for i_C at BP_1 is

$$i_C = \frac{0.01 V_{CC}}{R_L} \qquad (6\text{-}16)$$

From this and (6-14) we can calculate v_I. For the specified numerical values with the saturation voltage V_{BEL}' taken to be 0.7 V, the breakpoint (v_I, v_O) is found to be (0.599, 0.971).

BREAKPOINT 2

At this point transistor Q_4 has moved to the boundary of the active region, and the active-mode Ebers–Moll equation can be used to determine i_{CL} in terms of v_{BEL}. Also, from Fig. 6-6 we note that i_{CL} is $V_{CC} - v_{OL}'$ divided by nR_L, with the fan-in $n = 2$ and with v_{OL}' denoting the saturation value of v_{OL}. Accordingly,

$$i_{CL} = \frac{V_{CC} - v_{OL}'}{nR_L} = \alpha_F I_{ES} \exp\left(\frac{q v_{BEL}}{kT}\right) \qquad (6\text{-}17)$$

For values of v_I less than that at BP_2 transistor Q_4 is saturated with $i_{BL} > i_{CL}/\beta_F$ (see P6-4). However, at the breakpoint Q_4 enters the active region and i_{BL} is i_{CL}/β_F, or

$$i_{BL} = \frac{V_{CC} - v_{OL}'}{nR_L \beta_F} \qquad (6\text{-}18)$$

The output voltage v_O is the sum of $i_{BL} R_B$ and v_{BEL}. Utilizing (6-17) and (6-18), we find that

$$v_O = \frac{(V_{CC} - v_{OL}')R_B}{nR_L \beta_F} + \frac{kT}{q} \ln\left(\frac{V_{CC} - v_{OL}'}{nR_L \alpha_F I_{ES}}\right) \qquad (6\text{-}19)$$

The collector current i_C is $(V_{CC} - v_O)/R_L$ less the load current i_L, and i_L is mi_{BL}. With the aid of (6-18) we ascertain that

$$i_C = \frac{V_{CC} - v_O}{R_L} - \frac{m(V_{CC} - v_{OL}')}{nR_L \beta_F}$$ (6-20)

Assuming a saturation output v_{OL}' of 0.1 V, v_O and i_C can be calculated from (6-19) and (6-20), and v_I is found from (6-14). The breakpoint BP_2 (v_I, v_O) is (0.735, 0.716).

BREAKPOINT 3

Transistor Q_4 changes from active-mode operation to *off*. The collector current i_{CL} is very small and i_{BL} is negligible. Therefore, v_O equals v_{BEL}. A reasonable value of v_{BEL} is deduced from the active-mode Ebers–Moll equation with i_{CL} taken to be $0.01 V_{CC}/R_L$. Therefore,

$$v_O = \frac{kT}{q} \ln\left(\frac{0.01 V_{CC}}{R_L \alpha_F I_{ES}}\right)$$ (6-21)

Because i_L is negligible, i_C is

$$i_C = \frac{V_{CC} - v_O}{R_L}$$ (6-22)

From (6-21), (6-22), and (6-14) the breakpoint BP_3 (v_I, v_O) is found to be (0.741, 0.599).

BREAKPOINT 4

The input transistor becomes saturated, and the output voltage v_O is

$$v_O = v_O'$$ (6-23)

with v_O' denoting the saturation value of v_{CE}, which is approximately 0.1 V. The load current i_L is zero. It follows that i_C is again given by (6-22). The breakpoint (v_I, v_O) is found from (6-14), (6-22), and (6-23) to be (0.753, 0.1).

Comparison of the curves of Fig. 6-7 reveals that the only significant difference between the piecewise-linear plot and the true characteristic occurs in the vicinity of BP_4 where the actual characteristic has considerable curvature. The breakpoints have been expressed in terms of the network parameters, including the fan-out m and the load fan-in n. Thus it is easy to examine the effects of a change in one or more parameters (see P6-5). We now investigate the behavior of a transistor that is suddenly switched from one logical state to the other.

6-3. THE TRANSISTOR AS A SWITCH

It is most desirable that digital gates switch states quickly so that the overall system can operate at high speed. In order to change the operating region of a transistor or even to change the quiescent point, the charge stores in the neutral regions and within the space-charge layers must be changed, and this requires time. To illustrate the effect of these charges on the switching speed, we shall investigate an inverter that is switched by step-function inputs from *off* to *on* and from *on* to *off*.

CHARGE-CONTROL EQUATIONS

Let us examine the charge stores in the neutral regions of an NPN transistor. Suppose the base-collector voltage v_{BC} is zero. For this case the base current i_B is found from (6-8) to be

$$i_B = I_{ES}(1 - \alpha_F)(e^{qv_{BE}/kT} - 1) \tag{6-24}$$

Under static conditions the current consists of the flow of holes into the base to replace those lost by recombination in the neutral base and emitter regions. Actually, there is very little storage of charge in the emitter because of the heavy doping of this region, and the component of i_B that supplies hole flow across the junction to replace the holes lost in the emitter is negligible. If we let q_F denote the excess charge in these neutral regions when v_{BC} is zero, then i_B is q_F divided by the effective lifetime τ_{BF}, and τ_{BF} is the minority-carrier lifetime τ_b in the base. It follows that

$$\frac{q_F}{\tau_{BF}} = I_{ES}(1 - \alpha_F)(e^{qv_{BE}/kT} - 1) \tag{6-25}$$

Next, suppose $v_{BE} = 0$. In a similar manner we find that the ratio of the excess charge q_R to the lifetime τ_{BR} is

$$\frac{q_R}{\tau_{BR}} = I_{CS}(1 - \alpha_R)(e^{qv_{BC}/kT} - 1) \tag{6-26}$$

In this case the current i_B is the hole flow required to replace those lost by recombination in the neutral base and collector regions. There is substantial excess charge in the neutral collector when v_{BC} is positive, and the holes lost by recombination in this region are replaced by hole flow across the junction with the holes supplied by i_B. Because of the nonuniformity of the base and because τ_{BR} is determined partially by recombination within the collector, the lifetime τ_{BR} differs in value from τ_{BF}.

For the more general case in which neither v_{BC} nor v_{BE} is zero, the excess charge stores in the neutral regions are the sum of q_F and q_R. Also, i_B is the sum of the expressions of (6-25) and (6-26), and this sum is the Ebers–Moll equation (6-8).

Thus far we have examined only static equations, which must be modified for transistor operation under dynamic conditions. The modified equations for i_E, i_C, and i_B, referred to as the *charge-control equations*, are as follows:

$$i_E = \frac{\beta_R q_R}{\tau_{BR}} - \frac{\beta_F q_F}{\tau_{BF}} - \frac{q_F}{\tau_{BF}} - \frac{dq_F}{dt} - \frac{dq_{SE}}{dt} \tag{6-27}$$

$$i_C = \frac{\beta_F q_F}{\tau_{BF}} - \frac{\beta_R q_R}{\tau_{BR}} - \frac{q_R}{\tau_{BR}} - \frac{dq_R}{dt} - \frac{dq_{SC}}{dt} \tag{6-28}$$

$$i_B = \frac{q_F}{\tau_{BF}} + \frac{q_R}{\tau_{BR}} + \frac{d}{dt}(q_F + q_R + q_{SE} + q_{SC}) \tag{6-29}$$

The charge q_{SE} denotes that supplied to the emitter-junction space-charge layer when v_{BE} changes from zero, and q_{SC} is the corresponding charge of the collector-junction space-charge layer. The reverse beta, designated β_R, equals $\alpha_R/(1 - \alpha_R)$.

With the time derivatives set equal to zero, it is not difficult to show (see P6-7) that the charge-control equations are identical with the Ebers–Moll equations (6-5), (6-6), and (6-8). Thus we need only to consider the time-derivative terms. In (6-27) the term $-dq_F/dt$ is the component of i_E that supplies a change in the excess charge q_F of the neutral regions, and the term $-dq_{SE}/dt$ is the component supplying a change in the charge of the emitter-junction space-charge layer. These are capacitive currents, with the capacitances being nonlinear. Both q_F and q_{SE} are defined so that they increase when v_{BE} increases. However, although the forward electron flow increases, the conventional current i_E decreases, which accounts for the negative signs. The equation for i_C corresponds to that for i_E, and (6-29) for i_B is minus the sum of i_E and i_C.

For abrupt junctions the charges q_{SE} and q_{SC} are given by (1-53) with the appropriate junction voltage substituted for V. However, the junctions of an IC transistor more nearly approximate the graded junction examined in P1-16 and P2-26, and (2-45) and (2-46) apply. Accordingly, q_{SE} is given by

$$q_{SE} = a[\psi_o^{2/3} - (\psi_o - v_{BE})^{2/3}] \tag{6-30}$$

with a denoting a constant and with ψ_o denoting the emitter junction built-in voltage. A similar equation applies for q_{SC}.

INVERTER RELATIONS

The inverter to be analyzed is shown in Fig. 6-8, along with the input waveform and the parameter values. The parameters are those of the logic gate of Fig. 6-6, with the additional values needed for the charge-control model also given. For simplicity the inverter has no load. The input voltage v_I is initially zero with

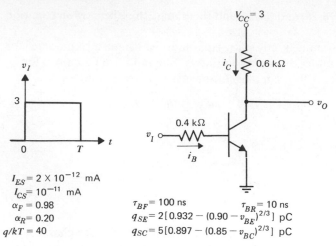

$I_{ES} = 2 \times 10^{-12}$ mA
$I_{CS} = 10^{-11}$ mA
$\alpha_F = 0.98$
$\alpha_R = 0.20$
$q/kT = 40$

$\tau_{BF} = 100$ ns
$q_{SE} = 2[0.932 - (0.90 - v_{BE})^{2/3}]$ pC

$\tau_{BR} = 10$ ns
$q_{SC} = 5[0.897 - (0.85 - v_{BC})^{2/3}]$ pC

Figure 6-8 BJT inverter, input-voltage waveform, and parameter values.

the gate *off*. At time zero the input changes to 3 V, remaining at this value until the gate has switched *on* and a steady state is reached. Then it drops abruptly to zero, turning the gate *off*. Using suitable approximations, we shall calculate the switching times. *Units of* kΩ, mA, *and* V *are used*. In addition, all charges are expressed in *picocoulombs* and time is given in *nanoseconds*. Note that pico-coulombs divided by nanoseconds yield milliamperes.

From the inverter circuitry of Fig. 6-8 we obtain the relations

$$v_I = 0.4i_B + v_{BE} \tag{6-31}$$

$$3 = 0.6i_C + v_{BE} - v_{BC} \tag{6-32}$$

Using the specified numerical values, which are shown in the figure, charge-control equations (6-28) and (6-29) become

$$i_C = 0.49q_F - 0.125q_R - \frac{d}{dt}(q_R + q_{SC}) \tag{6-33}$$

$$i_B = 0.01q_F + 0.10q_R + \frac{d}{dt}(q_F + q_R + q_{SE} + q_{SC}) \tag{6-34}$$

The excess charges q_F and q_R in picocoulombs are found from (6-25) and (6-26) to be

$$q_F = 4 \times 10^{-12}e^{40v_{BE}} \tag{6-35}$$

$$q_R = 80 \times 10^{-12}e^{40v_{BC}} \tag{6-36}$$

with the insignificant minus-one terms omitted. Also, the equations for q_{SE} and q_{SC}, which have the form of (6-30), are

$$q_{SE} = 2[0.932 - (0.90 - v_{BE})^{2/3}] \qquad (6\text{-}37)$$

$$q_{SC} = 5[0.897 - (0.85 - v_{BC})^{2/3}] \qquad (6\text{-}38)$$

with the charges in picocoulombs.

The charge-control equations (6-33) and (6-34) can be expressed in terms of the voltages v_I, v_{BE}, and v_{BC} by using the other equations to eliminate the currents and charges. With a little manipulation they can be arranged so that each has a single time derivative (see P6-8), with one having the form

$$\frac{dv_{BE}}{dt} = f(v_I, v_{BE}, v_{BC}) \qquad (6\text{-}39)$$

The other is similar, containing the derivative dv_{BC}/dt. With v_I specified as a function of time and with the initial values of v_{BE} and v_{BC} known, they can be solved with the aid of a digital computer using numerical integration techniques. This is suggested as an exercise for the interested student (P6-9). However, the procedure does not clearly reveal the causes of the switching delays that occur. Hence we shall use an approximate method. Although less accurate, it is simpler and has the advantage of relating the charge stores to the time delays. To avoid complicated equations the analysis utilizes numerical values rather than symbols representing network and transistor parameters. First, let us briefly examine the switching transient qualitatively.

THE SWITCHING TRANSIENT

Figure 6-9 shows the input voltage and the collector-current response of the inverter. The input pulse is applied at time zero. At this instant v_{BE} is zero and v_{BC} is -3. An initial current of 7.5 mA, equal to v_I/R_B, supplies charge to the emitter-junction space-charge layer, causing v_{BE} to rise from zero. During the first portion of the *delay time* t_d, defined as the time required for i_C to reach 10% of its maximum value, there is no appreciable change in v_{BC}. Thus the rise in v_{BE} causes the collector-emitter voltage to become greater than V_{CC} by as much as 0.5 V, and i_C is negative. Near the end of the delay time at the instant of zero i_C, enough charge has gone to the collector-junction space-charge-layer to cause its voltage to fall precisely the amount (about 0.6 V) required to offset the rise in v_{BE}, and v_{CE} equals V_{CC}. The delay time occurs because the nearly constant base current must supply considerable charge to the two space-charge layers and to the neutral base region before the excess charge q_F becomes sufficient to give appreciable collector current.

The *rise time* t_r is that required to increase i_C from 10 to 90% of its maximum value. In this interval v_{BE} changes very little, and i_B supplies charge to the neutral

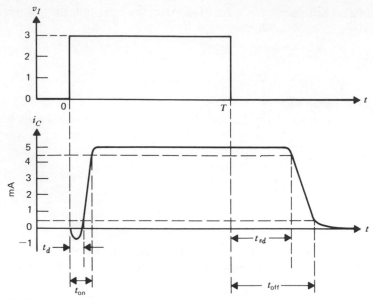

Figure 6-9   Inverter input and collector-current response.

base and to the collector-junction space-charge layer, increasing both q_F and q_{SC}. The output voltage drops rapidly to the saturation value. The sum of the delay time t_d and the rise time t_r is the *turn-on time* t_{on}. Shortly after turn-on the current i_C and the output voltage v_O become steady. However, current continues to supply charge to the neutral base and collector regions, increasing q_F and q_R until i_B, now about 5.7 mA, equals the charge recombination rate. At this point the transistor is deep in saturation with considerable charge stored in the neutral regions.

At time T the input drops to zero volts. The base current reverses with an initial value $-v_{BE}/R_B$ of about -1.8 mA. Excess charge decreases due to both i_B and recombination. However, not until the transistor leaves the saturation region does i_C begin to fall. The *storage-delay time* t_{sd} is the time required for i_C to drop 10% from its maximum value.

As charge is withdrawn, the transistor moves from saturation to active-mode operation to cutoff. *Turn-off time* t_{off} is the sum of the storage-delay time t_{sd} and the *fall time*, which is that required to drop i_C from 90 to 10% of its maximum. Let us now proceed to the calculations.

6-4. TURN-ON

As we shall see, throughout the turn-on time v_{BC} never exceeds $+0.4$ V, and consequently the excess charge q_R is negligible. Thus the charge-control equations (6-33) and (6-34) in units of mA, pC, and ns become

$$i_C = 0.49q_F - \frac{dq_{SC}}{dt} \tag{6-40}$$

$$i_B = 0.01q_F + \frac{d(q_F + q_{SE} + q_{SC})}{dt} \tag{6-41}$$

Initially, q_F is zero, remaining negligible until v_{BE} rises above 0.6 V. In this time interval i_C is negative as previously noted and as indicated by (6-40). In (6-41) the first term on the right is q_F/τ_{BF}, representing charge recombination. During turn-on the recombination current is almost negligible and hence a rough estimate of its effect on the turn-on time is satisfactory.

Because we have no idea of the value of dq_{SC}/dt in (6-40), let us *initially* neglect this term in order to obtain easily an approximate solution. Then an estimate of the derivative can be made and the process repeated. When the transistor is saturated with v_{CE} approximately 0.1 V, the collector current is 2.9/0.6, or 4.83 mA. At the end of turn-on the collector current is about 90% of this, or 4.4 mA. From (6-40) with the derivative neglected, q_F is found to be 9.0 pC, and v_{BE} is 0.711 from (6-35). With a collector current of 4.4 the collector-emitter voltage is 0.36. Hence, v_{BC} is 0.35. By (6-31) the base current is 5.72 mA, and q_R, q_{SE}, and q_{SC} are determined from (6-36) through (6-38) to be 0.0001, 1.21, and 1.34 pC, respectively.

To find the turn-on time we integrate (6-41) from time zero to $t = t_{on}$. We note that i_B drops from its initial value of 7.50 to 5.72 at the end of turn-on. Because this change is not large, a reasonable approximation is to treat i_B as a constant. The average of the extremes is 6.6. However, during most of the time v_{BE} is between 0.6 and 0.7 where i_B is close to its lower value. Thus we select 6 mA as an estimate. The integral of (6-41) becomes

$$6t_{on} = 0.01 \int_0^{t(on)} q_F \, dt + \Delta q_F + \Delta q_{SE} + \Delta q_{SC} \tag{6-42}$$

An estimate of the recombination integral of (6-42), *which is almost negligible*, is obtained by assuming q_F varies linearly with time from 0 to 9 pC at t_{on}. Thus q_F is $9t/t_{on}$. The terms Δq_F, Δq_{SE}, and Δq_{SC} are the increases in the charges, and the respective values are 9.0, 1.2, and 9.1 pC. These were determined from the calculated values of q_F, q_{SE}, and q_{SC} at t_{on} and the values at time zero, given in row 1 of Table 6-2. At time zero we note that v_{BC} is -3, and q_{SC} is -7.8 pC by (6-38). From (6-42) the initial estimate of t_{on} is calculated to be 3.2 ns.

The results show that q_{SC} varies from -7.8 at time zero to 1.3 at 3.2 ns. The average time rate of change is 2.8 pC/ns. At time t_{on} the base-emitter voltage is changing very little, because it is quite close to its steady-state value. However, the collector current is still rising rapidly, as indicated by the curve of Fig. 6-9.

Table 6-2   Inverter Calculated Values in ns, mA, V, and pC

	Time t	i_B	i_C	v_{BE}	v_{BC}	q_F	q_R	q_{SE}	q_{SC}
1	$t = 0$	7.50	0	0	-3	0	0	0	-7.8
2	$t_d = 1.3$ ns	5.76	0.5	0.698	-2.00	5.5	0	1.2	-5.6
3	$t_{on} = 4.2$ ns	5.70	4.4	0.723	0.360	14.7	0	1.2	1.4
4	Steady-state $t \approx 50$	5.66	4.9	0.735	0.681	23.9	54	1.3	3.0
5	$t_{sd} = 16.3$ ns	-1.76	4.4	0.702	0.342	6.4	0	1.2	1.3
6	$t_{off} = 27$ ns	-0.49	0.5	0.196	-2.50	0	0	0.3	-6.7
7	Steady state	0	0	0	-3	0	0	0	-7.8

Therefore, v_{CE} must be decreasing appreciably. With v_{BE} practically constant, v_{BC} and hence q_{SC} are rising at a rapid rate. As an approximation, we shall assume that at turn-on the charge q_{SC} is increasing at 2.8 pC/ns, which is the calculated average rate of change. This is undoubtedly an improvement over the value of zero that was used initially.

The preceding sequence of calculations is simply repeated, using 2.8 for dq_{SC}/dt in (6-40). This gives 14.7 for q_F at t_{on}, noting that i_C is 4.4 mA as before. The calculated values are given in row 3 of Table 6-2. The numerical values reveal that *the turn-on time is largely due to charges that must be supplied by* i$_B$ *to* q$_F$ *in the neutral base region and to the collector-junction space-charge layer so as to change* v$_{BC}$ *from* -3 *to* $+0.4$ V.

DELAY TIME

The delay time is determined similarly. From the data of rows one and three of Table 6-2 we note that q_{SC} changes 9.2 pC in 4.2 ns, with an average rate of change of 2.2. Using this as an estimate for dq_{SC}/dt in (6-40), the data of row two of Table 6-2 are calculated (see P6-10). The delay time is 1.3 ns, which is caused by changes in q_F, q_{SE}, and q_{SC}, with q_F having the dominant effect.

The difference between the turn-on and delay times is 3 ns, and this is the rise time. It is clear from the data of rows two and three of Table 6-2 that both q_F and q_{SC} are changing rapidly during this interval. Thus the rise time is largely the time required to store the necessary charge in the neutral base and also to charge the collector-junction space-charge layer so as to change its bias from reverse to forward.

STEADY STATE

After the collector-current rise is essentially completed, there is very little change in i_C, v_O, and the conditions of the space-charge layers. However, the base current continues to add charge to the neutral regions, driving the transistor deep into saturation. Both q_F and q_R increase until charge recombination balances i_B. This is the steady-state condition at which the time derivatives in the charge-control equations are zero. Using the steady-state form of the charge-control equations (6-33) and (6-34) to eliminate i_B and i_C in (6-31) and (6-32), we obtain

$$3 = 0.004q_F + 0.04q_R + v_{BE} \qquad (6\text{-}43)$$

$$3 = 0.294q_F - 0.075q_R + v_{BE} - v_{BC} \qquad (6\text{-}44)$$

These are easily solved in conjunction with (6-35) and (6-36) for the charges and voltages (see P6-11). Then (6-31) and (6-32) give the currents, and (6-37) and (6-38) give the space-layer charges. The results are shown in the fourth row of Table 6-2. As the steady state is approached, an increased percentage of i_B feeds recombination. Thus the approach is gradual. The process of charge build-up is essentially complete after about 50 ns. This is indicated by the results of a computer analysis (see P6-9).

6-5. TURN-OFF

STORAGE DELAY

At the end of the input pulse, indicated as time T in Fig. 6-9, the input voltage drops to zero. At this instant the collector current i_C is 4.9 mA, as shown in the fourth row of Table 6-2, and i_B is $-0.735/0.4$, or -1.84 mA. The charge-control equations in units of pC and ns are

$$4.9 = 0.49q_F - 0.125q_R - \frac{dq_R}{dt} \qquad (6\text{-}45)$$

$$-1.84 = 0.01q_F + 0.1q_R + \frac{d(q_F + q_R)}{dt} \qquad (6\text{-}46)$$

The charges q_{SE} and q_{SC} are not included because they change very little as long as the transistor is in saturation. Throughout the storage delay the transistor currents remain nearly constant at their initial values. Therefore, the charges q_F and q_R can be determined approximately in this time interval by solving (6-45) and (6-46) subject to the boundary conditions.

Solving (6-45) for q_F and substituting for q_F in (6-46), we obtain

$$\frac{d^2q_R}{dt^2} + 0.625\frac{dq_R}{dt} + 0.0503q_R = -0.951 \qquad (6\text{-}47)$$

For convenience let us refer to the initial time when v_I drops to zero as time zero. At this instant q_R is 54.3 as found in P6-11 for row four of Table 6-2, and dq_R/dt

is zero, which is obtained from (6-45) at time zero. The solution to (6-47) subject to these boundary conditions (P6-12) is

$$q_R = 89.2e^{-0.0949t} - 16e^{-0.530t} - 18.9 \text{ pC} \qquad (6\text{-}48)$$

From this and (6-45) the charge q_F is

$$q_F = 5.48e^{-0.0949t} + 13e^{-0.530t} + 5.2 \text{ pC} \qquad (6\text{-}49)$$

The excess charge q_R is zero when $t = 16.3$ ns, and this is the approximate storage-delay time. At this instant (6-49) gives 6.4 pC for q_F, and v_{BE} is found from (6-35) to be 0.702. As i_C is about 4.4 mA at the end of the storage delay, v_{CE} is 0.36, which gives 0.342 for v_{BC}. From (6-37) and (6-38) q_{SE} and q_{SC} are determined to be 1.2 and 1.3 pC, respectively. The base current is $-0.702/0.4$, or -1.76. These calculated values are those of the fifth row of Table 6-2.

Storage delay is the greater portion of the turn-off time and is considerably longer than the turn-on time. Thus every effort should be made to minimize this effect, which results from the extra charge that must be removed from the neutral regions of the saturated transistor. Both base current and internal recombination contribute to the removal of q_F and q_R, although in some switching circuits the transistor is turned off simply by opening the base lead. In such cases the excess charge is eliminated only by recombination, and storage delay is substantially increased. Transistors used in saturating logic circuits are frequently subjected to a gold-diffusion process during fabrication. This reduces the minority-carrier lifetime by aiding recombination.

SPEED-UP CAPACITOR

There are two common procedures for reducing the delay. One consists of turning the gate *off* with a negative v_I along with the smallest possible R_B. This increases the magnitude of the base current, thus providing a more rapid flow of charge out of the neutral regions. Negative base current during turn-off is referred to as turn-off *overdrive*. The other method is to shunt the base resistor R_B with a *speed-up capacitor*.

The inverter modified with the addition of a speed-up capacitor is shown in Fig. 6-10. In the *on*-state the voltage across C is $i_B R_B$, and the charge stored is $i_B R_B C$. Suppose C is chosen so that its charge is the same or greater than the total excess charge $(q_F + q_R)$ present in the *on*-state. Then when v_I drops to zero the capacitor draws an impulse current that quickly removes the excess charge. In the example considered, the *on*-state voltage across C is 2.3 and the total excess stored charge is found from Table 6-2 to be 78 pC. Accordingly, C should be 35 pF or more. Such a capacitance reduces the storage delay to about a nanosecond.

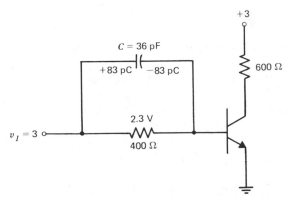

Figure 6-10 Inverter in the *on*-state with a speed-up capacitor.

FALL TIME

Throughout the fall time the charge q_R is negligible. The charge-control equations in units of mA, pC, and ns become

$$i_C = 0.49q_F - \frac{dq_{SC}}{dt} \tag{6-50}$$

$$i_B = 0.01q_F + \frac{d(q_F + q_{SE} + q_{SC})}{dt} \tag{6-51}$$

During this interval the collector current drops from 4.4 to 0.49. Nearly all the excess charge q_F is removed from the base, as well as much of that of the space-charge layers.

Let us consider conditions at the end of turn-off. As i_C is 0.49 mA and q_F is negligible, dq_{SC}/dt is -0.49 pC/ns in accordance with (6-50). In (6-51) both q_F and dq_F/dt are negligible. Furthermore, because q_F is nearly zero, the junction voltage v_{BE} is low, which indicates that q_{SE} is approaching zero. Thus we expect the derivative of q_{SE} in (6-51) to be quite small. Neglecting this term and those with q_F in (6-51), we obtain a value of -0.49 for i_B. For this i_B the junction voltage v_{BE} is found from (6-31) to be 0.196 V (see P6-13).

We can now calculate approximate values for the parameters at the turn-off time t_{off}. As i_C is 0.49, the collector-emitter voltage is 2.7, giving -2.5 for v_{BC}. Using this and 0.196 for v_{BE}, we find that both q_F and q_R are zero, and the respective values of q_{SE} and q_{SC} are 0.28 and -6.71 pC. The calculated results are entered in row six of Table 6-2.

The fall time t_f can be estimated by integrating (6-51) from the initial time t_{sd} to the turn-off time t_{off}. The recombination term $0.01q_F$ is negligible. Thus the fall time t_f is the ratio of the total charge removed to the magnitude of the average value of i_B. From the data of rows five and six of Table 6-2 the charge removed is found to be 15.3 pC. The base current changes from -1.76 to -0.49. However, it does not drop appreciably until v_{BE} falls below about 0.6 V, and this does not occur until nearly the end of the fall time when q_F becomes very small. Estimating the average base current to be 1.4 mA gives a fall time of 11 ns and a turn-off time of 27 ns.

The fall time is largely the time required to remove q_F from the neutral base and q_{SC} from the collector-junction space-charge layer. At the end of the fall time the base current is -0.49. It decreases to zero while removing the small charge stores present at the end of the turn-off time. The steady state is reached in about 6 ns.

The complete results of the numerical analysis based on the approximate procedures are summarized in Table 6-2. A computer procedure for solving Eqs. (6-31) through (6-38) is discussed in Appendix A. The rather close agreement between the calculations and the computer results justifies the approximations that have been made.

PROPAGATION DELAY

We have examined and analyzed the switching of a transistor inverter with a rectangular pulse applied to the input. For logic gates the input is not a rectangular pulse but is a waveform with finite rise and fall times, as shown in Fig. 6-11. The inverted output of the illustration has the same time scale as the input.

A common method of specifying the transient performance of a logic gate is to quote the *propagation delay* along with rise and fall times. The propagation delay t_{PHL} when the output is switched from the high to the low state is the time difference between the instant the input pulse reaches 50% of its final value and the instant the gate output has fallen to its 50% point. The propagation delay t_{PLH}

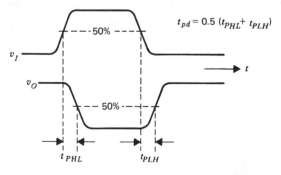

Figure 6-11   Propagation delay of a logic gate.

when the output is switched from low to high is similarly defined. The average of the two is the *propagation delay time* t_{pd} of the gate; that is,

$$t_{pd} = \tfrac{1}{2}(t_{PHL} + t_{PLH}) \tag{6-52}$$

Rise and fall times of either waveform of Fig. 6-11 are defined in the usual manner in terms of 10 and 90% points. Logic families are often compared with respect to speed by quoting for each a typical propagation delay time and the maximum frequency of a flip-flop.

The dc analysis previously considered assumed known values of the Ebers–Moll parameters I_{ES}, I_{CS}, α_F, and α_R. In addition, we have seen that a transient analysis requires values of the lifetimes τ_{BF} and τ_{BR} along with the constants necessary for expressing the charges of the depletion layers in terms of the junction voltages. The necessary quantities can either be estimated from data sheet information or measured.[1]

We have examined both the static and transient behavior of a transistor operated as a switch. The principles apply not only to saturated logic gates but also to control circuits utilizing transistors as switches. The main objective has been to relate the phenomena occurring within the device to the static response and also to the turn-on and turn-off times that limit the speed of operation.

REFERENCES

1. P. E. Gray and C. L. Searle, *Electronic Principles*, Wiley, New York, 1969.
2. C. S. Meyer, D. K. Lynn, and D. J. Hamilton, *Analysis and Design of Integrated Circuits*, McGraw-Hill, New York, 1968.
3. D. J. Hamilton and W. G. Howard, *Basic Integrated Circuit Engineering*, McGraw-Hill, 1975.

PROBLEMS

Section 6-1

6-1. For the basic logic gate of Fig. 6-2, calculate the output voltage v_O when each of the 3 inputs is 0.2 V. Assume v_{BE} of a saturated transistor is 0.8 V.

6-2. Deduce that $Y = \overline{A + B + \overline{C} + \overline{D} + E}$ in the configuration of Fig. 6-12, which has open-collector gates and wired-AND logic.

6-3. A MOSFET inverter has a transfer characteristic such that v_O is 10 for $v_I \leq 4$ and v_O is $-v_I^2 + 8v_I - 6$ for values of v_I between 4 and 6.7. For $v_I > 6.7$, the voltage is

$$v_O = v_I - 3.5 - (v_I^2 - 7v_I + 2.25)^{1/2}$$

The low-level output voltage is v_O when v_I is 10. From the unity-gain points, determine analytically the noise margins V_{NH} and V_{NL}. (*Ans:* $V_{NL} = 2.67$).

[1] See Reference 1, pages 781–782 and 824–832.

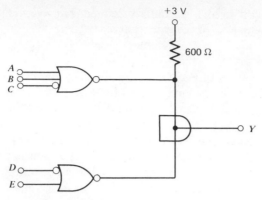

Figure 6-12   Network for P6-2.

Section 6-2

6-4. For the logic gate of Fig. 6-6 with $v_{OL} = 0.1$, calculate i_{CL} and the product $\beta_F i_{BL}$ in mA. From comparison of these determine the operating region of Q_4. Also, calculate v_I and v_O and from these values and the break points calculated in Sec. 6-2, verify that the operating region of Q_4 agrees with the appropriate entry of Table 6-1. (*Ans: $v_I = 0.715$*).

6-5. Suppose the logic gate of Fig. 6-6 is modified so that the fan-out is 2, and each load transistor has a fan-in of 1. Also, V_{CC} is changed from 3 to 3.6 V. Calculate the four breakpoints (v_I, v_O). Assume the saturation values V_{BEL}' and v_{OL}' are 0.7 and 0.1 V, respectively. (*Ans: BP_3 is 0.755, 0.604*).

6-6. For the logic gate of Fig. 6-6 calculate v_I and v_O when v_{OL} is $3 - 3 \times 10^{-11}$ and also when v_{OL} is 0.07658. (*Ans: $v_I = 3.09, 0.456$*).

Section 6-3

6-7. Deduce that (6-27) for i_E, in conjunction with (6-25) and (6-26) and when applied to static conditions, is identical to the Ebers–Moll equation (6-5). Also, deduce (6-28) from (6-27) and symmetry, and deduce (6-29) for i_B from (6-27) and (6-28).

6-8. With reference to (6-36) and (6-38), note that dq_R/dt is $40q_R \, dv_{BC}/dt$ and dq_{SC}/dt is $C_1 \, dv_{BC}/dt$ with C_1 equal to $3.33(0.85 - v_{BC})^{-1/3}$. Deduce that the charge-control equation (6-33) can be expressed as

$$\frac{dv_{BC}}{dt} = \frac{-5 + 1.67v_{BE} - 1.67v_{BC} + 0.49q_F - 0.125q_R}{40q_R + C_1}$$

with q_F and q_R given in terms of v_{BE} and v_{BC} by (6-35) and (6-36). Also, using this result, deduce in a similar manner the corresponding equation for dv_{BE}/dt.

6-9. Using a computer and a suitable numerical integration procedure, solve the charge-control equations of P6-8 for $v_I = 3$. The initial values of v_{BE} and v_{BC} are 0 and -3, respectively. Let time t vary from 0 to 30 ns and plot the results. Refer to appendix A.

Section 6-4

6-10. Calculate the data of row 2 of Table 6-2, which apply at the instant when the inverter of Fig. 6-8 has a collector current 10% of its maximum. Assume dq_{SC}/dt at this instant is the average rate of change throughout the turn-on time. Estimate the effect of the recombination term of (6-41) by assuming q_F varies linearly during the time delay and use the average value of the base current to calculate the delay time.

6-11. Excluding the time entry, calculate the data of the fourth row of Table 6-2. First, subtract (6-44) from (6-43). Eliminate q_F in the result by using (6-44) with $v_{BE} - v_{BC}$ replaced with an estimate of the saturation value (about 0.05 V) of v_{CE}. Solve for q_R in conjunction with (6-36) by trial and error. Use (6-35) to calculate v_{BE} after finding q_F from q_R.

Section 6-5

6-12. Deduce (6-47) from (6-45) and (6-46), and solve (6-47) subject to the boundary conditions $q_R = 54.3$ and $dq_R/dt = 0$ at time zero. From the solution and (6-45) find $q_F(t)$, and from these results calculate the data of the fifth row of Table 6-2. The storage delay is the time when $q_R = 0$, and i_C is 4.4 mA.

6-13. In calculating the fall time in Sec. 6-5, we assumed dq_{SE}/dt in (6-51) to be negligible at the end. However, the computer analysis of Appendix A gives a value of -0.172 pC/ns for this derivative. Using this and following the procedure of Sec. 6-5, calculate t_{off} and the corresponding values of v_{BE}, v_{BC}, q_{SE}, and q_{SC} in ns, pC, and V. Assume the same average base current of 1 4 mA used in Sec. 6-5 when integrating (6-51).

6-14. Using (6-31) through (6-38) with v_I zero, find the equations relating the junction voltages in the form of (6-39). Solve these with the aid of a computer, noting the initial conditions as given in Table 6-2. Let time t vary from 0 to 30 ns and plot the currents, voltages, and charges. Refer to Appendix A. Time zero corresponds to time T of Fig. 6-8.

Chapter Seven
BJT Gate Circuitry

In this chapter we emphasize the gate circuitry and performance characteristics of three very important bipolar logic families. These families are transistor-transistor logic (TTL), integrated injection logic (I^2L), and emitter-coupled logic (ECL). MOSFET gates are examined in the next chapter.

7-1. TRANSISTOR-TRANSISTOR LOGIC (TTL)

A popular saturating logic family characterized by high speed and low power dissipation is TTL, or T^2L. The integrated circuits are extremely versatile, and they are small, economical, and reliable. Available ICs include a wide variety of gate circuits and flip-flops in the small-scale-integration (SSI) category. There are MSI circuits consisting of decoders, memories, adders, counters, shift registers, multiplexers, and many others. The large number of gates and functional networks that are available provide for easier design implementation.

BASIC GATE

The basic gate is shown in Fig. 7-1. At the input is a multiple-emitter transistor. Suppose one or more of the inputs is a logical zero, say 0.2 V. The base of Q_1 is about 0.8 V above the lowest input voltage, which gives a base current of 1 mA. This current places Q_1 in the saturation mode. The collector voltage of

182

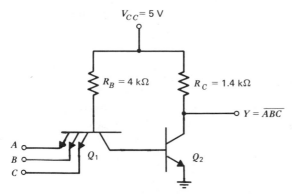

Figure 7-1   Basic TTL NAND gate.

Q_1, which is also the base voltage of Q_2, is too low to support conduction in the output transistor, and Q_2 is cut off. Thus the output is a logical 1. Because the collector of the input transistor is essentially an open circuit with $i_{C1} \approx 0$, the collector-junction of the saturated device is appreciably forward biased. This was discussed in Sec. 2-5 and is examined further in P7-1. The 1 mA base current divides between the emitters that are at low levels, with most of the current going to the emitter having the lowest voltage.

With all inputs at a high level, say 3.6 V, currents are driven into the emitters, opposite to the normal direction for NPN devices, and the input transistor operates in the inverse mode. The collector junction is again forward biased, with V_{BC1} approximately 0.7 V. The collector current of Q_1 feeds into the base of the output transistor, which becomes saturated with a logical-zero output of 0.2 V. With the base of Q_1 being two diode drops (about 1.5 V) above ground, the base current is 0.9 mA. Most of this goes to the collector and thence into the base of Q_2. The input transistor is designed so that the reverse alpha α_R associated with each emitter is very small, typically less than 0.02. Consequently, the reverse beta β_R is also small, with β_R defined as the ratio $\alpha_R/(1 - \alpha_R)$, which also equals the ratio i_E/i_B. For a reverse beta of 0.02 per emitter, the input current to each emitter is only 18 μA, or 2% of the base current.

Although we shall not do so here, a dc analysis of the TTL gate of Fig. 7-1 can be made following the procedure of Sec. 6-2. Equations (6-7) and (6-8) relate the voltages v_{BE} and v_{CE} and the currents i_B and i_C of each transistor. Noting that i_{C1} equals $-i_{B2}$ and v_O is v_{CE2}, the four Ebers–Moll relations combined with three loop equations are sufficient for plotting a static transfer characteristic, with the effect of the load current neglected. A piecewise-linear analysis is also possible.[1]

---

[1] Chapter 10 of Reference 1. (Recall that superscript numerals refer to References.)

For faster switching speed, higher fan-out capability, and improved noise immunity, logic gates should be designed so that their output impedances are small regardless of the logic state. Loads are usually somewhat capacitive. Whenever the logic state changes, the charge stored on a capacitive load must also change. For the gate to accomplish this quickly, it must be capable of both *supplying* and *sinking* large currents, and accordingly, a low output impedance is essential. An additional advantage arises when noise is coupled into a logic circuit. It is obvious that the noise voltage will be low provided the circuit impedance is low.

The basic gate of Fig. 7-1 has a low output impedance when the output is a logical 0, because the resistance between the collector of the saturated transistor Q_2 and ground is quite small. However, in the 1-state Q_2 is *off*, and current supplied to a capacitive load must pass through the pull-up resistor R_C. Thus the output impedance is 1.4 kΩ, which is excessive. Reducing R_C is not practical. This would increase the power dissipation and give an undesirable rise in the low-level output voltage due to increased current through Q_2.

STANDARD GATE

Shown in Fig. 7-2 is a practical version of a TTL gate. Transistor Q_3 replaces R_C, providing an *active pull-up* output. The combination of Q_3 and Q_4 is referred to as a *totem-pole* arrangement. It allows a high-capacitance load to be driven without seriously degrading the switching time. When Q_4 is switched *off*, transistor Q_3 in the active mode provides a low-impedance (70 ohms) driving

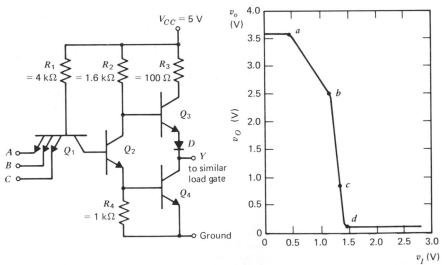

Figure 7-2   Standard TTL gate and transfer characteristic.

source that supplies a large current to the capacitive load, causing the output voltage to rise quickly from the low to the high level, with considerably improved turn-off time. When Q_4 is *on*, the output impedance of the gate is the low saturation resistance of this device, about 10 ohms.

TRANSFER CHARACTERISTIC

Let us examine the static transfer characteristic of Fig. 7-2. Whenever there is reference to the voltage of a terminal, it should be understood that this denotes the voltage drop from the specified terminal to ground. To simplify the discussion it is convenient to assume a drop of 0.7 V across a forward-biased "diode," a term used here and often employed to signify the conducting junction of a transistor as well as that of an actual diode. However, in the event the transistor is saturated, the drop is assumed to be 0.8 V. In addition, when the input transistor Q_1 is saturated along with a collector current very small compared with the base and emitter currents, we shall assume the collector voltage of Q_1 is 200 mV higher than the input v_I. About 100 mV are attributed to the difference between the forward-bias junction voltages. Although this difference is larger than normal for a transistor in saturation with an open collector, it is typical of transistors with very low reverse alphas, as shown in P7-1. The remainder of 100 mV is attributed to internal ohmic drops, largely in the active base region. The three input terminals are connected to the voltage v_I, and the output feeds into a load gate not illustrated.

With v_I zero, transistor Q_1 is saturated. Its collector voltage is 0.2, which is too low to drive Q_2. Thus Q_2 is *off*, the current through R_4 is zero, and Q_4 is *off*. The high-level output delivers a small current to the load gate, with the current supplied through transistor Q_3 and diode D. Although the base current of Q_3 passes through R_2, this current is very small, and the voltage drop is negligible. The load current flows through R_3, but both the current and the resistance are small enough so that the voltage across R_3 is also negligible. It follows that the base and the collector of Q_3 are each at about 5 V, and Q_3 is biased in the active mode with a base-collector voltage of zero. Because the output is two diode drops (v_{BE3} and V_D) below the base of Q_3, the output v_O is 3.6 V.

Conditions are unchanged for inputs from 0 to 0.5 V, which is breakpoint a of the transfer characteristic. At a the base of Q_2 is 0.7 V, which is sufficient to turn Q_2 on. Because the base of Q_1 is $v_I + 0.8$, or 1.3 V, i_{B1} is 0.9 mA and this current divides between the emitters.

From a to b, corresponding to values of v_I from 0.5 to 1.2, transistor Q_1 remains saturated with substantial base and emitter currents and practically no collector current. Both Q_2 and Q_3 are active, and Q_4 is *off*. The current of Q_2 flows through R_2, thereby lowering the base voltage of Q_3. Again, the output is *clamped* at two diode drops below the base, and v_O drops as the base voltage of Q_3 drops. Because v_{B2} is v_I plus 0.2, the emitter voltage is $v_I - 0.5$, and this voltage divided by R_4 gives the current through R_4. The current through R_2 is $(V_{CC} - 1.4 - v_O)/R_2$. With Q_2 in the active mode these currents are approximately equal. Equating them and differentiating the equation with respect to v_I gives

$$\frac{dv_O}{dv_I} = \frac{-R_2}{R_4} = -1.6 \tag{7-1}$$

This is an estimate of the slope of the transfer characteristic from a to b.

At breakpoint b transistor Q_4 is on the threshold of conduction with $v_{BE4} = 0.7$ V. The base currents of the totem-pole pair are still negligible. Thus the collector current of Q_2 is 0.7 mA, and the drop across R_2 is 1.1 V. Subtracting this and two diode drops from V_{CC} gives 2.5 V for v_O. Because v_{B2} is 1.4, the input v_I is 1.2 V. An additional increment in v_I increases i_{C2} and causes Q_4 to conduct; both the base and collector voltages of Q_3 drop. The current through R_3 increases substantially, and the base currents of the totem-pole transistors can no longer be neglected. Between breakpoints b and c transistor Q_1 is saturated, and Q_2, Q_3, and Q_4 are in the active mode.

At c the output transistor Q_4 approaches saturation with zero collector-base voltage. Thus v_{BE4} and v_O are each 0.8 V. Adding two diode drops to v_O gives 2.2 V for v_{B3}. Therefore, the current through R_2 is 1.8 mA, which supplies 0.8 mA to R_4 and 0.5 mA to each base terminal of the totem-pole pair. For a beta of 50 the collector currents are 25 mA. The base voltage of Q_2 is 1.5, v_I is 1.3, and v_{B1} is 2.1. Although the upper totem-pole transistor is active, its collector-emitter voltage is only 1 V.

Between breakpoints c and d the situation changes appreciably. A slight increase in v_I above 1.3 V to about 1.37 V raises the base voltage v_{B1} above 2.1 V. However, the base of Q_2 is fixed at about 1.5 V, which is the sum of the base-emitter voltages of Q_2 and Q_4. Consequently, the forward bias of the collector junction of Q_1 rises and the base current of this device begins to shift from the emitters to the collector. The collector-emitter voltage of Q_1 drops below 0.2 V. A consequence of the increased currents of Q_2 is a reduced collector voltage which turns the upper totem-pole transistor *off*. For a beta of 50, a base current of about 50 μA will saturate Q_2, because its collector current cannot exceed 2.5 mA. Thus Q_2 quickly saturates. The presence of diode D assures that Q_3 is *off*.

As v_I increases further, driving Q_2 into greater saturation, its base voltage rises to about 1.6 V and is clamped at this value. Nearly all the base current of Q_1 now flows to the collector, placing Q_1 in the inverse mode with a very small reverse beta. Thus at breakpoint d transistor Q_1 is in the inverse mode, Q_2 and Q_4 are saturated, Q_3 is *off*, and the output is 0.2 V. The lower totem-pole transistor functions as a *sink*, with current from the load gate flowing through its small saturation resistance of about 10 ohms.

Shown in Table 7-1 are the transistor operating modes with each corresponding range of values of v_I and v_O. During turn-on, when the input rises from 0.2 to

Table 7-1   **Transistor Operating Modes**

Region	v_I	v_O	Q_1	Q_2	Q_3	Q_4
to a	0 to 0.5	3.6	saturation	off	active	off
a to b	0.5 to 1.2	3.6 to 2.5	saturation	active	active	off
b to c	1.2 to 1.3	2.5 to 0.8	saturation	active	active	active
c to d	1.3 to 1.4	0.8 to 0.2	saturation	active–saturation	active–off	active–saturation
beyond d	> 1.4	0.2	inverse	saturation	off	saturation

3.6 V, the upper totem-pole transistor Q_3 does not saturate. As indicated in the analysis, Q_3 is cut off before it can store enough charge to saturate. Furthermore, the fast turn-on actually prevents the current of the totem-pole transistors from rising to the high value calculated in the static analysis. However, during turn-off there is a brief instant when both Q_3 and Q_4 are saturated, and there is a current spike. In most cases it reaches only about 15 mA for a few nanoseconds. For normal system speeds *with adequate power-supply bypassing*, the current spike that occurs during turn-off is a minor problem.

TTL has very high speed for a saturated logic system. During turn-off the output transistor Q_4 switches quickly from saturation to cutoff as charge flows from the base to ground through the 1 kΩ base resistor. Also, the low output impedances of both logic states significantly reduce the *RC* time constants. Standard medium-speed TTL can operate with a maximum clock rate of 20 MHz with a propagation delay of 12 nanoseconds. The power dissipation is typically 12 mW per gate. Fan-outs up to 10 are reasonable. At the input the low-level impedance is R_1, or 4 kΩ, and the approximate high-level impedance is 400 kΩ. The low-level and high-level output impedances are 10 and 70 ohms, respectively. The guaranteed noise margin of each logic state is usually given by the manufacturer to be 400 mV. However, noise voltages of 1 V or more can typically be tolerated without error.

Between each input terminal and ground of the gate of Fig. 7-2 is a *clamp diode*, not illustrated. These are connected so that they are reverse biased with no effect for normal voltages. However, for negative inputs they limit the voltage to a diode drop. The diodes serve to damp oscillations, referred to as *ringing*, that sometimes occur during switching operations due to LC resonant effects. The input voltage should be restricted to values not less than −0.5 V. Also, v_I should not exceed 5.5 V in order to avoid emitter-junction breakdown. Unused inputs should never be left open. This reduces noise immunity through stray pick-up and also reduces the speed because of input capacitance. An unused input can be tied to an active input although this increases loading. Also, it could be tied to the output of an unused gate or connected through a 1 kΩ resistor to the supply.

OPEN-COLLECTOR GATE

The totem-pole output configuration is unsuitable for the wire-AND logic connection. If the AND operation is required, the outputs can be fed into a NAND gate followed by an inverter. An alternate procedure is to use open-collector gates, discussed in Sec. 6-1. These are similar to the gate of Fig. 7-2 except that transistor Q_3, diode D, and resistor R_3 are omitted. The outputs of several open-collector gates can then be tied together with a single pull-up resistor added externally to the integrated circuits. Various TTL gates with open-collectors are commercially available, but speed is sacrificed.

A knowledge of the input characteristic is important for proper utilization of TTL gates, especially when these are to be interfaced with other logic families. A typical characteristic is shown in Fig. 7-3. With v_I equal to 0 the input current is -1 mA, and for v_I greater than 1.6 V the current is only a few microamperes. Clearly, *a device that drives a TTL gate must both source and sink current.*

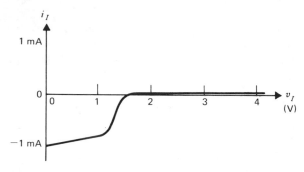

Figure 7-3   TTL input characteristic.

LOW-POWER TTL

In addition to the standard medium-speed TTL gate there is a *low-power* series. The resistors have higher values and the diode of Fig. 7-2 is eliminated. Typical values of R_1, R_2, R_3, and R_4 are 40, 20, 0.5, and 12 kΩ, respectively. The power dissipation per gate is reduced from about 12 mW to 1 mW, but the propagation delay is increased from 12 ns to 33 ns. This speed is roughly comparable to that of CMOS logic, discussed in Chapter 8. The output of a CMOS gate can sink a current up to 1 mA. When the input of a low-power TTL gate is connected to a logical zero, a current of about 0.18 mA flows to ground. Thus a

CMOS gate can directly drive up to five low-power TTL inputs, and this constitutes an important advantage of the low-power series. In order to drive standard TTL from a CMOS gate, a *buffer* circuit must be connected to the output of the CMOS gate to increase the output-drive capability.

HIGH-SPEED TTL

There is also a high-speed series. The gate is basically the same as that of Fig. 7-2. The resistors have lower values and the upper totem-pole transistor and diode are replaced with a *Darlington pair* of transistors. Such a pair is shown in Fig. 16-26a and discussed in Sec. 16-6. The propagation delay is reduced from 12 to 6 ns, but the power dissipation is nearly twice that of the standard series.

THE SCHOTTKY-CLAMPED GATE

Even higher speed is obtained from the *Schottky-clamped* TTL gate, which has a typical propagation delay of 3 ns and an average power dissipation of 20 mW per gate. The circuitry is quite similar to that of the high-speed gate, but Schottky transistors are used for those that normally saturate.

The Schottky barrier diode (SBD) consists of a contact between a suitable metal and an N-type semiconductor. *By the proper choice of metal a rectifying junction is obtained with a volt-ampere characteristic similar to that of a PN junction.* However, the forward diode drop is only about 0.4 V, with the precise value depending on both the metal and the dopant density of the semiconductor. Because there are no appreciable charge stores, the diode can switch *on-to-off* much faster and *off-to-on* somewhat faster than can a PN junction device. The SBD is represented by the symbol of Fig. 7-4a.

In Fig. 7-4b is shown the diagram of a Schottky transistor. It consists of a normal BJT with a Schottky barrier diode connected between the base and collector terminals as indicated. The symbol for such a configuration is that of Fig. 7-4c. When the BJT heads toward saturation, the SBD clamps the base-

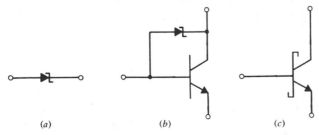

(a)   (b)   (c)

Figure 7-4 Schottky transistor. (a) Schottky diode. (b) Schottky transistor. (c) Transistor symbol.

collector voltage at 0.4 V, rather than the usual 0.8 V. Most of the charge store of a saturated transistor normally occurs in the vicinity of the collector junction because of its relatively large area and low collector-region dopant density. This is discussed further in Sec. 7-5. Because the excess charge is proportional to exp qv_{BC}/kT, the charge that is stored at 0.4 V is not nearly so great as that when the junction voltage is 0.8 V. Thus the BJT is effectively kept out of appreciable saturation (see P7-4). Consequently, *switching approaches the speed of unsaturated logic such as ECL*, which is examined later. A germanium diode substituted for the Schottky diode would serve a similar purpose. However, only slight additional processing is required to fabricate the SBD on the same chip with the transistor. Schottky TTL *with reduced power dissipation* is achieved by using resistances several times greater than those of standard TTL. This is referred to as *low-power Schottky TTL*.

7-2. INTEGRATED INJECTION LOGIC (I^2L)

The monolithic integrated circuit of the type discussed in Sec. 4-2 consists of "discrete" transistors, resistors, and capacitors, with the elements of the silicon chip interconnected through a surface metallization and set apart from one another by means of isolation islands. These islands are wasteful of surface area, thus limiting the amount of circuitry that can be placed on a single chip. This is a serious disadvantage relative to MOSFET circuitry with its natural isolation of elements. In both cases, design is quite similar to that of a circuit made from discrete components, although the designer of the IC must consider the restrictions imposed by the technology and the advantages offered by the ease of obtaining well-matched components.

Integrated injection logic (I^2L), also called *merged transistor logic* (MTL), is an integrated-circuit technology utilizing bipolar junction transistors that are merged into what is sometimes referred to as a superintegrated circuit. Because isolation islands are not required and because many of the interconnections between elements are accomplished *internally* through the P and N regions of the silicon chip, the circuit density that can be obtained exceeds even that of the MOSFET technology. Consequently, I^2L is especially suitable for large-scale integration (LSI), and thousands of gates can be fabricated on a single chip.

NOT GATE

To illustrate the basic operation of I^2L inversion let us examine the circuit diagram of the NOT gate of Fig. 7-5. The gate consists of an NPN transistor and a direct current source. A logical 0 is the saturation value of v_{CE}, and a logical 1 is the saturation value of v_{BE}. In I^2L gates these are typically 0.05 V and 0.75 V, respectively, providing a logic swing of 0.7 V.

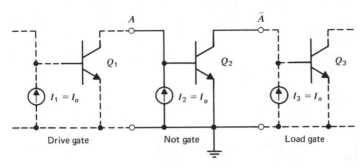

Figure 7-5   Circuit diagram of a NOT gate.

With Q_1 in saturation the input to the NOT gate is 0.05 V, and this low voltage places Q_2 in cutoff. The direct current I_O of the current source I_2 flows into the collector of Q_1, and I_O of the source I_3 enters the base of Q_3. Because the collector current of Q_3 is also I_O and because β_F is greater than unity, Q_3 is saturated with a base-emitter voltage of 0.75. Thus a logical 0 at the input of the NOT gate gives a logical 1 output.

Now suppose Q_1 is *off*. The current I_O into the base of Q_2 saturates this device, and its low output voltage turns Q_3 *off*. Consequently, both the base and collector currents of Q_2 equal I_O, and the logical 1 input gives a logical 0 output. Note that each transistor must have β_F greater than unity to ensure that a transistor with equal base and collector currents is in saturation.

NOR GATE

A two-input NOR gate is shown in Fig. 7-6. It consists of two inverters with wired-AND logic at the output. Suppose A and B are logical zeros. Both Q_1 and Q_2 are *off*, and the source currents at the inputs flow into the collectors of the driving gates not included in the sketch. In addition, Q_3 is saturated, and the output Y is a logical 1. Now suppose A is 1 and B is 0, making Q_1 saturated and Q_2 *off*. The output voltage is 0.05, and transistor Q_1 sinks the current I_O of the

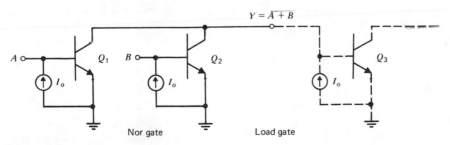

Figure 7-6   Circuit diagram of NOR gate.

output load gate. Clearly, the output of the gate is low-level if either or both of the inputs are high-level, and the NOR operation is accomplished. With both Q_1 and Q_2 on, the collector current of each is $0.5I_O$. Wired-AND logic is extensively used with I^2L.

MULTIPLE-OUTPUT GATE

The NPN transistor of an I^2L gate normally has multiple collectors. As we shall see, this is easily accomplished during fabrication, and it has the advantage of providing flexibility in the logic system. For example, the gate of Fig. 7-7 with inputs A and B has outputs $\overline{A}$, $\overline{B}$, $\overline{A + B}$, and $A + B$. Each transistor has three collectors, and two pairs of these are connected for wired-AND logic.

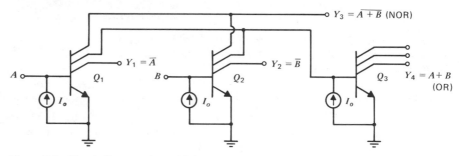

Figure 7-7 Circuit diagram of a multiple-output gate.

Suppose $A = B = 0$. Transistors Q_1 and Q_2 are *off*, and Q_3 is saturated. Outputs Y_1, Y_2, and Y_3 are logical 1's at 0.75 V, which is the base-emitter saturation voltage of the load gates at these outputs. Y_4 is 0.

Now let $A = 1$ and $B = 0$. Here, Q_1 is saturated *with each collector at* 0.05 V. The total collector current is $3I_O$. With a base current I_O it follows that Q_1 must have a β_F greater than 3 to ensure saturation. In general, an NPN transistor of an I^2L gate must have β_F greater than the number of its collectors. Because each collector provides one output, the number of collectors is the fan-out. Both Q_2 and Q_3 are *off. The collectors of* Q_2 *are not at the same voltage.* The bottom collector is at 0.75 V and the voltage of the other two collectors is 0.05. Of course, each collector current is zero. In this case, Y_1 and Y_3 are 0, and Y_2 and Y_4 are 1.

When both A and B are 1, Q_1 and Q_2 are saturated and Q_3 is *off*. Because the collectors in the wired-AND configuration now share the current of a source, the total collector current of Q_1 is $2I_O$, and that of Q_2 is the same.

The illustrated network of Fig. 7-7 contains three basic I^2L gates that are interconnected to perform desired logic. Each of these basic gates consists of a multiple-collector NPN transistor and its own direct current source. The base of the transistor is the input terminal, the collectors are the outputs, and the emitter is ground. All emitters of an I^2L network are common. Let us consider the current source.

THE CURRENT SOURCE

A very simple arrangement serves to supply the current I_o of a gate. Each gate contains a PNP transistor that delivers I_o from its collector terminal, which is connected to the base of the NPN device. The base of the PNP transistor goes to ground, and the emitter joins those of the other current sources. An external resistor connects the junction to a positive voltage, as shown in Fig. 7-8. It should be noted that one voltage supply and one resistor are adequate for biasing the current sources of all the gates of an integrated circuit, but each gate must have its own PNP current source. The following example illustrates the bias design.

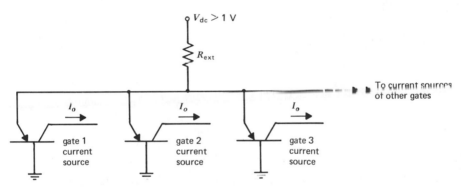

Figure 7-8   Power supply and gate current sources.

Example

An IC has 2000 basic I^2L gates, each of which is associated with an injection current I_o of 60 μA. The collector current of a source is 75% of the emitter current, and each v_{EB} is 0.8 V. Calculate the required value of the external resistor and the power dissipation of the chip. A 5 V supply is used.

Solution

Each emitter current should be 80 μA. Therefore, the 2000 current sources contribute 0.16 A to the external resistor. The voltage across this resistor is the supply voltage less v_{EB}, or 4.2. Thus the proper value for R_{ext} is 26 ohms. The power dissipated on the chip is the product of the total current and v_{EB}, which is 128 mW, or 64 μW/gate. This is several orders of magnitude less than the power dissipation (12 mW) of a standard TTL gate.

The power calculated in the example is not excessive. Silicon chips can normally dissipate up to 0.5 W without requiring specially designed heat sinks. The equal division of current among the emitters of the current sources is accomplished by careful design of the geometry of the chip. This is discussed later. There is no serious current-hogging problem.

In Fig. 7-9 is shown the circuit configuration of the basic NAND gate. Note the complete absence of resistors. The PNP transistor operates in the forward saturation mode. When its collector is at the saturation value of v_{CE} of an NPN transistor, the forward bias of the collector junction is so small that operation is equivalent to the active mode. At times the collector will be at the saturation value of v_{BE} of the NPN transistor, which may be as much as 0.8 V. However, the geometry is such that the saturation current of a diode equivalent to the base-emitter input of the NPN device is considerably larger than I_{CS} of the PNP transistor. This ensures that the forward bias of the emitter junction

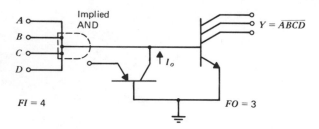

Figure 7-9   Basic I^2L gate with current-source transistor.

of a current source is greater than that of the collector junction by perhaps 50 mV or more, and the ratio of I_o to the input emitter current is approximately α_F (see P7-5). Again, operation corresponds closely with that of the active mode. The PNP transistor of a current source is usually a lateral transistor with a beta less than 10. Current sources associated with different transistors should not be connected in parallel. If they were, a current hogging problem might cause logic errors. Thus a connection between transistor base terminals is not allowed. *Each base terminal must go to a collector terminal.* This is why multiple collectors are so important.

POWER-DELAY PRODUCT

The propagation delay of a logic gate should be small so that the circuitry can perform logic functions at high speed. One way to reduce the delay is to increase current levels. Higher currents drive transistors into saturation and take them out of saturation more quickly. Unfortunately, this increases power dissipation, and special packaging and heat sinks are required when the power per chip exceeds about 0.5 W. The dissipation per gate may limit the number of gates, and thus the level of integration, that can be placed on a chip. In space applications power dissipation is usually a critical factor. A meaningful figure-of-merit of a logic gate is the product of the power dissipation and the propagation delay. Power multiplied by time gives energy, and the *power-delay product* is usually stated in picojoules (pJ). It is frequently called the *speed-power* product. Note that a *low* power-delay product signifies that the gate can operate at high speed at low power.

I²L has the lowest power-delay product of any technology in use. Values from 4 pJ to about 0.1 pJ are typical, and it is expected that values approaching 0.001 pJ will be obtained as processing techniques improve. In comparison, standard TTL with a power dissipation of 12 mW per gate and a delay of 12 ns has a power-delay product of 144 pJ. An especially interesting characteristic of the I²L gate is that its power-delay product is constant over a wide range of injection currents. In the example the injection current I_o was chosen to be 60 μA and the power dissipation per gate was 64 μW. Assuming the gate was designed with a power-delay product of 1 pJ, the propagation delay is 16 ns. However, if a delay this small is not required, the power can be lowered by reducing I_o. Suppose I_o is reduced to a level that gives a dissipation of only 100 nW per gate. The propagation delay becomes 10 μs, which is adequate for many applications.

Practical injection currents have a range covering six orders of magnitude, from about a nanoampere to a milliampere. Simply by adjusting the external resistor or voltage supply, one can control the operating speed. In fact, the supply current can be increased momentarily to provide fast operation at certain desired times, such as during reading or writing operations. As we have seen, TTL has several forms, each designed for a different speed and power. With I²L the speed and power are adjustable. For example, the same I²L structure used in the circuit of a watch, operating at slow speed at the microwatt power level, might also be used in a high-speed microprocessor at the milliwatt level.

7-3. I²L STRUCTURE AND CHARACTERISTICS

I²L fabrication is accomplished by the same bipolar process described in Secs. 4-1 and 4-2. A cross-section of a basic gate is shown in Fig. 7-10. The substrate is a heavily doped N region providing excellent conductivity as well as structural strength. On this substrate is grown an N-type epitaxial layer. All P regions of the integrated circuit are formed simultaneously by the diffusion process, and a second diffusion forms the heavily doped N regions that constitute the collectors. The formation of the integrated circuit is accomplished with only two diffusions and four masking steps. One mask is used for windows

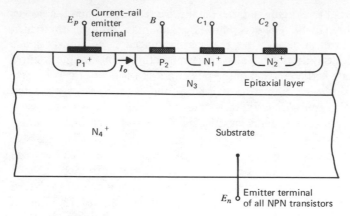

Figure 7-10   Cross-section of an I^2L gate.

in the oxide that are required for the P-type diffusion, and one is for windows for the N-type diffusion. A third provides windows for the terminal contacts, and a fourth is employed for the surface metallization pattern. In comparison, both TTL and ECL require four diffusions and seven masking steps. CMOS integrated circuits are made with three diffusions and six masking steps. Of the logic technologies available, only low-performing PMOS is simpler to fabricate, requiring one diffusion and four masking steps.

The emitter of the NPN transistor is the N_3 region of the epitaxial layer. *This region extends throughout the chip and serves as the emitter of each of the NPN transistors of the numerous gates.* The heavily doped substrate provides interconnections of the emitters, and a single external terminal, labeled E_n, is connected to the ground terminal of the voltage supply. Region P_2 is the lightly doped base of the NPN transistor. Both C_1 and C_2 are collector terminals. Because the gate operates with forward biased collector junctions, base-width modulation is negligible. Accordingly, the N-type collector regions are heavily doped so as to minimize the ohmic resistances. These resistances are so small that the saturation value of v_{CE} is typically about 50 mV.

The multicollector transistor is somewhat similar to the TTL multiemitter device operating in the inverse mode. There are important differences, however. When the multiemitter transistor operates in the inverse mode, the electrons injected from the N-type epitaxial layer are mostly lost by recombination in the base. Furthermore, the injection efficiency, defined as the ratio of the injected electron current to the total junction current, is low because the layer is lightly doped. Thus, α_R is very small. However, the NPN transistor of the I^2L gate operates so that most of the electrons injected from the epitaxial layer into the base are collected, and the heavy doping of the layer gives high injection efficiency. Alpha is reasonably large, perhaps 0.90 or more, and this is the forward alpha. The reverse alpha is about 0.99. The forward beta β_F is typically 10 and the reverse β_R, defined as $\alpha_R/(1 - \alpha_R)$, is typically 100. There are techniques that give larger values of β_F. The main advantage is reduced storage time. Also, as we have seen, the fan-out can be no greater than β_F.

The PNP current source is a lateral transistor having the $P_1{}^+$ region as its emitter. Its base is the N_3 region, which also serves as the emitter of the NPN device. Region P_2 is both the collector of the current source and the base of the NPN transistor. Note that the lateral PNP transistor and the vertical NPN transistor are *merged* into the same region of the silicon chip, giving a simplified structure with reduced interconnections. The name *merged transistor logic* (MTL) is often used in place of I²L. The only wiring is that between gates, which is minimal.

The $P_1{}^+$ region of Fig. 7-10 actually serves as the emitter for many current sources. It is a current-supply rail that runs the length of a large number of gates. The geometry is such that the current divides equally between the different P_2 regions of these gates. The I²L gate of the figure requires about the same surface area as a normal multiemitter transistor. Isolation between gates is not required, and the fabrication is fairly simple. Consequently, a packing density of 200 gates per square millimeter of surface area is typical, with high processing yields. This is about 10 times the density of TTL and from 4 to 6 times that of ECL and Schottky TTL designed for LSI. It even exceeds the density obtainable with MOS technology. *The large packing density permits a single I²L integrated circuit to perform the same logic as a substantial number of wired-together TTL circuits, with less space, reduced power, and lower cost.* MOS technology initially dominated large-scale integration. At least partly because it is well-established and available, it will continue to be important for years to come. However, I²L has advanced rapidly since it was introduced in the literature in 1972, and its future is bright.

In Fig. 7-11 is shown the cross-section of a NOR gate, along with the circuit diagram. The P-type current-supply rail injects a current I_o into the base of each vertical transistor, with equal current division assured by the geometry. Collectors C_1 and C_2 are wired together by means of surface metallization. The base terminals B_1 and B_2 are connected to collectors of driving gates, and the outputs Y_1 and Y_2 go to the bases of load gates. The second collector of Q_2 gives the NOT operation. In comparison with conventional integrated circuits having "discrete" elements in isolation islands, with the elements wired together, this is truly *superintegration.*

The extremes of the voltage swing of an I²L gate can be expressed in terms of the temperature, the injection current I_o, and the Ebers–Moll parameters of the NPN transistor of the gate. When the transistor is saturated, both the base and collector currents equal I_o. Applying this information to the Ebers–Moll equations for i_C and i_B, given as (6-6) and (6-8) of Sec. 6-2, we easily deduce

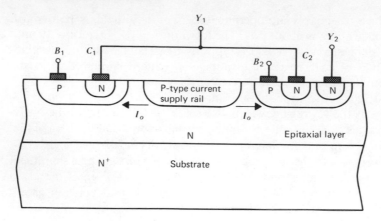

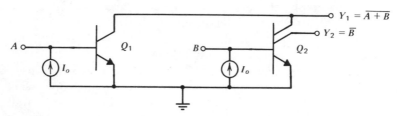

Figure 7-11   Cross-section of NOR gate and circuit diagram.

expressions for the saturated values of v_{CE} and v_{BE}, which are the low-level and high-level voltages (see P7-7). The results are

$$\text{low-level voltage} = \frac{kT}{q} \ln \frac{I_{CS}(2 - \alpha_R)}{I_{ES}(2\alpha_F - 1)} \tag{7-2}$$

$$\text{high-level voltage} = \frac{kT}{q} \ln \frac{I_o(2 - \alpha_R)}{I_{ES}(1 - \alpha_F \alpha_R)} \tag{7-3}$$

It is interesting that the expression of (7-2) for the low-level voltage does not include the injection current. However, practical gates have small ohmic drops dependent on I_o, and these add to the voltages of (7-2) and (7-3).

VOLTAGE-TRANSFER, TRANSCONDUCTANCE, AND INPUT CHARACTERISTICS

The voltage transfer characteristic of a gate is affected by its load. We shall consider the case in which a single output feeds into the base-emitter input of an identical gate, and a diode is employed to simulate this load. The network is that of Fig. 7-12. The saturation current I_s of the diode is selected so

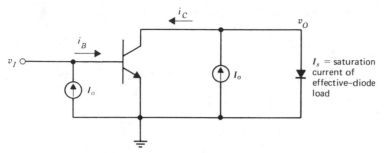

Figure 7-12 Gate with equivalent-diode load.

that the diode current is I_o when the voltage across it is the high-level value of (7-3). Accordingly, I_s is

$$I_s = \frac{I_{ES}(1 - \alpha_F \alpha_R)}{2 - \alpha_R} \qquad (7\text{-}4)$$

The transfer characteristic can be determined by equating the Ebers–Moll equation (6-6) for i_C to the difference between I_o and the forward diode current. We note that v_{BE} is v_I, and v_{BC} is $v_I - v_O$. With these substitutions, at a temperature of 290 K the equation becomes

$$I_o - I_s e^{40v_O} = \alpha_F I_{ES} e^{40v_I} - I_{CS} e^{40(v_I - v_O)} \qquad (7\text{-}5)$$

Being unimportant, the minus-one terms of both the diode equation and the Ebers–Moll relation have been omitted. Solving for exp $40v_I$ gives

$$\exp 40v_I = \frac{I_o - I_s \exp 40v_O}{\alpha_F I_{ES} - I_{CS} \exp(-40v_O)} \qquad (7\text{-}6)$$

For known parameters a static transfer characteristic can be plotted by assuming values for v_O and calculating the corresponding values of v_I from (7-6). For each set of values the currents can be determined from the relations:

$$i_C = I_o - I_s \exp 40v_O \qquad (7\text{-}7)$$

$$i_B = I_{ES}(1 - \alpha_F)\exp 40v_I + I_{CS}(1 - \alpha_R)\exp 40(v_I - v_O) \qquad (7\text{-}8)$$

In the Ebers–Moll equation for i_B the voltages v_{BE} and v_{CE} of (6-8) have been replaced with v_I and v_O, respectively,

The sketches of Fig. 7-13 represent voltage-transfer, transconductance, and input characteristics. These were calculated from the preceding equations using the following parameters:

$$
\begin{aligned}
I_o &= 10^{-4} \text{ A} & I_s &= 8.83 \times 10^{-18} \text{ A} \\
I_{ES} &= 10^{-16} \text{ A} & I_{CS} &= 0.93 \times 10^{-16} \text{ A} \\
\alpha_F &= 0.92 & \alpha_R &= 0.99
\end{aligned}
\qquad (7\text{-}9)
$$

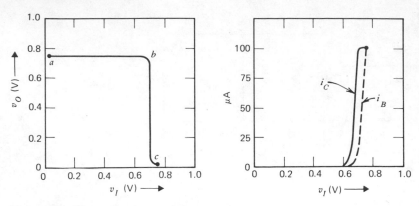

Figure 7-13 I^2L transfer characteristics.

The injection current was chosen to be 100 μA, and I_s was calculated from (7-4). The calculations reveal that *both junctions are forward biased at all times*. However, for values of v_I below 0.578 the collector current is less than 1% of I_o and the device is effectively cut off. *Throughout most of the transition region, corresponding to input voltage between 0.578 and 0.693, the forward bias on the collector junction is small enough so that the operating mode is essentially active, with i_C within 0.1% of $\beta_F i_B$* (see P7-8). Above 0.693 V the transistor is appreciably saturated. The voltage transfer curve begins at point a, breaks at b, and ends at c. The maximum v_I is slightly less than 0.752 V.

7-4. SCHOTTKY I^2L

Various techniques for improving the performance of I^2L have been developed. One of these is the use of ion implantation, which was discussed in Sec. 4-4. This method of forming the doped regions gives better quality integrated circuits with more precise geometry, and it increases the mobilities and lifetimes, thereby improving the betas. As noted in the previous section, most of the transition between states takes place in the active mode. The voltage transfer characteristic of Fig. 7-13 enters saturation with respective values of v_I and v_O of 0.693 and 0.176, as determined in P7-8. At this point the base current is 8.7 μA. *A larger β_F gives saturation at a lower base current, and switching speed is improved.* The propagation delay of ion-implanted I^2L gates is typically from 10 to 25 ns, and ion-implantation is becoming fairly common.

Another way to increase the speed is to employ some form of isolation that is designed to reduce the number of emitted electrons that are lost by lateral migration. One technique is to fabricate the integrated circuit with a recessed oxide sidewall, such as that shown in Fig. 7-16. The effect of the recessed oxide is to increase β_F. Other isolation methods are being developed. The increased

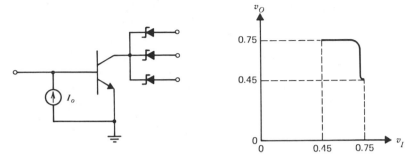

Figure 7-14   Schottky I²L gate and voltage characteristic.

processing costs and the larger surface area per gate are, of course, disadvantages of isolated I²L.

The circuit diagram of a Schottky I²L gate is shown in Fig. 7-14. The Schottky diodes do not keep the transistor out of saturation as in the case of Schottky TTL. *Their purpose is simply to reduce the logic swing.* Assuming the diodes are fabricated so that the forward voltage drop is 0.4 V, the output voltage is never less than 0.45 V. As indicated by the static characteristic of Fig. 7-14, the logic swing is from 0.45 to 0.75 V.

When Schottky diodes are used, there is no need to have multiple collectors. The advantage of multiple collectors is the natural isolation between them. As indicated clearly in Fig. 7-10, between the two collectors are back-to-back PN junctions. Thus no current can flow from one to the other as long as the base input is low level, which eliminates the injection current. Therefore, it is possible to have one collector at the high-level voltage with the other at the low-level voltage, depending on external connections. Of course if the transistor is saturated, both collector voltages must be low-level. *In Schottky I²L, gate isolation is provided by the rectifying diodes*, and a single N-type collector region is sufficient. The result is a smaller and simpler transistor. However, because multiple collectors have the advantage of reduced leakage current between the diodes, they are sometimes used.

Let us consider turn-on followed by turn-off of a *standard* I²L inverter, with input and output voltage waveforms similar to those of Fig. 6-11 of Sec. 6-5 used to define the propagation delay of a gate. Initially, v_I rises rather rapidly from 0.05 to 0.75 V, switching the output from the high to the low state with a propagation delay t_{PHL}. The interval of v_I between 0.05 and 0.45 contributes appreciably to this delay. The charge stores of the space-charge layers must be changed, and these actually dominate the delay when the injection current I_o is low. In addition, the active charge stores of the neutral base region also contribute to the delay. Although these change very little in this voltage range, the effect on the delay is significant because of the extremely small currents.

During turn-off when the output switches from low to high, the interval from 0.45 to 0.05 increases the delay t_{PLH} for the same reasons. When Schottky diodes are employed, as indicated in Fig. 7-14, the input voltage range from 0.05 to 0.45 V is eliminated. Consequently, Schottky I^2L is considerably faster than standard I^2L. For ion-implanted gates with oxide-isolation sidewalls and Schottky diodes, propagation delays from about 10 to 1 ns are possible.

The curves of Fig. 7-15 compare the performance of Schottky I^2L with the standard gate. At low current levels corresponding to a gate dissipation less than about 1 μW, the delay time is largely caused by space-charge-layer and parasitic capacitances. In the vicinity of the minimum point of a curve the delay is determined mainly by the charge stores of the neutral regions, and this delay is nearly independent of the power dissipation and the injection-current level. At high current and dissipation levels the effects of ohmic base resistance and high-level injection are detrimental to the delay. These effects determine the maximum injection current that should be used.

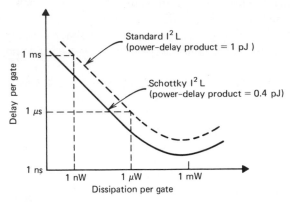

Figure 7-15 Typical curves of propagation delay versus power for Schottky and standard I^2L gates.

In Fig. 7-16 is a cross-sectional view of a Schottky I^2L gate with an oxide sidewall. The diodes are fabricated simply by depositing an appropriate metal, such as a compound of platinum and silicon (PtSi), in contact with the N-type collector region. This is then coated with aluminum. The diodes conduct in the forward direction, which is from the metal into the N-type collector, with a voltage drop between about 0.3 and 0.5 V. Conduction in the reverse direction is negligible, and there is no appreciable storage of charge in the diode.

The lowered logic swing of Schottky I^2L reduces the noise margin. Because there are no resistors between gates it is not possible for a noise voltage to produce a difference between the output voltage of one gate and the input voltage

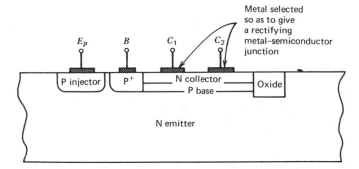

Figure 7-16   Cross-section of Schottky I²L gate with oxide sidewall.

of the next. Therefore, an absolute noise margin cannot be given in the usual manner. A standard gate that is *off* has $v_{BE} = 50$ mV. In order to switch it *on*, an injected noise current sufficient to change v_{BE} by nearly 700 mV is required, and this represents the high-level noise margin. For a Schottky gate with $v_{BE} = 450$ mV, the corresponding noise margin is 300 mV. For both types of gates a noise current that reduces v_{BE} from 750 mV to about 690 mV will switch an on-gate to the *off* state, and the difference of 60 mV represents the approximate low-level noise margin. Because all nodes of an I²L integrated circuit are at low impedance levels, the noise immunity is quite good.

By using a P-type substrate with an N-type epitaxial layer, I²L logic can be mixed with standard bipolar and MOS structures, both linear and digital, on the same chip. Of course, space-consuming isolation wells are required. The I²L circuitry can be placed within a single well and used in conjunction with other logic systems such as TTL, ECL, and MOS families as well as with linear circuits such as operational amplifiers. All of these can be fabricated on the same chip with interconnections provided by surface metallization. The processing procedure is that of the standard bipolar technology with four diffusions and seven masking steps. Circuit functions include those of logic arrays, microprocessors, read-only memories, memory decoders, frequency counters, frequency dividers, oscillators, many other digital systems, and all types of linear functions. A high degree of circuit integration is accomplished.

The number of collectors joined together and connected to the base of an I²L gate represents the fan-in and there is no limit. In most logic systems fan-outs of 5 are adequate. The fan-out of an I²L gate is limited by β_F, and fan-outs of 10 or more are possible. The output of an I²L gate is at a low voltage. There are two practical ways to raise this voltage to a higher level when necessary. One is to connect the outputs of several gates in series so that the voltages add, and the other is to connect a pull-up resistor between the output collector and a voltage source of 5 V or more. This latter method is especially suitable for driving

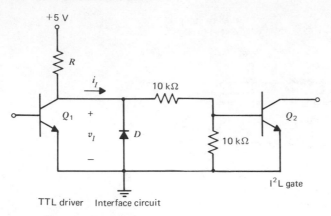

Figure 7-17   TTL to I^2L interface circuitry.

a TTL network. When this is done, the I^2L output is, in effect, treated as an open-collector TTL output. The high-level rise time is determined by the pull-up resistor. A typical I^2L output in saturation can sink about 40 mA without pulling out of saturation, and this is adequate for direct driving many TTL loads.

In order to drive an I^2L circuit from a TTL output, a simple interface arrangement such as that of Fig. 7-17 is adequate. It consists of a voltage divider to reduce the TTL output to the proper level. The input clamping diode limits negative excursions resulting from oscillations, referred to as ringing, that may occur from LC resonance effects. The interface, or buffer, circuit can be fabricated on the same chip with the I^2L gates.

7-5. EMITTER-COUPLED LOGIC

A typical integrated-circuit transistor, other than an I^2L device, has a collector-junction area considerably greater than that of the emitter junction, and the impurity level in the collector region near the junction is low compared with that of the base. Consequently, most of the excess charge of a transistor in saturation is stored in the neutral region of the collector, although considerable charge is also stored in the base. This is indicated in the sketch of Fig. 7-18. When the transistor is switched *off*, the excess charge must be removed before the output voltage can change, and the storage-delay time required to remove the charge is an appreciable portion of the propagation delay. Charge removal can be accomplished by means of reverse base current and also by recombination.

In the active mode a transistor has no significant charge stores in the collector region, and those of the base are small. In cutoff the charge stores of the neutral regions are negligible. Therefore, a gate designed so that the transistors operate

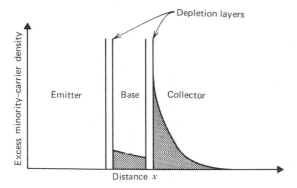

Figure 7-18   Excess minority-carrier densities of a transistor
in saturation.

only in the active and cutoff modes will have an extremely small storage-delay
time. As we have seen, Schottky transistors in TTL gates minimize saturation
effects.

An excellent example of a *nonsaturating* logic family is *emitter-coupled logic*
(ECL), also called *current mode logic* (CML). Not only is it nonsaturating but
the voltage swing is small, and these properties make ECL the fastest of the
technologies in use today.

Because of large variations that may occur in the current gain β_F of a transistor,
it is difficult to avoid saturation of a switching circuit by controlling the base
current in a production item. The ECL gate is designed so that the emitter
current is controlled, and operation is essentially independent of the base
current. This is accomplished by connecting the common-emitter terminal of
several input transistors, arranged with their collector-emitter outputs in
parallel, to the emitter of a control transistor with a base reference voltage.
The junction of the coupled emitters is supplied by a current source. Outside
the switching interval, the design is such that either all or none of the current
goes to the control transistor, so that this device is either active or cut off. If it
is *on*, the input transistors are *off*, and if it is *off*, one or more of the input transistors are active. Logic states change when current is switched, which is the basis
for the name *current-mode logic*. The name *emitter-coupled logic* is derived from
the coupled emitters of the circuit.

BASIC GATE

The basic gate is that of Fig. 7-19, and the number of inputs may be more or
less than the three that are shown. The control transistor is Q_4. Together,
V_{EE} and R_E constitute the effective current source, and the supplied current is
$v_E - V_{EE}$ divided by R_E, with v_E denoting the drop from the coupled emitters

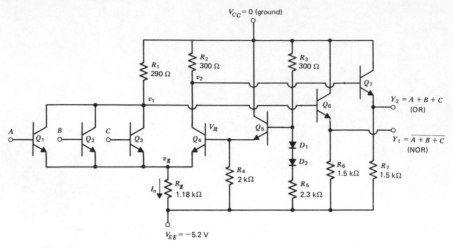

Figure 7-19 Basic ECL gate.

to ground. Usually V_{CC} is zero volts, or ground, as shown. Transistor Q_5, and its associated circuitry, supplies the reference voltage V_R to the base of the control transistor Q_4.

There are both NOR and OR outputs provided by transistors Q_6 and Q_7, respectively, and the symbol of the basic gate is that of Fig. 7-20. At output Y_1 the voltage is less than the output v_1 of the emitter-coupled stage by v_{BE6}. Also, at output Y_2 the voltage is less than v_2 by v_{BE7}. Thus each output is one diode drop V_D less than the respective output of the emitter-coupled stage. The output stages not only shift the voltage level downward by a diode drop but also, as we shall see, provide low output impedances.

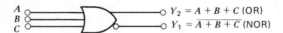

Figure 7-20 Basic ECL gate symbol.

INPUT STAGE

Let us examine the operation of the emitter-coupled input stage. The circuit is that of Fig. 7-21 with only one input shown. For this discussion the inputs to Q_1 and Q_2 of the network of Fig. 7-19 are specified to be sufficiently negative so that these devices are *off*. For positive logic the gate is designed for a high-level voltage of -0.75 and a low-level voltage of -1.60, with 0.75 V assumed for a diode drop V_D (see P7-10). The reference voltage V_R at the base of Q_4 is -1.175,

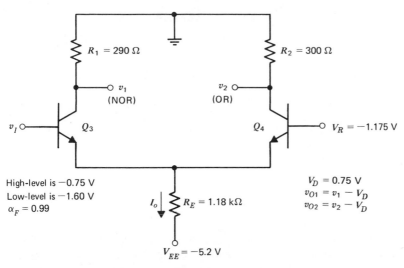

Figure 7-21   The emitter-coupled input stage.

which is precisely at the center of the logic swing. Each transistor is assigned a value of 0.99 for α_F.

Suppose v_I is high level. Then v_E is $v_I - V_D$, or -1.5 V, and the base-emitter voltage of the control transistor is too low to support conduction. Thus Q_4 is *off*. It cannot possibly be *on*, because this would fix v_E at a value making v_{BE3} greater than a volt, which is absurd. The emitter current of Q_3 is I_o of the current source, and this is

$$I_o = \frac{v_I - V_D - V_{EE}}{R_E} \qquad (7\text{-}10)$$

The voltage v_1 is $-\alpha_F I_o R_1$ and v_2 is 0. Recalling that the respective output voltages v_{O1} at Y_1 and v_{O2} at Y_2 are less than v_1 and v_2 by V_D, with the aid of (7-10) we find that

$$v_{O1} = -\frac{\alpha_F(v_I - V_D - V_{EE})R_1}{R_E} - V_D \qquad (7\text{-}11)$$

$$v_{O2} = -V_D \qquad (7\text{-}12)$$

For the numerical values of Fig. 7-21 with v_I equal to -0.75 V, calculations show that I_o is 3.14 mA, v_{O1} is -1.65 (low level), and v_{O2} is -0.75 (high level). For the case of multiple inputs the results are the same provided at least one of the inputs is at the high level. If two or more inputs are high level, the current I_o divides among the emitters of these inputs, but the current through R_1 is still the same. Accordingly, when one or more inputs is a logical 1, the output Y_1 is 0 and Y_2 is 1.

Now let us suppose v_I is low level at approximately -1.6 V. *The precise low-level value is not at all critical,* because any value near -1.6 is sufficient to place Q_3 in cutoff. The control transistor Q_4 is *on,* and v_E is $V_R - V_D$. Therefore, I_o is

$$I_o = \frac{V_R - V_D - V_{EE}}{R_E} \tag{7-13}$$

This is the emitter current of Q_4. The outputs v_{O1} and v_{O2} at Y_1 and Y_2 are found to be

$$v_{O1} = -V_D \tag{7-14}$$

$$v_{O2} = -\frac{\alpha_F(V_R - V_D - V_{EE})R_2}{R_E} - V_D \tag{7-15}$$

Calculations show that I_o is 2.78 mA, v_{O1} is -0.75 (high level), and v_{O2} is -1.57 (low level). With all inputs at the low level so that the input devices are cut *off,* the output at Y_1 is a logical 1 and that at Y_2 is 0. It is now evident that the output Y_1 provides NOR logic and its complement Y_2 provides OR logic.

The value of I_o is 3.14 mA when Q_3 is *on,* and it is 2.78 mA when Q_4 is *on.* The values chosen for R_1 and R_2 partially compensate for the difference in the currents. Other considerations are the effects of supply-voltage and temperature variations. From the equations that have been developed, we note that the output voltages depend on resistance ratios rather than on absolute resistance values. This is an important consideration in the design of integrated circuits, because resistance ratios can be accurately controlled. Furthermore, the transistor beta does not need to be precise, and values from 40 to 300 or more are acceptable. Collector-emitter voltages are low, allowing low breakdown voltages. The high-level output voltage is simply a diode drop below ground.

REFERENCE-VOLTAGE STAGE

The network that supplies the proper reference voltage V_R to the base of the control transistor Q_4 is shown in Fig. 7-22. With the small base current of Q_5 neglected, the current i_1 is found from a single loop equation. The base voltage of Q_5 is $-i_1 R_3$, and subtraction of a diode drop gives V_R. The result can be expressed as

$$V_R = \frac{R_3 V_{EE} - (R_5 - R_3)V_D}{R_5 + R_3} \tag{7-16}$$

Substitution of the numerical values gives -1.177 for V_R.

The values chosen for V_{EE}, R_3, and R_5, along with the two diodes, give the desired reference voltage V_R. In addition, the drop of about 0.4 V across R_3 provides a suitable collector-junction voltage. Resistor R_4 fixes the quiescent

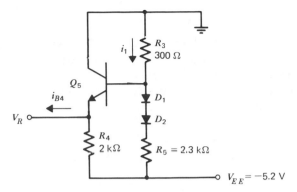

Figure 7-22 The reference-voltage network.

current of Q_5 at 2 mA, because the voltage across it is $V_R - V_{EE}$, or 4 V. The two diodes are used in place of a resistor in order to provide compensation for changes in both temperature and supply voltage V_{EE}. We shall now examine this in more detail.

The reference voltage V_R should be at the midpoint of the logic swing so as to give high-level and low-level noise margins that are equal. Let us consider the OR output. With v_I a logical 1, v_{O2} is $-V_D$, and when v_I is a logical 0, v_{O2} is that of (7-15). The desired V_R is 0.5 times the sum of these two output voltages, or

$$\text{desired } V_R = -\frac{\alpha_F(V_R - V_D - V_{EE})R_2}{2R_E} - V_D \qquad (7\text{-}17)$$

Solving for V_R and substituting numerical values for α_F and the resistances, we obtain

$$\text{desired } V_R = 0.112V_{EE} - 0.776V_D \qquad (7\text{-}18)$$

The actual V_R is that of (7-16). With numerical values substituted for the resistances, V_R becomes

$$V_R = 0.115V_{EE} - 0.769V_D \qquad (7\text{-}19)$$

Comparison of (7-18) and (7-19) reveals that *the actual V_R is approximately the same as the desired V_R for wide variations in V_{EE} and V_D.* A change in the supply voltage shifts the low logic level, but V_R is automatically adjusted so as to remain at the center of the logic swing. An increase in temperature causes all diode voltages, including each v_{BE}, to decrease about 2 mV/°C, and both high and low levels shift. However, V_R is again maintained at the proper value. Analysis of the NOR output reveals a similar situation (see P7-11).

7-6. ECL-GATE OUTPUT STAGES

Because the two output stages are identical except for a small difference in the base-to-ground resistors, we shall investigate only the one of Fig. 7-23. Between the base of Q_6 and ground are the resistor R_1 of the input stage and a current source I_1. This current is supplied by the collectors of the input transistors to the junction of R_1 and the base of Q_6. If at least one input is high-level, I_1 is $\alpha_F I_o$, or 3.1 mA, with I_o determined from (7-10). On the other hand, if all inputs are low-level, I_1 is zero. The base and emitter currents are expressed in terms of i_C for an assumed beta of 100.

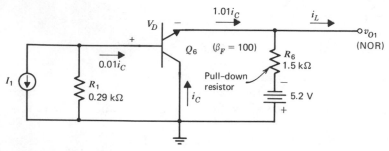

$I_1 = 0$ for logical 0 input.
$I_1 = 3.1$ mA for logical 1 input.

Figure 7-23　NOR output stage.

For $V_D = 0.75$ and i_L negligible, a single loop equation is sufficient for determining i_C. When I_1 is zero, i_C is 2.93 mA and v_{O1} is -0.76 V. When I_1 is 3.1, the collector current is 2.34 mA and v_{O1} is -1.66 V. When the current through R_6 decreases in magnitude, the output voltage drops, becoming more negative. Therefore, R_6 is referred to as a *pull-down* resistor. Note that a pull-down resistor is connected to a negative voltage, whereas a pull-up resistor connects to a positive voltage.

Let us now examine qualitatively the effect of a load. As long as Q_6 is out of saturation, the voltage across R_1 caused by the small base current i_B is almost inconsequential. It follows that *the base voltage of* Q_6 *is essentially independent of the load, and* v_{O1} *is lower. than this base voltage by a diode drop for all load currents.* As long as V_D holds its value of approximately 0.75 V, the output v_{O1} is at the proper value. We conclude that i_L has little effect, provided it is not excessive. This is true for both low-level and high-level outputs.

A change in i_L causes i_C to change so as to keep the voltage across R_6 fairly constant. If the change is such that i_C becomes large, the transistor may saturate, and the base current is no longer negligible. On the other hand, if i_C becomes very small the transistor moves into cutoff, and V_D drops (see P7-14). As long as the transistor remains in the active mode, we note that the emitter voltage *follows* the base voltage. Although it is less than the input by a diode drop, the emitter and base voltages rise simultaneously and fall simultaneously, with both changing the same amount. Because of this, the circuit of Fig. 7-23 is called an *emitter follower*. It is also known as a *common-collector* configuration because the collector is the common terminal to which both the input and output voltages are referred. An emitter-follower circuit is characterized by a high input impedance and a low output impedance. These properties are examined in more detail in later chapters.

To obtain quantitative results we need a relation between i_C and V_D. They are related by the active-mode form of the Ebers–Moll equation (6-6). Specifying $\alpha_F I_{ES}$ to be 2.5×10^{-13} mA at 290 K, we determine V_D to be

$$V_D = 0.025 \ln(4 \times 10^{12} i_C) \qquad (7\text{-}20)$$

with i_C in mA. The parameters were selected to make V_D about 0.75 V with a collector current of 3 mA.

With a specified load current i_L, a loop equation obtained from the network of Fig. 7-23 relates i_C and V_D. These can then be calculated with the aid of (7-20). When i_L changes from 0 to 0.4 mA with $I_1 = 0$, calculations show that the output voltage v_{O1} changes from -0.7608 to -0.7651. The ratio $|\Delta v_{O1}/\Delta i_L|$ of the incremental changes is 11 ohms (see P7-12). This is the output resistance of the emitter follower. It is not appreciably changed when the output is low-level (see P7-13).

Shown in Fig. 7-24 are equivalent output circuits of the NOR output stage. The previous analysis justifies the circuit of Fig. 7-24a, and that of (b) is based on the results of P7-13. Low output resistances, typically between 5 and 15 ohms, enable an ECL gate to drive a large number of identical gates. A fan-out of 25 is reasonable. However, there are capacitances associated with the input transistors of a gate, and these have to be charged and discharged through the output resistance when logic states change. Because an excessive fan-out increases the propagation delay, fan-out is usually restricted to about 10 or 15.

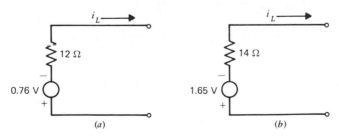

Figure 7-24   Equivalent output circuits of NOR output stage. (a) High-level output. (b) Low-level output.

7-7. ECL CHARACTERISTICS

Typical voltage transfer characteristics for both the NOR and OR outputs are shown in Fig. 7-25. Although these are easily deduced from Ebers–Moll equations, a piecewise linear approach is satisfactory and will be employed here.

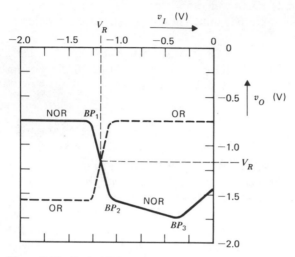

Figure 7-25   Typical ECL voltage transfer characteristics.

SPECIFICATIONS

The numerical values given on Fig. 7-26 will be used, and the approximate relationship between v_{BE} and i_C of both Q_3 and Q_4 is that of (7-20) with V_D representing v_{BE}. Inputs not shown on the network of Fig. 7-26 are logical 0's. When v_{BE} of a transistor is 0.65, i_C is 0.05 mA by (7-20), and the voltage across the collector resistor (R_1 or R_2) is 0.015, which is fairly negligible. Accordingly, we assume that *a transistor is off when its base-emitter voltage is less than* 0.65 V *and it is* on *when* v_{BE} *exceeds this value. When one transistor is off, the on-device has a base-emitter voltage of* 0.75 V. In addition, *a transistor is regarded as saturated when its collector-emitter voltage drops below* 0.1 V. *At this point* v_{BE} *is* 0.75 *and* v_{BC} *is* 0.65. *The NOR and OR outputs are each* 0.75 V *less than the corresponding output of the input stage.* These assumptions are necessary and sufficient for calculating the breakpoints of the transfer characteristic.

BREAKPOINT 1 ($v_I = -1.28$, $v_{O1} = -0.75$, $v_{O2} = -1.57$)

When v_I is -1.6, we know from previous analysis that Q_3 is *off* and Q_4 is active, making v_1 zero and v_2 equal to -0.82 V. The respective NOR and OR

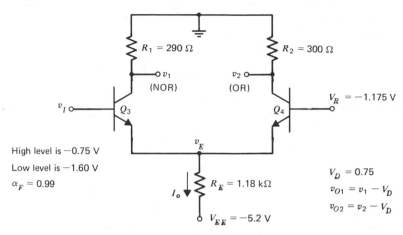

Figure 7-26   The emitter-coupled input stage.

outputs are -0.75 and -1.57 V. At BP_1 transistor Q_3 turns *on* with $v_{BE3} = 0.65$. At this point the emitter voltage v_E is $V_R - 0.75$, or -1.925. Adding 0.65 to this gives -1.275 for v_I. The NOR and OR outputs are the same as before.

BREAKPOINT 2 $(v_I = -1.08, v_{O1} = -1.57, v_{O2} = -0.75)$.

From BP_1 to BP_2 the current I_o shifts gradually from Q_4 to Q_3, and both devices are active. As the current shifts, the NOR output drops and the OR output rises. At BP_2 transistor Q_4 turns *off* with $v_{BE4} = 0.65$, which makes v_E equal to $V_R - 0.65$, or -1.825. Adding 0.75 to this gives -1.075 for v_I. The current I_o is $(v_E - V_{EE})/R_E$, and v_1 is $-\alpha_F I_o R_1$. From these the NOR output is calculated to be -1.57 V. Because Q_4 is *off*, the OR output is -0.75, remaining at this value as v_I is increased to zero.

BREAKPOINT 2 TO 3

The emitter voltage is now clamped at 0.75 V below v_I. From BP_2 to BP_3 the current I_o is $v_I - 0.75 - V_{EE}$ divided by R_E. The voltage v_1 is $-\alpha_F I_o R_1$, and the NOR output is a diode drop lower. We find that

$$v_{O1} = -0.243 v_I - 1.83 \qquad (7\text{-}21)$$

In this region the NOR characteristic has a slope of -0.243.

BREAKPOINT 3 $(v_I = -0.35, v_{O1} = -1.75, v_{O2} = -0.75)$.

Transistor Q_3 saturates with $v_{CE} = 0.1$ V at BP_3. At this point the current through R_1 is $\alpha_F I_o$. Kirchhoff's voltage law applied to the loop through R_1, from collector to emitter, and through R_E and V_{EE} gives 3.48 mA for I_o. The emitter voltage v_E is determined to be -1.10, and v_I is $v_E + 0.75$, or -0.35. The NOR output is found from (7-21) to be -1.75 V.

SATURATION REGION

With Q_3 saturated, v_E is $v_I - 0.75$ and v_1 is $v_E + 0.1$. Therefore, v_1 is $v_I - 0.65$. With v_I zero, the NOR output is -1.40 V. The saturation region is, of course, avoided in normal operation.

The characteristics of Fig. 7-25 are sketched from the calculations with curvatures added at the breakpoints. At the intersection of the NOR and OR characteristics the input and output voltages equal the reference voltage V_R. The noise margins are easily found from the unity-gain points of the NOR characteristic. As shown in (6-3) and (6-4) and Figs. 6-4 and 6-5 of Sec. 6-1, the high-level noise margin V_{NH} is $V_{OH(\text{min})}$ minus $V_{IH(\text{min})}$, and the low-level noise margin V_{NL} is $V_{IL(\text{max})}$ minus $V_{OL(\text{max})}$. With reference to the NOR characteristic the high-level output is -0.75, which is the value we shall use for $V_{OH(\text{min})}$. The minimum value of the high-level input is v_I at BP_2, giving -1.08 for $V_{IH(\text{min})}$. The difference of 330 mV is V_{NH}. The maximum value of the low-level input voltage is v_I at BP_1, or -1.28. The maximum value of the output low-level voltage is v_{O1} at BP_2, or -1.57. These give 290 mV for V_{NL}. The calculations have not considered differences between maximum and minimum characteristics, as shown in Fig. 6-4. When this is done, both noise margins become about 200 mV.

The ECL gate that has been considered has a propagation delay of about 4 ns and a power dissipation of 60 mW. The dissipation is the same for both logic states (see P7-15). The smooth power flow drawn from the power supply is desirable from the viewpoint of noise, and it minimizes the problem of power-supply bypassing. With four gates fabricated on a chip, with the gates using the same reference-voltage circuit, the dissipation is reduced to 47 mW/gate. In addition, if the emitter-followers providing the OR outputs are omitted, the dissipation is about 33 mW/gate. There are 1-ns and 2-ns versions of ECL, with increased power dissipation, and operation below a nanosecond is being utilized in custom-designed fast-computer ECL. The high speed, which results from the nonsaturating mode of operation and the small voltage swing, is the main advantage of ECL. In many main-frame computers, in high-speed counters and memories, and in various areas of instrumentation, digital communications, minicomputers, analog-to-digital conversion, and other applications the high speed is essential.

Although the power-delay product is moderate, having typical values from 50 to 200 pJ, the power dissipation is high. Thus heat sinks and cooling systems are required in LSI applications. The supply voltage is not critical, and in fact, it can be changed over a wide range. We have seen that V_R and the low-level output voltage track proportionately as the supply voltage is changed. Increasing the voltage gives a wider noise margin but also results in greater

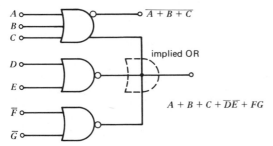

$\overline{A+B+C}$

implied OR

$A + B + C + \overline{DE} + FG$

Figure 7-27 Wired-OR logic.

dissipation and a larger logic swing (see P7-16 and P7-17). The choices of -5.2 for V_{EE} and zero for V_{CC} are compromises based on speed, noise margin, and power considerations.

In Sec. 6-1 we found that the implied-AND operation takes place when two or more outputs of RTL gates are wired together, and a similar effect occurs with TTL. The implied-AND operation gives a low-level output if one or more of the individual outputs are low-level. The outputs of two or more ECL gates can also be wired together, but in this case the OR operation is performed. Because of the low output resistances of emitter-followers, there is no need for external resistors as in RTL and TTL.

Suppose either the NOR or OR output Y_3 of a gate is wired to one of the outputs Y_4 of a second gate. If both were initially high-level or both were low-level, there is no interaction. However, in the event Y_3 was at -0.75 V and Y_4 was at -1.6 V, both become high-level at -0.75 V after the connection is made. The emitter-follower transistor giving Y_4 is simply cut off. Other transistors of the gates are not affected. The wired-OR logic is extensively used with ECL gates. An example is shown in Fig. 7-27.

Many gates are designed with multiple NOR and OR outputs, accomplished by adding additional emitter-follower stages. When this is done, the NOR outputs are independent of one another, as are the OR outputs. Along with implied-OR logic, such networks provide the designer with considerable flexibility, and the number of gates can be minimized. In Fig. 7-28 is shown the symbol of a gate with multiple outputs.

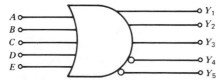

Figure 7-28 ECL gate with multiple outputs.

REFERENCES

1. C. S. Meyer, D. K. Lynn, and D. J. Hamilton, *Analysis and Design of Integrated Circuits*, McGraw-Hill, New York, 1968.
2. L. S. Garrett, *Integrated-Circuit Digital Logic Families*, Parts I, II, and III, IEEE Spectrum, Oct., Nov., Dec., 1970.
3. R. L. Horton, J. Englade, G. McGee, "I^2L Takes Bipolar Integration a Significant Step Forward," *Electronics*, **48**, 3, pp. 83–90, Feb. 6, 1975.
4. H. H. Berger and S. K. Wiedmann, "The Bipolar LSI Breakthrough," Parts 1 and 2, *Electronics*, **48**, 18 and 20, Sept. 4 and Oct. 2, 1975.
5. H. H. Berger and S. K. Wiedmann, "The Injection Model—A Structure-Oriented Model for Merged Transistor Logic," *J. Solid-State Circuits*, **SC-9**, 5, pp. 218–227, Oct. 1974.
6. IC Applications Staff of Texas Instruments, Inc., *Designing With TTL Integrated Circuits*, McGraw-Hill, New York, 1971.
7. *MECL Integrated Circuits Data Book*, Motorola Semiconductor Products, Inc., Phoenix, Ariz., 1972.

PROBLEMS

Section 7-1

7-1. The input transistor of a TTL gate has saturation currents I_{ES} and I_{CS} of 5.1×10^{-16} and 25×10^{-15} A, respectively, and the forward and reverse alphas are 0.98 and 0.02. At 290 K the junction hole currents in amperes are

$$I_{BE} = 10^{-18}(e^{40V_{BE}} - 1) \qquad I_{BC} = 2 \times 10^{-14}(e^{40V_{BC}} - 1)$$

The transistor is driven with a base current of 1 mA with the collector open-circuited. (a) Calculate the junction voltages V_{BE} and V_{BC}, and determine V_{CE} assuming internal IR drops are negligible. (b) Find the ratio $n_b(O)/n_b(W)$ of the free-electron densities at the extremities of the neutral base region. (c) Calculate the hole currents I_{BB}, I_{BE}, and I_{BC} in mA, with the sum of these being I_B. (d) Determine the electron currents in mA across each junction. Treat the multiple inputs as one emitter. [*Ans:* (d) 0.998, 0.8]

7-2. Suppose the transistor of P7-1 is in the inverse mode with a base current of 0.7 mA. The emitter junction is reverse biased by 1 V. (a) Calculate I_E in μA and I_C in mA. (b) Determine in mA both the hole and electron currents across the collector junction.

7-3. Suppose v_I at each input of the standard TTL gate of Fig. 7-2 is 1.37 V, which places Q_2 with a beta of 50 on the boundary between the active and saturation modes. Assume v_{BE2} is 0.72 and v_{BC2} is zero. Both v_{BE1} and v_{BE4} are 0.8 V. Calculate i_{E1}, i_{C1}, i_{B4}, and i_{E4} in mA, and also determine v_{CE1}. The output voltage is approximately 0.3 V, and the load is an *identical* gate with its inputs connected together. (*Ans:* $i_{E4} = -2.39$)

7-4. The respective saturation currents I_{ES} and I_{CS} of a certain transistor of a logic gate are 10^{-15} and 10^{-14} A, the respective forward and reverse alphas are 0.98 and 0.10, and the respective lifetimes τ_{BF} and τ_{BR} are 100 and 10 nanoseconds. The transistor is saturated with $v_{BE} = 0.8$ and $q/kT = 40$. The total excess charge stored in the neutral regions when a Schottky diode is used to clamp v_{BC} at 0.4 V is what percent of that when the diode is omitted, with v_{BC} at 0.75 V?

Section 7-2

7-5. At 290 K an I^2L current source supplies I_o to the base of an NPN transistor in saturation. Assume the load presented to the current source is approximately that of a diode, so that I_o is $I_s \exp 40 V_{BE}$ with I_s denoting the saturation current of the equivalent diode. Using the Ebers–Moll equations for the PNP transistor with $\alpha_F = \alpha_R = 0.8$ and $I_{ES} = I_{CS} = 0.01 I_s$, calculate in mV the difference $V_{EB} - V_{CB}$ between the forward-biased junction voltages and also determine the current ratio I_C / I_E. Note that I_C can be replaced with $-I_s \exp 40 V_{CB}$. First, deduce the relations

$$V_{EB} - V_{CB} = \frac{1}{40} \ln \frac{I_{CS} + I_s}{\alpha_F I_{ES}} \qquad \frac{I_C}{I_E} = \frac{-\alpha_F}{1 + (I_{CS}/I_s)(1 - \alpha_F \alpha_R)}$$

7-6. An I^2L network consists of 1500 basic gates, each with an injection current I_o of 100 nA. The collector current of a current source is 75% of the emitter current and each v_{EB} is 0.8 V. Each gate has a power-delay product of 0.5 pJ. The supply is 4 V. (a) Calculate the external resistance in kΩ, the chip dissipation in μW, and the propagation delay of a gate in μs. (b) For an external resistor of 80 ohms calculate the injection current I_o of a gate in μA, the chip dissipation in mW, and the gate propagation delay in ns.

Section 7-3

7-7. When the NPN transistor of an I^2L gate is saturated, both the base and collector currents equal I_o. At 290 K deduce (7-2) and (7-3) from the Ebers–Moll equations. Using the parameters of (7-9), calculate v_{BE}, v_{BC}, and v_{CE}.

7-8. Using the parameters of (7-9) at 290 K, calculate v_I and v_O of the I^2L inverter of Fig. 7-12 at the boundary of the cutoff region, with this boundary arbitrarily defined as that at which i_C is 1% of I_o. Also, calculate v_I and v_O at the boundary between the active and saturation modes with this arbitrarily defined as that at which i_C is 99.9% of $\beta_F i_B$ (Ans: 0.693, 0.176).

Section 7-4

7-9. Suppose R is 2 kΩ in the TTL to I^2L interface circuit of Fig. 7-17. Assume v_{CE1} is 0.1 V when Q_1 is saturated and v_{BE2} is 0.75 V when Q_2 is saturated. With Q_1 on, calculate i_I in mA, v_{BE2}, and the power P in mW supplied by the voltage source. Also, determine i_I, v_I, and P when Q_1 is off.

Section 7-5

7-10. In (7-12) and (7-15), which apply to the OR output of the ECL gate of Fig. 7-19, replace v_{O2} with the appropriate symbol V_L or V_H, representing the respective low-level and high-level voltages, and replace V_R with $0.5(V_L + V_H)$. (a) Calculate V_L and V_H using the numerical values of Sec. 7-5. (b) Repeat (a) with V_{EE} and V_D changed to −7 and 0.70 V, respectively.

7-11. With the three inputs of the ECL gate of Fig. 7-19 connected, the NOR output is the complement of the input, and (7-11) and (7-14) apply. (a) Replacing v_{O1} and v_I with the appropriate symbol V_L or V_H, representing the respective low-level and high-level voltages, and using the values for α_F, R_1, and R_E given in Sec. 7-5, determine V_L and V_H in terms of V_{EE} and V_D. From the result find the desired reference voltage V_R' in terms of V_{EE} and V_D, and compare with the actual V_R given by (7-19). (b) With $V_{EE} = -8$ V and $V_D = 0.6$ V calculate V_L, V_H, and the percent $[100 \times (V_R - V_R')/V_R']$ by which V_R deviates from the desired value. [Ans: (b) −3.29%]

Section 7-6

7-12. (a) For the output stage of Fig. 7-23 with $I_1 = 0$ and $i_L = 0.4$ mA, calculate i_C in mA, V_D, and v_{O1} to four significant figures. (b) With i_L zero, v_{O1} is -0.7608. Use this and the results of (a) to determine the output resistance in ohms. V_D is given by (7-20).

7-13. (a) For the output stage of Fig. 7-23 with $I_1 = 3.1$ mA and $i_L = -0.5$ mA, calculate i_C in mA, V_D, and v_{O1} to four significant figures. (b) With i_L zero, find v_{O1} to four significant figures, and use this and the data of (a) to determine the output resistance in ohms. V_D is given by (7-20).

7-14. In the NOR output stage of Fig. 7-23 with $I_1 = 3.1$ mA, determine the load current i_L in mA that causes V_D to drop to 0.6 V, with V_D given by (7-20).

Section 7-7

7-15. For the ECL gate of Fig. 7-19 calculate the total power dissipation in mW when all inputs are low-level and again when all inputs are high-level.

7-16. Using the assumptions of the *Specifications* of Sec. 7-7 for the ECL gate of Fig. 7-19, change V_{EE} to -7 V and calculate v_I and v_O at breakpoints 1, 2, and 3 of the NOR characteristic of Fig. 7-25. Also, determine from the results the high-level and low-level noise margins. Note that V_R changes. (*Ans:* noise margins, 0.54, 0.47)

7-17. For the ECL gate of Fig. 7-19 with V_{EE} changed to -7 V, calculate the total power dissipation in mW when all inputs are low-level. Note that V_R and the low-level voltage change.

Chapter Eight
MOSFET Logic
Gates

Large-scale integrated circuitry has initially been dominated by MOS technology. Complex logic functions having *minimum power dissipation* as well as *moderately high speed* can be built at *low cost* in *small packages* with *excellent reliability*. In addition, the technology is ideally suited for memory systems, shift registers, multiplexers, analog switching operations, and many other applications. Here we shall examine PMOS, NMOS, and CMOS logic gates. MOSFET flip-flops, memories, and other digital integrated circuits are presented in the next three chapters.

8-1. PMOS LOGIC

PMOS integrated circuits consist exclusively of p-channel enhancement-mode MOSFETs. There are no other elements, such as resistors or capacitors. The fabrication is relatively simple. Only a single diffusion into the N-type silicon wafer is required, and this forms the P regions that constitute the sources and drains of the transistors. There are four masking steps. The first is for the P-region diffusion, and this is followed by the mask that opens the windows for the gates so that the gate oxide can be deposited. The next two masks are necessary for the contact openings and for metal removal. The simplicity of fabrication allows a high circuit density with excellent processing yields. The small size and low cost of complex LSI circuits account for the popularity of PMOS logic.

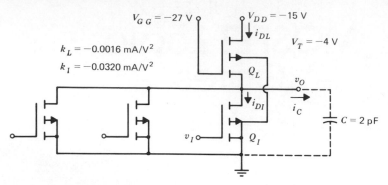

Figure 8-1   Three-input PMOS negative-logic NOR gate.

In Fig. 8-1 is shown the basic negative-logic NOR gate. It is customary in PMOS networks to employ *negative* logic, defined in Sec. 5-1, and we shall do so throughout this section. The numerical values depend on the specifications used for the design. Those of the figure are typical of an internal gate that drives similar gates of the chip. Each transistor is assumed to have a threshold voltage of -4 V. The gate-source voltage v_{GS} of the *load* device Q_L is $-27 - v_O$, and $v_{GS} - V_T$ is $-23 - v_O$. For all values of output voltage this is more negative than v_{DS}, which is $-15 - v_O$. Therefore, the load transistor is biased in the nonsaturation (linear) region. *It is essentially a resistor, but its resistance is somewhat nonlinear.*

The processing parameter $0.5\mu_h \varepsilon_{ox}/t_{ox}$ is 0.002 mA/V^2, which applies to all transistors of the chip. For each device the designer selects the channel width W and length l, and the product of the processing parameter and $-W/l$ gives k in accordance with (3-4) of Sec. 3-2. For the gate illustrated, the load transistor has a width W_L of 8 μm and a length l_L of 10 μm, giving -0.0016 mA/V^2 for k_L. Each input transistor has width W_I equal to 80 μm and length l_I equal to 5 μm, for a k_I of -0.032 mA/V^2. Five micrometers represent the minimum value that can be used for either W or l without special processing techniques, such as ion implantation. From these dimensions it is seen that each input transistor occupies considerably greater surface area than does the load device.

The capacitance C is the effective load. It consists of the capacitance that exists between the output node of the gate and ground plus the input capacitances of the various gates at the output, including stray capacitance. For a fan-out of three or four, which is somewhat greater than the average fan-out of a typical LSI network, the total capacitance C is about 2 pF as specified. Because the output is connected to insulated gates, the load resistance is infinite. *The transient response of the gate is determined approximately from the static behavior of the transistor and the effective load capacitance C.* Internal charge stores, other than those effectively included in C, introduce second-order effects which will be neglected. In this chapter we shall assume that the mathematical model of a MOSFET is adequately represented by equations (3-1) through (3-3) of Sec. 3-2. Actual IC design is often based on more complicated models.

BASIC EQUATIONS

The drain current i_{DL} of the unsaturated load transistor is given by (3-2) with v_{GS} replaced with $-27 - v_O$ and v_{DS} replaced with $-15 - v_O$. The current in milliamperes is

$$i_{DL} = -0.0016(31 + v_O)(15 + v_O) \tag{8-1}$$

The drain current i_{DI} of the input device Q_I is given by (3-1), (3-2), and (3-3) with $v_{GS} = v_I$ and $v_{DS} = v_O$. These become

$$i_{DI} = 0 \qquad\qquad \text{for } v_I > -4 \tag{8-2}$$

$$i_{DI} = -0.032[2(v_I + 4)v_O - v_O{}^2] \qquad \text{for } v_O \geq v_I + 4 \tag{8-3}$$

$$i_{DI} = -0.032(v_I + 4)^2 \qquad\qquad \text{for } v_O \leq v_I + 4 \tag{8-4}$$

with the currents in mA.

When the gate is *on*, the combination of the input devices gives a low resistance between the output node and ground. The highest resistance, and thus the worst case, occurs when only one transistor is *on*. Accordingly, the static and transient analyses to be made are based on the assumption that the input transistors other than the one labeled Q_I are *off* at all times. It follows that

$$i_{DL} = i_{DI} + i_C \tag{8-5}$$

STATIC TRANSFER CHARACTERISTIC

The voltage transfer characteristic with $i_C = 0$ is easily determined. For $v_I > -4$, the currents are zero, and v_O is V_{DD}, or -15 V. For $v_I \geq v_O - 4$, (8-1), (8-4), and (8-5) give

$$-0.0016(31 + v_O)(15 + v_O) = -0.032(v_I + 4)^2 \tag{8-6}$$

By replacing $v_I + 4$ with v_O and solving, we obtain -3.88 for v_O. Between this value and zero output voltage the input device is unsaturated, and from (8-1), (8-3), and (8-5) we find that

$$-0.0016(31 + v_O)(15 + v_O) = -0.032[2(v_I + 4)v_O - v_O{}^2] \tag{8-7}$$

From (8-6) and (8-7) we can determine v_I as a function of v_O. The results are

$$v_I = -4 - \sqrt{0.05(31 + v_O)(15 + v_O)} \qquad \text{for } -15 \leq v_O \leq -3.88 \tag{8-8}$$

$$v_I = 0.525v_O - 2.85 + \frac{11.63}{v_O} \qquad \text{for } -3.88 \leq v_O < 0 \tag{8-9}$$

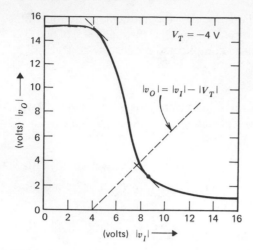

Figure 8-2 Static voltage transfer characteristic.

These two relations, along with our deduction that v_O is -15 for v_I between 0 and -4, were used to plot the transfer characteristic of Fig. 8-2. Because negative logic is used, the scales have been selected so as to relate the magnitudes of v_O and v_I. A logical 1 output corresponds to $|v_O| = 15$, and we shall refer to this as high-level. For $|v_O| \approx 1$, the output is a logical 0, or low-level, again ignoring the negative signs of the voltages.

When the gate is *off*, the input voltage is too small in magnitude to form a conducting channel in Q_I and the current is zero. Because the load transistor Q_L, which functions as a pull-down resistor, is in the nonsaturation region and has no current through its conductive channel, the magnitude of the output is 15 V. There is no power drawn from the voltage supplies.

When the gate is *on*, with -15 V giving a logical 1 at the input, both Q_L and Q_I have conducting channels in series. The channel resistances constitute a voltage-divider network with V_{DD} being the input. Because the magnitude of v_O should be no more than about 1 V for reasonable noise immunity, the input device must have a much lower channel resistance than that of the load device. Accordingly, k_I was chosen to be $20k_L$, accomplished by making the ratio W/l of Q_I twenty times that of Q_L. Examination of (8-6) and (8-7) shows that the transfer characteristic depends on the ratio k_I/k_L. For this reason the PMOS gate of Fig. 8-1 is referred to as a *ratio logic gate*. Increasing the ratio reduces the output voltage, thus improving the transfer characteristic and noise immunity. Unfortunately, the chip area of each input device also increases. Although the effective resistance of the input transistors is lowered when all are *on*, the design must be based on the worst case in which one and only one input is a logical 1. A ratioless logic gate is examined in Sec. 8-3.

There is appreciable power dissipation. In the *on* state with v_I at -15 V, the output v_O is -1. From (8-3) the drain current is found to be -0.67 mA. Multiplication by V_{DD} gives a dissipation of 10 mW. The average power of a gate depends on the percent of the time it is *on*.

Noise margins can be determined from either the transfer characteristics or (8-8) and (8-9). Using the equations, the unity-gain points are found (see P8-1) to be at (v_O, v_I) equal to $(-14.8, -4.41)$ and $(-2.76, -8.51)$. Thus V_{OH} is 14.8 and $V_{IH(min)}$ is 8.51, giving a high-level noise margin V_{NH} of 6.3 V. Also, V_{OL} is 2.76 and $V_{IL(max)}$ is 4.41 for a low-level noise margin V_{NL} of 1.7 V. The actual noise margins may be substantially lower because of variations in the supply voltages, the transistor parameters, and other effects. For example, although designed for a threshold voltage of -4, the processing may produce values of V_T anywhere between -3 and -5 V. Variations in k_I and k_L are also significant.

The transition region of the transfer characteristic of Fig. 8-2 is much broader than that typical of bipolar logic gates. The current of a BJT changes from nearly zero to its saturation value when v_{BE} changes from perhaps 0.65 to 0.75 V, a difference of only 0.1 V. However, the current of a MOSFET is not nearly as sensitive to changes in the gate-source voltage. As indicated by the transition region of Fig. 8-2, a change in v_{GS} from -4 to about -8 is required for the transition from cutoff to the input-device nonsaturation region, which corresponds to the BJT saturation mode. The change in v_{GS} is 40 times the corresponding change in v_{BE}.

8-2. TRANSIENT RESPONSE

There are switching delays during both turn-off and turn-on caused by the effective load capacitance C of the logic gate of Fig. 8-1. Because the greatest delay occurs during turn-off, we shall examine this first. Initially, the input transistor Q_I is *on*, with an input voltage v_I of -15, and the other inputs are assumed to be *off*. The output voltage that appears across C is -1 V.

TURN-OFF

At time zero v_I is abruptly switched to -1, turning Q_I *off*. The voltage across the capacitor must drop from -1 at time zero to -15 when the steady state is reached. During this interval, current through the load transistor is changing the charge stores and voltage of the capacitor. For convenience we shall use units of mA, V, kΩ, pF, and ns throughout this section. These constitute a consistent set. The capacitor current is $2\,dv_O/dt$ and this must equal i_{DL} at each instant. Using (8-1), we obtain the first-order differential equation:

$$\frac{2\,dv_O}{dt} = -0.0016(31 + v_O)(15 + v_O) \tag{8-10}$$

To determine $v_O(t)$ we need to solve (8-10) subject to the condition $v_O = -1$ at $t = 0$. Separating variables gives

$$dt = \frac{-1250 \, dv_O}{(31 + v_O)(15 + v_O)} \tag{8-11}$$

With the help of a table of integrals we integrate both sides of (8-11), using time limits 0 to t and voltage limits -1 to v_O. This gives a relation between t and v_O. Solving for v_O (see P8-3), with t in nanoseconds we determine that

$$v_O = -\frac{31 - 15 \exp 0.0128(t + 59.5)}{1 - \exp 0.0128(t + 59.5)} \tag{8-12}$$

A plot of the output voltage versus time, obtained from (8-12), is shown in Fig. 8-3. The time required for the voltage to rise from 1 V to 90% of its final value of 15 is 132 ns. This delay is large compared with that of other logic gates that have been considered, clearly indicating that PMOS logic is relatively slow.

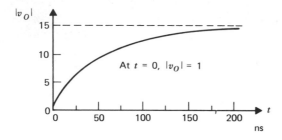

Figure 8-3   Output voltage versus time during turn-off.

The reason for the large delay is the considerable channel resistance R_L of the load transistor. Resistance R_L is v_{DS}/i_D with i_D given by (3-2). Replacing v_{GS} with $V_{GG} - v_O$ and v_{DS} with $V_{DD} - v_O$, we find that R_L is

$$R_L = \frac{1/|k_L|}{2|V_{GG}| - 2|V_T| - |V_{DD}| - |v_O|} \tag{8-13}$$

This is $625/(31 + v_O)$ kΩ. During cutoff the output drops from -1 to -15 V and R_L increases from 20.8 kΩ to 39.1 kΩ, with an average value of about 30 kΩ. Shown in Fig. 8-4a is an approximate equivalent circuit that can be used to calculate the turn-off time (see P8-4). In this circuit R_L is represented as a constant resistance of 30 kΩ. The 1 V battery accounts for the initial voltage of the capacitor.

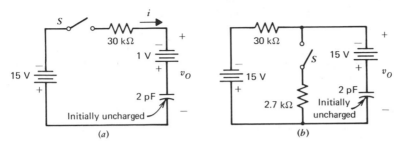

Figure 8-4   Approximate equivalent circuits. In each network the switch S is closed at time zero. (a) Turn-off. (b) Turn-on.

TURN-ON

Let us now briefly consider turn-on. With an output of -15 V, suppose v_I is abruptly switched from -1 to -15. The input transistor Q_I is turned *on* at time zero. With $v_I = -15$ the drain current i_{DI} in mA is found from (8-3) and (8-4) to be

$$i_{DI} = 0.032v_O(22 + v_O) \qquad \text{for } v_O \geq -11 \qquad (8\text{-}14)$$

$$i_{DI} = -3.87 \qquad \text{for } v_O \leq -11 \qquad (8\text{-}15)$$

Using (8-1), (8-5), (8-14), and (8-15) we can determine v_O as a function of time t following a procedure similar to that used in the turn-off analysis (see P8-5). The results show that the turn-on time is much less than the turn-off time. This is evident from consideration of the channel resistance R_I of the input transistor Q_I. For values of v_O from -15 to -11 the channel resistance in kilohms is determined from (8-15) to be $-v_O/3.87$ and for the values greater than -11 it is determined from (8-14) to be $31.25/(22 + v_O)$. From these we deduce that R_I decreases from 3.88 kΩ to 1.49 kΩ when v_O changes from -15 to -1. This is small compared with the resistance R_L of the load transistor. Consequently, the capacitor C discharges through R_I much faster than it charges through R_L when the gate turns *off*. An approximate equivalent circuit representing turn-on is shown in Fig. 8-4b. The 2.7 kΩ resistor represents the average value of R_I, and the 15 V battery in series with the capacitor accounts for the initial voltage. The turn-on time is less than 10% of the turn-off time (see P8-6).

SPEED CONSIDERATIONS

To increase the speed of the PMOS gate we want to minimize R_L. From (8-13) we deduce that V_{GG} should be large in magnitude. There is a limit, however, because values in excess of about 30 V are liable to damage the thin oxide insulation of the gate. The threshold voltage V_T should be small. This is accomplished by the low-threshold process, which reduces V_T with improvement in both speed and noise margin. The disadvantage is the increased cost because of ion implantation and other processing steps. These negate the advantageous PMOS property of simplicity. Also evident from (8-13) is the fact that a low voltage supply for V_{DD} reduces R_L, but of course, this adversely affects the noise margin. Performance is improved by using a silicon gate, which lowers V_T. This is discussed in Sec. 8-4.

A disadvantage of the gate of Fig. 8-1 is the requirement for two voltage supplies. One with a single supply is shown in Fig. 8-5. The top gate terminal is connected to the drain, which makes $V_{GG} = V_{DD}$ and $v_{GS} = v_{DS}$. From (3-3) it is evident that the load device is biased in the saturation mode for all values of input and output voltages. The drain current i_{DL} is given by (3-3), from which the load resistance R_L is found to be

$$R_L = \frac{V_{DD} - v_O}{k_L(V_{DD} - v_O - V_T)^2} \tag{8-16}$$

This is $625(15 + v_O)/(11 + v_O)^2$ kΩ for $V_{DD} = -15$ and $V_T = -4$ V.

For $v_O = -11$ the resistance is infinite and i_{DL} is zero. For this condition there is a 4 V drop across the pinched-off region of the channel. During turn-off the capacitor charges only to -11 V. This lower magnitude of the logical-1 output reduces the noise margin. Another serious disadvantage is the substantially increased turn-off time. As the voltage across C drops toward -11, the channel resistance R_L rises, becoming very large. The only advantage of the gate of Fig. 8-5 is the single supply, and the disadvantages are great.

It is possible to replace the load transistor with a diffused resistor. This is unreasonable, because the large resistance requires much more surface area than does the MOSFET load. Also, a depletion-mode device can be used for Q_L, and there are definite advantages which are discussed in Sec. 8-4 on NMOS

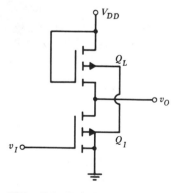

Figure 8-5   PMOS gate with saturated load transistor.

logic. The ion-implantation process permits fabrication of both enhancement and depletion-type MOS devices at the same time.

Although we have examined only the inverter and the negative-logic NOR gate, we know that complex logic can be implemented by interconnections of these. However, it is simpler to build logic systems from different types of basic gates. Several useful configurations are shown in Fig. 8-6.

Because each of the gates will have at times two conducting input devices in series, the physical size of these must be enlarged to keep the resistance low. For this reason it is desirable to minimize the number of series inputs, even though combinations such as those of the figure greatly increase design flexibility. The logic function of each gate is left as an exercise (see P8-7).

The numerical values of the PMOS gate of Fig. 8-1 are typical of an internal gate. An output gate that drives a load external to the chip may have a capacitive load of 20 pF or more. In order to maintain a reasonable propagation delay the transistors of the output gates are made with larger channel widths to give smaller resistances. Thus the output gates occupy greater chip area. A typical PMOS gate has a power dissipation of 6 mW and a propagation delay of 150 ns, for a power-delay product of 900 pJ. There are many applications in which the relatively slow speed is adequate. A familiar example is the pocket calculator, many of which contain several LSI circuits with thousands of PMOS logic gates.

In the section that follows we examine dynamic PMOS gates. These have increased speed, more compact circuitry, and reduced power dissipation.

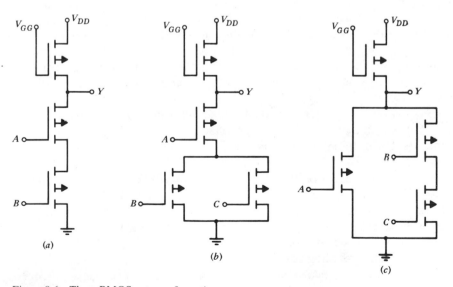

Figure 8-6   Three PMOS gate configurations.

8-3. DYNAMIC MOS GATES

The basic principle of a dynamic gate consists of utilizing the capacitances of transistors to store logic signals. These are transmitted through the gate from one capacitance to another, with timing pulses controlling the sequence. At least two sets of periodic pulses are required, with the pulses of each set occurring at different times from those of the other. The number of such sets is the *phase* of the network. Clock pulses are assumed to be sufficiently large to ensure transistor operation in the nonsaturation region.

TWO-PHASE GATE

Shown in Fig. 8-7 is a two-phase logic network, which is a cascade of inverters. During each clock period the inverters of a stage amplify and reshape the input signal representing the bit. Clock pulses are shown in Fig. 8-8, and the logic design is that of Fig. 8-9. The supply voltage V_{DD} is the logical 1 level. Each capacitor of the schematic diagram represents the total nodal capacitance, including interconnect capacitance and the intrinsic capacitance of the transistors connected to the node. Transistors are p-channel enhancement-mode devices.

Suppose the bit at the input is V_{DD} volts representing a logical 1. When pulse ϕ_1 appears, the voltage v_1 across C_1 becomes V_{DD} because Q_1 is *on*, and transistor Q_2 forms a channel. However, it does not conduct current because both Q_3 and Q_4 are *off*. The charge on C_1 is retained when ϕ_1 returns to zero. Then pulse ϕ_2 turns Q_3 and Q_4 *on*, and *capacitor* C_2 *is effectively short-circuited by the conducting channels of* Q_2 *and* Q_4. Most of the supply voltage V_{DD} appears across the higher resistance of the channel of Q_3. The very low voltage across C_2 ensures that Q_5 is *off*. When pulse ϕ_1 reappears, the conducting path through Q_6 and Q_7 fixes the output voltage across C_3 of the next stage at V_{DD}. Thus the input logical 1 bit has been transferred to the output with a time delay of one clock period.

The situation is simply reversed when the input bit is a logical 0. Pulse ϕ_1 fixes v_1 at zero, and pulse ϕ_2 causes C_2 to charge to V_{DD}, turning Q_5 *on*. When ϕ_1 reappears, the output voltage becomes zero because of conduction through Q_5 and Q_7.

The two-phase network that has been examined can easily be modified to accomplish complex logic. For example, input elements can be arranged in parallel, or in series, or in various series-parallel arrangements. Either or both of the inverters can be so modified. An example is the series-parallel logic gate of Fig. 8-10, which can be used for one of the stages. Note that each input has an associated coupling device.

The logic gates examined in this section operate in a sequence that is controlled by timing pulses supplied by a *clock*, or pulse generator. Clearly the pulse width

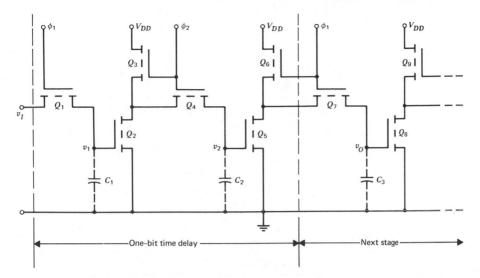

Figure 8-7   Two-phase dynamic shift register.

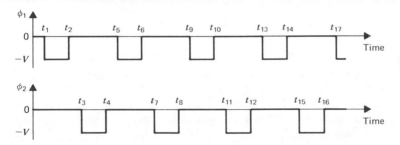

Figure 8-8   Waveforms of clock pulses.

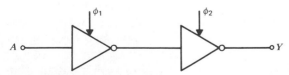

Figure 8-9   Logic diagram.

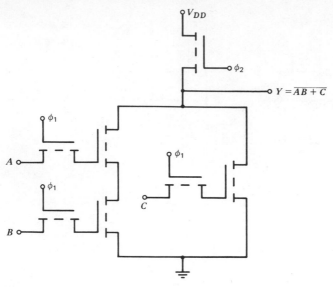

Figure 8-10 Series-parallel dynamic logic gate.

in nanoseconds must be greater than the turn-off time of a gate, and furthermore the interval between successive pulses of a phase ϕ must be greater than the pulse width. These requirements limit the maximum operating frequency to about 5 MHz, with the value depending on the design details.

In synchronous logic systems the binary output of a gate can change only at discrete instants of time. Gates operate in step, with the timing controlled by the clock pulse generator. Each bit is represented by the presence or absence of a pulse. Usually a pulse denotes a logical 1 and the absence of a pulse is a logical 0. Although timing is more critical than that of asynchronous networks in which one operation triggers the next, fewer and simpler circuits are required.

Frequently it is necessary to delay a bit by some interval. The two-inverter stage of Fig. 8-7 is useful for this purpose. We have seen that the output voltage is a replica of the input, except for a delay of a clock period which corresponds to a bit interval. A word consisting of a sequence of pulses or no pulses, fed into the input of the logic gate, appears at the output delayed by one bit interval. Thus the configuration serves as a *one-bit delay* circuit. With n such networks in cascade we have a circuit with an n-bit delay.

The configuration can also be used for storage of information. If precisely five bits are fed into a five-bit delay network and the input and clock pulses are then removed, the output voltage of the fifth stage is the first bit entered, the output of the fourth stage is the second bit entered, and so forth. Thus an n-bit delay network can be used for n-bit storage.

There is a problem associated with use of the delay network for memory. Because of leakage resistance, the capacitances of the gates cannot hold a charge very long. If the input is opened after a word is stored, the bits of the word are soon lost by leakage. A solution is to connect the output terminal to the input, which causes the bits to circulate around the loop (see P8-10). Each capacitance now has its charge either changed or refreshed before there is appreciable decay. A five-bit word is properly arranged in the memory at the end of each five clock pulses, and the word can be stored indefinitely. A *shift register* is a network, usually made of flip-flops, that shifts a set of digits one or more places to the right or left. Because the bits of the delay network used for memory advance around a loop, moving one position at each clock pulse, the network is called a *circulating shift register*, or a *circulating memory*.

Gates with logic states that are dependent on charge stores of capacitances which continually have to be refreshed are said to be *dynamic*. In contrast, the output of a static gate is constant as long as the gate inputs do not change.

A major advantage of dynamic logic is the reduced dissipation. Power is supplied only during a pulse that turns a load transistor *on*. Another advantage is the reduction in chip area because fewer and smaller transistors are required for most logic functions. Speed is increased, and in addition, system timing problems are simplified. Among the disadvantages are greater layout difficulties, increased loading of clock circuits, lower noise immunity, and reduced drive capability. Also, data are lost if the clock runs too slowly or is stopped.

A FOUR-PHASE RATIOLESS SHIFT REGISTER

The two-phase dynamic logic that has been considered is a ratio system. The low-level voltage at the output of an inverter depends on the ratio of the k parameters of two transistors that form a conducting path between V_{DD} and ground. Sometimes used are gates with transfer characteristics independent of the ratio k_I/k_L, and these are said to be *ratioless*. Such circuits are faster and have no dc power dissipation. However, they demand greater transient power from the clocks and more precise timing. There are many different configurations of multiphase logic, one of which is shown in Fig. 8-11. This is a four-phase ratioless dynamic logic network which employs the four illustrated pulse trains.

The operation of the four-phase network of Fig. 8-11 is as follows. Capacitance C_1 charges during pulse ϕ_1, regardless of the value of v_I. When pulse ϕ_2 appears, the voltage v_1 across C_1 becomes the complement of v_I. During ϕ_3 the capacitance C_2 is charged, and when ϕ_4 occurs, the output voltage v_O across C_2 becomes the complement of v_1. Accordingly, the logic state at the input is transferred to the output with a one-bit delay. Power is drawn from the clock sources only during the charging of a capacitor. Ratioless dynamic gates have a typical dissipation of 0.75 mW per gate and a delay of 70 ns for a power-delay product of 53 pJ.

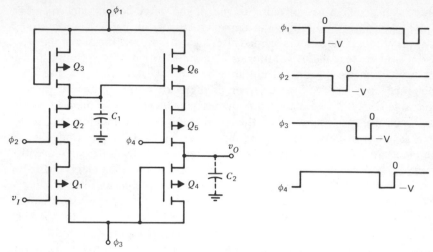

Figure 8-11   A four-phase dynamic 1-bit delay network.

In addition to a maximum operating frequency, dynamic shift registers and other logic configurations have a minimum allowed frequency, normally between 10 Hz and 10 kHz, depending on the rate of charge leakage from the intrinsic capacitances. The registers are quite often used to store bits of information in sequential memory systems, considered in Sec. 10-4. Although the discussion of this section assumed that the configurations consisted of p-channel devices, these can be replaced with *n*-channel transistors with the voltage polarities reversed. Indeed, dynamic NMOS shift registers are common, usually made of ratio-type circuits. In Chapter 10 we investigate several dynamic memory cells.

8-4. NMOS LOGIC

Numerous techniques can be employed to improve MOS logic performance. One of these is to use complementary circuits with both p-channel and n-channel devices, and this is the subject of the next section. Other techniques are the use of ion implantation, self-aligned gates, silicon gates, and depletion-mode devices for loads. Although these can be and are applied to PMOS technology, the main advantage of PMOS is its simplicity, and this advantage is partially compromised by the additional processing steps required.

NMOS integrated circuits consist exclusively of n-channel devices. We have learned that these are more difficult to fabricate than p-channel devices. Because one or more of the refinements that have been mentioned are usually employed, and also because of the greater electron mobility which is about 2.4 times that of the hole, the logic outperforms PMOS in both speed and power. The PMOS ratio circuits and dynamic ratioless circuits are equally applicable to NMOS, with n-channel enhancement-mode devices and positive voltage supplies. Here we examine only some of the special techniques commonly used in NMOS circuits and sometimes used with PMOS as well.

VOLTAGE TRANSFER CHARACTERISTIC

Throughout this section we employ units of V, mA, kΩ, pF, and ns. In Fig. 8-12 is shown a typical NMOS inverter. The input n-channel *enhancement-mode* MOSFET has a threshold voltage V_{TI} of 1 V and a k_I of 0.05 mA/V^2. The low-threshold voltage allows use of a 5 V supply. This lower voltage reduces power dissipation and simplifies interfacing with other logic systems, such as TTL. Providing *active pull-up* is the n-channel *depletion-mode* transistor Q_L with a threshold voltage V_{TL} of -2.5 and a k_L of 0.03 mA/V^2. Its I_{DSS} is $k_L V_{TL}^2$, or 0.188 mA. *The inclusion of a depletion-mode device eliminates the need for a second voltage supply.* Positive logic is used, and the logic levels are 0.5 and 5 V. Additional input devices in series and parallel arrangements can be included so as to provide complex logic.

The static relations are easily deduced. When v_I is lower than $+1$ V, the currents are zero and v_O is 5 V. For v_O lower than 5 and higher than 2.5, the load transistor is in the nonsaturation region, and (3-2) applies with v_{GS} zero. With the output between 2.5 and 0 the load device is saturated with a drain current of I_{DSS}. Thus with the currents in mA we have

$$i_{DL} = 0.03 v_O (5 - v_O) \qquad 5 > v_O \geq 2.5 \qquad (8\text{-}17)$$

$$i_{DL} = 0.1875 \qquad\qquad 2.5 \geq v_O > 0 \qquad (8\text{-}18)$$

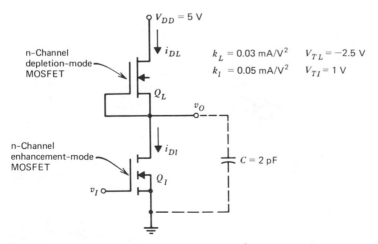

Figure 8-12   Typical NMOS inverter.

From (8-18) we note that the depletion-mode device is a *constant-current source*, supplying a current of 0.1875 mA as long as it is maintained in the saturation mode with an output less than 2.5 V. This constant current gives an increase in the switching speed. The drain current i_{DI} of the input transistor in mA is found from (3-2) and (3-3) to be

$$i_{DI} = 0.05v_O(2v_I - 2 - v_O) \qquad v_O \le v_I - 1 \qquad (8\text{-}19)$$

$$i_{DI} = 0.05(v_I - 1)^2 \qquad v_O \ge v_I - 1 \qquad (8\text{-}20)$$

The static transfer characteristic is shown in Fig. 8-13. Between BP_1 and BP_2 the load transistor is in the nonsaturation region and the input device is saturated. Accordingly, i_{DL} of (8-17) must equal i_{DI} of (8-20), which gives the relation

$$v_I = 1 + \sqrt{0.6v_O(5 - v_O)} \qquad BP_1 \text{ to } BP_2 \qquad (8\text{-}21)$$

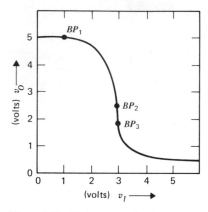

Figure 8-13   Voltage transfer characteristic.

Between BP_2 and BP_3 both devices are saturated. Equating the currents of (8-18) and (8-20) shows that v_I is

$$v_I = 2.94 \qquad BP_2 \text{ to } BP_3 \qquad (8\text{-}22)$$

For values of v_I greater than this, the load transistor is saturated and the input transistor is in the nonsaturation region. Using (8-18) and (8-19), we find that

$$v_I = 0.5v_O + 1 + \frac{1.875}{v_O} \qquad v_I > 2.94 \qquad (8\text{-}23)$$

Equations 8-21 through 8-23 were used to plot the transfer characteristic. The noise margins determined from the unity-gain points are greater than 1 V (see P8-12).

TURN-OFF

Let us suppose that v_O is a logical zero at 0.5 V and that the input voltage is switched from 5 to 0.5, abruptly turning off the input transistor at time zero. For $t > 0$, i_{DI} is zero, and i_{DL} equals C dv_O/dt. From (8-18) and (8-17), again using units of V, mA, pF, and ns, we deduce the relations

$$0.1875 = \frac{2\,dv_O}{dt} \qquad v_O < 2.5 \qquad (8\text{-}24)$$

$$0.03v_O(5 - v_O) = \frac{2\,dv_O}{dt} \qquad v_O > 2.5 \qquad (8\text{-}25)$$

By separating the variables of (8-24) and integrating time from 0 to t and voltage from 0.5 to v_O, we find $v_O(t)$ for values of output voltage from 0.5 to 2.5. From the result we find the time t_1 at which v_O is 2.5. Then by separating the variables of (8-25) and integrating time from t_1 to t and voltage from 2.5 to v_O, we find $v_O(t)$ for values of voltage from 2.5 to 5 (see P8-13). The result gives a delay of 51 ns as the time it takes the output to rise from 0.5 to 4.5 V, and this is the approximate turn-off time.

The turn-on time can be determined in a similar manner. As in the case of PMOS, this is substantially less than the turn-off time, attributable to the smaller resistance of the input transistor. Most of the propagation delay consists of the time t_{PLH} required for the output state to change from low to high. For the NMOS gate that has been analyzed, the propagation delay is approximately 60 ns, which is about half that of a typical PMOS gate. The power dissipation is 0.94 mW when the gate is *on* and zero when it is *off*. For a 50 % duty cycle the power-delay product is about 28 pJ.

SELF-ALIGNED SILICON GATES

The moderate threshold voltage (1 V) of the input transistor, the use of a depletion-mode load device for constant-current drive, and the reduced propagation delay indicate that the transistors of the inverter of Fig. 8-12 are probably fabricated with self-aligned silicon gates perhaps using ion implantation. The processing techniques are more complex. The gate area is about half the size of a typical PMOS gate, which is possible because of the higher carrier mobility, providing an appreciable increase in circuit density. This is an important advantage in LSI networks. As with PMOS, the elements are self-isolated because of the depletion regions surrounding the source, channel, and drain regions, and isolation wells are not required. Among the many applications of NMOS gates are memories, microprocessors, medium-speed data processing, analog-to-digital conversion, consumer controls, and high-speed calculators. Both dynamic and static logic systems are common.

Ion implantation is particularly advantageous in the fabrication of both depletion-mode and enhancement-mode transistors on the same chip. Smaller devices with shorter channels are possible. In addition to the increased circuit density that results, the input capacitance of a gate is reduced, and speed is increased. The normal MOS diffusion technology for fabricating n-channel transistors leads to depletion-mode devices. Ion implantation can be employed to add dopants of the desired type and density to the channel regions so as to give a threshold voltage of the proper value. This is one technique used to make n-channel enhancement-mode transistors. The threshold voltage must be greater than zero, but a reasonably low value permits use of a low supply voltage with reduced dissipation.

The metal gate of an enhancement-mode MOSFET must extend on the surface from the source region to the drain region. There can be no gap in the channel. Consequently, when the metal is applied after formation of the diffused regions, it must overlap these regions a small amount to allow for alignment errors. These overlaps increase the gate-drain and gate-source capacitances.

There are several procedures for eliminating the objectional overlaps. One of these consists of fabricating the transistor in the normal manner except that the gate metallization is deliberately made slightly shorter than the channel length, with small gaps at the ends. Ion implantation is then employed with the ion beam directed normal to the surface. The metal gate and the thick oxide outside the gate region prevent the ions from penetrating the silicon, but they do penetrate the thin oxide regions at both ends of the metal gate, forming doped regions just below this oxide. These doped regions extend the source and drain to the precise edges of the metal gate. There are now no gaps and no overlaps, and the gate is said to be *self-aligned*.

Another procedure that eliminates much of the overlap capacitance and also improves the threshold voltage consists of using a gate that is made of doped polycrystalline silicon rather than metal. The silicon gate (SG) is formed ahead of the diffusion, with the polycrystalline silicon on top of the thin oxide layer of the channel region. Windows are opened for the diffusion, and the source, drain, and polycrystalline silicon gate are doped simultaneously. Because there is some lateral diffusion under the gate oxide, overlap capacitances exist although they are quite small.

Self-aligned silicon-gate p-channel transistors have the advantage of low threshold voltages, typically from 1 to 3 V. Particularly attractive are n-channel SG devices. Depletion layers in the low threshold transistors are narrower, allowing shorter channel lengths. The smaller size leads to substantially higher circuit densities. A consequence of small size, low device capacitance, and high electron mobility is increased operating speed. In fabricating LSI circuits the polycrystalline silicon is often used for interconnections between terminals as well as for the gates.

8-5. COMPLIMENTARY SYMMETRY MOS LOGIC (CMOS)

One of the most important and popular logic families is *complementary symmetry MOS*, also called *CMOS* and *COS/MOS*. The gates employ pairs of p-channel and n-channel enhancement-mode transistors fabricated side-by-side on the silicon chip. The circuit is designed so that only one transistor of a pair is *on* at any given instant, and no power is drawn from the source in either static logic state. Dissipation occurs only when the logic state is switched. In fact, *the low power consumption is the major advantage of CMOS.*

The noise immunity is high, with noise margins that are about 30 to 45% of the supply voltage. The high noise immunity is a consequence of the sharp transfer characteristic. A wide range of voltages can be used, from 3 to 15 V, and a single supply is sufficient. The higher voltages give better noise immunity and increase the speed, but power dissipation is greater. The switching voltage tracks the supply voltage. Over 18 V can damage the circuit and less than 3 V provides inadequate noise margins and interface drive. For both logic states the output resistance is moderate, typically several hundred ohms. This is the channel resistance of the *on* device. Values are larger at low supply voltages.

The input capacitance of a CMOS gate is about 5 pF. With a fan-out of 3 giving a load capacitance of 15 pF, typical propagation delays are 25 ns for a gate with a 10 V supply and 35 ns for one with a 5 V supply. Each additional fan-out contributes an additional 5 pF of load capacitance and increases the delay by about 3 ns. Thus a 10 V gate with a fan-out of 50 would have a delay of 166 ns. Usually, the speed requirement restricts fan-out to 10 or less.

Shown in Fig. 8-14 is a CMOS inverter with typical values specified for the parameters. The n-channel transistor Q_I, which will be referred to as the *input* device, and the p-channel transistor Q_L, which will be called the *load* device, have their drains connected together; this node constitutes the output. A 15 pF load capacitor is assumed, corresponding to a fan-out of 3. The source of the load device is connected to the 10 V supply, and both gates are connected to the input voltage. For positive logic the voltage of the 0-state is zero and that of the 1-state is 10. Frequently, the threshold voltages V_{TL} and V_{TI} are more closely matched than indicated here.

The n-channel input device Q_I has $v_{GS} = v_I$ and $v_{DS} = v_O$. From (3-1) through (3-3) we find that

$$i_{DI} = 0 \qquad\qquad v_I < V_{TI} \qquad\qquad (8\text{-}26)$$

$$i_{DI} = k_I v_O(2v_I - 2V_{TI} - v_O) \qquad v_O \leq v_I - V_{TI} \qquad (8\text{-}27)$$

$$i_{DI} = k_I(v_I - V_{TI})^2 \qquad\qquad v_O \geq v_I - V_{TI} \qquad (8\text{-}28)$$

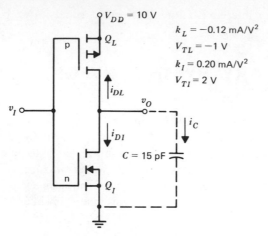

Figure 8-14   CMOS inverter.

The p-channel load device Q_L has v_{GS} equal to $v_I - V_{DD}$ and v_{DS} equal to $v_O - V_{DD}$. From (3-1) through (3-3) we deduce the relations:

$$i_{DL} = 0 \qquad\qquad\qquad\qquad\qquad\qquad v_I > V_{DD} + V_{TL} \quad (8\text{-}29)$$

$$i_{DL} = k_L(V_{DD} - v_O)(V_{DD} + 2V_{TL} + v_O - 2v_I) \qquad v_O \geq v_I - V_{TL} \quad (8\text{-}30)$$

$$i_{DL} = k_L(V_{DD} + V_{TL} - v_I)^2 \qquad\qquad\qquad v_O \leq v_I - V_{TL} \quad (8\text{-}31)$$

Note that k_L and V_{TL} are negative.

TRANSFER CHARACTERISTIC

The static voltage characteristic can be found from the specified numerical values and the preceding equations. However, the procedure is simplified and the results are informative if we assume symmetrical devices with

$$V_{TI} = -V_{TL} = V_T \qquad k_I = -k_L = k \qquad\qquad (8\text{-}32)$$

Both V_T and k are positive. The transfer characteristic is shown in Fig. 8-15.

Let us examine the transfer characteristic as v_I is increased from zero. Because i_C is zero, the current equation is

$$i_{DI} = -i_{DL} \qquad\qquad\qquad\qquad\qquad\qquad (8\text{-}33)$$

For values of v_I below V_T the input device is *off* and there is no current. The load transistor Q_L is biased in the nonsaturation region with v_{DS} zero. The output is V_{DD} as indicated from point a to b.

When v_I becomes equal to V_T, the input device turns *on* with a drain-source voltage of V_{DD}, which places this transistor in the saturation region. The

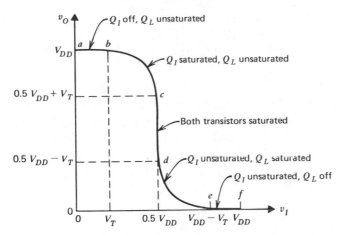

Figure 8-15   Voltage transfer characteristic.

characteristic from b to c is described by (8-33) with i_{DI} and i_{DL} given by (8-28) and (8-30). For the conditions specified by (8-32) it is not difficult to deduce that

$$v_I = \sqrt{2(V_{DD} - v_O)(V_{DD} - 2V_T)} - V_{DD} + V_T + v_O \qquad (8\text{-}34)$$

At point c the output voltage has dropped sufficiently to place the load transistor in saturation, and between c and d both devices are saturated. From (8-33) with the currents given by (8-28) and (8-31), subject to (8-32), we find that v_I is $0.5V_{DD}$ for values of v_O between $0.5V_{DD} + V_T$ and $0.5V_{DD} - V_T$. This is the vertical portion of the characteristic from c to d.

At point d the output has dropped sufficiently to place the input transistor in nonsaturation, and between d and e we use the currents of (8-27) and (8-31) in (8-33). The result can be expressed in the form

$$v_I = V_{DD} - V_T + v_O - \sqrt{2v_O(V_{DD} - 2V_T)} \qquad (8\text{-}35)$$

Finally, at point e the input voltage of $V_{DD} - V_T$ turns the load transistor *off* in accordance with (8-29). The currents are zero, the input device is in the nonsaturation region, and v_O is zero. The logic levels are zero and V_{DD}. From the unity-gain points, which are easily found from (8-34) and (8-35), the high-level and low-level noise margins are determined (see P8-14) to be

$$V_{NH} = V_{NL} = 0.25V_{DD} + 0.5V_T \qquad (8\text{-}36)$$

For $V_{DD} = 10$ and $V_T = 2$, the noise margins are 3.5 V. With matched transistors the high-level noise margin is equal to the low-level noise margin for all supply voltages, and each increases as V_{DD} is increased, in accordance with (8-36).

The equations that have been deduced and the transfer characteristic of Fig. 8-15 apply only to an inverter with symmetrical devices. However, the analysis of the inverter of Fig. 8-14 and the results are similar, in spite of the 1 V threshold mismatch. The calculated breakpoints (v_O, v_I) at b, c, d, and e of Fig. 8-15 are (10, 2), (6.1, 5.1), (3.1, 5.1) and (0, 9), respectively (see P8-15).

TRANSIENT ANALYSIS

The inverter of Fig. 8-14 with the specified parameters will be analyzed. First let us consider turn-off. Before switching occurs, the input is 10 V, the output is zero, and the load transistor is *off*. At time zero the input is switched abruptly to zero, turning the input transistor *off*. The load transistor is now *on* and saturated with a drain-source voltage of 10. Its drain current i_{DL} equals $-i_C$ with i_{DL} given by (8-31) with v_I zero. Accordingly,

$$k_L(V_{DD} + V_{TL})^2 = \frac{-C\,dv_0}{dt} \tag{8-37}$$

Separating variables and integrating with respect to time from 0 to t and with respect to voltage from 0 to v_O, we obtain

$$v_O = \left(\frac{-k_L}{C}\right)(V_{DD} + V_{TL})^2 t \tag{8-38}$$

This applies for values of v_O from 0 to $-V_{TL}$, by (8-31).

For values of v_O greater than $-V_{TL}$ the drain current is given by (8-30) with v_I zero, and the differential equation relating v_O and t is

$$k_L(V_{DD} - v_O)(V_{DD} + 2V_{TL} + v_O) = \frac{-C\,dv_O}{dt} \tag{8-39}$$

As before, we separate variables and integrate, using the integral given in P8-3. Voltage is integrated from $-V_{TL}$ to v_O and time is integrated from t_1 to t, with t_1 denoting the time when v_O is $-V_{TL}$. From (8-38) we determine t_1 to be $CV_{TL}/[k_L(V_{DD} + V_{TL})^2]$. The result can be expressed as

$$t = \frac{-0.5C}{k_L(V_{DD} + V_{TL})}\left[\frac{-2V_{TL}}{V_{DD} + V_{TL}} + \ln\left(\frac{V_{DD} + 2V_{TL} + v_O}{V_{DD} - v_O}\right)\right] \tag{8-40}$$

If we desire the output voltage as a function of time, we can solve (8-40) for v_O.

Let us define the turn-off time as the time required for v_O to rise from zero to 90% of V_{DD}. This is the time t of (8-40) with v_O replaced with $0.9V_{DD}$. For the numerical values given in Fig. 8-14 the turn-off time is calculated to be 21 ns.

The turn-on time is found in a similar manner. At time zero the output voltage is 10 and the input voltage is abruptly switched from 0 to V_{DD}. This places the input device in the saturation region. Analysis shows that

$$v_O = V_{DD} - \left(\frac{k_I}{C}\right)(V_{DD} - V_{TI})^2 t \tag{8-41}$$

for values of v_O between V_{DD} and $V_{DD} - V_{TI}$.

For smaller values of v_O the input device is in nonsaturation, and we find (see P8-17) that v_O and t are related by

$$t = \frac{0.5C}{k_I(V_{DD} - V_{TI})}\left[\frac{2V_{TI}}{V_{DD} - V_{TI}} + \ln\left(\frac{2V_{DD} - 2V_{TI} - v_O}{v_O}\right)\right] \tag{8-42}$$

From this we can determine v_O as a function of time for values of v_O less than $V_{DD} - V_{TI}$.

Let us define the turn-on time as that required for v_O to drop from V_{DD} to $0.1V_{DD}$. Replacing v_O in (8-42) with $0.1V_{DD}$ and using the numerical values of Fig. 8-14, we find the turn-on time to be 15 ns.

INVERTER POWER

In either logic state the dissipation is only a few nanowatts, which is a consequence of the very small leakage current of the *off* transistor. When the inverter switches from a low-level output to a high-level output, the capacitor voltage changes from zero to V_{DD}. During this switching interval power is dissipated in the channel resistance of the load transistor and energy is supplied to the capacitor. When the inverter switches back to the low level, the energy of the capacitor is dissipated in the channel resistance of the input transistor. No additional energy is supplied by the source.

Instantaneous power from the source is $-i_{DL}V_{DD}$. Because i_{DL} is zero during turn-on, the total energy supplied through a complete switching cycle from *off*-to-*on* and *on*-to-*off* can be found from consideration of turn-off only. This energy W is

$$W = -\int_0^{t_1} i_{DL}V_{DD}\,dt - \int_{t_1}^{\infty} i_{DL}V_{DD}\,dt \tag{8-43}$$

The load transistor is saturated from time 0 to t_1, and it is nonsaturated from t_1 to infinity.

Although the integrals of (8-43) give the correct result (see P8-18), there is an easier way to determine W. The charge on the capacitor rises from zero to its final value Q, which equals CV_{DD}. The total energy supplied by the source is QV_{DD}. It follows that W is CV_{DD}^2. Half of this is stored in the capacitor, with the other half lost as heat in the channel of the load transistor. When turn-on occurs, the energy of the capacitor is dissipated in the input device.

Let us suppose that a switching cycle from *on* to *off* and back to *on* is repeated periodically with period T and frequency f equal to $1/T$. For this case the average power dissipation P is the product of the frequency and the energy W supplied per cycle. Accordingly,

$$P = CV_{DD}^2 f \tag{8-44}$$

For the inverter of Fig. 8-14 operating at a frequency of 2 MHz the power is 3 mW.

A typical CMOS flip-flop, such as that of Fig. 9-18, consists of four or more inverters. Assuming each of four inverters is switched through a complete cycle in a period of two clock pulses, the power is twice that of (8-44) with f denoting the toggle frequency of the clock pulses. We note that *the power is directly proportional to the frequency, to the square of the supply voltage, and to the fan-out* which determines the capacitance C. At high toggle frequencies CMOS power dissipation is appreciable.

OUTPUT RESISTANCES

The high-level output resistance R_{OH} is simply the channel resistance of the load transistor when the input voltage is zero and the output voltage is V_{DD}. This resistance is the ratio of $V_{DD} - v_O$ to $-i_{DL}$, with i_{DL} given by (8-30). Replacing v_I with zero and v_O with V_{DD} we find R_{OH} to be

$$R_{OH} = \frac{-1}{2k_L(V_{DD} + V_{TL})} \tag{8-45}$$

The low-level output resistance R_{OL} is the channel resistance of the input transistor when v_I is V_{DD} and v_O is zero. This resistance is v_O/i_{DL} with the current given by (8-27). We find that

$$R_{OL} = \frac{1}{2k_I(V_{DD} - V_{TI})} \tag{8-46}$$

For the numerical values of Fig. 8-14 the output resistances R_{OH} and R_{OL} are found from (8-45) and (8-46) to be 463 and 313 ohms, respectively. Each of these is inversely proportional to the transistor k, and the resistances are higher at lower supply voltages.

Most of the analyses of this and preceding sections are based on theoretical MOSFET equations. These are, of course, only approximations. However, the deductions are typical of practical MOS logic circuits. CMOS gates and additional properties are examined in the following section.

8-6. CMOS GATES

STRUCTURE

There are usually at least several, and perhaps hundreds or more, gates on a CMOS silicon chip. They may be interconnected to perform complex logic or the gate terminals may be connected to external pins for flexible design. The simplest gate is the two-transistor inverter. Other gates require additional devices. Figure 8-16 is a cross-section of part of a CMOS integrated circuit, not drawn to scale. Shown are two p-channel and two n-channel transistors.

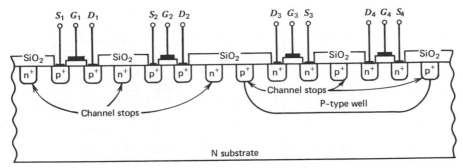

Figure 8-16   CMOS IC cross-section.

In the direction normal to the paper the gate of a p-channel device must be longer than that of an n-channel device because of lower hole mobility.

The structure has an N-type substrate with the n-channel devices within a P-type isolation well. Each p-channel device is surrounded by a continuous n$^+$ guard ring which prevents the formation of a channel between P regions of adjacent devices. These guard rings are referred to as *channel stops*. Also, each n-channel transistor is surrounded by heavily doped P-type channel stops. The necessity for P-type isolation wells and channel stops is a disadvantage, *and the average CMOS gate requires greater surface area than does a PMOS or an NMOS gate*. Some integrated circuits use a recessed oxide in place of the doped channel stops, reducing the surface area by about one-third. Not only are circuit densities greater but also the speed is increased. Processing is more complicated, however.

A structure with substantially improved performance is one using a sapphire substrate with silicon that is grown epitaxially within depressions on the substrate surface. The small silicon islands are insulated from one another by the sapphire, and transistors are made within the islands by conventional methods. MOSFET circuits using the silicon-on-sapphire (SOS) technology have greater circuit densities with reduced capacitances. There is less dissipation and an improvement in speed. SOS CMOS logic circuits fabricated with self-aligned silicon gates and employing ion implantation techniques can operate with a 10 V supply at propagation delays as small as 1 or 2 ns per gate. Insofar as performance is concerned, SOS is ideal for LSI networks. However, the high cost of the sapphire substrate is a serious disadvantage. Even so, SOS CMOS is growing in importance.

LOGIC GATES

The basic CMOS gates are the NOR and NAND gates, and these are shown in Fig. 8-17, each with three inputs. When the NOR gate turns off, the current that charges the load capacitance passes through the series-connected p-channels

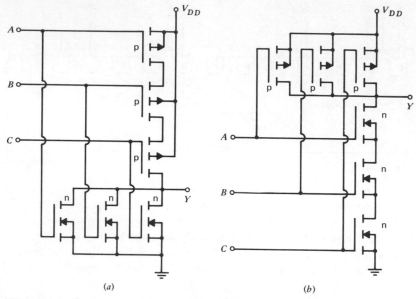

Figure 8-17 CMOS NOR and NAND gates. (*a*) Three-input NOR gate. (*b*) Three-input NAND gate.

of the three load transistors. Therefore, these devices should be fabricated with large W/l ratios. For the NAND gate the series connected n-channel transistors should have large W/l ratios to minimize turn-on time, and each of these is fabricated in a separate P-type well, with the substrate connected to the source. The p-channel transistors of each gate have a common substrate.

Available are many different types of CMOS logic gates in integrated-circuit form. An example is the CD4000 A integrated circuit of Fig. 8-18. It contains seven p-channel and seven n-channel transistors that provide two three-input NOR gates plus an inverter. The external pin labeled V_{SS} is the network ground, which is usually at zero volts. The recommended dc supply $V_{DD} - V_{SS}$ is from 3 to 15 V, and the input voltages of the gates should be between V_{SS} and V_{DD}. Input diode protection is provided. Another example is the CD4030A integrated circuit. It contains four independent Exclusive-OR gates on a silicon chip. Each gate consists of four n-channel and four p-channel enhancement-mode transistors, and all inputs and outputs are protected against electrostatic effects and high voltages. The gates can be connected to perform addition and subtraction, as well as many other logic functions. Speed is medium, with a typical gate propagation delay of 40 ns using a 10 V supply with a load of 15 pF. There are numerous other configurations. In fact, one of the major advantages of CMOS logic is the wide variety of gates and logic functions available at low cost.

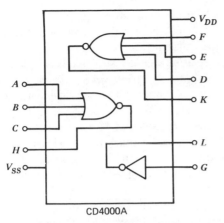

Figure 8-18   External pins and logic functions of a CMOS integrated circuit.

TRANSMISSION GATES

A CMOS gate extensively used in both digital and analog circuits is the *transmission gate* of Fig. 8-19a. It is an electronic switch with *off* and *on* states controlled by gate voltages ϕ and its complement $\bar{\phi}$, with the complement obtained from an inverter with input ϕ. The input voltage is assumed to be greater than v_O. Actually, the switch is *bilateral*, because current can be driven equally well in either direction. If v_I is less than v_O, the source and drain terminals are simply interchanged.

To illustrate the operation suppose a resistor is connected from the output terminal to ground, with ϕ zero and its complement equal to 10. With v_I zero,

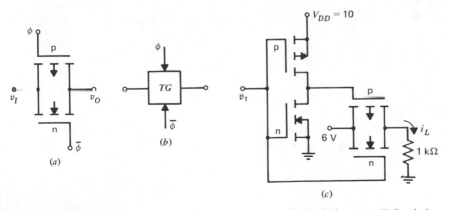

Figure 8-19   CMOS transmission gate and application. (*a*) Transmission gate. (*b*) Symbol. (*c*) Application as a switch.

the p-channel transistor is *off*, and the n-channel transistor is *on* in the non-saturation region with $v_{GD} = 10$. With v_1 changed to 10 V the n-channel device is *off*, but now the p-channel device is *on* and in the nonsaturation region with a gate-source voltage of -10. In fact, *for all positive and negative values of* v_1 *that do not exceed breakdown voltages at least one of the two transistors is on and non-saturated.* Note that the source of one device is connected to the drain of the other.

A typical channel resistance of a transmission gate that is *on* is 300 ohms. In most applications the load resistance is much larger than this, in which case the output voltage v_O is approximately equal to the input voltage v_I. The voltage transfer characteristic for this case is a straight line through the origin with a slope of unity.

For values of v_I between 0 and 10 the switch is *off* when ϕ is 10 and $\bar{\phi}$ is zero. The gate voltages have incorrect polarities for conduction. *Thus the switch is controlled by the voltage* ϕ. In many logic circuits ϕ consists of clock pulses having alternate values of 0 and 10.

In the elementary application of Fig. 8-19c the inverter provides the control voltages ϕ and $\bar{\phi}$. With $v_1 = 10$ the 6 V source drives a current i_L through the 1 kΩ resistor. When v_1 is changed to zero, the load current drops to zero (see P8-19).

In the dynamic logic networks of Sec. 8-3 PMOS transmission gates were used. Such a gate is shown in Fig. 8-20 along with its transfer characteristic for a large R_L, perhaps 10 kΩ or more. The device is saturated when $v_{DS} < v_{GS} - V_T$. As v_{DS} is $v_I - v_O$, and v_{GS} is $\phi - v_O$, it follows that the transistor is saturated when v_I is less than $\phi - V_T$, which is -6 V. In saturation with a drain current of zero, v_{GS} equals V_T, from the square law (3-3). Noting that v_{GS} is $\phi - v_O$, we find that v_O is constant at $\phi - V_T$ when v_I is equal to or less than this value. The transfer characteristic of Fig. 8-20 shows the saturation condition.

Now suppose the gate of Fig. 8-20 is replaced with an n-channel device, with $\phi \doteq 10$ and $V_T = 4$ V. For this case v_O saturates at $+6$ V when v_I is

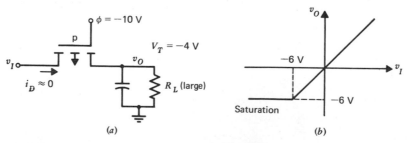

(a)   (b)

Figure 8-20   PMOS transmission gate. (a) PMOS gate circuit. (b) Transfer characteristic.

equal to or greater than 6. By using complementary devices the CMOS transmission gate avoids the restriction on v_O, for both positive and negative values of v_I. A disadvantage is the need for the complement of ϕ, which requires an inverter. In the dynamic networks of Sec. 8-3 saturation was avoided in the PMOS transmission gates by specifying a sufficiently large value of ϕ.

INTERFACING AND BUFFERING

Let us consider briefly some of the problems involved in connecting CMOS networks to other systems. A general weakness of MOS logic families is the lack of drive capability to handle many computer interface conditions. Often, much of the speed of a MOS system is lost at an interface. There is no particular difficulty in connecting CMOS to PMOS circuits. CMOS gates can operate with $V_{DD} = 0$ and with values of V_{SS} from -3 to -15 V, and there is no degradation of performance. However, both should use either positive logic or negative logic. CMOS can be interfaced directly with NMOS devices that operate at the same voltage level.

Integrated circuit CD4050A, classified as a Hex Buffer/Converter, has six double inverters that are useful in converting the logic levels to those of TTL and in providing sufficient drive for direct connection to two TTL loads. Each of the six individual gates has the circuitry of Fig. 8-21.

A buffer is a gate that isolates another gate from a load and supplies increased driving capability. It minimizes changes in switching thresholds caused by load currents, restores logic levels to optimum values, and improves noise immunity by providing a better transfer characteristic. The output gates of a buffer have large W/l ratios so as to lower the high-level and low-level output resistances. Thus the transistors occupy greater surface area. The low resistances enable the gate both to supply and to sink greater load currents without undue changes in the logic values. A *converter*, also called a *level shifter*, shifts the voltages of

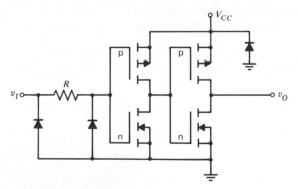

Figure 8-21   CMOS buffer network.

the logic levels. Both buffering and conversion are usually necessary when driving TTL gates from CMOS circuits.

The circuit of Fig. 8-21 is noninverting. An inverting buffer is obtained by omission of the second stage. Assuming V_{CC} is 5 V, a high-level input of 10 is shifted to 5 at the output. The zero level is unchanged. The resistance R and the diodes are for protection from static charges and excessive voltages.

With $V_{CC} = 5$ and a low-level output, the buffer can sink at least 3.75 mA, and typically 6 mA, while providing an output voltage less than 0.4 V. This is an adequate low-level voltage for TTL, and the current is more than twice that supplied by a TTL gate. The current from the load passes through the resistance of the n-channel transistor to ground.

With a high-level output the buffer can supply through the p-channel transistor a current from 1.25 to 2.5 mA to a load, while maintaining an output greater than 2.5 V. This is a reasonable 1-state voltage for TTL, and furthermore, the input TTL gate current is only a few microamperes. The gate delay is typically 65 ns with a supply of 5 V and a load capacitance of 15 pF. In general, considerable care must be exercised when interfacing one type of digital integrated circuit with another. Special attention should be given to the logic swing, the output drive capability, the noise immunity, and the speed. Buffers and level shifters are often required.

Although the inputs and outputs of MOS integrated circuits are usually protected with breakdown diodes, the circuits are nevertheless rather easily damaged by excessive static charges and voltages. An input must never be left floating, but each should be connected to the source or drain supply or to a device output, whichever is appropriate. Signals should not be applied to inputs when the supply is off. Test equipment should be grounded.

REFERENCES

1. Engineering Staff of American Micro-Systems, Inc., W. M. Penney, Editor, *MOS Integrated Circuits*, Van Nostrand Reinhold, New York, 1972.
2. *COS/MOS Digital Integrated Circuits*, SSD-203B, RCA Solid State Division, Somerville, New Jersey, 1973.
3. R. H. Crawford, *MOSFET in Circuit Design*, McGraw-Hill, New York, 1967.
4. D. J. Hamilton and W. G. Howard, *Basic Integrated Circuit Engineering*, McGraw-Hill, New York, 1975.

PROBLEMS

Section 8-1

8-1. By differentiation of (8-8) and (8-9) determine the unity-gain points (v_O, v_I) of the voltage transfer characteristic of Fig. 8-2.

8-2. With the three inputs of the NAND gate of Fig. 8-1 connected, determine the static voltage transfer equations corresponding to (8-8) and (8-9). Calculate the output voltage at which the input transistors enter the nonsaturation region, and find (v_O, v_I) at the unity-gain points.

Section 8-2

8-3. Deduce (8-12) for $v_O(t)$ during turn-off from (8-10), with the aid of the integral

$$\int \frac{dx}{(a + bx)(c + dx)} = \frac{1}{ad - cb} \ln \frac{c + dx}{a + bx}$$

Then calculate the time in ns it takes v_O to drop from -1 to -14 V.

8-4. Using the turn-off approximate equivalent circuit of Fig. 8-4a, find $v_O(t)$ and calculate the time in ns it takes v_O to drop from -1 to -13.5 V.

8-5. Using (8-1), (8-5), (8-14), and (8-15), determine v_O as a function of time during turn-on. The input voltage v_I switches from -1 to -15 V at time zero. Calculate the time it takes v_O to rise from -15 to -2 V. Refer to the integral of P8-3. (*Ans*: 10.2 ns)

8-6. Using the turn-on approximate equivalent circuit of Fig. 8-4b, calculate the time in ns required for v_O to rise from -15 to -2.5.

8-7. (a) Deduce that the gate of Fig. 8-6a is a negative-logic NAND gate and a positive-logic NOR gate. (b) With negative logic show that the output of the gate of Fig. 8-6b is $\bar{A} + \bar{B}C$. (c) With negative logic show that the output of the gate of Fig. 8-6c is $\bar{A}(\bar{B} + \bar{C})$. Also, determine the output for positive logic and compare the result with that of (b).

Section 8-3

8-8. At time zero suppose the voltages v_I, v_1, v_2, and v_O of the two-phase dynamic shift register of Fig. 8-7 with the clock pulses of Fig. 8-8 are -1, -1, -10, and -1 V, respectively. The network is designed to give a logical zero of -1 V and V_{DD} is -10. If v_I changes to -10 V at time t_3, determine the instants when v_1, v_2, and v_O change to -10, -1, and -10 V, respectively.

8-9. For the data of P8-8, suppose v_I changes to -10 at time t_3 and returns to -1 V at time t_{11}. Sketch the waveforms of v_I, v_1, v_2, and v_O on four time axes from 0 to t_{17}. Identify the instants at which each change occurs. Assume the capacitors charge and discharge in zero time.

8-10. Sketch the circuit of Fig. 8-7 with two complete stages and connect the output of the second stage to the input of the first stage. Label capacitances C_1, C_2, C_3, and C_4 having respective voltages v_1, v_2, v_3, and v_4. On time scales corresponding to those of Fig. 8-8, sketch the waveforms of v_1, v_2, v_3, and v_4 from time zero to t_{17}, assuming the voltages are -10, -1, -1, and -10 V, respectively, at time zero, and V_{DD} is -10 V. Note that the stored bits of C_1 and C_3 are unchanged after each two clock periods.

Section 8-4

8-11. From (3-2) and (3-3) derive (8-17) through (8-23) for the NMOS inverter of Fig. 8-12. Then deduce the values of v_O and v_I at BP_3 of the transfer characteristic of Fig. 8-13.

8-12. Using (8-21) and (8-23) to determine the unity-gain points of the static voltage transfer characteristic, calculate the high-level and low-level noise margins of the NMOS inverter of Fig. 8-12. (*Ans*: 1.24, 1.07)

8-13. With $v_O = 0.5$ at $t = 0$, determine $v_O(t)$ from (8-24) and (8-25) for values of v_O less than 5 and greater than 2.5. Calculate the time in ns it takes v_O to rise from 0.5 to 4.7 V. Refer to the integral of P8-3.

Section 8-5

8-14. For the CMOS transfer characteristic of Fig. 8-15, verify from the unity-gain points to be determined from (8-34) and (8-35) that both noise margins are those of (8-36).

8-15. In the *Transfer Characteristic* discussion of Sec. 8-5 the breakpoint numerical values for the inverter of Fig. 8-14 are given. Deduce these breakpoints from the equations, using the values of Fig. 8-14.

8-16. For the inverter of Fig. 8-14 with V_{DD} reduced to 5 V, calculate the turn-off and turn-on times in ns, with these defined as in Sec. 8-5. Then find the dissipation in mW at a toggle frequency of 2 MHz. Also, determine the high-level and low-level output resistances in ohms. Compare with the values for a 10 volt supply.

8-17. From the current equations, deduce the turn-on equations (8-41) and (8-42), and calculate the time in ns required for the output voltage to drop from 10 to 0.05 when the input voltage of the inverter of Fig. 8-14 is suddenly switched from 0 to 10 V. (*Ans*: 29)

8-18. Show that the integrals of (8-43) give $W = CV_{DD}^2$. First, change the variable of integration from time t to voltage v_O. Use (8-38) and (8-39) to find dt in terms of dv_O, and use (8-30) and (8-31) with v_I zero to replace i_{DL}. Integration limits become 0, $-V_{TL}$, and V_{DD}.

Section 8-6

8-19. For the transmission-gate circuit of Fig. 8-19c, assume each transistor has a threshold voltage of magnitude three and a k of magnitude 0.5 mA/V². Determine the load current i_L in mA when v_1 is 10 V. Use (3-2), noting that the n-channel device has $v_{GS} = 4 + v_{DS}$.

8-20. Repeat P8-19, except change the 6 V supply to 9 V.

Chapter Nine
Flip-Flops

A *flip-flop* is an electronic circuit that has two static states and the ability to switch states when an appropriate signal is applied. They are sometimes referred to as *bistable multivibrators*, *binaries*, or *toggles*, and they are widely used as building blocks in counter circuits, shift and storage registers, and control circuits. Among the many additional applications are various decoding, comparison, and timing functions, including both frequency division and generation of multiphase clock pulses for dynamic MOS networks.

The output of a flip-flop is related to the input that was applied at a certain time, and this logic signal can be stored temporarily or permanently even though the input may change. Control of data into and out of a flip-flop is accomplished by means of clock pulses and transmission gates. A flip-flop is an example of a *sequential circuit*, defined as one in which the logic signal at the output depends on some sequence of signals that occured earlier. Another example is the dynamic MOS configuration that was examined in Sec. 8-3. Clearly, sequential circuits have memory capability. On the other hand, the output of a combinational circuit consisting of combinations of static logic gates depends only on the inputs present at the moment, and there is no true memory.

9-1. THE RS LATCH

Basically, a *latch* is a bistable network with an output logic state Q determined by one or more inputs, with this state maintained indefinitely when the inputs are

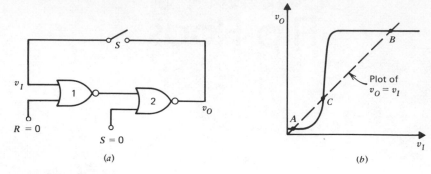

Figure 9-1 Two-stage NOR network and transfer characteristic. (a) Two-stage NOR network. (b) Static characteristic.

turned off in some allowed manner. Usually, a second output gives the comple-ment $\bar{Q}$ of the output. Operation is asynchronous, because Q changes whenever the input logic states have certain specified values. The RS latch has two inputs, with the output state Q depending on the possible combinations of the inputs.

In Fig. 9-1a is shown a two-stage NOR network. Because the inputs labeled R and S are zero, the network with switch S open is actually a two-stage inverter configuration. The output state is that of the input. The static voltage transfer characteristic of Fig. 9-1b is a combination of the transfer characteristics of two inverters in cascade. Now suppose the input signal is removed and the output terminal is connected to the input by closing the switch S. This makes $v_I = v_O$, and a plot of this equation is the dashed line of Fig. 9-1b. Clearly, the static operation point must be at one of the intercepts labeled A, B, and C.

Operating point C is unstable. Suppose operation is at C and an infinitesimal noise voltage causes the operating point to shift slightly. The circuit connections are such that v_O and v_I are equal at each instant of time. Therefore, the shift is along the straight line away from C, with both v_O and v_I changing precisely the same amount. Although operation is no longer on the static transfer character-istic, the fact that its slope dv_O/dv_I at C is greater than unity indicates that the change in v_I *tends* to produce a larger change in v_O. Accordingly, transients cause the instantaneous operating point to continue to move along the dashed line away from C until a stable condition is reached at A or B.

At either A or B any small transient that shifts the operating point is counter-acted by transients developed within the gates. This is because the slope of the transfer characteristic at each point is less than unity. Thus there are two stable outputs: the one at A is low-level and that at B is high-level. Which state exists is uncertain, depending on network parameters and initial transients.

In Fig. 9-2a is shown the network of Fig. 9-1a, rearranged and with the switch closed. Note the cross-coupling and the symmetry. We have determined that the

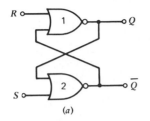

R	S	Q	$\bar{Q}$
0	0	no change	
0	1	1	0
1	0	0	1
1	1	not allowed	

(b)

Figure 9-2   RS NOR latch and truth table. (a) Network. (b) Truth table.

output state is ambiguous when R and S are both logical zeros. Positive control is essential, and this is the function of the RS inputs.

As indicated by the truth table of Fig. 9-2b, when the *SET* input S is 1 with R zero, the output Q of the latch is 1, and the latch is said to be *set*. When the *RESET* input R is 1 with S zero, Q is zero and the latch is *reset*. With both R and S zero, the inputs have no effect, just as though they were removed. For this case the output Q is unchanged from its previous state. With both inputs at 1, both outputs are zero. This contradicts the requirement that one output is the complement of the other. Therefore, this input condition is *not allowed*. When it cannot be avoided, a JK flip-flop should be used. Verification of the truth table is left as an exercise.

In Fig. 9-3 is an RS latch implemented with NAND gates in place of NOR gates. The use of bars over R and S signifies that an input must drop to a low level to give the desired output state. Inputs of 1 and 1 have no effect on the output, and inputs of 0 and 0 are not allowed because they give identical outputs of 1.

Some form of latch is used in every flip-flop. An integrated circuit called a latch is one designed primarily for bit storage. A typical application is that of a holding register in a multiregister system, where the data are stored temporarily before transfer to another register. A clocked D type latch has only a single input with input data controlled by clock pulses. Because the output is coupled directly to the input, the usual one-bit delay of a flip-flop is absent. Its use is mainly for temporary storage (see P9-2).

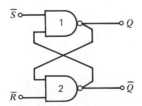

$\bar{R}$	$\bar{S}$	Q	$\bar{Q}$
0	0	not allowed	
0	1	0	1
1	0	1	0
1	1	no change	

Figure 9-3   RS NAND latch and truth table.

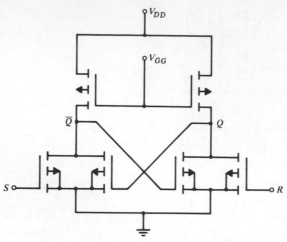

Figure 9-4   RS PMOS latch.

The RS latches can be made by interconnecting properly two NOR or two NAND gates. For example, integrated-circuit SN7402 is a TTL quad two-input NOR network that can be used to make two RS NOR latches of the form of Fig. 9-2. Perhaps the simplest of all latch configurations is the PMOS RS latch of Fig. 9-4. It has six p-channel enhancement-mode MOSFETs and no resistors or capacitors.

Integrated circuit CD4043A is a CMOS quad tri-state RS latch. Because a latch stores one bit, the IC provides independent storage for four bits. The logic diagram and truth table are shown in Fig. 9-5. Each NOR gate is a two-input CMOS network similar to that of Fig. 8-17a, and the inverters are identical

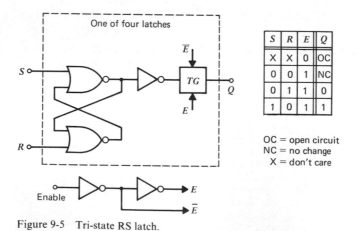

S	R	E	Q
X	X	0	OC
0	0	1	NC
0	1	1	0
1	0	1	1

OC = open circuit
NC = no change
 X = don't care

Figure 9-5   Tri-state RS latch.

to that of Fig. 8-14. The inverter at the latch output provides buffering. Following this inverter is a transmission gate. A logical 1 at the *enable* terminal E connects the latch state to the output. The same signal is supplied to the other transmission gates, and the four bits are delivered simultaneously to four data bus lines. A logical 0 disconnects the latches, giving open-circuited outputs. Consequently, there is a *tri-state* output, with possible states of 0, 1, and the open-circuit condition.

The purpose of the enable input is to allow the same four output bus lines to be used by other identical integrated circuits. A typical application consists of a combination of four of these circuits employed to store four words of four bits each. The outputs can be connected to common busses without interaction by supplying enabling signals at different times.

SWITCH BOUNCE

In many systems an operator pushes a button to initiate a sequence of events. After momentary contact the button is released, springing back to its original position, and an electrical pulse is generated. However, sometimes the switch after making the initial contact momentarily loses the connection, and then reestablishes it. In fact, switch vibration can cause this to happen a number of times in a fraction of a second. This bounce effect generates multiple pulses that can be troublesome.

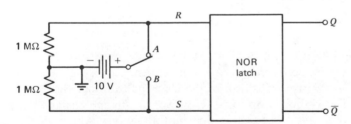

Figure 9-6   Switch-bounce eliminator.

The latch shown in Fig. 9-6 is used to eliminate switch bounce. Suppose the switch is pushed from A to B and returned to A by a spring. Let us assume that multiple pulses are generated at B. However, these do not occur at the output terminals of the latch. At the instant of initial contact at B, inputs R and S are 0 and 1, respectively, and the corresponding outputs Q and $\bar{Q}$ are 1 and 0. The bouncing at B causes the logic state of S to fluctuate, but R remains at 0. Because of its inherent memory, the output of the latch remains fixed when both R and S are zero. A NAND latch can be substituted for the NOR latch provided the polarity of the battery is reversed.

9-2. THE RS FLIP-FLOP

The RS flip-flop has asynchronous reset and set inputs, similar to the latch. The difference is the clock input, as shown in the logic diagram of Fig. 9-7. The *clear* and *preset* controls are optional. They can be used to fix the outputs in desired states regardless of the RS inputs, and there is no effect when both are zero as indicated. We note that the inputs are admitted only when a clock pulse occurs, assuming a pulse denotes a logical 1. In the absence of a clock pulse the RS inputs have no effect, and the outputs of the flip-flop are constant. In digital systems a *clock* is a pulse generator used to synchronize the timing of switching circuits.

When the clock pulse rises from low to high, the RS inputs are fed into the latch, and the output Q may or may not change depending on the logic states

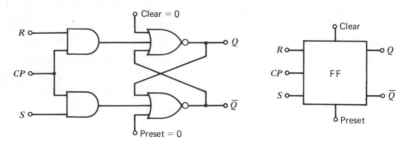

Figure 9-7   RS flip-flop and symbol.

of R and S. When the clock pulse drops back to the low-level, there is no change, because the latch inputs are both zero. Designs are frequently such that flip-flop operation occurs at a negative transition of the clock pulse rather than at a positive transition. The choice is optional. However, *a flip-flop cannot shift states at both edges of a pulse*. The design does not allow this. Although the actual configuration of Fig. 9-7 is simply one of many possible arrangements, the functional operation of all RS flip-flops made with a single latch is similar. Clearly, the AND gates of Fig. 9-7 could be replaced with clock-operated transmission gates.

Numerous applications require a cascade of flip-flops designed so that stored bits are transferred from one to the next in a specified time interval. However, a "race" problem occurs when flip-flops such as that of Fig. 9-7 are cascaded. To illustrate, suppose four flip-flops having a common clock are in cascade, each with an output Q of zero which supplies the R-input of the succeeding stage, with $CP = 0$. At the input of the cascade assume R is 0 and S is 1 when a long clock pulse is initiated at time zero. Furthermore, let us assume that all stages change state during a time interval t_1. Then at t_1 the flip-flops have switched and each Q output is 1. However, the S inputs of the last three stages are now logical zeros. Therefore, at time $2t_1$ these have switched a second time, giving respective outputs of 1, 0, 0, and 0. By time $3t_1$ the last two stages have again switched, with the outputs now being 1, 0, 1, and 1. Finally, the last stage is switched at $4t_1$ to a logical 0, and the steady state is reached.

The preceding analysis has been idealized. The individual stages do not switch states simultaneously, and also, the clock pulse may terminate before the steady state is reached. Thus there is uncertainty in the logic states. Clearly, we must isolate the flip-flops so that changes in their states occur in an orderly and controlled manner.

MASTER-SLAVE FF

On way to accomplish the desired result is to form a *single* flip-flop by cascading two flip-flops of the form of Fig. 9-7 with one important change. The clock input of the second stage is now the complement $\overline{CP}$ of the first-stage clock-pulse input CP. Such a configuration is shown in Fig. 9-8. The first cross-coupled latch is the *master*, and the second is the *slave*. The *characteristic table* of Fig. 9-8 relates the results at the output, after a clock pulse, to the inputs prior to the pulse.

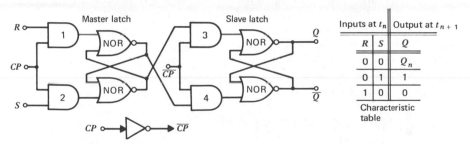

Figure 9-8   RS master-slave flip-flop.

The master and slave latches cannot interact. The circuit is designed so that when a clock pulse *begins* its rise, the master is disconnected from its slave. Then near the top of the leading edge of the pulse, the input AND gates connect the RS inputs to the master latch. When the pulse *begins* to drop, the RS inputs are disconnected, and near the bottom the AND gates 3 and 4 connect the two latches together. Thus RS data enter the master when the clock pulse rises from low-level to high-level. They are transferred to the slave when the pulse drops from high-level to low-level. Only one latch is enabled at any given instant, and the outputs are totally isolated from the data inputs. In many designs the operations are reversed with respect to the clock-pulse transitions.

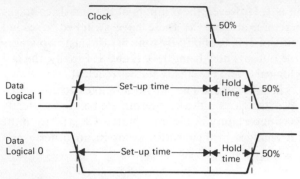

Figure 9-9   Set-up and hold times.

SET-UP AND HOLD TIMES

For all flip-flops it is important to observe the required *set-up* and *hold* times. The set-up time is defined as the time interval between the beginning of a data pulse and the time at which the clock pulse *activates the outputs*. This is illustrated in Fig. 9-9 for a clock pulse *assumed to activate the outputs on a negative transition*. Also shown is the hold time, defined as the interval between the time at which the clock pulse activates the output and the end of the data pulse. Note that the times are measured between the midpoints of the pulse edges. For example, with a clock pulse of 50 ns width, a data sheet might specify that a logical bit must be present at least 20 ns prior to the clock-pulse edge that activates the output. This is the set-up time required for proper operation of the flip-flop. The time that the logical bit must be held after the clock-pulse edge is the minimum hold time. It is frequently specified as zero.

9-3. THE JK FLIP-FLOP

A disadvantage of the RS flip-flop is the not-allowed condition of both inputs equal to 1. The JK flip-flop eliminates this problem. A possible configuration is shown in Fig. 9-10. The two connections from the outputs to the inputs, sketched with heavy lines, convert the RS flip-flop into a JK type. With one exception the performance is identical with that of the RS flip-flop. When both J and K are 1 at the beginning of a clock pulse, *the output switches state* at the end of the pulse, as does its complement.

It is possible to make a JK flip-flop with a single latch. However, the "race" problem described in the last section with regard to cascaded single-latch flip-flops is especially troublesome. Connections from output to input, make a "race-around" likely. There. are various ways to eliminate this situation, such as the use of delay lines, but the master-slave concept is the one usually employed in integrated circuits.

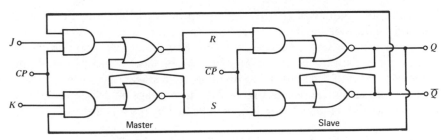

Figure 9-10   JK master-slave flip-flop.

The standard symbol and characteristic table are shown in Fig. 9-11. In the table the eight possible combinations of logic states for J, K, and Q_n at the beginning of a clock pulse at time t_n are given along with the output Q_{n+1} occurring immediately after the pulse. Preset and clear inputs are often provided as indicated on the symbol.

FREQUENCY DIVISION

One of the many applications of the JK flip-flop is frequency division. Suppose we wish to generate a pulse train having a frequency one-half that of the periodic clock pulses. This is easily accomplished with a JK flip-flop. The JK inputs are connected to the dc voltage of a logical 1, and the periodic pulses are supplied to the clock input. During each pulse the output switches state, because both J and K are 1. The change occurs on either a positive or negative transition but not both. Thus the output changes only half as fast as the clock input, and the frequency is divided precisely by two. This is illustrated by the top two waveforms of Fig. 9-12.

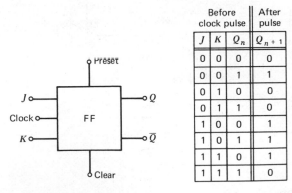

Before clock pulse			After pulse
J	K	Q_n	Q_{n+1}
0	0	0	0
0	0	1	1
0	1	0	0
0	1	1	0
1	0	0	1
1	0	1	1
1	1	0	1
1	1	1	0

Figure 9-11   JK flip-flop symbol and characteristic table.

By feeding the output of the divide-by-two network into the clock input of a second JK flip-flop having J and K equal to 1, the frequency is again halved. Any number of such stages can be cascaded. For example, a popular crystal frequency for wristwatches is 32.768 kHz. When clock pulses having this frequency are fed into a cascade of 15 divide-by-two stages, the output consists of one pulse per second. Shown in Fig. 9-12 is a three-stage binary ripple counter that can be used for frequency division, as indicated by the waveforms.

THE RIPPLE COUNTER AND T FLIP-FLOPS

A counter is a network that provides outputs indicating the number of input pulses received. Counters are essential to the operation of most digital equipment. A common and elementary type is the *ripple counter* of Fig. 9-12. Let us assume that all outputs are initially zero and that only a negative-slope pulse edge can trigger a flip-flop. As shown by the waveforms, each time the clock input drops, the state of Q_1 changes. Each time the Q_1-pulse drops, the state of Q_2 changes, and finally, each time the Q_2-pulse drops, the state of Q_3 changes. The flip-flop transitions ripple through from one flip-flop to the next in sequence.

Suppose there is only a single clock pulse. After this, the respective states of Q_1, Q_2, and Q_3 are 1, 0, and 0. If there are five pulses, the states become 1, 0, and 1, as indicated by the waveforms of the illustration. With seven pulses the states are all logical ones. Thus the output states indicate the number of pulses, and there is no ambiguity from 0 to 7 (see P9-9). After an eighth pulse, the states are zero. Because this condition is the same as for no pulses, the counter cannot

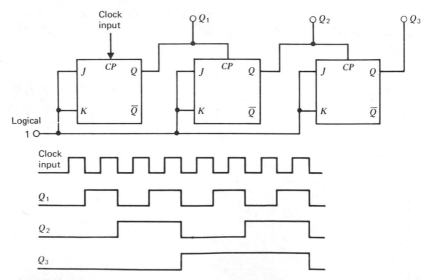

Figure 9-12   Three-stage binary ripple counter.

read this or any larger number. By cascading many stages much larger counts are possible. An *n*-stage configuration can count up to 2^n pulses, and this number is over two million for 21 stages.

Each JK flip-flop of the ripple counter of Fig. 9-12 is connected so that it operates as a *T type* flip-flop. This configuration has a single input T in addition to the clock terminal, and the flip-flop *changes its output state at the proper clock-pulse edge whenever the* T *input is a logical* 1. No switching occurs with an input of 0. The T, or *trigger*, input is the combination of the JK terminals connected together. When periodic pulses are supplied to the clock terminal with a logical 1 present at the input, the back-and-forth switching of the output logic states is called *toggling* (see P9-11). The T flip-flop has output Q and frequently $\bar{Q}$. Its truth table is that of the JK flip-flop with J and K equal. Clear-preset controls are optional.

The ripple counter is asynchronous, because the flip-flops are controlled by different clocks. There are various kinds of synchronous counters. These can operate faster because of the elimination of the flip-flop delays of the ripple counter. Most can be used for frequency division. Some are designed for division by N, with the number N programmed by the user, and N may typically be any integer from 2 to 10. Flip-flops change states synchronously with the clock pulse.

REGISTERS

When a number of flip-flops are grouped together and treated as an entity, a *register* is formed. The design is often such that each stored bit shifts one place to the right whenever a clock pulse is applied, and this type of network is known as a *shift register*. Shown in Fig. 9-13 are four-stage JK and RS shift registers.

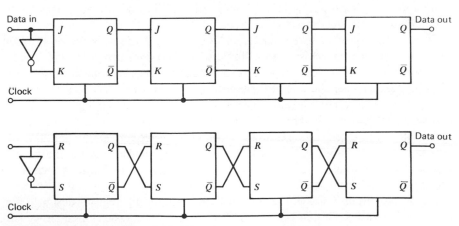

Figure 9-13   Four-stage JK and RS shift registers.

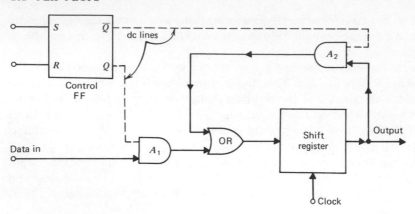

Figure 9-14   Circulating register.

The inverter at each input provides signals at the flip-flop terminals that are complements of one another. Each register can store four bits of a data word.

In Fig. 9-14 is the diagram of a *circulating* register. Suppose the output Q of the control FF is a logical 0 with both R and S zero. The output of AND gate A_1 is zero, and AND gate A_2 and the OR gate act as conducting transmission gates. Each stored bit of the register shifts to the right through one flip-flop when a clock pulse occurs, with the bits at the output returned to the input. The bits circulate around the loop indefinitely.

To enter data, a pulse is applied to the SET input of the control latch and the output state Q is switched to a logical 1. Both R and S are again zero. However, data now enter through the AND gate A_1 with the loop opened at gate A_2. As soon as the proper number of bits have been stored in the register, a pulse applied to the reset input switches the output Q of the control latch back to its original state, and the data circulate (see P9-8).

Data normally enter a register in *serial* form, with the data pulses in time sequence fed into the register on a single wire. The data can be taken serially from the output. However, some registers have connections at the output of each flip-flop that transfer the bits of the stored word *simultaneously* to appropriate decoders. The conversion is said to be *serial-to-parallel*. Desired logic operations can be performed on the bits simultaneously. Parallel operation is faster than serial, but more circuitry is required. *Parallel-to-serial* conversion can also be accomplished by a shift register with the individual bits fed into the individual flip-flops at the same instant by means of preset inputs. Clock pulses then shift the word out in serial form.

Integrated circuit CD4035A is a CMOS four-stage shift register designed for parallel-in and parallel-out operation. The stored bits can be shifted either left or right on clock pulses. Serial inputs are possible via $J\bar{K}$ logic, which signifies that the $\bar{K}$ input is inverted ahead of the K terminal. This feature is useful in counting and sequence-generation applications. When the $J\bar{K}$ inputs are connected together, the first stage is a D flip-flop, which is examined in the next section. All flip-flops are master–slave (MS) types with *static* operation, in contrast to the *dynamic* mode of multiphase configurations such as that of Fig. 8-7 of Sec. 8-3.

9-4. CMOS FLIP-FLOP CIRCUITS

D FLIP-FLOP

RS, JK, and T flip-flops have been examined. Another configuration of importance is the *D type*, with D representing *delay*. The output after a clock pulse equals the input before the pulse. In Fig. 9-15 are shown the symbol and characteristic table. Optional are the clear–preset inputs and the complement $\bar{Q}$ of the output. A D flip-flop can be made from a $J\bar{K}$ flip-flop by connecting the J and $\bar{K}$ inputs, with the connection serving as the data input. When periodic pulses are applied to the clock input, the output is that of the input delayed by one clock pulse.

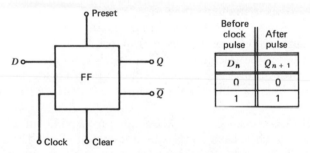

Before clock pulse		After pulse
D_n		Q_{n+1}
0		0
1		1

Figure 9-15   D flip-flop and characteristic table.

A clocked *D latch* differs from a D flip-flop in that the one-bit delay is eliminated. The network is designed so that when the clock pulse triggers the gate, the output is coupled directly to the input D, and Q equals D. The output is then held, or *latched*, in this state until the next pulse triggers the gate. The clock simply acts as an enable input to the latch. It has important applications in registers, especially for temporary data storage.

The logic diagram of a CMOS clocked D latch is shown in Fig. 9-16. Transmission gate TG_2 turns *off*, and TG_1 then turns *on*, during a clock-pulse rise from low to high. The reason TG_2 turns *off* is to prevent the output at Q_2 from interacting with the data input. When TG_1 turns *on*, the input bit enters

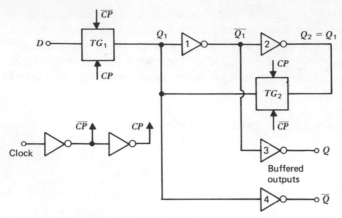

Figure 9-16   CMOS clocked D latch.

the latch. This stored bit appears at the buffered output terminal Q with very little delay. The propagation delays of the inverters are small compared with the clock period.

When the clock pulse drops from high to low, TG_1 cuts *off*, TG_2 then turns *on*, and the bit remains stored until the next pulse appears. The reason for the inclusion of inverter 2 and TG_2 is to maintain the proper stored charge on the insulated gate terminals of inverters 1 and 4. If they were eliminated, any charge stored on these insulated gates would soon be lost by leakage. With TG_2 *on*, inverters 1 and 2 constitute a cross-coupled latch. Transmission gates are used instead of NOR gates to control the operation. The use of two inverters for the clock circuitry provides buffering to reduce the loading of the clock and to improve the pulse waveforms.

Integrated circuit CD4042A is classified as a CMOS quad clocked D latch. It consists of four separate D latches, each strobed by a common clock. The configuration is that of Fig. 9-16. A *polarity* circuit of two cascaded inverters can be used to program the pulse transition, either positive or negative, that switches the output. The gate propagation delay is typically 50 ns with a 10 V supply and a load capacitance of 15 pF, corresponding to a fan-out of three. In the low state the gate can sink about 2 mA while maintaining an output less than 0.5 V, and in the high state it can supply 2 mA with the voltage held above 9.5. A toggle frequency up to about 8 MHz is reasonable. Typical applications include buffer storage and use as a holding register in digital systems.

A D type master-slave (MS) flip-flop can be made simply by cascading two D latches of the form of Fig. 9-16, with the transmission gates clocked so that only one latch receives data at a time. By replacing inverters 1 and 2 of each latch with NOR gates, preset and clear controls can be added, which are often referred to as *set-reset* controls. Such a configuration is shown in Fig. 9-17, along

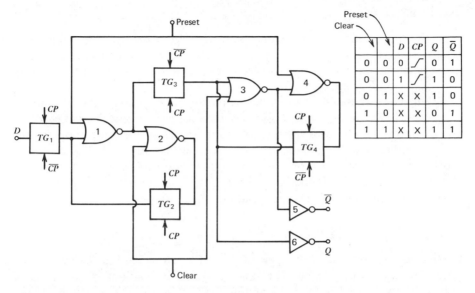

		D	CP	Q	$\bar{Q}$
0	0	0	⌐	0	1
0	0	1	⌐	1	0
0	1	X	X	1	0
1	0	X	X	0	1
1	1	X	X	1	1

Figure 9-17 Logic diagram of D-type master-slave flip-flop.

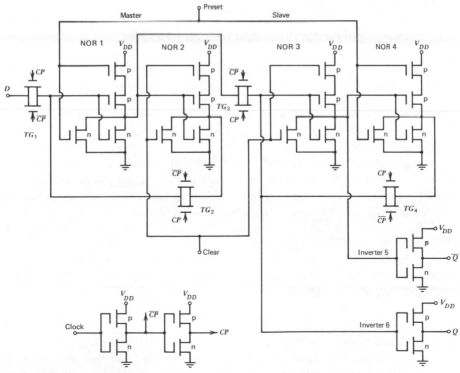

Figure 9-18 D-type master-slave flip-flop circuitry.

265

with the truth table. When a clock pulse rises from low to high, which is a positive transition, the logic level present at the D input becomes the Q output. Data enter the master on negative transitions and are transferred to the slave on positive transitions.

The first two rows of the truth table are those of a D flip-flop, with the symbols under the clock column indicating the level change at which the D input becomes the Q output. The bottom three rows simply represent the truth table for the case in which one or both of the preset–clear inputs is a logical 1. The states and clock transitions marked X have no effect on the output. These are referred to as *don't-care* conditions. When a logical 1 is present at a preset or clear input, the output is independent of the data input and the clock pulses.

CMOS FLIP-FLOPS

Circuitry accomplishing the logic of Fig. 9-17 is shown in Fig. 9-18 with each gate identified. Both the numbers and the relative positions of the gates of Fig. 9-18 correspond with those of Fig. 9-17. Note the symbol used for the transmission gate. Because transmission through a gate is possible in both directions, the position of the gate terminal is centered. All transistors are enhancement-mode devices.

Integrated circuit CD4013A consists of dual D type flip-flops. Each of the two identical flip-flops has the circuitry of Fig. 9-18. Operation is static, rather than dynamic, with the state of the flip-flop retained indefinitely when the clock input is constant at either a high-level or low-level voltage. A toggle rate of about 8 MHz is typical with a 10 V supply, and the respective high-level and low-level output impedances are typically 400 and 200 ohms. The dc supply V_{DD} should be between 3 and 15 V. By connecting the $\bar{Q}$ output to the D input the flip-flop toggles at each clock pulse. Applications include shift registers, counters, and control circuits.

A D flip-flop can be converted to a JK configuration in a number of ways, one of which is shown in Fig. 9-19. In addition to the logic arrangement, the figure includes the characteristic table, assuming zero preset and clear inputs. Let us consider the first row of the table. With the present state of Q equal to 0 and the input at J a logical 1, the outputs of gates 1 and 2 are logical zeros regardless of K, which is a don't-care state. Therefore, the output of NOR gate 3 is 1. As this is the D input, Q becomes 1 after the positive pulse transition. Verification of the other rows is left as an exercise. The process is simplified by recognizing that the output D of NOR gate 3 is given by

$$D = \overline{KQ + \overline{J + Q}} \tag{9-1}$$

with Q denoting the present state. Output D is the next state of Q.

Figure 9-20 shows suitable circuitry that performs the logic of (9-1), along with the proper connections to the D flip-flop. All p-channel transistors have a

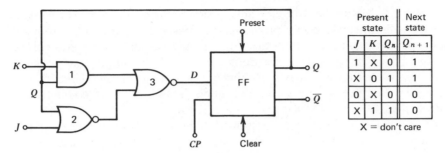

Present state			Next state
J	K	Q_n	Q_{n+1}
1	X	0	1
X	0	1	1
0	X	0	0
X	1	1	0
	X = don't care		

Figure 9-19   Conversion of a flip-flop from D-type to JK.

common substrate connected to V_{DD}, and the substrates of the n-devices are connected to the respective sources. Transistors Q_1 through Q_4 comprise a NOR gate with inputs Q and J, corresponding to gate 2 of the logic diagram of Fig. 9-19. The rest of the circuit implements the logic of gates 1 and 3 of Fig. 9-19, although the individual AND and NOR gates are not identifiable.

Integrated circuit CD4027A consists of dual CMOS JK master-slave flip-flops with preset–clear capability. The detailed circuitry of each of the flip-flops

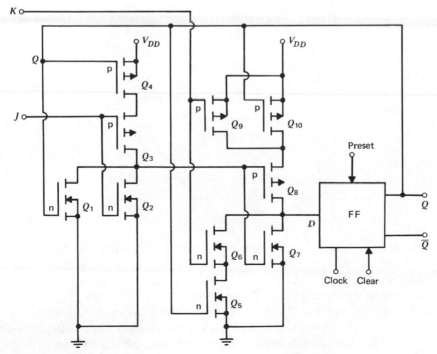

Figure 9-20   CMOS circuitry for flip-flop conversion from D to JK.

is that of Fig. 9-20 with the D flip-flop having the circuitry of Fig. 9-18. In addition, there are buffering inverters for providing the clock pulses CP and $\overline{CP}$ from the clock input. The connection from the Q output to the input is taken ahead of the inverting buffer that gives $\overline{Q}$. Operation is static, with the output state retained indefinitely when the clock is fixed at either high level or low level. The characteristic table is that given in Fig. 9-19. Applications include registers, counters, and control circuits.

The circuits that have been presented here are representative of those commonly used in CMOS latches and flip-flops. BJT flip-flops are generally made by interconnections of logic gates that are somewhat standard. I^2L is quite different in many ways, and these flip-flops are examined after a brief comparison of the major logic families.

The speed of a logic system depends on both the clock rate for flip-flop operation and the average propagation delay per gate. Shown in Table 9-1

Table 9-1   Comparison of Logic Families

Logic system	Clock rate (MHz)	Propagation delay per gate (ns)	Power dissipation per gate (mW)
Standard TTL	15–30	12	12
High-speed TTL	30–60	6	22
Schottky TTL	200	2	15
4-ns ECL	60–120	4	40
2-ns ECL	200	2	60
1-ns ECL	400	1	60
I^2L	4–120	5–100	0.005–0.01
PMOS	2	300	0.2–10
(SG) NMOS	15	20	0.2–1
CMOS	10	30	1 mW at 1 MHz

are some typical values for the common logic systems. The flip-flop clock rate is usually less than one-half the frequency listed, which is a typical value of the maximum toggle rate that can be used. The numbers are only approximations, because the actual values depend on the voltages, the average fanout, the loading conditions, and the switching factor. In addition, technology has a large effect on the performance characteristics of the MOS systems, with the use of silicon gates (SG), ion implantation, depletion loads, and silicon-on-sapphire (SOS) being particularly important.

9-5. I²L LATCHES AND FLIP-FLOPS

Integrated-circuit I²L latches are attractive for use in memory arrays and flip-flops. For memory applications the densities are comparable with those obtainable from dynamic MOSFET configurations, and performance is superior with regard to power dissipation and speed. The equivalent circuit of an RS latch is presented in Fig. 9-21.

The latch outputs most likely will go to base inputs of transistors not shown. Let us recall that a direct connection between any two base terminals is prohibited, because this would place current sources in parallel with associated current-hogging problems. Accordingly, *it is not possible to use the cross-coupled outputs for additional connections to base inputs.* Multiple collectors eliminate any problem in this regard.

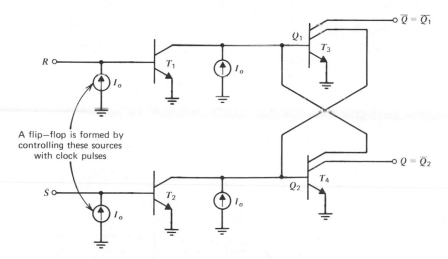

Figure 9-21   An I²L RS latch.

Each base-collector junction acts as an inverter and each junction of two or more collectors constitutes an implied-AND logic function. The collectors of transistors T_1 and T_4 are joined at the junction labeled Q_1. Therefore,

$$Q_1 = \overline{R}\overline{Q}_2 = \overline{R + Q_2} \tag{9-2}$$

In a similar manner we find that

$$Q_2 = \overline{S}\overline{Q}_1 = \overline{S + Q_1} \tag{9-3}$$

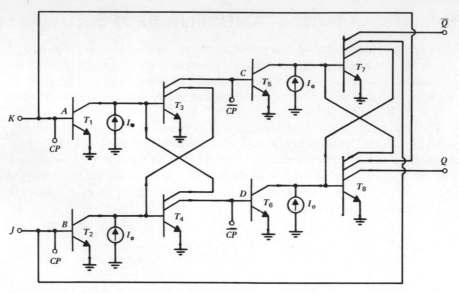

Figure 9-22   JK I²L flip-flop.

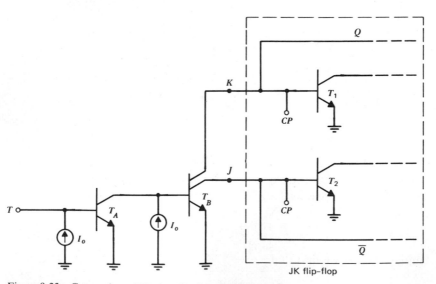

Figure 9-23   Conversion of JK to T flip-flop.

Either from these logic relations or from examination of the circuitry, the truth table is easily determined to be that of the RS cross-coupled NOR latch given in Fig. 9-2, noting that Q is the complement of Q_2 and $\bar{Q}$ is the complement of Q_1. The logic configuration equivalent to Fig. 9-21 consists of two inverters followed by two cross-coupled NAND gates (see P9-16).

In Fig. 9-7 a latch is converted to a flip-flop by the addition of a clock and two AND gates. We accomplish this in Fig. 9-21 simply by converting the injection currents at the inputs to clock current pulses. The required AND logic is implemented by the wired junctions. By cascading two such stages with the clock pulses of one being the complements of those of the other, an RS master-slave flip-flop is formed. The logic diagram of Fig. 9-10 indicates a way to convert an RS flip-flop to JK. This is applicable to I^2L, and the input AND gates can be replaced with wired-AND connections. Shown in Fig. 9-22 is a JK master-slave flip-flop made in this manner. Note that each output transistor must have three collectors.

A trigger flip-flop is formed from JK by connecting the inputs so that $J = K$, and the junction of the terminals is the T input. However, we cannot directly connect J to K in Fig. 9-22, because the outputs Q and $\bar{Q}$ and also the base terminals of T_1 and T_2 would be joined, and neither is acceptable. A scheme for surmounting this problem is shown in Fig. 9-23. Note that once again multiple collectors allow a connection but avoid a junction of base terminals. If the connection from the top collector of transistor T_B to K is replaced with one from a second collector of T_A to K, the T flip-flop becomes a D type (see P9-18).

Another way to make a D type is to connect an inverter across the inputs of an RS flip-flop so that $R = \bar{S}$ with $S = D$. With reference to Fig. 9-21 this is accomplished simply by adding a second collector to transistor T_2 and connecting it to terminal R. The S terminal is used for data input, giving a D type flip-flop. It may be desirable to add additional input circuitry to provide buffering for isolating the data input from the clock pulses. Such a network using transistors has been fabricated with a chip area of only 18.4 mils^2, or about 0.012 mm^2. The six transistors are formed in a row alongside a current rail which provides I_o to four of the devices. In addition to the D input there are two inputs for the clock pulses. Surface metallization provides the necessary interconnections.

9-6. SCHMITT TRIGGER

At various points within a flip-flop the voltages are determined not only by the inputs but also by the outputs. This is a consequence of the fact that the outputs are connected in some manner either to the inputs or to certain terminals inside the network. Hence there is *feedback*, and the type used in the flip-flops we have examined is *regenerative*, or *positive*, feedback. It is the regenerative feedback that enables the flip-flop to exhibit memory. There are other forms of regenerative switching circuits having two stable output states. One of particular importance in both digital and analog circuits is the *Schmitt trigger*.

Such a circuit is shown in Fig. 9-24a with numerical values specified for the components. The static voltage transfer characteristic is that of Fig. 9-24b. With reference to these let us examine qualitatively the behavior as v_I is increased from zero. The resistors have been selected so that transistor Q_2 is biased in the active region with v_I zero. Because of the positive voltage v_E, the base-emitter junction of Q_1 is reverse biased, and Q_1 is off. Analysis reveals that

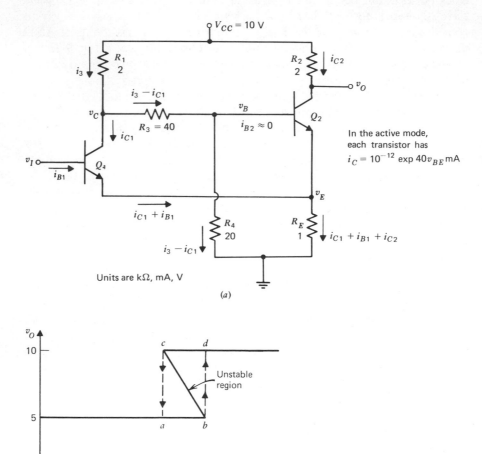

(a)

(b)

Figure 9-24   (a) A Schmitt trigger circuit. (b) Static voltage transfer characteristic.

the output is 5 V. The respective node voltages v_B and v_E are 3.2 and 2.5, which gives 0.7 V for v_{BE2}.

As v_I is increased from zero, the currents of the network remain unchanged until v_I becomes 0.6 V greater than v_E, or 3.1 V. This is point b on the characteristic. Transistor Q_1 is on the threshold of conduction with $v_{BE1} = 0.6$, and Q_2 is conducting in the active mode with $v_{BE2} = 0.7$.

Suppose the input voltage is increased about 0.01 V from its value at b and *held constant*. The small collector current that now flows through Q_1 increases the voltage across R_1, reducing v_C. Because v_B is one-third of v_C, it also drops, causing i_{C2} to decrease. The current through R_E consists of an increased i_{C1} and a decreased i_{C2}. However, the net current is slightly lower, and the node voltage v_E drops. As usual in this trigger circuit, the parameters have been chosen so that v_B declines at a faster rate than does v_E, assuring that v_{BE2} and i_{C2} do indeed drop as assumed. The reduction in v_E increases v_{BE1}, which equals the constant input voltage v_I less v_E. Consequently, i_{C1} rises even more, and this in turn further reduces v_{BE2} and i_{C2}. The process continues until i_{C2} becomes zero. Node voltage v_E decreases only slightly, because it is clamped at a diode drop below v_I.

The transition occurs rapidly. The instantaneous operating point (v_I, v_O) moves directly along the vertical path from b to d during the transient. Switching is produced by the regenerative feedback through R_E that causes v_{BE1} to rise as i_{C2} falls. Throughout the interval, transistor Q_1 remains in the active mode and Q_2 moves from active to *off*. The drop that occurs in v_E is only about 0.1 V, but this small change raises v_{BE1} from effective cutoff at about 0.6 V to active-mode operation at 0.7 V. Node voltage v_B drops considerably more, from 3.2 to 1.7 V, and the emitter junction of Q_2 is reverse biased by 0.7 V at the end of the transient (see P9-21). If v_I is increased to 4 V, the input transistor becomes saturated. The output remains constant at 10 V.

Now suppose v_I is reduced. As v_I drops, v_E also drops because it is clamped one diode drop below v_I. The node voltage v_E drops faster than does v_B. At point c with $v_I = 2.35$ the base-emitter voltage of Q_2 reaches the threshold of conduction. Suppose v_I is decreased perhaps 0.01 V below its value at c. The increased v_{BE2} gives a current i_{C2} which increases v_E. This lowers v_{BE1} and hence i_{C1}. The reduced i_{C1} produces an increase in v_B. The cycle continues with v_B rising faster than v_E until Q_1 is cut off. The instantaneous operating point moves along the vertical path from c to a. As before, the rapid switching is caused by regenerative feedback through R_E. Analysis shows that during the transient the node voltage v_B rises 1.1 V, v_E rises 0.9, v_{BE2} increases from 0.52 to 0.71, and v_{BE1} drops from 0.70 to -0.16 V (see P9-22).

Let us now examine the static transfer characteristic. In the transition region of Fig. 9-24b between points b and c both transistors are in the active mode. Only at each end point is a device at the boundary between the active mode and cutoff. For simplicity we assume identical transistors with sufficiently large values of β_F to justify neglecting the active-mode base currents. Also, we assume that the active-mode Ebers–Moll equations are

$$i_{C1} = 10^{-12}e^{40v_{BE1}} \qquad i_{C2} = 10^{-12}e^{40v_{BE2}} \qquad (9\text{-}4)$$

with the currents in mA. Units of mA, kΩ, and V are used in this analysis.

With reference to the network of Fig. 9-24 we find from a single loop equation through V_{CC}, R_1, R_3, and R_4 that

$$i_3 = 0.1613 + 0.9677i_{C1} \qquad (9\text{-}5)$$

By writing two additional loop equations and utilizing (9-5) to eliminate i_3, we determine the emitter-junction voltages to be

$$v_{BE1} = v_I - i_{C1} - i_{C2} \qquad (9\text{-}6)$$

$$v_{BE2} = 3.226 - 1.645i_{C1} - i_{C2} \qquad (9\text{-}7)$$

The four equations of relations (9-4), (9-6), and (9-7) have five variables including v_I. It is not difficult to eliminate the base-emitter voltages and i_{C2} (see P9-23). We eliminate i_{C2} by means of the relation $2i_{C2} = 10 - v_O$ and use the developed equations to eliminate i_{C1}. Solving for v_I gives

$$v_I = 4.20 - 0.196v_O - 0.0152 \ln(10 - v_O)$$
$$+ 0.025 \ln[0.304v_O - 1.49 - 0.0152 \ln(10 - v_O)] \qquad (9\text{-}8)$$

Any desired value of v_O between 5 and 10 V can be selected, and (9-8) can be used to calculate the corresponding input voltage v_I. This procedure was followed in determining the points used in plotting the static characteristic of Fig. 9-24b. Although the transfer characteristic indicates three possible static output states corresponding to a value of v_I between the thresholds, the intermediate state is unstable. Accordingly, the Schmitt trigger is bistable in this range. The difference between the threshold values of v_I is known as *hysteresis*. For the circuit of Fig. 9-24 the hysteresis voltage is 0.75.

In designing the network a bias current of 2.5 mA was selected for transistor Q_2. A larger current increases power dissipation, whereas a smaller value requires larger resistors objectional in integrated circuits. For a low-level output of 5 V, R_2 must be 2 kΩ. The selection of R_E determines the value of v_I that switches the output from low-level to high-level. For a 3 V threshold, v_E should be 2.4. We chose R_E to be 1 kΩ. Subtracting the drops across R_2 and R_E from V_{CC} gives 2.5 for the collector-emitter voltage of Q_2, which is adequate for active-mode operation.

Resistors R_3 and R_4 reduce the supply voltage to the proper value at the base of Q_2. This voltage v_B is the sum of v_{BE2} and v_E, or about 3.2 V when Q_2 is conducting. Accordingly, R_4 must be approximately one-half of R_3. Choosing large values for these resistors minimizes power dissipation, and 40 and 20 kΩ are suitable. Finally, R_1 is selected to give the desired hysteresis. When v_I is reduced with operation in the high-level state, the collector current i_{C1} drops, and when the voltage across R_1 drops sufficiently, Q_2 switches on. Because a large resistance gives switching at a low value of i_{C1}, the input threshold voltage of the high-level state is lowered when R_1 is increased and raised when it is decreased (see P9-24). Resistance R_1 was selected for a threshold input voltage of 2.35.

CMOS integrated circuit CD4093B consists of quad two-input NAND Schmitt triggers. Each of the four circuits is represented by the logic symbol of Fig. 9-25. Also shown is the transfer characteristic. The output is inverted from that of Fig. 9-24b, and the low state is the voltage of the V_{SS} supply, assumed to be zero. The high-level output is the V_{DD} supply. Each circuit can be operated as a two-input NAND gate. When used for triggering, one input is connected to the supply voltage, and the other is for the trigger voltage v_I. With a 10 V supply the respective threshold voltages V_N and V_P of Fig. 9-25 are 4 and 6 V, giving a hysteresis of 2 V. When v_I rises from a low value, switching occurs at V_P with the subscript indicating the *positive* slope of the input. Switching on a *negative* slope occurs at V_N. A buffered output allows sourcing and sinking 1.8 mA with the output voltage held within 0.5 V of the zero and 10 V states.

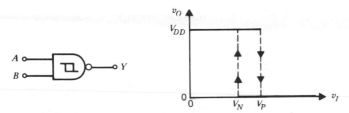

Figure 9-25   Logic symbol and hysteresis characteristic of two-input NAND Schmitt trigger.

Applications of the Schmitt trigger are numerous. It is used in various timing circuits to mark the instant when an input voltage reaches the trigger level. A sinusoidal input voltage is converted into a pulse train at the output. External circuitry can be added to form either a monostable or astable multivibrator, which are examined in Sec. 22-4. Especially important applications involve wave shaping. Poorly shaped pulses supplied to the input are much improved at the output. This is illustrated in Fig. 9-26. The peak of each input pulse must exceed the threshold voltage V_P. The hysteresis is advantageous. It prevents multiple switching caused by superimposed noise, as indicated in the figure. Without hysteresis, the Schmitt trigger would be a comparator of the type discussed in Sec. 23-2. Schmitt triggers made from operational amplifiers, along with applications in waveform generation, are examined in Sec. 23-6.

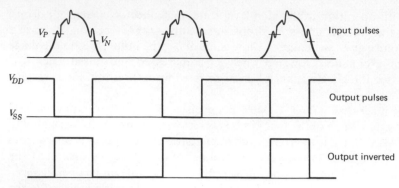

Figure 9-26   Pulse shaping.

9-7. SEQUENTIAL NETWORK DESIGN

In a sequential circuit the output depends on both the present inputs and the previous history of the inputs. Such circuits usually consist of an interconnection of flip-flops and logic gates. The flip-flop outputs often feed into the inputs of the combinational circuits that drive the flip-flops. The gates and the bits stored in the flip-flops before a clock pulse determine the external outputs after the pulse.

FF EXCITATION TABLES

We have learned that the characteristic tables of flip-flops specify the next state when the inputs and the present state are known. In the design process we often know the required transition from the present to the next state, and we need to find the flip-flop inputs that give the transition. For a given change of state the table listing the proper input bits is called a *flip-flop excitation table*.

In Fig. 9-27 are the excitation tables for the four basic types of flip-flops. The columns on the left side of each table show the output $Q(t)$ before a pulse and the output $Q(t + 1)$ after the pulse. On the right side are columns giving the required inputs for each specified transition, with the data obtained from the FF characteristic tables. Note the don't-care inputs. Each X can be either 0 or 1, as convenient.

STATE-EXCITATION TABLES

The operation of a sequential circuit can be described by a *state table*, or *transition table*, which relates the outputs and the next states to the inputs and the present states. All possible transitions are included. In clocked circuits the transition from the present state to the next state occurs during a clock pulse.

$Q(t)$	$Q(t+1)$	S	R
0	0	0	X
0	1	1	0
1	0	0	1
1	1	X	0

(a)

$Q(t)$	$Q(t+1)$	J	K
0	0	0	X
0	1	1	X
1	0	X	1
1	1	X	0

(b)

$Q(t)$	$Q(t+1)$	D
0	0	0
0	1	1
1	0	0
1	1	1

(c)

$Q(t)$	$Q(t+1)$	T
0	0	0
0	1	1
1	0	1
1	1	0

(d)

Figure 9-27   FF excitation tables. (a) RS flip-flop.
(b) JK flip-flop. (c) D flip-flop. (d) T flip-flop.

The form of a state table is shown in Fig. 9-28, which is applicable to a sequential circuit consisting of a combinational network, two flip-flops A and B, one external input X_1, and one external output Y_1. Flip-flop A has outputs A and $\bar{A}$, and the outputs of flip-flop B are B and $\bar{B}$. Data entered in the left three columns represent the eight distinct combinations of present states A and B and external inputs X_1. These represent the possible inputs to the associated combinational network. With one flip-flop and two external inputs X_1 and X_2 the number of rows is also eight. The table for a network with three flip-flops and two external inputs would have 2^5, or 32, rows. Data entered in the columns for the present output Y_1 and the next states A and B depend on specified information about the particular network. Sometimes there are two outputs Y_1 and Y_2, or more.

Present states A B	Input X_1	Present output Y_1	Next states A B
0 0	0	0	0 0
0 0	1	0	1 0
0 1	0	1	0 1
0 1	1	1	1 1
1 0	0	0	0 0
1 0	1	0	1 1
1 1	0	1	0 0
1 1	1	1	0 0

Figure 9-28   State table.

Present states		Input	Present output	Next states		FF input conditions			
A	B	X_1	Y_1	A	B	J_A	K_A	J_B	K_B
0	0	0	0	0	0	0	1	0	0
0	0	1	0	1	0	1	0	0	0
0	1	0	1	0	1	0	1	0	0
0	1	1	1	1	1	1	1	0	0
1	0	0	0	0	0	0	1	0	1
1	0	1	0	1	1	1	0	1	1
1	1	0	1	0	0	0	1	0	1
1	1	1	1	0	0	1	1	1	1

Figure 9-29   State-excitation table.

In many circuits the outputs are those of the flip-flops, in which case the present-output column Y_1 is often omitted.

An extension of the state table is the sequential-circuit *excitation table* of Fig. 9-29, often called a *state-excitation table*, or a *transition-excitation table*. It consists of the state table with the required flip-flop input conditions added to it. Symbols J_A and K_A refer to flip-flop A, and they represent JK inputs *that give the specified transition from the present state A to the next state A*. Data for columns J_A and K_A are determined from the columns for present-state A and next-state A, with the aid of the FF excitation table of Fig. 9-27b. The terms J_B and K_B are similarly defined.

STATE DIAGRAMS

The information of a state table may be represented graphically in a *state diagram*, which contains precisely the same information. In Fig. 9-30 is the state diagram corresponding to the state table of Fig. 9-28. There are four possible flip-flop AB states. These are 00, 01, 10, and 11, and each is represented on the state diagram by appropriate binary numbers within a circle. The directed lines between the circles indicate possible transitions between states that can occur during clock pulses, and each line is labeled with two binary numbers separated by a slash. The first of these is the value of input X_1 that causes the state transition, and the number after the slash is the present output Y_1 before the transition occurs. Output Y_1 is omitted if the external outputs are those of the flip-flops.

A directed line connecting a circle with itself indicates that the state does not change. For example, suppose the state is 00. Then A and B are each 0 and $\bar{A}$ and $\bar{B}$ are each 1. From Fig. 9-28 we observe that for $X_1/Y_1 = 0/0$ the next state is also 00. Therefore, the directed line 0/0 properly connects the circle with itself. For $X_1/Y_1 = 1/0$ the next state is shown in the table of Fig. 9-28 to be 10. Thus the 1/0 directed line goes to the 10 state.

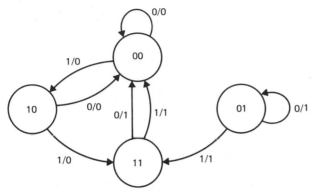

Figure 9-30 State diagram corresponding to table of Fig. 9-28.

The operation of the circuit is described graphically by the state diagram. Suppose the initial state of the diagram of Fig. 9-30 is 00 and the input $X_1 = 1$. The first clock pulse produces a transition to state 10, the second pulse changes the state to 11, and the third returns the state to 00. The cycle repeats itself continuously. There is no way to reach the 01 state, although the circuit could possibly be in this state initially. If the initial state is either 10 or 11 with $X_1 = 0$, the circuit returns to state 00 at the first clock pulse and remains there as long as X_1 is 0.

The design of a sequential network consists of finding a suitable circuit from a specified state diagram. This is examined shortly. In analysis the network is given, and the objective is to deduce the state diagram. It is usually best to find the excitation table first and then to determine the state diagram from the table. The technique is illustrated by the following example.

Example 1

Find the state diagram that applies to the sequential network of Fig. 9-31.

Solution

From the network we deduce that

$$J_A = X_1 \qquad K_A = \overline{X_1}\,\overline{B}$$
$$J_B = X_1 A \qquad K_B = A \qquad Y_1 = B \tag{9-9}$$

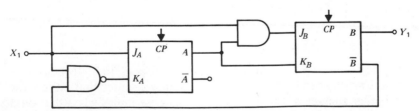

Figure 9-31 Sequential network.

The excitation table is that of Fig. 9-29. The three columns on the left give the eight possible combinations of A, B, and X_1. The column for Y_1 is identical with that of B by (9-9). Next, the FF input conditions are found from (9-9), and these four columns are completed. The next state A is determined from present-state A and inputs J_A and K_A, using the FF excitation table of Fig. 9-27b, and the next state B is found similarly. From the excitation table the state diagram of Fig. 9-30 is deduced.

NETWORK DESIGN

The design procedure is the inverse of the analysis technique. The state diagram must be determined from the specifications. Once the state diagram is known, the design is accomplished in a routine manner as follows:

1. Determine the number of flip-flops to be used from the number of bits inside the circles of the diagram. Also, determine the number of required external inputs X_1, X_2, and so forth. These are specified along the directed lines between the circles.
2. Select flip-flop types and define the FF inputs. For RS flip-flops use R_A, S_A, R_B, S_B, and so on, for the inputs. For other types use J, K, D, and T.
3. From the state diagram write the state table.
4. Expand the state table into an excitation table. The FF input conditions are determined from the present states, the next states, and the FF excitation tables of Fig. 9-27.
5. The combinational-circuit truth table is contained within the state-excitation table. Columns for the FF present-states and the external inputs are the inputs of the combinational circuit. The FF input columns are the outputs of the combinational circuit. Form a Karnaugh map for each FF input, with the variables being the FF present states and the external inputs.
6. From the Karnaugh maps design the combinational network that supplies the inputs to the flip-flops.
7. If there are external outputs other than the states of the flip-flops, develop a map for each output, with the variables being the FF present-states and the external inputs. Use data from the excitation table. Design suitable combinational networks. The following examples illustrate the procedure.

Example 2

An n-bit binary counter is a network that counts pulses up to 2^n. It then resets and counts again from zero. Design a two-bit binary counter that counts the input clock pulses. Include a control input X_1 that determines when the count begins. Design the circuit so that the output state is 00 whenever X_1 is 0.

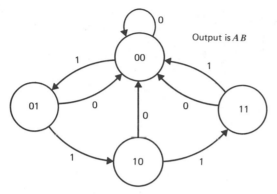

Figure 9-32   State diagram for two-bit binary counter.

Solution

A two-bit counter counts to 4. Therefore, only two flip-flops are required. With $X_1 = 1$, the binary states AB must sequence through 00, 01, 10, and 11 as clock pulses are received, with the process then repeated. Whenever X_1 becomes 0, the state must return to 00 and remain there until X_1 again becomes 1.

The state diagram based on the specifications is shown in Fig. 9-32. The number beside each directed line is X_1, and the output is the binary number AB, with A and B denoting the FF outputs. Although any of the four types of flip-flops can be used (see P9-25), we select the JK type because it is the most flexible with two input terminals and four don't-care states in its excitation table.

The state table is found from the state diagram. It consists of the five left-hand columns of the table of Fig. 9-33. The entries of the next-state columns A and B are read from the state diagram. Then the table is expanded into an excitation table by adding to it the FF input conditions. The inputs J_A and K_A are found from the present-state A, the next-state A, and the JK excitation table of Fig. 9-27. The columns for J_B and K_B are obtained similarly.

Also shown in Fig. 9-33 are the Karnaugh maps for J_A, K_A, J_B, and K_B, along with the simplest Boolean functions. From these the network of Fig. 9-34 is deduced.

Example 3

Design a JK flip-flop from a D flip-flop and NAND gates, using the state-diagram approach.

Solution

The external inputs are J and K, and the single flip-flop with input D has states 0 and 1. The state diagram is deduced from the JK characteristic table of Fig. 9-11. Shown in Fig. 9-35 are the state diagram, the excitation table, the map for the input D of the flip-flop, and the network. The present output Y_1

Present states		Input	Next states		FF input conditions			
A	B	X_1	A	B	J_A	K_A	J_B	K_B
0	0	0	0	0	0	X	0	X
0	0	1	0	1	0	X	1	X
0	1	0	0	0	0	X	X	1
0	1	1	1	0	1	X	X	1
1	0	0	0	0	X	1	0	X
1	0	1	1	1	X	0	1	X
1	1	0	0	0	X	1	X	1
1	1	1	0	0	X	1	X	1

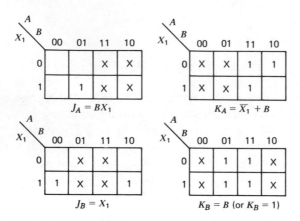

Figure 9-33   Excitation table and maps for Example 2.

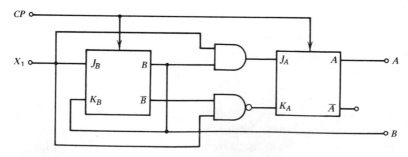

Figure 9-34   Two-bit binary counter.

282

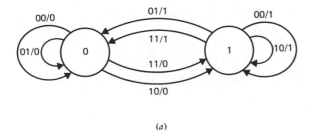

(a)

Present state	Inputs		Present output	Next state	FF input
A	J	K	Y_1	A	D
0	0	0	0	0	0
0	0	1	0	0	0
0	1	0	0	1	1
0	1	1	0	1	1
1	0	0	1	1	1
1	0	1	1	0	0
1	1	0	1	1	1
1	1	1	1	0	0

(b)

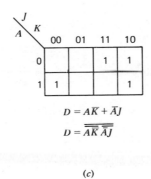

$$D = A\overline{K} + \overline{A}J$$

$$D = \overline{\overline{A\overline{K}}\ \overline{\overline{A}J}}$$

(c)

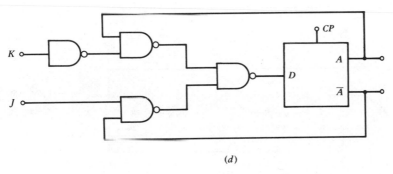

(d)

Figure 9-35 Design of JK flip-flop from D-type and NAND gates. (a) State diagram. (b) Excitation table. (c) Map and logic. (d) JK flip-flop.

shown on the state diagram and also in the excitation table is simply the present state A of the D flip-flop. The next state A of the excitation table is found from the state diagram. Of course, the FF input D is identical to that of the next-state A, as required by the D type excitation table of Fig. 9-27. The logic expression for D, which is found from the Karnaugh map, is used to deduce the network of the figure.

REFERENCES

1. IC Applications Staff of Texas Instruments, Inc., *Designing with TTL Integrated Circuits*, McGraw-Hill, New York, 1971.
2. *RCA COS/MOS Integrated Circuits Databook*, SSD-203C, Solid State Division, Somerville, N.J. 1975.
3. V. H. Grinich and H. G. Jackson, *Introduction to Integrated Circuits*, McGraw-Hill, New York, 1975.

PROBLEMS

Section 9-1

9-1. Shown in Fig. 9-36 is a bistable network with inputs v_S and v_R. Assume a transistor in saturation has v_{BE} and v_{CE} equal to 0.8 and 0.2 V, respectively, and each transistor has a beta of 100. (a) Determine v_{O1} and v_{O2} in volts when v_S is zero and v_R is 4 V. (b) If v_R is now reduced to zero, what are the outputs? (c) Find the outputs with $v_S = 4$ and v_R zero. (d) Repeat for v_S and v_R both equal to 4 V.

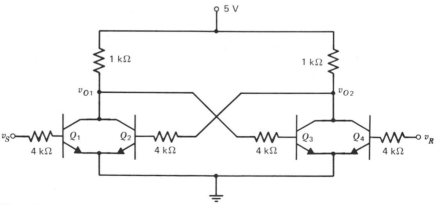

Figure 9-36   Bistable network for P9-1.

9-2. In Fig. 9-37 is the logic diagram of a clocked D-type latch used for data storage. During a clock pulse, which is a logical 1, TG_1 is an open circuit and TG_2 is a low resistance. (a) Give the truth table relating the data input D, the clock state CP, and the outputs Q_1 and Q_2. (b) Sketch the complete circuit diagram using CMOS, with a single clock input terminal.

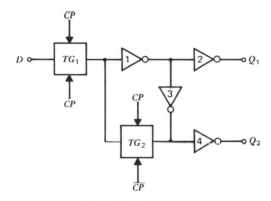

Figure 9-37 D-latch for P9-2.

9-3. Sketch the circuitry of a PMOS RS latch using two cross-coupled positive-logic NAND gates. Include the truth table. Let V_{DD} and V_{GG} be -10 and -20 V, respectively.

Section 9-2

9-4. The RS flip-flop of Fig. 9-7 has positive clock pulses at a clock frequency of 5 MHz. Selecting time zero as the instant a particular clock pulse begins, suppose the set S input is a logical 0 at 0 V until t is 650 ns and a logical 1 at 10 V thereafter. The reset R input is 10 V from 150 to 350 ns and zero at all other times. The output at Q, which is initially 10 V, can change only during a clock pulse. Assuming rectangular pulses and negligible set-up and hold times, find and plot the output voltage as a function of time.

9-5. Repeat P9-4 for the RS master–slave flip-flop of Fig. 9-8. Note that the output can change only on the complement of a clock pulse.

9-6. Suppose the master–slave flip-flop of Fig. 9-8 is CMOS with $V_{DD} = 10$ and a clock pulse frequency of 5 MHz. Each latch feeds into a load of 10 pF. Assume the flip-flop changes state during each clock pulse and dissipation in the AND gates is negligible. Calculate the power in mW. (*Ans*: 10)

9-7. In the RS flip-flop configuration of Fig. 9-7, replace the AND gates with NAND gates, change the preset and clear inputs to logical 1's, reverse the outputs Q and $\bar{Q}$, and change the cross-coupled NOR latch to a cross-coupled NAND latch. Sketch the result. Also, give the characteristic table relating Q at time t_n^+, just after a clock pulse is applied, to the R and S inputs at t_n immediately before the pulse is applied.

Section 9-3

9-8. The circulating register of Fig. 9-14 has eight JK flip-flops in cascade, similar to the configuration of Fig. 9-13. A dc voltage representing a logical 1 is present at the data input. Just before a clock pulse the Q outputs of the flip-flops from left to right are 11000110. Determine the respective outputs after five clock pulses if (a) $S = 0, R = 1$, and (b) $S = 1, R = 0$.

9-9. A four-stage binary ripple counter similar to that of Fig. 9-12 has an input of n pulses with n less than 16. The respective outputs left to right of the flip-flops are Q_1, Q_2, Q_3, and Q_4, and each is initially a logical 0. Prepare a table showing the four-bit binary number $Q_4 Q_3 Q_2 Q_1$ corresponding to values of n from 0 to 15 inclusive, and give the decimal equivalents.

9-10. For the JK MS flip-flop of Fig. 9-10, connect the input terminal J to K through an inverter. Join together the $\bar{Q}$ and J inputs to the AND gate, forming a *toggle* circuit. On corresponding time scales sketch the first five periodic clock pulses and the output Q, assuming Q is zero just before the first clock pulse. Assume rectangular pulses and neglect the gate propagation delay.

9-11. For the JK flip-flop of Fig. 9-10 the input J is a high-level dc voltage representing a logical 1. The clock frequency is 1 MHz, and prior to the first pulse the output Q is 0. On corresponding time scales sketch the first five clock pulses and the output Q, assuming the terminal J is connected (a) directly to K and (b) through an inverter to K. Neglect propagation delay and capacitance effects.

9-12. When data pulses at the input of a JK flip-flop begin or end during clock pulses, the characteristic table alone is inadequate for determining the output. With values in ns, suppose each periodic clock pulse has a width of 100. The time interval between pulses is 100, and the first clock pulse begins at $t = 100$. The J input is a logical 1 from 125 to 175, 250 to 450, and 550 to infinity. It is zero at other times. The K input is zero until 475 and a logical 1 thereafter. Both Q and S are initially zero. With reference to Fig. 9-10, carefully sketch on corresponding scales the clock pulses, J, K, S, and Q for values of time from 0 to 1050.

Section 9-4

9-13. From (9-1) deduce that $D = (J + Q)(\bar{K} + \bar{Q})$. Use this result to develop a logic diagram that converts a D type FF to JK, using two OR gates, two inverters, and one AND gate. Show the complete logic configuration similar to that of Fig. 9-19. The complement of the output is not available.

9-14. Directly from the CMOS circuitry of Fig. 9-20 determine the truth table relating D to J, K, and Q. Then deduce that this is also the truth table for (9-1).

9-15. The logic diagram of a D type master–slave flip-flop using NOR gates is shown in Fig. 9-17. Redraw the diagram using NAND gates in place of the NOR gates and interchange the clear–preset inputs. Give the truth table similar to that of Fig. 9-17 with each don't-care state indicated with an X. Sketch the CMOS circuit of one NAND gate, including substrate connections.

Section 9-5

9-16. Using two inverters followed by cross-coupled NAND gates, sketch the logic configuration equivalent to the RS latch of Fig. 9-21. Replace the output $\bar{Q}$ with Q', which is not restricted to values equal to the complement of Q, and deduce the truth table relating the inputs and the outputs. Also, show that $Q = \overline{SQ'}$ and $Q' = \overline{RQ}$.

9-17. With the current sources included, sketch a four-transistor I^2L network with inputs A and B and outputs Q_1 and Q_2 that are related by the expressions

$$Q_1 = \bar{Q}_2 + A \qquad Q_2 = \bar{Q}_1 + B$$

The outputs must be from separate collectors. Give the truth table relating the input and output logic states.

9-18. (a) From Fig. 9-23 determine J and K in terms of T. (b) Then sketch the complete circuit of an I^2L JK flip-flop converted to D type and determine J and K in terms of D.

9-19. With all current sources included, sketch an I^2L network using no more than seven NPN transistors that performs the logic

$$Y = \overline{A + B} + \overline{C + D}$$

9-20. For the JK I²L flip-flop of Fig. 9-22 prepare a characteristic table with the eight possible combinations of J, K, and the output Q just before a clock pulse. Include the states A and B immediately after the positive clock transition and the states C, D, and Q after the negative clock transition.

Section 9-6

9-21. The Schmitt trigger circuit of Fig. 9-24 switches from low level to high level at an input voltage of 3.1. Calculate the node voltages v_B, v_E, and v_O in the low state with Q_1 *off* and also in the high state with the input at 3.1 V. Use (9-4) and assume β_F is infinite.

9-22. Switching of the Schmitt trigger of Fig. 9-24 from high level to low level occurs at $v_I = 2.35$. Calculate v_{BE1} and v_{BE2} in the high state with Q_2 *off* and also in the low state at this value of input voltage. Use (9-4) and assume β_F is infinite.

9-23. For the Schmitt trigger circuit of Fig. 9-24 deduce (9-8) from (9-4), (9-6), and (9-7). First, solve (9-4) for the base-emitter voltages and use the results to eliminate these in (9-6) and (9-7). Solve the modified (9-7) for i_{C1} and substitute in (9-6). Replace the current i_{C2} with $5 - 0.5v_O$. Also, use (9-8) to calculate approximate values of v_I for v_O precisely equal to 4.98198, 6.5, and $10 - 10^{-18}$.

9-24. It is sometimes desirable to minimize hysteresis in a Schmitt trigger circuit. With reference to the network of Fig. 9-24, reducing R_1 helps to keep v_B from dropping rapidly during the transition. With other parameters constant, determine the value of R_1 in ohms that reduces hysteresis to zero. Assume the base-emitter voltage of a transistor in the active mode is 0.7 V and it is 0.6 V at the boundary between the active mode and cutoff. Do not use (9-4).

Section 9-7

9-25. Design the two-bit binary counter of Example 2 of Sec. 9-7 using T flip-flops, 2 two-input AND gates, 2 two-input OR gates, and one inverter. On the network label X_1, CP, and the outputs.

9-26. Redesign the two-bit binary counter of Example 2 of Sec. 9-7 so that the output state AB remains fixed when X_1 changes from 1 to 0, rather than returning to 00. When X_1 changes from 0 to 1, the count should commence from the fixed state. Use JK flip-flops and a two-input AND gate. On the network label X_1, CP, and the outputs.

9-27. Design a four-bit binary counter using T flip-flops and 2 two-input AND gates, with no external inputs other than the clock pulses to be counted. On the network label CP and the outputs.

9-28. Repeat P9-27 using JK flip-flops. In the excitation table label the flip-flop states D, C, B, and A in this order from left to right.

9-29. Using an RS flip-flop, design the circuit specified by the state diagram of Fig. 9-38. The numbers on the directed lines are the inputs X_1 and X_2.

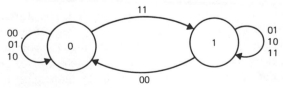

Figure 9-38   Figure for P9-29.

Chapter Ten
Semiconductor Memories

Digital systems normally require some way to store logical bits. Punched cards, paper tape, magnetic tape, and magnetic disk serve as permanent memory. For temporary storage the random-access memory systems of computers are employed, and for many years magnetic cores have been extensively used as storage elements. However, the use of semiconductors in such systems, both large and small, is widespread and increasing rapidly. We confine our attention to semiconductor memories.

A *memory cell* is a device or circuit that stores a bit of information, and an orderly cell arrangement constitutes a *memory*. In addition to cells, an integrated-circuit memory includes associated networks such as those for address selection, amplifiers, and control. Three categories are particularly important. One is the *random access memory (RAM)*, in which the cells can be individually addressed in any desired sequence with similar access times. Although the term can be applied to either read-only or read/write memories, it is often understood to refer to the latter. The *read/write* memory allows the stored bit of an addressed cell to be read at an output, and in addition, it permits a bit at a data input terminal to be written into the cell for storage.

A second category is the *read-only memory (ROM)*. This type has stored bits that are fixed during fabrication and cannot be altered. It is often referred to as *fixed-program storage*. There are various ways to permit the user to program the memory once and only once, in which case it is called a *programmable* read-only memory (PROM). If the data content can be altered more than once, it is a *reprogrammable* ROM.

In the third basic category, known as *serial-access memory*, the bits of a word are entered sequentially at a single input and the stored bits are read sequentially from a single output. A cascade of flip-flops is a common arrangement. RAMs and serial-access memories are examined in this chapter, along with various applications, and ROMs are presented in the chapter that follows.

10-1. RANDOM ACCESS MEMORY (RAM)

MEMORY ORGANIZATION

We consider here the read/write type of RAM. Memories are organized by words and bits. To illustrate, a 1024-bit memory might be designed so as to have 256 words of 4 bits each, referred to as 256 × 4. Although 256 words must be addressable, the number of input address terminals on the IC package is easily reduced to eight by means of *decoders*. Four pin terminals are required for data input and output. In some designs input and output terminals are separate, in which case eight pins are needed. The four bits of a word are entered simultaneously at the inputs or taken simultaneously from the outputs. This is known as *parallel access*.

The 1024-bit memory requires 1024 cells. These are arranged on the silicon chip in a matrix, perhaps having 32 rows and 32 columns with each row consisting of 8 words. The matrix is not necessarily square. There could be 64 rows and 16 columns, or some other configuration. Frequently, the memories are designed with only one bit per word. A 16,384 × 1 memory has 16,384 words of one bit each. A single input-output data terminal is sufficient with only one bit entered or retrieved at a time. Any one of the words can be addressed for reading or writing. It is often called a 16 K memory, with K denoting 1024 cells.

Shown in Fig. 10-1 is a 256 × 4 RAM cell array in a 32 by 32 matrix. There are 32 row-select lines and 8 column-select lines. The proper voltage applied, for example, to line X_2 and line Y_1 addresses the four-bit word at the X_2–Y_1

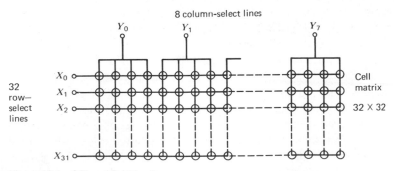

Figure 10-1   256 × 4 RAM cell array.

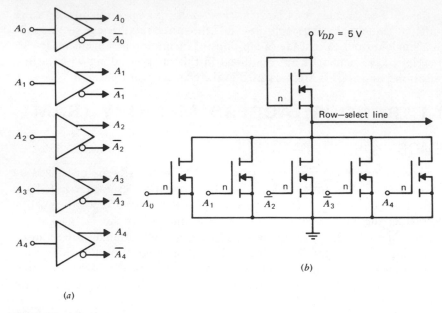

Figure 10-2   MOS row-select circuitry for 32-row cell matrix, using n-channel enhancement-mode devices. (*a*) Input buffer-inverters. (*b*) One of 32 identical row-decode circuits.

location for either reading or writing. Both row-select and column-select decoders are normally used with this configuration. The purpose is to reduce the number of package pins required for address, as well as the number of input address signals, to a reasonable number. In this case the reduction is from 40 to 8.

A row-select network that can be used with a MOS 256×4 memory having a 32×32 matrix is indicated in Fig. 10-2. There are five pin terminals designated A_0, A_1, A_2, A_3, and A_4. These are connected to five buffer-inverter amplifiers, shown in Fig. 10-2a. The five inputs and their complements provide 2^5 or 32 combinations of address words. The address word at the five-terminal input of the row-decode circuit of Fig. 10-2b is $A_0 A_1 \bar{A}_2 \bar{A}_3 A_4$. For any binary levels of the five inputs, only one of the 32 combinations consists of five zeros, and this is the one that activates a row. Let us suppose that the word $A_0 A_1 \bar{A}_2 \bar{A}_3 A_4$ is 00000. The NOR gate of Fig. 10-2b is *off*, feeding a logical 1 to the row-select line. This line is enabled for reading or writing, and all other rows are disabled.

To complete the address, one of the eight four-bit words of the enabled row must be selected. Circuitry similar to that of Fig. 10-2 can be used. Because there are eight column lines, eight NOR gates are required, along with three input buffer-inverter amplifiers. The logic configuration is shown in Fig. 10-3. It is referred to as a 3-to-8 decoder. We note that the 256×4 memory requires

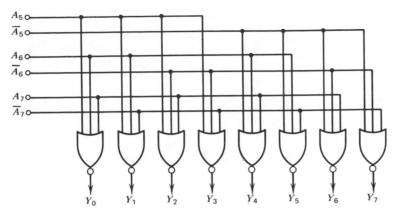

Figure 10-3   A 3-to-8 decoder for column select.

a total of eight address pins connected to eight inverters. Five of these are part of a 5-to-32 decoder for row select, and three are part of a 3-to-8 decoder for column select. The decoder is a combinational network with n inputs and 2^n outputs. For each binary input combination of ones and zeros there is a single output with a value of 1.

In addition to the cell array and decoders, an integrated-circuit memory contains sense (read) amplifiers, write amplifiers, and control circuitry. The sense and write amplifiers are simply logic gates that provide proper logic levels and adequate buffering. In Fig. 10-4 is a functional block diagram of a 256-word by four-bit static RAM. There are four sense and four write amplifiers for amplifying the bits of a word. The two AND gates with inverting inputs as shown provide control. A bar over an input logic control symbol is often used to indicate that the control is active when the input is low-level, whereas the absence of a bar implies that the control is active with a high-level input. Accordingly, the chip is enabled when the chip-enable ($\overline{CE}$) input is low-level. However, when the $\overline{CE}$ input is 1, the outputs of both AND gates are 0, and the sense and write amplifiers are *off*. Thus the input/output (I/O) circuits are inhibited, or disabled.

Let us suppose the chip is enabled with a logical 0 on the CE input. A logical 1 on the read/write ($R/\overline{W}$) control gives a zero output on AND gate 1, which disables the write amplifiers. The logical-1 output of AND gate 2 enables the sense amplifiers, and the stored bits of the addressed word appear at the outputs. On the other hand, a logical 0 on the $R/\overline{W}$ input reverses the outputs of the AND gates, enabling the write amplifiers and disabling the sense amplifiers. The bits at the input terminals are now stored in the addressed word.

The IC package contains 16 pins. Fourteen of these are identified on the diagram of Fig. 10-4. The other two are for the supply voltage.

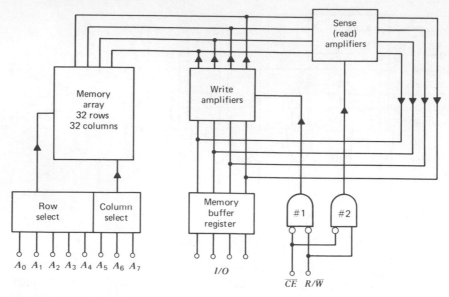

Figure 10-4   Functional block diagram of 256 × 4 RAM.

VOLATILITY

Readout from semiconductor memories is usually *nondestructive*, which signifies that the operation does not affect the storage. In contrast, the stored bits of magnetic cores are removed when reading occurs. This destructive readout normally requires that each READ operation be immediately followed by a WRITE cycle which reenters the bits into the magnetic cores. The decreased operating speed and the additional circuitry are disadvantages. An undesirable feature of semiconductor memories, which is not shared by magnetic cores, is *volatility*. A volatile memory is one in which the data are lost when power is removed.

Semiconductor memories can be made nonvolatile by adding circuitry that switches the voltage supply terminals to a battery whenever power is interrupted. An example of one that is easily adapted to nonvolatile use is the Intel 5101 CMOS 1024-bit static RAM organized as 256 words by four bits.[10] The CMOS devices are made with silicon gates using ion-implantation technology. Because the standby power-supply current is typically an ultra-low 0.2 μA and the supply voltage need be only 2 V or more, two small button-type primary battery cells mounted on the printed-circuit board are often adequate. Of course, it is necessary to include circuitry designed to switch operation to the battery when power fails. The IC includes an output disable pin that places the internal data buffers in a high impedance state. This tri-state output is particularly useful in systems

having a common data bus. The memory can be expanded by properly interconnecting a number of the integrated circuits. For example, 16 memories in a
4×4 matrix can provide a system having 1024 words of 16 bits each.

CMOS CELL

The CMOS cell used in the Intel 5101 is shown in Fig. 10-5. Transistors T_1
through T_4 constitute a cross-coupled latch that draws a current of about a
nanoampere in the steady state. Transistors T_5 and T_6 are gating devices that

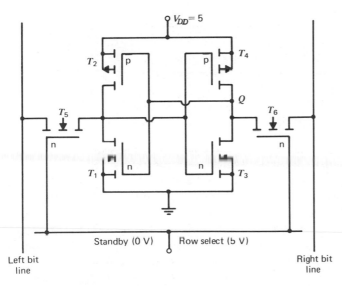

Figure 10-5   CMOS memory cell.

couple the bit lines to the latch when the row-select line is high-level. The
output Q is a logical 1 when T_3 is *off* and T_4 is *on*, and it is zero when these states
are reversed. Reading and writing are accomplished through the bit lines.
Access time is less than 650 ns.

A six-transistor cell of a 4096-bit 5 V static NMOS memory is illustrated in
Fig. 10-6. Silicon-gate technology is employed, and the cross-coupled latch has
depletion-mode load devices. The 4-kilobit integrated circuit is available in an
18-pin package having a typical access time of 150 ns and a power dissipation of
350 mW. Organization is either 1024×4 or 4096×1, with a cell matrix of 64
rows and 64 columns.

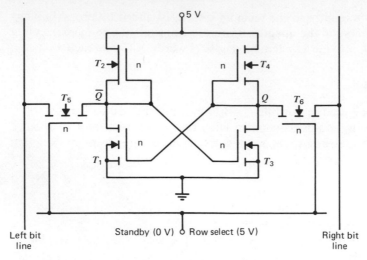

Figure 10-6 NMOS six-transistor static cell.

TIMING REQUIREMENTS

Data sheets specify certain timing constraints attributable to delays that occur within the chip. To illustrate typical requirements, specified data for a TMS 4043 NMOS 256 × 4 static RAM will be utilized. In Fig. 10-7 are shown the waveforms applicable to the READ cycle. The address is at the top, with the sketch indicating that the input at each of the address pins may be high or low. Note the READ cycle time t_{RC}. This must be at least 1000 ns. Below the address is the chip enable $\overline{CE}$ logic, and the chip is enabled when this input is low-level. At the bottom is the time interval of the valid output data.

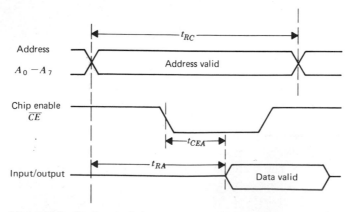

Figure 10-7 Read-cycle timing.

An important parameter is the READ *access time* t_{RA}. As indicated, it is the READ-operation delay from the instant the address logic is applied to the input pins to the instant the bits appear at the data output pins. This parameter is specified as 1000 ns maximum. Also shown is the access time t_{CEA} from the chip enable, which is 800 ns maximum. Accordingly, the data at the output are not valid for 1000 ns after addressing and for 800 ns after enabling the chip.

The WRITE-cycle timing is indicated in Fig. 10-8. A minimum of 1000 ns is required for the cycle time t_{WC}. *Set-up time* t_S is defined as the time a logic level at one input is set up before a transition occurs at another input. A certain minimum time between these transitions may be required to avoid logic errors. Three set-up times are identified on the timing waveforms. The *address* set-up time t_{AS} is the time the address is set up before the WRITE pulse is applied. To avoid the possibility of writing erroneous data into the memory, this interval must be at least 150 ns. The *chip-enable* set-up time t_{CES} has a specified minimum of zero. Then there is the *data* set-up time t_{DS} with a minimum of 600 ns, which is the interval between the application of input data and the disabling of the WRITE line.

Finally, let us consider *hold* time t_H, defined as the time during which a logic level is held at an input after a transition takes place at another input. Identified on the WRITE waveforms is the *address* hold time t_{AH}, which has a specified minimum of 50 ns. For this interval after the WRITE line is disabled, the address must not change in order to avoid writing data into an incorrect cell. The *chip-enable* hold time t_{CEH} is specified as zero minimum, and the *data* hold time t_{DH} must be at least 100 ns.

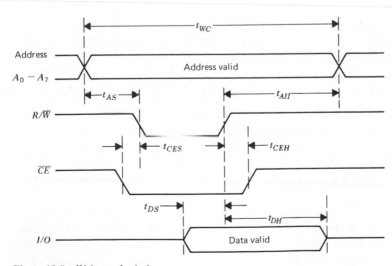

Figure 10-8   Write-cycle timing.

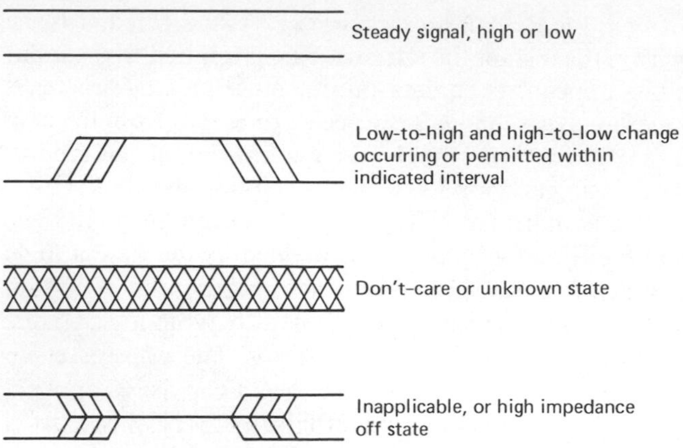

Steady signal, high or low

Low-to-high and high-to-low change occurring or permitted within indicated interval

Don't-care or unknown state

Inapplicable, or high impedance off state

Figure 10-9   Timing-diagram symbols.

In designing a memory system it is good procedure to draw a timing diagram with a known pulse train taken as reference. Other signals should be sketched relative to it, with waveforms similar to those of Figs. 10-7 and 10-8. There are always uncertain transition intervals, possibly don't-care or unknown logic levels, and perhaps inapplicable or high-impedance *off* states. The symbols of Fig. 10-9 are convenient and conventional. Care must be taken in the design to stay within specified time limits.

10-2. BIPOLAR MEMORIES

Random access memories using bipolar junction transistors are characterized by exceptionally high speed, with access times from about 5 to 200 ns. We consider TTL, ECL, and I^2L configurations.

TTL

Figure 10-10 is a schematic diagram of a TTL cell, which consists of a cross-coupled latch with one transistor in saturation and the other in cutoff. In the standby condition the voltage of the word line is 0.3 and that of each bit line is 0.5 V. Therefore, the current of the *on* transistor flows into the word line, and the current sensed by the bit lines is negligible. The standby power is about 0.9 mW.

To read the cell the word line is raised from 0.3 to 3.0 V, which transfers the current from the word line to the bit line connected to the conducting transistor. The readout current is about 0.15 mA (see P10-4). To write a logical 1 into the cell, both the word line and the right bit line are raised momentarily to 3 V, with the left bit line remaining at 0.5 V. This places transistor T_1 in the *on* state and T_2

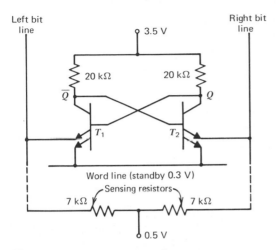

Figure 10-10   TTL multiple-emitter cell.

in the *off* state. These remain fixed when the 3 V pulse is removed. The procedure for writing a zero into the cell is the same except that the left bit line is employed. The WRITE cycle of a single cell requires about 20 ns, which can be substantially reduced by using Schottky transistors. There are several variations of this basic circuit that give reduced power dissipation and faster speed.

Integrated circuit SN54S201 is a TTL random access memory organized with 256 words of one bit each.[4] The transistors are Schottky-clamped for high speed. A three-state output makes it possible to connect the output and other similar outputs to a common bus. This is especially convenient when a number of the integrated circuits are interconnected to form an expanded memory system. A version with an open-collector output is also available, offering the capability of direct interface with a data line with a passive pull-up. The access time is typically 42 ns, the WRITE cycle requires 100 ns, and the power dissipation is about 500 mW.

The package has 16 pins. Eight are for address, two for the 5 V supply, and one each for data input, data output, and read/write control. The other three are chip-enable pins connected to a NOR gate. The chip is disabled if any one of these is high level, and the output is the high-impedance *off* state. Because the cells are arranged in a 16 × 16 matrix, there are two 4-to-16 line decoders, with one providing row select and the other providing column select.

The data input feeds into a WRITE amplifier which gives adequate buffering and suitable logic levels for writing into the cells. Both the input and output circuits are shown in Fig. 10-11. The input configuration utilizes a PNP transistor with a negative base current, which keeps the low-level input current

---

[4] Reference 4, page 190. (Recall that superscript numerals refer to References.)

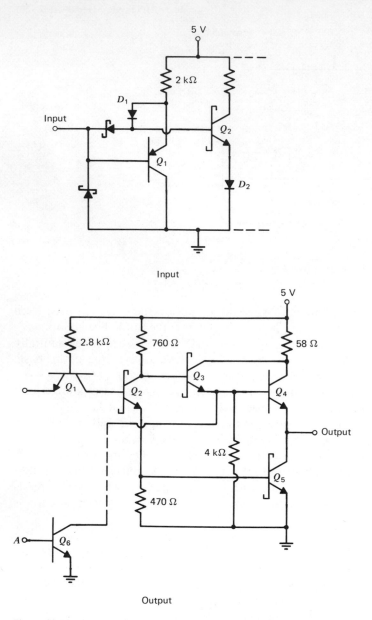

Input

Output

Figure 10-11   Input and output circuits.

298

around -0.2 mA (see P10-5). The high-level input current consists of a few microamperes of leakage through the diodes.

The illustrated output circuit is the conventional high-speed TTL logic gate, having a typical propagation delay of 6 ns. Active pull-up is provided by the totem-pole arrangement. Switching is quite fast because of the Darlington pair Q_3 and Q_4 and the Schottky clamps on transistors that would otherwise saturate. The output impedance is only a few ohms in both logic states. The high-impedance output suitable for bussing is obtained by driving transistor Q_6 into saturation, thereby turning off the two output transistors. The actual circuitry associated with Q_6 is modified somewhat from that shown.

ECL

Emitter-coupled-logic memories are often used when ultra-high speed is required. The schematic diagram of an ECL cell is shown in Fig. 10-12, along with sense-write circuits. Because each transistor operates in either the active

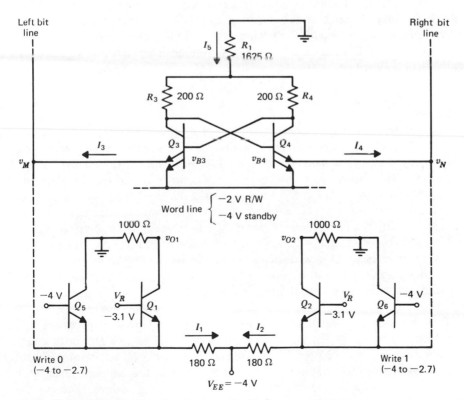

Figure 10-12   ECL memory cell and sense-write circuits.

or cutoff mode, all base currents are very small and will be ignored in the discussion here. Also, we shall assume all diode drops are 0.75 V. The storage cell consists of a cross-coupled latch with multiple-emitter transistors. When the word line is at -4 V, the cell is in standby and effectively isolated from the bit lines. Address for reading or writing is accomplished by raising the word-line voltage to -2.

Suppose the cell is in standby with transistor Q_4 conducting. This condition is defined as the logical-zero state. Transistor Q_3 is *off*. The base voltage v_{B4} of Q_4 is -3.25, which is 0.75 V above that of the word line. Because the current through R_3 is negligible, the voltage across the 1625 ohm resistor is v_{B4}. Hence I_5 is 2 mA, giving a cell dissipation of 8 mW. The base voltage v_{B3} is found to be -3.65, providing a base-emitter bias of 0.35 V. This is too low for conduction, and Q_3 is indeed *off* as assumed. Although conducting transistor Q_4 has a forward bias on the collector junction equal to $I_5 R_4$, or 0.4 V, this is sufficiently small to justify treating Q_4 as operating in the active mode. The purpose of the 1625 ohm resistor R_1 is to limit the current so as to ensure that the product $I_5 R_4$ is small enough to avoid appreciable saturation.

Let us consider the sense-write circuits in standby. We note that the emitters of Q_2, Q_4, and Q_6 are coupled together. Consequently, *only the emitter connected to the transistor with the highest base voltage can conduct*. Because v_{B2} is -3.1, v_{B4} is -3.25, and v_{B6} is -4, the base voltage of Q_2 is the highest, and the current I_2 is supplied by Q_2. The emitter of Q_2 is 0.75 V below the reference voltage V_R of -3.1. Thus I_2 is 0.83 mA and the node voltage v_N is -3.85. Of course, I_1 equals I_2 and v_M equals v_N. The output voltages v_{O1} and v_{O2} are -0.83 V.

To read the cell the word line is raised to -2 V, which raises the base voltages of Q_3 and Q_4. The network is designed so that v_{B4} is now higher than the reference voltage V_R, but v_{B3} remains lower than V_R. Accordingly, the current of Q_4 transfers to the sense circuit, with $I_2 = I_4 = I_5$. Both Q_2 and Q_6 are *off*. From a loop equation through R_1, R_3, and v_{BE4} to V_{EE} the current is found to be 1.80 mA. Conditions in the left bit line and its sense circuit are unchanged from the standby condition. The respective base voltages v_{B3} and v_{B4} of the latch transistors are found to be -3.29 and -2.93. These values confirm the assumption that the bit-line emitter of Q_3 is *off* and the current of Q_4 has shifted to its bit line. The respective outputs v_{O1} and v_{O2} are -0.83 and 0. These are supplied to a differential amplifier, such as that discussed in Sec. 14-1, which is designed to give a logical-1 output.

To write a logical one the cell is placed in the READ mode by accessing the word line with -2 V. Then a pulse is applied to the base of Q_6, raising the voltage from -4 to -2.7 for a brief interval. This turns Q_4 *off*, causing Q_3 to conduct, and a logical 1 is stored (see P10-6). A zero is written similarly, using Q_5.

There are variations in the circuitry and parameter values. The power dissipation of about 8 mW per bit is large and restricts the number of bits that can be arranged on a chip. Increasing the resistances decreases the dissipation but also reduces the speed. ECL random access memories usually have 64, 128, or 256 bits.

An example is the SN10142 ECL memory with a 64-word-by-one-bit organization, available in a 16-pin dual-in-line package. Included are row and column decoders, sense/write amplifiers, input and output buffers, and two enable inputs for READ, WRITE, and chip-disable control. The emitter-follower output can be connected to similar outputs to achieve wired-OR word expansion. Typical high and low logic levels are -0.9 and -1.8 V. The access time is less than 10 ns, which is the outstanding feature of the memory. With a supply voltage of -5.2 the dissipation is about 450 mW, or 7 mW/bit.

I^2L

This type of memory is characterized by high speed, low dissipation, and large circuit densities. Thousands of cells can be placed on a single chip. Although impractical in size, the eight-word-by-one-bit array of Fig. 10-13 serves to

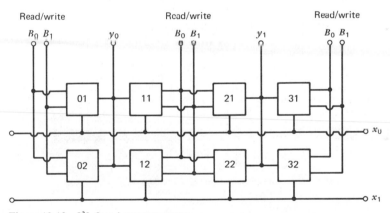

Figure 10-13 I^2L 8×1 memory array.

illustrate a static I^2L random access memory. The row-address lines are x_0 and x_1 and the column-address lines are y_0 and y_1. Each y line addresses two columns. The bit lines labeled B_0 read and write logical zeros and those labeled B_1 read and write logical 1's.[5]

Shown in Fig. 10-14 is the schematic diagram of cell 11, with values of currents and voltages given for the standby condition. Transistors T_1 and T_2 are current-source loads for the NPN transistors of the cross-coupled latch, and transistors

[5] Reference 5, pages 332–337.

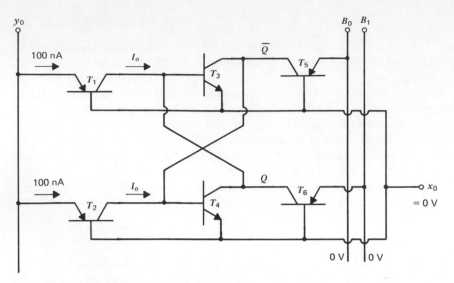

Figure 10-14   Schematic of cell 11 in standby.

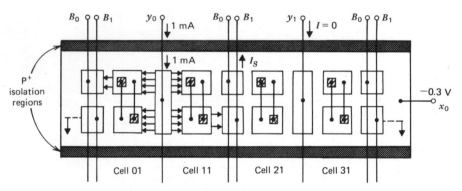

Read operation, word 1, top view, cell 21 addressed.

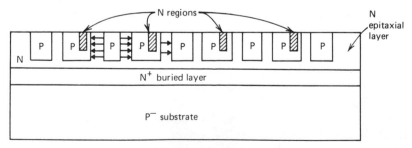

Read operation, word 1, cross sectional view through
lower portion of top view, cell 11 addressed.

Figure 10-15   Memory structure.

T_5 and T_6 couple the latch to the bit lines. The current-source and coupling transistors are PNP lateral devices. The y_0 address line is a P-type current rail which also serves as the emitters of current-source transistors. It supplies 100 nA to each device. Thus the standby dissipation is 150 nW per bit, which is much less than the 8 mW per bit typical of an ECL cell. The row address x_0 and the bit lines are connected to ground.

Suppose cell 11 in standby has transistor T_3 in saturation and T_4 in cutoff. The node voltage at output Q is 0.75, which is the saturation voltage v_{BE3}. That at output $\bar{Q}$ is 0.05 V, which is the saturation value of v_{CE3}. Therefore, a logical 1 is stored in the cell, with $Q = 1$ and $\bar{Q} = 0$. Because both T_4 and T_5 are *off*, all the current I_o from the collector of T_2 flows into the collector of the saturated transistor T_3. As we shall see, the geometry is such that most of the current from the collector of T_1 enters the base of T_3, although a fraction of this current feeds through T_6 into bit line B_1. Actually, the bit line current is negligible compared with the READ current.

The top and cross-sectional views of the memory structure are shown in Fig. 10-15. Each x-line has an N^+ buried layer which connects the N-type emitters and bases together. This layer is connected to the x address terminal. To read from or write into a cell of line x_0, the line voltage is lowered to -0.3 V. Because the voltage of line x_1 must be kept at zero, P-type isolation regions are added to the structure. These are shown in the top view of Fig. 10-15. All devices of a row are truly merged, but the individual rows are isolated from one another. The geometrical arrangement of the devices of cell 11 in the schematic diagram of Fig. 10-14 corresponds with the cell structure shown in the top view of Fig. 10-15. Note the latch is merged between the current rail on the left and the coupling transistors on the right. Shaded boxes denote N-type collectors.

Cell 11 is addressed for reading by applying -0.3 V to line x_0 and supplying a current of 1 mA to line y_0. Simultaneously, the current of line y_1 is reduced to zero. The base-emitter voltages of transistors T_1 and T_2 of Fig. 10-14 are 0.75 V, which clamps line y_0 at 0.45 V. Because this is the base-emitter voltage of the current-source transistors of the cells of the other row, all injection currents to the transistors of this row are zero. In fact, only cells 01 and 11 receive current. As indicated in the top view of Fig. 10-15, the rail supplies 0.25 mA to each of four current sources. Assuming a logical 1 is stored, transistors T_4 and T_5 are *off*, and no current is supplied to bit line B_0. However, a portion of the current from current-source T_1 is fed into the collector of T_6 and from there to bit line 1. Note that T_6 operates in the inverse mode. An amplifier in line B_1 senses the current I_s. The currents are shown in Fig. 10-15.

When cell 11 is read, cell 01 is also addressed. As indicated by the currents of Fig. 10-15, cell 01 has a logical-zero bit. The bit lines of cell 01 can be disabled when reading cell 11. The reading operation takes place quickly during brief pulses, and within the read cycle the injection currents to all but two cells are zero. However, the bits do not vanish. The charge stores are in the space-charge layers, and they do not diminish appreciably before the read cycle terminates and the injection currents are restored. Reading is nondestructive.

Now suppose a logical 0 is to be written into cell 11. With the x_0 line accessed, the current of line y_0 is reduced to zero and a current of 0.5 mA is driven into line B_0, as shown in Fig. 10-16. Transistor T_3 is *off*, because there is no way for any current to flow into its base. This is evident from examination of Fig. 10-14, noting that the only source of current is that from the collector of T_5. This current cannot flow through T_3. Therefore, it crosses over to the base of T_4 through the surface metallization, as indicated in Fig. 10-16. Transistor T_4 saturates with zero collector current and a very low collector-emitter voltage. When conditions are switched back to standby, T_4 remains saturated and T_3 is *off*. Thus Q is 0 and $\bar{Q}$ is 1.

To avoid changing the state of cell 21 during the WRITE cycle, a current of 1 mA is driven into line y_1 immediately before and during the interval of the 0.5 mA current applied to bit line B_0. This ensures that current injected by bit line B_0 into cell 21, as shown in Fig. 10-16, will not switch the latch.

It is anticipated that a 4-kilobit static I^2L memory with cells somewhat similar to the one described will soon be fabricated on a monolithic chip in a 16-pin dual-in-line package. The cell density should be at least as great as that which has been obtained with MOS technology, and the I^2L memory should have greater speed and lower power dissipation. An access time of about 50 ns appears to be a reasonable expectation, allowing extremely fast operation. A 16 kilobit I^2L memory is likely sometime in the near future. In the next section we examine a dynamic I^2L random access memory which is soon to be in production. The chip stores 4096 bits, has an access time less than 100 ns, and requires only 80 mW in the standby mode. The performance characteristics of I^2L memories are indeed outstanding. The only serious questions at this time relate to processing yield and cost.

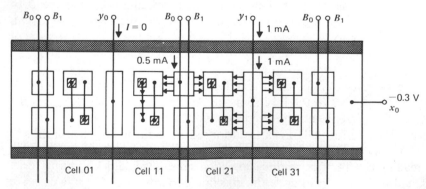

Figure 10-16   Write operation. Logical 0 fed into cell 11.

10-3.   DYNAMIC MEMORIES

Two factors largely determine the number of cells that can be placed on a single monolithic chip. One of these is the power dissipation and the other is the required real estate per cell. In comparison with static cells, dynamic cells have lower dissipation, and because they are made with fewer transistors, each cell occupies less surface area of the chip. Also, there is reduced capacitance because of the smaller number of transistors per cell, which allows faster operating speed. Consequently, integrated-circuit random access memories with more than 4 kilobits of storage capacity are usually dynamic. In large memory systems the cost per bit must be minimized, and this requires maximizing the number of bits per chip.

In our study of dynamic MOS gates we learned that device capacitances are utilized for temporary charge storage, and a dynamic system is clocked so that charge leakage from each capacitor with a stored logic level is periodically refreshed. This principle is employed in dynamic memories. Clearly, the timed voltage or current pulses which must be applied periodically to refresh the charge stores constitute a disadvantage. The interval between pulses must be small enough to avoid losing a logic level. Refresh of each cell of a MOS dynamic memory is usually performed every two milliseconds or less.

FOUR-TRANSISTOR CELL

To illustrate the principle let us examine the four-transistor dynamic NMOS cell of Fig. 10-17, along with some associated circuitry.[1] The reference voltage V_R is 5, and the gates of control transistors T_3 through T_{10} are either high-level at 10 V or low-level at zero, with the level of each determined by timing and control circuits. The cell is similar to the six-transistor static cell of Fig. 10-6. However, transistors T_3 and T_4 serve dual functions as both load and coupling devices, thus reducing the number of transistors from six to four. This is possible only because the parasitic capacitances C_1 and C_2 are used for temporary charge storage.

In the network of Fig. 10-17 let us assume that the voltages of the row, chip, and column select terminals are zero, making transistors T_3 through T_8 nonconducting. Also, assume the reset terminal is at 10 V, thereby fixing the potentials of the two refresh lines at 5 V. Furthermore, let us suppose that at time zero the illustrated cell has output $Q = 1$, corresponding to 5 V. The complement $\bar{Q}$ is zero, indicating a voltage of zero. These conditions imply that T_1 is *on* with capacitance C_1 charged to 5 V, and T_2 is *off* with no charge on C_2.

Unfortunately, because the cell is isolated from the bit and refresh lines, the charge on C_1 decays. This is largely caused by leakage through the reverse-biased PN junction separating the N-type drain of T_2 from the P-type substrate.

---

[1] Reference 1, page 120.

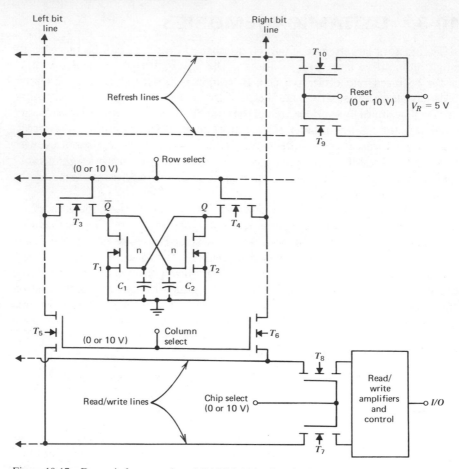

Figure 10-17   Dynamic four-transistor NMOS RAM cell and related circuitry.

In perhaps 20 ms the voltage of C_1 drops from its initial value of 5 to a value below the threshold voltage, both transistors of the latch are now *off*, and the stored bit is lost. However, if the row-select line is accessed with a 10 V pulse before the bit is lost, with the column-select voltage remaining at zero, the latch is connected to the 5 V refresh lines with coupling transistors T_3 and T_4 serving as unsaturated loads. Because T_1 conducts, the voltage across C_2 is maintained at a low value, and T_2 is *off*. Charge flows into C_1 to replace that lost by leakage, and the cell is *refreshed*.

Suppose the cell matrix is 32×32, organized as a 1024-word memory with one bit per word. Assume a maximum interval of 2 ms between refreshing cycles is specified. With the column and chip selects at zero and the reset at 10 V, the 32 cells of a row are refreshed simultaneously whenever the row is accessed. A common procedure is to select the rows in sequence. This can be accomplished by connecting the outputs of a five-stage binary ripple counter, similar to that of Fig. 9-12 of Sec. 9-3, to the five row addresses. The binary counter cycles through the addresses, and the row decoder supplies pulses to the rows in succession (see P10-9) at a speed determined by the clock pulses to the counter. The entire refresh cycle, during which all cells are refreshed, may require about 20 μs, which is only 1 % of the duty cycle. If desired, refresh can be accomplished simultaneously on all rows in one interrupt.

To access the cell for reading or writing, both row and column selects are raised to 10 V. Note that only a single memory cell is connected to the read/write lines. After the minimum address set-up time the reset is switched from low level to high level. Then the chip is enabled by applying 10 V to the chip-select terminal.

In the READ cycle the polarity of the differential voltage between the bit lines is sensed, and the READ amplifier converts this into the correct logic at the output. In the illustrated circuit the current drawn by a bit line causes the voltage of a refresh line to drop excessively. This is avoided by including a *refresh amplifier* in each column, and these amplifiers provide buffering. Readout is nondestructive. To write a bit into the cell, a pulse from the WRITE amplifier places one bit line at a higher voltage than the other.

The TMS 4062 large-scale integrated-circuit PMOS memory, organized 1024×1, uses the four-transistor dynamic storage cell. Three voltage supplies are required. With respect to V_{SS} the reference voltage V_R is -13, the substrate voltage V_{BB} is 2.5, and V_{DD} is -20, which supplies the amplifiers and also serves as the enabling voltage for the reset, access, and chip-select inputs. The positive substrate voltage increases the reverse bias of the PN junction, which enlarges the depletion region, and the main effect is to increase the magnitude of the threshold voltage by about 0.5 V. Each cell must be refreshed at least once in every 2 ms period by cycling through the row addresses or by addressing each row for reading or writing or refreshing at least once in that period. The access time is less than 130 ns, and the read and write cycles require only 200 ns. Power dissipation is low, typically 120 mW in normal operation and 2 mW in standby. Memory expansion can be accomplished with the chip-select.

ONE-TRANSISTOR CELL

The four-transistor dynamic cell occupies less real estate, consumes less power, and operates faster than does the six-transistor static cell. Additional improvements are possible by reducing further the number of transistors per cell. There are various arrangements of three-transistor, two-transistor, and one-transistor cells employed in dynamic MOS memories.[1] Particularly important is the one-transistor cell having a row-select line, one data line, and a fabricated capacitor. A simplified schematic diagram including related circuitry is shown in Fig. 10-18.

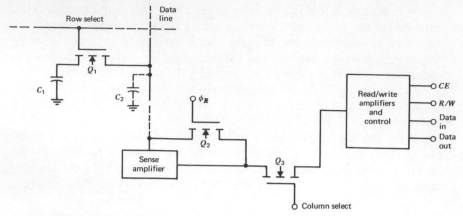

Figure 10-18   One-transistor RAM cell and related circuitry.

The basic cell consists of transistor Q_1 and capacitor C_1 which stores the bit. A charge store is a logical 1 and no charge is logical 0. The capacitor is a polysilicon gate separated from an inversion layer by a thin oxide, with a typical value of 0.1 pF. The inversion layer serves as an electrode. Connected to the cell is a row-select line and a data line. The parasitic capacitance C_2 of the data line is typically 10 times or more greater than C_1 and its effect is detrimental.

To read the cell the row-select line is accessed. The logic state of C_1 is sensed by the sense amplifier connected to the data line and fed through transistor Q_3 and the READ amplifier to the output. Because the readout is destructive, the bit withdrawn from the cell must be resupplied. This is accomplished by turning Q_3 *off* and Q_2 *on*, causing the bit held at the output of the sense amplifier to be rewritten into the cell. In a similar manner the other cells of the accessed row are refreshed during this operation. In fact, whenever a row is accessed, the cells of the row are refreshed regardless of whether the operation is READ, WRITE, or simply REFRESH.

Prior to reading, the data line is usually precharged to a voltage midway between the high-level and low-level cell voltages, which are typically 10 and 0, respectively. An individual cell is accessed by enabling the row-select and column-select lines. The voltage that appears on the data line is only about 200 mV above or below the precharged value, depending on the stored logic. Charge sharing between C_1 and C_2, which occurs when they are connected in parallel by means of transistor Q_1, results in a rather small change in the voltage across C_2. This is a consequence of the much larger value of C_2, and a contributing factor is the charge leakage that has occurred since the last refresh. The sense amplifier must detect the small 200 mV signals. Furthermore, it must be designed for low power dissipation because there are as many sense amplifiers as there are columns.

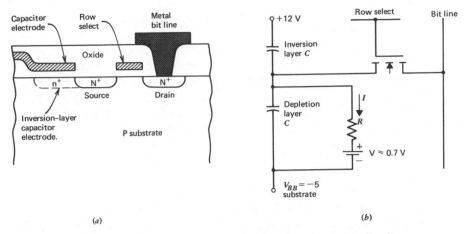

Figure 10-19   One-transistor cell. (a) Structure. (b) Approximate equivalent circuit.

In Fig. 10-19 is shown the structure and approximate equivalent circuit of a practical one-transistor cell.[5] A polycrystalline doped silicon gate is connected to the row-select line. The top capacitor electrode, also polysilicon, is connected to a 12 V supply; the bottom electrode of the capacitor is the inversion layer, which joins the source. Added to the approximate equivalent circuit is the depletion-layer capacitance. This is shunted by a large resistance R in series with a 0.7 V source representing the equilibrium voltage across the depletion layer. Although much oversimplified, this arrangement accounts for currents through the layer under nonequilibrium conditions.

Let us briefly consider the inversion layer. Suppose the voltages of $+12$ and -5 are suddenly applied at time zero. Prior to this the substrate is in equilibrium, with no inversion layer. The 17 V between the top capacitor electrode and the substrate connection supply an electric field to the substrate, which pulls mobile electrons toward the surface and repels holes. Only a few nanoseconds are required to form a strong inversion layer with a depletion region beneath it. The P substrate, the depletion layer, and the n^+ inversion layer are similar to a PN junction having its depletion layer entirely within the P region. In equilibrium there are equal-and-opposite drift and diffusion currents. Thus the current through the resistor R of Fig. 10-19b is zero, the substrate-to-source voltage V_{BS} is -0.7, and that across the inversion-layer capacitance is 16.3 V.

Now suppose the row is accessed, forming a channel, with the bit line at $+10$ V. Assuming a threshold voltage of 2, all mobile electrons of the inversion layer are drawn into the source and through the channel, leaving the surface region at a potential of 10 V. Row select is then disabled. The voltage across the

[5] Reference 5, pages 305–310.

inversion-layer capacitance is 2 and that across the depletion layer is 15. This is a nonequilibrium situation.

An inversion layer does not reappear immediately. The electric field is confined to the depletion region, which is devoid of mobile charge carriers. Electrons enter this region by chance as a consequence of their random motion within the neutral substrate region and also from thermal generation within the layer. Gradually a strong inversion layer develops, but the time required to reach equilibrium is tens of milliseconds at room temperature and perhaps as much as 10 ms at 75°C. The leakage current through the depletion layer is represented in Fig. 10-19b by the current I. For electron flow upward into the inversion layer, I is positive.

Thus when the row is accessed with the bit line at 10 V, electrons flow out of the inversion layer into the bit line with the surface potential rising from -4.3 to $+10$. This is equivalent to a positive current into the storage capacitor, with positive charge stored. The cell stores a logical 1. With the bit line at zero the inversion-layer potential changes from -4.3 to 0, and the charge flow is much less. In normal operation the charge store corresponds to either a logical 1 or a logical 0, and equilibrium does not exist. Refreshing is accomplished at 2 ms intervals to replace charge lost by leakage through the depletion layer.

An example of a dynamic RAM using the one-transistor silicon-gate NMOS cell is the TMS4030-2 four-kilobit integrated circuit. Organization is 4096×1 with a 64×64 cell matrix. For a memory of this capacity, operation is fast with an access time less than 200 ns. Dissipation is typically 400 mW in operation and only 0.2 mW in standby. Refreshing must be accomplished at 2 ms intervals. A single external clock is required. The three-state output buffer provides a high-level voltage of 2.5, a low-level voltage of 0.4, and a high impedance state. Two standard TTL gates can be driven.

Supply voltages V_{DD}, V_{CC}, and V_{BB} are 12, 5, and -3, respectively. As in this case, negative substrate voltages are common in NMOS integrated circuits. Normal processing may give a threshold voltage that is marginal for an enhancement-mode device, such as 0.4 V, and a negative substrate-to-source voltage raises this to a reasonable value, perhaps 1.5 V. Also, the depletion layers around the transistors are enlarged because of the increased reverse bias, resulting in improved isolation between devices.

In Fig. 10-20 is a compact version of the one-transistor cell.[6] Referred to as a *surface-charge cell*, it is less than half the size of the cell of Fig. 10-19. Although the source diffusion is omitted, the inversion layer serves as both a capacitor electrode and the MOSFET source. Note the double level of polysilicon in the region above the inversion layer. The fact that the drains of all transistors of a column are common with the bit line is exploited by using a single diffused region for these. The cell is used in the Intel 2116 n-channel RAM with a storage

[6] Reference 6, page 128.

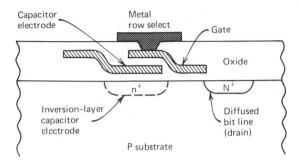

Figure 10-20   Compact one-transistor RAM cell.

capacity of 16 kilobits. Maximum access time is about 250 ns, operating power is 700 mW, and standby power is 12 mW. The 16-pin package requires supply voltages of 12, 5, and -5.

Organized as 16,384 words of one bit each, the memory has four identical cell arrays of 32 × 128. A total of 256 sense amplifiers provide sensing, rewriting, and refreshing, which must be done every 2 ms. Each amplifier is a cross-coupled latch designed for low power consumption. Clock signals are generated internally. A high-impedance output state enables simplified memory expansion. It is possible to arrange 64 of these 16-kilobit memories, along with the required address multiplexer, decoders, drivers, and buffers, on a memory board about 15 × 30 cm, thereby obtaining a *one-megabit* memory system. This might be organized in the form of 64-kilobit words of 16 bits each.

I²L CELL

Bipolar transistors are also used in dynamic memories. Because of its considerable importance, an I²L configuration is selected as an example. Shown in Figs. 10-21*a* and *b* are a cross-sectional view of a dynamic I²L cell and its schematic diagram with standby voltages.[6] With reference to these, line W_P is connected to the P-type injector, the row-select line is the collector terminal of the NPN transistor, and the heavily doped buried N^+ region serves as the column-select line. Presence or absence of charge stored in the emitter-junction depletion-layer capacitance C_1 of the NPN device represents the logic, with a junction voltage of 3 denoting a logical 1 and a voltage of zero denoting a logical 0. In the standby state of Fig. 10-21*b*, transistor Q_1 is *off* and capacitance C_1 can discharge only by means of leakage current through the reverse-biased junctions. Refreshing at regular intervals prevents appreciable decay of any stored charge.[9]

[9] Reference 9, page 280.

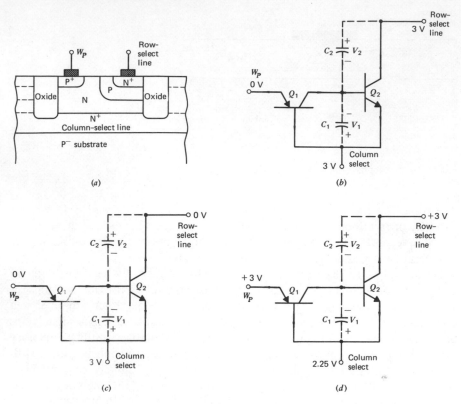

Figure 10-21   Dynamic I^2L cell and READ/WRITE operations. (a) I^2L dynamic cell. (b) Cell schematic with standby voltages ($V_1 = 3$ or 0). (c) WRITE 1, and READ. (d) WRITE 0.

Except for the common N^+ buried layer which serves as the column-select line, the cells of a column are separated from one another by recessed-oxide isolation walls. The columns are similarly isolated, using deeper oxide walls. The technique gives a cell no larger than the one-transistor MOS cell. In order to write a logical 1 into the cell, the row-select line is lowered to a voltage of zero, as shown in Fig. 10-21c. Transistor Q_2 conducts briefly in the inverse mode, charging C_1 to about 3 V. In equilibrium the base current is zero, which requires a small forward bias across the collector junction, but the negative voltage V_2 is very nearly zero. When the cell is returned to standby, the capacitances C_1 and C_2 are in parallel with equal voltages. Because of the relatively large emitter junction, C_1 is substantially larger than C_2, and the voltage across C_1 drops only a small amount below 3 V. The *net* negative charge on the electrodes of C_1 and C_2 connected to the base is trapped, representing a logical 1. To write a zero, the column-select line is lowered to about 2.25 V and the row line W_P is raised to 3 V. Transistor Q_1 drives Q_2 into active-mode operation with V_1 equal to -0.75 and V_2 zero. When standby is resumed, C_1 discharges through Q_2, making V_1 zero.

The procedure for writing a one is also used for reading; that is, the voltage of the row-select line is lowered to zero. If a logical 1 is stored, there is no appreciable transient current in the column-select line. Capacitor C_2 discharges through Q_2, and the voltage across C_1 remains almost constant at 3 V. However, if a logical 0 is stored, it is sensed by the current required to charge C_1, and because the cell now stores a one, the zero is rewritten into it. Refreshing is accomplished by performing a READ operation, followed by a rewrite if the stored bit is sensed as a logical zero. Although the write-zero operation has no significant effect on unaddressed cells, the write-one operation affects all cells of the addressed row. Column currents are sensed to detect stored zeros in cells of the row other than the cell addressed, and these zeros are rewritten.

A cell similar to the one described is employed in a 4-kilobit dynamic random access memory. The integrated circuit requires a single 5 V supply. Access time is 140 ns, and power dissipation is 400 mW in normal operation and 80 mW in standby. The cell matrix is 32×128. The 128 cells of a row are refreshed simultaneously.

There are variations of the cells that have been presented here, as well as new ideas with considerable promise. Undoubtedly, research and development will lead to improved cells with increased densities, higher speeds, and reduced dissipation. The only safe prediction for the future of memory cells is that the rapid change of the past will continue into the foreseeable future. However, the basic principles presented here will certainly be important for years to come.

10-4. SEQUENTIALLY ACCESSED STORAGE

In this chapter we have thus far considered only memories that store bits in a somewhat random manner, with approximately the same access time for each bit. However, there are important memories, known as *sequentially accessed memories* or *serial-access memories*, that store data in serial form. Data are entered in time sequence, with each bit having a unique time delay from a reference. To access a location the stored bit must be moved in sequence with the others until it reaches the output. The time required to access a particular bit is the *latency time*, and this depends considerably on the relative position of the bit in the serial data train.

EXAMPLES OF SERIAL-ACCESS MEMORIES

A familiar example of a sequentially accessed memory is magnetic tape. Mechanical motion of the tape moves the stored bits in sequence to an output for data processing. Both magnetic tape and magnetic disk store sequential data and both are mechanically accessed. Although their speed is quite slow compared with that of semiconductor RAMs, they offer very low cost per bit. Along with paper tape and punched cards they dominate the sequentially accessed mass-memory market. It appears likely, however, that certain forms of highly integrated serial-access semiconductor memories will achieve similar low cost per bit, thus enabling them to complement and compete with these types. One of these semiconductor memories, referred to as *CCD*, will be examined shortly.

A cascade of electronic shift registers is a serial-access memory. Examples are the four-step JK and RS shift registers and the circulating register of Figs. 9-13 and 9-14 of Sec. 9-3. The stored bits are shifted in synchronism from one stage to the next during a clock period. There is always an output from the last stage, and outputs from the other stages are sometimes provided. In a static register the clock rate can be reduced to zero and the bits remain stored, whereas a dynamic register has a minimum allowed shift rate determined by charge leakage from storage capacitors. An example of a dynamic shift register is a cascade of one-bit delay networks of the form of the two-phase logic MOS configuration of Fig. 8-7 or the four-phase configuration of Fig. 8-11, both of which were discussed in Sec. 8-3.

Shift registers are extensively utilized for temporary data storage. An example is the storage associated with on-chip peripheral circuitry that refreshes the rows of a dynamic RAM. They serve as buffer input/output memories in many types of integrated circuits and are widely used in calculators. Several applications related to counting, serial-to-parallel and parallel-to-serial data conversion, and frequency division were mentioned in Sec. 9-3.

There are many serial-access integrated-circuit memories of different capacities and arrangements. One is the TMS 3140 PMOS memory consisting of nine static shift registers of 133 stages each. Depletion-mode P-channel load devices made with ion-implantation technology are employed. Each register has on-chip logic to provide circulation of data. During circulation, data are available at the output and the data input is inhibited. A clock frequency between zero and 1.5 MHz is allowed. At 1 MHz the typical power dissipation is 330 mW.

An example of a dynamic sequential memory is the TMS 3409 Quad PMOS integrated circuit consisting of four 80-bit shift registers with independent inputs, outputs, and recirculate controls for each register. P-channel enhancement-mode devices with low-threshold processing and self-aligned gates are used. Necessary clock phases are generated on the chip from an external clock signal with a frequency between 10 kHz and 5 MHz. Typical power dissipation at 1 MHz is 285 mW.

A serial memory of increasing importance today is that which uses a *charge-coupled device* and is referred to as a *CCD* memory.[7] Although latency times are much greater than the access times of RAMs, typically from 100 microseconds to several milliseconds, the memories provide large bit capacity at low cost. Storage of 64 kilobits or more, along with peripheral circuits, can be placed on a silicon chip with an area less than half a square centimeter. Such high-capacity, low-cost memories provide the potential for replacing magnetic drum and disk memories of digital equipment with solid-state circuits. After a brief study of the principle of the charge-coupled device, we examine the characteristics of a CCD memory.

A CHARGE-COUPLED DEVICE

Figure 10-22 is a simplified representation of a limited portion of the cross-section of a charge-coupled device with electrons stored. Shown are three metal plates above a thin oxide layer on the surface of a P substrate. MOS technology is employed. With respect to the neutral substrate the electrode voltages are 5, 10, and 5. Assuming a threshold voltage V_T of 2, an inversion layer is certainly present under equilibrium conditions.

As we shall see, in normal operation electrons clustered at the surface are swept from left to right by moving fields and when they reach the end they can be eliminated by a logic gate. In fact, whenever a WRITE operation is performed, the first step is to remove inversion-layer electrons. Let us assume that the free electrons of the inversion layer of Fig. 10-22 have been removed, leaving a net charge of zero in the surface region between the depletion layer and the oxide. Also, we assume that the flow of electrons into the surface region from the depletion layer is negligible. Because a new inversion layer can be formed by raising slightly the voltages of the metal plates, it follows that the voltages at points of the surface are less than the respective electrode voltages by V_T, or 2 V.

In Fig. 10-22 the dashed line represents the *surface potential* ϕ_s, which is 3 V at points under the 5 V plates and 8 V at points beneath the 10 V plate. Note that the potential axis is directed downward.

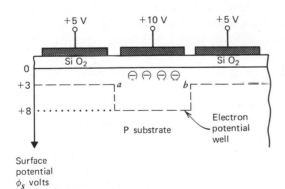

Figure 10-22   Charge storage in a charge-coupled device.

Now suppose a few free electrons are injected into the surface region between the depletion layer and the oxide. (There are various ways to accomplish this; several are described later.) These electrons are attracted most strongly to the surface region directly beneath the 10 V plate and held there, because this is the region with the highest voltage. Here the surface potential is 8 V, and the electron potential energy is a minimum at -8 eV. The electrons are said to be trapped within the illustrated *potential well*, which lies between points a and b. This is an expedient for expressing the fact that they are strongly attracted to the 10 V plate. The potential well stores the electron charge.

The flow of electrons into the surface region from the depletion layer has been neglected. Electrons are constantly being thermally generated within this layer, and the intense field normally present quickly pulls such electrons to the surface. However, once depleted of electrons the surface remains reasonably depleted for a substantial period of time. With processing techniques designed to reduce the densities of carrier-generation centers to a minimum, thereby giving a large minority-carrier lifetime, this period is at least tens of microseconds and may be many milliseconds. The thermally generated carriers that do accumulate are eliminated by frequent refreshing.

In the illustration the electrons are properly shown at the surface. The potential well drawn below the electrons visually portrays the fact that they are held in a region of minimum potential energy by the most positive electrode above the oxide. To transport the charge we need only to shift the position of the potential well. This is accomplished by applying two or more pulse trains to the electrodes with the pulses timed so as to move the minimum potential from left to right.

CCD MEMORY

A three-phase system uses three pulse trains to transfer the charge, as indicated in Fig. 10-23. Using a logical 1 for electron store and zero for no charge, the bits of the potential wells of the top figure are 1, 0, and 1 from left to right at time t_0. Then the voltage applied to electrodes 2, 5, and 8 is raised to 15, and the bits are stored below these at time t_1 as shown in the middle configuration. The voltages are again changed as indicated in the bottom figure. The potential wells and their stored bits have shifted to the right through a distance of one electrode during the sequence of pulses occurring between t_0 and t_2.

At the input of a charge-coupled device, bits are introduced in various ways. One procedure is to pulse the electrode at the input end of the device with a sufficiently high voltage to produce a momentary avalanche breakdown. Breakdown occurs in the depletion region beneath the electrode, supplying a packet of electrons to the surface. Timing must be such that a potential well is present and commencing its trip from input to output. The absence of a breakdown pulse stores a zero in the well. Holes generated during breakdown flow into the neutral substrate.

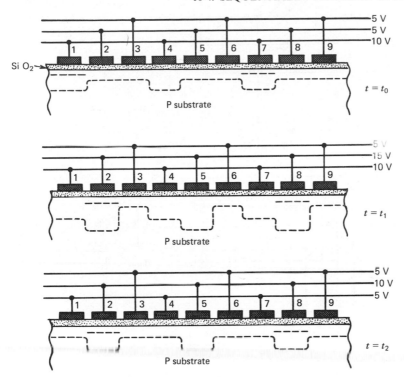

Figure 10-23   Charge transfer in a three-phase CDD.

An alternate technique consists of applying a pulse of light to the surface region below the input electrode. Hole-electron pairs are created within the depletion layer, and the electrons flow to the surface. The principle can also be used to sense light and to store optic signals serially. CCD *image sensors* have many applications, among which are TV cameras, missile-guidance systems, and satellite surveillance.

A third method often employed to supply charge to a potential well utilizes a PN junction A heavily doped N region at the input side of the surface is forward biased with respect to the P substrate. Electrons are injected from the N region, and those that flow into the portion of the depletion layer located between the neutral substrate and the surface constitute the packet which is stored as a bit. This method and a somewhat similar method for detecting the stored bits at the output are illustrated in Fig. 10-25, which will be examined shortly.

A practical configuration is the Intel 2416 CCD 16-kilobit serial memory, organized in the form of 64 recirculating shift registers of 256 bits each and available in a standard 18-pin package. A simplified block diagram is shown in Fig. 10-24. The 18 pins provide for the 15 inputs of the figure plus one each for

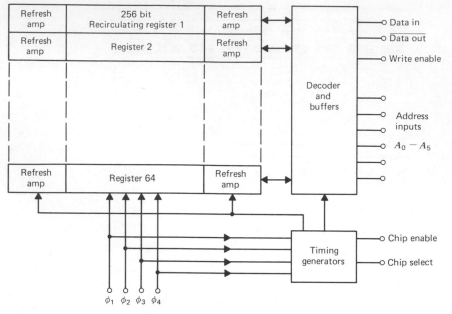

Figure 10-24   Block diagram of 16,384-bit CCD memory.

V_{DD}, V_{BB}, and V_{SS}. These supplies are normally 12, -5, and 0 V, respectively. Because 255 shift operations are necessary to move a bit stored at the input end of the memory to the output and a shift cycle requires 750 ns, the maximum latency time is about 200 μs. The electrodes are silicon gates.[8]

The shift registers recirculate data as long as the four-phase clocks ϕ_1, ϕ_2, ϕ_3, and ϕ_4 are continuously applied and write-enable is low-level. When either ϕ_2 or ϕ_4 makes a low-to-high transition, a one-bit shift occurs in each of the 64 registers. Readout is nondestructive. Access to a recirculating register is accomplished randomly between shift operations by means of the six-bit address, which selects the desired register for reading or writing. Because of substrate leakage currents, operation is dynamic. A shift operation must be performed every 9 μs, and refreshing is accomplished by amplifiers positioned at each end of each recirculating register.

REFRESH AMPLIFIER

A simplified diagram of a refresh amplifier is shown in Fig. 10-25. Pulses ϕ_P, ϕ_R, ϕ_1', and ϕ_2' are generated on the chip, with a high level of 12 and a low level of 0. Let us assume each transistor has a threshold voltage of 2, and the

[8] Reference 8, pages 10–18.

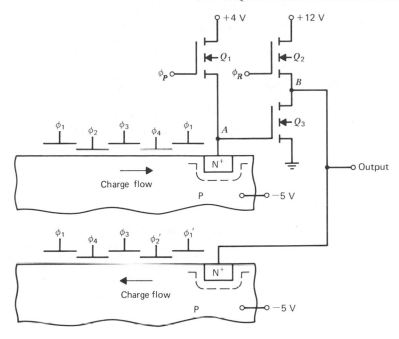

Figure 10-25 Simplified diagram of refresh amplifier.

charge packets of the top channel have charges of -0.03 pC and -0.20 pC for respective logic states of 0 and 1. Also, node A has a capacitance of 0.07 pF. The sequence of events is as follows:

1. Node A is preset to 4 V by pulse ϕ_P which goes from 0 to 12 to 0. The node stores $+0.28$ pC, equal to the product of capacitance C and voltage V.
2. Next, pulse ϕ_2' goes high, creating a deep potential well under its electrode of the bottom channel.
3. Transistor Q_3 is *on* with its gate at 4 V. Hence the voltage of node B is zero. Pulse ϕ_1' goes from low to high, establishing a positive surface potential beneath the electrode. Electrons are injected from the N^+ region of the bottom channel into the surface region, flowing into the potential well under electrode ϕ_2'. The charge store under electrode ϕ_2' becomes a logical 1, with the potential well completely filled with electrons.
4. The timing is such that negative charge from the top channel is now dumped onto node A and its voltage falls. A logical zero causes the voltage of node A to drop to 3.6, but transistor Q_3 remains *on*. A logical 1 lowers the voltage to 1.1 V and Q_3 turns *off* (see P10-12).
5. Transistor Q_2 is pulsed *on* and *off* by ϕ_R. After this pulse, the voltage of node B is 0 if Q_3 is *on* and it is 10 V $(V_{DD} - V_T)$ if Q_3 is *off*.

6. For the case of zero voltage at node B, the charge stored under electrode $\phi_2{'}$ remains, and the packet stores a logical 1. However, if node B is at 10 V, the potential of the N^+ region of the bottom channel is slightly greater than the surface potentials under the adjacent electrodes. Thus charge is withdrawn, and the packet stores a zero.

As indicated by the preceding analysis, a logical 0 received at the output of the top channel is transformed into a 1 at the input of the bottom channel, and vice versa. Thus the refresh amplifier inverts the data. Because there are two refresh amplifiers in each recirculating loop, data consistency is maintained.

CHARGE-COUPLED RAM CELL

The charge-coupled principle can be applied to random-access memories as well as to serial-access memories. In Fig. 10-26 is shown the structure of a charge-coupled RAM cell.[8] It is simpler and smaller than both the one-transistor cell of Fig. 10-19 and the surface-charge RAM cell of Fig. 10-20. Note the P-type ions implanted in the silicon just below the oxide layer. As a

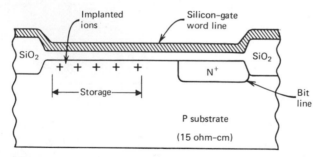

Figure 10-26   Charge-coupled RAM cell.

consequence, with the word-line voltage at zero the potential under the implants is larger than that on either side, and a potential well capable of holding electrons is formed. The doped N^+ region is the bit line. When the word line is raised to 10 V, a forward bias on the bit line injects electrons into the potential well, thus storing a logical 1. To store a zero the bit-line voltage reverse biases the junction, and any stored electrons are removed. With the word-line voltage at zero, the potential well is isolated from the bit line. The READ operation is the same as the WRITE operation, with the bit line sensing current.

[8] Reference 8, pages 58–63.

REFERENCES

1. G. Luecke, J. P. Mise, and W. N. Carr, *Semiconductor Memory Design and Application*, McGraw-Hill, New York, 1973.
2. L. L. Vadasz, H. T. Chua, and A. S. Grove, "Semiconductor Random-Access Memories," *IEEE Spectrum*, **8**, May 1971.
3. J. Springer, "Designers' Guide to Semiconductor Memory Systems," *EDN*, **19**, Sept. 5, 1974. R. J. Frankenberg, "Designers' Guide to Semiconductor Memories—Part 2," *EDN*, **20**, Aug. 20, 1975.
4. *The Semiconductor Memory Data Book*, Texas Instruments, Inc., Dallas, Texas, 1975.
5. *IEEE Journal of Solid-State Circuits (JSSC)*, **SC-8**, 5, Oct. 1973, Special Issue on Semiconductor Memory and Logic.
6. IEEE 1976 International Solid State Circuits Conference, *Digest of Technical Papers*, **XIX**, IEEE Cat. No. 76CH1046–2.
7. W. S. Boyle and G. E. Smith, "Charge-Coupled Devices—A New Approach to MIS Device Structures," *IEEE Spectrum*, **8**, July 1971.
8. *IEEE JSSC*, **SC-11**, 1, Feb. 1976, Special Issue on Charge-Transfer Devices.
9. *IEEE JSSC*, **SC-6**, 5, Oct. 1971, Special Issue on Semiconductor Memories and Digital Circuits.
10. J. Oliphant, "Designing Non-Volatile Memory Systems with Intel's 5101 RAM," Application Note AP-12, Intel Corp., Santa Clara, Calif., 1975.

PROBLEMS

Section 10-1

10-1. In the row-decode circuit of Fig. 10-2b, assume that k_I of each input transistor is 10 times k_L of the load device, each transistor has a threshold voltage of unity, and the current of the row-select line is negligible. Determine the row-select voltage if the respective logic states of A_0, A_1, A_2, A_3, and A_4 are (a) 0, 0, 1, 1, 0 and (b) 1, 0, 0, 0, 0. A logical 1 is 5 V. (*Ans*: 4, 0.065)

10-2. For the 3-to-8 decoder of Fig. 10-3, construct a table relating the logic states of A_5, A_6, and A_7 to each high-level column select.

10-3. Suppose a memory with WRITE-cycle timing of Fig. 10-8 has actual times in ns as follows: WRITE cycle 800, address set-up 200, $\overline{CE}$ set-up 40, data set-up 350, address hold 100, $\overline{CE}$ hold 30, and data hold 50, All times are at least as great as specified minima. With time zero taken as the instant at which the address commences, determine in ns the instants at the beginning and end of the WRITE pulse, the $\overline{CE}$ pulse, and the valid data. The timing points are the 50% points.

Section 10-2

10-4. Suppose transistors T_1 and T_2 in the TTL cell of Fig. 10-10 have respective base-emitter and collector-emitter saturation values of 0.8 and 0.2 V. Calculate the standby cell dissipation in mW and the readout current in mA.

10-5. With reference to the input circuit of Fig. 10-11, assume a conducting transistor has a base-emitter voltage of 0.75, PN-diode forward drops are 0.75 V, Schottky-diode forward drops are 0.45 V, and the PNP lateral transistor has a dc beta of 10. In mA find the input current when the input voltage is zero, and also find the base current of the Schottky transistor Q_2 when the input voltage is 3.

10-6. In the ECL memory of Fig. 10-12, suppose the word line is accessed with -2 V and the base voltage of transistor Q_6 is raised from -4 to -2.7. Assume Q_4 is *off* and calculate the currents $I_1, I_2, I_3, I_4,$ and I_5 in mA, the base-emitter voltages v_{BE4} (both emitters), and the total power dissipation in mW. Neglect base currents and use 0.75 V for diode drops. (*Ans:* P $= 19.4$)

10-7. With the ECL memory cell of Fig. 10-12 in standby with transistor Q_3 conducting, assume each of the six transistors has a dc beta of 40 and a base-emitter voltage of 0.70 when conducting. In mA calculate the currents $I_1, I_2, I_3, I_4,$ and I_5, the base-emitter voltages v_{BE4} (both emitters), and the power dissipation in mW of the storage cell only. (*Ans:* P $= 8.10$)

10-8. Repeat P10-7, except let the word line be at -2 V, which gives the READ operation. (*Ans:* P $= 6.70$)

Section 10-3

10-9. In a form similar to that of Fig. 9-12 of Sec. 9-3, sketch a five-stage binary ripple counter. Show and label outputs $A_4, A_3, A_2, A_1,$ and A_0 and also their complements, with A_4 denoting the output of the first flip-flop. During refresh, these outputs supply the five row addresses of a 32-row dynamic RAM. The 10 outputs are connected to the 160 inputs of 32 five-input NOR gates, which supply positive pulses in sequence from row 0 to row 31. Sketch connections from the ripple counter to the NOR gates of row 0, row X, and row 31, which have respective outputs $\bar{A}_0\bar{A}_1\bar{A}_2\bar{A}_3\bar{A}_4$, $A_0\bar{A}_1\bar{A}_2 A_3\bar{A}_4$, and $A_0 A_1 A_2 A_3 A_4$. What is the number of row X?

10-10. In the one-transistor cell of Fig. 10-18 suppose the storage capacitor is charged to $+2$ V representing a logical 0. The data line is precharged to $+5$ V, and its total impedance is represented by the capacitance C_2 which equals $10C_1$. When transistor Q_1 turns *on* so that the cell can be read, determine the drop in the data-line voltage in mV. Also, find the rise in this voltage if the storage capacitor is charged to 7 V representing a logical 1.

10-11. Suppose the I²L cell of Fig. 10-21 is in standby, storing a logical 1 with $V_1 = 2.5$ V at time zero. For respective emitter-junction and collector-junction capacitances of 0.8 and 0.2 pF and a total leakage current of 0.4 nA across the reverse-biased PN emitter and collector junctions, calculate the time at which refreshing must occur to prevent the voltage from dropping below 1.7 V. Assume the leakage current and the capacitances are constant throughout the voltage range.

Section 10-4

10-12. In the CCD refresh amplifier of Fig. 10-25 the capacitance of node A is 0.07 pF, and it is precharged by pulse ϕ_p which goes from 0 to 12 to 0. A logical-0 charge packet of -0.03 pC is detected initially. With node A again precharged, this is followed by a logical-1 packet of -0.20 pC. Calculate the respective voltages of node A immediately after the packets are received. From the results deduce that transistor Q_3 is first *on* and then *off*, with $V_{DD} = 12$ and $V_T = 2$. Also, deduce that the data transmitted on the lower channel are inverted. In addition, find the number of electrons in a logical-0 packet.

10-13. A one-megabit storage card contains 64 Intel 2416 CCD memories. Suppose an external BJT circuit, which is used to drive one of the four clock phases of the storage card, supplies to the chip a constant current I during each 50 ns rise interval. Each bit presents a capacitance of 0.04 pF to this clock phase, and the clock pulse is 13 V. Calculate I in amperes. (*Ans*: 10.9)

10-14. An LSI dynamic shift register is organized four words by 80 bits. If the clock repetition rate is 1 MHz, calculate in microseconds the maximum and minimum latency times and also the time it takes a bit to recirculate once.

Chapter Eleven
ROMs, Multiplexers, and Microprocessors

Read-only memories, multiplexers, and demultiplexers implement combinational logic. In the first four sections of this chapter we examine the ROM, the programmable read-only memory (PROM), the erasable programmable read-only memory (EPROM), the programmable logic array (PLA), and multiplexers and demultiplexers. A few of the many diverse applications are investigated, including character generation, code conversion, look-up tables, counters, decoders, and data selection. These are followed by an introduction to both the hardware and software of the important integrated circuit known as the *microprocessor*. Included is a brief discussion of the associated system called the *microcomputer*.

11-1. READ-ONLY MEMORY (ROM)

A read-only memory is a network that stores a fixed program. *Writing is not possible*. Because the digital access word provides a unique preprogrammed word at the output, an ROM is basically a code converter. Each output bit can be expressed in terms of Boolean logic operations performed on the input variables. *Thus a read-only memory is equivalent to an array of suitably selected combinational logic networks*. However, there are usually significant differences. The ROM is fabricated normally as a matrix array of fixed logical 1's and 0's, whereas the equivalent combinational network would consist of interconnected logic gates. Each static or dynamic cell of an ROM can be made from a single MOSFET or BJT, a feature that allows a very dense memory structure. Furthermore, *the bits of an integrated-circuit* ROM *can often be programmed initially by the user*, although once established, they cannot be altered. In contrast, there is no way to change the logic of an IC combinational network without rewiring.

Access is random, with the various locations having approximately the same access times. With the WRITE operation unavailable, ROMs are not true memories. Indeed, they are combinational networks, whereas all true memories are sequential. The outputs of a static ROM are easily fed into a synchronous system by utilizing storage registers of clocked flip-flops. Reading is non-destructive and the memory is nonvolatile.

The ability to initially program a standard ROM integrated circuit provides the digital designer with considerable flexibility. Combinational logic needed for a particular digital function is easily obtained, quite often at very low cost. On the other hand, design and fabrication of equivalent combinational networks are likely to be expensive and may require an intolerable amount of time. In general, such LSI custom-wired gates are economical only when large quantities are needed and when there is sufficient time for design and development. However, most LSI chips with custom-wired gates are substantially faster than LSI read-only memories.

To illustrate certain basic principles of read-only memories let us refer to the NMOS static ROM of Fig. 11-1, recognizing that it is much too small to be practical. There are transistors at 10 cell locations of the 4×4 matrix array, and none are present at the other six. Inputs A, B, and C provide the address for the eight words of two bits each. Transistors Q_3 through Q_6, along with saturated load devices Q_1 and Q_2, constitute the column-select circuitry. Suppose A, B, and C are each zero. Cells 31 and 33 are addressed, because Q_4 and Q_6 are *on* and row $\overline{B}\overline{C}$ is enabled. Because transistors Q_4 and Q_{31} have very small channel resistances, there is a logical 0 at output D_o, and the absence of a transistor at cell 33 gives a logical 1 at output D_1. *The stored bit of each location of the* ROM *is represented by the presence or absence of a transistor*. In the illustrated circuit a transistor denotes a zero. With inverters added at the outputs, a transistor denotes a logical 1.

The Karnaugh map relating the outputs D_o and D_1 to the inputs A, B, and C is easily established directly from the network (see P11-1). The map can then be used to verify that the combinational logic performed by the ROM is represented by the logical expressions:

$$D_o = ABC + A\overline{B}\overline{C} + \overline{A}B\overline{C}$$
$$D_1 = AC + \overline{A}\overline{B}\overline{C} \tag{11-1}$$

In Fig. 11-1 the row-select circuitry can be designed as a second on-chip ROM (see P11-2).

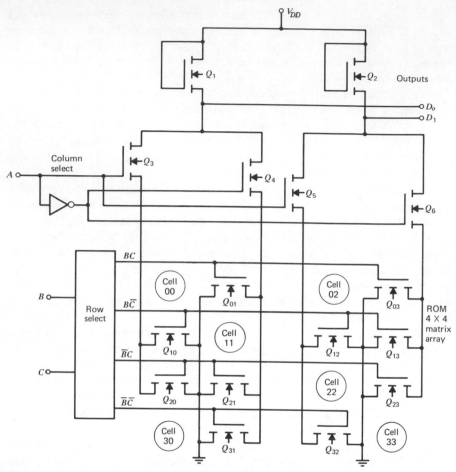

Figure 11-1   NMOS 8-word-by-2-bit static ROM.

Fabrication consists of forming partial transistors at each matrix location. The devices are without gates when data describing the specific ROM organization and characteristics are published. The customer supplies the manufacturer with programming instructions, frequently in the form of a data deck that identifies the desired words and bits. This is the software package. The term *software* is used to describe programs and instructions for digital equipment, in contrast to *hardware* which denotes the physical components. The manufacturer then prepares an appropriate mask which is used to open windows for deposition of the thin oxides of the gates. The mask is designed so that those cells that are to have no transistors retain their thick oxide. After the gate-oxide formation, the metallization, testing, and packaging are completed, and the product is sent to the customer. The difference between the cell structures representing the different bits is simply the oxide thickness.

The MCM6590 is a mask-programmable static 16-kilobit ROM fabricated with N-channel metal-gate technology. Organization is 2048 × 8. Tri-state outputs are provided for easy memory expansion. Access time is less than 800 ns, and typical power dissipation is about 400 mW. In addition to the 128 × 128 matrix array, the chip contains row-select circuitry having seven input pins for addressing the 128 rows, column-select circuitry having four pins for addressing 16 columns of eight bits each, and output buffers with eight pins for data output. There is a pin for chip-select, and the remaining four of the 24-pin dual-in-line package are for the voltage supplies.

A companion integrated circuit is the MCM6591 which is preprogrammed. Half the memory consists of six different character generation codes useful in certain types of character displays. The other half has its 1024 words grouped in sequences of eight, providing 128 sets. Each set contains eight words and each word has eight bits. A set is programmed so that it can be used to identify a character on a visual display. The characters consist of numerals, alphabetic letters, and various special symbols. Using a cathode-ray tube (CRT), for example, and a parallel-to-serial circulating shift register, the bits of a word can be supplied serially to the video input of the CRT. A logical 0 blanks the electron beam and a logical 1 unblanks it. By adjusting the vertical sweep so that the beam moves up the face of the scope during the time interval of the serial word, with the actual displacement in synchronism with the system clock, the character bits are sequentially displayed in a vertical line. Dots appear at locations corresponding to logical 1's. Shown in Fig. 11-2 is a vertical-deflection waveform alongside an 8 × 8 character.

The horizontal deflection is shifted one space to the right when a new word of the set appears, with the shift initiated by a counter controlled by the system clock. In this manner a character of eight columns with eight bits per column is displayed on the scope by the set. Careful timing is essential. The vertical and horizontal step generators must be properly synchronized with the circulating register and also with the address counters that regulate the flow of words into the register. Provision must be made for adequate refreshing of the CRT.

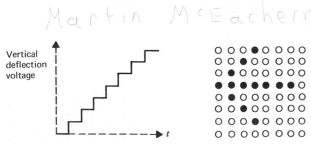

Figure 11-2   Vertical-deflection voltage and 8 × 8 character.

We have found that an ROM cell of a MOS memory consists of a single transistor. It can also be fabricated as a BJT, a diode, a resistor, or a capacitor, and there are other possibilities. In a BJT memory each transistor is arranged with its base terminal either connected to a row-select line or left open-circuited. An example is the SN54S370. This is a static TTL 2-kilobit mask-programmable ROM that employs Schottky-clamped transistors for high-speed operation. Access time is less than 95 ns and power dissipation is less than 770 mW.

PROGRAMMABLE READ-ONLY MEMORY (PROM)

Because mask programming of an ROM is performed during the final phase of semiconductor fabrication, it must be done by the manufacturer. However, there are electrically *programmable read-only memories* (PROM), and the packaged devices are programmed by the user. Figure 11-3 shows the schematic

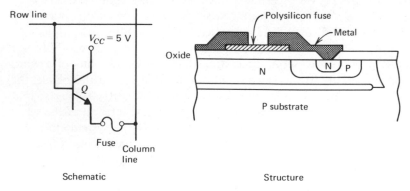

Figure 11-3   Bipolar PROM cell.

diagram and cross-section of a PROM cell. A fuse is in series with the emitter. With the fuse intact the cell represents a logical 1. If a zero is desired at the location, the cell is addressed and driven with a current pulse sufficient to blow the fuse. Programming the zeros at the proper locations is accomplished one cell at a time in accordance with instructions.

The illustrated structure is that of an NPN transistor with a fuse consisting of a notched strip of polycrystalline silicon. The resistivity of the fuse is controlled by doping, and the value is fixed so that a current of about 25 mA opens the fuse with the temperature at the notch near 1400°C. This causes the silicon to oxidize, thus forming an insulating material that eliminates any possibility of reconnection. Because the memory cells have common collectors, many are fabricated within the same isolation well. The Intel 3604 bipolar 4K PROM, organized 512 × 8, has polysilicon-fused cells of the form of the illustration. Access time is less than 90 ns and power dissipation is less than 900 mW.[1]

Metal fuses are also employed in electrically programmable read-only memories. An example is the SN10139 256-bit ECL integrated circuit. This PROM is organized as 32 words of 8 bits each, with an access time less than 20 ns. The low access time is characteristic of ECL memories.

ERASABLE PROGRAMMABLE READ-ONLY MEMORY (EPROM)

Although PROMs can be programmed initially, they then become unalterable. Another type of memory, referred to as *erasable programmable read-only memory* (EPROM), can be reprogrammed. An example is the *F*loating-gate *A*valanche-junction *MOS* charge-storage device (FAMOS). A cross-section of a p-channel device is shown in Fig. 11-4 along with its symbol. There is no connection to the polysilicon gate. The floating gate is isolated from the substrate by a thin oxide layer.[3]

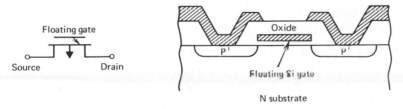

Figure 11-4   EPROM memory cell.

Suppose a 5 ms drain-to-source pulse of -50 V is applied. With a substrate resistivity from 5 to 8 ohm-cm this is sufficient to produce avalanche breakdown in the depletion layer of the drain-substrate junction. Electrons from this layer are accelerated by the intense field toward the P-type drain and some are directed toward the surface region adjacent to the drain. Many of these pass through the thin oxide into the silicon gate at a rate giving a current of about 1 nA. A negative charge accumulates on the gate, and when the drain voltage is removed, this trapped charge is sufficient to induce an inversion layer between source and drain. At least 70% of this charge is present after 10 years or more. The presence or absence of charge can be sensed by means of the conductance between the source and drain.

The procedure that has been described is useful for programming the ROM. However, it is sometimes desirable to be able to erase the program and to develop a new one. Erasure of the stored bits of a packaged device is accomplished

[3] Reference 3, pages 301–306.

by exposure to X-radiation. This results in a flow of electrons from gate to substrate, thus eliminating the charge of the floating gate. The memory can then be reprogrammed, although the number of times this can be done is limited because of radiation damage.[1]

The addition of a metal gate to the surface of the oxide, above the floating gate of the structure of Fig. 11-4, provides an electrical means of erasure. During programming, a positive voltage applied to this gate increases the rate at which electrons accumulate on the floating gate. The storage is erased by again avalanching the PN junction, but with a negative voltage on the second gate. This draws holes to the floating gate, and these recombine with the stored electrons. After programming, the channel is gone.

In addition to the address-decode combinational network and the memory array for bit storage, IC ROMs may have buffers on the bit-line outputs. These circuits provide suitable drive for capacitive loads and are quite common. We have considered here only some of the techniques used to make programmable and reprogrammable read-only memories, and new methods are being developed. ROMs can be interconnected to form programmable logic arrays. These are examined next.

11-2. PROGRAMMABLE LOGIC ARRAYS (PLA)

A simplified manner of representing the logic of an ROM is desirable in order that large arrays can easily be illustrated. In Fig. 11-5a is shown an eight-cell PMOS ROM with four row-select lines and two outputs. For *positive* logic, the circuitry is equivalent to the arrangement of the NAND gates of Fig. 11-5b. Let us consider the output $\overline{A}\overline{B}$. If either A or $\overline{B}$ is at -10 V, corresponding to a logical 0, at least one of the two transistors connected to the output line is *on*, and the output voltage is near zero, which is a logical 1. The symbolic representation is that of Fig. 11-5c. Whenever this form is used, we shall understand that *the output gives the* NAND *operation on all those inputs joining the output line at the dotted junctions.* Unless specifically stated otherwise, this is assumed to be valid regardless of the type of transistor or device employed as cells of the ROM.

It is possible to extend the complexity of ROM logic by using more than one cell matrix. Suppose two matrices are formed on the same silicon chip, with the outputs of one fed directly into the inputs of the other. The array can be mask-programmed during production, just like a conventional ROM, and the combination is referred to as a *programmable logic array*, or *PLA*. The advantage is that complicated logic can be realized with substantially fewer cells than required by an equivalent ROM.[2]

---

[2] Reference 2, Chapter 6.

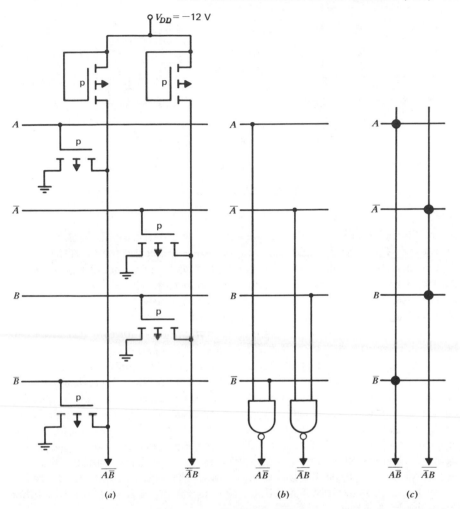

Figure 11-5   (*a*) ROM matrix. (*b*) Positive-logic equivalent diagram. (*c*) Symbolic representation.

Figure 11-6 shows a PLA configuration, although too small to be practical. The outputs of the top ROM are $\overline{A\overline{B}}$, $\overline{\overline{A}B}$, and $\overline{\overline{A}\overline{B}}$. These are the inputs of the bottom ROM, which also performs the NAND operation. With the aid of DeMorgan's laws the outputs Q_1 and Q_2 are found to be

$$Q_1 = A\overline{B} + \overline{A}B + AB = A\overline{B} + B$$
$$Q_2 = A\overline{B} + \overline{A}B \qquad\qquad (11\text{-}2)$$

We note that each output has the form of a sum of products, and the logic is functionally equivalent to that of a collection of AND gates followed by OR

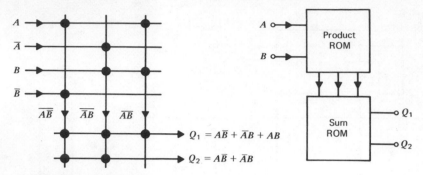

Figure 11-6   PLA configuration consisting of two ROMs.

gates. The product bits are determined by the top matrix, which is referred to as a *product* (AND) ROM, and the sums are determined by the bottom, or *sum* (OR), matrix. For example, the two dots of the bottom row signify that Q_2 is the sum of two product terms. Projecting upward from these dots into the top matrix, the corresponding dots of this matrix identify the product terms. By feeding the outputs Q_1 and Q_2 through flip-flops or some other form of time-delay circuitry into the A and B inputs, *sequential logic is possible.* Flip-flops can be included on the chip of a PLA. *There are no decoders.*

Because large-scale integrated PLAs are available at low cost, substantial savings often are possible by using them in place of random-logic designs made of interconnected gates. Furthermore, the designer can more quickly implement complex multilevel logic, with programming accomplished in a single mask step. The speed of a PLA is almost independent of the complexity of the logic equation realized. In contrast, the speed of an equivalent random-logic design depends strongly on the total number of gates and especially on the number that are in series. Logic systems consisting of only a few series-connected gates are likely to be considerably faster than the equivalent PLA, but in many other cases the PLA is faster.

FIELD-PROGRAMMABLE LOGIC ARRAY (FPLA)

Some PLAs are designed so that the packaged device can be programmed electrically by the user. Such a device is referred to as a *field-programmable logic array (FPLA)*. It is also possible to make reprogrammable arrays. Various methods are employed, and these are similar to those applied to PROMs and EPROMs.

An example is the IM5200 FPLA, available in a 24-pin dual-in-line package and having a logic diagram of the form of that of Fig. 11-6. There are 14 inputs and 8 outputs, and the remaining two pins are for the 5 V supply. Associated with each input are four inverters which provide buffering and inversion. The 28 outputs of the buffers are the 14 inputs and their complements. These connect to 28 row lines, and each row line of the product array has 48 cells.

The 48 outputs of the product array feed into the 8 × 48 sum array, which has eight buffered outputs. Each one consists of from 0 to 48 sum terms with each term made up of a product of various inputs and their complements. The logic depends, of course, on the programming. This is accomplished electrically. The cells are NPN transistors with open bases, and they act as open circuits. By forcing a high current through a transistor from emitter to collector, the emitter junction breaks down, and the high temperature that results causes aluminum to short-circuit the junction. This forms a diode. Thus each cell is either an open-circuit or a conducting diode.

MODULO-10 COUNTER

Applications of PLAs are numerous. They can provide random combinational logic, code conversion, microprogramming, and control of sequential circuits, counters, registers, and RAMs. They also find use as look-up tables, character generators, decoders, and encoders. Some of these are discussed in the next section. Here we examine an important application, which is that of a modulo-10 counter.[2]

The function of the counter is to activate on a seven-segment display a decimal digit from 0 through 9, representing the number of clock pulses that have occurred. After the ninth pulse the counter is to repeat the count, beginning with zero. The PLA must be programmed to count in sequence, and the binary number must supply the proper pulses to the circuitry driving the display unit. The sequence from 0 to 9 is shown in Table 11-1. The binary number $ABCD$ denotes the present states of the flip-flops of the counter and $A'B'C'D'$ of the state table represents the next FF states. Ths most significant bit (MSB) is A, and D denotes the least significant bit (LSB). The number $A'B'C'D'$ of Table 11-1, which corresponds to pulse n, is the same as $ABCD$ corresponding to pulse $n + 1$. After the ninth pulse, the next state is 0000 because the sequence repeats itself. The design is easily accomplished with Karnaugh maps (see P11-5), and the equations that apply are

$$A' = BCD + A\bar{D}$$
$$B' = B\bar{C} + B\bar{D} + \bar{B}CD$$
$$C' = \bar{A}CD + C\bar{D} \qquad (11\text{-}3)$$
$$D' = \bar{D}$$

We note that (11-3) satisfies the sequence of Table 11-1. For example, after pulse number 5 the binary number $ABCD$ is 0101 by Table 11-1. The equations of (11-3) give $A'B'C'D'$ to be 0110, which is the binary number after pulse 6.

---

[2] Reference 2, Chapter 6.

Table 11-1   State Table of Modulo-10 Counter

Pulse Number	Present FF States A B C D				Next FF States A′ B′ C′ D′			
0	0	0	0	0	0	0	0	1
1	0	0	0	1	0	0	1	0
2	0	0	1	0	0	0	1	1
3	0	0	1	1	0	1	0	0
4	0	1	0	0	0	1	0	1
5	0	1	0	1	0	1	1	0
6	0	1	1	0	0	1	1	1
7	0	1	1	1	1	0	0	0
8	1	0	0	0	1	0	0	1
9	1	0	0	1	0	0	0	0

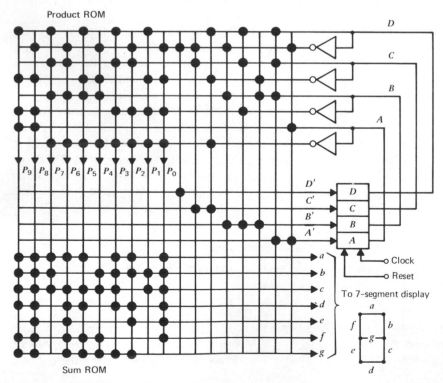

Figure 11-7   Modulo-10 counter using PLA.

334

Now let us examine the modulo-10 counter of Fig. 11-7. It consists of a product ROM, a sum ROM, and four D-type flip-flops. The clock pulses to be counted control these one-bit delay flip-flops. At the flip-flop outputs is the binary number $ABCD$. The eight columns on the right side of the two ROMs provide the logic of (11-3). Consequently, the input word $ABCD$ of the product ROM changes with each clock pulse in accordance with the sequence of Table 11-1.

The outputs of the lower ROM labeled a through g feed into the circuits driving the respective segments of the display. Suppose the word $ABCD$ is 0000. For this input the product ROM is designed so that P_0 is a logical zero and outputs P_1 through P_9 are ones. Consequently, outputs a through f are logical 1's and g is 0. The segments corresponding to the 1's properly display a zero. Each time a clock pulse occurs, the seven-segment display advances one decimal digit.

Let us view the logic configuration from another viewpoint. From the dots of the ROMs of the PLA, we find that output e has the logic

$$e = A\bar{B}\bar{C}\bar{D} + \bar{A}BC\bar{D} + \bar{A}B\bar{C}\bar{D} + \bar{A}\bar{B}C\bar{D} \tag{11-4}$$

From this and the data of Table 11-1 we find that segment e is activated for pulse numbers 0, 2, 6, and 8. A little reflection based on the preceding discussion reveals that the complete programming of the ROMs is readily deduced from the stated requirements and (11-3).

11-3. ROM APPLICATIONS

We have learned that the read-only memory is basically a combinational logic network with a specified function corresponding to each word address. This feature is especially useful in code conversion, program storage, and character generation such as that considered in Sec. 11-1. By incorporating feedback from one ROM to another through a one-bit delay network, sequential logic is possible, and the modulo-10 counter is an example. ROMs are extensively employed as *look-up tables* for many different mathematical functions. These include x^2, $\sqrt{x}$, trigonometric, exponential, and logarithmic functions, and others. The memories can perform arithmetic logic, and the operation of many control systems is governed by the fixed program storage of an ROM. Here we briefly examine only a few applications.

CODE CONVERSION

To illustrate code conversion, let us consider the procedure for converting from a binary code to a Gray code. The codes are related by the truth table of Fig. 11-8. Although there are different Gray codes, each has the characteristic

Decimal	Binary ABCD	Gray abcd
0	0 0 0 0	0 0 0 0
1	0 0 0 1	0 0 0 1
2	0 0 1 0	0 0 1 1
3	0 0 1 1	0 0 1 0
4	0 1 0 0	0 1 1 0
5	0 1 0 1	0 1 1 1
6	0 1 1 0	0 1 0 1
7	0 1 1 1	0 1 0 0
8	1 0 0 0	1 1 0 0
9	1 0 0 1	1 1 0 1
10	1 0 1 0	1 1 1 1
11	1 0 1 1	1 1 1 0
12	1 1 0 0	1 0 1 0
13	1 1 0 1	1 0 1 1
14	1 1 1 0	1 0 0 1
15	1 1 1 1	1 0 0 0

(a)

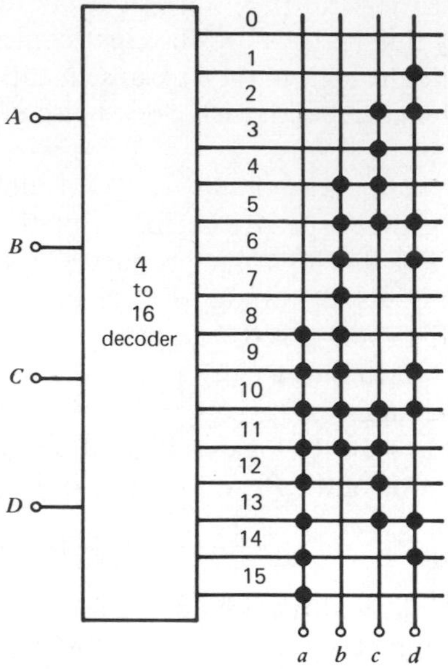

(b)

Figure 11-8   Code conversion, binary to Gray. (a) Truth table. (b) ROM.

that only a single bit changes when the corresponding decimal number changes by one. This is especially important when electromechanical systems are controlled by a computer. For example, suppose the speed of a motor is regulated by a binary digital signal. When the sequence changes from decimal 7 to 8, four bits of the binary number change as shown in the table of Fig. 11-8. Because the changes in these bits are not exactly synchronized, the motion of the motor is momentarily erratic. This problem is avoided with the Gray code.

The ROM organization is that of Fig. 11-8b. Assuming PMOS with positive true logic, the dot matrix performs NAND logic. A row is accessed with a logical zero corresponding to a negative voltage. The 4-to-16 decoder is designed so that row X is accessed by the binary number corresponding to the decimal number X. For example, the input binary number 0111 supplies a logical zero to row 7 and a logical one to all other rows. Utilizing NAND logic, the matrix of Fig. 11-8 reveals that the output word abcd is 0100. Note that row 7 stores the Gray code corresponding to the binary number that represents 7. The conversion is extended to larger words by employing ROMs with more cells.

SINE LOOK-UP TABLE

Suppose a 1024-bit ROM, organized 128 × 8, is designed as a look-up table for the function sin θ, with θ restricted to values from 0 to 90°. The first row is programmed for $\theta = 0$, and the remaining 127 rows are programmed for multiples of 90/127 degrees, or 0.709 degrees. The seven-bit address word with a 7-to-128 decoder addresses the row corresponding to θ in accordance with the data of Table 11-2. For example, with $\theta = 28.35°$ the input address is 0101000 which supplies a logical zero to row 40, and all other rows are disabled with logical 1's. The eight-bit binary output is 01111001, representing the decimal 121. Because the maximum binary number at the output is 11111111, representing the decimal number 255 corresponding to the sine of 90°, which is unity, the decimal number 121 denotes 121/255, or 0.4745. This is the approximate value of the sine of 28.35°.

Table 11-2 Sine Look-Up Table

Address	Degrees	Programmed Binary Output
0000000	0°	00000000
0000001	0.71°	00000011
0000010	1.42°	00000110
—	—	—
0101000	28.35°	01111001
—	—	—
1111111	90°	11111111

Each row of the look-up table is programmed to give the sine function of the corresponding angle (see P11-11). Greater accuracy can be obtained by using a larger matrix. Suppose a θ accuracy of 11 bits and a sin θ accuracy of 12 bits are required. The ROM must have 2048 rows of 12 bits each, for a total of over 24 kilobits.

An alternate procedure, which reduces the circuitry considerably, is to employ two ROMs arranged so that their outputs are added. The first ROM, organized 256 × 12, is addressed with the eight most significant bits of θ, and these bits define θ_M. Because the maximum angle of 90° is divided into 255 increments, the angles of successive rows differ by 0.353°. The second ROM, organized 256 × 5, is addressed with the eight least significant bits of θ. These bits define θ_L, which has values between 0.353° and zero. The angle θ is the sum of θ_M and θ_L.

The interpolation approach is based on the identity

$$\sin \theta = \sin(\theta_M + \theta_L) = \sin \theta_M \cos \theta_L + \cos \theta_M \sin \theta_L \qquad (11\text{-}5)$$

In (11-5) the term $\cos \theta_L$ can be dropped, because its values are between 0.99998 and 1. Programming is accomplished so that the first ROM gives $\sin \theta_M$ with 12-bit accuracy, and the second gives the product $\cos \theta_M \sin \theta_L$ with 5-bit accuracy. With the outputs fed into adders, $\sin \theta$ is obtained with 12-bit accuracy. This procedure requires two ROMs and adders, but the total bit capacity is reduced from 24 kilobits to 4352 bits. There are other techniques that further increase the accuracy while keeping the number of cells at a minimum. It should be noted that the basic principle of a look-up table is identical to that of a code converter.

GENERATION OF MULTIPHASE PULSE TRAINS

Read-only memories are often used to generate clock phases for multiphase systems, such as those required for driving dynamic random-access and CCD memories. Table 11-3 illustrates the design principles. With T denoting the period of clock pulses fed to a counter, the binary address of the table is sequenced by the counter as indicated, and the ROM is programmed to give the desired outputs ϕ_1 through ϕ_4. The complete cycle requires an interval of $8T$. Each output represents a pulse train with the timing determined by the clock period and the ROM program (see P11-12). Normally, the outputs are supplied to clock-controlled D-type latches, which eliminate unwanted ROM transients from the four-phase lines.

Applications are indeed varied and essentially unlimited. The network designer should certainly consider the use of an ROM whenever combinational logic on a large scale is required.

Table 11-3   ROM Design for a Four-Phase Clock System

Time	Binary Address	ϕ_1	ϕ_2	ϕ_3	ϕ_4
0	000	0	0	0	1
T	001	0	1	0	1
$2T$	010	1	1	0	1
$3T$	011	1	1	0	0
$4T$	100	0	1	0	0
$5T$	101	0	1	0	1
$6T$	110	0	1	1	1
$7T$	111	0	0	1	1
$8T$	000	0	0	0	1

11-4. MULTIPLEXERS

A generally useful integrated circuit is the digital *multiplexer*, also called a *data selector*. It has many applications associated with semiconductor memories. In addition, it is often employed to generate various logic functions. Basically, the network selects one of several input words, supplying the selected word to an output. An example is the four-input multiplexer of Fig. 11-9. The four one-bit input words are designated D_0, D_1, D_2, and D_3. One of these is supplied to the Y output, with the address inputs A and B determining the selection. A low-level voltage at the strobe input enables the multiplexer, whereas a high-level signal disables it, providing a zero output at Y regardless of the data inputs. With the strobe enabled, the network performs combinational logic in accordance with the relation

$$Y = \bar{A}\bar{B}D_0 + \bar{A}BD_1 + A\bar{B}D_2 + ABD_3 \qquad (11\text{-}6)$$

An eight-input multiplexer has three address pins. Thus the logic corresponding to (11-6) contains a sum of eight terms with each consisting of a product of four variables. The input word of a multiplexer may have more than one bit. For example, a multiplexer designed for four-bit words would have four outputs, with the address selecting the desired word. Frequently, the address is sequenced with a counter. Suppose the AB inputs of (11-6) are shifted by a counter through the sequence 00, 01, 10, and 11, with the sequence then repeated. The output would be a serial data train with D_0 appearing first, followed by D_1, D_2, and D_3. This technique is useful in converting data that is in parallel form to series form. It is an example of *time multiplexing*, with the output line transmitting the inputs in time sequence.

There are also *analog multiplexers*. The term *multiplexing* denotes the combining of a number of data lines into a single channel. It is extensively employed in both digital and analog communications systems to reduce the required wiring. The analog multiplexer selects one analog input out of a number of multiple inputs, switching the selected signal to the output.

A *demultiplexer* is the opposite of a multiplexer in that it directs information from a single input to one of several outputs in a sequence determined by information applied to control inputs. We have examined the decoder, in which the binary input address determines which one of the multiple outputs goes low, or perhaps high. By utilizing an enable (strobe) input for data input, the decoder becomes a demultiplexer. For example, in the 3-to-8 decoder of Fig. 10-3 of Sec. 10-1 suppose a fourth input is added to each of the NOR gates, with these inputs connected together through an inverter to an external pin of the IC package. This pin can be used for enabling or disabling the decoder. However, with input data supplied to this pin, the address selects the output line that receives the data. Furthermore, the data can be shifted from one line to another by changing the address (see P11-14).

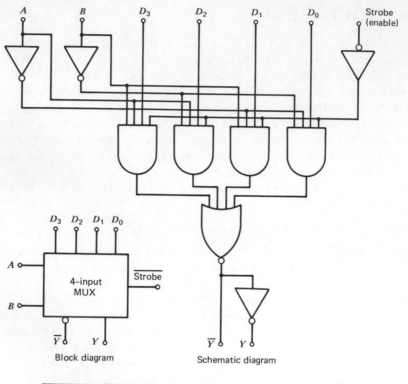

Inputs							Outputs	
A	B	Strobe	D_0	D_1	D_2	D_3	Y	$\overline{Y}$
X	X	1	X	X	X	X	0	1
0	0	0	0	X	X	X	0	1
0	0	0	1	X	X	X	1	0
0	1	0	X	0	X	X	0	1
0	1	0	X	1	X	X	1	0
1	0	0	X	X	0	X	0	1
1	0	0	X	X	1	X	1	0
1	1	0	X	X	X	0	0	1
1	1	0	X	X	X	1	1	0

Truth table

Figure 11-9. Four-input multiplexer.

340

11-5. MICROPROCESSORS

Of special importance to the electronics engineer is a large-scale integrated circuit known as the *microprocessor*. It provides the designer of electronic systems with digital-computing capability which can be used where desired at low cost with minimum turn-around time. Present applications in the industrial, military, and consumer fields are numerous and varied, and future potential applications are indeed unlimited.

Both large and small digital computers can be represented functionally by the block diagram of Fig. 11-10. Input devices are utilized to feed data into the *central processing unit* (CPU). Examples are analog-to-digital (A/D) converters, tape readers, card readers, keyboards, disk drives, and optical readers. The computer processes the input data and delivers results to circuits and devices such as digital-to-analog (D/A) converters, card and paper-tape punches, printers, displays, plotters, and communication lines. Input/output devices are called *peripheral* elements.

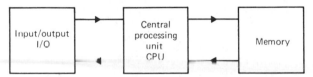

Figure 11-10   Simplified block diagram of a computer.

Computer programs and data are stored in the memory. This usually consists of chips of both ROMs and RAMs, either bipolar or MOS. Timing and general operation are controlled by the CPU, which also performs the necessary arithmetic and logical operations. Synchronous operation is governed by clock pulses.

A microprocessor is an integrated circuit designed for use as the central processing unit of a computer. In some cases it consists of more than one chip. A computer having a *microprocessor* as its CPU is called a *microcomputer*. In contrast with the minicomputer which is designed and sold as a complete operating system, the microcomputer is assembled by the electronics designer who purchases and interconnects various integrated-circuit components. This is done in a manner that satisfies specific requirements of a desired application.

Although a microprocessor is often referred to as a "computer on a chip," it is actually a component intended for interconnection with other integrated circuits to form a computer system. In comparison with a minicomputer the amount of hardware is substantially less, resulting in reduced computing flexibility. Consequently, software instruction programs are more complicated and critical, and greater attention in programming must be given to system hardware components. Software considerations are treated in the next section.

MICROPROCESSOR ARCHITECTURE

The organizational structure has the same basic architecture as the CPU of a larger computer. The simplified block diagram of Fig. 11-11 indicates a number of control inputs and outputs. An external clock provides pulse trains, shown as having two phases. The actual number of inputs and outputs, and their functions, depend on the specific microprocessor.

The instruction word usually has either 4, 8, or 16 bits, and these are processed in parallel. An eight-bit word referred to as a *byte* is common. The *bus* lines in Fig. 11-11 have a wire for each bit, and these busses allow the exchange of words between the blocks. A microprocessor is often classified as either a *CPU* or a *slice*. The types referred to as slices are designed so that they can be connected in parallel, thus increasing the bit capacity. For example, the TI SBP0400 is a four-bit slice made with I^2L technology. It can be operated directly as a four-bit microprocessor, or if desired, four of these can be connected in parallel to give a 16-bit processor. The Motorola 10800 is an extremely fast ECL four-bit slice. An especially popular processor is the Intel 8080 NMOS eight-bit CPU with a

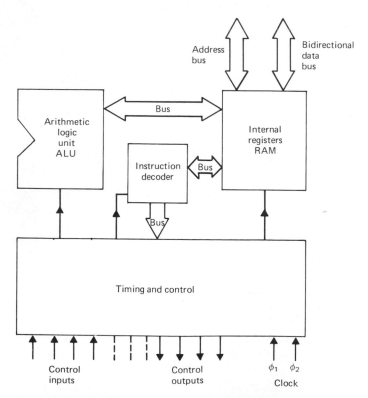

Figure 11-11   Simplified block diagram of a microprocessor.

16-bit address bus. There are many others, and improved devices are being developed. Most use NMOS, silicon-gate CMOS, or Schottky bipolar technology.

ARITHMETIC/LOGIC UNIT (ALU)

As implied by the name, the ALU performs both arithmetic operations and logic operations. Only two words, or *operands*, are operated on at a time. Word A always comes from the internal register known as the *accumulator*, or *A register*. This is a shift register with the number of flip-flops equal to the number of bits in the operand and having parallel-in and parallel-out capability. Word B is either that stored in another internal register or one brought by the data bus from the main memory.

For arithmetic operations the word is interpreted as representing a binary number with the leftmost bit being a *sign* bit. This bit denotes a plus sign $(+)$ when 0 and a minus sign $(-)$ when 1. The twos-complement representation of negative numbers is widely used in microcomputers. In this system a negative number is obtained from the positive number by complementing each bit including the sign bit, and adding 1 to the result. For example, the four-bit word denoting $+3$ is 0011, and -3 becomes 1101. Their sum is 0000 with the leftmost 1 dropped. To perform subtraction an adder simply adds the negative number and drops the leftmost bit in the result. In conjunction with shifting networks, adders can also perform multiplication and division.

Logical operations, such as OR, operate on the corresponding bits of the two words. For example, with A and B represented by 0110 and 1100, the OR operation gives 1110. Upon completion of an arithmetic or logical operation the result is placed in the accumulator.

INTERNAL REGISTERS

We have considered the accumulator register, which always stores one of the words to be processed and accepts new data after processing. Among the other registers is the *instruction register*. It stores the next instruction to be executed. At the proper time this stored word is fed into an internal ROM or other logic network which decodes the instruction and supplies the proper signals to the control section. Depending on the particular bits of the word, the signals may signify that the word located in the main memory at the address stored in a specific internal register is to be incremented by 1 with the result placed in the accumulator. When execution is accomplished, the new word in the instruction register is processed.

Another internal register of particular importance is the *program counter* (PC). It specifies the address of the next instruction to be *fetched* from memory and executed. The program controlling the entire operation is stored in an external

ROM, and instructions are normally withdrawn in order, with each new address incremented by 1. However, the stored program can instruct the CPU to *jump* to some other address.

There is a *memory address* register. It holds the address of the memory location being accessed. *Scratch-pad* registers are often included. These are random-access general-purpose registers. They provide work space, permitting calculations without transferring data to and from the main memory. Intermediate results of ALU operations can be stored here temporarily. One of these may serve as a *pointer*, which is a register holding the memory address of an operand to be used by an instruction. The register "points" to the memory location of the data or other word. Possibly there are *index registers* containing words used to modify an instruction address. Some of the registers may be unavailable for use in the program, being reserved for internal processing of instructions. Transfer of eight-bit bytes between the internal registers through the busses is accomplished by a register-select decoder.

The Intel 8080 has a register array with six 16-bit registers plus an accumulator. an instruction register, and several others reserved for special purposes. In addition to the register-decode network, there is a multiplexer. It is used to transfer contents of the various registers to an internal data bus in a time-multiplexed operation when double-precision operation with 16-bit words is desired.

TIMING AND CONTROL

The control circuitry maintains the proper sequence of events for processing. When an instruction is fetched and decoded, the control circuits provide the appropriate signals to both internal and external units. There are external input and output control terminals. Signals on these can request an *interrupt* or a *wait*. The *interrupt* request temporarily interrupts the main program execution, jumps to a special routine to service the interrupting device, and finally returns to the main program. A *wait* request may come from an external peripheral element that operates slower than the processor. The CPU circuitry is idled until the element is ready.

Timing of the orderly sequence is controlled by the external clock, often with more than one phase. In the Intel 8080 the *clock generator* consists of a 20-MHz crystal oscillator, a four-bit counter, and gating circuits. It is available as an integrated circuit. The proper waveforms are fed from the counter to the timing and control unit. The interval of an *instruction cycle* is the time required to fetch and execute the simplest instruction. The instruction is extracted from memory, deposited in the instruction register, decoded, and executed. Most instructions require from 1 to 5 cycles of time. The smallest interval of processing activity, which is the portion of a cycle identified with a clearly defined process, is known as a *state*. From 5 to 15 states per instruction are typical.

The processor uses the external bus for reading, writing, and other purposes. However, there are intervals when only internal operations are being performed. During such times, which are indicated by signals supplied by the processor to a control terminal, the external bus can be used for external operations such as a transfer of data from a peripheral device to memory. This *cycle stealing* increases system efficiency. The direct transfer of data from a peripheral device to memory or in the reverse direction, without processor control, is known as *direct memory access* (DMA).

MICROCOMPUTER SYSTEMS

The organization of a typical system is illustrated in Fig. 11-12. Because the Motorola M6800 NMOS microprocessor has depletion-mode load devices, only a single 5 V supply is needed. There are three external busses. The program that controls the microprocessor is stored in a look-up table of the ROM. Software implemented by an ROM is often referred to as *firmware*.

A memory stack can be organized in the RAM by means of the microprocessor *stack pointer* (SP). The word stored in the pointer is the address of the memory location at the top of the stack. Words are withdrawn from the top and also entered at the top. This last-in/first-out (LIFO) stack is very useful for sub-routine nesting in the program and for other purposes. The size of the stack is

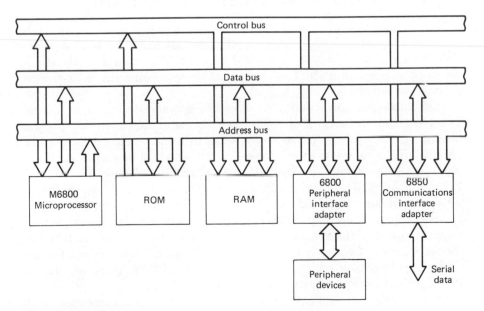

Fig. 11-12   Microprocessor system.

limited only by the address capability of the pointer, which may have as many as 16 bits.

Systems typically have from 1 to 16 external ROMs for storing programs and data tables and up to 16 or more RAMs for storing instructions and data used in operations. The read-only memories may be mask-programmable types, or they may be PROMs or EPROMs. Erasable read-only memories allow the processor program to be modified. A simple microcomputer can be made from a single microprocessor with one ROM.

The *peripheral interface adapter* of the system allows utilization of any I/O device compatible with TTL logic. The *communications interface adapter* is an LSI chip that uses clocked shift registers to convert the bit-parallel data stream from the processor or memory into serial form for transmission along a communication channel. Control and synchronization signals are supplied by the control bus. Transmission in the reverse direction from the remote device to the microprocessor is also possible. The address bus supplies bits to the decoders associated with the various devices, thereby selecting the desired device or memory location.

Available are single chips containing all elements of a minicomputer. For example, the Intel NMOS eight-bit 8048 monolithic chip has a CPU, a 1 kilobyte EPROM, a 512-bit static RAM for scratch-pad data handling, and I/O channels. External memory can be added if desired.

11-6. MICROCOMPUTER SOFTWARE

Microcomputer operation is controlled by a program stored in an ROM. The computer accepts only 0's and 1's. An eight-bit instruction byte consists of eight binary bits, as does an eight-bit data word. The program must identify the word as either instruction or data, and the microcomputer must properly process the information.

The sequence of instructions controlling the operation is known as *software*. The term also includes design aids such as instruction manuals, program libraries, codes, and various languages, along with equipment used to translate the prepared program into the 0's and 1's of the machine language. The cost of microprocessor systems is reduced by substituting software for hardware. For example, random-logic hardware can often be replaced with lower-cost software programs. Instead of using two hardware components for an operation, an instruction may accomplish the desired result with multiple operations on a single component. If a system requires modification to perform new functions, the software can frequently be changed without affecting hardware. The user simply reprograms or replaces the stored-program ROM. Trade-offs between hardware and software depend on the specific application.

Many instruction routines are so common that they are encoded in the instruction decoder of the microprocessor. The set of instructions selects data paths and their sequence of activation. In a *fixed-instruction* computer these instructions are fixed by the wiring designed by the manufacturer, and application programs are prepared for conformity. Some microprocessors are organized so that the user can program the instruction set into an internal ROM, or perhaps a PLA, thereby providing flexibility. These processors are said to be *micro-programmable*. Often the ROM and its control circuits are separate, with the microprocessor consisting of a processing chip and a microprogram chip.

PROGRAM LANGUAGES

Each microprocessor has an associated table, or dictionary, specifying the binary word corresponding to each possible instruction. There are data transfer instructions, arithmetic and logic instructions, and branch instructions such as JUMP, MOVE, or HALT. Each instruction has a *mnemonic* code given in the dictionary. For example, the code *ADD* 011 may represent the byte 10000011, with the five leftmost bits denoting the *operator* ADD and the next three signifying that the *operand* to be added to the contents of the accumulator is located in the register identified by 011.

The *machine language*, or *object language*, to be stored in the program ROM is a sequence of *n*-bit binary words representing instructions and data. A typical program has hundreds of such words. With the aid of the instruction-binary dictionary, a program can be prepared directly in machine language. As this is a most tedious process, it is rarely done.

In developing a microcomputer system the I/O devices must be selected, memory must be organized, and the types and frequency of required instructions need to be determined. These factors should be considered in choosing a suitable microprocessor. After the system has been clearly defined, a *flow chart* is prepared, indicating which input signals must be read, what processing and computations are required, and what output signals must be produced. From the flow chart and the functional block diagram of the selected microprocessor, a *source program* is written on a properly designed program form. This is often coded in an *assembly language*, which employs the symbolic mnemonic code provided by the instruction-set dictionary of the specific processor. For each operation required by the program, the appropriate mnemonic is selected. In general, an instruction statement begins with a symbolic name such as VOLTAGE, X5, or COUNT, and the name represents the actual binary address of the assigned location of the instruction in the ROM. This part of the instruction can be omitted if the instruction is performed in sequence, always following the preceding one. Next is an operator and this is followed by the operand. Remarks can be added as a convenience to the programmer.

The source program in assembly language must be converted into an *object program* in machine language before it can be stored in the ROM. This is accomplished by an *assembler*, which is a program designed to translate assembly language into machine language for the type microprocessor being used. It also assigns machine addresses to the symbolic address names of the source program. The assembler program performing the translation can be one prepared for execution by the microcomputer, with the assembler program read into the main memory. Often it is written for a different and larger computer, in which case it is called a *cross-assembler*.

Typically, a keyboard is used to punch an assembly-language program on paper tape. This is read into a large computer. Also read into the computer is a suitable cross-assembler program. Translation is executed, and the machine-language program is punched out on paper tape. This can be used for mask-programming of an ROM by the manufacturer. An alternate procedure is to employ a PROM and a *PROM programmer*. This is hardware designed to transfer the program to a PROM.

Source programs from the teletype keyboard can be modified by special computer programs known as *editors*. Without an editor, an error would require retyping the entire program.

Programs are easier to prepare in high-level languages. These use problem-oriented or function-oriented statements. One functional statement translates into a number of instructions or subroutines in machine language, and the number of instruction statements is greatly reduced. However, there are disadvantages. In general, assembly language tailored for a particular application uses memory more efficiently and gives higher operating speed. These considerations are sometimes critical in a small microcomputer.

General-computer languages such as FORTRAN, COBOL, and BASIC can be used. Then there are programming languages developed especially for microcomputers. Many of these are derived from PL/1, a problem-solving language developed in the 1960s for business and scientific calculations. High-level language is converted into machine language by a *compiler* program.

REFERENCES

1. B. Greene and D. House, "Designing with Intel PROMs and ROMs," Application Note AP-6, Intel Corp., Santa Clara, Calif., April 1975.
2. G. Luecke, J. P. Mise, and W. N. Carr, *Semiconductor Memory Design and Application*, McGraw-Hill, New York, 1973.
3. *IEEE JSSC*, **SC-6**, 5, Oct. 1971, Special Issue on Semiconductor Memories and Digital Circuits.
4. T. R. Blakeslee, *Digital Design with Standard MSI and LSI* (especially Chapters 3 and 4 on combinational logic, multiplexers, demultiplexers, and ROMs), Wiley, New York, 1975.
5. D. R. McGlynn, *Microprocessors*, Wiley, New York, 1976.
6. A Barna and D. I. Porat, *Introduction to Microcomputers and Microprocessors*, Wiley, New York, 1976.
7. B. Souček, *Microprocessors and Microcomputers*, Wiley, New York, 1976.

PROBLEMS

Section 11-1

11-1. For the ROM of Fig. 11-1 establish the Karnaugh maps relating the inputs A, B, and C and the outputs D_0 and D_1. From the maps determine that the combinational logic is that given by (11-1).

11-2. Design a 2-to-4 NMOS decoder using an ROM with a 4×4 cell matrix, which can be used to replace the row-select block of Fig. 11-1. Use two inverters to provide respective row inputs of B, $\bar{B}$, C, and $\bar{C}$ from top to bottom. Connect each of the four vertical column lines through a saturated load device to V_{DD}, and arrange the eight transistors of the cell matrix so that the respective column outputs from left to right are BC, $B\bar{C}$, $\bar{B}C$, and $\bar{B}\bar{C}$.

11-3. The FAMOS memory cell of Fig. 11-4 is programmed with a -50 V drain-to-source 5 ms pulse that produces an avalanche current from gate to substrate of 0.6 A/m^2. The gate area is 10^{-9} m^2. At the end of the pulse the gate-to-substrate voltage is 10. Calculate the number of stored electrons and the capacitance in pF between the gate and the inversion layer.

Section 11-2

11-4. Design a PLA configuration in the form of that of Fig. 11-6 with two outputs such that Q_1 is $AB + \bar{A}\bar{B}$ and Q_2 is $A\bar{B} + \bar{A}B$. Then sketch the complete circuit using positive-logic PMOS similar to that of Fig. 11-5. Label the outputs of both the product and sum ROMs in terms of A and B.

11-5. From Karnaugh maps for A', B', C', and D' of Table 11-1, deduce that the minimal sum-of-products expressions are those of (11-3). Use X to represent a don't-care state and treat each X as a logical 0 or 1, as convenient.

11-6. For the modulo-10 counter of Fig. 11-7 express the output a in terms of the inputs A, B, C, and D. From this result determine the $ABCD$ words of Table 11-1 that supply a logical 1 to output a, and sketch the corresponding decimal seven-segment displays corresponding to each of these words.

11-7. Repeat P11-6 for output g.

Section 11-3

11-8. Using Karnaugh maps, for the data of Table 11-3 determine the minimal sum-of-products logic expressions relating each of the phases ϕ_1, ϕ_2, ϕ_3, and ϕ_4 to A, B, and C of the binary address ABC. Give all possible minimal expressions.

11-9. Using Karnaugh maps, for the ROM of Fig. 11-8b determine the minimal sum-of-products logic expressions relating each of the outputs a, b, c, and d to the inputs A, B, C, and D. Design a network that implements output b, using *only* NAND gates. Complements of the inputs are not available.

11-10. Redesign the ROM of Fig. 11-8b so that it converts the Gray code of Fig. 11-8a to the binary code. Specify logic expressions for the 16 outputs of the decoder in terms of the inputs a, b, c, d, and their complements. Assume PMOS positive logic.

11-11. Determine the address and binary-output words for the sine look-up table of Table 11-2 for the angle corresponding to address 80 of the ROM. Also, calculate the percent error of the eight-bit output for this angle. (*Ans*: 0.053%)

11-12. (a) In the form of Fig. 11-8b, sketch the ROM design for the four-phase clock system of Table 11-3. Label the binary addresses of the eight rows, beginning with 000. (b) Sketch the outputs ϕ_1 through ϕ_4 for values of time from 0 through 8T. Label the axes, showing the instants when transitions occur. Assume PMOS positive logic.

Section 11-4

11-13. Design the logic configuration of a multiplexer with two input words of four bits each. Use eight three-input AND gates, four two-input NOR gates, and necessary inverters. Label the select input S, the enable input $\bar{E}$, and the words at the inputs and output.

11-14. For the 3-to-8 decoder of Fig. 10-3 of Sec. 10-1 modify the network so that it can be used as a demultiplexer. Use an inverter at the data input to avoid data inversion at the output. Sketch the complete logic diagram. Which output receives the data when address word $A_5 A_6 A_7$ is 010 and also when it is 101?

11-15. If the four-input multiplexer of Fig. 11-9 has $D_0 = 0$, $D_1 = 1$, and D_2 and D_3 each equal to D, deduce that Y is $\bar{A}B + AD$. Also, with D_0 changed to 1, show that Y is $\bar{A} + AD$.

Sections 11-5 and 11-6

11-16. In the twos-complement system, express the decimal numbers 7 and -5 as four-bit binary numbers including the sign bit, and determine the difference $7 - 5$ in this system. Also determine $5 - 7$ in the twos-complement system and show that the result is -2.

11-17. Identify the acronyms CPU, ALU, A/D, D/A, and DMA. Also, with reference to a microcomputer define the terms *peripheral element, fetch, slice, accumulator, program counter, scratch-pad register, instruction cycle, state,* and *stack pointer.*

11-18. With reference to a microcomputer define the terms *microprogram, mnemonic, operator, operand, source language, object language, assembler, cross-assembler, editor,* and *compiler.*

PART THREE
Analog Circuits

Chapter Twelve
Transistor Small-Signal Models

The nonlinear Ebers–Moll equations, which relate the static voltages and currents of a BJT, are valid for time-changing variables if the changes are not too rapid. Many transistor circuits process signals so small that the transistor behaves *linearly* insofar as these signals are concerned. Although the Ebers–Moll equations are still applicable, assuming sufficiently low frequencies, a linear model is needed in order that the powerful tools of linear circuit theory can be utilized. The main purpose of this chapter is to develop the small-signal hybrid-π model and to investigate the properties of the circuit parameters and their experimental determination. Included is a brief introduction to two-port network theory.

12-1. AN ELEMENTARY AMPLIFIER

Our study of amplifiers begins with an analysis of the circuit of Fig. 12-1, containing an NPN transistor biased in the active region. Although the base width W is a function of the collector voltage, this is a second-order effect that will be ignored in order to keep this discussion as simple as possible. Consequently, I_{ES} and α_F are treated as constants.

NOTATION

The standard notation for designating currents and voltages is employed here and throughout this book. The current i_B is

$$i_B = I_B + i_b \tag{12-1}$$

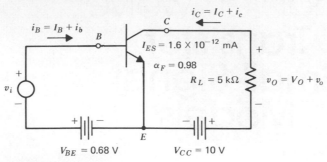

Figure 12-1   Amplifier.

With v_i zero, the circuit of Fig. 12-1 is said to be *quiescent*, or at rest, and the quiescent base current is I_B. When v_i is not zero, the total current i_B differs from its quiescent value by i_b. The symbol i_b denotes the *incremental* current, also called the *signal-component* of i_B. Note that the conventional reference directions for i_B, I_B, and i_b are into the terminal B of the device.

The voltage drop from the base B to the emitter E is designated v_{BE}, which is expressed similarly as the sum of a quiescent voltage V_{BE} and an incremental voltage v_{be}. In the circuit of Fig. 12-1, v_{be} is v_i. Summarizing, lowercase letters with capital subscripts denote total currents and voltages. Capital letters with capital subscripts designate quiescent quantities. Lowercase letters with lowercase subscripts are used for the incremental variables. Current reference directions are taken into the device, unless specifically stated otherwise. Voltage reference directions are designated by double subscripts, or by plus and minus signs as for v_O in Fig. 12-1. The transistor is said to be biased at the *operating point*, or *Q-point* (for quiescent), and the Q-point voltages and currents are the quiescent quantities.

APPROXIMATE ANALYSIS

The signal voltage v_i may be positive or negative, static or time-varying. However, any variations with time are assumed here to be sufficiently slow to justify neglecting the reactive currents of the junction capacitances. Furthermore, the magnitude of v_i is restricted so that operation is maintained in low-level injection in the active mode.

We know the emitter and collector currents are nearly equal to the electron diffusion current in the base. The electron gradient is $n_b(0)/W$ with W assumed to be constant. Therefore, i_E and i_C are proportional to $n_b(0)$ and hence to $\exp qv_{BE}/kT$ by the boundary condition (1-44) of Sec. 1-6. Both components I_{BB} and I_{BE} of the base current i_B are also proportional to $n_b(0)$. *Clearly the terminal currents are linearly related to one another.* However, they are exponentially related to v_{BE}, and the exponential function is quite nonlinear.

An increase in v_{BE} reduces the field in the emitter space-charge layer, allowing greater electron injection into the base. A small percentage change in v_{BE} produces a relatively large change in each of the terminal currents. Hence, v_{BE} acts somewhat like a valve that controls the currents. For example, an incremental voltage v_i of 0.0025 V, representing an increase in v_{BE} of only about 0.4 %, makes $40v_{be} = 0.1$, and increases $n_b(0)$ by the factor exp 0.1, or 1.1. This 10 % increase in $n_b(0)$ gives 10 % increases in the terminal currents. As will be shown shortly, I_C and I_B of the amplifier of Fig. 12-1 are approximately 1 mA and 0.02 mA, respectively. Therefore, the incremental currents i_c and i_b become 0.1 mA and 0.002 mA, respectively. The current gain A_i of the amplifier, defined as $-i_c/i_b$, is -50.

The incremental output voltage v_o is $-i_c R_L$, or -0.5 V. This gives a voltage gain A_v, defined as v_o/v_i, of -200. The negative voltage gain simply means that the sign of the output voltage is opposite that of the input. The output signal power is $-v_o i_c$, or 0.05 mW, and the input power is $v_i i_b$, or 5×10^{-6} mW. Hence the power gain G is 10^4. *Signal power to the load resistor is four orders of magnitude greater than that supplied by the input signal*, with the additional power furnished by the batteries.

The preceding approximate analysis showed that the amplifier has appreciable current gain, voltage gain, and power gain. Although the currents are linearly related to one another, there is nonlinearity because of the dependence of $n_b(0)$ on the exponential of the input voltage. The nonlinearity problem will now be investigated.

A SMALL-SIGNAL MODEL

In the active mode with $q/kT = 40$ the Ebers–Moll equations (2-10) and (2-11) can be simplified by neglecting the exponential of $40v_{BC}$ and also the small saturation currents. The results show that i_E is $-I_{ES} \exp 40v_{BE}$ and i_C is $-\alpha_F i_E$. For convenience let us understand units of mA, kΩ, and volts, which constitute a consistent set. Then i_E is $-1.6 \times 10^{-12} \exp 40v_{BE}$. Multiplication by $-\alpha_F$ gives i_C, and i_B is $-i_E - i_C$. Thus

$$i_C = 1.57 \times 10^{-12} e^{40v_{BE}} \text{ mA} \tag{12-2}$$

$$i_B = 0.03 \times 10^{-12} e^{40v_{BE}} \text{ mA} \tag{12-3}$$

When v_i is zero, v_{BE} is 0.68. With the aid of (12-2) and (12-3) the quiescent values are found to be

$$I_C = 1.0 \text{ mA} \qquad I_B = 0.02 \text{ mA} \qquad V_O = -5.0 \text{ V} \tag{12-4}$$

Let us replace exp $40v_{BE}$ in (12-2) and (12-3) with exp $40(V_{BE} + v_i)$, which equals $6.5 \times 10^{11} \exp 40v_i$. The incremental variables are found by subtracting the quiescent values of I_C and I_B from the total variables of (12-2) and (12-3). With the currents in mA, the results are

$$i_c = e^{40v_i} - 1 \qquad i_b = 0.02(e^{40v_i} - 1) \tag{12-5}$$

These equations show that i_c is proportional to i_b, with the current gain $-i_c/i_b$ being -50, which agrees with the earlier estimate. The relationship between each current and v_i involves a nonlinear exponential function. The incremental output voltage v_o is $-5i_c$, with i_c given by (12-5). Figure 12-2 shows two plots of the voltage transfer characteristic v_o versus v_i, each to a different scale.

Suppose $v_i = 0.015 \cos \omega t$. From Fig. 12-2a we note that v_o varies between the limits of $+2.3$ and -4.1 V, and the variation of v_o with time is not at all sinusoidal. The nonlinearity of the transfer characteristic of Fig. 12-2a has introduced considerable *nonlinear distortion*. Now let us suppose that v_i is $0.001 \cos \omega t$. From Fig. 12-2b we observe that v_o varies between the limits of ± 0.20 V. In this range the transfer characteristic of Fig. 12-2 is nearly a straight line, and nonlinear distortion is quite small. In fact, the curve of v_o versus v_i is fairly linear for values of v_i between ± 0.005 V (see P12-1).

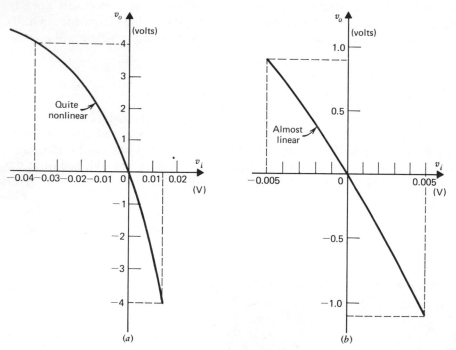

Figure 12-2 Plots of v_o versus v_i, with $v_o = -5(e^{40v_i} - 1)$. (a) $-0.040 < v_i < 0.015$. (b) $-0.005 < v_i < 0.005$

Of considerable significance is our observation that, *for sufficiently small signals, the BJT is a linear device insofar as the incremental variables are concerned.* The transfer characteristic is curved, as in Fig. 12-2a, but a small region of this curve, as in Fig. 12-2b is approximately a straight line. A large and important area of electronics involves the amplification and processing of signals that are sufficiently small that nonlinear distortion is negligible. This area is often referred to as *linear electronics.*

The graphical result can, of course, be verified analytically. The Taylor's series expansion of exp x about the point $x = 0$ is

$$e^x = 1 + x + \frac{x^2}{2!} + \frac{x^3}{3!} + \cdots \tag{12-6}$$

From this it is easily deduced that

$$e^{40v_i} - 1 = 40v_i(1 + 20v_i + 267v_i^2 + 2667v_i^3 + \cdots) \tag{12-7}$$

If v_i is very small, say less than 0.005, (12-7) becomes

$$e^{40v_i} - 1 \approx 40v_i \tag{12-8}$$

Using (12-8), the incremental variables of (12-5) are found in mA to be

$$i_c = 40v_i \qquad i_b = 0.8v_i \tag{12-9}$$

The incremental output voltage is $-5i_c$, or $-200v_i$, giving a voltage gain of -200. This agrees with the estimate obtained earlier. The power gain G is $A_i A_v$, or 10,000.

In terms of v_{be}, (12-9) can be written

$$i_c = g_m v_{be} \qquad i_b = g_\pi v_{be} \tag{12-10}$$

with $g_m = 40$ mmhos and $g_\pi = 0.8$ mmho. If we had performed the analysis with appropriate symbols in place of numerical values, we would have found that

$$g_m = \left(\frac{q}{kT}\right) I_C \qquad g_\pi = \frac{g_m}{\beta_F} \tag{12-11}$$

with $\beta_F = \alpha_F/(1 - \alpha_F) = 49$.

Equations 12-10 represent a linear mathematical model of the transistor, one that properly relates the small-signal incremental variables. The incremental circuit model of the amplifier, based on (12-10), is given in Fig. 12-3.

The incremental circuit gives no information about the quiescent currents and voltages, and *no dc sources are included*; the quiescent quantities are determined from the nonlinear Ebers–Moll model. For the incremental variables the linear circuit is an excellent approximation for a sufficiently small signal, and linear circuit theory can be applied to determine the incremental response to a specified input signal.

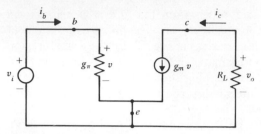

Figure 12-3   Incremental circuit model ($v = v_{bc}$).

Although the model of Fig. 12-3 is sufficient for many purposes, it is over-simplified. For example, if R_L is increased without limit, the signal power to R_L also increases without limit. An actual transistor is incapable of such perform-ance. The assumption that base-width modulation is negligible is not always reasonable. If v_{CE} varies with time, the base width W also varies, as do α_F and I_{ES}. Other effects have also been neglected, and a better model is needed. In the next section a more general approach is undertaken, with base-width-modula-tion effects included.

12-2. A LOW-FREQUENCY MODEL

The static characteristics of the transistor of Fig. 12-4 can be described mathematically in terms of two current variables and two voltage variables. One choice is to express i_B and i_C as functions of v_{BE} and v_{CE}, as indicated by the relations

$$i_B(v_{BE}, v_{CE}) \qquad i_C(v_{BE}, v_{CE}) \tag{12-12}$$

There are numerous other possibilities but this one is convenient. The third current (i_E) depends on the currents of (12-12), and the third voltage v_{BC} depends on the voltages of (12-12).

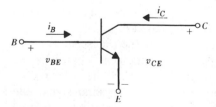

Figure 12-4   Transistor.

TAYLOR'S SERIES EXPANSION

The currents i_B and i_C can be expanded into a Taylor's series. Each current is a function of two variables. Let us first review the expansion of a function $f(x)$ of a single variable. The infinite series is

$$f(x) = f(a) + (x - a)f'(a) + \frac{(x - a)^2}{2!} f''(a) + \frac{(x - a)^3}{3!} f'''(a) + \cdots$$

(12-13)

with $f(a)$ denoting the value of $f(x)$ at $x = a$, with $f'(a)$ denoting the first derivative evaluated at the same point, and so forth. The selection of the point $x = a$ is arbitrary.

Suppose that i_B is a function of the single variable v_{BE}. Then $i_B(v_{BE})$ corresponds to $f(x)$. For an expansion about the Q-point, the point $v_{BE} = V_{BE}$ corresponds to $x = a$, and the series of (12-13) becomes

$$i_B(v_{BE}) = I_B + (v_{BE} - V_{BE}) \frac{di_B}{dv_{BE}} + \frac{(v_{BE} - V_{BE})^2}{2!} \frac{d^2 i_B}{dv_{BE}^2} + \cdots \quad (12\text{-}14)$$

with $i_B(V_{BE})$ replaced with I_B and with the derivatives evaluated at the Q-point. From the relations $i_b = i_B - I_B$ and $v_{be} = v_{BE} - V_{BE}$, (12-14) can be expressed as

$$i_b = v_{be}\left(\frac{di_B}{dv_{BE}}\right) + \frac{v_{be}^2}{2!}\left(\frac{d^2 i_B}{dv_{BE}^2}\right) + \frac{v_{be}^2}{3!}\left(\frac{d^3 i_B}{dv_{BE}^3}\right) + \cdots \quad (12\text{-}15)$$

with the derivatives evaluated at the Q-point.

In the previous section i_B was assumed to be a function of the single variable v_{BE}, and the exponential relationship was given as (12-3). Using (12-3), the derivatives of (12-15) can be found, giving i_b in terms of v_{be}. The result is that found in the previous section, expressed as (12-5) and (12-7). (See P12-5.)

The base current of (12-12) is a function of two variables. The expansion by Taylor's series, in the form that corresponds to (12-15), is

$$i_b = \left(v_{be}\frac{\partial}{\partial v_{BE}} + v_{ce}\frac{\partial}{\partial v_{CE}}\right)i_B + \frac{1}{2!}\left(v_{be}\frac{\partial}{\partial v_{BE}} + v_{ce}\frac{\partial}{\partial v_{CE}}\right)^2 i_B + \cdots \quad (12\text{-}16)$$

with the partial derivatives of i_B evaluated at the Q-point. Terms such as $\partial/\partial v_{BE}$ are operators that operate on i_B. An operator raised to a power n implies that the operation is performed n times; for example, $(\partial/\partial v_{BE})^2 i_B$ is $\partial^2 i_B/\partial v_{BE}^2$.

For a sufficiently small input signal, v_{be} and v_{ce} are small, and the only significant terms on the right side of (12-16) are those proportional to v_{be} and v_{ce}; that is,

$$i_b(v_{be}, v_{ce}) = \frac{\partial i_B}{\partial v_{BE}}\bigg|_Q v_{be} + \frac{\partial i_B}{\partial v_{CE}}\bigg|_Q v_{ce} \quad (12\text{-}17)$$

Similarly, the current i_c can be expressed as

$$i_c(v_{be}, v_{ce}) = \frac{\partial i_C}{\partial v_{BE}}\bigg|_Q v_{be} + \frac{\partial i_C}{\partial v_{CE}}\bigg|_Q v_{ce} \qquad (12\text{-}18)$$

It is convenient to introduce a set of g-parameters:

$$g_\pi = \frac{\partial i_B}{\partial v_{BE}}\bigg|_Q \qquad g_\mu = -\frac{\partial i_B}{\partial v_{CE}}\bigg|_Q$$

$$\qquad (12\text{-}19)$$

$$g_m = \frac{\partial i_C}{\partial v_{BE}}\bigg|_Q \qquad g_o = \frac{\partial i_C}{\partial v_{CE}}\bigg|_Q$$

Each of these is defined as the ratio of a differential current to a differential voltage. Accordingly, the parameters are called *incremental conductances*, with the SI unit *mho*. Equations 12-17 and 12-18 can be written

$$i_b = g_\pi v_{be} - g_\mu v_{ce}$$

$$\qquad (12\text{-}20)$$

$$i_c = g_m v_{be} + g_o v_{ce}$$

This set of equations represents a *linear mathematical model* that applies to the small-signal incremental variables of a transistor.

The analysis leading to the linear model of (12-20), with the conductance parameters defined by (12-19), is applicable to any three-terminal device (see P12-10). In a later chapter the equations are used as the basis of a suitable model for the field-effect transistor with the subscripts changed to conform with FET terminology. The model is valid only for small signals. Just how small the signal must be depends on the characteristics of the particular device. The incremental variables must be small enough to justify neglecting the higher-order terms of (12-16) and the higher-order terms of the corresponding equation for i_c. In the previous section the linear model was found to be at least a fair approximation for the BJT incremental variables *provided* v_{be} *is less than 5 millivolts.* This value is not changed appreciably by the second-order effects of base-width modulation that have been included in this section.

In order to find the g-parameters of a BJT in terms of the quiescent currents, the temperature, and basic transistor parameters, suitable equations for i_C and i_B are needed. The active-mode current i_C is $\alpha_F I_{ES} \exp qv_{BE}/kT$, in accordance with the Ebers–Moll equation (2-11) of Sec. 2-2, neglecting the small saturation currents. The product $\alpha_F I_{ES}$ is $qAD_b n_{bo}/W$, by (2-14) with W/L_b small. Therefore,

$$i_C = \frac{qAD_b n_{bo}}{W} e^{qv_{BE}/kT} \qquad (12\text{-}21)$$

The active-mode current i_B is the sum of the base recombination current I_{BB} and the emitter-junction hole current I_{BE}. From (2-27) and (2-30) of Sec. 2-3 we find that

$$i_B = qA\left(\frac{Wn_{bo}}{2\tau_b} + \frac{D_e p_{eo}}{L_e}\right)e^{qv_{BE}/kT} \qquad (12\text{-}22)$$

The parameters g_m and g_π are found by differentiating (12-21) and (12-22) with respect to v_{BE}, with v_{CE} constant, and then substituting V_{BE} for v_{BE}. Although W is a function of the voltages, the change in W due to an increment Δv_{BE}, with v_{CE} constant, is negligible compared with the change in exp qv_{BE}/kT. Accordingly, W is treated as a constant, and the results are

$$g_m = \left(\frac{q}{kT}\right)I_C \qquad g_\pi = \left(\frac{q}{kT}\right)I_B \qquad (12\text{-}23)$$

with I_C and I_B denoting the quiescent currents.

The conductance g_o is found by differentiating i_C with respect to v_{CE}. Although v_{CE} does not appear explicitly in (12-21), the base width depends on this voltage. Hence the partial derivative of i_C is the product of $\partial i_C/\partial W$ and $\partial W/\partial v_{CE}$. The derivative $\partial i_C/\partial W$ is determined from (12-21) to be $-I_C/W$ at the Q-point. The expression $\partial W/\partial v_{CE}$ equals $-\partial W/\partial v_{BC}$, because $dv_{CE} = -dv_{BC}$ when v_{BE} is constant. Therefore, g_o is the product of $-I_C/W$ and $-\partial W/\partial v_{BC}$ at the Q-point. Clearly, this product gives $g_o = g_m\eta$, with the *base-width-modulation factor* η given in (12-25).

The parameter g_μ is defined as the negative of the derivative of i_B with respect to v_{CE}. The derivative $-\partial i_B/\partial W$ is found from (12-22) to be $-I_{BB}/W$, with I_{BB} given by (2-30). As before, $\partial W/\partial v_{CE}$ can be replaced with $-\partial W/\partial v_{BC}$, and g_μ can be expressed as $(q/kT)I_{BB}\eta$. Because I_{BB} is quite a bit larger than I_{BE} in the active mode in low-level injection, it is satisfactory to replace I_{BB} with I_B, giving $g_\mu = g_\pi\eta$.

As $I_B \approx I_C/\beta_F$, in accordance with (2-36), the g-parameters can be expressed in the form

$$g_m = \left(\frac{q}{kT}\right)I_C \qquad g_o = g_m\eta$$

$$g_\pi = \frac{g_m}{\beta_F} \qquad g_\mu = g_\pi\eta \qquad (12\text{-}24)$$

with $\beta_F = \alpha_F/(1 - \alpha_F)$ and with the base-width-modulation factor defined by

$$\eta = \frac{kT}{q}\frac{1}{W}\frac{dW}{dV_{BC}} \qquad (12\text{-}25)$$

The term dW/dV_{BC} signifies the derivative evaluated at the Q-point, with the other junction voltage maintained constant.

It should be noted that *each of the incremental conductances is directly proportional to the quiescent collector current*, and they are functions of temperature and collector voltage (see P12-7 and P12-8). Also, because β_F is normally greater than 20 and η is much smaller than $1/\beta_F$, it follows that

$$g_m \gg g_\pi \gg g_o \gg g_\mu \tag{12-26}$$

Were it not for base-width modulation, g_o and g_μ would be zero; these parameters did not appear in the discussion of the previous section because W was treated as a constant. Let us consider an example.

Example

A silicon ($\varepsilon = 10^{-10}$) NPN transistor at 290 K has $V_{BE} = 0.65$ V, $V_{BC} = -4$ V, $I_C = 2$ mA, and $\beta_F = 40$. The respective majority-carrier densities n_{eo}, p_{bo}, and n_{co} are 10^{24}, 10^{22}, and 10^{21} m^{-3}, and the distance between the abrupt metallurgical junctions is 2×10^{-6} m. Calculate the incremental parameters.

Solution

As q/kT is 40 and I_C is 2 mA, the transconductance g_m is 80 mmhos, and g_π is g_m/β_F, or 2 mmhos. The built-in junction voltages ψ_{oe} and ψ_{oc} are determined from (1-40), noting that the dopant densities equal the majority-carrier densities, to be 0.829 and 0.656, respectively. From (2-21) with W_B equal to 2×10^{-6} m, the base width W is found in terms of V_{BC} to be

$$W = (1.85 - 0.107\sqrt{0.656 - V_{BC}})10^{-6} \text{ m}$$

This is used to find dW/dV_{BC}, which is 2.48×10^{-8} at -4 V. The value of W is calculated to be 1.62×10^{-6} m. These values substituted into (12-25) give a base-width modulation factor η of 0.000383. The conductance g_o is $g_m\eta$, and g_μ is $g_\pi\eta$. In mmhos the results are

$$g_m = 80 \qquad g_o = 0.031$$
$$g_\pi = 2 \qquad g_\mu = 0.00077$$

The inequality of (12-26) is satisfied quite well.

LINEAR CIRCUIT MODEL

The mathematical model of (12-20) is represented by the linear circuit of Fig. 12-5a. With respect to reference node e, the equations of nodes b and c are (12-20). Because of inequality (12-26), the model of Fig. 12-5b is equivalent to that of Fig. 12-5a. The $g_m v$ current source supplies a current proportional to the voltage drop from b to e; thus it is a *voltage-dependent current source*. Although derived for an NPN transistor, *the g-parameter definitions and relations, and also the incremental circuit model, apply to PNP devices provided the magnitude of* I_C is used in (12-24) and V_{BC} in (12-25) is replaced with V_{CB}.

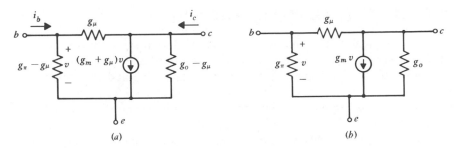

Figure 12-5   Linear incremental circuit models. (*a*) Model based on (12-20). (*b*) Approximate model.

The incremental conductance g_m is called the *mutual conductance*, or *transconductance*, of the transistor. The subscript *m* stands for mutual. Because g_m is defined as the ratio of a small change in i_C to the corresponding change in v_{BE} with v_{CE} constant, it is a measure of the effectiveness of the emitter-junction voltage in controlling the collector current. Large values of g_m are desired for large voltage gain, and *an outstanding feature of the BJT is its relatively large mutual conductance compared with that of most other electronic devices.* This is a consequence of the sensitive exponential relationship between $n_b(0)$ and v_{BE}. At room temperature g_m is approximately $40I_C$, and a current of 1 mA gives a g_m of 40 mmhos.

The conductance g_π is the ratio of a small change in i_B to the corresponding change in v_{BE}, with v_{CE} constant. An increase in v_{BE} produces a rise in the charge store in the base, and i_B increases. For a g_m of 40 mmhos and a β_F of 40, g_π is 1 mmho; the incremental resistance r_π is 1 kΩ.

The *incremental output conductance* g_o, so-called because the collector-emitter terminals are usually the output terminals, is the ratio of a small change in i_C to the corresponding change in v_{CE}, with constant base-emitter voltage. A change in v_{CE} changes the base width W, which in turn affects the minority-carrier diffusion current in the base. The output conductance equals $g_m\eta$, this being a consequence of the fact that the emitter voltage is approximately $1/\eta$ times as effective as the collector voltage in controlling i_C. The modulation factor η is typically between 10^{-3} and 10^{-5}. For an operating current of 1 mA, values of g_o between 0.0400 and 0.0004 mmho are likely, and the corresponding values of the incremental output resistance r_o are from 25 kΩ to 2500 kΩ.

Also dependent on base-width modulation is g_μ, which is the ratio $|\Delta i_B/\Delta v_{CE}|$ with constant emitter voltage. A change in v_{CE} changes W, and this produces a change in the electron store in the base, which affects i_B. Because the effect on i_B is slight, g_μ is very small. For $g_\pi = 1$ mmho, values of g_μ (or $g_\pi\eta$) between 10^{-3} and 10^{-5} mmho are to be expected; the corresponding values of r_μ are from 1 to 100 megohms.

12-3. THE HYBRID-π MODEL

The low-frequency circuit model that has been developed is improved by the addition of a resistance r_x in series with the base terminal, as shown in Fig. 12-6. Also shown are the incremental junction capacitances C_π and C_μ, which were discussed in Sec. 2-7. These capacitances extend the useful frequency range of the model to radio frequencies. The circuit of Fig. 12-6 is known as the *hybrid-π model*,[5] because it consists of a π-configuration to which r_x is added. It is extensively used in the ånalysis of the incremental behavior of transistor circuits and *it applies to both NPN and PNP transistors*. Let us now examine the justification for the resistance r_x.

The base resistance is a second-order effect that was not considered in the development of the Ebers–Moll equations. The base current I_B consists largely of the flow of majority carriers into the base to replace those lost by recombination with the minority carriers. This flow is a drift process in a region with appreciable resistance because of light doping. As shown in Fig. 12-7, the majority carriers

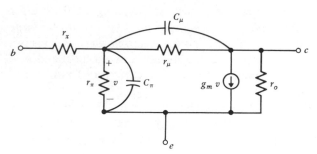

Figure 12-6   Hybrid-π model of BJT.

drift in the base in a direction transverse to the normal transistor current. Because the voltage drops along the path of the majority-carrier flow, the actual emitter-junction voltage is less than the terminal voltage v_{BE}. By adding a suitable resistance r_x to the model, the effect of this voltage is included.

Examination of Fig. 12-7 indicates that the emitter-junction voltage is less at the midpoint of the junction than at the extremities, because of the ohmic drop between these points. The higher forward bias near the edges of the emitter region results in a higher current density near these edges. This is known as *current crowding*, or *pinch-out*. It is evident that there is a different resistance between terminal B and each point of the junction; consequently, treating the base resistance as a single lumped element r_x is an arbitrary and rather crude approximation. Fortunately, this is adequate for most situations.

The resistance r_x has values that usually range from a few ohms to 100 ohms and more. At very low current levels at low frequencies the effect of r_x is negligible. At the high current levels commonly encountered in high-power transistors the

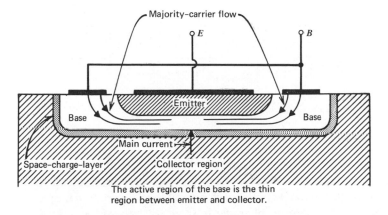

The active region of the base is the thin
region between emitter and collector.

Figure 12-7   Majority-carrier flow in base.

current crowding (pinch-out) that results from the base resistance is especially
deterimental, limiting the maximum current. *At high frequencies* r_x *is particularly
important*, because the base current that charges the capacitances C_π and C_μ
must pass through this resistance. In noise studies r_x must always be included
in the model. It is an ohmic resistance that produces thermal noise which appears
amplified in the collector-emitter circuit.

An increase in the reverse bias on the collector junction causes the base width
W to decrease. This reduces the effective cross-sectional area normal to the
majority-carrier flow, which increases r_x. Hence r_x increases as $|V_{BC}|$ increases.

An increase in I_C causes r_x to decrease. There are two reasons for this effect.
First, a larger current results in a greater percentage of this current occurring
in the region near the extremities of the emitter junction; this effect results from
the exponential dependence of the current on the junction voltage. Consequently,
the effective resistance of the active region of the base is lowered. Second, as I_C
increases, injection changes from low-level to high-level, causing the majority-
carrier density in the base to increase; therefore, the base resistivity drops. These
effects are second-order, and r_x is usually treated as a constant in the typical
operating region.

Numerical values of the elements of the hybrid-π model are obtained from
measurements, which involve parameters from two-port network theory. In the
next two sections this theory is considered, and some of the results are applied in
Sec. 12-6 to the problem of finding the hybrid-π parameters.

12-4. TWO-PORT NETWORK THEORY

Our study of small-signal BJT circuits will utilize the hybrid-π model that has
been developed and also the basic concepts of two-port network theory. This
theory will now be investigated.

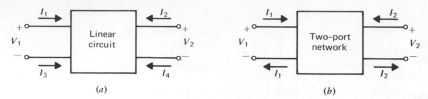

(a)   (b)

Figure 12-8   (a) Four-terminal network. (b) Two-port network.

Suppose the box of Fig. 12-8a contains a linear circuit with four terminals, referred to as a four-terminal network. The four currents are assumed to be excited by one or more *external* sources, each having the same radian frequency ω, and the capital letters denote phasors. If $I_3 = -I_1$ and $I_4 = -I_2$ as in Fig. 12-8b, the arrangement is called a *two-port network*. By convention, the terminals on the left constitute the input port and those on the right the output port, regardless of the external connections.

Our discussion is restricted to the usual case in which the circuit enclosed within the box contains no independent sources at frequency ω. An independent source is a generator with an open-circuit voltage or a short-circuit current not dependent on other circuit variables. Dependent sources such as the $g_m v$ generator of Fig. 12-6 are allowed, and there may be independent sources at other frequencies, such as batteries ($\omega = 0$). Sources capable of supplying positive time-average power are said to be *active*. A circuit with at least one such source is called an active network; otherwise it is passive. The theory presented here applies to sinusoidally excited two-port linear networks, either active or passive, but free of independent sources at the excitation frequency ω.

Let us suppose that external circuitry, not shown in Fig. 12-8b, produces the indicated voltages and currents. If the phasor voltages V_1 and V_2 are calculated or measured, the external circuitry could be replaced with ideal voltage sources V_1 and V_2, as shown in Fig. 12-9, without affecting the currents I_1 and I_2. The only independent sources at frequency ω are V_1 and V_2. Because the circuit is linear, we can employ superposition. The current I_1' due to V_1 (with $V_2 = 0$) is $y_i V_1$, and the current I_1'' due to V_2 (with $V_1 = 0$) is $y_r V_2$, with y_i and y_r denoting

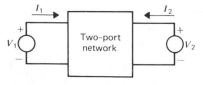

Figure   12-9   Network   with   voltage
sources.

complex constants having dimensions of admittance. The total current I_1 is the sum of $y_i V_1$ and $y_r V_2$. Current I_2 can be found similarly, with the results:

$$I_1 = y_i V_1 + y_r V_2$$
$$I_2 = y_f V_1 + y_o V_2 \qquad (12\text{-}27)$$

Equations 12-27 state that the terminal currents are linear functions of the terminal voltages. The admittance parameters are functions of ω and depend on the properties of the enclosed network.

The two-port network has four terminal variables, and (12-27) is not the only way of relating them. The equations can be solved for V_1 and V_2, expressing these variables in terms of the currents I_1 and I_2 and certain impedance parameters. In fact, any two of the four variables can be expressed as functions of the other two. A little reflection reveals that there are six possibilities, and four of these are presented in matrix form as (12-28).

$$\begin{bmatrix} V_1 \\ V_2 \end{bmatrix} = \begin{bmatrix} z_i & z_r \\ z_f & z_o \end{bmatrix}\begin{bmatrix} I_1 \\ I_2 \end{bmatrix} \qquad \begin{bmatrix} I_1 \\ I_2 \end{bmatrix} = \begin{bmatrix} y_i & y_r \\ y_f & y_o \end{bmatrix}\begin{bmatrix} V_1 \\ V_2 \end{bmatrix}$$

$$\begin{bmatrix} V_1 \\ I_2 \end{bmatrix} = \begin{bmatrix} h_i & h_r \\ h_f & h_o \end{bmatrix}\begin{bmatrix} I_1 \\ V_2 \end{bmatrix} \qquad \begin{bmatrix} I_1 \\ V_2 \end{bmatrix} = \begin{bmatrix} g_i & g_r \\ g_f & g_o \end{bmatrix}\begin{bmatrix} V_1 \\ I_2 \end{bmatrix} \qquad (12\text{-}28)$$

The significance of the matrix form is apparent by comparing each expression of (12-27) with its matrix equivalent.

MATRIX PARAMETERS

The parameters z, y, h, and g are interrelated. For example, (12-27) can be solved for V_1 and I_2, giving

$$V_1 = \left(\frac{1}{y_i}\right)I_1 - \left(\frac{y_r}{y_i}\right)V_2$$

$$I_2 = \left(\frac{y_f}{y_i}\right)I_1 + \left(\frac{y_i y_o - y_f y_r}{y_i}\right)V_2 \qquad (12\text{-}29)$$

Comparison with the h-parameter equation of (12-28) shows that h_i equals $1/y_i$, h_r equals $-y_r/y_i$, and so forth. Clearly, from a known set of parameters, the other sets are easily determined. For convenience the interrelations are presented in Appendix B.

Inspection of (12-28) reveals that z_i, y_i, h_i, and g_i represent either the input impedance (V_1/I_1) or the input admittance (I_1/V_1) with the output either shorted ($V_2 = 0$) or opened ($I_2 = 0$). For example, z_i is V_1/I_1 with $I_2 = 0$. The subscript i stands for input.

The parameters z_o, y_o, h_o, and g_o are either the output impedance (V_2/I_2) or admittance (I_2/V_2) with the input either shorted or opened, and the subscript o

stands for *output*. For example, y_o is the output admittance with the input short-circuited $(V_1 = 0)$.

Each of the *forward* matrix parameters, having the subscript f, is the ratio of an output variable (V_2 or I_2) to an input variable (V_1 or I_1) with the other output variable equal to zero. For example, h_f is the forward current gain I_2/I_1 with the output shorted $(V_2 = 0)$. Measurement requires driving the network in the *forward* direction, from the input port to the output port.

The *reverse* parameters, with the subscript r, are ratios of an input variable (V_1 or I_1) to an output variable, with the other input variable equal to zero. An example is h_r, which is the reverse voltage gain V_1/V_2 with the input port open $(I_1 = 0)$; its measurement is made by driving the network in the *reverse* direction.

We have learned that each of the matrix parameters is defined by (12-28) as a ratio of two variables from the set (V_1, V_2, I_1, I_2), with one of the ports either short-circuited or open-circuited. Because the z parameters have dimensions of ohms and are defined with one port open, they are called *open-circuit impedance parameters*. The y parameters, or *short-circuit admittance* parameters, have dimensions of mhos and are defined with one port shorted. One h parameter (h_i) is an impedance, two (h_f, h_r) are dimensionless, and one (h_o) is an admittance; consequently, they are known as *hybrid* parameters. The g parameters, which also have hybrid dimensions, have no other name.

COMMON-EMITTER HYBRID PARAMETERS

A three-terminal linear network or transistor can always be regarded as a two-port network. This is done by selecting arbitrarily a *common terminal*, and designating the currents of the other two terminals as I_1 and I_2. The derivation of (12-27) for the three-terminal network is the same as for the two-port of Fig. 12-8b. An example of a three-terminal linear device is the BJT biased in the active region, provided the input signal is sufficiently small.

In the next section we investigate ways to find the hybrid-π parameters of a transistor, using measured h parameters with the emitter selected as the common terminal. The common-emitter (CE) h parameters are designated h_{ie}, h_{re}, h_{fe},

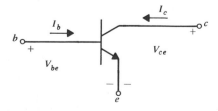

Figure 12-10   BJT as a two-port.

and h_{oe}, with the second subscript identifying the common terminal. From Fig. 12-10 and (12-28) the two-port equations are found to be

$$V_{be} = h_{ie}I_b + h_{re}V_{ce}$$
$$I_c = h_{fe}I_b + h_{oe}V_{ce} \tag{12-30}$$

The transistor of Fig. 12-10 is assumed to be biased in the active region, and (12-30) relate the incremental terminal variables, regardless of the external circuitry. Of course, the parameters depend on the Q-point and on the frequency.

In accordance with (12-30), we have:

$$h_{ie} = \frac{V_{be}}{I_b}\bigg|_{V_{ce}=0} = \text{CE short-circuit input impedance} \tag{12-31a}$$

$$h_{re} = \frac{V_{be}}{V_{ce}}\bigg|_{I_b=0} = \text{CE open-circuit reverse voltage gain} \tag{12-31b}$$

$$h_{fe} = \frac{I_c}{I_b}\bigg|_{V_{ce}=0} = \text{CE short-circuit forward current gain} \tag{12-31c}$$

$$h_{oe} = \frac{I_c}{V_{ce}}\bigg|_{I_b=0} = \text{CE open-circuit output admittance} \tag{12-31d}$$

We observe that h_{ie} and h_{fe} are determined with the phasor voltage $V_{ce} = 0$, which implies that the total voltage v_{CE} equals the quiescent voltage V_{CE}. The output terminals are shorted incrementally. The parameters h_{re} and h_{oe} are found with $I_b = 0$, or $i_B = I_B$. Although the output terminals are incrementally open-circuited, the proper Q-point must be maintained during measurements.

Figure 12-11 represents a CE h-parameter model of a transistor. The terminal voltages and currents are properly related in accordance with (12-30). It is a valid incremental circuit, no matter what external connections are made to the three terminals, provided the sinusoidal signal is sufficiently small so that the transistor can be treated as a linear device.

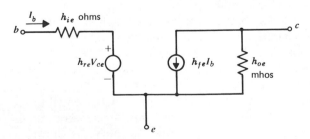

Figure 12-11　CE h-parameter model of BJT.

12-5. MEASUREMENTS OF HYBRID PARAMETERS

At low frequencies the reactances of the incremental junction capacitances of a BJT are so large that they are, in effect, open circuits. For such low frequencies the matrix parameters are real numbers and independent of frequency. Measurements of any four independent parameters can be used for calculating any or all of the others.

PARAMETERS h_{ie} AND h_{fe}

A circuit such as that of Fig. 12-12 is employed to measure the low-frequency values of h_{ie} and h_{fe}. The selected Q-point is 1 mA and 20 V, with these values of I_C and V_{CE} obtained by adjustment of V_{EE} and V_{CC}. Approximate values of V_{EE} and V_{CC} are found by assuming $\beta_F = 50$, giving a base current of 0.02 mA. This makes the base potential -2 V, and for a silicon device the emitter potential is about -2.7 V. This dictates values of 53.7 V for V_{EE} and 17.4 V for V_{CC}.

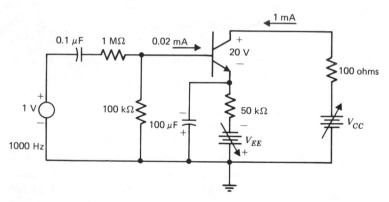

Figure 12-12   Circuit for measuring h_{ie} and h_{fe}.

A frequency of 1000 Hz is selected, because this is low enough to treat C_π and C_μ as open circuits and high enough to justify ignoring the small reactances of the μF-capacitances of Fig. 12-12. Utilizing the model of Fig. 12-11 leads to the incremental circuit of Fig. 12-13. The 100 kΩ resistor is omitted, *because it is shunted by the much smaller input resistance of the transistor.*

The input resistance V_{be}/I_b of a CE transistor is approximately equal to r_π, which is typically 1 or 2 kΩ for a transistor biased at 1 mA. A generator output of 1 V gives an I_b of 1 μA, and this makes V_{be} about 1 or 2 mV. As shown in Sec. 12-1, this is sufficiently small to justify the linear model. The voltage V_{be}

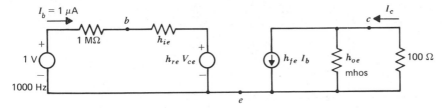

Figure 12-13   Incremental circuit.

is measured, and I_b is determined by measuring the voltage across the 1 megohm resistor, using a high-impedance voltmeter.

In the output circuit the resistance $1/h_{oe}$ is large (usually 50 kΩ or more), and it can be omitted. For an h_{fe} of 50, the output current is 50 μA, and V_{ce} is -5 mV. Power supply ripple is very low because of the small signals. The output current I_c is calculated from the measured ac voltage across the 100 ohm resistor. It should be noted that the 100 ohm resistance is so small that it has no appreciable effect on I_b, I_c, and V_{be}, which are the variables used to find h_{ie} and h_{fe}. Hence the 100 ohm resistor, placed in the circuit for the purpose of determining I_c, is an effective incremental short circuit. For assumed experimental values of $I_b = 1$ μA, $I_c = 50$ μA, and $V_{bv} = 1$ mV, (12-31a) and (c) give $h_{ie} = 1$ kΩ and $h_{fe} = 50$.

PARAMETERS h_{re} AND h_{oe}

To measure the low-frequency values of h_{re} and h_{oe} of the same transistor at the same Q-point, a circuit such as that shown in Fig. 12-14 is used. In the incremental circuit of Fig. 12-15 the 1 megohm resistor is omitted because it is practically an open-circuit. A high-impedance ac voltmeter is used to measure V_{be}, V_{ce}, and the voltage across the 1 kΩ resistor, from which I_c is calculated. Assuming these are found to be 0.2 mV, 1 V, and 20 μA, respectively, we employ

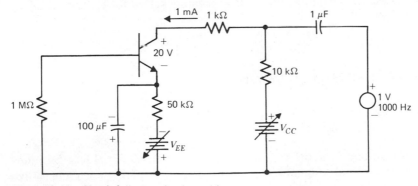

Figure 12-14   Circuit for measuring h_{re} and h_{oe}.

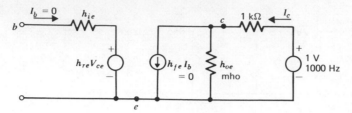

Figure 12-15 Incremental circuit.

(12-31b,d) to find that $h_{re} = 2 \times 10^{-4}$ and $h_{oe} = 2 \times 10^{-5}$ mho. From these results it is easy to justify the omission of the one-megohm resistor from the incremental model.

COMMON-BASE (CB) AND COMMON-COLLECTOR (CC) HYBRID PARAMETERS

The CE h parameters have been considered in some detail, because they are the easiest to measure at low frequencies. However, the transistor can also be represented by an incremental model using common-base (CB) or common-collector (CC) h parameters. These are easily determined from the CE parameters.

To illustrate the procedure, we write the equations that define the CC h parameters:

$$V_{bc} = h_{ic}I_b + h_{rc}V_{ec}$$
$$I_e = h_{fc}I_b + h_{oc}V_{ec} \tag{12-32}$$

The CE parameters are defined by (12-30). In (12-30) we can replace the CE terminal variables V_{be}, V_{ce}, and I_c with the equivalent CC terminal variables $V_{bc} - V_{ec}$, $-V_{ec}$, and $-I_b - I_e$, respectively, noting that I_b is the input current for both cases. The resulting equations are readily solved for V_{bc} and I_e, giving

$$V_{bc} = h_{ie}I_b + (1 - h_{re})V_{ec}$$
$$I_e = -(1 + h_{fe})I_b + h_{oe}V_{ec} \tag{12-33}$$

Comparison of (12-32) and (12-33) shows that

$$
\begin{array}{ll}
h_{ic} = h_{ie} & h_{rc} = 1 - h_{re} \\
h_{fc} = -1 - h_{fe} & h_{oc} = h_{oe}
\end{array} \tag{12-34}
$$

The CC h-parameter model of a BJT is obtained from Fig. 12-11 by interchanging the terminal designations c and e, replacing V_{ce} in the dependent source with V_{ec}, and using CC h parameters.

PARAMETER DETERMINATION FROM STATIC CHARACTERISTICS

Each low-frequency matrix parameter can be expressed as a partial derivative representing a ratio of differential changes in appropriate total variables. For example,

$$h_{oe} = \frac{I_c}{V_{ce}}\bigg|_{I_b=0} = \frac{\partial i_C}{\partial v_{CE}}\bigg|_{i_B=I_B} \tag{12-35}$$

with the partial-derivative form valid only when ω approaches zero. This is a consequence of the fact that the phasor ratio I_c/V_{ce} equals the ratio i_c/v_{ce} of the instantaneous quantities when $\omega \approx 0$. Because i_c denotes the incremental change Δi_C in the total current, and v_{ce} is Δv_{CE}, the ratio i_c/v_{ce} is $\Delta i_C/\Delta v_{CE}$, or $\partial i_C/\partial v_{CE}$. The partial derivative implies that the variations from the Q-point are small and that i_B is maintained constant at its quiescent value I_B. In other words, $I_b = 0$.

A family of collector characteristics, showing curves of i_C versus v_{CE} with i_B constant, is convenient for finding graphically the low-frequency h_{oe}. It is clear from (12-35) that h_{oe} is the slope of the characteristic curve at the Q-point. The collector characteristics are also suitable for determining h_{fe}, but a suitable set of input characteristics is needed for evaluation of h_{ie} and h_{re}. Because h_{re} is so small (10^{-3} to 10^{-5}), it cannot be determined accurately from such curves.

Incremental models based on the z, y, or g parameters in CE, CB, and CC configurations are also important. Of particular importance is the CE y-parameter model, being especially convenient for circuit analysis. It is treated in detail in the next chapter.

12-6. DETERMINATION OF HYBRID-π PARAMETERS

In order to make calculations with the hybrid-π model, numerical values of the transistor elements are needed. Measurements are required, by either the manufacturer or circuit designer, or perhaps both. The transconductance g_m is readily found by multiplying the quiescent current I_C by q/kT. Then the incremental resistances are calculated from g_m and the measured low-frequency CE h parameters. Suitable equations will now be developed.

In Fig. 12-16 is the low-frequency hybrid-π circuit. Current sources at the input and output ports supply the phasor currents I_b and I_c. Selecting the emitter as the reference node, the three node equations are

$$0 = -g_x V_{be} + (g_x + g_\pi + g_\mu)V - g_\mu V_{ce} \tag{12-36}$$

$$I_b = g_x V_{be} - g_x V \tag{12-37}$$

$$I_c = (g_o + g_\mu)V_{ce} + (g_m - g_\mu)V \tag{12-38}$$

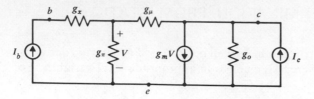

Figure 12-16 Low-frequency hybrid-π circuit.

Adding (12-36) and (12-37) and solving for V, we obtain

$$V = \frac{I_b + g_\mu V_{ce}}{g_\pi + g_\mu} \qquad (12\text{-}39)$$

Let us substitute for V into (12-37) and (12-38). Upon rearranging terms, the respective equations become

$$V_{be} = \left[\frac{1}{g_x} + \frac{1}{g_\pi + g_\mu}\right]I_b + \left[\frac{g_\mu}{g_\pi + g_\mu}\right]V_{ce} \qquad (12\text{-}40)$$

$$I_c = \left[\frac{g_m - g_\mu}{g_\pi + g_\mu}\right]I_b + \left[g_o + \frac{g_\mu(g_m + g_\pi)}{g_\pi + g_\mu}\right]V_{ce} \qquad (12\text{-}41)$$

Clearly, the terms in brackets are the CE h parameters.

We know that $g_m \gg g_\pi \gg g_o \gg g_\mu$. Therefore, the h parameters are approximately

$$h_{ie} = r_x + r_\pi \qquad h_{re} = \frac{r_\pi}{r_\mu}$$

$$h_{fe} = g_m r_\pi \qquad h_{oe} = g_o + g_m h_{re} \qquad (12\text{-}42)$$

Accordingly, the elements of the low-frequency hybrid-π can be determined for both NPN and PNP transistors from the following relations:

$$g_m = \left(\frac{q}{kT}\right)|I_c| \qquad r_o = \frac{1}{h_{oe} - g_m h_{re}}$$

$$r_\pi = \frac{h_{fe}}{g_m} \qquad r_\mu = \frac{r_\pi}{h_{re}} \qquad (12\text{-}43)$$

$$r_x = h_{ie} - r_\pi$$

The low-frequency h parameters are usually measured at 1000 Hz. It is clear from (12-24) and (12-42) that $h_{re} \approx \eta$ and $h_{oe} \approx 2\eta g_m$, with η defined by (12-25).

CAPACITANCES C_μ AND C_π

Measurement of the magnitude of the admittance between the base-collector terminals with the emitter incrementally open is sufficient for determining the capacitance C_μ, provided the frequency is properly chosen. The incremental circuit is shown in Fig. 12-17, with y_μ and y_π representing the respective admittances $g_\mu + j\omega C_\mu$ and $g_\pi + j\omega C_\pi$. A generator connected to terminals bc produces a current through r_o equal to $(g_m + y_\pi)V$. Therefore, the voltage V_1 across y_μ is $(1 + g_m r_o + y_\pi r_o)V$. The current I_1 is $y_\pi V$, and the admittance I_1/V_1 in parallel with y_μ is $y_\pi/(1 + g_m r_o + y_\pi r_o)$. At the measurement frequency this admittance is small compared with y_μ and can be neglected (see P12-22). By selecting a frequency such that $g_x \gg \omega C_\mu \gg g_\mu$, the magnitude of the input admittance is ωC_μ. For example, a frequency of 1 MHz could be used for measuring a capacitance of 2 pF, giving $\omega C_\mu \approx 10^{-5}$ mho. The capacitance found in this manner is called the CB output capacitance C_{ob}, which approximately equals C_μ. It is given on transistor data sheets if this parameter is important in the intended use of the device.

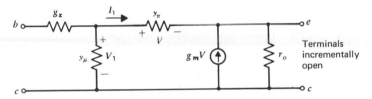

Figure 12-17   Incremental circuit for determining C_μ.

The parameter h_{fe} is $(g_m - y_\mu)/(y_\pi + y_\mu)$, obtained from (12-41) by replacing g_μ with y_μ and g_π with y_π. Because $g_\pi \gg g_\mu$, $g_m \gg g_\mu$, and $g_m \gg \omega C_\mu$ at even the highest frequency for which the hybrid-π is a reasonable model, h_{fe} can be expressed as

$$h_{fe} = \frac{(h_{fe})_o}{1 + j\omega r_\pi(C_\pi + C_\mu)} \tag{12-44}$$

where $(h_{fe})_o = g_m r_\pi = $ zero-frequency h_{fe}.

Equation 12-44 is employed to find C_π. It is necessary to measure the magnitude of h_{fe} at a frequency sufficiently high to make the imaginary term of (12-44) significant. A convenient choice is the frequency, designated ω_β, that makes the j-term unity and $|h_{fe}| = 0.707(h_{fe})_o$. In terms of ω_β, C_π becomes

$$C_\pi = \frac{g_\pi}{\omega_\beta} - C_\mu \tag{12-45}$$

The CE short-circuit forward current gain h_{fe} is quite frequently referred to as β, and $(h_{fe})_o$ is β_o. Accordingly, the frequency at which $|\beta|$ equals $0.707\beta_o$ is called the *beta cutoff frequency*. From (12-45) it follows that ω_β is $g_\pi/(C_\pi + C_\mu)$.

THE f_T FREQUENCY

To introduce another frequency of particular importance let us substitute $r_\pi g_m$ for $(h_{fe})_o$ in (12-44) and consider the range of radian frequencies greater than $10\omega_\beta$, which makes the unity term in the denominator negligible. The magnitude of h_{fe}, or β, becomes $g_m/\omega(C_\pi + C_\mu)$, and taking common logarithms gives the relation

$$\log_{10}|\beta| = \log_{10}\left(\frac{g_m}{C_\pi + C_\mu}\right) - \log_{10}\omega \qquad (12\text{-}46)$$

A plot of $\log|\beta|$ versus $\log\omega$ is a straight line with a slope of -1, and the product $\omega|\beta|$ equals the constant $g_m/(C_\pi + C_\mu)$.

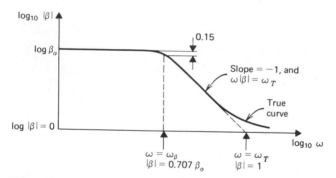

Figure 12-18   Typical plot of $\log|\beta|$ versus $\log\omega$.

Figure 12-18 shows a typical plot of $\log|\beta|$ as a function of $\log\omega$ from very low to very high frequencies. In the low-frequency region β is the constant β_o. At the beta cutoff frequency, $|\beta|$ is $0.707\beta_o$, and $\log|\beta|$ is less than the low-frequency value by $|\log 0.707|$, or 0.15. At higher frequencies the slope becomes approximately -1 in accordance with (12-46), except for the decrease in the magnitude of the slope at the extreme right which is not predicted by the hybrid-π model. The g-parameters of the model were deduced from the Ebers–Moll equations based on the dc diffusion equation. Because of this and various high-frequency effects, the frequency range of the hybrid-π model is restricted. The dashed line in Fig. 12-18 is a projection of the straight portion of the curve to the axis, where $\log|\beta|$ is zero and $|\beta|$ is unity. At the intercept we have the *transition frequency* ω_T, defined as that frequency at which $|\beta|$ *extrapolates to unity.* At any point on the straight line with the slope -1 the product $\omega|\beta|$ is constant, and this constant is ω_T. It should be noted that $|\beta|$ is actually greater than one at ω_T.

In (12-46) $\omega = \omega_T$ when $|\beta| = 1$. Accordingly, ω_T becomes $g_m/(C_\pi + C_\mu)$, and f_T is

$$f_T = \frac{g_m}{2\pi(C_\pi + C_\mu)} \tag{12-47}$$

For comparison, f_β is

$$f_\beta = \frac{g_\pi}{2\pi(C_\pi + C_\mu)} \tag{12-48}$$

Clearly, $f_T = \beta_o f_\beta$, with $\beta_o = r_\pi g_m$.

Transistor data sheets usually specify the minimum value of f_T and perhaps a typical value, along with a curve showing the variation of f_T with I_C. This is discussed further in Sec. 15-2. Assuming g_m and C_μ have been determined, C_π can be calculated from a specified f_T with the aid of (12-47). The hybrid-π model is not valid at frequencies greater than about one-third of f_T.

DISCUSSION

Measurements of C_μ and C_π are at high frequencies and must be made with extra care. Instruments such as electronic voltmeters and cathode-ray-oscilloscopes have input capacitances that may be tens of picofarads. DC milliammeters and power supplies are likely to present considerable ac impedance at high frequencies, and they should be bypassed with capacitors.

Data-sheet information is often used to find the hybrid-π parameters. If the specified Q-point does not correspond to the one in the circuit under consideration, corrections should be made. We have learned that g_m, g_π, g_o, g_μ, and C_π are nearly proportional to the quiescent current. The parameters g_x and C_μ are relatively independent of current, but r_x decreases at high-current levels for the reasons given in Sec. 12-3. Except for g_m, the conductances are functions of base-width W and hence V_{CE}, and C_π and C_μ also depend on V_{CE}.

In the hybrid π circuit there are five branches. These are the conductances g_x and g_o, the admittances y_π and y_μ, and the dependent generator $g_m V$. On the other hand a two-port model is characterized by only four complex parameters. The great advantage of the hybrid-π model is that its elements are reasonably independent of frequency. Thus it is especially useful for nonsinusoidal excitations and for sinusoidal signals with arbitrary frequency. Furthermore, the hybrid-π parameters and their variations with temperature and operating point are simply related to the physics of the device.

The two-port models are convenient at low frequencies, where the parameters are real, and for any case in which the excitation is sinusoidal provided the parameters are known at the signal frequency. At very high frequencies where the hybrid-π is no longer an accurate representation of the device, the two-port models using complex parameters are invaluable.

12-7. FURTHER DISCUSSION OF THE BASE RESISTANCE

When the base resistance r_x is found by subtracting r_π from h_{ie}, as indicated in (12-43), the result may include an appreciable component attributable to the ohmic resistance r_e' of the emitter region. This is true even though the emitter is so heavily doped that its resistance is quite low, normally ranging from about 0.1 ohm to 2 ohms. In Fig. 12-19 is shown a simplified hybrid-π with the emitter resistance added and the collector-emitter terminals shorted. The omission of r_o and r_μ introduces no appreciable error when the output is shorted, which is discussed in detail in Sec. 13-2. The actual base ohmic resistance is labeled r_x'. Note that $g_m V$ of the dependent current source equals $\beta_o I_1$, and the current through r_e' is $(\beta_o + 1)I_1$.

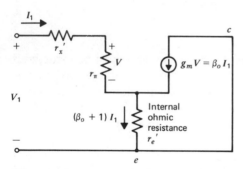

Figure 12-19 Simplified hybrid-π with emitter resistance.

A loop equation around the input circuit of Fig. 12-19 is sufficient for determining that the ratio V_1/I_1 is the sum of r_x', r_π, and $(\beta_o + 1)r_e'$, which is the measured h_{ie}. Subtraction of r_π from h_{ie} gives r_x to be

$$r_x = r_x' + (\beta_o + 1)r_e' \qquad (12\text{-}49)$$

To illustrate, suppose a transistor has an ohmic base resistance r_x' of 50 ohms, an ohmic emitter resistance of 0.2 ohm, and a beta of 100. For this example the difference $h_{ie} - r_\pi$, or r_x, is 70.2 ohms.

We shall henceforth ignore the ohmic resistance of the emitter, as well as that of the collector. However, the approximate effect of r_e' on the incremental performance of the transistor is included in r_x, which is calculated from the measured value of h_{ie}.

REFERENCES

1. P. E. Gray and C. L. Searle, *Electronic Principles*, Wiley, New York, 1969.
2. R. P. Nanavati, *Semiconductor Devices*, Intext Education Publishers, New York, 1975.
3. P. M. Chirlian, *Electronic Circuits: Physical Principles, Analysis, and Design*, McGraw-Hill, New York, 1971.
4. J. Lindmayer and C. Y. Wrigley, *Fundamentals of Semiconductor Devices*, D. Van Nostrand, New York, 1965.
5. J. L. Giacoletto, "Study of p-n-p Alloy Junction Transistor from d-c through Medium Frequencies," *RCA Review*, **15**, pp. 506–552, 1959.

PROBLEMS

Section 12-1

12-1. As indicated by (12-5), the incremental collector current i_c of a BJT is approximately proportional to $(e^{40v_{be}} - 1)$ at 290 K. (a) Using the first three terms of the exponential series of (12-6), deduce that i_c is proportional to $(v_{be} + 20v_{be}^2)$. Then let v_{be} be $\sqrt{2}\,V_{be}\cos\omega t$, and show that

$$i_c = C_1 V_{be}(10\sqrt{2} + \cos\omega t + 10\sqrt{2}\,V_{be}\cos 2\omega t)$$

with C_1 a constant. (b) From this result, calculate the percent distortion D for rms base-emitter signal voltages of 1, 5, and 10 mV, with D defined as the ratio of the rms value of the second harmonic (2ω) component of i_c to the rms value of the fundamental (ω) component, multiplied by 100%.

12-2. For the amplifier of Fig. 12-1, the true value of i_b is given by (12-5) and the approximate value for small signals is given by (12-9). With $v_i = 0.005$ V, what is the percent error in i_b introduced by the approximate equation? Repeat for $v_i = 0.0005$ V. (*Ans*: 9.67, 0.99)

12-3. For the amplifier of Fig. 12-1 with $v_i = 10^{-5}$ V, calculate the currents i_c, i_b, and i_e in μA, the voltages v_{be}, v_{ce}, and v_{bc} in mV, and the gains A_i, A_v, and G.

12-4. Using the incremental model of Fig. 12-3 of Sec. 12-1, find the voltage gain v_o/v_i, the current gain $-i_c/i_b$, and G as functions of g_m, r_π, and R_L; $(r_\pi = 1/g_\pi)$. Calculate A_i, A_v, and G for $g_m = 40$ mmhos, $r_\pi = 1$ kΩ, and $R_L = 4$ kΩ.

Section 12-2

12-5. Using (12-3) and (12-4) of Sec. 12-1, converted to SI units, find in terms of v_{be} the first four terms of the series of (12-15) of Sec. 12-2. Also, substitute the series of (12-7) into (12-5) for i_b, converted to SI units. Compare results. Note that $d^n i_B/dv_{BE}^n$ at the Q-point is $40^n I_B$.

12-6. At 290 K $(q/kT = 40)$ the current I_C of a silicon NPN transistor is 2.5 mA with $V_{BE} = 0.70$ and $V_{BC} = -15.3$ V. Assume that the electron diffusion length in the base is 5×10^{-5} m and that the base width W is

$$W = 10^{-6}(11.68 - \sqrt{0.86 - V_{BE}} - 0.32\sqrt{0.7 - V_{BC}})\ \text{m}$$

Calculate g_m, r_π, r_o, and r_μ. (*Ans*: 100 mmhos, 0.49 kΩ, 100 kΩ, 4900 kΩ)

12-7. Repeat P12-6, but with $|V_{BC}|$ increased to 24.3 V. From the results, comment on the variations of the parameters with respect to $|V_{BC}|$.

12-8. At 350 K ($q/kT = 33.2$) the silicon transistor of P12-6 has $\psi_{oe} = 0.75$ (instead of 0.86) and $\psi_{oc} = 0.60$. The expression for W must be corrected. The Q-point (I_C, V_{CE}) is the same, but V_{BE} is lower, being 0.66 V. Because of the longer excess-carrier lifetime in the base, the diffusion length L_b is greater, being 6×10^{-5} m. Find $g_m, r_\pi, r_o,$ and r_μ, and compare with the values given in P12-6 at 290 K.

12-9. (a) Calculate $r_\pi, r_o,$ and r_μ in kΩ for $\beta_F = 100, \eta = 10^{-4}$, and an operating current of 1 mA at 290 K. (b) Sketch the incremental model of Fig. 12-5, showing numerical values of the conductances on the circuit, using mA, mmhos, and volts. (c) Write two node equations that give i_b and i_c in terms of v_{be} and v_{ce}.

12-10. Shown in Fig. 12-20 is a three-terminal device (not a BJT) with terminals labeled B, C, and E. The currents and voltages of the device are related by the equations

$$i_B = 0.1v_{BE} - 0.2 \qquad i_C = (v_{BE} - 1)^2 + 0.01v_{CE}{}^2$$

with v_{CE} restricted to positive values and with i_B and i_C in mA. Find the quiescent

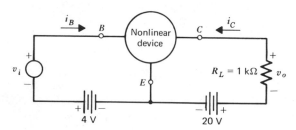

Figure 12-20 Amplifier.

terminal voltages and currents, use (12-19) to calculate the g parameters at the Q-point, and employ the small-signal model of Fig. 12-5a to determine the voltage gain v_o/v_i and the current gain $-i_c/i_b$. (Ans: $A_v = -5, A_i = -50$)

Section 12-3

12-11. Shown in Fig. 12-21 is the low-frequency hybrid-π with sources that supply i_b and i_c. Units are in mA, mmhos, and volts. (a) Using these units, write the node equations for the nodes labeled v, v_{be}, and v_{ce}. Solve for i_b and i_c as functions of v_{be} and v_{ce}.

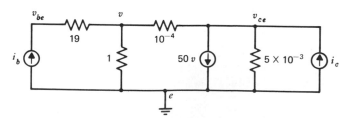

Figure 12-21 Circuit for P12-11 and P12-12.

(b) Change g_x from 19 mmhos to infinity (r_x from 52.6 ohms to zero), making $v_{be} = v$. Solve the network for i_b and i_c as functions of v_{be} and v_{ce}. Compare results with those of (a), noting that r_x has little effect.

Section 12-4

12-12. Calculate the CE y parameters in mmhos of the two-port network of Fig. 12-21, which has element values in units of mA, mmhos, and V. Then find the CE h parameters in SI units.

12-13. Superposition was applied to the network of Fig. 12-9 of Sec. 12-4 to derive the y-parameter equations (12-27). Replace the voltage sources V_1 and V_2 with current sources I_1 and I_2. Then apply superposition, and deduce the z-parameter equations of (12-28).

12-14. Define the CE z parameters of a transistor in the form of (12-31), which define the CE h parameters.

12-15. By either equation manipulation or matrix inversion, find the z parameters of (12-28) as functions of the y parameters.

12-16. The tee-network of Fig. 12-22 is an equivalent circuit of a two-port network provided the elements are properly chosen. Find Z_1, Z_2, Z_3, and Z_4 as functions of the matrix z parameters.

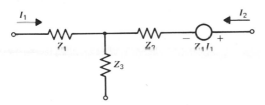

Figure 12-22   Tee equivalent circuit.

12-17. In terms of g_m, g_π, g_μ, and g_o, find the CE y parameters of the circuit of Fig. 12-5a.

Section 12-5

12-18. For the circuit of Fig. 12-14 of Sec. 12-5, determine V_{EE} and V_{CC} at the selected Q-point, assuming $\beta_F = 50$ and $V_{BE} = 0.7$ V.

12-19. (a) Write the low-frequency parameter h_{fe} as a partial derivative, in the form of (12-35). (b) The respective CE hybrid parameters h_{ie}, h_{fe}, h_{re}, and h_{oe} of a certain transistor are 1000, 50, 0.0002, and 0.00002 in SI units. Calculate the corresponding CC hybrid parameters, and using these, sketch the incremental model.

12-20. Find the CB h parameters in terms of the CE h parameters. Also, sketch the incremental model of a BJT, using CB h parameters.

Section 12-6

12-21. In Fig. 12-16 of Sec. 12-6, replace the current source at the output with a short-circuit and solve for h_{ie} and h_{fe}. Then remove the current source at the input, apply a signal to the output port, and solve for h_{re} and h_{oe}. Calculate these parameters for values of g_x, g_m, g_π, g_o, and g_μ respectively equal to 20, 40, 1, 4×10^{-3}, and 10^{-4} mmhos. Give answers in SI units.

12-22. In Fig. 12-17 of Sec. 12-6, a generator is connected to the input port with $\omega = 10^7$ radians/second. When V_{bc} is 1 V, $|I_b|$ is 20 μA. The values of the g parameters are those given in P12-21. Assuming C_π is 100 pF, calculate the admittance in parallel with y_μ and compare with y_μ. Also, calculate C_μ. Use SI units.

12-23. If $\beta_o = 100$ and $|\beta| = 50$ at 4 MHz, calculate f_β and f_T in MHz. Also find C_π, assuming $g_m = 50$ mmhos and $C_\mu = 4$ pF. (*Ans*: 30.5 pF)

12-24. Measurements of the low-frequency h parameters h_{ie}, h_{re}, h_{fe}, and h_{oe} at a quiescent current of 0.5 mA yield the respective SI values of 2050, 5 $\times$ 10^{-4}, 40, and 2 $\times$ 10^{-5}. The CB output capacitance C_{ob} is 3 pF and $|h_{fe}|$ at 1 MHz is 0.707 times its low-frequency value. At 290 K calculate g_m, r_x, r_π, r_o, r_μ, C_π, and C_μ.

Chapter Thirteen
Basic Amplifier Configurations

We have found that the transistor is an active device capable of gain. Although the currents are nonlinear functions of the voltages, the distortion is negligible when the BJT is biased in the active region and fed with small signals, which permits the use of a linear incremental circuit. Most applications employ the CE configuration, in which the base and emitter terminals constitute the input port and the collector-emitter terminals are the output. Also important, however, are the common-collector (CC) and common-base (CB) configurations. This chapter presents certain basic properties of small-signal amplifiers, and compares the performances of the three configurations.

A signal supplied to an amplifier may come from a microphone, an antenna, a measuring instrument, the output of another amplifier, or some other device or circuit. In any event the source can be represented by its Thévenin equivalent circuit. Amplifier loads are the input impedances of other amplifiers, antennas, speakers, measuring instruments, and other devices and circuits. Each of these loads can be represented by an impedance.

Usually the amplifier should supply the load with signal power that is large compared with the available power of the source. This must be accomplished over the desired range of frequencies with minimum distortion. Noise introduced by the amplifier should be small compared with the signal. The bias circuit must not be wasteful of power, and the amplifier should be small, low in cost, and capable of operating under various environmental conditions with long life. Designing the amplifier for easy construction as an integrated circuit may be essential. Of course the relative importance of the various requirements depends on the intended use.

13-1. A TYPICAL CE AMPLIFIER

A configuration frequently used to amplify small signals is shown in Fig. 13-1. Connected to the input port is the Thévenin equivalent circuit of the source, assumed to be sinusoidal and having an internal impedance that is resistive. The open-circuit rms phasor voltage is denoted V_s and R_s is the internal resistance. Direct currents that may be present in the circuitry of the source have no effect on the amplifier because of the blocking capacitor C_1, and furthermore, direct currents of the amplifier cannot pass through the source. Similarly, C_2 is a blocking capacitor, providing dc isolation between the amplifier and the circuitry represented by the load resistance R_L. This resistance may be the input resistance of an amplifier stage in cascade with the one in Fig. 13-1, or some other load.

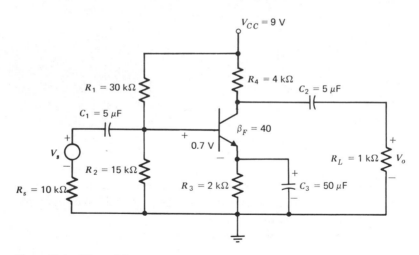

Figure 13-1   CE amplifier with NPN silicon transistor.

The combination of a resistor and a capacitor to permit transfer of signal energy from one stage to another, while blocking the direct currents, is referred to as *RC coupling*. A transformer can be used for the same purpose, though this is not often done. Many amplifiers are directly coupled, and there is interaction between their dc circuits.

Located between the point identified by the ground symbol and the terminal labeled V_{CC} is the dc power supply, which maintains the voltage V_{CC}. The electric potential of the ground is selected as zero, and the V_{CC} terminal has potential V_{CC}.

In some electronic diagrams two or more points may have ground symbols, in which case these points are understood to be connected electrically. The ground symbol does not indicate a connection to earth. At the signal frequencies for which the amplifier is designed, capacitors C_1, C_2, and C_3 have reactances so small that they are, in effect, ac short circuits. The incremental model of Fig. 13-2 applies to signal frequencies high enough to justify treating the circuit microfarad capacitors C_1, C_2, and C_3 as shorts, but low enough to justify treating the transistor picofarad capacitances C_π and C_μ as opens. It is assumed that the internal impedance of the power supply is zero, which places the V_{CC} terminal and the ground point at the same potential *insofar as the signals are concerned*.

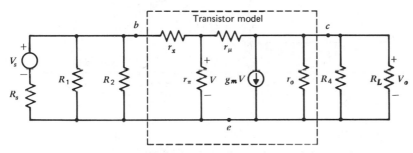

Figure 13-2   Midfrequency incremental model of amplifier of Fig. 13-1.

In Fig. 13-2 the emitter terminal is common to both the input and output ports of the amplifier. Accordingly, the configuration is called a CE amplifier, even though at very low frequencies the amplifier is not common emitter in the strict sense. When R_3 is very small, C_3 is sometimes omitted, and the signal currents pass through R_3. In this case the amplifier is approximately common emitter. As we shall see from an analysis in the next section, elimination of C_3 reduces the voltage gain.

THE BIAS CIRCUIT

The dc circuit Fig. 13-1 is shown in Fig. 13-3. A voltage-divider circuit is formed by R_1 and R_2. They are selected so that the voltage across R_2 gives the proper values for V_{BE} and the IR drop across R_3. Because R_1 and R_2 are across the input in the incremental circuit of Fig. 13-2, their presence reduces the current gain of the amplifier. This undesirable effect is minimized by choosing values for the resistances that are large compared with the input resistance of the amplifier. However, large values of R_1 and R_2 cause V_{BE} to be sensitive to changes in the quiescent base current resulting from temperature variations or other effects. Because this is also undesirable, the selection of values of R_1 and R_2 involves a compromise.

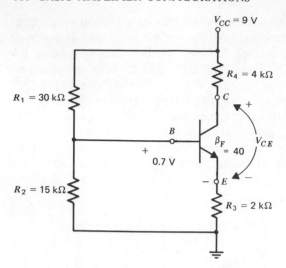

Figure 13-3   Bias circuit of amplifier of Fig. 13-1.

Stabilization of the quiescent collector current is accomplished by R_3. The voltage across R_2 less that across R_3 is V_{BE}. A small increase in I_C, due perhaps to a change in temperature or a change in a circuit parameter, gives an increased voltage across R_3, thereby decreasing V_{BE}. This tends to reduce I_C. Hence, a change in I_C causes V_{BE} to change so as to oppose the initial change in I_C. The resistance R_3 is said to introduce dc *negative feedback* from the collector-emitter circuit into the base-emitter circuit. Good stabilization is accomplished by using a large R_3.

There are three reasons, however, for keeping R_3 as small as possible. First, for a specified quiescent current, the larger one makes R_3, the greater is the voltage across it and the electrolytic capacitor C_3. The dc voltage rating of C_3 must be sufficiently large. Furthermore, if V_{CC} is limited, it may be necessary to reduce V_{CE} below the optimum value. Second, dc power dissipated in R_3 is wasted as heat. This is especially objectionable in battery-operated equipment. Third, the *impedance* between the emitter and ground, caused by the parallel $R_3 C_3$ combination, *should be small* in order to prevent an excessive amount of negative feedback of the signal with consequent reduction of the gain. *A sufficiently small* R_3 *allows the selection of* C_3 *with fewer microfarads.* In fact, in some circuits C_3 is omitted. Clearly, the selection of R_3 involves compromise.

As far as the bias circuit is concerned, the resistor R_4 is unnecessary. Because it consumes voltage and dissipates power, just like R_3, it should be as small as possible. It cannot be too small, however, because in the incremental circuit of Fig. 13-2 it is in parallel with the load resistance. Signal current through R_4 means less signal current through R_L, with reduced gain. From the viewpoint of the incremental circuit, R_4 should be as large as possible. Again, the designer must compromise, and the value selected for R_4 is usually from 2 to 5 times R_L.

THE TWO-BATTERY EQUIVALENT CIRCUIT

Any bias circuit consisting of resistors, batteries, and possibly direct-current sources is a three-terminal network as viewed from the terminals B, E, and C of the transistor. By selecting one of the terminals as common, the bias circuit becomes a linear active two-port network. This network differs from the two-ports considered in Chapter 12 because it has internal independent sources at the frequency ($\omega = 0$) of interest. It is easy to show (see P13-5) that such a network can always be reduced to an equivalent tee containing two batteries, as shown in Fig. 13-4, in which the transistor is connected to the terminals of the tee. Because the common terminal was chosen to be the emitter, the batteries appear in the base and collector leads.

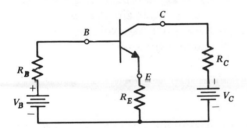

$V_B = 3\,\text{V}, \; V_C = 9\,\text{V}$
$R_B = 10\,\text{k}\Omega, \; R_E = 2\,\text{k}\Omega, \; R_C = 4\,\text{k}\Omega$
$V_{BE} = 0.7, \; \beta_F = 40$

Figure 13-4 Two-battery equivalent circuit.

The resistances R_B, R_E, and R_C are found by letting the sources be inactive and reducing the three-terminal resistive network to a tee, recalling that a zero voltage source is a short and a zero current source is an open. With the transistor removed and the terminals B, E, and C open, the voltages V_B and V_C are the terminal voltages V_{BE} and V_{CE}, respectively. Both the analysis and design of single-stage bias circuits are usually expedited by working with the circuit of Fig. 13-4.

For the bias circuit of Fig. 13-3 the equivalent tee is easily obtained. Removal of the transistor, leaving terminals B, E, and C open, reveals that the terminal voltages are

$$V_B = \frac{R_2}{R_1 + R_2} V_{CC} \qquad V_C = V_{CC} \qquad (13\text{-}1)$$

Replacing the battery with a short circuit puts R_1 and R_2 in parallel, and the resistances of the tee are observed to be

$$R_B = R_1 \| R_2 \qquad R_E = R_3 \qquad R_C = R_4 \qquad (13\text{-}2)$$

The symbolism $R_1 \| R_2$ denotes the equivalent resistance of the parallel combination. Using the numerical values of Fig. 13-1 gives the tee-parameter values shown in Fig. 13-4.

The quiescent currents are determined by applying Kirchhoff's voltage law to the emitter-base loop of Fig. 13-4. Because the silicon transistor is in the active mode at room temperature, I_{CO} is negligible, and $I_B = I_C/\beta_F$. As β_F is 40, I_B is $0.025I_C$ and I_E is $-1.025I_C$. Here, V_{BE} is specified to be 0.7 V. Using numerical values with units of mA, kΩ, and V, the loop equation in terms of I_C is

$$3 = 10(0.025I_C) + 0.7 + 2(1.025I_C) \tag{13-3}$$

Solving gives $I_C = 1.0$, $I_B = 0.025$, and $I_E = -1.025$ mA. By writing a loop equation around the collector-emitter circuit, the collector voltage V_{CE} is found to be 3 V.

In the preceding analysis the specified numerical values for V_{BE} and β_F were used. In Chapter 15 it is shown that the Q-point in a well-designed bias circuit is not very sensitive to changes in V_{BE} and β_F. If V_{BE} and β_F are unknown, the assumption of reasonable values enables one to ascertain the approximate quiescent conditions. At room temperature the emitter-base voltage of a silicon transistor in the active mode is about 0.7 V, and for germanium it is about 0.2 V. A typical value of β_F is 50.

AN EASY DESIGN PROCEDURE

Values selected for the elements of most bias circuits are not critical, and the design is straightforward. Normally R_L and V_{CC} are specified, a suitable quiescent current is chosen, and the design consists of selecting R_B, R_E, R_C, and V_B, from which the resistors R_1 and R_2 are determined. As previously mentioned R_4 (or R_C) is often chosen to be about 2 to 5 times R_L. The precise value is a compromise depending on the available voltage and also on the relative importance of the power dissipation in R_4 and the sacrifice in gain that results from R_4 being too small. There is a simple and usually satisfactory way to select the remaining resistors.

Because $I_E = -I_C(\beta_F + 1)/\beta_F$ and $I_B = I_C/\beta_F$, the loop equation for the base-emitter circuit of Fig. 13-4 can be written as

$$V_B = I_C\left[\frac{R_E(\beta_F + 1)}{\beta_F} + \frac{R_B}{\beta_F}\right] + V_{BE} \tag{13-4}$$

with V_{BE} equal to 0.7 V for Si and 0.2 for Ge at room temperature. As shown in Fig. 1-15 of Sec. 1-7, V_{BE} decreases about 2 mV per degree Celsius increase in temperature, and β_F also varies with temperature. Furthermore, data sheets specify the minimum β_F, and sometimes the maximum, but the precise value of beta is most often unknown.

A simple design procedure that is usually satisfactory provided temperature variations are not extreme is to select R_E and R_B as follows:

$$R_E = \frac{1}{I_{C(min)}} \qquad R_B \leq 0.3 R_E \beta_{F(min)} \qquad (13\text{-}5)$$

$I_{C(min)}$ is the minimum value of the quiescent current, with this value selected by the designer, and $\beta_{F(min)}$ is the minimum beta as specified on the transistor data sheet. In order to avoid excessive signal degradation, R_B should be at least several times r_π, which is assured if the maximum R_B permitted by (13-5) is chosen (see P13-4). The restriction on the maximum R_B given in (13-5) makes R_B/β_F in (13-4) less than $0.3 R_E$, thereby ensuring that the quiescent collector current is rather insensitive to the actual value of β_F. Note that the selection of R_E in (13-5) prescribes an emitter-resistance voltage of about 1 V when the collector current is a minimum. Accordingly, as indicated by (13-4), moderate changes in V_{BE} have but slight effect on I_C.

We calculate V_B from (13-4). The resistances R_1 and R_2 are easily shown from (13-1) and (13-2) to be

$$R_1 = \frac{R_B V_{CC}}{V_B} \qquad R_2 = \frac{R_B V_{CC}}{V_{CC} - V_B} \qquad (13\text{-}6)$$

Determination of R_1, R_2, R_E, and R_B by the method suggested here is adequate for many cases in which an optimum bias circuit is not required. A more careful design procedure is treated in Chapter 15. The small-signal performance of a CE amplifier is investigated in the next section.

13-2. CE AMPLIFIER ANALYSIS

Figure 13-5 shows an amplifier, represented as a two-port network, connected between a source and a load. The complex voltage gain A_v, the complex current gain A_i, and the input impedance Z_i are defined in terms of the rms phasor voltages and currents as follows:

$$A_v = \frac{V_o}{V_i} \qquad A_i = \frac{I_o}{I_i} \qquad Z_i = \frac{V_i}{I_i} \qquad (13\text{-}7)$$

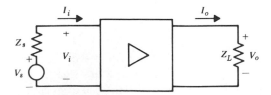

Figure 13-5 Amplifier between source and load.

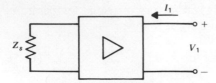

Figure 13-6 Z_o is V_1/I_1. Note that Z_L is removed.

Also of interest is the output impedance Z_o. This is the impedance looking back into the amplifier from the load, with the source replaced with its internal impedance. Referring to the network of Fig. 13-6, Z_o is

$$Z_o = \frac{V_1}{I_1} \tag{13-8}$$

With Z_L in Fig. 13-5 removed, the amplifier as viewed from the output terminals can be replaced by a Thévenin equivalent circuit. In this circuit, Z_o is the impedance that would be used, just as Z_s is the impedance of the Thévenin equivalent circuit of the source.

The output power P_o is the power supplied to the load impedance Z_L, and the input power P_i is that supplied to the input terminals of the amplifier. Their ratio is the power gain G; that is,

$$G = \frac{P_o}{P_i} \tag{13-9}$$

With R_i and R_L denoting the resistive parts of Z_i and Z_L, respectively, P_o is $|I_o|^2 R_L$ and P_i is $|I_i|^2 R_i$. Hence,

$$G = |A_i|^2 \frac{R_L}{R_i} = \left| A_i A_v \left(\frac{Z_i}{R_i} \right) \left(\frac{R_L}{Z_L} \right) \right| \tag{13-10}$$

If Z_i and Z_L are pure resistances, A_i and A_v are real numbers with the same sign, and G is $A_i A_v$.

The power gain is often expressed in *decibels*, or dB, with the dB-gain defined as

$$G = 10 \log_{10} \frac{P_o}{P_i} \text{ dB} \tag{13-11}$$

There are several reasons for the use of the decibel. The response of the ear is logarithmic; numerical values of gains typically encountered are smaller and easier to work with mathematically when expressed in dB; and the total gain of several stages in cascade is found simply by adding the dB gains of the individual stages, because log ab is the sum of log a and log b. For a single-stage amplifier the power gain is typically from 10 to 10 000, or from 10 dB to 40 dB.

We shall analyze the CE amplifier considered in the previous section, at frequencies sufficiently high to justify treating the microfarad coupling and bypass capacitors as short circuits but sufficiently low so that the picofarad capacitances C_π and C_μ are essentially open circuits. This frequency region is called the *mid-frequency band*. Both Z_s and Z_L are assumed to be pure resistances. Hence, the incremental circuit is a resistive model, and all currents and voltages are real quantities.

In the next two sections the CC and CB amplifiers are analyzed. The objective is to determine and compare the basic properties of the three different configurations. This objective is not compromised and simplicity is achieved by ignoring the effect of the bias resistors. These resistors downgrade the performance, but their effects are of secondary importance (see P13-6).

Equations 12-24 give $g_m = (q/kT)I_C$, $g_\pi = g_m/\beta_F$, $g_o = \eta g_m$, and $g_\mu = g_o/\beta_F$. These are approximations, because they were deduced from the Ebers–Moll equations derived for an idealized transistor. However, they are assumed to be valid for the analysis of this and the following sections. For the amplifier of Fig. 13-1 the quiescent current is 1 mA. At room temperature q/kT is 40, and β_F was given as 40. We need to specify the base-width modulation factor η and the incremental ohmic conductance g_x of the base. Assuming η is 10^{-4} and g_x is 19 mmhos, the hybrid-π parameters in mmhos become

$$g_x = 19 \qquad g_m = 40 \qquad g_\pi = 1 \qquad g_o = 0.004 \qquad g_\mu = 10^{-4} \qquad (13\text{-}12)$$

The corresponding resistances are $r_x = 52.6\ \Omega$, $r_m = 25\ \Omega$, $r_\pi = 1000\ \Omega$, $r_o = 250\ \text{k}\Omega$, and $r_\mu = 10\ \text{M}\Omega$. These are reasonable values for a diffused planar transistor biased at 1 mA. For convenience the units of volts, mA, mmhos, and kΩ are employed in the remainder of this chapter.

INCREMENTAL CIRCUIT ANALYSIS

The CE incremental circuit is given in Fig. 13-7. The values of R_s and R_L are chosen to be 10 kΩ and 1 kΩ, respectively. Because r_μ is so large, it appears that approximate results can be obtained by regarding r_μ as an open circuit, which gives the simplified circuit of Fig. 13-8. The justification for this will be examined after the analysis. In Fig. 13-8 the dependent source supplies the current $g_m V$, which is $g_m r_\pi I_i$, or $\beta_o I_i$. Because R_L is much smaller than r_o, the output current $I_o \approx -\beta_o I_i$. From this relation and a careful examination of the circuit, we can find the voltage gain, the current gain, and the input resistance, with the results:

$$A_v \approx -g_m R_L\left(\frac{r_\pi}{r_\pi + r_x}\right) \qquad A_i \approx -\beta_o \qquad R_i \approx r_x + r_\pi \qquad (13\text{-}13)$$

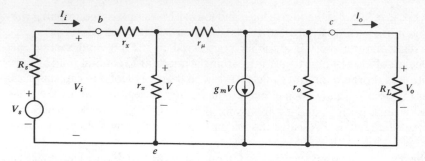

Figure 13-7   CE incremental model. $R_s = 10 \text{ k}\Omega$, $R_L = 1 \text{ k}\Omega$.

Calculations using (13-13) give the incremental values of A_v, A_i, and R_i to be -38, -40, and $1.053 \text{ k}\Omega$, respectively. The negative signs signify that the output variables are $180°$ out of phase with the input variables, and this $180°$ phase shift in the midfrequency band is an important property of the CE amplifier. The power gain G is $A_i A_v$, or 1520, which is 31.8 dB.

To justify the deductions made from the approximate model, we need to show that these results do indeed lead to a negligible current in r_μ in the circuit of Fig. 13-7. The voltage $V_o - V$ across r_μ is approximately V_o, because V_o is much greater than V_i and V is very slightly less than V_i. Hence the current I_μ through r_μ is about V_o/r_μ. A rough approximation for I_i is V_i/r_π, and the ratio I_μ/I_i becomes $(r_\pi/r_\mu)A_v$, or ηA_v. In this example the magnitude of ηA_v is 0.0038. Clearly, I_μ is negligible compared with I_i and even more insignificant in comparison with I_o, which justifies the approximate model.

As long as ηA_v is small, say less than 0.1, the simplified model is reasonable. Because η equals $1/(g_m r_o)$, the requirement is that A_v be less than $0.1 g_m r_o$. From (13-13) A_v is only slightly less than $g_m R_L$. Accordingly, the requirement is substantially satisfied provided

$$R_L < 0.1 r_o \qquad (13\text{-}14)$$

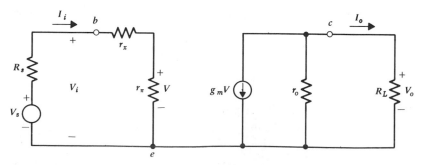

Figure 13-8   Simplified model for $R_L < 0.1 r_o$.

Most CE amplifiers have load resistances less than 10% of r_o*, and the resistance* r_μ
can be omitted. This simplifies the analysis.

THE OUTPUT RESISTANCE

To find R_o the circuit of Fig. 13-9 is used. The symbol G_s' denotes $1/(R_s + r_x)$.
For convenience in writing the node equations, a current source is connected to
the output terminals. Of course the ratio V_1/I_1 is independent of the type of
source that drives the output. The node equations are

$$I_1 = (g_o + g_\mu)V_1 + (g_m - g_\mu)V$$
$$0 = -g_\mu V_1 + (G_s' + g_\pi + g_\mu)V \tag{13-15}$$

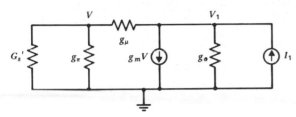

Figure 13-9 R_o is V_1/I_1. $G_s' - 1/(R_s + r_x)$.

Solving for I_1/V_1 gives the output conductance G_o to be

$$G_o = g_o + g_\mu\left(\frac{g_m + g_\pi + G_s'}{g_\mu + g_\pi + G_s'}\right) \tag{13-16}$$

This can be simplified by neglecting g_π in the numerator and g_μ in the denominator, and the output resistance R_o can be expressed in the form

$$R_o = r_o \left\| \left[\frac{r_\mu}{\beta_o}\left(\frac{1 + r_\pi G_s'}{1 + r_m G_s'}\right)\right] \right. \tag{13-17}$$

with $G_s' = 1/(R_s + r_x)$. From (13-17) R_o is determined to be 130 kΩ, or about
$\frac{1}{2}r_o$. It should be noted that R_o depends on the base-width-modulation parameters r_o and r_μ. Without base-width modulation R_o would be infinite. Both R_i and
R_o decrease as the quiescent current I_C is increased, because r_π, r_o, and r_μ in
(13-13) and (13-17) decrease. Consequently, *large-signal amplifiers with large
quiescent currents usually have lower input and output resistances than do small-
signal amplifiers.*

BETA CUTOFF AND TRANSITION FREQUENCIES

In Sec. 12-6 it was found that at high frequencies the CE short-circuit current gain β, also called h_{fe}, is less than the low-frequency value β_o because of C_π and C_μ. The beta cutoff radian frequency was shown to be

$$\omega_\beta = \frac{g_\pi}{C_\pi + C_\mu} \tag{13-18}$$

Assuming C_π is 98 pF and C_μ is 2 pF, ω_β for the transistor is 10^7 radians/s, and f_β is 1.59 MHz. The transition frequency f_T is $\beta_o f_\beta$, or 63.7 MHz.

β_F AND β_o

At low frequencies the short-circuit current gain β_o is $r_\pi g_m$. We know that the product $r_\pi g_m$ is also approximately β_F, by (12-24). However, there is an important distinction. Transistor data sheets give values of β_F that are based on the active-mode Ebers–Moll equation (2-32) of Sec. 2-4. From this equation it is clear that

$$\beta_F = \frac{I_C - I_{CO}}{I_B + I_{CO}} \approx \frac{I_C}{I_B} \tag{13-19}$$

Data sheets usually refer to β_F as h_{FE}. The parameter β_o is the low-frequency h_{fe}, and using the notation of (12-35) of Sec. 12-5, we obtain

$$\beta_o = \left.\frac{\partial i_C}{\partial i_B}\right|_{v_{CE} = V_{CE}} = \left.\frac{I_c}{I_b}\right|_{V_{ce} = 0} \tag{13-20}$$

In (13-19), I_C and I_B are quiescent currents, and I_c and I_b in (13-20) are the phasor currents. Although these equations show the distinction, in most cases $\beta_F \approx \beta_o$.

UNBYPASSED EMITTER RESISTANCE

Sometimes R_3 (or R_E) in a circuit such as that of Fig. 13-1 is quite small and the bypass capacitor is omitted. For this case the incremental circuit with r_μ and r_o neglected can be put into the form of Fig. 13-10. By inspection the current gain is $-\beta_o$ as before, but analysis reveals that

$$R_i = r_x + r_\pi + (\beta_o + 1)R_E \tag{13-21}$$

$$A_v = -g_m R_L \left(\frac{r_\pi}{r_x + r_\pi + (\beta_o + 1)R_E} \right) \tag{13-22}$$

Omission of the emitter bypass capacitor gives an input resistance increased by the factor $(\beta_o + 1)R_E$. The current gain is not appreciably affected, but the voltage gain is reduced. For the amplifier analyzed in this section, with R_E equal to 2 kΩ, removal of the capacitor causes R_i to increase from 1 kΩ to 83 kΩ and A_v to change from -38 to -0.5, an unacceptable reduction.

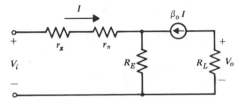

Figure 13-10   Model with R_E.

13-3. THE CC AMPLIFIER

Another amplifier configuration of considerable importance is the one of Fig. 13-11. The power supply V_{CC} and the three capacitors are short circuits to the incremental variables, provided the frequency is not too low. Consequently, insofar as the signal is concerned, the circuit may be redrawn in the form shown

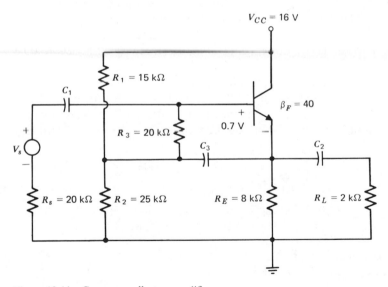

Figure 13-11   Common-collector amplifier.

in Fig. 13-12. Because the collector of Fig. 13-12 is common to both input and output, this is a *common-collector amplifier*, also called an *emitter follower*.

The resistor R_3 has little effect on the performance of the incremental circuit, and $R_1, R_2,$ and R_E reduce A_i, R_i, and R_o, with negligible effect on A_v (see P13-9). In the analysis here these bias resistors are eliminated from consideration.

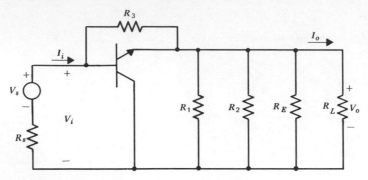

Figure 13-12   Incremental circuit.

DC AND INCREMENTAL CIRCUITS

The dc circuit is that of Fig. 13-13. With reference to the two-battery equivalent circuit of Fig. 13-4, it is evident that R_B is 29.4 kΩ, this being the sum of R_3 and $R_1 \| R_2$. Equation 13-1 gives V_B to be 10 V, and I_C is found from (13-4) to be 1 mA.

In designing the bias circuit R_E was chosen to be four times R_L to minimize its shunting effect. Of course, a large R_E is needed to stabilize the selected quiescent current. Because $\beta_F R_E$ is 320 kΩ, R_B should be less than 96 kΩ in accordance with (13-5). If this were accomplished with R_3 of Fig. 13-13 equal to zero, as was done in the CE circuit, the parallel combination of R_1 and R_2 would shunt the input with a resistance less than 96 kΩ. This would be fine for a CE amplifier. However, as we shall see, an important property of the CC amplifier is its inherently large input impedance. Shunting the input with bias resistors is usually

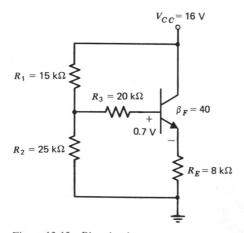

Figure 13-13   Bias circuit.

intolerable. For this reason the $R_3 C_3$ combination is added as shown in Fig. 13-11, and the arrangement places R_1 and R_2 in parallel with the load rather than the input.

The incremental base-emitter resistance consisting of the sum of r_x and r_π is shunted by R_3. To minimize its effect on the incremental performance R_3 was made 20 kΩ, a value 20 times greater than r_π. Then, R_1 and R_2 were selected to give V_B and R_B values providing a quiescent current of 1 mA while keeping R_B less than 96 kΩ.

Replacement of the transistor in Fig. 13-12 with its hybrid-π model and elimination of the bias resistors give the incremental circuit of Fig. 13-14. The CC amplifier is normally used with load resistances that are very small compared with

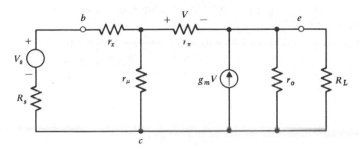

Figure 13-14   Hybrid-π circuit.

r_o. Therefore, r_o can be omitted from the circuit. As we shall see, the effective resistance in parallel with r_μ, due to that part of the circuit of Fig. 13-14 to the right of r_μ, is approximately the sum of r_π and $\beta_o R_L$. This sum is nearly always much smaller than r_μ, and hence r_μ is also negligible. The simplified model with the base-width-modulation terms omitted is shown in Fig. 13-15. For convenience the current source $g_m V$ has been replaced with the source $\beta_o I_i$.

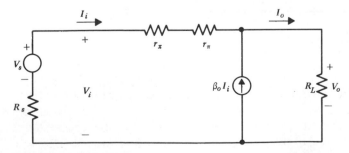

Figure 13-15   Simplified model.

The output current I_o is $(\beta_o + 1)I_i$, giving a current gain of $\beta_o + 1$. The input resistance is the sum of r_x, r_π, and $(\beta_o + 1)R_L$, which approximately equals $(\beta_o + 1)R_L$ provided R_L is not exceedingly small. The voltage gain V_o/V_i is $A_i R_L/R_i$, or $(\beta_o + 1)R_L/R_i$. Because $(\beta_o + 1)R_L$ and R_i are nearly equal, A_v is only slightly less than unity, again assuming R_L is not too small. The results are

$$R_i = r_x + r_\pi + (\beta_o + 1)R_L \approx (\beta_o + 1)R_L$$

$$A_i = \beta_o + 1 \qquad A_v = \frac{(\beta_o + 1)R_L}{R_i} \approx 1$$

(13-23)

Using the hybrid-π parameters of (13-12) and the specified load resistance of 2 kΩ, the numerical results are $R_i = 83$ kΩ, $A_i = 41$, and $A_v = 0.988$. The power gain $A_i A_v$ is 40.4, or 16 dB. We note that R_i is large, and A_i is about the same as that of the CE configuration. The voltage gain is slightly less than unity. This implies that the incremental emitter-collector and base-collector voltages are approximately equal and in time phase, a consequence of the fact that the incremental base-emitter voltage is quite small in comparison. The name *emitter follower* comes from the fact that the output emitter potential "follows" the input base potential, both rising and falling simultaneously, with a difference of about 0.7 V dc at room temperature.

Referring back to Fig. 13-11, we see that the end points of R_3 have potentials that rise and fall in unison. Because of the obvious analogy between these potentials and the straps of a boot, the bias arrangement of R_3 and C_3 is called a *bootstrap* circuit. In this case, R_3 provides a path for the dc base current, but its effect on the incremental circuit is negligible.

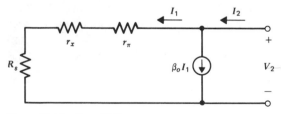

Figure 13-16 $R_o = V_2/I_2$.

To find R_o let us refer to the circuit of Fig. 13-16. If placed in the circuit, r_μ would appear in parallel with $R_s + r_x$, and r_o would be across the output terminals. Because r_μ is usually large compared with R_s, and r_o is large compared with the output resistance R_o, these base-width-modulation parameters are again omitted. Note the reversal of the reference direction of the current source from that of Fig. 13-15, resulting from the selected reference direction of the current

I_1. Here, I_2 is $(\beta_o + 1)I_1$ and V_2 is $(r_x + r_\pi + R_s)I_1$. Consequently, the output resistance is

$$R_o = \frac{r_x + r_\pi + R_s}{\beta_o + 1} \tag{13-24}$$

For the specified R_s of 20 kΩ, R_o is 513 Ω. This is much smaller than R_i and also small compared with the output resistance of the CE amplifier.

13-4. THE CB AMPLIFIER

The third and last configuration to be examined in this chapter is that of Fig. 13-17. Except at very low frequencies, the bypass capacitor C_1 is an effective ac short circuit. This makes the base common to both input and output, giving a common-base amplifier. Between the amplifier and the load is a 1:1 transformer, assumed to be ideal. The 1:1 ideal transformer offers zero resistance to the flow of direct current from the battery to the collector, while presenting an incremental load resistance between the collector and ground equal to the actual load R_L. For a brief discussion of the ideal transformer, refer to Sec. 16-2, particularly to Fig. 16-7 and (16-10) through (16-12). Perhaps the most important application of the CB configuration is the amplification of high-frequency signals, and the direct current of a radio-frequency (RF) amplifier is supplied through an inductor. Large-signal audio power amplifiers sometimes are common-base amplifiers with transformer coupling as in Fig. 13-17.

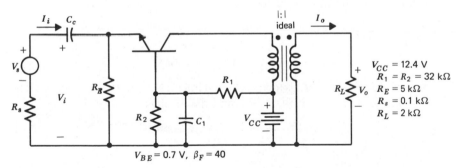

Figure 13-17   Common-base amplifier.

Figure 13-18 shows the bias circuit. By inspection, the two-battery equivalent circuit has an R_B of 16 kΩ and a V_B of 6.2 V. Using (13-4), the collector current is determined to be 1 mA. Although R_E shunts the input, its effect on the signal is negligible, because the input resistance of a CB amplifier is seldom more than a few tens of ohms. The 5 V drop across R_E may be more than needed for bias-point stabilization. However, $\beta_F R_E$ is 200 kΩ, and hence R_B should be less than 60 kΩ, which it is. A much smaller R_E requires a smaller R_B, and the reduced

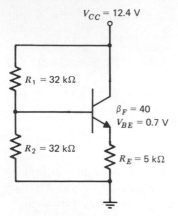

$V_{CC} = 12.4$ V

$R_1 = 32$ kΩ

$\beta_F = 40$
$V_{BE} = 0.7$ V

$R_2 = 32$ kΩ

$R_E = 5$ kΩ

Figure 13-18   DC circuit.

values of R_1 and R_2 would dissipate more power from the V_{CC} supply. Resistors R_1 and R_2 do not appear in the midfrequency incremental circuit, because they are effectively shorted out by the capacitor C_1 and the ideal battery.

INCREMENTAL ANALYSIS

In Fig. 13-19 is the incremental circuit with a current source at the input. Numerical values of the parameters are those of (13-12), and the units are volts, mA, and mmhos. The three node equations are

$$I_i = 41V_1 - 41V_2 - 0.004V_3$$
$$0 = -V_1 + 20V_2 - 0.0001V_3$$
$$0 = -40V_1 + 40V_2 + 0.5041V_3$$

These can be solved for V_1/I_i, which is R_i, and for V_3/V_1, which is A_v. The current gain can be determined from $A_v R_i/R_L$, and the results are

$$R_i = 26 \ \Omega \qquad A_v = 75.4 \qquad A_i = 0.975$$

The product $A_v A_i$ is the power gain G, which is 73.5, or 18.7 dB.

The network of Fig. 13-19 can be redrawn in the form of Fig. 13-20. Let us suppose that r_μ is so large that we can neglect it. Because I_i and I_o are nearly equal and in time phase, the current through the very small base resistance r_x is rather insignificant, and r_μ is approximately in parallel with R_L. Practical load resistances are always much smaller than r_μ and our assumption of negligible r_μ is justified.

The current through r_o is $(\beta_o + 1)I_o - \beta_o I_1$, and the mesh equations are

$$V_1 = (r_\pi + r_x)I_1 - (r_\pi + r_x)I_o$$
$$0 = -(r_\pi + r_x + \beta_o r_o)I_1 + (r_\pi + r_x + \beta_o r_o + r_o + R_L)I_o$$

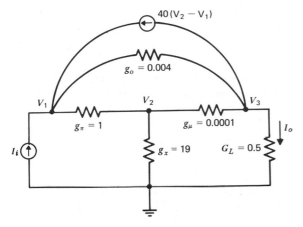

Figure 13-19   Incremental CB circuit (volts, mA, mmhos).

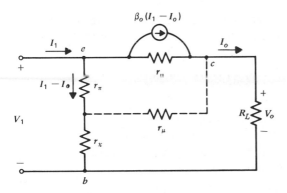

Figure 13-20   CB model.

Solving for V_1/I_1 and I_o/I_1 (noting that $r_\pi + r_x \ll \beta_o r_o$), we obtain

$$R_i = \frac{(r_\pi + r_x)(r_o + R_L)}{(\beta_o + 1)r_o + R_L} \qquad A_i = \frac{\beta_o r_o}{(\beta_o + 1)r_o + R_L} \qquad (13\text{-}25)$$

In practical amplifiers R_L is much less than $\beta_o r_o$, and the term R_L in each denominator of (13-25) can be omitted. With this done and with A_v found from $A_i R_L/R_i$, the results are

$$R_i = \left(\frac{r_\pi + r_x}{\beta_o + 1}\right)\left(1 + \frac{R_L}{r_o}\right) \qquad A_i = \frac{\beta_o}{\beta_o + 1}$$

$$A_v = g_m R_L\left(\frac{r_\pi}{r_\pi + r_x}\right)\left(\frac{r_o}{r_o + R_L}\right) \approx g_m R_L\left(\frac{r_\pi}{r_\pi + r_x}\right) \qquad (13\text{-}26)$$

Calculations using (13-26) give values for R_i, A_i, and A_v within 1% of those found from the exact model of Fig. 13-19. *An important property of the CB amplifier is its small input resistance.* For $R_L \ll r_o$, the input resistance is that of the CE amplifier divided by $\beta_o + 1$, as indicated by (13-26). This implies that I_e is $\beta_o + 1$ times I_b, which is consistent with our knowledge that the quiescent current I_E is $\beta_F + 1$ times I_B. *The current gain is slightly less than unity*, and this is expected in view of the fact that the quiescent current I_C is $\beta_F/(\beta_F + 1)$ times I_E. With identical transistor parameters and loads, *both the CB and CE amplifiers have about the same voltage gain*, except that *the CB amplifier has no phase inversion*. We could have deduced the equivalence of the voltage gains without the aid of the mathematical analysis by noting that both configurations have the same pair of input terminals and their output voltages V_{cb} and V_{ce} differ only by the much smaller V_{be}. Because the current gain is nearly unity, A_v is approximately the ratio R_L/R_i.

THE OUTPUT RESISTANCE

The circuit of Fig. 13-21 can be used to find R_o. For the present let us neglect r_μ. Because the voltage across r_x is much smaller than V_o, r_μ is approximately in parallel with the output. Two loop equations are

$$V_o = r_o(I_o + \beta_o I_1) + (r_\pi + r_x)I_1$$
$$0 = -R_s(I_o - I_1) + (r_\pi + r_x)I_1$$

Recalling that the sum $r_\pi + r_x$ is small compared with $\beta_o r_o$, it is easy to show that the ratio V_o/I_o is the sum of r_o and $\beta_o r_o R_s/(r_\pi + r_x + R_s)$. The output resistance R_o is approximately the parallel equivalent of this ratio and r_μ, or

$$R_o = r_\mu \left\| \left[r_o \left(1 + \frac{\beta_o R_s}{r_\pi + r_x + R_s}\right)\right]\right. \tag{13-27}$$

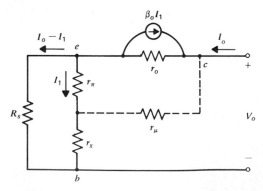

Figure 13-21 CB model for determination of R_o.

Using the hybrid-π parameters and the source resistance of the CB amplifier of Fig. 13-17, R_o is calculated from (13-27) to be one megohm. The actual value is 0.96 MΩ as determined from analysis of the exact hybrid-π model with r_μ included (see P13-14). The error introduced by (13-27) is small. *An important characteristic of the CB amplifier is its large output resistance.* For moderate values of load impedance, the output current is nearly independent of the load. As shown by (13-27), R_o is finite only because of base-width modulation.

THE ALPHA CUTOFF FREQUENCY

A parameter of considerable importance is the CB forward short-circuit current gain α, defined as $-h_{fb}$. In Fig. 13-22 the capacitances C_π and C_μ are included in the impedances and the output is shorted. Alpha is the ratio I_o/I_i, and this ratio is a function of frequency. The current $I_i - I_o$ through r_x is considerably smaller than I_o. Because the impedance z_μ is in parallel with r_x

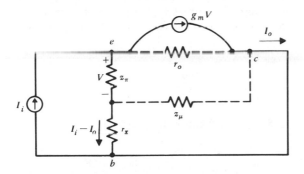

Figure 13-22   Network for determination of α.

and because the magnitude of this impedance is much greater than r_x even at the highest frequencies of interest here, the current through z_μ is also very small compared with I_o. Hence we can neglect z_μ. The large resistance r_o is in parallel with $z_\pi + r_x$. Thus the current through r_o is also negligible, and $I_o = g_m V$. At the emitter node $I_i - I_o$ equals $y_\pi V$, or $\cdot y_\pi I_o/g_m$. From this it is easy to deduce that the ratio I_o/I_i, or α, is

$$\alpha = \frac{g_m}{g_m + y_\pi} = \frac{\beta_o}{\beta_o + 1 + j\omega r_\pi C_\pi} \tag{13-28}$$

The expression on the right was obtained by replacing y_π with $g_\pi + j\omega C_\pi$ and multiplying both numerator and denominator by r_π.

At low frequencies $\alpha = \alpha_o = \beta_o/(\beta_o + 1)$ by (13-28). In terms of α_o, (13-28) can be expressed in the form

$$\alpha = \frac{\alpha_o}{1 + j\omega/\omega_\alpha} \tag{13-29}$$

with

$$\alpha_o = \frac{\beta_o}{\beta_o + 1} \quad \text{and} \quad \omega_\alpha = \frac{(\beta_o + 1)g_\pi}{C_\pi} \tag{13-30}$$

The value of β_o is 40 and α_o is 0.975. The conductance g_π is 1 mmho. Assuming C_π is 98 pF, which is the same value used in Sec. 13-2 to calculate ω_β, we find that ω_α is 42×10^7, and f_α is 67 MHz. The frequency f_α is called the *alpha cutoff frequency*. At this frequency the CB short-circuit current gain is 0.707 times its low-frequency value α_o, and f_α gives the approximate useful frequency range of the amplifier. For the same hybrid-π parameters the CB alpha cutoff frequency of 67 MHz is considerably larger than the CE beta cutoff frequency, which was found in Sec. 13-2 to be 1.59 MHz, and is about the same as f_T, which is 64 MHz.

At low frequencies α_o is $\beta_o/(\beta_o + 1)$, and the dc parameter α_F is $\beta_F/(\beta_F + 1)$. The distinction between the small-signal parameter β_o and the dc parameter β_F was discussed in Sec. 13-2. There is a somewhat similar distinction between α_o and α_F. From the active-mode Ebers–Moll equation (2-31) we can express α_F, also referred to as $-h_{FB}$, in the form

$$\alpha_F = -h_{FB} = -\frac{I_C}{I_E} \tag{13-31}$$

assuming I_{CO} is negligible. As α_o is $-h_{fb}$, we can write

$$\alpha_o = -h_{fb} = -\frac{\partial i_C}{\partial i_E}\bigg|_{v_{CB}=V_{CB}} = -\frac{I_c}{I_e}\bigg|_{v_{cb}=0} \tag{13-32}$$

Both I_C and I_E in (13-31) are quiescent currents, and I_c and I_e in (13-32) are incremental phasor currents.

13-5. THE ADMITTANCE PARAMETERS

The incremental analyses of the preceding sections can be nicely handled using the matrix y parameters introduced in Sec. 12-4. For the two-port network of Fig. 13-23 the y parameters are defined by the equations

$$\begin{aligned} I_1 &= y_i V_1 + y_r V_2 \\ I_2 &= y_f V_1 + y_o V_2 \end{aligned} \tag{13-33}$$

In matrix form these become

$$\begin{bmatrix} I_1 \\ I_2 \end{bmatrix} = [y]\begin{bmatrix} V_1 \\ V_2 \end{bmatrix} \quad \text{with } [y] = \begin{bmatrix} y_i & y_r \\ y_f & y_o \end{bmatrix} \tag{13-34}$$

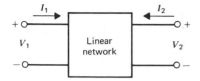

Figure 13-23   Two-port network.

It is evident that:

$$y_i = \frac{I_1}{V_1}\bigg|_{V_2=0} = \text{short-circuit } \textit{input} \text{ admittance} \qquad (13\text{-}35)$$

$$y_r = \frac{I_1}{V_2}\bigg|_{V_1=0} = \text{short-circuit } \textit{reverse} \text{ transfer admittance} \qquad (13\text{-}36)$$

$$y_f = \frac{I_2}{V_1}\bigg|_{V_2=0} = \text{short-circuit } \textit{forward} \text{ transfer admittance} \qquad (13\text{-}37)$$

$$y_o = \frac{I_2}{V_2}\bigg|_{V_1=0} = \text{short-circuit } \textit{output} \text{ admittance} \qquad (13\text{-}38)$$

These are the most useful matrix parameters for transistor circuit analysis. There are several reasons for this. Perhaps most important is that y parameters are read directly on most instruments designed to measure transistor parameters at high frequencies. They have consistent units, whereas the matrix h and g parameters do not. In addition, the y-parameter circuit models are best suited for nodal analysis, which is often the easiest method for analyzing electronic circuits.

y-PARAMETER CIRCUITS

In Fig. 13-24 is shown a two-generator circuit that can be used to represent the network of Fig. 13-23. This model constitutes an equivalent circuit, because the relations between the terminal currents and voltages of the model are

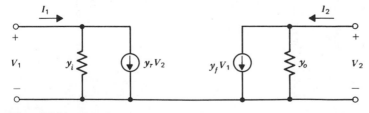

Figure 13-24   Two-generator y-parameter model.

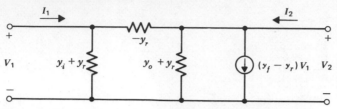

Figure 13-25   One-generator y-parameter model.

precisely the same as those of the network of Fig. 13-23, provided the y-param-
eters are accurately determined. A one-generator model is shown in Fig. 13-25.
Equations 13-33 are those of the two upper nodes, a fact that validates the
configuration. If the network of Fig. 13-23 is passive, the reciprocity theorem
requires that $y_f = y_r$. However, we shall most often be analyzing active networks.

CE PARAMETERS

The CE y parameters of a BJT are easily determined in terms of the hybrid-π
parameters. In Fig. 13-26 is shown the CE configuration with current sources
at the input and output terminals. The equations at the nodes labeled V_1, V,
and V_2 are

$$I_1 = g_x V_1 - g_x V \tag{13-39}$$

$$0 = -g_x V_1 + (g_x + y_\pi + y_\mu)V - y_\mu V_2 \tag{13-40}$$

$$I_2 = (g_m - y_\mu)V + (g_o + y_\mu)V_2 \tag{13-41}$$

From (13-40), $V = (g_x V_1 + y_\mu V_2)/Y$, with $Y = g_x + y_\pi + y_\mu$. Substituting
for V into (13-39) and (13-41) gives I_1 and I_2 as functions of V_1 and V_2. Comparison with the y-parameter equations of (13-33) yields the desired relations.

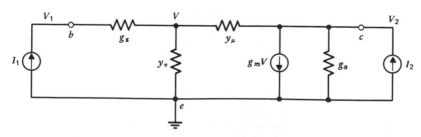

Figure 13-26   Circuit for determining CE y parameters.

A similar procedure can be employed to determine the CC and CB y parameters. The results are

$$[y_e] = \frac{1}{Y}\begin{bmatrix} g_x(y_\pi + y_\mu) & -g_x y_\mu \\ g_x(g_m - y_\mu) & g_o Y + y_\mu(g_m + g_x + y_\pi) \end{bmatrix} \quad (13\text{-}42)$$

$$[y_c] = \frac{1}{Y}\begin{bmatrix} g_x(y_\pi + y_\mu) & -g_x y_\pi \\ -g_x(g_m + y_\pi) & g_o Y + (g_m + y_\pi)(g_x + y_\mu) \end{bmatrix} \quad (13\text{-}43)$$

$$[y_b] = g_o\begin{bmatrix} 1 & -1 \\ -1 & 1 \end{bmatrix} + \frac{1}{Y}\begin{bmatrix} (g_m + y_\pi)(g_x + y_\mu) & -y_\mu(g_m + y_\pi) \\ -g_m g_x - g_m y_\mu - y_\pi y_\mu & y_\mu(g_m + g_x + y_\pi) \end{bmatrix}$$
$$(13\text{-}44)$$

with

$$Y = g_x + y_\pi + y_\mu \quad (13\text{-}45)$$

The subscripts e, c, and b on the symbol y in the preceding equations signify the common terminal.

It is easy to show (see P13-16) that the determinant of the y matrix is precisely the same for each of the the three configurations. This determinant is often represented by the symbol Δ, and Δ is

$$\Delta = y_i y_o - y_r y_f = \frac{g_x[g_o(y_\pi + y_u) + y_u(g_m + y_\pi)]}{Y} \quad (13\text{-}46)$$

CC AND CB PARAMETERS

The expressions for the CC and CB y parameters in terms of the CE parameters are readily found by the procedure described in Sec. 12-5 (see P13-17 and P13-18). The relations are

$$[y_c] = \begin{bmatrix} y_{ie} & -y_{ie} - y_{re} \\ -y_{ie} - y_{fe} & y_{ie} + y_{re} + y_{fe} + y_{oe} \end{bmatrix} \quad (13\text{-}47)$$

$$[y_b] = \begin{bmatrix} y_{ie} + y_{re} + y_{fe} + y_{oe} & -y_{re} - y_{oe} \\ -y_{fe} - y_{oe} & y_{oe} \end{bmatrix} \quad (13\text{-}48)$$

In Fig. 13-27 is a circuit with a two-port network between a source and a load. There are four terminal variables V_1, V_2, I_1, and I_2, and these are related by the two y-parameter equations (13-33). The relation $V_2 = -I_2 Z_L$ and (13-33) constitute three linear equations with four unknowns. Therefore, the ratio of any two of the four variables can be found from the equations. In this manner expressions for the voltage gain V_2/V_1, the current gain $-I_2/I_1$, and the input impedance V_1/I_1 are determined (see P13-19). The expression for the output

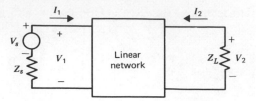

Figure 13-27 Two-port network between source and load.

impedance V_2/I_2 is developed from (13-33) and the relation $V_1 = -I_1 Z_s$, which applies when V_s is zero. We find that

$$A_v = \frac{-y_f}{y_o + Y_L} \qquad A_i = \frac{-y_f Y_L}{\Delta + y_i Y_L} \qquad (13\text{-}49)$$

$$Z_i = \frac{y_o + Y_L}{\Delta + y_i Y_L} \qquad Z_o = \frac{y_i + Y_s}{\Delta + y_o Y_s} \qquad (13\text{-}50)$$

with $A_i = -I_2/I_1$ and $\Delta = y_i y_o - y_r y_f$. In Appendix B expressions for A_v, A_i, Z_i, and Z_o are presented in terms of each of the four sets (z, y, h, g) of matrix parameters.

The equations developed here are applied in the next section to the amplifiers of the previous sections. The procedure is exact, and the calculated results are used to compare further the three configurations.

13-6. COMPARISON OF CE, CC, AND CB CONFIGURATIONS

Each of the basic amplifiers has been analyzed using the same hybrid-π parameters. The y parameters and the determinant of these parameters are calculated using (13-42) through (13-46), and the results for each configuration are given in Table 13-1.

Table 13-1. The Low-Frequency y Parameters of a BJT with $g_x = 19$, $g_\pi = 1$, $g_m = 40$, $g_o = 0.004$, and $g_\mu = 0.0001$ mmho.

millimhos		CE		CC		CB	
y_i	y_r	0.95	-0.95×10^{-4}	0.95	-0.95	39	-0.0042
y_f	y_o	38	0.0043	-39	39	-38	0.0043

$$\Delta = y_i y_o - y_r y_f = 0.0077 \text{ mmho}^2$$

Using the numerical values of Table 13-1 and employing (13-49) and (13-50) for A_v, A_i, Z_i, and Z_o, these quantities are readily evaluated for any specified source and load resistances. Calculated results are presented in Figs. 13-28, 13-29, 13-30, and 13-31 for values of R_s and R_L between 1 ohm and 1000 MΩ, using log scales. Approximate expressions given for the limiting values are based on the assumptions that r_π is considerably larger than r_x, β_o is large, and $r_\mu \approx \beta_o r_o$.

Figures 13-28 and 13-29 clearly show that the CE amplifier has appreciable voltage and current gain for moderate values of R_L, whereas the CC configuration has $A_v \approx 1$, and the CB configuration has $A_i \approx 1$. The voltage gain A_v of

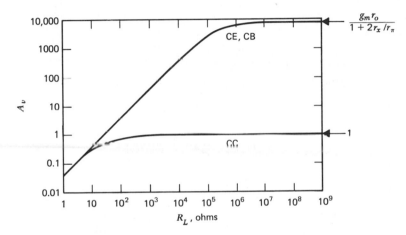

Figure 13-28 A_v versus R_L.

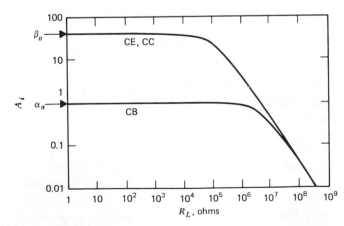

Figure 13-29 A_i versus R_L.

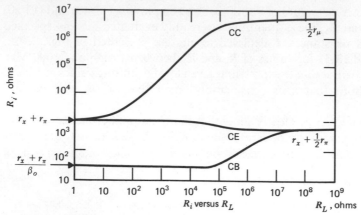

Figure 13-30 R_i versus R_L.

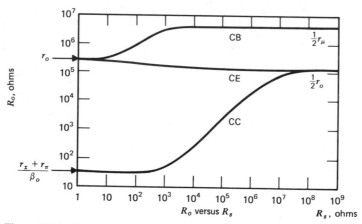

Figure 13-31 R_o versus R_s.

the CB connection is nearly identical to that of the CE connection, and the CC current gain approximately equals the CE current gain. Not shown in the plots is the 180° phase reversal of the CE amplifier.

From Figs. 13-30 and 13-31 it is evident that both the input and output resistances of the CE amplifier are intermediate in relation to the corresponding resistances of the CC and CB configurations. Furthermore, the variations of R_i and R_o with R_L and R_s are considerably less for the CE amplifier. In the midfrequency range, CC amplifiers are characterized by large input and small output resistances, whereas the reverse is true for CB amplifiers. A cascade of CC amplifiers, with the load of one stage being the input resistance of the next, would be grossly mismatched at the interconnections, as would be a cascade of CB amplifiers. The impedance mismatches are considerably less in a CE multistage amplifier.

13-7. TRANSDUCER GAIN

When an amplifier is inserted between a source and a load, the power to the load should be and usually is much greater than that which would be obtained by connecting the load directly to the source. An elementary example is presented here to show that the determination of the power gain G fails to indicate whether or not this is accomplished. In Fig. 13-32a is the network to be considered. With

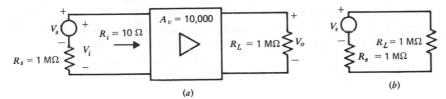

Figure 13-32 (a) Source, amplifier, load. (b) Source, load.

the 1-megohm load across the output, the amplifier is assumed to have an input resistance R_i of 10 ohms and a voltage gain of 10,000. The input voltage V_i is $V_s R_i/(R_s + R_i)$, or $10^{-5}V_s$, and the power supplied to the input terminals of the amplifier is $10^{-11}V_s^2$. Since V_o is $10^4 V_i$, the output voltage is $0.1V_s$, giving a power of $10^{-8}V_s^2$ to the load. The power gain G is 1000, or 30 dB, which is indeed substantial. In Fig. 13-32b the amplifier has been removed, and the load is connected directly to the source. The power to the load is now $25 \times 10^{-8}V_s^2$, which is 25 times greater than before. In this example the amplifier *reduces* the power to the load by a factor of 25, even though the gain G is 30 dB. Clearly, we need a more informative measure of gain.

THE AVAILABLE POWER OF A SOURCE

The source of Fig. 13-33 has a variable load impedance Z_L across its terminals. For any fixed R_L, the current I is maximized by adjusting X_L so as to cancel the reactance of the source, and the power to R_L is $|I|^2R_L$. For this condition I is $V_s/(R_s + R_L)$, and the power becomes

$$P = \frac{|V_s|^2 R_L}{(R_s + R_L)^2} = \frac{|V_s|^2}{4R_s}\left[\frac{4R_L/R_s}{(1 + R_L/R_s)^2}\right] \qquad (13\text{-}51)$$

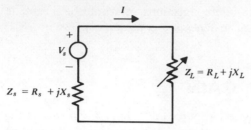

Figure 13-33   Source with variable load.

With R_s constant, the quantity in brackets has a maximum value of unity, occurring when $R_L = R_s$, and the maximum power to R_L is $|V_s|^2/4R_s$. This is illustrated in Fig. 13-34.

The peak of the curve represents all the power that is available from the source. The source may supply a load with less power than $|V_s|^2/4R_s$, but never more. Accordingly, we shall refer to this as the *available power* P_a of the source, with

$$P_a = \frac{|V_s|^2}{4R_s} \tag{13-52}$$

Here, V_s is the open-circuit rms phasor voltage and R_s is the resistive part of the impedance of the source.

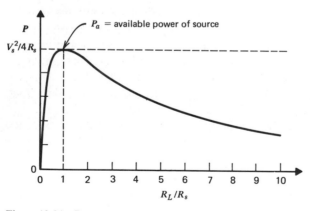

Figure 13-34   Power versus R_L/R_s, with R_s constant.

TRANSDUCER GAIN

The curve of Fig. 13-34 indicates why the amplifier of Fig. 13-32 is so poor. The input resistance R_i is the load connected to the source, and R_i/R_s is 10^{-5}. Because of this enormous mismatch, the power supplied to the input terminals of the amplifier is much less than P_a. The effect of a mismatch at the input is taken into consideration by defining the *transducer gain* G_T of an amplifier as the ratio of the output power to the available power of the source. Hence,

$$G_T = \frac{P_o}{P_a} \qquad (13\text{-}53)$$

For the amplifier of Fig. 13-32, P_o is $10^{-8}V_s^2$ and P_a is $25 \times 10^{-8}V_s^2$, giving a transducer gain of 0.04. If a lossless impedance-matching network, such as an ideal transformer, is connected between the source and the amplifier, the input power to the amplifier is P_a, and the transducer gain G_T equals the power gain G, which is 1000.

An expression for G_T is easily deduced with reference to the network of Fig. 13-35. The output power P_o is $|I_o|^2R_L$, and P_a is $|V_s|^2/4R_s$, or $|I_i|^2|Z_s + Z_i|^2/4R_s$. The transducer gain can be expressed in the form

$$G_T = \frac{4R_sR_L}{|Z_s + Z_i|^2}|A_i|^2 \qquad (13\text{-}54)$$

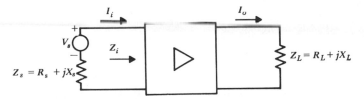

Figure 13-35   Source, amplifier, and load.

In the remainder of this section the discussion is restricted to resistive networks. For such cases (13-54) becomes

$$G_T = \frac{4R_sR_L}{(R_s + R_i)^2} A_i^2 \qquad (13\text{-}55)$$

Both R_i and A_i are functions of the y parameters and R_L, as given by (13-49) and (13-50). Using these equations to replace R_i and A_i in (13-55) gives, after some manipulation, the expression

$$G_T = \frac{4R_sR_Ly_f^2}{(R_sR_L\Delta + y_iR_s + y_oR_L + 1)^2} \qquad (13\text{-}56)$$

with $\Delta = y_iy_o - y_ry_f$. From this equation we can calculate G_T from specified source and load resistances and known y parameters.

MAXIMUM GAIN

Let us now investigate the maximum gain G_M that can be obtained from an amplifier. We know that G and G_T are equal when R_i equals R_s, which gives maximum transfer of power from the source to the amplifier. Also, of course, the gain is maximized by selecting R_L equal to R_o. Shown in Fig. 13-36 is an amplifier between a source and a load, with ideal matching transformers at both input and output. The term R_s' denotes the resistance R_s transformed to the secondary of the input transformer, and R_L' is the transformed load resistance R_L, as indicated on the figure.

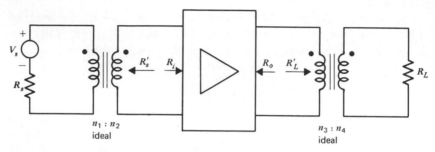

Figure 13-36 Amplifier with matching transformers.

Resistance R_i depends on R_L', and R_o depends on R_s'. When the terminal resistances R_L' and R_s' are adjusted by means of the turns ratios so that impedances are matched, the input resistance R_i is called the *image input resistance* R_{ii} and the output resistance R_o is called the *image output resistance* R_{oi}. Hence,

$$R_s' = R_{ii} \quad \text{and} \quad R_L' = R_{oi} \tag{13-57}$$

Also, from (13-50) we have

$$R_{ii} = \frac{y_o R_L' + 1}{R_L'\Delta + y_i} \qquad R_{oi} = \frac{y_i R_s' + 1}{R_s'\Delta + y_o} \tag{13-58}$$

Using (13-57) to eliminate R_s' and R_L' in (13-58) yields two equations relating R_{ii} and R_{oi}. Solving these equations gives the image resistances, which are

$$R_{ii} = \sqrt{\frac{y_o}{y_i \Delta}} \qquad R_{oi} = \sqrt{\frac{y_i}{y_o \Delta}} \tag{13-59}$$

These are applicable to networks with complex y parameters, in which case R_{ii} and R_{oi} become Z_{ii} and Z_{oi}.

Every two-port network has image input and output impedances that are functions of the y parameters according to (13-59). For any source and load, maximum power transfer occurs if lossless matching networks are designed so that the transformed R_s equals R_{ii} and the transformed R_L equals R_{oi}. Clearly, the proper turns ratios of the transformer of Fig. 13-36 are

$$\frac{n_1}{n_2} = \sqrt{\frac{R_s}{R_{ii}}} \qquad \frac{n_3}{n_4} = \sqrt{\frac{R_{oi}}{R_L}} \qquad\qquad (13\text{-}60)$$

A useful expression for the maximum gain is found by replacing R_s and R_L in (13-56) with the respective y-parameter expressions of (13-59). The result can be put in the form

$$G_M = \frac{y_f{}^2}{(\sqrt{y_i y_o} + \sqrt{\Delta})^2} \qquad\qquad (13\text{-}61)$$

For the hybrid-π model considered in this chapter, having the y parameters given in Table 13-1 of Sec. 13-6, the image resistances and the maximum gain are found from (13-59) and (13-61), and the results are tabulated in Table 13-2.

Table 13-2. R_{ii}, R_{oi}, and G_M.

	CE	CC	CB
R_{ii}, ohms	767	73,000	120
R_{oi}, ohms	169,000	1,780	1,090,000
G_M, dB	48.0	16.0	37.7

COMPARISON OF CE, CC, AND CB AMPLIFIERS

In Fig. 13-37 are shown plots of the transducer gain G_T as a function of R_L for three different values of R_s. The curves of the top figure were plotted from calculations that were made from (13-56) with R_s equal to the common-emitter R_{ii}, or 767 ohms. In the center figure R_s is the common-collector R_{ii}, or 73 kΩ, and R_s in the bottom figure is the common-base R_{ii}, or 120 Ω. It should be noted that in each case *the CE transducer gain is either approximately equal to or greater than the CC and CB gains for all values of* R_L. This is always true, which can be shown analytically with the aid of (13-56); (see P13-29 and P13-30).

In the uppermost plot of Fig. 13-37 the CE transducer gain has the maximum value given in Table 13-2. The CE gain is much greater than those of the CC and CB connections except at the extremes of R_L. The center figure has the source resistance that is ideal for the CC configuration. Even so, the CE curve is above the CC curve except at low values of R_L where the gains are about the same. The

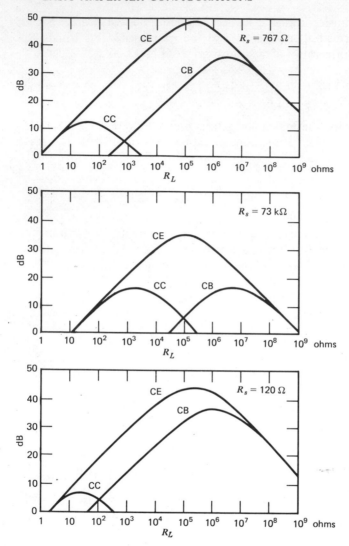

Figure 13-37   Transducer gain versus R_L.

CC maximum gain is that of Table 13-2. The source resistance used in the calculations for the bottom figure is ideal for the CB connection, and the CB maximum gain is that of Table 13-2. Once again the CE amplifier is the best with respect to gain, except at the extremes.

From the results we conclude that in order to get the most power to a specified load from a specified source, in the midfrequency band, the CE amplifier should be used. It is superior to the other two configurations, or at least approximately as good, *regardless of the values of* R_s *and* R_L. Consequently, the CE configuration is much more widely used than the others.

The CB alpha cut-off frequency is considerably greater than the CE beta cut-off frequency. In Secs. 13-2 and 13-4 we found these to be 67 MHz and 1.59 MHz, respectively, for the transistor analyzed in this chapter. Because of this, there are important applications of the CB connection in high-frequency oscillators and tuned amplifiers. Another advantage of the CB configuration is the large output voltage swing that can be obtained with very little distortion. The magnitude of the voltage is limited by avalanche breakdown, whereas that of a CE amplifier is limited by the lower sustaining voltage. These breakdown conditions were investigated in Sec. 2-6.

The allowed *large voltage swing* combined with *low input impedance* and *wide frequency range* are features that make the CB configuration ideally suited for use in the *cascode* amplifier, which is examined in Sec. 19-3. We have learned that the emitter and collector currents of a transistor are nearly equal, and these are the respective input and output currents of a CB amplifier. Thus the relationship is linear. Because this is most desirable in audio power amplifiers, the CB configuration is attractive for such use. However, the reduced distortion is at the expense of power gain. This is discussed further in Sec. 16-3. There are many specialized applications of the CB amplifier in which the low input impedance or the high output impedance or both are utilized advantageously, and some of these will be encountered later.

The common-collector amplifier, or emitter follower, is extensively used in all types of integrated circuits, in operational amplifiers, as drivers of CE power amplifiers, and as power amplifiers. Most amplifiers feed into loads that can be represented approximately by an RC parallel combination. In order to keep the capacitance from reducing the output voltage excessively at moderate frequencies, *it is essential that the output impedance of the amplifier be low*. Emitter-follower outputs are commonly used for this purpose. They are also employed as *buffer* stages to prevent two amplifier stages from interacting with one another. The very large input impedance of a CC stage between two amplifying stages provides isolation. In addition, the CC amplifier is often used as a power amplifier.

A transducer supplying a signal into the input of a multistage amplifier may have a frequency response or nonlinear characteristic that depends on the impedance into which it operates. If a low impedance is required, the CB circuit is advantageous, and the CC circuit is best when the transducer must operate into high impedance. In both cases gain is sacrificed. Noise considerations are certainly important. However, each of the three basic considerations gives about the same signal-to-noise ratio. Most applications employ the CE connection to obtain high gain, often utilizing negative feedback to increase the frequency range, reduce nonlinear distortion, change the input and output impedances, and generally improve the performance.

REFERENCES

1. V. H. Grinnich and H. G. Jackson, *Introduction to Integrated Circuits*, McGraw-Hill, New York, 1975.
2. J. Millman and C. Halkias, *Integrated Electronics*, McGraw-Hill, New York, 1972.
3. D. L. Schilling and C. Belove, *Electronic Circuits: Discrete and Integrated*, McGraw-Hill, New York, 1968.

PROBLEMS

Section 13-1

13-1. Find the values of R_1, R_2, and R_3 that should be used in the circuit of Fig. 13-1 to give a quiescent collector current of 0.5 mA. Use (13-5), selecting the maximum value of R_B. (*Ans*: 107, 31, 2 kΩ)

13-2. In the amplifier of Fig. 13-1 the quiescent collector current is 1 mA. Suppose a rise in temperature causes β_F to increase to 200 and V_{BE} to decrease to 0.4 V. Assuming the other parameters of the circuit are unchanged, find I_C.

13-3. In the circuit of Fig. 13-1 suppose the upper terminal of R_1 is connected to the collector rather than to the power-supply terminal. Find the resistances of the two-battery equivalent circuit (using a delta-to-wye, or pi-to-tee, transformation). Also, find V_B and V_C, and calculate I_C.

13-4. In a CE amplifier with a silicon transistor at room temperature ($q/kT = 40$), suppose R_E and R_B are chosen in accordance with (13-5), using the maximum R_B. Assuming $\beta_F \approx \beta_o$, deduce that R_B is 12 times r_π. What is the ratio R_B/r_π if the actual β_F is twice that used to calculate R_B?

13-5. By successive wye-delta transformations any two-port network containing only resistors can be reduced to a tee. (a) Sketch this tee, labeling V_1, V_2, I_1, I_2, and the resistances R_A, R_B, and R_C. (b) From the network find V_1 and V_2 as functions of the currents and the resistances. (c) If independent sources are added to the two-port, the equations of (b) must be modified by adding V_A to the expression for V_1 and V_B to the expression for V_2. These voltages represent the respective components of V_1 and V_2 due to all internal sources, with $I_1 = I_2 = 0$, and their addition to the equations is justified by the principle of superposition. Add V_A and V_B to the equations and deduce from them an equivalent circuit consisting of a tee with two batteries. The result verifies the two-battery bias circuit of Fig. 13-4.

Section 13-2

13-6. Draw the midfrequency incremental circuit of Fig. 13-1, including R_1, R_2, and R_4. The hybrid-π parameters are given in (13-12) and r_μ and r_o can be neglected. Place numerical values on the model, using kilohms, mA, and volts. Calculate A_v, A_i, and R_i. Compare the results with those of the analysis of Sec. 13-2, which neglected the effects of the bias circuit.

13-7. Suppose the bias circuit of the amplifier of Fig. 13-1 is changed so as to make I_C equal to 0.5 mA with $q/kT \doteq 40$. Assume that the transistor has $\beta_F = \beta_o = 100$, $\eta = 0.0005$, and $r_x = 100$ ohms. Neglecting the effect of the bias resistors R_1, R_2, and R_4, calculate the midfrequency values of A_v, A_i, R_i, R_o, and G in decibels. (*Ans. $R_o =$ 59.6 kΩ; 32.83 dB*)

13-8. At very low frequencies C_3 in the amplifier of Fig. 13-1 is no longer an effective bypass capacitor and the impedance Z_E, consisting of the parallel combination of R_3 and C_3, must be added to the incremental model. Assuming C_1 and C_2 are still short circuits, the model is that of Fig. 13-10 with R_E changed to Z_E. Using this one-capacitor model, find the *lower half-power frequency* f_1, defined as that frequency at which $|A_v|$ equals 0.707 times the midfrequency value of A_v. The hybrid-π parameters are those of (13-12). [*Ans:* 125 Hz (With C_1, C_2, and bias resistors included, f_1 is 156 Hz)].

Section 13-3

13-9. For the emitter follower of Fig. 13-11 calculate A_v, A_i, R_i, R_o, and G in decibels in the midfrequency region. Neglect r_μ but include all other resistances. Compare the results with those of the analysis in Sec. 13-3, which neglected the bias resistors. Use the hybrid-π parameters of (13-12). (*Ans:* 0.982, 26.5, 54 kΩ, 478 Ω, 14.2 dB)

13-10. With C_3 omitted in the emitter follower of Fig. 13-11, find A_v, A_i, R_i, and G. Neglect r_μ. Compare the results with those of P13-9. Use the parameters of (13-12).

13-11. What values of R_1 and R_2 should be used in the amplifier of Fig. 13-11 in order to give a quiescent collector current of 0.5 mA and to give $R_B = 30$ kΩ? The other circuit parameters are as specified.

13-12. In the CC amplifier of Fig. 13-11, R_3 should be large compared with r_π. To show the effect of a small R_3, suppose R_3 is reduced to a value equal to $r_x + r_\pi$, with R_1 and R_2 changed so as to maintain the same R_B and I_C as before. Add R_3 to the incremental circuits of Figs. 13-15 and 13-16 and calculate A_v, A_i, R_i, R_o, and G in decibels. Use the hybrid-π parameters of (13-12). Compare the results with those of the analysis in Sec. 13-3. (*Ans:* 42.5 kΩ, 977 Ω)

Section 13-4

13-13. Suppose the bias circuit of the CB amplifier of Fig. 13-17 is changed so as to make $I_C = 0.5$ mA with $q/kT = 40$. Assume that the transistor has $\beta_F = \beta_o = 100$, $\eta = 0.0005$, and $g_x = 19.8$ mmhos. Neglecting the effect of the bias circuit, calculate the midfrequency values of A_v, A_i, R_i, R_o, and G.

13-14. Find the output resistance R_o of the CB amplifier of Fig. 13-17. To do this use the hybrid-π model with the parameters of (13-12). With $V_s = 0$ and R_L replaced with a current source, write three node equations, and from these determine R_o. Compare the result with the value of 1 MΩ obtained from the approximate equation (13-27).

13-15. A transistor has $\beta_o = 50$, $I_C = 1$ mA, $C_\pi = 50$ pF, and $C_\mu = 5$ pF. Calculate α_0, f_α, and f_β. Assume $q/kT = 40$.

Section 13-5

13-16. Using (13-47) and (13-48), deduce that the determinants of the CE, CC, and CB y parameters are identical.

13-17. Derive (13-47), which gives the CC y parameters in terms of the CE y parameters.

13-18. Derive (13-48), which gives the CB y parameters in terms of the CE y parameters.

13-19. Derive (13-49) and (13-50), which give A_v, A_i, Z_i, and Z_o as functions of the y parameters and Y_L and Y_s.

13-20. Draw the CC hybrid-π circuit with ideal current sources I_1 and I_2 feeding the input and output ports. Write three node equations and solve for I_1 and I_2 in terms of V_1, V_2, and the admittances. Show that the CC y parameters are those given in (13-43).

13-21. Repeat P13-20 for the CB circuit, deducing that the CB y parameters are those of (13-44).

13-22. Find the y parameters of the tee of Fig. 13-38. Then sketch the equivalent y-parameter model in the form of Fig. 13-25, showing on the sketch the numerical values of the resistances in ohms. Note that a wye-delta (or tee-pi) transformation has been accomplished.

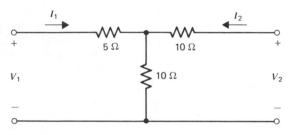

Figure 13-38   Tee network for P13-22.

Section 13-6

13-23. For an amplifier with $R_s = 10$ kΩ, $R_L = 1$ kΩ, and the CE y parameters of Table 13-1 of Sec. 13-6, calculate A_v, A_i, R_i, and R_o. Compare the results with those found in the analysis of Sec. 13-2.

13-24. The CE parameters of a BJT are $y_{ie} = 0.5$, $y_{re} = -0.0002$, $y_{fe} = 20$, and $y_{oe} = 0.01$, all in mmhos. The transistor constitutes a common-collector amplifier with source and load resistances of 100 kΩ and 1 kΩ, respectively. Calculate A_v, A_i, R_i, R_o, and G in decibels.

13-25. If the transistor of P13-24 constitutes a common-base amplifier with source and load resistances of 100 ohms and 100 kΩ, respectively, calculate A_v, A_i, R_i, R_o, and G in decibels. (*Ans:* $R_o = 280$ kΩ, $G = 29.71$ dB)

Section 13-7

13-26. In Fig. 13-33 suppose $V_s = 10$, $Z_s = 4 + j3$, and Z_L is a variable resistor R_L. If R_L is adjusted for maximum power, what is the power to R_L?

13-27. Assume the network of Fig. 13-36 is a CB amplifier with $y_i = 20$, $y_r = -0.01$, $y_f = -19$, and $y_o = 0.01$, all in mmhos. R_s is 5 kΩ and R_L is 2 kΩ. The turns ratios are such that the transducer gain is a maximum. Calculate n_1/n_2, n_3/n_4, and G_M.

13-28. Derive the expressions for the image resistances of (13-59) from (13-57) and (13-58).

13-29. In Fig. 13-37 the CB transducer gain is shown to be considerably less than the CE gain except at large values of R_L. Using (13-56), show that the ratio of the CB gain to the CE gain, as R_L approaches infinity, is $(y_{fb}/y_{fe})^2$ for all values of R_s. Deduce that this ratio is slightly greater than unity, using (13-42) and (13-44).

13-30. In Fig. 13-37 the CC transducer gain is clearly much less than the CE gain except at small values of R_L. Using (13-56), show that the ratio of the CC gain to the CE gain, as R_L approaches zero, is $(y_{fc}/y_{fe})^2$ for all values of R_s. Then show that this ratio is the square of $(\beta_o + 1)/\beta_o$, or approximately unity.

13-31. Deduce that $R_{ii} = \sqrt{h_i \Delta^h / h_o}$, $R_{oi} = \sqrt{h_i / h_o \Delta^h}$, and

$$G_M = \frac{h_f^2}{(\sqrt{h_i h_o} + \sqrt{\Delta^h})^2}$$

Chapter Fourteen
The Operational Amplifier

In addition to the CE, CC, and CB circuits discussed in the preceding chapter, there is another basic configuration of particular importance. This is the *differential amplifier*. It has inputs for two signal voltages and an output proportional to their difference. Frequently, one input voltage is a fraction of the output, taken from a voltage-divider network providing negative feedback, and sometimes an input is simply grounded. In either case the amplifier becomes *single-ended*, having one input and one output.

As we shall see, the differential amplifier can handle relatively large signals without excessive nonlinear distortion, and this greater dynamic range is one of its many attributes. Because the input impedance is moderate-to-high for small bias currents, the loading of the signal source is not excessive. Operation at low frequencies, including dc, is possible. The circuitry is especially well-suited for IC fabrication, and most linear integrated circuits contain one or more differential-amplifier stages. Examples of such circuits are analog-computer networks, monolithic voltage regulators, video amplifiers, analog comparators, and operational amplifiers. The versatility and popularity of the operational amplifier justifies the emphasis given to it in this and later chapters.

The *operational amplifier* is a multistage configuration with a differential input stage and characterized by high voltage gain, high input impedance, and low output impedance. It is widely used in many different types of linear and nonlinear circuits. There are applications associated with instrumentation circuits, special-purpose linear amplifiers, oscillators, active filters, and others. In fact, whenever inexpensive voltage amplification is required, the operational amplifier should be considered.

In this chapter some of the basic features of the operational amplifier are examined. Included are a number of applications, and others are presented in later chapters. The circuits of three different OP AMPs are shown and briefly discussed in Chapters 15 and 17. Let us begin with a study of the differential amplifier.

14-1. THE DIFFERENTIAL AMPLIFIER

EMITTER-COUPLED AMPLIFIER

There are various types of differential amplifiers. A common configuration is one in which two bipolar junction transistors are arranged *so that their emitters are incrementally in series.* An example is shown in Fig. 14-1a. In this

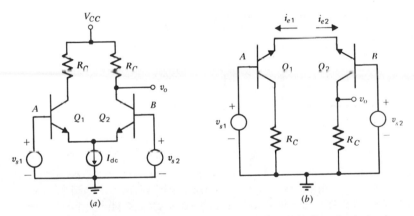

Figure 14-1   An emitter-coupled differential amplifier. (*a*) Complete circuit. (*b*) Incremental model.

circuit the ideal current source supplies a *constant* direct current I_{dc}. Therefore, it is an *open circuit* to the incremental components of the emitter currents, and the incremental circuit has the form of Fig. 14-1b. The ideal dc voltage and current sources have been replaced with a short-circuit and an open circuit, respectively. Clearly, $i_{e1} = -i_{e2}$. Amplifiers having emitters in series insofar as the incremental currents are concerned, such as the emitters of Fig. 14-1b, are called *emitter-coupled amplifiers.*

It is important that transistors Q_1 and Q_2 be closely matched so that their characteristics are nearly identical. With matched transistors and zero input voltages the collector currents of the two transistors are identical. This is evident from the symmetry of the circuit. Even if the collector resistor of Q_1 is omitted, the currents are almost equal, because the collector current of a BJT in the active mode is nearly independent of V_{CE}. Thus this resistor is sometimes omitted. Its purpose is to improve the dc balance. In the incremental model the resistor has negligible effect, being in series with the high-impedance collector which acts very much like a current source.

The input terminals at A and B are the base terminals of Q_1 and Q_2. For small signals the incremental model is linear, and the superposition principle applies. Hence we can treat the inputs one at a time. Let $v_{s2} = 0$. For this case the incremental circuit can be drawn in the form of a two-stage amplifier as shown in Fig. 14-2. As noted, R_C in the first stage has negligible effect. It follows that this stage is approximately a common-collector configuration, and the second stage is common base. The input impedance of this cascade is twice that of a CE amplifier having the same values of r_x and r_π.

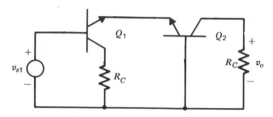

Figure 14-2   Incremental model.

In Secs. 13-3 and 13-4 we learned that the resistances r_μ and r_o can be neglected in both CC and CB amplifiers when transmission is in the forward direction, provided the load resistance is small compared with r_o. The requirement is obviously satisfied for the CC stage, which has as its load the small input resistance of the CB stage. It is also satisfied for the CB stage because the load resistance R_C is normally much less than r_o in practical circuits. The situation is similar when v_{s2} is applied with $v_{s1} = 0$, and the output voltage can be taken from either collector. In fact, as we shall see, the incremental voltages across the collector resistors are equal in magnitude and opposite in sign. Hence our analysis will neglect the parameters that depend on base-width modulation.

The midband incremental model with source resistances R_s added to the circuit is that of Fig. 14-3. The transistors are assumed to be matched, with identical quiescent points and incremental parameters. Each collector resistor R_C is in series with the incremental current source of the collector. Consequently, the currents of the circuit are unaffected by these resistors. Furthermore, if one resistor is eliminated, the voltage across the other is unchanged. A node

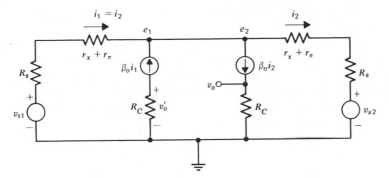

Figure 14-3   Midband model.

equation at the emitter node shows that $i_1 = i_2$. Clearly the output voltages are equal in magnitude and opposite in sign.

DIFFERENCE-MODE AND COMMON-MODE VOLTAGE GAINS

From a loop equation around the outside circuit of Fig. 14-3, we find that

$$v_{s1} - v_{s2} = 2i_2(r_x + r_\pi + R_s) \tag{14-1}$$

The output voltage v_o equals $\beta_o R_C i_2$, and the *difference-mode voltage gain* A_d, defined as the ratio of the output voltage to the difference between the two input voltages, is

$$A_d = \frac{v_o}{v_{s1} - v_{s2}} = \frac{\beta_o R_C}{2(r_x + r_\pi + R_s)} \tag{14-2}$$

The output v_o' is the negative of v_o, which permits the amplifier to be used as a phase-splitter, as discussed in Sec. 16-5. Each output is proportional to the difference $v_{s1} - v_{s2}$ of the inputs. This difference is called a *difference-mode input voltage* v_d, or

$$v_d = v_{s1} - v_{s2} \tag{14-3}$$

The average value of the two input voltages is $\frac{1}{2}(v_{s1} + v_{s2})$ and represents the *common-mode input voltage* v_c. Hence,

$$v_c = \frac{1}{2}(v_{s1} + v_{s2}) \tag{14-4}$$

The ratio of the output voltage to the common-mode input voltage, when the two input voltages are equal to one another, is the *common-mode gain* A_c, or

$$A_c = \frac{v_o}{v_c} \quad \text{with } v_{s1} = v_{s2} \tag{14-5}$$

For the network of Fig. 14-3 the output voltage is zero when the two inputs are equal, and the common-mode gain is zero.

As long as the simplified model of Fig. 14-3 applies, the common-mode voltage gain is zero even for unmatched transistors. For this case the currents i_1 and i_2 are proportional to one another, as evidenced by a node equation, and both must be zero when the two input voltages are equal. However, if the base-width modulation resistances r_o and r_μ are included, it is easy to show that A_c is not normally zero for unmatched transistors. Consequently, in evaluating A_c both r_o and r_μ should be included in the models even though these resistances have a negligible effect on the differential gain. The model of Fig. 14-3 does not include the incremental resistance of the practical source that supplies the quiescent emitter currents. The resistance is usually so large that it can be ignored in determining the differential gain, but it may be important in determining A_c. Even for matched transistors the presence of this resistance in the model leads to a nonzero value for A_c, and this value may be either positive or negative at low frequencies.

Because an output that is proportional to the difference between the inputs is desired, the output should be extremely small when the inputs are equal. Ideally, A_c should be zero. An important figure-of-merit is the *common-mode rejection ratio* (CMRR), which is defined as

$$\text{CMRR} = \left| \frac{A_d}{A_c} \right| = 20 \log_{10} \left| \frac{A_d}{A_c} \right| \text{ dB} \qquad (14\text{-}6)$$

Large values of CMRR are obtained by the use of a perfectly symmetrical circuit with matched transistors and with a direct-current source having a very high incremental resistance. The CMRR is usually found experimentally, and values greater than 10^5, or 100 dB, are possible. From known values of CMRR and A_d the magnitude of the common-mode gain A_c can be found, but not the sign.

Example

Find the common-mode rejection ratio for the differential amplifier of Fig. 14-1 if the direct-current source has an incremental resistance of 100 kΩ. Assume $R_C = 1.39$ kΩ, $r_\pi = 2$ kΩ, $r_x = 200$ ohms, and $\beta_o = 80$. Add source resistances of 800 ohms at each input.

Solution

The effect of the 100 kΩ resistor on the differential gain is negligible, and A_d is 18.53 from (14-2). The model for finding the common-mode gain is shown in Fig. 14-4. Two loop equations are sufficient for determining that $i_1 = i_2 = 6.172 \times 10^{-5}$ mA. The output voltage v_o is -0.00686, and the common-mode input voltage v_c is 1, giving a common-mode gain A_c of -0.00686. The ratio A_d/A_c is -2700, and the CMRR is 69 dB.

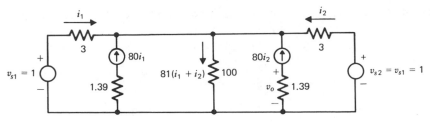

Figure 14-4   Incremental model (kΩ, mA, V).

THE DYNAMIC RANGE

Several advantages of the differential amplifier are revealed by the dynamic transfer characteristic, which will now be determined and examined. In the amplifier of Fig. 14-5 the base ohmic resistances have been placed external to the transistors, and hence the Ebers–Moll equations apply to the idealized devices that remain. Because emitter resistors are sometimes used in differential amplifiers to provide both dc and ac feedback, although their presence reduces the midband gain (but increases the input impedance), they are included here for generality. Usually, R_e is 50 ohms or less. Transistors Q_1 and Q_2 are assumed to be identical with the same quiescent points.

In the active mode at a temperature of 290 K the Ebers–Moll equation (2-10) indicates that each total instantaneous emitter current i_E is proportional to $\exp 40v_{BE}$. From this, and noting that the difference $v_{BE1} - v_{BE2}$ equals $v_{be1} - v_{be2}$ because the quiescent base-emitter voltages are equal, we find that the ratio of the emitter currents is

$$\frac{i_{E1}}{i_{E2}} = e^{40(v_{BE1} - v_{BE2})} = {}^{40(v_{be1} - v_{be2})} \tag{14-7}$$

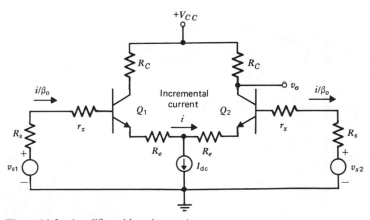

Figure 14-5   Amplifier with emitter resistors.

As shown in Fig. 14-5, the incremental emitter currents are the current i and the incremental base currents are approximately i/β_o. A loop equation obtained from the circuit containing the two inputs relates the incremental variables as follows:

$$v_{be1} - v_{be2} = v_{s1} - v_{s2} - iR = v_d - iR \tag{14-8}$$

with

$$R = 2\left[R_e + \frac{R_s + r_x}{\beta_o}\right] \tag{14-9}$$

From (14-7) and (14-8) we obtain

$$\frac{i_{E1}}{i_{E2}} = e^{40(v_d - iR)} \tag{14-10}$$

With reference to Fig. 14-5 we note that i_{E1} equals $-\frac{1}{2}I_{dc} - i$ and i_{E2} equals $-\frac{1}{2}I_{dc} + i$. Making these substitutions in (14-10), with a little manipulation we find that

$$i = \frac{1}{2}I_{dc}\left[\frac{e^{40(v_d - iR)} - 1}{e^{40(v_d - iR)} + 1}\right] = \frac{1}{2}I_{dc} \tanh 20(v_d - iR) \tag{14-11}$$

Shown in Fig. 14-6 is a plot of the dynamic transfer characteristic i/I_{dc} versus v_d, obtained from (14-11) for the case in which the iR terms are negligible. Because the output voltage v_o is proportional to i, a plot of v_o versus v_d is similar. We note that, when v_d is changed from a positive value to an equal negative value, the normalized current i/I_{dc} simply changes sign. This type of symmetry indicates that *the incremental current* i *has no even harmonics* when the input difference

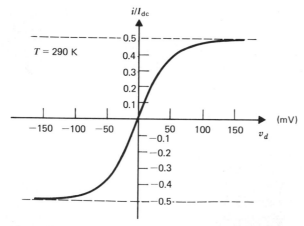

Figure 14-6   Dynamic transfer characteristic with $R = 0$.

voltage v_d is sinusoidal with time. Even harmonics generated by the nonlinear transistors represent common-mode signals that cancel in the output. In contrast, we found in Sec. 12-1 and P12-1 that the distortion in an amplifier with a single BJT is dominated by the second harmonic component of the incremental output current.

The elimination of even harmonics in the output leads to a substantially greater dynamic range for the differential amplifier. In P12-1 it was shown that an incremental base-emitter voltage of 5 mV rms produces 7.1 % distortion in a single-ended BJT circuit. From (14-11) with R negligible it is easy to show (see P14-4) that the distortion in a differential amplifier is essentially the same value when the differential input is 30 mV rms. Thus *the dynamic range of a differential amplifier is about* 30 mV *compared with about* 5 mV *for the single-ended BJT amplifier.* This statement is reaffirmed by examination of the transfer characteristic of Fig. 14-6.

For a fixed difference-mode voltage v_d with $R = 0$, it is evident from (14-11) that i is proportional to I_{dc}, and hence v_o is also proportional to I_{dc}. This property is sometimes exploited in automatic-gain-control (AGC) circuits that utilize I_{dc} to control the gain. We note also from Fig. 14-6 that $|i|$ has a maximum value of $\frac{1}{2}I_{dc}$, with $|i|$ being approximately equal to this maximum for values of $|v_d|$ greater than 100 mV. Thus the differential amplifier can be used as a *limiter*, which prevents the output voltage from exceeding a certain value on both positive and negative voltage swings.

The effect of the iR terms in (14-11) can be easily included by solving (14-11) for v_d. The result is

$$v_d = iR + 0.025 \ln \frac{I_{dc} + 2i}{I_{dc} - 2i} \qquad (14\text{-}12)$$

For a specified I_{dc} and a value of R found from (14-9), Equation 14-12 can be used to plot the transfer characteristic i versus v_d. The effect of the linear resistance R is to decrease somewhat the nonlinearity of the curve, thus extending the dynamic range (see P14-5).

The input resistance R_i of the emitter-coupled amplifier, defined either as the ratio v_d/i_2 or as that seen at either input with the other input terminal grounded, assuming both source resistances are present, is

$$R_i = 2(r_x + r_\pi + R_s) \qquad (14\text{-}13)$$

It is twice that of a CE amplifier having the same transistor parameters and R_s, and is made large by selecting a small quiescent collector current and high-beta transistors. In OP-AMP integrated circuits the input differential stage typically operates at a few microamperes of collector current.

From the input of Fig. 14-1 labeled A the single-ended voltage gain is positive and that from input B is negative. Accordingly, terminal A is the *noninverting* input and *B* is the *inverting* input. Because the output is proportional to the difference of the two input voltages, the configuration is often referred to as a *difference* amplifier.

The direct-current source of the differential amplifier can be obtained by using a large resistor in series with a large dc voltage. However, a more practical current source consists of a transistor biased so as to provide the desired direct current. Transistor current sources are common in integrated circuits, and such sources are examined in Secs. 15-7 and 19-1.

14-2. MONOLITHIC OPERATIONAL AMPLIFIERS

Whenever possible, standard monolithic integrated circuits should be used to provide a desired electronic function because of the low cost per circuit that accompanies large production runs. This economic consideration has led to standard linear ICs that are designed for wide flexibility in their applications, and such circuits normally require the addition of discrete components, with the selection depending on the particular application. An example is the operational amplifier that requires external feedback and compensation networks. Sometimes external circuitry is added because of limitations of the technology. For example, intermediate-frequency IC amplifier stages use external tuned circuits, because LC tuned circuits are not easily integrated into a chip.

The use of capacitors in integrated circuits should be minimized, with values over 100 pF avoided whenever possible. In fact, a capacitance of only 100 pF may require as much as 10 times the surface area of a monolithic chip as does a transistor. Thus direct-coupled stages are the rule, and circuits are designed so that bypass capacitors are usually unnecessary.

Because large resistors occupy more chip area than small ones, circuits are designed with the smallest possible resistances, preferably below 10 kΩ. A current source is used in place of a large resistor in series with a voltage source. *Matched components are easily obtained, and circuits depending on matched transistors and resistance ratios are preferred*, such as the emitter-coupled amplifier with its balanced circuitry. Matched elements are placed close together on the same chip, and they are at nearly the same temperature and have almost identical temperature variations. Balanced monolithic circuitry is relatively drift-free over the useful temperature range.

Low voltages, preferably below 15 V, are desired to avoid transistor breakdown and also for minimum power dissipation. Voltages greater than 100 V present serious breakdown problems in monolithic design. Power is dissipated within a tiny chip, and the packaging must provide a sufficiently low thermal impedance to the environment to prevent excessive temperature rise. Heat sinks are sometimes used. Most monolithic ICs dissipate about 100 mW or less, but amplifiers supplying and dissipating watts of power are available. The small size of the circuit reduces unwanted electrical pickup, which allows the transistors to operate at somewhat smaller signal levels with lower bias voltages and power levels.

Perhaps the most important of the monolithic linear integrated circuits is the operational amplifier. Basically, *this is a high-gain amplifier with high-input and low-output impedances*, and normally there are two inputs—one inverting and one noninverting. In most applications *OP AMPs are used with negative feedback applied externally*. The performance of a typical operational amplifier depends on the matched transistors and resistors of the balanced circuitry of the emitter-coupled configuration. Stages are direct-coupled, capacitances greater than a few tens of picofarads are avoided, supply voltages are usually less than ± 20 V, most resistors have values that are moderate or low, and transistors with common collectors are often included. Thus *the operational amplifier is ideally suited for monolithic ICs*.

OP AMP 741

A popular general-purpose operational amplifier is one known as the 741. Its input differential stage feeds into a CE stage, which uses a *Darlington* pair of transistors for high input impedance to avoid seriously reducing the voltage gain of the input differential amplifier. This stage is followed by a *complementary* emitter-follower output stage made up of an NPN and a PNP substrate transistor. The use of an emitter-follower output circuit provides the desired low output impedance. A Darlington-pair "composite" transistor, or "compound pair," is shown in Fig. 14-7. It is equivalent to a single transistor with a very large beta. In Fig. 14-8 is a complementary emitter-follower output stage. Both Darlington pairs and complementary outputs are examined in Secs. 16-5 and 16-6.

On the monolithic chip are 13 NPN and 7 PNP transistors, some of which serve to provide the proper bias currents to the active devices. There are 11

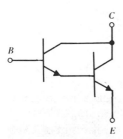

Figure 14-7 Darlington-pair composite transistor.

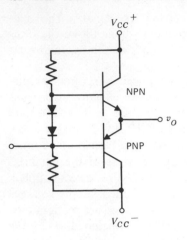

Figure 14-8   A complementary emit-ter-follower output stage.

resistors ranging in value from 50 kΩ to 25 ohms, along with one 30 pF capacitor. The detailed circuitry is shown in Fig. 15-18 and discussed in Sec. 15-7. The package has seven connected pins. Two of these are the inverting and non-inverting inputs, and one is the output. Two are for application of $+15$ and -15 V. The remaining two are for optional connection of a 10-kilohm external potentiometer. With the variable tap of the potentiometer connected to the negative supply of -15 V, proper adjustment nulls the output dc voltage. The network has no ground. In fact, many OP AMPs do not include a ground connection, and their schematic diagrams show no ground symbol. However, one is established by the midpoint of the voltage supplies. *This serves as the zero reference point for the signals.*

DC PARAMETERS

Although the transistors of the input differential stage are closely matched and located adjacent to one another on the chip, they may well have values of V_{BE} that differ by several millivolts for equal collector currents. For the 741 the data sheet specifies a value of 1 mV as a typical *input offset voltage*, defined as the dc voltage that must be applied between the input terminals to reduce the quiescent output voltage to zero. With a differential voltage amplification A_{VD} of 200 000, which is the normal gain expected, a 1 mV input would give a dc output of 200 V assuming linearity. However, the assumption is absurd because one or more transistors are driven out of the active region by a dc output exceeding the supply voltage, typically ±15 V.

The calculation clearly indicates that *the operational amplifier must be used in conjunction with a significant amount of dc feedback to reduce the effect of small dc imbalances in the input stages, or else a balancing network must be used at the input and critically adjusted to null the dc output offset voltage.* The required reduction in the output offset depends on the particular application. For example, when used as a true dc amplifier, the output offset must be much less than the output signal (see P14-6).

In addition to the input offset voltage, data sheets specify the *input offset current*. It is defined similarly, being the difference between the direct currents into the two input terminals required for nulling the output. For the 741 the input offset current I_{IO} is typically 20 nA. In normal operation the output is approximately nulled either by negative feedback or by means of a balancing network, but I_{IO} passing through a source impedance gives a difference-mode voltage that is amplified. Thus the input offset current is especially important when source impedances are very large. This is discussed further in Example 2 of the next section.

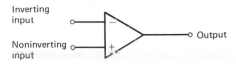

Figure 14-9   Operational amplifier symbol.

Also specified is the *supply voltage sensitivity*, which is the ratio of the change in the input offset voltage to the change in the supply voltages producing it, with both positive and negative supply voltages varied symmetrically. For the 741, a 1 V change in V_{CC} typically produces a 30 μV change in the input offset voltage. The corresponding supply voltage sensitivity is 30 μV/V. This is of particular importance in battery-operated operational amplifiers. With ac power lines employed, dc voltage regulators are normally used, and these provide a more stable supply voltage than do batteries.

The conventional symbol for an operational amplifier is shown in Fig. 14-9. The two input voltages and the output are referenced to ground.

INCREMENTAL PARAMETERS

We have learned that an important feature of an operational amplifier is high voltage gain at frequencies down to and including direct current. The gain may be from about 1000 to 10^7 or higher. The corresponding decibel gain ($20 \log A_V$) is 60 to 140 dB. Internal or external *compensation* is used to control the high-frequency response so that a large amount of negative feedback can be added without causing oscillations, and the feedback must include direct current. The 741 is internally compensated.

A characteristic of operational amplifiers is high input impedance. It is usually in the range from 100 kΩ to 10 MΩ, but substantially lower and higher values are not uncommon. For the 741 the input impedance is typically that of a 2 MΩ resistor shunted by a 1.4 pF capacitor. High values are obtained by using input-stage quiescent collector currents of tens of microamperes. In such cases the collector resistors of the input-stage differential amplifier should be large, perhaps 25 kΩ or more. As in the 741, these resistors are often replaced with transistors that provide high-impedance *active* loads. The input resistance of the second stage should be high to avoid excessive shunting of a collector resistor. This is accomplished in the 741 by using a Darlington pair in the second stage. Many high-performance operational amplifiers have extremely high input resistances, which are obtained by using either FETs or emitter followers at the input. These arrangements also have the advantage of drawing very little direct current from a source, and this is essential for some types of sources. The use of Darlington pairs at the input is generally not desirable because of the difficulty of maintaining a low percentage difference in the input bias currents over a wide temperature range.

Most operational amplifiers have emitter-follower (CC) outputs for low output impedance. Ideally, this impedance should be zero so that the output voltage will remain constant when the load draws current. Typical values of output impedances vary from 50 to 300 ohms, and that of the 741 is about 75 ohms.

The specified *input voltage range* V_I is the range of voltage which, if exceeded at either input terminal, will cause the amplifier to cease functioning properly. For the 741 with supply voltages of ± 15 V, this range is ± 13 V. Also important is the maximum peak-to-peak output voltage swing V_{OPP}, which is the output that can be obtained without waveform clipping when the quiescent dc output offset voltage is zero. With ± 15 V supplies V_{OPP} is 28 V for the 741.

The common-mode rejection ratio (CMRR) has been defined as the ratio of the differential voltage gain to the common-mode gain. Typical values for operational amplifiers range from 60 dB to 110 dB, and for the 741 it is 90 dB. Sometimes circuits are included internally that increase the CMRR by means of negative feedback of the common-mode signal, with the circuits designed so as to have little effect on the difference-mode gain.

The *unity-gain bandwidth* of the 741 is about 1 MHz. This is defined as the frequency at which the voltage amplification without feedback is unity. Most applications are restricted to frequencies from 0 to about 10 kHz.

Figure 14-10 is the schematic of a transient-response test circuit for the 741 operational amplifier. As we shall see in the next section, this circuit has unity gain, with $v_O = v_I$. At the right side are shown the input and output voltage waveforms. With $V_I = 20$ mV the *rise time* t_r is typically 300 ns, which is the time it takes v_O to rise from 10% to 90% of its final value V_O, and the *overshoot* is typically 5% of V_O. The total time required for the output to settle within a

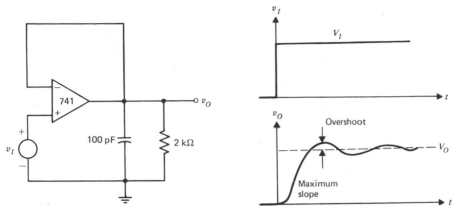

Figure 14-10   OP AMP test circuit and waveforms.

specified percentage of its final value is known as the *settling time t_s*. A fast settling time is particularly desirable for amplifiers used in switching circuits.

Of special importance in operational amplifiers is the *slew rate SR*, defined as *the maximum time rate of change of the closed-loop amplifier output voltage for a step-function input*. In the network of Fig. 14-10 it equals the maximum slope of the output voltage waveform. With V_I and V_O equal to 10 V, the slew rate is typically 0.5 V per microsecond, or 5×10^5 V/s. It is normally measured with a square-wave input under large-signal conditions.

Suppose the input and output of the OP AMP are sinusoidal, with v_O equal to $V_m \sin \omega t$. The maximum rate of change of v_O is ωV_m, and this must be less than the slew rate SR if observable distortion is to be avoided. Therefore, $\omega V_m < SR$ and $f < SR/(2\pi V_m)$. The maximum frequency that can be used without undue distortion resulting from slew-rate limiting is the *full-power bandwidth f_p*. Hence,

$$f_p < \frac{SR}{2\pi V_m} \tag{14-14}$$

With the peak value V_m of the output maintained constant, the full-power bandwidth is measured by increasing the frequency until distortion becomes apparent. The inequality of (14-14) provides an upper bound on f_p. It is useful in estimating the maximum frequency that can be used at a specified output voltage level.

14-3. THE IDEAL OPERATIONAL AMPLIFIER

Operational amplifiers with typical inverting and noninverting feedback configurations are shown in Fig. 14-11. In each case Z_1 denotes the impedance

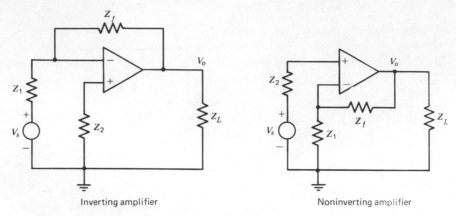

Inverting amplifier   Noninverting amplifier

Figure 14-11   Typical feedback configurations.

connected to the inverting input and Z_2 is that which is in series with the input impedance Z_i of the operational amplifier. To analyze the networks we need an appropriate model for the operational amplifier, and that of Fig. 14-12 is usually adequate. Between the inverting and noninverting inputs is the large differential impedance Z_i. Note that the reference direction for the differential input voltage V_i across Z_i corresponds with the $\pm$ input terminals. The gain A_v is the ratio of the *open-circuit* output voltage to the input V_i. Because the output impedance Z_o is small, A_v approximately equals the differential gain V_o/V_i, and at low frequencies A_v is a positive real number.

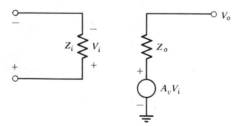

Figure 14-12   Operational amplifier model.

The model has two inputs, one being from the inverting input to ground and the other from the noninverting input to ground. For equal input voltages, V_i is zero. Thus the model gives no output for these common-mode voltages. However, imbalances present in practical amplifiers cause a common-mode input to produce small input currents and output voltage. A better model includes impedances from each input to ground. These *common-mode input impedances* are typically much larger than Z_i and can be neglected in most cases without appreciable error. We shall do so here.

Numerical values associated with practical configurations normally satisfy the following inequalities:

$$A_v \gg 1 \qquad\qquad A_v Z_1 \gg Z_f \qquad\qquad (14\text{-}15)$$

$$Z_i \gg Z_1, Z_2, \text{ and } \frac{Z_f}{A_v} \qquad\qquad (14\text{-}16)$$

$$Z_o \ll Z_f \text{ and } Z_L \qquad\qquad (14\text{-}17)$$

The magnitude is understood in the event a quantity is complex. Inequality (14-15) is a consequence of the large voltage gain that is characteristic of operational amplifiers, that of (14-16) is the result of the large input impedance Z_i, and (14-17) results from the low output impedance. We shall assume the inequalities are applicable throughout this section.

THE INVERTING AMPLIFIER

Appropriate models for the inverting configuration of Fig. 14-11 are shown in Fig. 14-13. Feedback takes place through the impedance Z_f connected between the input and output nodes. In the network of Fig. 14-13a, the operational amplifier has been replaced with its model of Fig. 14-12. From this network we can find the gain and the input and output impedances without utilizing the inequalities that have been presented.

In the approximate network of Fig. 14-13b the impedances Z_2, Z_o, and Z_L are omitted. Because Z_2 is in series with Z_i, its omission is justified by inequality (14-16). Application of Norton's theorem to the branch with Z_o and the dependent source places Z_o in parallel with Z_L, and thus Z_L can be eliminated, justified by (14-17). This places Z_o in series with Z_f, which is much larger than Z_o by

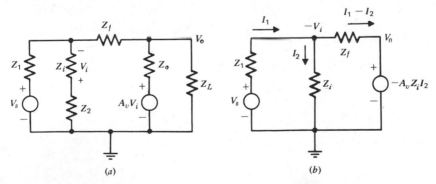

Figure 14-13   Inverting-configuration models. (a) Model. (b) Approximate model.

(14-17). Hence the voltage across Z_o is small compared with that across Z_f. Because the voltage across Z_f is the sum of V_o and V_i, that across Z_o is very small compared with $A_v V_i$, and Z_o can be replaced with a short-circuit. This gives the approximate model.

From the approximate model it is easy to deduce that

$$Z_{in} = Z_1 + \frac{Z_i Z_f}{A_v Z_i + Z_i + Z_f} \approx Z_1 \tag{14-18}$$

$$A_{Vs} = \frac{-A_v Z_i Z_f}{Z_1 (A_v Z_i + Z_i + Z_f) + Z_i Z_f} \approx -\frac{Z_f}{Z_1} \tag{14-19}$$

$$\frac{-V_i}{V_s} = \frac{-A_{Vs}}{A_v} \approx \frac{Z_f}{A_v Z_1} \ll 1 \tag{14-20}$$

with $Z_{in} = V_s/I_1$ and $A_{Vs} = V_o/V_s$. The final expressions of the equations were derived with the aid of the inequalities. The symbol A_{Vs} denotes the ratio of the output voltage to the open-circuit voltage of the source. *The feedback reduces the input impedance at the inverting input terminal of the amplifier to a value so low that the input impedance seen by the source V_s is simply Z_1*, as indicated by (14-18).

By (14-19), A_{Vs} is $-Z_f/Z_1$. With precision resistors replacing the impedances, the amplifier becomes a *scale changer*. The output voltage V_o is $K V_s$ with the scale factor $K = -R_f/R_1$. For equal resistances K is -1, and the OP AMP is an *inverter* having an output voltage equal in magnitude but opposite in sign to the input. The use of complex impedances for Z_f and Z_1 with equal magnitudes but different angles converts the amplifier into a *phase shifter*. The phase of a sinusoidal input can be shifted by any value between 0 and $\pm 180°$ (see P14-10).

The output impedance is found from the network of Fig. 14-14, deduced from that of Fig. 14-13a looking back from the load Z_L with $V_s = 0$. Omitted is the

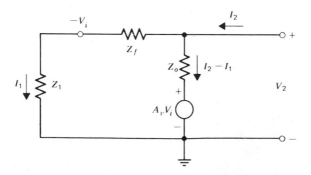

Figure 14-14   Network for determining Z_{out}.

branch containing Z_i, which is in parallel with Z_1. This is justified by the assumption that Z_1 is much smaller than Z_i, according to (14-16). Analysis reveals that

$$Z_{out} = \frac{Z_o(Z_1 + Z_f)}{A_v Z_1 + Z_1 + Z_f + Z_o} \approx \frac{Z_o(Z_1 + Z_f)}{A_v Z_1} \qquad (14\text{-}21)$$

Because Z_o is small and A_v is large, Z_{out} *is usually no more than a fraction of an ohm.*

THE NONINVERTING AMPLIFIER

Let us briefly examine the noninverting configuration of Fig. 14-11. The analysis is performed by replacing the operational amplifier with its model, shown in Fig. 14-12. For transmission in the forward direction the impedances Z_2 and Z_o are negligibly small and Z_L can be treated as an open circuit, again assuming the inequalities of (14-15) through (14-17) apply. We find that (see P14-9)

$$Z_{in} \approx Z_i\left(1 + \frac{A_v}{1 + Z_f/Z_1}\right) \approx \frac{Z_i A_v}{A_{Vs}} \qquad (14\text{-}22)$$

$$A_{Vs} \approx \frac{Z_1 + Z_f}{Z_1} = \frac{V_o}{V_s} \qquad (14\text{-}23)$$

$$\frac{V_i}{V_s} \approx \frac{Z_1 + Z_f}{A_v Z_1} \ll 1 \qquad (14\text{-}24)$$

with A_v denoting the open-loop voltage gain of the operational amplifier. The network for determining the output impedance can be arranged precisely in the form of that of Fig. 14-14, provided the inequalities of (14-15) through (14-17) apply. Therefore, (14-21) for Z_{out} applies to the noninverting amplifier as well as to the inverting amplifier.

THE IDEAL OP AMP

This is one having the model of Fig. 14-12 *with the inequalities of* (14-15) *through* (14-17) *being completely valid.* Ideally, both A_v and Z_i are so large that making either of them larger has no appreciable effect on the performance of the amplifier. In addition, Z_o is so small that a lesser value likewise has no appreciable effect. In other words, *in an ideal operational amplifier* A_v *and* Z_i *are infinite and* Z_o *is zero.*

With Z_o negligible, V_o equals $A_v V_i$, and the gain A_{Vs} becomes $A_v V_i/V_s$. It follows that V_i/V_s equals A_{Vs}/A_v, as indicated by (14-20). Because V_s and A_{Vs} are limited, V_i approaches zero as A_v approaches infinity. Therefore, *between the inverting and noninverting input terminals of an ideal operational amplifier is an infinite impedance with zero voltage.* Because the voltage V_i is zero, the branch is referred to as a short-circuit, and because the infinite impedance requires that the input current I_i also be zero, the term *virtual short-circuit* is used. Unlike a normal short-circuit, a virtual short-circuit carries no current. The model shown in Fig. 14-15 defines an ideal operational amplifier. Although the output voltage of the model is indeterminate, it is fixed by the external feedback circuitry.

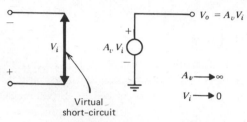

$$V_o = A_v V_i$$

$$A_v \longrightarrow \infty$$

$$V_i \longrightarrow 0$$

Virtual short-circuit

Figure 14-15　Ideal operational-amplifier model.

To illustrate the use of the ideal operational amplifier, let us consider the noninverting amplifier of Fig. 14-11. An operational amplifier can be regarded as ideal provided the feedback is appreciable, both Z_i and A_v are so large that they can be regarded as infinite, and Z_o is so small that it can be assumed zero without appreciable error. It is understood, of course, that the rated output must not be exceeded, and the frequency must not be too high. For values normally encountered in many practical circuits, these requirements are satisfied. Accordingly, the network of Fig. 14-16 is applicable. Because no current passes through the virtual short-circuit, the voltage across Z_2 is zero. Therefore, the voltage across Z_1 is V_s, the current through Z_1 is I_1, and $V_s = Z_1 I_1$. Also, $V_o = (Z_1 + Z_f)I_1$, and the ratio V_o/V_s becomes $(Z_1 + Z_f)/Z_1$, which is (14-23). The model gives an input impedance of infinity and an output impedance of zero.

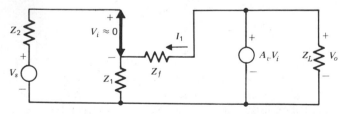

Figure 14-16　Noninverting amplifier model.

Example 1
　Suppose the noninverting OP AMP of Fig. 14-16 has $Z_1 = 1\ \text{k}\Omega$, $Z_f = 4\ \text{k}\Omega$, and $V_s = 1$ V dc. (a) Calculate the output voltage V_o assuming an ideal amplifier. (b) If the OP AMP is nonideal, having an open-loop gain $A_v = 10^5$ and a CMRR

of 80 dB, calculate V_o and the voltage V_i across the input terminals of the OP AMP. Assume the input impedance Z_i is infinite, the output impedance Z_o is zero, and the common-mode voltage gain is positive.

Solution

(a) For the ideal amplifier the output voltage is 5 V dc by (14-23). (b) For the nonideal case the voltage across Z_1 is $0.2V_o$, with V_o to be determined, because the input impedance of the OP AMP is infinite. The voltage at the noninverting input is precisely 1 V. Therefore, the difference-mode input voltage V_i is $1 - 0.2V_o$ by (14-3), and the common-mode input voltage V_c is $0.5 + 0.1V_o$ by (14-4). The difference-mode gain A_d is the specified open-loop gain A_v of 10^5. With a CMRR of 80 dB, or 10^4, the common-mode voltage gain A_c is found from (14-6) to be 10. The sum of $V_i A_v$ and $V_c A_c$ gives V_o to be

$$V_o = (1 - 0.2V_o)10^5 + (0.5 + 0.1V_o)10$$

From this, V_o is found to be precisely 5.00025 V, and the input voltage V_i is $1 - 0.2V_o$, or -5×10^{-5} V.

It is interesting to observe that $V_i A_v$ is -5 and $V_c A_c$ is $+10.00025$, with the sum being approximately 5 V. In spite of the large effect of the common-mode gain, the output voltage is nearly the same as that of an ideal OP AMP with a common-mode gain of zero. In both cases the external resistors determine the voltage gain of the feedback amplifier. A negative common-mode gain leads to similar results (see P14-12).

Example 2

The input bias currents I_A and I_B of an inverting OP AMP with no input signal are represented by the current sources of Fig. 14-17. Except for these sources the OP AMP is ideal. (a) Find V_O as a function of the bias currents and resistances. (b) For perfectly matched differential inputs ($I_A = I_B$), determine R_2 such that the output is zero. (c) For this R_2 find V_O in terms of R_1, A_{Vs}, and the input offset

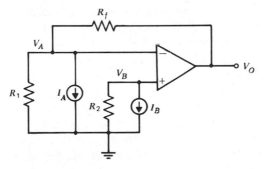

Figure 14-17   OP AMP with input bias currents I_A and I_B.

current I_{IO}, defined as $I_A - I_B$. Calculate V_O for $R_1 = 50 \text{ k}\Omega$, $A_{Vs} = -4$, $I_A = 400 \text{ nA}$, and $I_B = 350 \text{ nA}$. (d) With $R_2 = 0$, and using the values of (c), determine V_O.

Solution

(a) The virtual short-circuit of Fig. 14-17 has zero current and zero voltage. Therefore, V_B is $-R_2 I_B$, and this is also V_A. From a nodal equation at node V_A we find that

$$V_O = R_f\left[I_A - \frac{I_B R_2}{R_1 \| R_f}\right] \tag{14-25}$$

(b) For matched inputs R_2 should equal $R_1 \| R_f$ in order to give zero output voltage. (c) With this R_2 the output becomes $R_f I_{IO}$, or $-R_1 A_{Vs} I_{IO}$, with $A_{Vs} = -R_f/R_1$. For $R_1 = 50 \text{ k}\Omega$, $A_{Vs} = -4$, and $I_{IO} = 50 \text{ nA}$, the output is 10 mV. (d) With $R_2 = 0$, the output is $R_f I_A$, or 80 mV for a bias current of 400 nA. The results clearly show that, for a specified voltage gain, the source resistance should be small so as to minimize the effect of I_{IO}. Also, the advantage of using identical dc resistances between each input terminal and ground is apparent.

SUMMARY AND DISCUSSION

Networks containing operational amplifiers are usually quite easy to analyze or design provided the ideal model of Fig. 14-15 can be used. It is important to remember that *both the voltage and current of the virtual short-circuit are zero*, and this virtual short appears directly between the inverting and noninverting terminals.

There are two basic feedback configurations. Assuming the ideal model is applicable, the voltage gains depend on external impedances and are independent of the parameters of the operational amplifier. These gains are $-Z_f/Z_1$ and $(Z_1 + Z_f)/Z_1$ for the inverting and noninverting configurations, respectively. *The input impedance of the inverting amplifier is simply the external impedance in series with the source, whereas that of the noninverting amplifier is extremely large, or infinite, in accordance with* (14-22). *Both configurations have very low output impedances, normally less than an ohm.*

In some applications the inequalities of (14-15) through (14-17) are not wholly justified. For these cases one must use the OP AMP model of Fig. 14-12 (see P14-11). Equations 14-18 through 14-24 are frequently quite useful.

As a consequence of the virtual short-circuit, the inverting configuration of Fig. 14-11 has both OP AMP inputs at approximately zero volts, and the current fed back through Z_f is V_o/Z_f. The output voltage V_o is referred to as the *sampled quantity*. The current is injected into the input node where it is, in effect, *compared* with that from the source. The difference between these nearly equal currents is the extremely small current flowing into the inverting terminal of the

OP AMP. This feedback is often called *voltage-sample current-sum*, and in Chapter 20 we learn that it reduces the input and output impedances and also the overall voltage gain V_o/V_s. These effects have been observed here.

In the noninverting amplifier of Fig. 14-11 the feedback network consisting of Z_1 and Z_f introduces a voltage into the input circuit, which is proportional to V_o. The feedback is called *voltage-sample voltage-sum*. Again, in Chapter 20 we learn that this type raises the input impedance, lowers the output impedance, and reduces the voltage gain. These effects have also been observed here. Further discussion of feedback concepts is deferred to Chapters 20 and 21, which investigate the beneficial properties of feedback and examine the design and analysis of feedback amplifiers.

14-4. APPLICATIONS

The design of a particular operational amplifier is oftentimes directed at rather specific applications. Some characteristics are inevitably compromised in order to achieve the best performance for the intended use. For example, certain amplifiers are designed for unusually high input impedance at the expense of voltage gain, or perhaps gain is emphasized with a sacrifice in bandwidth, and so forth. For each application examined here we shall assume the network has been designed so that it satisfies approximately the requirements of an ideal OP AMP, provided there is appreciable feedback. However, in practical situations it is important to ascertain that the assumption is reasonable for the actual operational amplifier selected for use, and care must be taken to avoid exceeding specified limitations.

VOLTAGE FOLLOWER

Figure 14-18 is the schematic diagram of a noninverting unity-gain amplifier known as a *voltage follower*. Because of the virtual short-circuit across the input terminals of the OP AMP, the output voltage V_o equals V_s. Some operational amplifiers are specifically designed for this application, with the connection

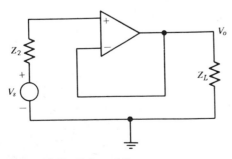

Figure 14-18   Voltage follower.

between the inverting input and the output made internally, along with internal compensation.

The amplifier has a very high input impedance and delivers the output from a low impedance. It is used as a buffer, allowing a source with a high internal impedance to be connected to a network or instrument requiring driving current. A typical application might consist of transforming 3 V peak-to-peak from a 100 kΩ source to a 500 ohm load without appreciable change in the voltage. The circuit of Fig. 14-18 is the noninverting voltage-sample voltage-sum feedback amplifier of Fig. 14-11 with infinite Z_1 and zero Z_f. The full output voltage is fed back into the input circuit. A low value of input dc bias current is important, because this current flows through the high resistance of the source. A blocking capacitor must not, of course, be placed in series with the source.

ADDITION AND SUBTRACTION

A general form of a summing amplifier, or *adder*, is shown in Fig. 14-19. The subscripts I and N refer to the inverting and noninverting inputs, respectively. Let R_A denote the parallel equivalent of all resistors connected to node A and let R_B denote the parallel equivalent of all resistors connected to node B. Mathematically,

$$R_A = R_f \| R_{I0} \| R_{I1} \| R_{I2}$$
$$R_B = R_{N0} \| R_{N1} \| R_{N2} \tag{14-26}$$

Because the current through the virtual short-circuit at the input of the OP AMP is zero, the node equations at nodes A and B are

$$\frac{v_O}{R_f} = -\frac{v_{I1}}{R_{I1}} - \frac{v_{I2}}{R_{I2}} + \frac{v_A}{R_A} \tag{14-27}$$

$$\frac{v_B}{R_B} = \frac{v_{N1}}{R_{N1}} + \frac{v_{N2}}{R_{N2}} \tag{14-28}$$

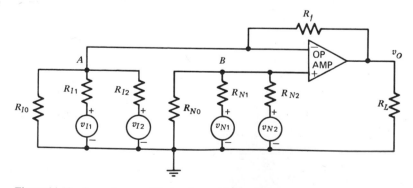

Figure 14-19   Summing amplifier.

The virtual short-circuit makes v_A equal to v_B. Therefore, (14-28) can be used to eliminate v_A in (14-27), and the result is

$$v_O = R_f \left[-\frac{v_{I1}}{R_{I1}} - \frac{v_{I2}}{R_{I2}} + \frac{R_B}{R_A}\left(\frac{v_{N1}}{R_{N1}} + \frac{v_{N2}}{R_{N2}} \right) \right] \qquad (14\text{-}29)$$

The output is the sum of the inputs, with each voltage at the inverting input multiplied by a negative scale factor and each voltage at the noninverting input multiplied by a positive scale factor. If the number of inputs at the inverting or noninverting terminals is more or less than two, (14-29) is modified appropriately.

The resistors are often selected so that the ratio R_B/R_A is unity, and either R_{I0} or R_{N0} is frequently infinite. For example, with $R_f = 100$ kΩ, suppose there is a single inverting input with $R_{I1} = 10$ kΩ and three noninverting inputs with R_{N1}, R_{N2}, and R_{N3} being 25, 50, and 50 kΩ, respectively. By choosing R_{N0} to be 33.3 kΩ with R_{I0} infinite, R_A and R_B have equal values of 9.1 kΩ, and (14-29) suitably modified becomes

$$v_O = -10v_{I1} + 4v_{N1} + 2v_{N2} + 2v_{N3}$$

DIGITAL-TO-ANALOG CONVERSION

It is frequently necessary to convert a digital word having each bit represented by a logical 0 or 1 into an analog signal. For example, a voice communication that has been converted to digital form for transmission over telephone cables to a receiver must be converted back to the original analog signal at or ahead of the receiver. The summing amplifier in conjunction with electronic switches controlled by the bits of the input digital signal can be employed for this purpose.

Figure 14-20 is the schematic diagram of a digital-to-analog, or D/A, converter. The network converts a four-bit binary word into a voltage that is proportional to the decimal number represented by the coded word. Each bit

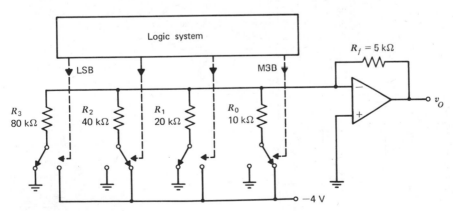

Figure 14-20   D/A converter.

electronically sets a switch, with the most significant bit (MSB) controlling the switch connected to the 10 kΩ resistor and the least significant bit (LSB) controlling the switch to the 80 kΩ resistor. A logical 1 connects the resistor to the reference voltage of -4, and a logical 0 connects it to ground. The resistors are *weighted* as are the bits of the digital word, with the MSB controlling the switch of the smallest resistor. Resistance R_n is the product of R_0 and 2^n. The system is easily expanded for conversion of words with more bits. For example, with an eight-bit word the largest resistor R_7 is $2^7 R_0$, or 1.28 MΩ, and its associated switch is controlled by the LSB.

The switches of the D/A converter of Fig. 14-20 are positioned for the binary word 1010, corresponding to the decimal 10. This coding system is explained in Sec. 5-1. Equation 14-29 applies with $R_B = 0$ and with each input voltage equal to either -4 or 0, depending on the switches. From the figure we find that

$$v_O = 5\left(\frac{4}{10} + \frac{0}{20} + \frac{4}{40} + \frac{0}{80}\right) = 2.5$$

Suppose the binary word is changed to 0101, corresponding to the decimal number 5. The position of each switch of the figure is now reversed, and the output voltage is found from (14-29) to be 1.25 V, precisely half of the previous result.

A possible circuit for the electronic switch is that of Fig. 14-21. When the bit line is -10 V corresponding to a logical 1, the D-type latch turns Q_1 *on* and Q_2 *off*. This is reversed with the bit line voltage at zero corresponding to a logical 0.

There are modifications of the circuit considered here, and various types are available in IC form. Typical is an eight-bit D/A converter which includes the

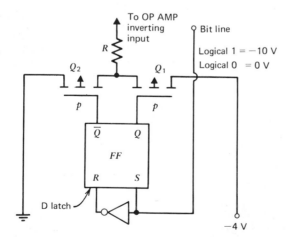

Figure 14-21 Electronic switch for D/A converter.

operational amplifier, the reference voltage, the resistor network, and the switches. The inverse operation of analog-to-digital (A/D) conversion is examined in Sec. 23-4.

DIFFERENTIAL INSTRUMENTATION AMPLIFIER

Although an OP AMP has a differential input, when it is used as a linear amplifier one input must be connected to the feedback network, and the amplifier is single-ended. Single-input amplifiers are not always adequate for measuring low-level signals commonly encountered in various types of instrumentation. The amplifier must accept signals from a few microvolts to several volts, while rejecting noise at frequencies from dc to hundreds of megahertz. By using a differential configuration in place of a single-ended operational amplifier, noise that is introduced on the input lines is largely rejected. However, the input leads should be twisted together to ensure that the introduced noise is largely a common-mode signal. To minimize the introduction of noise from ground connections, the impedance between each input terminal and ground should be high.

In Fig. 14-22 is a common configuration used as an instrumentation amplifier. At each input is an SN72310 voltage follower. This IC is internally connected as a unity-gain noninverting amplifier and it utilizes Darlington-pair inputs to an emitter-coupled stage. The arrangement gives an extremely low input bias current, typically 2 nA. The high input resistance of about 10^{12} ohms provides excellent buffering between the signal supply and the amplifier. Shown is a

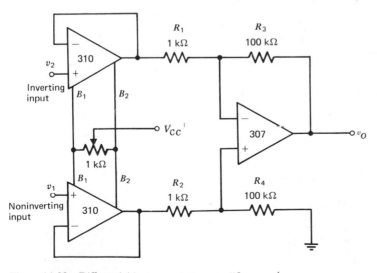

Figure 14-22   Differential instrumentation amplifier.

potentiometer for balancing the offset voltage. It is adjusted to give an output voltage of zero when the inputs are grounded.

The voltage followers at the inputs can be and frequently are omitted. When this is done, any external resistance in series with R_1 should be nearly equal to that in series with R_2, or otherwise there may be significant error. It is important that the ratio R_3/R_1 be equal to R_4/R_2, with R_1 and R_2 including any series-connected resistances. By using superposition and treating each OP AMP input as a virtual short-circuit, it is evident that the output voltage v_O is the product of the ratio R_3/R_1 and the difference-mode voltage $v_1 - v_2$. The resistors should be accurate to within 0.1 %.

The differential instrumentation amplifier has high performance, and it is versatile and inexpensive. Among its many applications are the amplification of dc or slowly varying low-level signals from transducers which transform a physical quantity into an electrical signal. Examples are strain-gauge bridges and thermocouples.

VOLTAGE-REFERENCE NETWORKS

Voltage regulators, comparators, D/A converters, and many other networks require a stable reference voltage for proper operation. Zener diodes and mercury cells are often used, but the voltage may have to be converted to a different value. An adjustable voltage-reference circuit is shown in Fig. 14-23.

The voltage V_A of pin 3 of the 741 OP AMP is also the voltage of pin 2 because of the virtual short-circuit. Therefore, V_R equals $V_z + V_A$. As there is no current

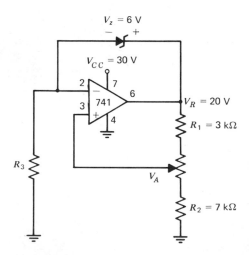

Figure 14-23   Voltage-reference circuit $(V_R > V_z)$.

through the sliding tap of the potentiometer, V_R also equals the product of V_A and the ratio $(R_1 + R_2)/R_2$. From these relations we find that

$$V_R = \frac{V_z(R_1 + R_2)}{R_1} \tag{14-30}$$

For the specified parameters of the figure, V_R is 20 V.

Pins 7 and 4 of the OP AMP are $V_{CC}{}^+$ and $V_{CC}{}^-$, respectively. In the figure these voltages are shown as $+30$ and 0. This is permissible, because the 741 has no internal ground point. The output voltage is restricted to values within this range, which eliminates any possibility of the diode conducting in the forward direction. If we desire a negative output voltage, we simply reverse the connections to the diode, connect pin 7 to ground, and apply -30 V to pin 4. With these changes, (14-30) still applies, but V_z is negative. The output is now -20 V.

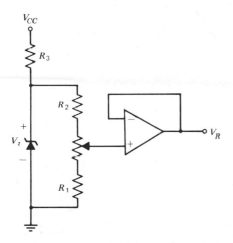

Figure 14-24   Voltage-reference circuit ($V_R < V_z$).

A stable voltage source having V_R less than the breakdown voltage V_z of a Zener diode is shown in Fig. 14-24. The diode voltage is reasonably independent of V_{CC} and the resistor values. Because the output impedance of the voltage follower is very low, it maintains a constant voltage across a variable load. Furthermore, it does not draw a significant current from the voltage-divider network. The output reference voltage is adjustable to any value between V_z and zero. There are, of course, many other voltage-reference networks designed with operational amplifiers. They have various special features, and some are exceedingly stable.

DIODE AND CLAMP NETWORKS

A semiconductor diode will not rectify very small signals because a silicon diode does not conduct appreciably below about 0.6 V. In Fig. 14-25 is a network designed to rectify voltages of the order of microvolts. In the equivalent circuit the OP AMP is replaced with the model of Fig. 14-12. When v_I is negative, the diode is an open circuit and the high impedance Z_i isolates the output from the input. However, when v_I is positive with a load connected to the output, the diode conducts with a forward drop V_D.

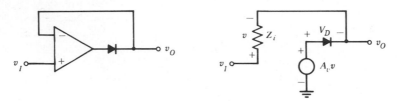

Figure 14-25   Precision diode network and equivalent circuit.

From the circuit we deduce that v_O equals $v_I - v$ and it also equals $A_v v - V_D$. Noting that the open-loop gain A_v is very large, we find from these relations that

$$v_O = v_I - \frac{V_D}{A_v} \quad \text{(for } v_I > 0) \quad (14\text{-}31)$$

With appreciable conduction V_D is 0.7 V; for $A_v = 10^5$ the output is less than v_I by only 7 μV. Let us now consider a *clamp* circuit, or *limiter*.

The precision clamp of Fig. 14-26 is somewhat similar in form to the diode network of Fig. 14-25. Its purpose is to prevent the output from going more positive than the reference voltage V_R. When v_I is less than V_R, diode D_1 is reverse biased and acts as an open-circuit. However, when v_I is greater than

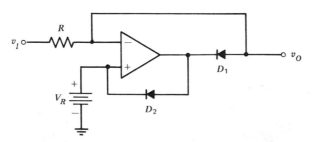

Figure 14-26   A precision clamp.

V_R, diode D_1 conducts, the feedback is effective, and the virtual short-circuit clamps the output at the reference voltage V_R.

Diode D_2 has no effect on the basic operation. It is reverse biased when the output is clamped with D_1 conducting. When v_I is less than V_R, diode D_1 is *off* and D_2 conducts. In this case it clamps the output a diode drop above V_R, thereby preventing the transistors of the OP AMP from becoming saturated. Thus D_2 is an output clamp which keeps the OP AMP out of saturation when D_1 is reverse biased. The result is faster response.

REFERENCES

1. E. R. Hnatek, *Applications of Linear Integrated Circuits*, Wiley, New York, 1975.
2. *Linear and Interface Circuits Applications*, Texas Instruments, Inc., Dallas, Texas, 1973.
3. J. K. Roberge, *Operational Amplifiers*, Wiley, New York, 1975.
4. *The Linear and Interface Circuits Data Book*, Texas Instruments, Inc., Dallas, Texas, 1973.

PROBLEMS

Section 14-1

14-1. (a) Each transistor of the emitter-coupled amplifier of Fig. 14-1 has $r_x = 200$ ohms, $r_\pi = 2$ kΩ, and $\beta_o = 80$. R_C is 2 kΩ. Add source resistances of 800 ohms in series with each base and calculate the difference-mode voltage gain and the input resistance in kilohms. (b) With the source resistances included, repeat the calculations of (a) with the addition of 25 ohms in series with each emitter. (*Ans*: 26.7, 6; 15.9, 10.1)

14-2. Suppose the transistors of the differential amplifier of Fig. 14-1 are mismatched, with Q_1 having $r_{x1} = 200$ ohms, $r_{\pi1} = 1$ kΩ, and $\beta_{o1} = 40$, and with Q_2 having $r_{x2} = 200$ ohms, $r_{\pi2} = 4$ kΩ, and $\beta_{o2} = 160$. R_C is 1.39 kΩ, and source resistances of 800 ohms must be added in series with each base. Assume the direct-current source has an incremental resistance of 100 kΩ, and calculate the CMRR in dB.

14-3. Add 800 ohms source resistances in series with each base of the emitter-coupled amplifier of Fig. 14-1. R_C is 1.5 kΩ, V_{CC} is 6 V, and I_{dc} is 2.02 mA. The transistors have identical dc betas of 100 but the transfer characteristics I_C-versus-V_{BE} are slightly mismatched. As a consequence, I_{C1} and I_{C2} are 0.8 and 1.2 mA, respectively. The input voltages are zero. Assume V_{BE1} is 0.7 V, and calculate V_{CE1}, V_{CE2}, the difference $V_{BE1} - V_{BE2}$, and the power in mW supplied by the voltage source and also that supplied by the current source.

14-4. Using (14-11) with $R = 0$ and $v_d = 0.03\sqrt{2}\cos \omega t$, calculate the percent harmonic distortion for a BJT differential amplifier. To do this, approximate the hyperbolic tangent (tanh x) in (14-11) with the first two terms $(x - x^3/3)$ of its infinite series. $(\cos^3 x = 0.25 \cos 3x + 0.75 \cos x)$. Refer to (16-19).

14-5. For the differential amplifier of Fig. 14-5, assume $R_s + r_x = 1000$ ohms, $R_e = 40$ ohms, $\beta_o = 100$, and $I_{dc} = 1$ mA. (a) Using (14-12), calculate v_d for $i = 0, \pm 0.1, \pm 0.2, \pm 0.3$, and ± 0.4 mA. (b) Repeat with $R = 0$ in (14-12). (c) On the same axes plot i versus v_d for both cases. Use the same current scale, but let the voltage scale for the curve of (a) be twice that of part (b) so that the two curves nearly coincide. Note the increased linearity due to R.

Section 14-2

14-6. In the network of Fig. 14-27 the OP AMP input currents flow through equal external resistances and these resistances are small. Consequently, any output voltage must be due to the input offset voltage V_{IO}, which is represented in the figure by the voltage source. If the output voltage is measured to be 0.35 V, calculate V_{IO}. Assume the input impedance of the OP AMP is infinite, and the voltage across this infinite impedance is zero. (*Ans*: 3.47 mV)

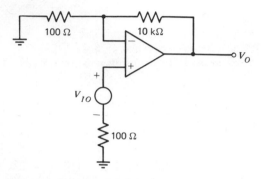

Figure 14-27   Network for P14-6.

14-7. In the measurement circuit of Fig. 14-28 with $R_3 = 0$ and R_L infinite, the 50 kΩ potentiometer is adjusted to make v_o zero when v_I is zero, and R_1 is determined to be 32 kΩ. Then v_I is adjusted to -80 mV, giving an output of 10 V. When 200 kΩ resistors are inserted for R_3 the output drops to 6 V. Next, resistors R_3 are removed and a 100 ohm load resistor is added. The output is now 4 V. For the OP AMP determine the input offset voltage V_{IO} in mV, the open-loop gain A_v, R_i in kΩ, and R_o in ohms.

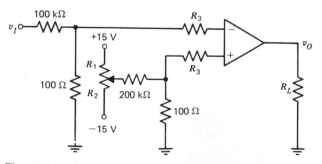

Figure 14-28   Measurement circuit for P14-7 ($R_1 + R_2 = 50$ kΩ).

14-8. An operational amplifier has a slewing rate of 1 V/μs. Estimate the full-power bandwidth in kHz for a peak-to-peak output swing of 20 V. Also, estimate the maximum frequency that can be used with a sinusoidal output having a peak value of 4 V.

Section 14-3

14-9. From the noninverting amplifier of Fig. 14-11 and the inequalities of (14-15) through (14-17), derive equations (14-22) through (14-24). Justify all approximations.

14-10. For the inverting amplifier of Fig. 14-11 with $Z_f = 1000$ ohms, determine Z_1 in ohms to give (a) an inverter, (b) a scale changer with $K = -5$, and (c) a phase shifter that makes V_o lag V_s by $120°$.

14-11. The noninverting amplifier of Fig. 14-11 has Z_1 infinite, $Z_2 = 500$ kΩ, $Z_f = 0$, and $Z_L = 1$ kΩ. The OP AMP can be represented by the model of Fig. 14-12 with $Z_i = 10$ kΩ, $Z_o = 200$ ohms, and $A_r = 10^4$. Calculate the input resistance in MΩ as seen by the source V_s, the output resistance in ohms (not including Z_L), and the voltage gain V_o/V_s. (*Ans*: 83.8, 1.01, 0.994)

14-12. Repeat Example 1 of Sec. 14-3 assuming a *negative* common-mode gain.

Section 14-4

14-13. For the voltage follower of Fig. 14-18, use the OP AMP model of Fig. 14-12 and deduce that the respective input and output impedances are approximately $A_r Z_i$ and Z_o/A_v. State approximations used. Calculate these impedances in ohms for the 741, assuming Z_i is 2 MΩ, Z_o is 75 ohms, and A_r is 200,000.

14-14. For a summing amplifier similar to that of Fig. 14-19, with $R_f = 50$ kΩ, $R_A = R_B$, and either R_{IO} or R_{NO} infinite, determine in kilohms the six resistors required to obtain an output given by

$$v_0 = 2v_1 + 10v_2 + 25v_3 - 25v_4 - 5v_5$$

14-15. Add appropriate circuitry to Fig. 14-20, making it an eight-bit D/A converter with the MSB controlling the electronic switch to the 10 kΩ resistor. Sketch the complete circuit, showing all parameter values. Set the switches for the input word 11010010, corresponding to 210, and calculate v_o. Also, determine v_o when the input word is 01010100. What is the corresponding decimal number (*Ans*: 84)?

14-16. For the differential amplifier of Fig. 14-22 connect the inverting input through 4 kΩ and $+0.12$ V to ground and the noninverting input through 4 kΩ and $+0.08$ V to ground. Calculate v_o. Then repeat the calculation with the voltage followers eliminated.

14-17. Eliminate the voltage-followers of the differential amplifier of Fig. 14-22. Then connect the input terminal at R_1 through 4 kΩ to a common-mode voltage of 1 V and connect the input at R_2 through 3 kΩ to the same voltage. Calculate v_o. (*Ans*: 0.192)

14-18. In the voltage-reference circuit of Fig. 14-23 reverse the connections to the Zener diode, connect pin 7 of the OP AMP to ground, apply -30 V to pin 4, and adjust the 10 kΩ potentiometer so that R_1 is 6 kΩ. Sketch the circuit and calculate the output reference voltage V_R.

14-19. In the inverting amplifier of Fig. 14-11 with 1 kΩ resistors used for Z_1 and Z_2, replace Z_f with two series-connected back-to-back diodes. Each diode has a breakdown voltage of 5.1 and a forward drop of 0.7 V. With $V_s = 1$ determine the output voltage and also the diode current in mA. Repeat with $V_s = -2$ V.

Chapter Fifteen
Bias Circuit
Design

Transistors made simultaneously in the same batch and assigned the same identification number by the manufacturer often have widely differing characteristics. Control of the excess-carrier lifetime τ is especially difficult, and the important parameter β_F is approximately proportional to τ, as shown in Sec. 2-4. Hence, it is not uncommon for β_F to vary over a range of 1 to 5, from perhaps 40 to 200, among transistors of a particular type. The range of this variation is frequently unspecified by the manufacturer. To complicate matters further, both τ and β_F change with temperature in a rather unpredictable manner, and both I_{CO} and the volt-ampere characteristic of the forward-biased emitter junction are quite temperature sensitive. Still, the Q-point must be maintained within suitable limits for an amplifier using any of the transistors of a particular type operating over a specified temperature range.

We shall investigate carefully the design of the dc circuitry of small-signal single-stage CE amplifiers. The design criteria are then extended to the CC and CB configurations, and bias concepts of direct-coupled stages are introduced. Also examined are *current mirrors*, which are extensively used in integrated circuits for biasing. In the next chapter many of the principles are applied to audio power amplifiers.

15-1. DC AND AC LOAD LINES

The CE amplifier of Sec. 13-1 is shown again in Fig. 15-1 and the equivalent two-battery dc circuit is sketched in Fig. 15-2. The units are kΩ, mA and V,

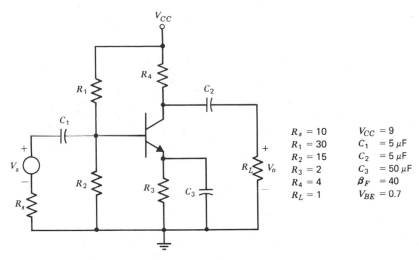

Figure 15-1   CE amplifier (units: kΩ, mA, V).

The component values shown in the figure:

$R_s = 10$	$V_{CC} = 9$
$R_1 = 30$	$C_1 = 5\,\mu F$
$R_2 = 15$	$C_2 = 5\,\mu F$
$R_3 = 2$	$C_3 = 50\,\mu F$
$R_4 = 4$	$\beta_F = 40$
$R_L = 1$	$V_{BE} = 0.7$

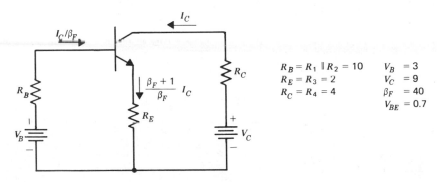

Figure 15-2   Two-battery equivalent circuit (kΩ, mA, V).

The component values shown in the figure:

$R_B = R_1 \parallel R_2 = 10$	$V_B = 3$
$R_E = R_3 = 2$	$V_C = 9$
$R_C = R_4 = 4$	$\beta_F = 40$
	$V_{BE} = 0.7$

and these are employed throughout this section. Because the silicon NPN transistor is biased in the active mode at moderate temperatures, I_{CO} is negligible, and $I_B = I_C/\beta_F$.

THE DC LOAD LINE

Application of Kirchhoff's voltage law to the base-emitter circuit of Fig. 15-2 gives the relation

$$V_B = V_{BE} + I_C\left[\frac{R_B}{\beta_F} + \frac{R_E(\beta_F + 1)}{\beta_F}\right] \qquad (15\text{-}1)$$

Using the specified numerical values, I_C is found from (15-1) to be 1 mA. The collector-emitter voltage can be expressed as

$$V_{CE} = V_C - I_C \left[R_C + \frac{R_E(\beta_F + 1)}{\beta_F} \right] \qquad (15\text{-}2)$$

For the specified values, with the quiescent collector current of 1 mA, V_{CE} is 3 V. Equation 15-2, which relates the quiescent collector current and the quiescent collector-emitter voltage, is referred to as the *dc load-line equation*, and the

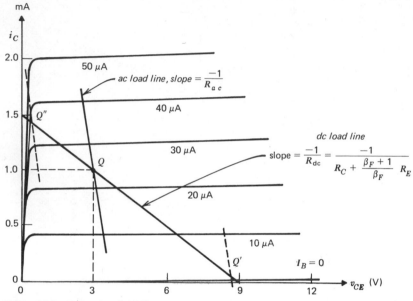

Figure 15-3   DC and ac load lines.

plot of I_C versus V_{CE} is a straight line. This dc load line is sketched on the collector characteristics of Fig. 15-3. For all values of network parameters that maintain operation in the active region, the Q-point must lie on the dc load line.

THE AC LOAD LINE

When a time-varying signal is applied, the instantaneous values of v_{CE} and i_C do not fall on the dc load line. Assuming that the coupling and bypass capacitors are effective ac shorts, the circuit configuration for the incremental components v_{ce} and i_c of v_{CE} and i_C is quite different from the dc circuit. Whereas all the quiescent current I_C must pass through R_4, much of the incremental current is shunted through the load resistor R_L. Also, this current is bypassed around R_3. With R_{ac} denoting the ac resistance of the external circuit between the collector and emitter terminals, the incremental components are related by the equation $v_{ce} = -i_c R_{ac}$. This can be written in the form

$$v_{CE} - V_{CE} = -(i_C - I_C)R_{ac} \qquad (15\text{-}3)$$

For the circuit of Fig. 15-1, R_{ac} is the parallel equivalent of R_4 and R_L, because the battery is an ac short-circuit and R_3 is bypassed. Hence, R_{ac} is 0.8 kΩ, and V_{CE} and I_C have been determined to be 3 V and 1 mA. Equation 15-3 becomes $v_{CE} = 3.8 - 0.8i_C$.

A plot of i_C versus v_{CE} is a straight line passing through the Q-point and having the slope $-1/R_{ac}$ in accordance with (15-3). This *ac load line* is shown on Fig. 15-3. *The instantaneous operating point* (i_C, v_{CE}) *is always on the ac load line.*

MAXIMUM AND MINIMUM COLLECTOR CURRENTS

The Q-point of Fig. 15-3 is on the dc load line at the quiescent collector current I_C of 1 mA. From (15-1) it is clear that I_C depends on V_B, R_B, R_E, β_F, and V_{BE}, and I_C also depends on I_{CO} if this parameter is important. Because there is usually a rather considerable uncertainty as to the value of β_F, there is some uncertainty as to the location of the Q-point. Furthermore, because β_F, V_{BE}, and I_{CO} are functions of temperature, the Q-point shifts its position on the load line whenever the ambient temperature changes. With an improperly designed bias circuit there is the distinct possibility that the Q-point may at times be in the saturation region, or perhaps in the cutoff region, and either of these situations is intolerable. In fact, if the signal is large enough to give a significant excursion along the ac load line, we must ensure that the *instantaneous* operating point along this line is kept out of both saturation and cutoff.

Suppose, for example, that the quiescent condition is in the active region but very near saturation, as indicated by the point labeled Q'' on the load line of Fig. 15-3. Now, if the sinusoidal source voltage v_s is sufficiently large to cause the instantaneous operating point to move into saturation on the positive half cycles of v_s, the waveforms of v_s, i_c, and v_o (or v_{ce}) that would be observed on an oscilloscope are somewhat as illustrated in Fig. 15-4a. Note the phase reversal from v_s to v_o. For the case in which the quiescent condition is near cutoff, as shown by Q' in Fig. 15-3, the waveforms are those of Fig. 15-4b. Such gross distortion is intolerable.

To find suitable mathematical expressions for the maximum and minimum values of I_C that ensure operation in the active region, let us refer to Fig. 15-5. The symbol V_m denotes the maximum deviation of v_{CE} from its quiescent value V_{CE}. For a bypassed emitter resistor, this is simply the amplitude of the output voltage. However, if R_E is unbypassed, V_m is the amplitude of the output voltage multiplied by the ratio $R_{ac}/(R_C \| R_L)$, with R_{ac} approximately equal to the sum of R_E and $(R_C \| R_L)$.

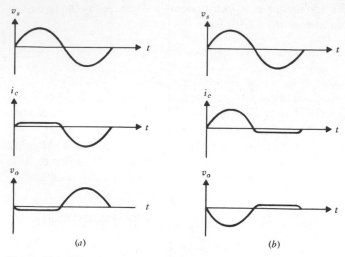

Figure 15-4 Waveforms showing large-signal distortion. (*a*) Q-point near saturation. (*b*) Q-point near cutoff.

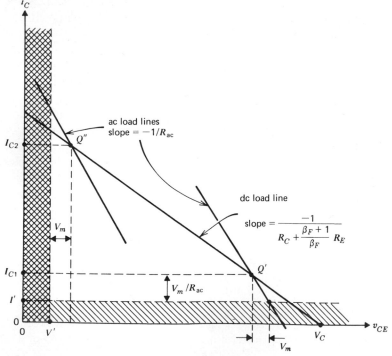

Figure 15-5 Maximum and minimum values of I_C.

The shaded region of Fig. 15-5 to the left of the voltage V' is selected so as to include the saturation region with some margin of safety. The instantaneous voltage v_{CE} must never be less than V'. The current I' is the minimum instantaneous collector current to be allowed. This could be nearly zero, but some safety margin is prudent to ensure that penetration into cutoff does not occur. For a quiescent current of a few milliamperes, reasonable values of V' and I' are 0.4 V and 0.1 mA, but larger values can be selected if desired.

Let I_{C1} and I_{C2} denote the respective minimum and maximum values of I_C, and let β_{F1} and β_{F2} be the respective minimum and maximum values of β_F. Minimum values of β_F, which is usually referred to as h_{FE} on data sheets, are given by the manufacturer, and sometimes maximum values are also specified. Because the excess-carrier lifetime τ increases with temperature, so does β_F. Therefore, β_{F1} should be the minimum possible β_F at the lowest temperature T_1, and β_{F2} should be the maximum possible β_F at the highest temperature T_2. If the maximum β_F is unknown, we shall assume the worst possible case of a very large value and take β_{F2} to be infinite.

Careful examination of the geometry of Fig. 15-5 reveals that within the desired active-mode region the minimum and maximum currents are

$$I_{C1} > I' + \frac{V_m}{R_{ac}} \tag{15-4}$$

$$I_{C2} \le \frac{V_C - V' - V_m}{R_C + R_E} \tag{15-5}$$

In the denominator of (15-5) the term $(\beta_F + 1)/\beta_F$ associated with R_E has been dropped. Because the maximum current I_{C2} exists when β_F is its maximum value β_{F2} and because β_{F2} is always large, treating the ratio as unity is justified. If the minimum quiescent current I_{C1} satisfies inequality (15-4), the instantaneous collector current will always be greater than I'. If the maximum current I_{C2} is always less than the right side of (15-5), the instantaneous voltage v_{CE} will always be greater than V'. *In many applications the designer may restrict* I_{C1} *and* I_{C2} *to a much narrower range than that indicated in Fig. 15-5, by substantially increasing* I' *and* V'. Because the transistor parameters depend on I_C, so do the voltage gain, the current gain, the input and output impedances, and the bandwidth. Noise is also a function of I_C. Having selected an optimum quiescent current, the designer usually chooses I' and V' to maintain this current within more restricted limits than those indicated on Fig. 15-5, but inequalities (15-4) and (15-5) still apply. In most small-signal amplifiers V_m is nearly zero and can be neglected.

15-2. TRANSISTOR SELECTION

Now that we have learned the procedure for finding the Q-point of a single-stage amplifier and have acquired an understanding of the restraints on the location of this point on the load line, we are ready to approach the more challenging problem of design. Assuming specifications have been determined, two major decisions that must be made initially are the choice of a suitable transistor and the selection of an appropriate quiescent point. There are many factors to be considered by the engineer who is to choose the transistor and its bias point, and it is the purpose of this section to investigate briefly the more important of these.

THE DC BETA

Let us first take a closer look at β_F, or h_{FE}, one of the most significant transistor parameters. Usually it is approximately the same as β_o, or h_{fe}. In the active mode β_F is the ratio I_C/I_B, assuming I_{CO} is negligible. It was defined as the ratio $\alpha_F/(1 - \alpha_F)$, which was shown in Sec. 2-4 to be $2\tau_b D_b/W^2$ to a first crude approximation. In accordance with this, we note that β_F increases with temperature T, primarily because the excess-carrier lifetime τ_b in the base increases with temperature, except at very high current levels. Although there is no equation that properly relates β_F to T, the curves of Fig. 15-6, which show β_F versus I_C at three different temperatures, are somewhat typical. Because of the term W^2 in the denominator of the expression for β_F, it is clear that base-width modulation will cause β_F, and also β_o, to increase as the collector-emitter voltage is increased, but this effect is slight. Of considerably greater importance to us, however, is the variation of β_F with I_C, as indicated by the curves of Fig.

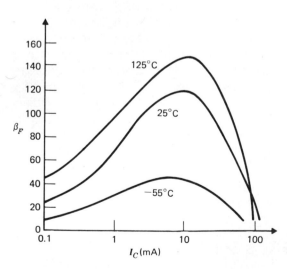

Figure 15-6 Typical variation of β_F with I_C and T. Note the logarithmic scale for I_C.

15-6. This variation is not predicted by the theoretical expression that was determined from the analysis of the idealized transistor of Chapter 2.

The 25°C curve of Fig. 15-6 gives β_F values of 25, 120, and 20 at currents of 0.1 10, and 100 mA, respectively. The corresponding values of α_F are 0.962, 0.992, and 0.952, respectively. The percentage changes in α_F are quite small compared with those of β_F. The relatively low value (0.962) of α_F at 0.1 mA is attributed to the fact that some of the majority carriers that constitute I_E are lost by recombination within the emitter-junction space-charge layer, and hence fail to reach the neutral base region. This is, of course, detrimental to the ratio I_C/I_E, and α_F and β_F. The effect, which was ignored in the theoretical analysis of Chapter 2, is especially pronounced at very low currents. As a consequence, *most BJTs must be biased at currents that are at least a few microamperes.* This restriction does not apply to field-effect transistors.

At the 1 mA point of the curve both α_F and β_F are higher. The recombination current within the emitter-junction space-charge layer is probably an inconsequential percentage of the total at this higher level, but another effect is beginning to appreciably influence alpha. The slight imbalance of the positive and negative excess-carrier densities in the base region is now large enough to give a small electric field that is sufficient to produce a significant minority-carrier drift current. This portion of the curve is a transition region between low-level and high-level injection. The field enables the carriers to move more quickly through the base, thus reducing recombination within this region and increasing α_F and β_F.

As the current is further increased, α_F and β_F continue to rise. At 10 mA α_F is 0.992 and the curve of β_F is near its maximum. At this point the transistor is well into high-level injection, and the majority-carrier density in the base is considerably greater than its equilibrium value. Accordingly, injection from the base into the emitter is substantial, giving an increased I_E but not I_C. Lowered is the *emitter efficiency*, defined as the ratio of the current resulting from injection from the emitter into the base to the total emitter current. At 100 mA the reduced α_F (0.952) is explained by the low emitter efficiency. This was mentioned earlier, in Sec. 2-4. The maximum β_F for BJTs designed for low-to-moderate bias currents, from 10 μA to several milliamperes, commonly occurs between 1 and 5 mA for small IC transistors and between 5 and 20 mA for discrete devices. Data sheets usually give a typical curve of h_{FE} versus I_C, often at several different temperatures.

THE TRANSITION FREQUENCY

A second parameter that may affect the choice of a transistor and its Q-point is f_T. This was defined in Sec. 12-6 as the product of $|h_{fe}|$ and the frequency f of measurement when this frequency is sufficiently high so that $|h_{fe}|$ is proportional to $1/f$ (refer to Fig. 12-18). It is also the product $\beta_o f_\beta$, with f_β denoting the CE beta cutoff frequency, and we know from (12-47) that $2\pi f_T$ is approximately equal to $g_m/(C_\pi + C_\mu)$. Transistors for low-level low-frequency amplifiers have f_T values typically in the range from 20–150 MHz. The range is from about 50 to several hundred megahertz for general-purpose BJTs and from about 300 to 3000 MHz for high-frequency devices.

The transition frequency is a function of I_C, and the relationship is given on data sheets in the form of the representative curve of Fig. 15-7. Because f_T equals $\beta_o f_\beta$, and $\beta_o \approx \beta_F$, the variation of f_T with I_C is in large part due to the same effects that cause β_F to vary with I_C. However, ω_β equals $g_\pi/(C_\pi + C_\mu)$, according to (12-48). As both g_π and the diffusion-capacitance component of C_π are proportional to I_C, the frequency f_β increases as I_C increases, although to a much lesser extent, but this also affects f_T. The maximum value of f_T occurs at a quiescent current that is usually about the same as for the maximum

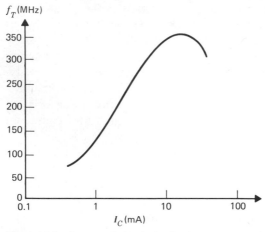

Figure 15-7 f_T versus I_C (typical values).

β_F. It has been mentioned that the value of β_F is not easily controlled in the manufacturing process, because of the troublesome excess-carrier lifetime τ_b. Consequently, there is always an uncertainty in the value of f_T. In addition to the curve of Fig. 15-7, transistor data sheets normally specify a minimum f_T as well as a typical f_T, which is most likely several times greater than the minimum because of beta variations.

NOISE

A third parameter of special interest to the designer is the noise figure. In a transistor each type of noise occurs because each individual charge carrier, which has a discrete charge, moves in an erratic way; quantities of such carriers flowing randomly through various cross-sections of minute volumes of semiconducting material give small instantaneous currents and i^2R power, even when the current is supposedly zero.

Three major sources of noise in a BJT are $1/f$, *thermal*, and *shot* noise. The first of these, $1/f$ or flicker noise, is attributed to the sporadic nature of the thermal generation and recombination processes. Because it is nearly inversely proportional to the frequency f, it is dominant at very low frequencies. *If signal amplification at low frequencies is not essential, the input and output coupling capacitors of an amplifier should be selected so as to block this objectional noise.* Thermal, or Johnson, noise is generated by the random movement of the charge carriers of an ohmic region, and *this noise is uniformly distributed over the radio-frequency spectrum.* Every resistor has thermal noise, including the transistor ohmic regions, especially the base. Shot noise in a BJT occurs in the space-charge layers. Some carriers that enter a layer are repelled back by the electric field and others are able to pass through. *The noise produced by this process is dominant in a BJT except at very low frequencies or very low currents.*

Specifying the noise applied to the input terminals of an amplifier to be the thermal noise of the Thévenin-equivalent source resistance at 25°C (or other stated reference temperature), the *average noise figure* $\bar{F}$ within a designated frequency band is the ratio of the total output noise power to that which would exist if the amplifier were noiseless. In decibels it is 10 times the common logarithm of the power ratio. The *spot noise figure* F at a single frequency is the value of $\bar{F}$ when the specified bandwidth approaches zero at the frequency. Clearly, the noise figure is always greater than unity, or 0 dB, and is a function of both R_s and f. For a BJT it also depends on I_C, and to a much lesser extent on V_{CE}. Data sheets give information in various forms, an example being that shown in Fig. 15-8.

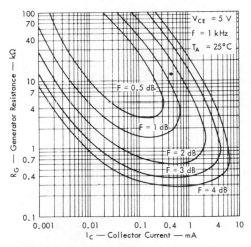

Figure 15-8 Contours of constant F at 1 kHz.

ADDITIONAL CONSIDERATIONS

Other data sheet parameters that must be considered when choosing a transistor and its bias point are the breakdown voltage $V_{(BR)CEO}$ (discussed in Sec. 2-6), the capacitance C_μ (or C_{ob}, C_{cb}), the allowable power dissipation, the allowable collector current, typical values of the CE h-parameters, and many others. Design engineers should refer to a suitable transistor data book with transistors classified under PNP and NPN headings such as general-purpose, low-level, high-frequency, high-voltage, power, switches, choppers, matched duals. They should select the appropriate classification for their intended use and examine carefully the specified characteristics of the various devices within the desired category.

Let us confine our discussion to small-signal amplifiers and hence to BJTs characterized as either general-purpose, low-level, or high-frequency. *General-purpose* low-power transistors have useful betas over a wide current range, perhaps from a few microamperes to tens of milliamperes. The minimum f_T is generally in the range 15 to 300 MHz, the breakdown voltage $V_{(BR)CEO}$ has values from about 15 to 80 volts, and the allowable power dissipation ranges from 300 to 600 mW at a free-air temperature of 25°C. The *low-level* devices are designed to have high betas at quite low quiescent currents, perhaps as low as 1 μA. Also, in general they have minimum f_T values from 15 to 100 MHz, rather low breakdown voltages (10–50 V), an allowable power dissipation of about 300 mW at 25°C free-air temperature, and low noise. All *high-frequency* transistors have high transition frequencies, with minimum values in the range from 250 to 2000 MHz, and very small incremental collector-base capacitances (C_μ), usually less than 1 or 2 pF. Some are designed for unusually low noise at high frequencies and some have a β_F that is fairly constant over a wide current range.

Transistor selection depends on the properties that have been discussed here, on the particular application, and perhaps on the transistors available on the shelf or in the laboratory. In the next section we consider the choice of transistors and bias points for each BJT of a multistage amplifier.

15-3. SELECTION OF THE Q-POINT

Let us suppose we are faced with the problem of designing a three-stage amplifier, such as that shown in Fig. 15-9. The output current I_o is $A_{i1} A_{i2} A_{i3} I_s$, with the current gains denoting those of the respective stages, and V_o is $I_o R_L$. Because the open-circuit source-voltage V_s is $I_s(R_s + R_{in})$, the voltage gain V_o/V_s becomes

$$\frac{V_o}{V_s} = \frac{A_{i1} A_{i2} A_{i3} R_L}{R_s + R_{in}} \tag{15-6}$$

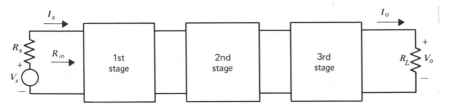

Figure 15-9   Three-stage amplifier between source and load.

The transducer gain G_T, defined by (13-53), is easily determined to be the product of $4R_s/R_L$ and the square of the voltage gain of (15-6). In nearly all cases the second and third stages have very little effect on R_{in}, unless the input stage is a CC amplifier, as indicated by the curves of Fig. 13-30. For large output voltage and transducer gain it is clear from (15-6) that *these stages should be designed for large current gain.* This criterion eliminates the CB configuration, because its current gain is less than unity. For the same load resistance the CE and CC configurations have approximately the same current gain in accordance with the curves of Fig. 13-29. These curves also show that A_i is reduced at the larger values of R_L, and hence the large input resistance of the CC configuration would act to reduce the current gain of the preceding stage. Accordingly, to achieve the highest possible output voltage and power, the second and third stages should be common-emitter amplifiers.

For the input stage both its current gain and its input resistance appear in (15-6). Obviously, if maximum output power is the only consideration, we desire a large A_i and a small R_{in}. It was shown in Sec. 13-7 that transducer gain is maximized by the use of the CE connection. Therefore, this should be our choice for the input stage unless the source is a transducer whose characteristics are such that it must feed into either a very high or a very low impedance in order to avoid excessive distortion. We shall assume that is not the case here. As each of the three basic configurations has about the same signal-to-noise ratio at the output, we select a CE amplifier for the first stage, as well as for the others.

In order to proceed with the design intelligently, we need more detailed specifications. Let us suppose that the source resistance is 500 ohms and that the amplifier must feed into a load resistance of 50 ohms, with a voltage gain V_o/V_s of at least 2000 and a bandwidth of 2 MHz or more. To achieve the specified gain and bandwidth we need transistors with moderately high values of f_T, perhaps 200 MHz or more. For convenience each transistor will be the same type, and we shall assume that the β_F, f_T, and noise-figure F curves of Figs. 15-6, 15-7, and 15-8 apply. The midfrequency incremental circuit configuration can be drawn in the form shown in Fig. 15-10.

From Figs. 15-6 and 15-7 we find that the maximum β_F at 25°C occurs at

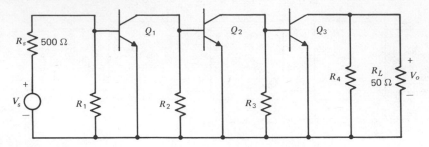

Figure 15-10 Midfrequency incremental circuit of three-stage CE amplifier.

10 mA and the maximum f_T is at 20 mA. For the input stage a low input resistance is desired, according to (15-6). This resistance is $r_{x1} + r_{\pi 1}$, neglecting the effect of R_1 of Fig. 15-10. We learned in Sec. 12-3 that r_x decreases as I_C increases, and of course, r_π is inversely proportional to I_C. Accordingly, to obtain a low input resistance and large β_F and f_T we might choose $I_{C1} = 20$ mA, which gives $\beta_{F1} = 110$, $f_{T1} = 350$ MHz, $g_{m1} = 0.8$ mho, and $r_{\pi 1} = 110/0.8$, or 140 ohms. Such a choice would most likely be completely unacceptable from the viewpoint of noise. We note from Fig. 15-8 that minimum noise typically occurs at about 0.6 mA for a source resistance of 500 ohms. Minimum noise normally occurs at quiescent currents that are considerably lower than those for maximum β_F and f_T. Because first-stage noise is amplified in the second and third stages, we must minimize this noise. Choosing 1 mA for I_{C1} gives a noise figure near the minimum. At this current β_F is 70, f_T is only 125 MHz, and $r_{\pi 1}$ is calculated to be 1.75 kΩ. The larger resistance gives reduced loading of the source.

Noise in the third stage is not so important, and I_{C3} should be chosen for large current gain and bandwidth. At 20 mA the transition frequency is a maximum at 350 MHz. However, at 10 mA f_T is fairly close to its maximum, being 330 MHz, the power consumption in the transistor and its bias resistors is substantially less, the noise figure is tolerable for this output stage, and β_F is a maximum at 120. Consequently, let us select 10 mA, assuming this current is greater than V_m/R_{ac} as required by inequality (15-4). For the 50-ohm load this choice of I_{C3} restricts V_m to a value less than 0.5 V corresponding to an output power less than 2.5 mW.

Bias considerations favor selecting an intermediate current for the second stage. According to (15-6) we want to maximize the current gain of each stage. Therefore, the base-to-emitter input resistance of a stage, which is almost inversely proportional to I_C, should be small compared with the collector resistance R_C of the preceding stage. Because the three stages operate off the same voltage supply, larger values of R_C can be used with lower quiescent currents. If Q_2 in Fig. 15-10 were biased at 10 mA, it might be impossible to select

R_3 large enough to prevent excessive shunting of signal current, and if the bias current is 1 mA, the shunting effect of R_2 at the input might become intolerable. *The effect of the bias circuitry on the gain of a CE cascade is minimized by selecting quiescent currents that increase gradually from input to output*, with collector resistances R_C and input resistances that decrease gradually (see P15-6). Another consideration is noise. Although noise introduced in the second stage is not nearly as detrimental as that of the first, it is still of significance, and the quiescent current should not be unnecessarily high. Let us choose an I_{C2} of 5 mA, at which current β_{F2} is 110 and f_T is 275 MHz.

We have selected quiescent currents of 1, 5, and 10 mA for the first, second, and third stages, respectively. If low power consumption is particularly important, these currents might be reduced, but the result would be a lower gain-bandwidth product. The dc betas at the selected currents are 70, 110, and 120. If the bias resistors R_1, R_2, R_3, and R_4 of Fig. 15-10 shunt negligible currents, then the overall current gain $A_{i1} A_{i2} A_{i3}$ is approximately the product of these betas, or 924 000. Assuming R_{in} is $r_{\pi1}$, (15-6) gives a voltage gain of 20 500, substantially greater than the specified 2000.

An analysis of the high-frequency circuit will not be undertaken here, but the result would reveal a bandwidth less than the required 2 MHz. There are two convenient ways to increase the bandwidth. One of these is simply *to reduce the values of the resistors* R_1, R_2, R_3, *and* R_4 *in Fig. 15-10, and the second way is to employ negative feedback*. Both reduce the voltage gain, a necessary penalty for increasing the bandwidth. We shall investigate these methods in Secs. 19-1 and 21-4.

A few comments on the selection of V_{CE} are in order. In certain cases V_{CE} may be as low as 0.5 V. At low and moderate currents the collector characteristic curves enter the normal region at values of V_{CE} from about 0.2 to 0.3 V, and a quiescent collector-emitter voltage of 0.5 V gives operation in the active mode provided V_m is less than about 0.1 V. *A disadvantage of this low bias voltage is the fact that* C_μ *is larger for the smaller values of collector voltage*, as shown by (2-48). At the other extreme V_{CE} may be nearly as large as $V_{(BR)CEO}$, and C_μ is quite small at these voltages. Usually, V_{CE} is less than one-half of $V_{(BR)CEO}$. The value selected depends on the available voltage supply, the circuit configuration, the importance of a low C_μ, power consumption considerations, and the required maximum value V_m of the output voltage. Some of these factors are examined in the remainder of this chapter.

The example that has been analyzed was undertaken to present some of the considerations involved in selecting the quiescent current. Each individual electronic circuit design has its own special requirements, many of which may influence the choice of I_C. There is no one current that is ideal. The choice is always a compromise between conflicting requirements.

Many sources have considerable internal resistance along with characteristics that change when appreciable current is supplied to the input of an amplifier.

Furthermore, in various applications the output voltage of an amplifier should be independent of the internal impedance of the source. Thus *a typical design objective of many amplifiers is the attainment of a high input impedance*. This is an important reason why input stages are frequently biased at low current levels and use high-beta transistors. Resistances of the incremental model of a transistor are quite large when bias currents are low. With a quiescent collector current of 10 μA, the parameter r_π is typically 100 kΩ or more. Emitter followers and FET inputs are often used.

At the output, loads are usually capacitive, and such a load can be represented by a parallel arrangement of a resistor and a capacitor. At high frequencies the load impedance becomes small, causing the output voltage to drop. *The effect is minimized by using an amplifier with a low output impedance, and this is another typical design objective.* Emitter-follower outputs biased at high current levels are common.

Power availability is an obvious factor that may influence the selection of the Q-points. In particular, battery-operated equipment should be designed for low power consumption.

15-4. BIAS PRINCIPLES

Figure 15-11 shows the two-battery equivalent circuit of an amplifier whose bias circuit consists of resistors and independent sources. It applies to a single stage of a cascade provided RC coupling is employed at both input and output, making the dc circuit independent of the others of the cascade.

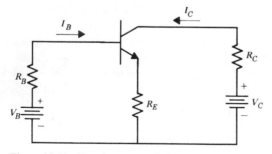

Figure 15-11   Two-battery equivalent circuit.

THE MAXIMUM COLLECTOR CURRENT

The active-mode base current I_B of an NPN transistor is found from (2-32) to be

$$I_B = \frac{I_C}{\beta_F} - \frac{I_{CO}(\beta_F + 1)}{\beta_F} \tag{15-7}$$

The open-circuit saturation current I_{CO} is positive. Let I_{C1} denote the minimum value of I_C, corresponding to the minimum β_{F1} at the lowest junction temperature T_1 that will be encountered, and let I_{CO1} and V_{BE1} be the values of these quantities at T_1. The maximum value of I_C is I_{C2}, corresponding to the maximum β_{F2} at the highest junction temperature T_2, and I_{CO2} and V_{BE2} are the values of these quantities at T_2. The voltage sources and resistors of Fig. 15-11 are assumed to be constants independent of temperature.

It is convenient and reasonable to assume that

$$I_{CO1} \approx 0 \qquad \frac{\beta_{F2} + 1}{\beta_{F2}} \approx 1 \qquad (15\text{-}8)$$

At the lowest temperature I_{CO} is nearly always negligible, and the maximum β_F at T_2 is much greater than unity. At temperatures T_1 and T_2 we find from (15-7) that

$$I_{B1} = \frac{I_{C1}}{\beta_{F1}} \qquad I_{B2} = \frac{I_{C2}}{\beta_{F2}} - I_{CO2} \qquad (15\text{-}9)$$

Application of Kirchhoff's voltage law to the base-emitter circuit of Fig. 15-11 at these two temperatures, utilizing Eqs. (15-9), gives the two relations:

$$V_B = \frac{R_B I_{C1}}{\beta_{F1}} + \frac{R_E I_{C1}(\beta_{F1} + 1)}{\beta_{F1}} + V_{BE1} \qquad (15\text{-}10)$$

$$V_B = R_B\left(\frac{I_{C2}}{\beta_{F2}} - I_{CO2}\right) + R_E(I_{C2} - I_{CO2}) + V_{BE2} \qquad (15\text{-}11)$$

The voltage V_B is eliminated by subtracting (15-11) from (15-10). Solving the resulting equation for I_{C2}, we obtain

$$I_{C2} = \frac{[R_E(\beta_{F1} + 1)/\beta_{F1} + R_B/\beta_{F1}]I_{C1} + \Delta V_{BE} + (R_E + R_B)I_{CO2}}{R_E + R_B/\beta_{F2}} \qquad (15\text{-}12)$$

with $\Delta V_{BE} = V_{BE1} - V_{BE2}$, a positive quantity. The minimum quiescent current I_{C1} corresponds to the minimum β_{F1} at the lowest junction temperature T_1. For selected resistors R_E and R_B, (15-12) gives the maximum quiescent current I_{C2} corresponding to the maximum β_{F2} at T_2. Of course the equation applies only if the transistor is in the active mode.

THE COLLECTOR CURRENT WITH $R_E = 0$

Examination of (15-12) with $R_E = 0$ clearly reveals why an emitter resistance is usually desirable. The equation becomes

$$I_{C2} = \left(\frac{\beta_{F2}}{\beta_{F1}}\right)I_{C1} + \frac{\beta_{F2}\Delta V_{BE}}{R_B} + \beta_{F2}I_{CO2} \qquad (15\text{-}13)$$

The first term on the right represents the increase in the current caused by the increase in beta from β_{F1} to β_{F2}, and this increase can easily be sufficient to give an I_{C2} in the saturation region. The second term is the increase resulting from the change in V_{BE}. As the temperature rises, V_{BE} decreases, which gives an increased voltage $V_B - V_{BE}$ across R_B, causing I_B and I_C to rise. This effect is minimized by using a large R_B and a large voltage V_B, which tend to maintain I_B constant. The third term is the increase in I_C resulting from the increase in I_{co} from I_{co1} (assumed negligible) to I_{co2}, and the change is multiplied by the large number β_{F2}.

When a bias circuit is employed without an emitter resistor, the base resistance should be very large so as to minimize the second term on the right of (15-13). This large R_B stabilizes the base current. The designer must ascertain that β_{F2} is sufficiently restricted to prevent I_{C2} from reaching the maximum allowable value given by (15-5). Shown in Fig. 15-12 is a circuit with zero emitter resistance that utilizes the largest possible R_B for a specified supply V_{CC}. In the two-battery equivalent circuit both V_B and V_C equal V_{CC}.

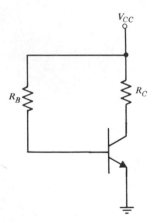

Figure 15-12   A simple bias circuit.

The bias circuit of Fig. 15-12 is avoided in commercial equipment, because it tends to stabilize the base current, allowing the collector current to change with temperature and β_F, which usually has an uncertain value. However, it is sometimes quite convenient for laboratory use when the transistor beta is measured. It is a simple circuit, easy to design. For example, suppose a collector current of 1 mA is desired with a supply of 12 V, and β_F is determined by measurement to be 100. Then I_B is 10 μA, the voltage across R_B is 11.3 V, assuming V_{BE} is 0.7, and a resistance of 1.13 MΩ is chosen for R_B.

EFFECTS OF A LARGE R_E

Having considered the case of zero emitter resistance and having found this very undesirable for most applications, let us now examine (15-12) with R_E large. For R_E much greater than R_B/β_{F1}, the equation becomes

$$I_{C2} = \left(\frac{\beta_{F1} + 1}{\beta_{F1}}\right)I_{C1} + \frac{\Delta V_{BE}}{R_E} + \left(1 + \frac{R_B}{R_E}\right)I_{CO2} \qquad (15\text{-}14)$$

Comparison with (15-13) shows the improvement. The first term on the right is approximately I_{C1}, rather than β_{F2}/β_{F1} times I_{C1}. The second term is also reduced, because R_E is much greater than R_B/β_{F2}. In fact, this term is negligible if we make $\Delta V_{BE}/R_E$ very small compared with I_{C1}, or $I_{C1}R_E$ much greater than ΔV_{BE}. In the third term I_{CO2} is multiplied by a number substantially less than the β_{F2} of (15-13).

The results show that *the quiescent collector current is well-stabilized if* R_E *is much greater than* R_B/β_{F1} *and if the voltage across* R_E *is much greater than the expected variation* ΔV_{BE}. The value of V_{BE} changes about 2 mV per degree Celsius change in temperature. From the viewpoint of good bias stability we want to make R_B equal to zero. This is simply not practical. With RC coupling at the input of a CE amplifier, R_B shunts the base-emitter terminals of the transistor, reducing the power transfer from the source to the device. This detrimental effect is not serious as long as R_B is several times greater than r_π, but it must also be substantially less than $\beta_{F1}R_E$. The choice is a compromise.

SIMPLIFIED DESIGN EQUATIONS

The arbitrary but simplified design equations presented earlier as (13-5) are repeated here for closer examination:

$$R_E = \frac{1}{I_{C1}} \qquad R_B \le 0.3\beta_{F1}R_E \qquad (15\text{-}15)$$

The first of these ensures that the *minimum* voltage across R_E is 1 V. In accordance with (15-12) this is sufficient to keep I_C from increasing more than 10% due to a drop in V_{BE} resulting from a 40°C rise in temperature. The second relation makes R_B fairly small compared with $\beta_F R_E$ for all values of β_F. If R_B is chosen to be the maximum value allowed by (15-15), with $q/kT = 40$ the combination of the equations gives

$$R_B = \frac{0.3\beta_{F1}}{I_{C1}} \approx \frac{0.3\beta_o}{I_{C1}} \approx 12r_{\pi 1} \qquad (15\text{-}16)$$

with $r_{\pi 1}$ denoting the minimum value of r_π. Because r_π is proportional to β for a reasonably constant g_m, this becomes $3r_{\pi 2}$ for a β_{F2} that is four times β_{F1}, still a satisfactory value for most purposes. A design based on this procedure should always be checked by calculating I_{C2}, using (15-12). The result must, of course, be less than the maximum allowable I_{C2} as given by (15-5).

DETERMINATION OF β_{F2}, ΔV_{BE}, AND I_{CO}

The manufacturer frequently does not specify β_{F2}, in which case it should be taken as infinity in (15-12), this being the worst case. Because V_{BE} decreases about 2 to 2.5 mV per degree Celsius, for conservative design we find ΔV_{BE} by multiplying 2.5 mV by the maximum temperature variation, unless a curve of V_{BE} versus T is given on the data sheet.

Transistor data sheets normally specify I_{CBO}, which is the saturation current of the reverse-biased collector junction with the emitter open. This is not the same current as I_{CO}. In fact, the current I_{CO} used in the Ebers-Moll equations and models is typically a fraction of a picoampere for a device rated in milliamperes, whereas I_{CBO} is several orders of magnitude greater. The difference occurs because I_{CBO} includes a thermal generation component resulting from carrier generation within the collector junction space-charge layer, as well as a surface leakage component, as discussed in Secs. 1-9 and 2-4. Because the collector junction is reverse biased, the symbol I_{CO} in (15-7) should include the thermal-generation and leakage components. Accordingly, *when the transistor is biased in the active mode, the symbol* $\mathrm{I_{CO}}$ *should be interpreted as* $\mathrm{I_{CBO}}$. Although I_{CO} is proportional to $n_i{}^2$, the current I_{CBO} does not vary nearly so rapidly with temperature. Data-sheet information on the variation of I_{CBO} with temperature should be used. If such information is not given, it is usually reasonable to assume that I_{CBO} doubles for each 10°C rise in temperature, for both silicon and germanium small-signal transistors. For most silicon devices I_{CBO} is negligible below about 125°C provided the quiescent current is above the microampere range and operation is low level.

The simple bias design based on (15-15) is sometimes unsatisfactory, particularly for designs based on military specifications. Values selected for R_E and R_B may give an I_{C2} in the saturation region. *The problem is to choose the three bias resistors* $\mathrm{R_E}$, $\mathrm{R_B}$, *and* $\mathrm{R_C}$ *so that* $\mathrm{I_{C2}}$ *is maintained below the maximum allowable, without reducing the gain of the ac circuit an intolerable amount.* A trial-and-error procedure, using (15-12) and (15-5), can be employed but this is tedious. *A major problem results from the fact that increasing* $\mathrm{R_E}$ *to improve stability simultaneously decreases the maximum allowable* $\mathrm{I_{C2}}$. In the section that follows, a logical procedure for selecting the bias resistors is presented.

Example

Design the bias circuit of a CE amplifier with the configuration of Fig. 15-1, with an 8 V supply and a load resistance R_L of 500 ohms. The transistor is silicon with a minimum β_F of 60 at 25°C. The output voltage is no more than a few millivolts. The quiescent current is chosen to be 2 mA minimum, V_{CE} should not be less than 0.3 V, and the temperature range is from 25°C to 125°C.

Solution

Here, I_{C1} is 2 mA. Using (15-15), we select $R_E = 0.5$ kΩ and $R_B = 9$ kΩ. This R_B is $12r_\pi$ at I_{C1}. The collector resistance R_C should be several times R_L. A reasonable value is 2 kΩ, which is $4R_L$, and this gives a V_{CE} of 3 V at I_{C1}. In (15-12), ΔV_{BE} is 0.25 V, I_{CO2} is negligible, and β_{F2} is infinite, giving $I_{C2} = 3.13$ mA. The maximum allowable I_{C2} is 7.7/(2.5), or 3.08 mA, by (15-5). Although marginal, this is probably satisfactory in view of the fact that β_{F2} is actually finite. However, there is no appreciable safety margin for voltage-supply fluctuations or resistance deviations, and it might be wise to reduce either R_B or R_C or both.

Using the calculated values, we find V_B from (15-10) to be 2.02, assuming V_{BE1} is 0.7 V. Equations 13-6 yield the values of 35.6 and 12.0 kΩ for R_1 and R_2, respectively.

In many cases the possible increase in I_C of 50% is unacceptable. The variation in I_C can be reduced by increasing R_E to perhaps 1 kΩ and reducing R_C the same amount to 1.5 kΩ, with R_B remaining at 9 kΩ. Proper values of V_B, R_1, and R_2 are calculated as before. The changes do not affect the dc load line and provide a more stable amplifier, but the smaller R_C lowers the voltage gain.

15-5. BIAS CIRCUIT DESIGN

For the present let us restrict our attention to the design of the bias circuit of a one-stage CE amplifier. Solving (15-12) for R_B gives

$$R_B = \beta_{F1}\left[\frac{(I_{C2} - bI_{C1} - I_{CO2})R_E - \Delta V_{BE}}{I_{C1} + \beta_{F1}I_{CO2} - aI_{C2}}\right] \qquad (15\text{-}17)$$

with

$$a = \frac{\beta_{F1}}{\beta_{F2}} \quad \text{and} \quad b = \frac{\beta_{F1} + 1}{\beta_{F1}} \qquad (15\text{-}18)$$

For $R_E = 0$ the numerator of (15-17) is negative, and the required value of R_B is positive only if the denominator is negative. However, the denominator is usually positive because a is small, and the stabilizing effect of an emitter resistance is essential.

Let us choose I_{C2} in (15-17) equal to the maximum that is allowed by (15-5), or

$$I_{C2} = \frac{V}{R_C + R_E} \quad \text{with} \quad V = V_C - V' - V_m \qquad (15\text{-}19)$$

With no emitter resistance, $I_{C2} = V/R_C$, and the denominator of (15-17) is positive provided

$$(I_{C1} + \beta_{F1}I_{CO2})R_C > aV \qquad (15\text{-}20)$$

In the event that the selected and specified values of the parameters are such that (15-20) is not satisfied, either β_{F2} or V' should be arbitrarily increased, reducing a or V and further restricting the maximum collector current. Often β_{F2} is unknown, in which case we choose a to be zero corresponding to the worst case of an infinite β_{F2}. *The design equations of this section are based on the assumption that inequality (15-20) applies.*

INTERRELATIONS AMONG BIAS RESISTANCES

The resistors R_B, R_E, and R_C are related by (15-17) and (15-19). Shown in Fig. 15-13 are plots of R_B versus R_E for three selected values of R_C, with parameters as specified. Each of the curves begins on the left at a negative R_B, rises to a maximum, and then falls. As R_E is increased from zero, the numerator of (15-17) increases and becomes positive. The allowed value of R_B rises, but not indefinitely because larger values of R_E reduce the maximum allowable current I_{C2} according to (15-19). This is also indicated in Fig. 15-5 by noting that the magnitude of the slope of the dc load line decreases as R_E is increased. To the right of the maximum point of a curve the term $(I_{C2} - bI_{C1} - I_{CO2})$ in (15-17) is decreasing faster than R_E is increasing.

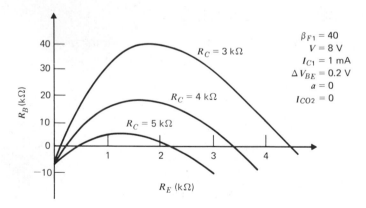

Figure 15-13   Plots of R_E versus R_B.

Each point of a curve fixes the resistances and establishes the corresponding I_{C2} at the maximum allowed value, but the difference between I_{C2} and I_{C1} is smaller for the larger values of R_E. For a selected R_E *a point on the curve gives the maximum* R_B *that can be used*. Reducing R_B improves stability, of course, making I_{C2} less than the maximum. The curves clearly show the trade-off between R_B and R_C. As R_C *is increased, the maximum* R_B *is decreased, and vice versa*. For high gain both R_B and R_C should be large. If suitably large values of both are unobtainable, the voltage V must be increased if possible, or I_{C1} decreased. A larger voltage allows a larger value of I_{C2}, moving the curves of Fig. 15-13 upward.

THE MAXIMUM R_B

The maximum points of the curves are of particular interest. For any specified R_C the value of R_E permitting the maximum R_B is found by setting dR_B/dR_E equal to zero and solving the result for R_E. Using (15-17) and (15-19), the procedure for the special but important case in which both a and I_{CO2} are zero is simple (see P15-10), and the expression for R_E at the maximum is

$$R_E = \sqrt{\frac{VR_C}{bI_{C1}}} - R_C \tag{15-21}$$

A direct relationship between R_B and R_C can now be found. We use (15-19) to eliminate I_{C2} in (15-17) and then substitute for R_E, obtaining

$$R_B = (\beta_{F1} + 1)\left(R_C - 2\sqrt{\frac{VR_C}{bI_{C1}}} + \frac{V - \Delta V_{BE}}{bI_{C1}} \right) \tag{15-22}$$

For any selected value of R_C, (15-22) gives the maximum permissible R_B, provided R_E is selected in accordance with (15-21). The equations apply only if β_{F2} is assumed to be infinite and I_{CO2} is zero.

Example 1

An RC coupled CE amplifier with a bypassed R_E feeds into a second stage known to have an input resistance between 500 and 1500 ohms, depending on the β_F of the second transistor. The source is 4 V, V' is chosen to be 0.5 V, and V_m is 2 mV. The minimum beta of the BJT is 40, and the temperature range is from 25°C to 125°C. As the transistor is silicon, I_{CO} is negligible. Determine suitable values for R_B, R_C, and R_E, for $I_{C1} = 0.5$ mA.

Solution

We find that $V = 3.5$, $a = 0$, $b = 1.025$, and ΔV_{BE} is 0.25 V assuming V_{BE} decreases 2.5 mV/°C. From (15-22) with resistances in kΩ, we get

$$R_B = 41(R_C - 5.23\sqrt{R_C} + 6.34) \tag{15-23}$$

In order to select R_C and R_B we need approximate values of the input resistance and the load resistance. With $I_{C1} = 0.5$ mA, the corresponding g_{m1} is 20 mmhos, and $r_{\pi1}$ is 2 kΩ. The maximum β_F is unknown, but we might reasonably expect the input resistance to be between 2 and 8 kΩ, and this is shunted by R_B. The resistance R_C shunts the load, which is between 0.5 and 1.5 kΩ. A few trial substitutions in (15-23) yields the reasonable values of 2.5 kΩ for R_C and 23.4 kΩ for R_B. This R_C is five times the minimum R_L and 1.7 times the maximum. The value of R_B is about 12 times the minimum r_π and three times the expected maximum. Equation 15-21 gives $R_E = 1.63$ kΩ. The maximum I_C is determined from (15-12) or (15-19) to be 0.85 mA.

The simplified design equations (15-15) give $R_E = 2$ and $R_B = 24$ kΩ. From (15-12) we find that I_{C2} is 0.79 mA, and (15-19) shows that R_C must be less than 2.44 kΩ for these values of R_E and R_B.

Perhaps it should be noted that any acceptable design can, if desired, be modified by decreasing R_C and increasing R_E exactly the same amount. This decreases I_{C2} as given by (15-12), without affecting the maximum allowable I_{C2} as specified by (15-19). The procedure increases the stability of the bias circuit by reducing the possible variation of I_C. The disadvantage, of course, is the increased shunting effect of R_C. For any current I_C, neither V_{CE} nor the total power dissipation is affected by such a change, but the voltage across the emitter bypass capacitor C_3 is increased.

The effects of a finite β_{F2} and a nonzero I_{CO2} can be included, but the results are a little more complicated. The equations that correspond to (15-21) and (15-22) are

$$R_E = \sqrt{V\left(R_C - \frac{aV}{f}\right)\left(\frac{1}{d} - \frac{a}{f}\right) + \frac{aV\Delta V_{BE}}{df}} - R_C + \frac{aV}{f} \qquad (15\text{-}24)$$

$$R_B = \frac{\beta_{F1}d}{f}\left[R_C - \frac{2aV}{f} - 2\sqrt{V\left(R_C - \frac{aV}{f}\right)\left(\frac{1}{d} - \frac{a}{f}\right) + \frac{aV\Delta V_{BE}}{df}}\right]$$

$$+ \frac{\beta_{F1}}{f}(V - \Delta V_{BE}) \qquad (15\text{-}25)$$

with $d = bI_{C1} + I_{CO2}$, $f = I_{C1} + \beta_{F1}I_{CO2}$, and a, b, and V defined by (15-18) and (15-19).

Example 2
A CE amplifier has a supply voltage of 8 V. The minimum V_{CE} is to be 1 V, and the maximum ac output voltage has $V_m = 1$ V. The value of β_{F1} is 50, β_{F2} is 250, and ΔV_{BE} is 0.3 V. The minimum quiescent current I_{C1} is selected as 1 mA. Find the maximum R_B that can be used corresponding to a selected R_C.

Solution
In this example, $V = 6$, $a = 0.2$, $b = 1.02$. From (15-25) with the resistances in kilohms, we find

$$R_B = 51R_C - 221\sqrt{R_C - 1.125} + 163$$

Some corresponding values of R_C and R_B are

R_C (kΩ)	1.125	2	3	3.5
R_B (kΩ)	220	58.3	13.4	0.92

If we select 3 kΩ for R_C, the maximum allowed R_B is 13.4 kΩ, provided R_E is 1.16 kΩ as determined from (15-24). Using these values, I_{C2} is calculated from either (15-12) or (15-19) to be 1.44 mA.

R_1 AND R_2 CALCULATIONS

Once the resistances R_B, R_E, and R_C have been selected, it is a simple matter to find R_1 and R_2 in the circuit of Fig. 15-1. First V_B is calculated from (15-10), which is the loop equation around the base-emitter circuit of Fig. 15-2. Then R_1 and R_2 are found from V_B and R_B. Refer to (13-6).

Although the equations of this chapter have been presented for an NPN transistor, they are equally applicable for PNP devices by using the magnitude of any negative quantity encountered, such as I_C, I_B, V_{CE}, and V_{BE}. Of course the batteries V_B and V_C in the two-battery equivalent circuit should be reversed.

JUNCTION TEMPERATURE

Most of the power dissipated in a transistor occurs at the reverse-biased collector junction, and the temperatures T_1 and T_2 referred to earlier are the minimum and maximum junction temperatures. The relation between the junction temperature T and the power P dissipated is

$$T - T_A = R_\theta P \qquad (15\text{-}26)$$

with T_A denoting the ambient temperature and R_θ the *thermal resistance*, in degrees Celsius per watt. To find T_1 we use the minimum ambient temperature and the lowest power dissipation, and T_2 is obtained from the highest values of T_A and P. The thermal resistance (or its reciprocal, or equivalent information) is given on the data sheet. Typical values are from 1 to 500°C/W, with the lower values applying to power transistors. In most low-level circuits the term $R_\theta P$ in (15-26) is negligible, and T equals T_A.

Example 3
The specifications state that an amplifier shall operate over an ambient temperature range from 0°C to 100°C. The transistor is biased at 2 mA, 5 V. The data sheet specifies a maximum allowable dissipation (at or below 25°C free-air temperature) of 0.4 W, with the transistor to be derated linearly to zero power at an ambient temperature of 175°C. Determine T_1 and T_2

Solution
The thermal resistance R_θ is $(175 - 25)/0.4$, or 375°C/W. The reciprocal of this is 2.67 mW/°C, which is the derating factor. At the bias point the dissipation is 10 mW, and the junction is 3.75°C above the ambient temperature. Hence T_1 is about 4°C and T_2 is 104°C. Most low-power transistors are rated at 0.3 W or more, and the junction temperature is only a few degrees above the ambient for typical Q points.

Example 4

Using the conventional circuitry of Fig. 15-1, design a CE amplifier with the following specifications: $V_{CC} = 6$ V, $R_L = 100$ Ω, $V_m \leq 0.6$ V, ambient temperature range 25°C to 75°C. At 25°C the *germanium PNP* transistor to be used has $V_{EB} = 0.2$ V, $I_{CBO} < 7$ μA, a derating factor $1/R_\theta = 5$ mW/°C, a maximum allowable current of 25 mA, and $\beta_F > 80$ with the maximum beta unspecified. Select $V' = 0.4$ V and $I' = 0.5$ mA.

Solution

In (15-4) R_{ac} is the parallel equivalent of R_C and R_L. Assuming the selected value of R_C will be at least twice R_L, R_{ac} is 67 ohms or more, and (15-4) gives a minimum allowed I_C of 9.5 mA. Let us select this value for I_{C1}. The product $V_C I_{C1}$ is 57 mW. Estimating that about half of this, or 29 mW, will be dissipated at the collector junction, we find from the derating factor that the junction temperature is about 6° above the ambient. We shall use 30°C for T_1 and 80°C for T_2.

The value of V is 5 and ΔV_{EB} is 0.125 over the 50°C range from T_1 to T_2. As I_{CBO} can be as large as 7 μA at 25°C, approximately doubling for each 10°C rise in temperature, I_{CO2} is $(2)^{5.5}$ times 0.007 mA, or 0.32 mA at 80°C. The constant b is 81/80, d is 9.94 mA, and f is 35.1 mA. Substituting into (15-25), with units of kΩ, mA, and volts, we obtain

$$R_B = 22.65 R_C - 32.13\sqrt{R_C} + 11.11$$

A value of R_C that appears suitable is 0.20 kΩ, which gives an R_B of 1.271 kΩ. From (15-24) R_E is 0.1172 kΩ, and I_{C2} is found from either (15-12) or (15-19) to be 15.8 mA. At T_1 with $I_C = 9.5$ mA, g_m is 364 mmhos. At the minimum beta of 80, r_π is 0.22 kΩ. The value selected for R_B is nearly six times r_π. At the higher values of β_F that are possible, r_π is larger. We can reasonably expect R_B to be twice as great as the input resistance or more, and R_C is twice R_L. We would like to make both R_B and R_C larger, but this cannot be done without modifying the specifications.

For the selected values of R_C and R_E, V_{CE} is 3 V when I_C is 9.5 mA, and the junction dissipation is 28.5 mW. Hence our value assumed for T_1 is satisfactory. At the higher currents V_{CE} is reduced sufficiently to give a reduction in the junction dissipation, and T_2 is a few degrees lower than initially assumed.

To complete the design we need to find R_1 and R_2. At T_1 the voltage V_{EB} is 0.1875, and from (15-10) we find V_B to be 1.47 V. As R_B is 1.27 kΩ, the resistances R_1 and R_2 are determined from (13-6) to be 5.18 and 1.68 kΩ, respectively.

The bias stability is improved if R_E is increased. Suppose R_E is increased by 0.05 kΩ to 0.1672 kΩ, with R_C decreased from 0.20 to 0.15 kΩ. The maximum collector current is found from (15-12) to be 14.0 mA, which is an improvement. However the gain suffers.

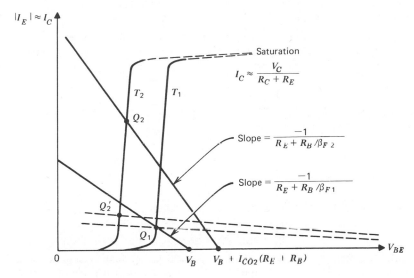

Figure 15-14   Input characteristics and load lines.

We have I_C and V_{BE} related at T_1 by (15-10) and at T_2 by (15-11). Typical plots of these load lines, superimposed on plots of the diode characteristics of the emitter-base junction, are shown in Fig. 15-14. The dashed load lines apply for a much larger R_E (and V_B), showing the improved bias stability. Clearly, a large R_E minimizes the effects of changes in β_F and I_{CO}.

The basic principles of bias-point stabilization apply to all amplifier configurations. The requirements on the dc circuitry are the same regardless of which terminals are used for signal input and output, and (15-12) applies to all cases. In the CC configuration there is no need for the resistance R_C and it is eliminated. No special problem is presented by R_E, which is in parallel with the load. Because R_B must not be unduly large for good bias stability, a bootstrap circuit is often used. Such a circuit was discussed in Sec. 13-3. In the CB amplifier R_B shunts neither the input nor the output and consequently is made as small as possible. Resistance R_E is in parallel with the very small input resistance. *For all three configurations the bias resistors must be selected so that* I_{C2} *as given by (15-12) is in the active region and less than the maximum allowable value as specified by (15-5).* Preferably I_{C2} should not be much greater than I_{C1}. In circuits with small values of external resistance in the base-emitter loop the ohmic base resistance r_x must be considered, because it is indeed a portion of the total resistance R_B.

There is no one design that is best for all bias circuits. The variables are numerous, the specifications vary from one situation to another, and trade-offs and compromises are necessary. Gain, bandwidth, noise, power consumption, and other factors are involved, as well as the variation of these with temperature, and the particular application determines which of these must be given special importance. Good judgment is required, based on a thorough understanding of the transistor characteristics and the fundamental concepts of bias-circuit design.

15-6. BIAS CIRCUIT PRINCIPLES FOR DIRECT-COUPLED AMPLIFIERS

In RC coupled amplifiers, operating-point selection and dc circuit design is accomplished stage by stage. The bias circuit of one stage is independent of the others and is usually designed to affect the signal only slightly. However, when stages are direct coupled, the dc circuits of the various stages interact, and the signal and bias circuits are closely interrelated. This leads to a considerable sacrifice in design freedom, but there are compensating advantages. *Direct coupling requires fewer resistors and capacitors, resulting in smaller, simpler, and less expensive circuits.* These advantages are especially significant in integrated circuits, wherein complex circuits are constructed on a single chip of silicon. When amplification at very low frequencies is required, the cost and physical size of coupling and bypass capacitors become objectional, and direct coupling is desirable. For amplification of dc signals, these capacitors must be avoided.

There are special problems associated with high-gain dc amplifiers. One of these is that large amounts of feedback cannot be used to stabilize the operating points of the transistors, because the feedback would greatly reduce the dc gain. Other problems are the temperature dependence of β_F and V_{BE}. Variation of these quantities with temperature may produce a dc output voltage that is unrelated to the input signal, and consequently, temperature-compensation schemes are essential. We shall exclude dc amplifiers from consideration here.

In a direct-coupled ac amplifier, operating-point stabilization is usually accomplished by utilizing a large amount of dc negative feedback. The feedback circuit is designed to sample the quiescent collector current I_C of the output stage and to introduce into the base circuit of the input stage a current proportional to I_C. The current that is fed back must have a direction such that it tends to counteract any change in I_C. This is the purpose of the emitter resistor of a single stage, and this feedback concept was discussed briefly in Sec. 13-1. In a direct-coupled amplifier the principle is the same, only the current is fed back from the output of the last stage to the input of the first stage. When the first stage is an FET amplifier, the quantity fed back is a voltage. This is discussed in Sec. 17-3.

AMPLIFIER DESIGN

In order to illustrate the feedback principle and a typical design procedure, let us consider the direct-coupled CE-CC amplifier of Fig. 15-15. The dc

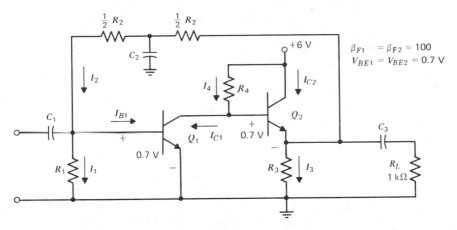

Figure 15-15   Direct-coupled CE-CC amplifier.

feedback occurs through R_2, which is connected from the second-stage emitter output to the first-stage base input. Capacitor C_2 eliminates ac feedback. Suppose we choose $I_{C1} = 0.1$ and $I_{C2} = 1$ mA. Assuming β_{F1} and β_{F2} are each 100, the respective base currents are 0.001 and 0.01 mA.

The feedback current I_2 increases when I_{C2} increases, and I_2 decreases when I_{C2} decreases. *For good stabilization we want a small percentage change in I_2 to produce a much larger percentage change in I_{B1}*, which equals $I_2 - I_1$. Accordingly, both I_2 and I_1 must be considerably larger than I_{B1}. Let us select R_1 to be 60 kΩ, which makes I_1 equal to 0.012 mA. Then I_2 is 0.013, which is 13 times I_{B1}. The current I_3 is found from I_2 and the emitter current of Q_2 to be approximately 1 mA. A reasonable choice for R_3 is 3 kΩ, which gives 3.7 V for V_{CE1} and 3 V for V_{CE2}. The voltage across R_2 is $I_3 R_3$ minus V_{BE1}, or 2.3 V, and this requires that R_2 be 180 kΩ. The voltage across R_4 is also 2.3, and I_4 is the sum of I_{C1} and I_{B2}, or 0.11 mA. Hence, R_4 is 20.9 kΩ. This completes the design.

In the incremental circuit, R_1 and $\frac{1}{2}R_2$ shunt the input, and R_3 and $\frac{1}{2}R_2$ shunt the load. These effects are rather negligible. However, R_4 shunts the input of the CC stage, thus reducing the voltage gain of the CE stage. A larger value of R_4 could be used if R_3 were increased, but this would reduce V_{CE2} and increase $C_{\mu 2}$.

Now let us suppose that β_{F2} decreases, causing I_{C2} to decrease. The voltage across R_3 decreases, causing I_2, I_{B1}, and I_{C1} to decrease, and I_4 increases. As I_{B2} equals $I_4 - I_{C1}$, clearly I_{B2} increases, which tends to counteract the decrease in I_{C2}. For example, if β_{F2} is reduced from 100 to 50, with the other parameters unchanged, analysis shows that I_{B2} increases from 0.01 to nearly 0.02 mA, whereas I_{C2} changes from 1 to 0.982 mA, a drop of only 1.8% (see

P15-14). The stabilization is improved by connecting a negative source of several volts between R_1 and ground, with R_1 appropriately increased. This minimizes the effect of a change in V_{BE1}.

A DIRECT-COUPLED CE–CE AMPLIFIER

A second example is the network of Fig. 15-16a with the dc circuit displayed alongside. Note that the battery, R_1, and R_2 have been replaced with a Thévenin equivalent circuit in Fig. 15-16b. The two-stage CE amplifier employs both dc and ac feedback. Although the emitter of Q_1 is not truly common to the signal input and output because of the 100-ohm unbypassed emitter resistor R_4, the stage is still referred to as CE. Resistors R_5 and R_6 constitute the dc feedback network. Because this combination would give far too much signal feedback, the small resistor R_4 is placed in parallel with R_6 for ac conditions. The incremental circuit is analyzed in P20-26.

With reference to the dc circuit of Fig. 15-16b, suppose β_{F2} decreases from 100 to 50. This causes $|I_{C2}|$ to decrease, and also I_6. Because the voltage across R_B is 2.3 minus $I_6 R_6$, this voltage increases, causing I_{B1} and I_{C1} to rise. I_3 is constant at 0.7 mA, and $|I_{B2}|$ is $I_{C1} - 0.7$. The larger I_{C1} gives a larger $|I_{B2}|$

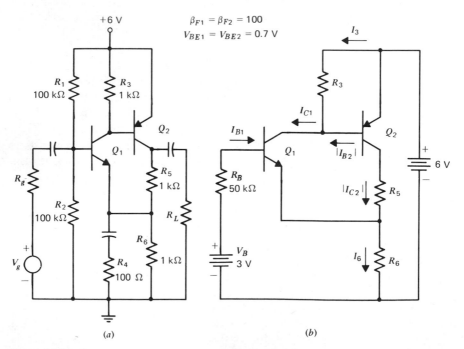

Figure 15-16   Two-stage CE amplifier. (a) Complete circuit. (b) DC circuit.

which tends to counteract the original decrease in $|I_{C2}|$. The numerical values are selected so that small changes in $|I_{C2}|$ and I_6 give a large percentage change in I_{B1}, and furthermore, a small change in I_{C1} has a large effect on I_{B2}. This is the feedback principle.

The dc circuit is easily analyzed. The value of I_{C1} is the sum of I_3, which is 0.7 mA, and $0.01|I_{C2}|$; I_6 is the sum of $1.01I_{C1}$ and $|I_{C2}|$. Along with a loop equation around the base-emitter circuit of Q_1, these relations are sufficient for determining the following results, with values in mA:

$$I_{C1} = 0.712 \qquad I_{C2} = -1.22 \qquad I_3 = 0.700 \qquad I_6 = 1.94$$

It should be noted that $(V_{BE1} + I_6 R_6)$ is only slightly less than V_B, thus making I_{B1} sensitive to changes in I_6. Also, I_{C1} and I_3 are nearly the same, which makes I_{B2} sensitive to changes in I_{C1}.

15-7. DIODE-TRANSISTOR CURRENT MIRRORS

Operational amplifiers sometimes use two matched transistors, with one connected as a diode, to transform an input direct current to a desired output direct current. An example of such a circuit is shown in Fig. 15-17. Here Q_2 is connected as a diode, and as we shall see, the ratio of the currents I_1 and I_2 is determined by the dc beta, the temperature, and the voltage across the resistor R.

From the Ebers–Moll equation (2-10) applied to the active mode, which is assumed here, I_3 equals $I_{ES} \exp q V_{BE1}/kT$ and I_4 equals $I_{ES} \exp q V_{BE2}/kT$. The matched transistors are assumed to have identical saturation currents I_{ES}, as

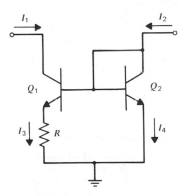

Figure 15-17   Diode-transistor current transformer.

well as the same beta and temperature. Noting that $V_{BE2} - V_{BE1}$ equals $I_3 R$, the ratio of the currents is

$$\frac{I_4}{I_3} = \exp\frac{I_3 R q}{kT} \tag{15-27}$$

$I_3 = I_1(\beta + 1)/\beta$, and $I_4 = I_2 - I_1/\beta$. These give the sum of I_3 and I_4 equal to the sum of I_1 and I_2 as required. By replacing I_3 and I_4 in (15-27), the result can be expressed as

$$\frac{I_2}{I_1} = \frac{1}{\beta} + \frac{\beta + 1}{\beta} \exp\left[\left(\frac{q}{kT}\right)\left(\frac{\beta + 1}{\beta}\right)(I_1 R)\right] \tag{15-28}$$

It is evident from (15-28) that I_2 is always greater than I_1. An important special case is the one with $R = 0$, which gives $I_2/I_1 = (\beta + 2)/\beta$. For a typical beta of 50 this ratio is 1.04. The network of Fig. 15-17 is referred to as a *current mirror*. Either current, I_1 or I_2, may be used as the input current that drives the current mirror.

Shown in Fig. 15-18 is the schematic diagram[7] of the popular 741 general-purpose monolithic operational amplifier. Although presented here primarily for the purpose of illustrating current mirrors, other important aspects of the circuitry will also be discussed.

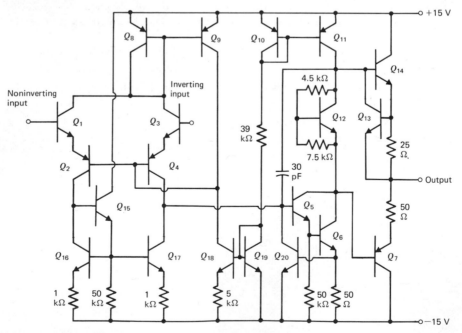

Figure 15-18   741 OP AMP circuit.

DC ANALYSIS OF 741 OP AMP

For the dc analysis each input is grounded, each V_{BE} is taken to be 0.7 V, and q/kT is 40. All PNP transistors except the output transistor Q_7 are lateral devices with low betas, assumed to be 4. Transistor Q_7 is a substrate type with a beta of approximately 25. Both lateral and substrate PNP transistors were examined in Sec. 4-2. The input transistors are designed with high betas, and a value of 200 is used in the analysis. Other betas are assumed to be 50. Units employed are mA, kΩ, and V.

Inspection of the circuit reveals that the voltage across the 39-kΩ resistor in the middle portion of the diagram is 30 less two diode drops, or 28.6 V, giving a current of 0.73 mA. Transistors Q_{18} and Q_{19} constitute a current mirror. We can apply (15-28) with $I_2 = 0.73$ mA, $R = 5$ kΩ, $\beta = 50$, and $q/kT = 40$. The value of I_1 is found to be 0.018 mA, which is I_{C18}.

For matched input transistors we find from a node equation at the collector of Q_{18} that

$$I_{C18} = -I_{C9} - 2I_{B2} = 0.018 \text{ mA} \tag{15-29}$$

We can express I_{B2} in terms of the collector current of Q_1. This base current equals $-I_{E2}/(\beta + 1)$, or $-0.2I_{E2}$ for a beta of 4. Because of the large beta of 200 for Q_1, I_{E2} is approximately equal to I_{C1}, and I_{B2} is $-0.2I_{C1}$. With this substitution, (15-29) becomes

$$-I_{C9} + 0.4I_{C1} = 0.018 \text{ mA} \tag{15-30}$$

In order to determine the input bias current we need a second equation relating I_{C1} and I_{C9}. It is found from the current mirror formed by Q_8 and Q_9. For this mirror the respective currents I_1 and I_2 of Fig. 15-17 are I_{C9} and $-2I_{C1}$, R is zero, and β is 4. Accordingly, from (15-28) we find that

$$\frac{-2I_{C1}}{I_{C9}} = 1.5 \tag{15-31}$$

From (15-30) and (15-31) the collector current I_{C1} of Q_1 is calculated to be 0.010 mA. Division by 200 gives 50 nA for I_{B1}. This is the input bias current I_{IB}. The data sheet gives a typical value of 80 nA and a maximum value of 500 nA.

The collector current of Q_{11} provides the quiescent current for the Darlington pair Q_5 and Q_6. It is easily found from the current mirror consisting of Q_{10} and Q_{11}. From (15-28) with $R = 0$ the current I_{C11} is determined to be -0.49 mA.

Transistors Q_7 and Q_{14} constitute a class AB (see Sec. 16-2) complementary emitter-follower output stage (see Sec. 16-5) with low output impedance, typically 75 ohms. They are biased near cutoff, with a small bias current that minimizes crossover distortion (see Fig. 16-21). The base-emitter voltages are supplied by transistor Q_{12}. With an input current from Q_{11} of 0.49 mA, the circuitry of Q_{12} is designed to give the required dc voltage of about 1.2 V while presenting a low incremental resistance of about 190 ohms (see P15-19). Because its collector-emitter voltage V_{CE} is approximately equal to the product kV_D, with k determined from the two resistors according to (16-30) and with V_D denoting the base-emitter voltage, the configuration is referred to as a V_D *multiplier circuit*. For output transistors with similar input characteristics, the output voltage is zero and the 1.2 V of Q_{12} are divided equally between the base-emitters of Q_7 and Q_{14}. Thus each of these devices is biased near cutoff with a collector current that is a fraction of a milliampere, and each collector-emitter voltage is 15.

Protection in the event of a short-circuit at the output is offered by Q_{13} and Q_{20}, which are normally *off*. For example, Q_{13} becomes biased in the active region, with $V_{BE} = 0.7$ V, when the emitter current of Q_{14} reaches about 28 mA, and Q_{13} prevents this current from rising appreciably higher by shunting current from the base of Q_{14}. An excessive current through Q_7 raises the collector current of Q_6, thereby turning Q_{20} *on*; this shunts current from the Darlington-pair input.

Transistors Q_{16} and Q_{17} constitute a current mirror, with Q_{16} connected as a diode by the base-emitter of Q_{15}. An advantage of this arrangement is that the base currents of Q_{16} and Q_{17} do not flow through Q_2. These transistors help in maintaining equal bias currents in each half of the differential input stage. Because Q_{16} is connected as a diode, the effective resistance in series with the collector of Q_2 is low, but the *active load* provided by Q_{17} for Q_4 is of the order of megohms. The output signal of the differential input stage is taken from the collector of Q_{17} and fed into the Darlington pair Q_5 and Q_6. Thus transistors Q_{16} and Q_{17} provide an efficient means for converting the input differential stage to single-ended drive. The high input resistance of the Darlington stage avoids excessive loading of the input differential stage, thereby maintaining high voltage gain. Note that the bases of Q_{15} and Q_5 are about two diode drops above the negative supply voltage, fixing the collectors of Q_{16} and Q_{17} at approximately -13.6 V.

INCREMENTAL CHARACTERISTICS

Let us briefly examine the incremental properties of the 741 OP AMP. For $I_{C1} = 10$ μA the input transistor Q_1 has $g_{m1} = 0.40$ mmho, $\beta_1 = 200$, and $r_{\pi1} = 500$ kΩ. These values also apply to Q_3. Transistor Q_2 has an emitter current of 10 μA and a collector current of about 8 μA for a beta of 4. Therefore, g_{m2} is 0.32 mmho and $r_{\pi2}$ is 13 kΩ, with the same values applicable to Q_4.

We can now estimate the input resistance. Suppose a signal is applied to the noninverting input with the inverting input grounded. The incremental input current I_{b1} flows through $r_{\pi1}$ to the emitter of Q_1. The emitter current of Q_1 is $I_{b1}(\beta_1 + 1)$, which is also the incremental emitter current of Q_2. Therefore, the base current of Q_2 is $I_{b1}(\beta_1 + 1)/(\beta_2 + 1)$, and this flows through $r_{\pi2}$ of Q_2.

It follows that the resistance from the input terminal to the base of Q_2 is the sum of $r_{\pi 1}$ and $r_{\pi 2}(\beta_1 + 1)/(\beta_2 + 1)$, or 1.0 MΩ for $r_{\pi 1} = 500$ kΩ, $r_{\pi 2} = 13$ kΩ, $\beta_1 = 200$, and $\beta_2 = 4$. Because an equal resistance exists between the base of Q_2 and the base of Q_3, the total input resistance is twice the value calculated, or 2.0 MΩ. Based on the assumed transistor parameters, the calculations are only approximate because of neglect of the base-width-modulation parameters, which have a second-order effect. Also, at room temperature the g_m of a lateral transistor is closer to $25I_C$ than $40I_C$ as assumed. However, the result agrees precisely with the typical input resistance of 2 MΩ specified by the manufacturer.

The input transistor Q_3 has the CC configuration. Its collector is connected to ac ground through Q_8, which is an integrated circuit diode. The output at the emitter of Q_3 feeds into the emitter of Q_4, with the output of Q_4 taken from its collector. Thus Q_4 approximates a CB amplifier. Transistors Q_1 through Q_4 give a differential input stage with each input consisting of a CC-CB arrangement. The arrangement is designed to give a high CMRR and a high common-mode input resistance. It also permits a large range of input voltage, typically ± 13 V for supply voltages of ± 15 V. The PNP lateral transistors, such as Q_2 and Q_4, have higher base-emitter breakdown voltages than do NPN vertical devices.

Major consideration in the design of the input stage was given to avoiding an effect known as *latch-up*. Occasionally an integrated-circuit NPN transistor driven into saturation remains saturated when the driving signal is removed. This results because of positive feedback from a parasitic substrate-type PNP transistor, and the interaction is similar to that of a flip-flop. The latched state is eliminated only by removing the power momentarily. Careful design, such as that employed in the 741, has nearly eliminated the latch-up problem in most integrated circuits.

The output of the differential stage is amplified by the Darlington pair consisting of Q_5 and Q_6 and connected as a CE amplifier. Because of the high input impedance at the base of Q_5, the input differential stage is able to contribute a large voltage gain, from 100 to 200. Furthermore, because of the high impedance of the active load provided by transistor Q_{11} which maintains a constant bias current, along with the high input resistance of the complementary emitter-follower output stage, the interstage CE amplifier produces an even larger voltage gain, from 1000 to 2000. Its output feeds into the bases of the output stage. The V_D multiplier of Q_{12} connects the bases incrementally while providing the proper dc bias.

Because of the small incremental resistance of the V_D multiplier, the 30 pF capacitor is effectively connected between the base and collector terminals of the Darlington pair. Its purpose is to modify the high-frequency characteristics and thus prevent oscillations. This internal compensation is discussed in more detail in Sec. 21-9.

REFERENCES

1. R. F. Shea (Ed.), *Transistor Circuit Engineering*, Wiley, 1967.
2. F. K. Manasse, J. A. Ekiss, and C. R. Gray, *Modern Transistor Electronics Analysis and Design*, Prentice-Hall, Englewood Cliffs, New Jersey, 1967.
3. P. E. Gray and C. L. Searle, *Electronic Principles*, Wiley, New York, 1969.
4. H. E. Stewart, *Engineering Electronics*, Allyn and Bacon, Inc., Boston, Mass., 1969.
5. D. L. Schilling and C. Belove, *Electronic Circuits: Discrete and Integrated*, McGraw-Hill, New York, 1968.
6. R. H. Mattson, *Electronics*, Wiley, New York, 1966.
7. *The Linear and Interface Circuits Data Book*, Texas Instruments Inc., Dallas, Texas, 1973.

PROBLEMS

Section 15-1

15-1. For $\beta_F = 40$ the Q-point (I_C, V_{CE}) of the transistor of Fig. 15-1 is 1 mA, 3 V. What is the Q-point if β_F, instead of being 40, is so large that we can assume it to be infinite? (*Ans*: 1.15 mA, 2.1 V)

15-2. Suppose the Q-point of the amplifier of Fig. 15-1 is moved along the dc load line by adjusting R_2, with all other parameters as specified in the figure. For a sinusoidal output signal of 0.4 V peak-to-peak, what are the maximum and minimum allowable quiescent collector currents, assuming v_{CE} is never to be less than 0.4 V and i_C is never to be less than 0.1 mA?

15-3. Suppose the amplifier of Fig. 15-1 has the following parameters, in units of kΩ and volts:

$$R_s = 5 \quad R_1 = 40 \quad R_2 = 10 \quad R_3 = 0.5 \quad R_4 = 2 \quad R_L = 0.5$$

$$V_{CC} = 10 \qquad\qquad V_{BE} = 0.7 \qquad\qquad \beta_F = 50$$

(a) Determine I_C, I_B, and I_E in mA. (b) Find V_{CE}. (c) Find the dc and ac load-line equations. Using suitable scales, plot these load lines in the manner shown in Fig. 15-3. Use mA, kΩ, and V. (d) For a sinusoidal input, what is the maximum amplitude V_m of the output voltage that is possible without encountering gross distortion of the type shown in Fig. 15-4 (*Ans*: 0.78 V)?

Section 15-2

15-4. The curves of Fig. 15-8 describe the spot noise figure F of a certain transistor at 25°C, 1 kHz, and 5 V. (a) For minimum noise what are the optimum values of source resistance and quiescent collector current? (b) If the source resistance is 3 kΩ, what bias current gives minimum noise? (c) For a bias current of 2 mA, determine the source resistance for minimum noise.

Section 15-3

15-5. For the three-stage amplifier of Fig. 15-9 find the voltage gain V_o/V_s in terms of A_{v1}, A_{v2}, A_{v3}, R_s, and R_{in}. The advantage of the form of (15-6) over the result obtained here is that the current gains of the three basic amplifier configurations are nearly independent of the load presented by a following stage, as indicated by Fig. 13-29 of Sec. 13-6.

15-6. For quiescent currents of 1, 5, and 10 mA selected for the transistors of Fig. 15-10 the corresponding dc betas are 70, 110, and 120. Assume that R_3 is selected for a 5 V drop, this being the maximum voltage that can be spared for this resistor. (a) Considering the shunting effect of R_3, calculate the current gain A_{i2}, assuming q/kT is 40 and r_{x3} is 100 ohms. The input current is I_{b2}. (b) Repeat the calculation for bias currents of 1, 10, and 10 mA, noting the reduction in A_{i2} even though β_{F2} is higher (Ans: -66.7).

Section 15-4

15-7. (a) In the simple bias circuit of Fig. 15-12 with $R_B = 415$ kΩ, $R_C = 4$ kΩ, $\beta_F = 50$, and $V_{CC} = 9$, calculate I_C and V_{CE}. Then repeat the calculations for $\beta_F = 150$. Assume $V_{CE} = 0.2$ V in saturation, and $V_{BE} = 0.7$ V. (b) Now suppose the bias circuit is changed to the form of Fig. 15-1, with R_1, R_2, R_3, and R_4 equal to 71, 21, 1, and 4 kΩ, respectively, and V_{CC} and V_{BE} as given in (a). Find I_C and V_{CE} for $\beta_F = 50$ and again for $\beta_F - 150$. Compare the results with those of part (a), noting the improvement.

15-8. The amplifier of Fig. 15-1 is to be used in ambient temperatures ranging from 25°C to 125°C. At the lower temperature the minimum β_F is 40, as specified on the figure, the corresponding collector current is 1 mA, and V_{BE} is 0.7 V. At the higher temperature the maximum β_F is 400, I_{CO2} is 5 μA, and V_{BE} is 0.45 V. (a) Calculate the maximum collector current I_{C2} at 125°C. (b) Selecting $I' = 0.1$ mA and $V' = 0.5$ V, determine the maximum V_m that can be used without exceeding the active-mode limitations of (15-4) and (15-5). [Ans: (b) 0.72 V]

15-9. Design the bias circuit of a CE amplifier with the configuration of Fig. 15-1, with a 10 V supply and a load resistance R_L of 200 ohms. Use (15-15), selecting the maximum R_B. The transistor is silicon with a minimum β_F of 30 at 25°C. The output voltage is no more than a few millivolts. The temperature range is from 25°C to 125°C. The maximum beta is unknown, and I_{CO} is negligible. At room temperature with the minimum beta the Q-point (I_C, V_{CE}) should be (5 mA, 5 V). Calculate I_{C2} and the corresponding V_{CE2} (Ans: 7.92 mA, 2.14 V). Sketch the circuit.

Section 15-5

15-10. Deduce (15-21) and (15-22) directly from (15-17) and (15-19). First, substitute for I_{C2} in (15-17) with β_{F2} infinite and $I_{CO2} = 0$. Then equate dR_B/dR_E to zero and solve for R_E. Substitute the result into (15-17).

15-11. Using (15-21) and (15-22) with $R_C - 800$ ohms and $V' - 2$ V, design the bias circuit of a CE amplifier with the configuration of Fig. 15-1. Assume $V_{CC} = 10$ V, $V_{BE} = 0.7$ V, $R_L = 200$ ohms, $\beta_{F1} = 30$, V_m is negligible, $\Delta V_{BE} = 0.25$ V, and $I_{C1} = 5$ mA. Find the Q-points (I_C, V_{CE}) at T_1 and β_{F1} and also at T_2 with beta infinite. Sketch the circuit. The specifications are essentially the same as those of P15-9. Explain why this design is superior to that of P15-9. (Ans: Q_2 is at 7.19 mA, 1.92 V)

15-12. A CE amplifier of the form of Fig. 15-1 is to be constructed in a lab, where the room temperature is constant at 25°C. The load resistance is 400 ohms, and R_4 is selected to be 4 kΩ. The data sheet specifies that β_F is between 30 and 100 at 2 mA, 6 V, 25°C. For the minimum beta we want I_C to be 2.00 mA with $V_{CE} = 6$, and for the maximum beta I_C is to be 2.15 mA. Resistance $R_B = 10$ kΩ. (a) Calculate R_E and V_{CC}. Assume

V_{BE} is 0.7 V and do not use approximate Eq. (15-17). (b) Determine R_1 and R_2. (c) Calculate V_{CE} at the maximum I_C (*Ans*: 4.95). (d) What is the required voltage rating for the electrolytic capacitor C_3?

15-13. A CE amplifier having the form of Fig. 15-1 has $V_{CC} = 10$, $R_L = 3$ kΩ, $V_{BE} = 0.7$ V at 25°C, and a minimum β_F of 40 at 25°C, 1 mA, 3 V. The temperature range is 25°C to 125°C, and V_{BE} decreases 2.5 mV/°C. The maximum value of V_m is 0.6 V, and the instantaneous v_{CE} must never be less than 0.4 V. I_{CO} is negligible, and the maximum beta is unknown. The minimum quiescent current is selected to be 1 mA. (a) For $R_4 = 6$ kΩ, which is twice R_L, calculate the maximum R_B that can be used. (b) Repeat for $R_4 = 5$ kΩ, and for this case also determine R_1, R_2, and R_3. (c) For the design of (b) calculate I_C at 125°C assuming that β_F is 250 at this current and temperature, and compare I_C with the maximum allowable.

Section 15-6

15-14. In the direct coupled CE–CC amplifier of Fig. 15-15, with $R_1 = 60$, $R_2 = 180$, $R_3 = 3$, and $R_4 = 20.9$ kΩ, we found in Sec. 15-6 that I_{C2} is 1 mA. If β_{F2} decreases to 50, with all other parameters unchanged, find I_{C2}. To do this, first write four equations relating I_{C1}, I_{C2}, I_2, and I_3. These should be node equations at nodes B_1 and E_2 and two loop equations—one around the feedback loop and the other around the loop with V_{CC}, R_3, and R_4, with I_4 expressed in terms of the two collector currents. What is the percent decrease in I_{C2}?

15-15. For the CE–CC amplifier of Fig. 15-15 determine I_{C1} and I_{C2} assuming V_{CC} is changed from 6 to 7 V and both β_{F1} and β_{F2} are increased from 100 to 200. Refer to P15-14 for data. (*Ans*: 0.156, 0.988 mA)

15-16. For the two-stage CE amplifier of Fig. 15-16 determine I_{C1} and I_{C2} with both β_{F1} and β_{F2} changed from 100 to 500.

Section 15-7

15-17. For the current mirror of Fig. 15-17 with $\beta = 50$, $q/kT = 38$, and $I_2 = 1$ mA, calculate the resistance R in kΩ required to make $I_1 = 50$ μA.

15-18. For the current mirror of Fig. 15-17 with $\beta = 50$, $q/kT = 38$, $R = 3$ kΩ, and $I_2 = 1$ mA, determine I_1 in μA.

15-19. In the operational amplifier of Fig. 15-18 the current source Q_{11} supplies 0.49 mA to the network of transistor Q_{12}. Assume the dc and ac betas are 50, $V_{BE} = 0.7$ V, and $q/kT = 40$. Calculate the dc voltage V_{CE} across transistor Q_{12} and find the incremental resistance in ohms.

15-20. The schematic diagram of the 702 OP AMP is shown in Fig. 15-19. The quiescent currents of the matched transistors Q_1 and Q_2 are identical, as are those of Q_3 and Q_4. Transistors Q_7, Q_8, and Q_9 constitute a current mirror. Ground the two inputs, use 0.7 V for all base-emitter voltages, and assume all base currents are zero. Use units of kΩ, mA, and V. (a) Determine I_{C8} from a loop equation from the negative supply through Q_8 to ground. (b) From a second loop equation that includes the base-emitter terminals of both Q_7 and Q_8, find I_{C7} and then I_{C1} and I_{C2}. (c) Find I_A through the 8-kΩ resistor from a loop equation from the positive supply to the base of Q_3 to ground. Then find I_{C3} and I_{C4}. (d) Noting that $I_{C5} = I_{C9}$ for negligible base currents, determine I_{C5}, I_{C6}, and I_{C9} from two additional loop equations. (*Ans*: $I_{C6} = 1.876$)

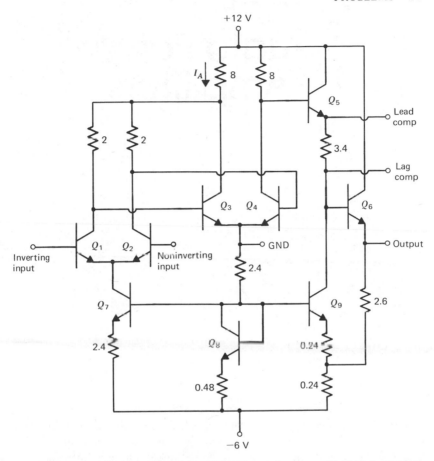

+12 V

I_A 8 8

Q_5

Lead comp

2 2

3.4

Lag comp

Q_3 Q_4

Q_6

Q_1 Q_2

Inverting input

Noninverting input

o GND

2.4

Output

Q_7

Q_9 2.6

2.4

Q_8

0.24

0.48

0.24

−6 V

Figure 15-19   Schematic diagram of the SN72702 general-purpose OP AMP, for P15-20, P15-21, and P15-22. Resistances are in kilohms.

15-21. The collector currents found in P15-20 are only approximate because the betas were taken as infinite. Assume now that all betas are 50. From the collector currents we find the approximate values of I_{B1} through I_{B9} to be 0.0037, 0.0037, 0.02, 0.02, 0.018, 0.038, 0.0074, 0.037, and 0.018 mA, respectively. Using these and treating the collector currents as unknowns, repeat the procedure of P15-20 and calculate I_{C1} through I_{C9} in mA. Although this method is approximate, for the assumed data the results are accurate within 1 %. (*Ans:* 0.175, 0.175, 0.974, 0.974, 0.858, 1.89, 0.357, 1.78, 0.837)

15-22. Using the results of P15-21, calculate the dc output voltage and also the total power dissipation in mW of the SN72702 OP AMP. The inputs are grounded and all betas are 50.

Chapter Sixteen
Audio Power
Amplifiers

We have learned techniques for analysis of amplifiers that operate with small signals. The term *small-signal* implies signals so small that distortion is negligible, even though the Ebers–Moll equations of a BJT are nonlinear. The hybrid-π and *h*-parameter models are examples of linear incremental models used to replace transistors. However, in Sec. 12-2 we learned that the use of such linear models requires, at least for accurate results, *that the instantaneous incremental base-emitter voltage be less than about* 5 mV.

In many electronic systems one or more stages are large-signal amplifiers that supply power to a load with very little distortion. Because the signal is large ($v_{be} > 5$ mV) and because the circuits are designed so that the signal voltage across the load is nearly undistorted (although v_{be} is distorted), such operation is often classified as *large-signal linear*. However, the linear incremental models are invalid, and careful analysis requires graphical procedures. A common example of a large-signal linear amplifier is the audio power amplifier that supplies speech or music signals to a speaker. Although this chapter is oriented toward the audio power amplifier, the basic concepts and circuits apply to all large-signal linear amplifiers, including those used in control, instrumentation, analog-computer, and other systems. Let us begin with a discussion of the power transistor.

16-1. THE POWER BJT

In comparison with a small-signal BJT the power transistor operates at relatively high current and voltage levels and dissipates more power. Consequently, the device has some special requirements. First, it must be capable of providing useful current gain at large quiescent currents. The curves of Fig. 15-6 of Sec. 15-2 show the rather sharp reduction in β_F at the higher currents. This is attributed to the lowered emitter efficiency caused by high-level injection, which is aggravated by current crowding in the base as discussed in Sec. 12-3. Because the emitter efficiency depends on the density of the current, high-current devices must have larger cross-sectional areas. Power transistors with useful betas have rated collector currents that range from a few hundred milliamperes to 100 amperes or more.

A second requirement is a high breakdown voltage V_{CEO}. In Fig. 2-16 of Sec. 2-6 the avalanche region between the sustaining voltage V_s and the avalanche voltage V_a is shown to be one of considerable nonlinearity, and power transistors are usually operated with the maximum v_{CB} less than V_s. The corresponding breakdown value of v_{CE} is referred to on data sheets as $V_{(BR)CEO}$. The sustaining voltage is that voltage at which the product $M\alpha_F$ is unity, with the multiplication factor M dependent on V_a in accordance with (2-44). A large sustaining voltage requires a large V_a and a small α_F (see P16-4). Of course, a small α_F implies a small β_F, which is undesirable, and the result is a compromise. Breakdown voltages (V_{CEO}) are usually in the range from 30 to 100 V, although devices are available with breakdown voltages of 500 V or more. Typical values of β_F are from 20 to 80. Data sheets nearly always specify both minimum and maximum values of β_F. The minimum may be as low as 5, and the maximum will nearly always be below 250. Many transistor applications, especially battery-operated and automobile equipment, use low supply voltages (6 to 12 V). For these, a low breakdown voltage is satisfactory, and devices with higher betas can be used.

At high currents the resistance r_π becomes very small, and the input resistance of a CE amplifier is approximately equal to the intrinsic base resistance r_x. Because V_o/V_s is $A_i R_L/(R_s + R_{in})$ by (15-6), the output voltage, current, and power to a specified R_L are greatest when r_x is a minimum. Hence power transistors are designed so as to minimize r_x. Using a large base diameter is helpful. Increasing the base width W reduces r_x but has a detrimental effect on β_F, and increasing the base dopant density lowers the emitter efficiency as well as r_x.

THERMAL RESISTANCE

Perhaps the most important attribute of a good power transistor is its ability to transfer heat rapidly from the collector junction to the environment. To expedite this heat flow the collector is usually in direct contact with the metal case.

This gives the lowest possible junction-to-case thermal resistance R_θ, defined by the equation

$$R_{\theta JC} = \frac{T_J - T_C}{P_D} \qquad (16\text{-}1)$$

The junction-to-ambient thermal resistance is

$$R_{\theta JA} = \frac{T_J - T_A}{P_D} \qquad (16\text{-}2)$$

The case temperature T_C is intermediate to the junction temperature T_J and the ambient temperature T_A. It is measured at a point on the case specified on the dimension sketch of the transistor given on the data sheet.

The maximum allowable collector-junction temperature of a silicon power transistor is from 150° to 200°C, and for a germanium device it is usually 100°C. Power dissipated at the collector junction produces heat, which causes the junction temperature T_J to rise. The temperature gradients from the junction to the case and from the case to the ambient cause this heat to flow through the total thermal resistance to the environment. A low thermal resistance allows a high power dissipation without exceeding the maximum allowable junction temperature. Data sheets specify the operating junction-temperature range and also both the junction-to-case and the junction-to-ambient thermal resistances, from which it is easy to determine the power capability of a device.

Example 1

Illustrated is a 2N5301 NPN silicon power transistor. The maximum dimension is about 1.6 inches. The data sheet states "200 W at 25°C case temperature" and specifies an operating collector-junction temperature range from -65°C to 200°C. The maximum values of the junction-to-case and junction-to-ambient

thermal resistances are given as 0.875°C/W and 35°C/W, respectively. For a free-air temperature of 25°C find the maximum allowable device dissipation (a) assuming no provision for cooling, (b) assuming a cooling system maintains the case temperature at 100°C, and (c) for $T_C = 25$°C.

Solution
(a) $P_D = (200 - 25)/35 = 5$ W.
(b) $P_D = (200 - 100)/0.875 = 114$ W.
(c) $P_D = (200 - 25)/0.875 = 200$ W.

In the example we note the very substantial increase in the power capability of the device when some means of cooling is provided. A common technique is to bolt the transistor tightly to the chassis or some extended metal surface. Such a surface is referred to as a *heat sink*. The metal conducts heat, thereby lowering the junction-to-ambient thermal resistance. Commercially available heat sinks often have fins to assist cooling by convection. For a required sink-to-ambient thermal resistance less than 1°C/W, forced-air or water cooling is necessary. Because the collector is connected to the metal case, the case is charged by the collector voltage and must be insulated from the heat sink. This is accomplished by using a thin mica washer between the case and the metal plate along with plastic bushings to insulate the bolts. The thermal resistance of the insulating washer is typically about 1°C/W. Shown in Fig. 16-1 is a curve of thermal resistance plotted as a function of the side length of a one-eighth-inch thick square sheet of bright aluminum, which is often used as a heat sink.

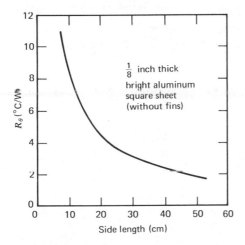

Figure 16-1 Thermal resistance of aluminum sheet.

Example 2

Suppose the power transistor of Example 1 is to be used in a circuit designed so that the maximum power dissipation $V_{CE}I_C$ is 30 W. Using Fig. 16-1 and allowing 0.5°C/W for the mica washer, determine the heat sink required. Also, at maximum power calculate the case temperature T_C and the *heat sink temperature* T_{HS} adjacent to the mica washer.

Solution

The total allowable thermal resistance is (200 − 25)/30, or 5.83°C/W, which equals the sum of the thermal resistances shown in Fig. 16-2. Accordingly,

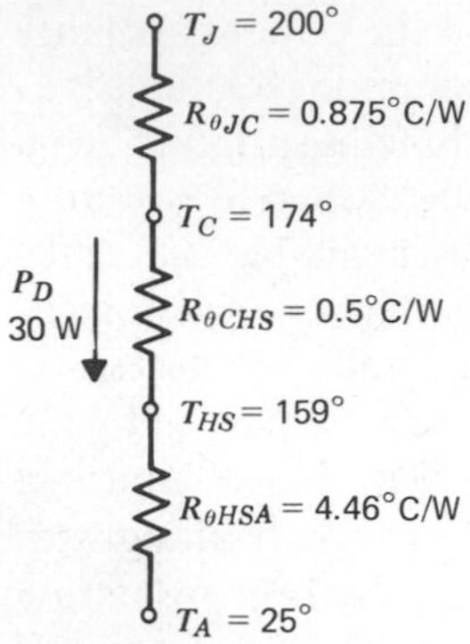

$T_J = 200°$

$R_{\theta JC} = 0.875°C/W$

$T_C = 174°$

P_D 30 W

$R_{\theta CHS} = 0.5°C/W$

$T_{HS} = 159°$

$R_{\theta HSA} = 4.46°C/W$

$T_A = 25°$

Figure 16-2 Schematic diagram.

the heat-sink thermal resistance must not be greater than 4.46°C/W. With reference to Fig. 16-1 a suitable heat sink is a one-eighth-inch thick bright aluminum square sheet of side length 20 cm or more. The schematic diagram of Fig. 16-2, with T analogous to voltage and P analogous to current, is used to calculate the temperatures shown on the figure.

SAFE OPERATING REGION

Every transistor has what is referred to as a "maximum safe continuous operating region." An example of such a region is the shaded part of Fig. 16-3.

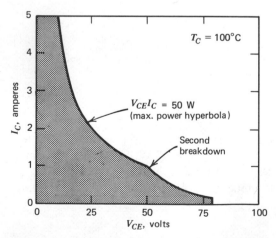

Figure 16-3 Maximum safe continuous operating region.

The transistor has a maximum allowable collector current of 5 A. Above this current the beta may be quite low, and there is the possiblity of damage to the leads and connections. The maximum voltage is 80, which is the minimum breakdown value of V_{CEO}. Although higher voltages can sometimes be used, thermal problems that are especially acute make this inadvisable unless great care is exercised. The curve, $V_{CE}I_C = 50$ W, is the *maximum power hyperbola*. At any point on this hyperbola the power dissipation is 50 W, and for a case temperature of 100°C, the junction temperature is the maximum allowable. From 50 to 80 V the maximum current is determined by a phenomenon known as *second breakdown*. Nonuniform thermal properties in the emitter may result in localized heating and "hot spots" at the junction. The temperature dependence of the junction current is such that the densities are greater at the hot spots than at other points of the emitter-junction cross-section. This concentration of current tends further to increase the temperature. The process can become cumulative, leading to thermal runaway and transistor failure. Second breakdown is avoided by confining operation to the shaded region. Power-transistor data sheets show the safe operating region, though the scales are usually logarithmic.

The dc load lines of small-signal devices are nearly always wholly within the safe operating region. However, such is not the case with power transistors. Let us suppose that the dc load line of the transistor of Fig. 16-3 is vertical at 25 V, with the Q-point at 1.5 A, 25 V. The vertical load line implies that the sum of R_C and R_E is zero, which is unrealistic, but dc load lines are frequently very nearly vertical. If the ambient temperature rises causing T_C to increase, the transistor becomes hotter. Consequently, V_{BE} decreases, and β_F and I_{CO} increase. Each effect tends to increase I_C in accordance with (15-12). A larger I_C produces more power dissipation and a higher temperature. The process can continue until the Q-point crosses the maximum power hyperbola, destroying the device. This is referred to as *thermal runaway*. Although the effect is avoided by the use of an adequate heat sink and by proper stabilization of the Q-point, clearly the design of the dc circuit is much more critical for power amplifiers than for the small-signal amplifiers analyzed in Chapter 15 (see P16-3).

POWER TRANSISTOR PARAMETERS

As indicated in Sec. 15-4, the active-mode collector cutoff current I_{CO} is essentially the same as I_{CBO}. In power transistors this current is usually of the order of tens of microamperes, though it may be as high as a milliampere in those devices with especially high current ratings. We have learned that I_{CO} approximately doubles with each 10°C rise in temperature for small-signal silicon and germanium transistors. This is also true for the alloy-junction germanium power BJT. However, for silicon power transistors the surface-leakage component of I_{CO} is a fairly large percent of the total, and I_{CO} does not increase so rapidly with temperature. A good approximation for I_{CO2} at temperature T_2 is, for silicon power transistors,

$$I_{CO2} = I_{CO1}(2^{\Delta T/15}) \tag{16-3}$$

which assumes that I_{CO} doubles each 15°C. For the mesa-type device the variation with temperature is substantially less than that indicated by (16-3). Unless R_B is very large, I_{CO} is usually negligible in amplifiers using silicon devices. Data sheets sometimes, but not always, give I_{CBO}. When I_{CBO} is not given, it is unacceptable to use either I_{CEO} or I_{CES} for I_{CO}. These currents are much larger than I_{CO}, perhaps by an order of magnitude or more.

In Fig. 15-6 of Sec. 15-2 the value of β_F at a given temperature is shown to increase as I_C increases, then dropping at the higher currents where high-level injection gives a lowered emitter efficiency. Whereas small-signal amplifiers are biased at quiescent currents to the left of the maximum, power amplifiers are frequently biased at currents to the right of the maximum. This gives a reduced gain but allows greater output power, as we shall see in the next section. In this region of high-level injection, the minority-carrier current through the base consists of both diffusion and drift components, and it can be shown that these are nearly equal. The theoretical development of Chapter 2 was based on the minority-carrier diffusion equation, which was developed in Sec. 1-4 by assuming that minority-carrier drift currents were negligible, a deduction based on low-level injection. However, the equations are sometimes applied to the case of high-level injection with the minority-carrier diffusion constant D_b doubled. This gives a diffusion current that is twice its true value, which is an artificial way of accounting for the equal drift current.

Another important high-current effect leads to modification of the expression used to find the transconductance g_m. To determine this modification let us first reexamine the diode equation. For the nonuniform dopant densities normally encountered in PN junctions it can be shown[6] that the forward current is more accurately represented as

$$I = I_s e^{qV/nkT} \tag{16-4}$$

At low currents n is unity, but it is typically 2 at the current levels of power devices. Therefore, at high currents the incremental diode resistance dV/dI equals $2kT/qI$. A similar relation applies to the *dynamic resistance* r_e of the emitter-base junction of a transistor biased in the active mode, with r_e defined as the magnitude of $\partial v_{BE}/\partial i_E$ with v_{CE} constant. Because I_C is $-I_E\beta_F/(\beta_F + 1)$, it is easy to show that the transconductance g_m is the product of $1/r_e$ and $\beta_F/(\beta_F + 1)$, or $g_m \approx 1/r_e$.

In (12-24) the transconductance is expressed as qI_C/kT. According to the preceding discussion, in power transistors we must replace kT with $2kT$, giving

$$g_m = \frac{qI_C}{2kT} \tag{16-5}$$

At room temperature g_m is $20I_C$. A consequence of the reduction in g_m by a factor of 2 is the doubling of r_π. For approximate results the equations previously developed for small-signal devices in low-level injection can be applied to power transistors in high-level injection, *provided we use (16-5) for* g_m *and use a value for* D_b *that is twice its actual value.*

The same considerations that lead to lower values of β_F in power transistors also lead to smaller transition frequencies. Typically, f_T has values from 1 to 50 MHz at the maximum point of the f_T-versus-I_C curve. The relatively large cross-section results in a large C_μ, with typical values ranging from 100 to 2000 pF. Of course, C_π is much greater. However, there are power transistors designed especially for use at radio frequencies that can deliver tens of watts at hundreds of megahertz.

Although the hybrid-π model of a transistor applies only for small signals, the model and its parameters are useful in deducing rough, approximate results. The numerical values of the parameters are quite different from their low-current values, as the following example shows.

Example 3

A CE power amplifier has a 20 ohm load in series with the collector. The transistor, biased at 1 A, 20 V, has $\beta_F = 20$, $f_T = 5$ MHz, $C_\mu - 500$ pF, and $r_x = 10$ ohms. At the junction temperature of $175°C$ the internal emitter-junction voltage is 0.4 V. Determine g_m, r_π, C_π, the terminal voltage V_{BE}, and the small-signal R_{in}, A_i, A_v, and G, assuming β_o equals β_F.

Solution

At the specified junction temperature, $q/2kT$ is 12.9, and g_m is 12.9 mhos. As β_o is 20, r_π is 1.55 ohms. From (12-47) C_π is found to be 0.41 μF. The value of V_{BE} is the sum of 0.4 and $I_B r_x$. The base current is 50 mA, and V_{BE} is 0.9 V.

The input resistance ($r_x + r_\pi$) is 11.6 ohms, 86% of which is r_x. The current gain equals β_o, or 20, and A_v is $A_i R_L/R_{in}$, or 34. The product $A_i A_v$ is the power gain, which is 680, or 28.3 dB. The calculated results are not very accurate because injection is high-level.

Most power amplifiers use silicon transistors. The small I_{CO} simplifies the design of the bias circuit, and the high allowed junction temperature is a distinct advantage. However, the germanium BJT has a lower saturation voltage, which permits operation at increased efficiency, particularly when operating from low-voltage supplies. The lower saturation voltage is caused by the thin heavily doped collector, which is characteristic of alloy-junction devices. Also, the higher mobilities of the carriers give germanium an advantage at high frequencies.

Many power amplifiers use power FETs. These devices have various other applications, and they are examined in Sec. 17-4 of the next chapter.

16-2. CLASS A AMPLIFICATION

The operating point of a transistor determines whether an amplifier is characterized as class A, AB, B, or C. For a sinusoidal input signal, the active element in class A operation conducts for the entire input cycle, and this is the case which we have considered exclusively thus far. In class B operation, the transistor is biased at cutoff, conducting for only half of each input cycle. Class AB is in between A and B, with collector current flowing for less than a full cycle but more than a half cycle. The transistor operating class C is biased well into the cutoff region, a procedure used mainly in RF tuned amplifiers. Only classes A, AB, and B are capable of reasonably linear operation. The general formulation of class A audio power amplifiers is the subject of this section.

AMPLIFIER WITH DIRECT-COUPLED LOAD

The amplifier of Fig. 16-4 has identical dc and ac load lines. The location of the Q-point on the load line is determined by the emitter-base circuit. Wherever this point is, the power P_D dissipated by the BJT is $V_{CE}I_C$ provided the input signal is zero, and hence P_D is $V_{CE}(V_{CC} - V_{CE})/R_L$. By differentiating P_D with respect to V_{CE} and equating the result to zero, we find that the maximum transistor dissipation occurs with the Q-point located at $V_{CE} = \frac{1}{2}V_{CC}$, as shown on the figure. The

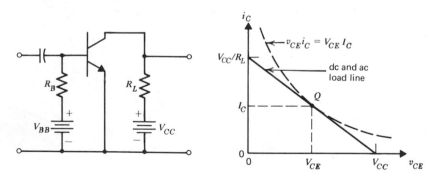

Figure 16-4   A class A amplifier and its load line.

power hyperbola $(v_{CE}i_C = V_{CE}I_C)$ is also illustrated. Because this hyperbola must coincide with the load line at the midpoint, and because no other point on the load line represents this much dissipation, it is evident that the power hyperbola is tangent to the load line at its midpoint. We know that an increase in temperature increases β_F and I_{CO} and decreases V_{BE}, with the result that the Q-point moves to higher current levels. It is interesting to note that, for V_{CE} between V_{CC} and $\frac{1}{2}V_{CC}$, the Q-point movement produces greater power dissipation, which tends to increase the temperature even more. On the other hand, for V_{CE} between $\frac{1}{2}V_{CC}$ and zero, the movement gives reduced dissipation, tending to oppose the temperature rise.

With distortion assumed negligible, a sinusoidal input signal gives a sinusoidal output, and the instantaneous operating point moves back and forth along the load line with equal maximum distances on the two sides of the Q-point. Let (V_{max}, V_{min}) and (I_{max}, I_{min}) denote the maximum and minimum values of v_{CE} and i_C, respectively. Then the rms values of the ac components of v_{CE} and i_C are

$$V_{ce} = \frac{1}{2\sqrt{2}}(V_{max} - V_{min}) \qquad I_c = \frac{1}{2\sqrt{2}}(I_{max} - I_{min}) \qquad (16\text{-}6)$$

The *ac output power* P_o, defined as the average power to the load due to the ac components of v_{CE} and i_C, is

$$P_o = V_{ce}I_c = I_c^2 R_L = \frac{V_{ce}^2}{R_L} \qquad (16\text{-}7)$$

The *collector-circuit efficiency* η is

$$\eta = \frac{P_o}{P_{CC}} \times 100\% \qquad (16\text{-}8)$$

with $P_{CC} = V_{CC}I_C$, which is the power supplied to the collector circuit by the battery. For a sinusoidal input the maximum P_o occurs when the variation in v_{CE} is over the entire load line of Fig. 16-4, from $v_{CE} = 0$ to $v_{CE} = V_{CC}$. Using (16-6) and (16-7) with $V_{max} = V_{CC}$ and $V_{min} = 0$, the maximum P_o is $V_{CC}^2/8R_L$. The quiescent current at the midpoint is $V_{CC}/2R_L$, and P_{CC} is V_{CC} times this current. Equation 16-8 becomes

$$\eta = \frac{V_{CC}^2/8R_L}{V_{CC}^2/2R_L} \times 100\% = 25\% \qquad (16\text{-}9)$$

This is the maximum theoretical conversion efficiency of a class A amplifier with a load resistor in series with the collector, assuming a sinusoidal input.

Shown in Fig. 16-5 is a power hyperbola and two load lines, each of which is tangent to the hyperbola. The load lines have different values of V_{CC} and R_L. The power supplied by the battery is $V_{CC1}I_{C1}$, or $2V_{CE1}I_{C1}$, for the line with Q_1, and it is $V_{CC2}I_{C2}$, or $2V_{CE2}I_{C2}$, for the line with Q_2. These battery powers are equal, because the product $V_{CE1}I_{C1}$ equals $V_{CE2}I_{C2}$. We know from (16-9) that the maximum collector-circuit efficiency is 25%, independent of V_{CC} and R_L. It follows that the maximum possible ac output power P_o of an amplifier is the same for each combination of V_{CC} and R_L that yields a load line tangent to the same power hyperbola. Consequently, if the designer is free to choose R_L, it is usually selected to give the most voltage gain, accomplished by using the maximum voltage V_{CC}. Then R_L is chosen so as to make the load line tangent to the

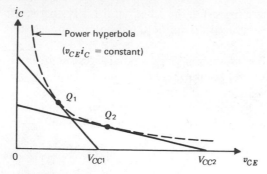

Figure 16-5   Power hyperbola and load lines.

selected power hyperbola (see P16-5). This gives the largest R_L, which is still smaller than the output impedance of the CE amplifier. *For the same maximum power hyperbola high-voltage class A amplifiers have greater voltage gain than do low-voltage amplifiers.*

The instantaneous power dissipated by the transistor is $v_{CE}i_C$, and this is $V_{CE}I_C$ when the input signal is zero. With the Q-point at the midpoint of the load line and with a signal present, the instantaneous power to the transistor is less than $V_{CE}I_C$, because the load line is below the power hyperbola everywhere except at the point of tangency. Clearly the transistor cools somewhat when a signal is applied to a class A amplifier (see P16-6).

The actual collector-circuit efficiency is usually no more than 15 to 20% for a practical class A amplifier with a resistive load. To avoid excessive distortion the voltage swing cannot go to the extremes of V_{CC} and zero, because both the saturation and cutoff regions must be carefully avoided. The load might be RC coupled, and losses occur in R_C. Also, power is dissipated in any elements added for bias stabilization, such as an emitter resistance. A marked improvement in the efficiency is obtained by using transformer coupling, which eliminates the dc power lost in R_L. Let us now consider the transformer-coupled class A audio power amplifier, such as the one illustrated in Fig. 16-6.

AMPLIFIER WITH TRANSFORMER-COUPLED LOAD

The iron-core transformer of the amplifier of Fig. 16-6 is designed to operate over a specified frequency range with a specified load resistance. As long as these requirements are reasonably satisfied, the device can be modeled approximately by lumped resistances R_p and R_s placed in series with the primary and secondary coils, respectively, of an *ideal transformer*. In the equivalent circuit of Fig. 16-7 the ideal transformer is a circuit element that is defined by the equations

$$\frac{v_1}{v_2} = \frac{n_1}{n_2} \qquad \frac{i_1}{i_2} = \frac{n_2}{n_1} \tag{16-10}$$

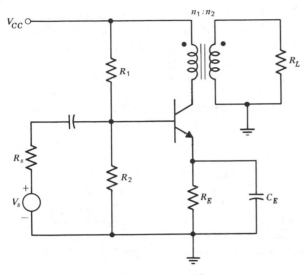

Figure 16-6   Transformer-coupled amplifier.

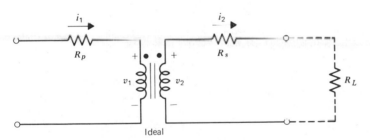

Figure 16-7   Approximate equivalent circuit.

with n_1/n_2 called the *turns ratio*. For the indicated reference directions relative to the transformer dots, v_1 and v_2 have the same sign and i_1 and i_2 also have the same sign, as required by (16-10). From the equations we find that

$$\frac{v_1}{i_1} = \left(\frac{n_1}{n_2}\right)^2 \left(\frac{v_2}{i_2}\right) = \left(\frac{n_1}{n_2}\right)^2 (R_L + R_s) \qquad (16\text{-}11)$$

The ratio $(n_1/n_2)^2$ is referred to as the *impedance ratio* of the transformer. For example, a transformer designed to transform an 8 ohm load into a resistance of 156 ohms has an impedance ratio of 156/8 and a turns ratio of 4.42.

Although the dc resistance of the primary is simply R_p, the ac resistance R_L' is the sum of R_p and the ratio v_1/i_1, or

$$R_L' = R_p + \left(\frac{n_1}{n_2}\right)^2 (R_L + R_s) \qquad (16\text{-}12)$$

By (16-10) the power $v_1 i_1$ into the primary of the ideal transformer equals the power $v_2 i_2$ out of the secondary. Thus the ideal transformer is 100% efficient. However, the actual transformer efficiency is typically about 75%. Some of the power supplied to R_L' is consumed by R_p and R_s, and there are significant iron-core losses as well, though these are not included in the equivalent circuit of Fig. 16-7.

For the amplifier of Fig. 16-6 the load lines are drawn in Fig. 16-8, along with the power hyperbola that passes through the Q-point. The dc collector-circuit resistance is the sum of R_E and R_p, usually no more than a few ohms. The ac load line is tangent to the power hyperbola through the Q-point, and the reciprocal of its slope determines R_L' and the transformer impedance ratio. To find the maximum theoretical collector-circuit efficiency η we assume that R_E, R_p, and R_s are zero. For this case $V_{CE} = V_{CC}$ and $I_C = V_{CC}/R_L'$. The power P_{CC} supplied to the collector circuit by the battery is $V_{CC} I_C$, or V_{CC}^2/R_L'. Assuming a sinusoidal signal, the maximum output power P_o is found from (16-6) and (16-7) to be $V_{CC}^2/2R_L'$ and (16-8) becomes

$$\eta = \frac{V_{CC}^2/2R_L'}{V_{CC}^2/R_L'} \times 100\% = 50\% \tag{16-13}$$

With P_o a maximum, half the power supplied by the battery is converted into output signal power, with the other half converted into heat within the transistor. In practical circuits, efficiencies of 35 to 40% are easily obtained.

To get the most output power with high voltage gain the designer should sketch the maximum power hyperbola of the selected transistor and draw the ac load line from the maximum allowed v_{CE} so that the line is tangent to the hyperbola. After calculating R_L', a suitable choice of a transformer is made with the aid of (16-12), and V_{CC} is found from the dc load line. The geometry is that of Fig. 16-8. For most waveforms and frequencies each point of the ac load line

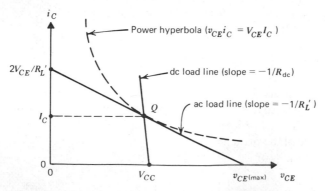

Figure 16-8 Load lines of transformer-coupled amplifier.

must be within the safe operating region as defined on the data sheet. In many cases the supply voltage V_{CC} is specified, and the preceding procedure must then be reversed, with the intersection of the dc load line and the power hyperbola determining the Q-point.

Because the dc load line of a transformer-coupled amplifier crosses the maximum power hyperbola, special care must be taken to ensure that the bias point is properly stabilized. In the amplifier of Fig. 16-6 the stabilization is by means of the emitter resistance R_E bypassed by C_E, an arrangement that is quite common. The proper value of R_E is found from the equation

$$R_E = \frac{(I_{C1}/\beta_{F1} - I_{C2}/\beta_{F2} + \Delta I_{CO})R_B + \Delta V_{BE}}{I_{C2} - bI_{C1} - \Delta I_{CO}} \tag{16-14}$$

with $b = (\beta_{F1} + 1)/\beta_{F1}$ and $\Delta I_{CO} = I_{CO2} - I_{CO1}$ (see P16-8). Equation 16-14 is (15-17) solved for R_F, except that ΔI_{CO} replaces I_{CO2}. In deriving (15-17) we assumed that I_{CO1} was negligible. This is not always true for power transistors, because the lowest junction temperature T_1 can be quite high.

When the quiescent current of a power transistor is large, the input and output impedances are low and R_E may be about one ohm or even less. The impedance from the emitter to ground should be much smaller than this at the lowest frequency of interest. The result is that the required C_E often becomes so large as to be economically impractical. A possible solution is to increase R_E and to replace C_E with a diode connected so that it is forward biased. With this arrangement the resistance to I_E is the dc diode resistance in parallel with R_E. For the signal component of the emitter current the diode presents a much smaller incremental resistance. A procedure that is generally more satisfactory is the utilization of multistage dc feedback for bias stabilization, similar to the methods discussed in Sec. 15-6. Signal feedback is also employed to improve performance and to minimize distortion.

16-3. NONLINEAR DISTORTION

Nonlinear distortion is always present in large-signal amplifiers. An example that we studied earlier is the voltage transfer characteristic of Fig. 12-2 of Sec. 12-1. In a CE amplifier there is input-circuit distortion owing to the nonlinearity of the emitter-base diode characteristic. There is also output-circuit distortion resulting from the dependence of β_F on collector current, which gives a nonlinear relationship between i_C and i_B. Let us first examine the input circuit, with RC coupling as utilized in Fig. 16-6. The incremental base-emitter circuit, with the input coupling capacitor and the emitter bypass capacitor treated as short-circuits, is shown in Fig. 16-9.

THE DYNAMIC INPUT CHARACTERISTIC AND INPUT LOAD LINE

From the circuit of Fig. 16-9 it is easy to deduce that

$$v_{be} = \left(\frac{R_s R_B}{R_s + R_B}\right)\left(\frac{v_s}{R_s} - i_b\right) \tag{16-15}$$

With v_{be} and i_b replaced with $(v_{BE} - V_{BE})$ and $(i_B - I_B)$, respectively, (16-15) becomes

$$v_{BE} = C_1 + v_g - i_B R_g \tag{16-16}$$

with $R_g = R_s \| R_B$, $v_g = v_s(R_g/R_s)$, and $C_1 = V_{BE} + I_B R_g$. Equation 16-16 is the generator-load-line equation, with R_g denoting the effective generator resistance, v_g the effective voltage, and C_1 a constant. At any fixed instant of time a plot of i_B versus v_{BE} is a straight line with the slope $(-1/R_g)$. Its intercept on the v_{BE} axis is

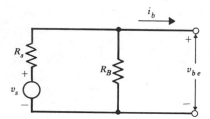

Figure 16-9   Incremental input circuit.

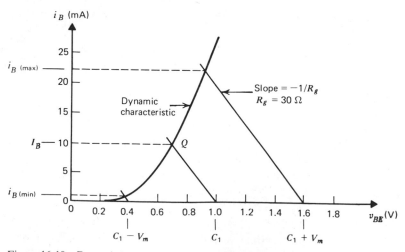

Figure 16-10   Dynamic input characteristic and load lines.

the sum of C_1 and v_g, and for sinusoidal time variations this intercept moves from $C_1 - V_m$ to $C_1 + V_m$, with V_m denoting the maximum value of v_g. Figure 16-10 is a sketch of a typical input characteristic of a CE amplifier, giving plots of the generator load line for v_g equal to $-V_m$, 0, and $+V_m$. The characteristic is obtained from the actual circuit. It is referred to as a *dynamic* characteristic to distinguish it from a static characteristic that has a constant v_{CE}.

Close examination of Fig. 16-10 reveals that the variation of i_B about its quiescent value I_B is not a pure sine function. Very small values of R_g give nearly vertical load lines. This eliminates the distortion of v_{BE} but greatly increases the base-current distortion. On the other hand, as R_g approaches infinity, the load lines become horizontal, and the input-characteristic curvature has no effect on i_B, although v_{BE} is now distorted. These two conditions are illustrated in Fig. 16-11.

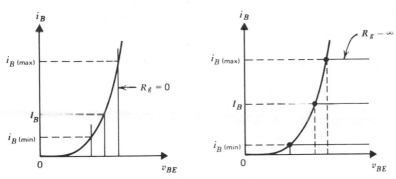

Figure 16-11   Generator load lines for $R_g = 0$ and $R_g = \infty$.

THE DYNAMIC TRANSFER CHARACTERISTIC

Next, let us examine the dynamic characteristic relating i_C and i_B. From the ac load line drawn on the collector characteristics, a curve such as that shown in Fig. 16-12 is obtained. Both Figs. 16-10 and 16-12 apply to the same amplifier, and their numerical values correspond. As i_C rises from a low level, the slope of the curve increases slightly because of the increase in β_F. At higher levels β_F drops because of the reduced emitter efficiency, and the slope decreases appreciably. When the ac load line enters the saturation region, i_C saturates with a slope that is nearly zero.

Figures 16-10 and 16-12 could be used to obtain a plot of i_C versus $(C_1 + v_g)$. This dynamic transfer characteristic could then be employed to sketch the instantaneous i_C corresponding to a sinusoidal v_g, but this will not be done here. Mathematically, the transfer characteristic can be expressed as an infinite power series in the form

$$i_C = A_1(C_1 + v_g) + A_2(C_1 + v_g)^2 + A_3(C_1 + v_g)^3 + \cdots \qquad (16\text{-}17)$$

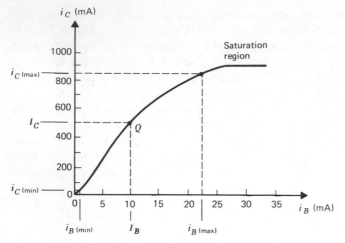

Figure 16-12   Dynamic transfer characteristic.

For $v_g = V_m \cos \omega t$, by means of appropriate trigonometric identities we can convert (16-17) into the cosine series

$$i_C = I_C + I_0 + I_1 \cos \omega t + I_2 \cos 2\omega t + I_3 \cos 3\omega t + \cdots \qquad (16\text{-}18)$$

PERCENT HARMONIC DISTORTION

In (16-18) I_C is the quiescent current, and the sum of I_C and I_0 is the average, or dc, collector current. The magnitude of the fundamental, or first harmonic, is I_1. This has the frequency of the input signal and is the desired output. The second and higher harmonics are distortion terms. The *percent distortion D* is defined by

$$D = \frac{\sqrt{I_2{}^2 + I_3{}^2 + I_4{}^2 + \cdots}}{I_1} \times 100\% \qquad (16\text{-}19)$$

The *second harmonic distortion D_2* is $(I_2/I_1) \times 100\%$.

A useful first-order approximation is to drop harmonics above the second in (16-18). With reference to Fig. 16-12, at $\omega t = 0$, $v_g = V_m$, and $i_C = i_{C(\max)}$. At $\omega t = \pi/2$, $i_C = I_C$. Finally, at $\omega t = \pi$, $i_C = i_{C(\min)}$. Putting these values into (16-18) with the third and higher harmonics eliminated, we obtain the equations

$$\begin{aligned}
i_{C(\max)} &= I_C + I_0 + I_1 + I_2 \\
I_C &= I_C + I_0 - I_2 \\
i_{C(\min)} &= I_C + I_0 - I_1 + I_2
\end{aligned} \qquad (16\text{-}20)$$

These can be solved for I_0, I_1, and I_2. From the results the second harmonic distortion is determined to be

$$D_2 = \frac{i_{C(\max)} + i_{C(\min)} - 2I_C}{2i_{C(\max)} - 2i_{C(\min)}} \times 100\% \tag{16-21}$$

From this we note that D_2 is zero if the quiescent current is midway between the maximum and minimum values.

Let us now reexamine Figs. 16-10 and 16-12. In Fig. 16-10 I_B is below the midpoint between the maximum and minimum values. In Fig. 16-12 the reduced slope at the higher values of i_C tends to compensate for the input circuit distortion. Clearly, the proper selection of the generator resistance R_g is important in minimizing distortion. *In CE class A amplifiers the curvatures of the input and current transfer characteristics are usually such that distortion is minimized when the generator resistance is about equal to the incremental input resistance at the Q-point.* In Fig. 16-10 the 30 ohm resistance of the generator is somewhat too large for minimum distortion (see P16-9). Using values of 850, 500, and 30 mA for the respective maximum, quiescent, and minimum collector currents, with these data taken from Fig. 16-12, the second harmonic distortion is determined from (16-21) to be 7.3 %.

CE class A audio power amplifiers have input resistances from a few ohms to perhaps several hundred ohms, depending on the bias current and the transistor. A CC driver stage is especially suitable. Because the load on the driver is the low input resistance of the power stage, both the CE and CC configurations are equally satisfactory from the viewpoint of transducer gain, as indicated in Fig. 13-37 of Sec. 13-7. However, the CC amplifier with its moderately low output impedance presents a more suitable value of R_g, resulting in lower distortion in the CE output stage. Also, in some applications the input to the driver comes from a detector or other network that should not be loaded appreciably, and the high input impedance of the CC amplifier is advantageous. CE amplifiers are sometimes used as drivers. In order to give a small, effective generator resistance for low distortion, and for good power transfer, transformer coupling between the CE stages is often employed.

CLASS A COMMON-BASE AND EMITTER-FOLLOWER AMPLIFIERS

Our attention thus far has been on the CE power amplifier, but both CB and CC configurations have certain advantages, as well as disadvantages. In Fig. 16-13 is a CB amplifier with transformer coupling at both input and output. The transfer characteristic relating the output collector current to the input emitter current is approximately a straight line through the origin. The nonlinearity would not be apparent, because i_C and i_E are nearly equal. By selecting the turns ratio so that the effective generator resistance R_g is large compared with the very

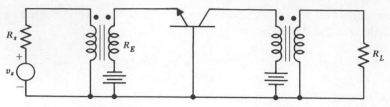

Figure 16-13   CB amplifier.

small input resistance of the CB amplifier, a sinusoidal signal drives a sinusoidal input current, and the output distortion is quite low, usually below 2%. However, the power gain suffers from the mismatch. With a turns ratio that matches the impedances at the input, the input diode characteristic produces excessive distortion. The designer must compromise.

The CB collector characteristics, consisting of plots of i_C versus v_{CB} for constant values of i_E, are nearly equally spaced horizontal lines for all values of v_{CB} greater than zero. This was shown in the curves of Fig. 2-9f of Sec. 2-4. Active-mode operation extends to zero v_{CB}. The output voltage can be driven from zero volts to $2V_{CC}$, with increased collector-circuit efficiency. *The main disadvantage of the CB amplifier is the low transducer gain*, as evidenced by the curves of Fig. 13-37.

Figure 16-14 illustrates the circuit of a CC power amplifier. Resistance R_2 is the dc resistance of the secondary of the input transformer, and the voltage V_B of

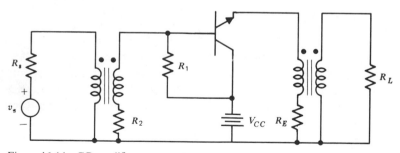

Figure 16-14   CC amplifier.

the two-battery equivalent circuit is developed from R_1, R_2, and V_{CC}. Resistance R_E is the dc resistance of the primary of the output transformer. This arrangement avoids the necessity of a separate emitter resistor along with its bypass capacitor, at the same time conserving both voltage and power. The low output resistance of the CC amplifier is conducive to direct coupling with an electrodynamic speaker having a typical impedance at 500 Hz of 3.2, 4, 8, or 16 ohms. With direct coupling, the load resistance becomes the stabilizing resistance R_E.

Distortion is negligible provided the effective generator resistance is small compared with the input resistance. This gives a base-collector input voltage equal to the signal voltage, and the emitter-collector output voltage is nearly identical to the base-collector input voltage. On the other hand, a large R_g makes the input base current proportional to the signal voltage. Because the output current i_E is about the same as i_C, the current transfer characteristic i_E-versus-i_B is similar to that of the CE configuration with similar distortion.

The transformed load resistance R_L' is determined from the ac load line drawn tangent to the maximum power hyperbola. This resistance is small when V_{CC} is small. Unless this is the case, the gain is much less than that of the CE configuration, as shown by the curves of Fig. 13-37. *Thus the CC class A power amplifier is rarely used unless the available voltage supply is quite small, in which case the CC amplifier has considerable merit.*

Although both the CB and CC amplifiers have certain advantages, especially with respect to linearity, most class A audio power amplifiers are common emitter. Signal feedback is frequently employed to reduce distortion, and this is discussed in Sec. 20-2. Typically, the power to the load is less than 2 W with a distortion of about 3%. A 2 W amplifier that delivers power to an electro-dynamic speaker having an efficiency of 2% produces 40 mW of acoustical power. Such amplifiers are widely used in many low-cost power-line-operated AM and FM receivers and most automobile radios, as well as TV sets.

16-4. CLASS A CE POWER AMPLIFIER DESIGN

SPECIFICATIONS

Most of the concepts introduced thus far in this chapter can be brought together and illustrated through the design of a low-cost audio amplifier. The amplifier is required to deliver up to 500 mW with less than 5% harmonic distortion to a transformer-coupled 4 ohm speaker. The input impedance should be greater than 4 kΩ in order to avoid undue loading of the detector that supplies the signal, and this signal voltage is a maximum of 110 mV rms. The ambient temperature range is from 25 to 60°C, and the available supply voltages are 12 and 90 V. These specifications apply to the audio section of many small-screen TV sets.

THE POWER TRANSISTOR AND HEAT SINK

In order to keep the cost low, the power stage should have high voltage gain, which will make it possible to meet the specifications with only two stages— a driver and a power output stage. High gain requires the use of a large effective load resistance R_L', and this in turn requires a transistor with a high breakdown

V_{CEO} voltage. The 90 V supply is not all available for V_{CE}, because there will be a few volts taken by R_E and the dc resistance of the transformer. However, we might expect V_{CE} to be between 80 and 85 V, and the breakdown V_{CEO} voltage must be at least twice this, or 170 V. The maximum theoretical efficiency is 50%, but the losses in R_E and the transformer will reduce this to around 35%. Thus we should select a transistor and a heat sink so that the maximum power hyperbola at the minimum quiescent current is at least 1.5 W.

The selected silicon transistor that will meet the specifications has an allowable junction temperature of 200°C, a junction-to-case thermal resistance of 35°C/W, a minimum breakdown voltage of 175 V, and a maximum I_{CBO} of 50 μA at 150°C. Allowing a margin of safety, we select 190°C as the maximum junction temperature T_2, with the environment at 60°C. For a 2 W maximum power hyperbola, the total thermal resistance must be no more than the 130°C temperature difference divided by 2, or 65°C/W. A commercial heat sink with a case-to-ambient thermal resistance of 30°C/W will be used.

THE Q-POINT AND THE EFFECTIVE LOAD RESISTANCE

At the minimum ambient temperature of 25°C with β_F having its lowest value β_{F1}, the Q-point on the nearly vertical dc load line must be selected well below the 2 W hyperbola. A reasonable choice at the minimum quiescent current I_{C1} is a 1.6 W hyperbola, which makes this current about 20% less than I_{C2}. From the thermal resistance we find the junction temperature T_1 to be 129°C, with $T_A = 25°C$. The minimum β_{F1} at 129°C is 50 and the maximum β_{F2} at 190°C is 200, with these values obtained from the data sheet.

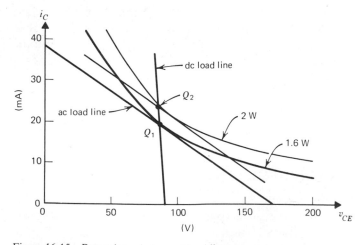

Figure 16-15   Power hyperbolas and load lines.

The quiescent point Q_1 corresponding to I_{C1} is the intersection of the dc load line ($v_{CE} = 90 - i_C R_{dc}$) with the 1.6 W hyperbola ($v_{CE} i_C = 1.6$). Estimating R_{dc} at 300 ohms, an estimate that can later be corrected, the quiescent point is found to be at $I_{C1} = 19$ mA and $V_{CE1} = 84.3$ V. This determination is made either analytically from the equations or graphically from the curves of Fig. 16-15. Perhaps it is well to note that the case is charged to 84 V, presenting a shock hazard as well as a burn hazard. The ac load line passes through the Q-point, with intercepts on the i_C and v_{CE} axes of 38 mA and 168.6 V, which is $2V_{CE1}$. From these values the effective load resistance R_L' is found to be 4440 ohms.

OUTPUT POWER AND VOLTAGE GAIN

The data sheet reveals that the saturation region is avoided as long as v_{CE} is greater than 1 V. Thus the maximum output power occurs for a voltage swing from 1 V to 167.6 V, giving an rms value of 58.9 V. As R_L' is 4440 ohms, the maximum power to this effective load is 781 mW. For a transformer efficiency of 75%, the power to the load is 586 mW. The battery power is $90I_{C1}$, or 1710 mW, and the collector-circuit efficiency is 34.3%.

The voltage gain from the input of the CE power stage to R_L' is approximately $\beta_o R_L'/R_{in}$. At T_1, $q/2kT$ is 14 and g_m is $14I_{C1}$, or 0.27 mho. For an ac beta range from 50 to 200, r_π is between 190 and 740 ohms. These are only approximations, but from them we deduce that the input resistance is probably in the range from about 300 to 850 ohms, allowing 110 ohms for base ohmic resistance. Using the minimum ac beta of 50 and the corresponding R_{in} of 300 ohms, we find the

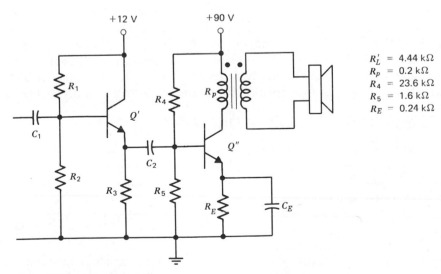

+12 V   +90 V

R_L'	= 4.44 kΩ
R_p	= 0.2 kΩ
R_4	= 23.6 kΩ
R_5	= 1.6 kΩ
R_E	= 0.24 kΩ

Figure 16-16   500-mW amplifier.

voltage gain to be 740. For the rms voltage of 58.9 across R_L', the required input voltage is 80 mV. From this estimate it is reasonable to expect that no additional voltage gain is needed, and a CC driver can be used. The low output impedance provides a suitable effective generator resistance for driving the power stage with low distortion. In addition, the high input impedance is essential for satisfying the specifications. The CC stage is used as an *impedance transformer* that supplies both current and power gain. RC coupling is employed for this design, as shown in Fig. 16-16.

BIAS RESISTORS

For a 4 ohm speaker at the output, the desired impedance ratio of the transformer is 4440/4. The dc resistance of the primary is assumed to be 200 ohms. We need to determine the bias resistors R_4, R_5, and R_E. With reference to the *output stage*, I_{C1} is 19 mA, and I_{C2} is $I_{C1}/0.8$, or 23.8 mA. This value can also be determined graphically, with I_{C2} read from the intersection of the dc load line with the ac load line drawn tangent to the 2 W hyperbola of Fig. 16-15. The minimum and maximum betas have been found to be 50 and 200. The term ΔV_{BE} is the product of $(T_2 - T_1)$ and 0.0025, or 0.153 V. As I_{CBO} is specified to be 50 μA maximum at 150°C, I_{CO1} and I_{CO2} are calculated from (16-3). Their values are 19 and 317 μA, and ΔI_{CO} is 0.298 mA. A suitable choice for R_B is 1.5 kΩ, which is nearly twice the largest expected input resistance of 850 Ω. Substituting the proper numerical values into (16-14) gives 241 ohms for the minimum value of R_E. At 129°C the data sheet specifies 0.5 V for V_{BE}. From (15-10), modified with the aid of (15-7) to include I_{CO1}, we find that V_B is 5.71 V, and R_4 and R_5 are determined from (13-6) to be 23.6 and 1.6 kΩ, respectively.

Unfortunately, our original estimate of the dc resistance of the collector circuit was rather low. The sum of R_p and R_E is 441 ohms, whereas our estimate was 300 ohms. For the selected Q-point (19 mA, 84.3 V), with a dc resistance of 441 ohms the voltage V_{CC} would have to be 92.7 V. To complete the design of the power stage, the procedure should be repeated using 441 ohms in place of 300. This is left as an exercise (see P16-12). The results are not significantly changed.

THE EMITTER-FOLLOWER DRIVER

The input voltage of the driver stage may be as large as 100 mV. However, if the voltage gain is greater than 0.95, the incremental base-emitter voltage is less than 5 mV, and the driver is a linear small-signal amplifier. From (13-23) the input resistance R_i and the voltage gain A_v are

$$R_i = r_x + r_\pi + (\beta_o + 1)R_L \qquad A_v = \frac{(\beta_o + 1)R_L}{R_i} \qquad (16\text{-}22)$$

with R_L denoting the effective load resistance, which is the parallel equivalent of R_3 and the input resistance of the output stage. Clearly, we desire a large β_o to make R_i large and A_v nearly unity. A transistor with a minimum beta of 50 is adequate. The bias point should be chosen so that beta is large, V_{CE} is adequate, and R_3 is at least several kilohms. After completion of the design of the driver, the amplifier should be built and tested to ascertain that the distortion requirements and other specifications are satisfied. Standard components should be selected to correspond closely with the design values.

One objection to the amplifier is the coupling capacitor C_2, which must be large because of the low values of the coupled impedances. Capacitor C_2 can be eliminated, along with two resistors, by coupling the stages directly as shown in Fig. 16-17. Except for the selection of R_E, the design and performance of the power stage is unchanged. However, its stability is now affected by the resistors of the driver, as well as by R_E.

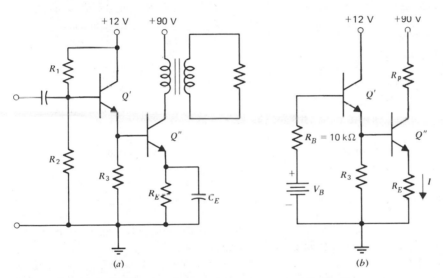

Figure 16-17   Direct-coupled amplifier. (a) Actual circuit. (b) DC circuit.

DIRECT-COUPLED AMPLIFIER DESIGN

The design will employ the same notation as before with respect to subscripts 1 and 2. A single prime is used to denote a quantity referring to the driver transistor Q', with a double prime for the power device Q''. The resistance R_B is detrimental to bias stability and should be as small as possible. However, it shunts the input so we select $R_B = 10 \text{ k}\Omega$. The input impedance must be maintained above 4 $\text{k}\Omega$ according to the specifications.

The driver BJT dissipates only a few milliwatts and thus its junction temperature nearly equals that of the environment. Accordingly, its V_{BE1}' is 0.7 V,

$\Delta V_{BE}'$ is 0.088 V, and I_{CO}' is negligible. The minimum and maximum values of β_F' are 50 and 200 at about 2 mA. The ambient temperature varies from 25 to 60°C.

In the discussion that follows, units of kΩ, mA, and V are understood. From Fig. 16-17b we get

$$IR_E = V_B - V_{BE}' - V_{BE}'' - \frac{10I_C'}{\beta_F'} \qquad (16\text{-}23)$$

We know that I is nearly the same current as I_C''. It is well-stabilized if we make the constant voltage V_B large compared with the changes in the other three terms on the right side of (16-23). The value of V_B is only one or two volts greater than IR_E. Clearly, the larger the value selected for R_E, the greater will be V_B, with improved bias stability. If we select R_E for a 4 V drop, V_B will be about 5 or 6 V, which should be sufficient to keep I from changing more than 20%. In (16-23) the maximum changes in the emitter-base voltages are 0.088 and 0.153. The change in the term $10I_C'/\beta_F'$ is $0.20I_{C1}'$ minus $0.05I_{C2}'$, which is easily kept under 0.5 V by choosing I_C' between 2 and 3 mA. Accordingly, let us tentatively select 0.2 kΩ for R_E.

The design of the remainder of the circuit will be at the minimum junction temperatures of 25°C and 129°C with each beta at its minimum of 50 and with $I_{CO}'' = 0.019$ mA. The value of I_{C1}'' is 19 mA, I_{B1}'' is 0.36 mA including the effect of I_{CO}, and I is 19.4 mA. This makes $V_{E1}'' = 3.88$ V. Adding V_{BE1}'' to this gives $V_{E1}' = 4.38$ V. The resistance R_3 must now be selected. It should be large compared with the input resistance of the power stage, at the same time giving a suitable bias current for the driver BJT. Let us select $R_3 = 2$ kΩ, which gives a current of 2.19 mA in this resistor. The value of I_{E1}' is this current plus I_{B1}'', or 2.55 mA; I_{C1}' becomes 2.50 mA, and I_{B1}' is 0.050 mA. Adding 0.7 to V_{E1}' gives $V_{B1}' = 5.08$ V. The sum of V_{B1}' and $10I_{B1}'$ is V_B, and V_B is 5.58 V. From V_B and R_B, using (13-6), we get $R_1 = 21.5$ and $R_2 = 18.7$ kΩ. The complete circuit is shown in Fig. 16-18, with standard resistance values.

In order to verify that I_{C2}'' does not exceed 23.8 mA, which is the value at quiescent point Q_2 of Fig. 16-15, we must now solve for I_{C2}'' using the circuit of Fig. 16-18. The procedure is similar. We begin with the unknown current I_{C2}'' instead of the known 19 mA that is I_{C1}''. We determine I_{B2}' and V_{B2}' in terms of I_{C2}'', which is then found by means of a loop equation around the V_B-R_B circuit. The analysis is simplified with negligible effect on the result by taking the maximum betas to be infinite, which is the worst case. Accordingly, I_{B2}' is zero, and V_{B2}' is V_B. We need only to find V_{B2}' in terms of I_{C2}'' and equate the result to V_B. We assume that V_{BE}' is 0.7 V at 25°C and V_{BE}'' is 0.5 V at 129°C.

By (16-23), V_B is the sum of $0.2I_{C2}''$ and the two base-emitter voltages. At 60°C V_{BE2}' is 0.612, and at 190°C V_{BE2}'' is 0.347 V. For the values of Fig. 16-18, V_B is 5.4 V. We now determine I_{C2}'' to be 22.2 mA, which is nearly 7% less than the maximum allowed value, and we see that the design is adequate. In fact, R_E could be reduced slightly.

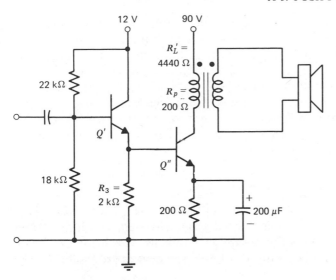

Figure 16-18   500-mW direct-coupled amplifier.

It is interesting to note that I_{CO} *has practically no effect on the bias stability of the direct-coupled power stage.* This current does not appear in (16-23) at any temperature except as a rather inconsequential part of the emitter current I. In the RC-coupled amplifier I_{CO} was forced to flow through the effective R_B of the power stage. In the direct-coupled amplifier of Fig. 16-18 the current through R_3 is well-stabilized by the bias circuitry of the driver stage. Any change in I_{CO} appears as a change in the emitter current of Q_2, and this current is not at all critical.

The amplifier of Fig. 16-18 has been tested and found satisfactory. For most applications the gain is more than adequate, and some signal feedback can be used. This is easily accomplished by inserting 2 or 3 ohms of unbypassed resistance in series with the 200 ohm emitter resistance of Fig. 16-18. The result is better performance and reduced distortion in exchange for a small sacrifice in gain.

The design of a *low-voltage* class A power amplifier is similar to the example considered. However, the effective ac load resistance R_L' will be much smaller, with reduced voltage gain. As a consequence, if a CC driver is used, it is usually necessary to precede it with a CE stage.

16-5. PUSH-PULL AMPLIFIERS

An efficiency substantially greater than that of a class A circuit is obtained by using two transistors in a power stage operating class AB. The devices are connected in some form of a *push-pull* arrangement that enables one transistor to amplify the positive part of the input signal with the other amplifying the negative part. The greater efficiency not only conserves power but also permits the use of relatively small and inexpensive transistors. In addition, the configuration is such that *even harmonics are suppressed with lower distortion*. When an output transformer is used, the push-pull circuit has dc currents that are of equal magnitude and in opposite directions in the two halves of the primary coil. Thus, there is no dc magnetization of the iron core, which reduces the cost and size of the transformer. Most power amplifiers with outputs of more than a few watts, as well as many low-power amplifiers, are push-pull.

A CLASS B PUSH-PULL CIRCUIT

Let us now consider the widely used circuit of Fig. 16-19. The transistors are in the CE configuration, with the inputs fed into the base-emitter terminals and the outputs taken from the collector-emitter terminals. This is the usual case when both transistors are NPN or both PNP. The devices are assumed to be

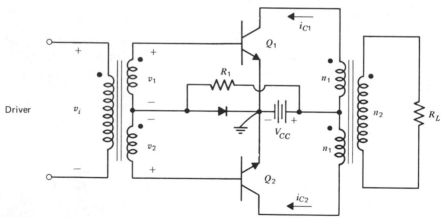

Figure 16-19   Push-pull amplifier.

identical, with each biased near cutoff on the knee of the input diode characteristic. Because this characteristic is temperature sensitive, the base-emitter bias voltage is taken from a forward-biased diode that is thermally bonded to the same heat sink as the output units.

For convenience let us idealize the situation somewhat by assuming that the diode voltage is 0.6 V and that each transistor is cut off when its v_{BE} is at and below 0.6 V. Suppose a sinusoidal signal v_i is supplied to the primary of the input transformer. Voltages v_1 and v_2 appear across the two halves of the secondary, and for the specified reference directions, v_1 and v_2 are equal in magnitude but opposite in sign. When the input v_i is positive, making v_1 positive, v_{BE1} is greater than 0.6 V and transistor Q_1 conducts. At the same time the negative v_2 maintains Q_2 in cutoff. In the next half-cycle, Q_1 is cut off and Q_2 conducts. Each transistor is *on* 50% of the time and *off* the other 50%, which is class B operation.

In the collector circuits when v_i is positive, i_{C1} is a positive half-cycle of a sine wave and i_{C2} is zero. In the next half-cycle when v_i is negative, i_{C1} is zero and i_{C2} is a positive half-cycle of a sine wave. These alternate half-cycles of currents through the n_1 turns of half the primary of the output transformer produce a sinusoidal magnetic flux in the iron core, which generates a sinusoidal voltage across R_L. Thus, the amplification is linear.

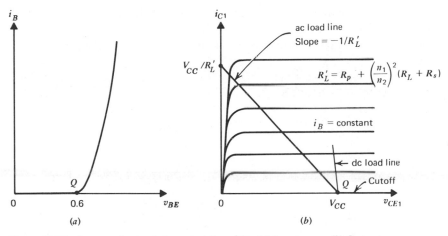

Figure 16-20   Input and output characteristics of Q_1. (*a*) Input curve. (*b*) Output curves.

Shown in Fig. 16-20 are assumed input and output characteristics of Q_1, along with the ac load line drawn through the Q-point of the output circuit selected at cutoff for class B operation. During the positive sinusoidal half-cycles of i_{C1}, the current i_{C2} is zero and hence the n_1 turns that are in series with the collector of Q_2 have no effect on i_{C1}. Consequently, the effective load resistance R_L' across that half of the primary carrying the current i_{C1} is given by (16-12) and indicated on the figure, with the understanding that n_1 denotes the turns of half the primary and R_p is the dc resistance of the n_1 turns.

COLLECTOR-CIRCUIT EFFICIENCY AND TRANSISTOR DISSIPATION

The current i_{C1} is a half-cycle sine wave when the input voltage v_i is positive. If this current were a full sine wave, the power to R_L' would be $\frac{1}{2}I_{max}^2 R_L'$, but because the current is zero half the time, the power is only one-half of this. However, i_{C2} supplies an equal wattage, and accordingly, the expression $\frac{1}{2}I_{max}^2 R_L'$ truly represents the total power into the primary of the transformer. With a maximum signal that drives v_{CE1} to zero, I_{max} is V_{CC}/R_L', which is evident from Fig. 16-20b, and the power can be expressed as $\frac{1}{2}I_{max} V_{CC}$. The current through the battery is a rectified sine wave, being the sum of i_{C1} and i_{C2}. The average value of this current is $(2/\pi)I_{max}$, which gives a battery power of $(2/\pi)I_{max} V_{CC}$. With maximum signal, the ratio of the power to the transformer to the battery power yields the maximum theoretical collector-circuit efficiency η for a class B amplifier, and the result is

$$\eta = \left(\frac{\pi}{4}\right) \times 100\% = 78.5\% \tag{16-24}$$

Practical efficiencies range from 55% to 65%. There are losses in the transformer and bias resistors. Furthermore, v_{CE} cannot be driven into the saturation region without excessive distortion, and as we shall see, a bias point at cutoff is impractical.

The difference between the battery power and that supplied to the transformer represents the total dissipation $2P_D$ in the two transistors. Therefore,

$$2P_D = \frac{2}{\pi} I_{max} V_{CC} - \frac{1}{2}I_{max}^2 R_L' \tag{16-25}$$

Setting the derivative of P_D with respect to I_{max} equal to zero, we find the maximum transistor dissipation occurs when I_{max} is $(2/\pi)V_{CC}/R_L'$, which gives

$$P_{D(max)} = \frac{V_{CC}^2}{\pi^2 R_L'} \tag{16-26}$$

The maximum theoretical output power $P_{o(max)}$ is $\frac{1}{2}I_{max}^2 R_L'$ with I_{max} equal to V_{CC}/R_L'. From this and (16-26) we obtain the ratio

$$\frac{P_{D(max)}}{P_{o(max)}} = \frac{2}{\pi^2} \approx 0.2 \tag{16-27}$$

According to (16-27), a class B amplifier designed to deliver 50 W to its output transformer has a maximum power dissipation per transistor of 10 W. In contrast, the BJT of a 50 W class A amplifier with a theoretical conversion efficiency of 50% would dissipate 100 W with zero signal, with an enormous heat-sink problem. While these results illustrate a major advantage of the class B amplifier, it should be kept in mind that they are based on sinusoidal signals and theoretical conditions. Another significant feature of class B operation is the very low drain on the battery when there is no signal. This is especially important in certain applications, such as hearing aids.

CROSSOVER DISTORTION

The input characteristic of Fig. 16-20a is represented as being at cutoff at 0.6 V. Actually, cutoff is more gradual, as shown in Fig. 16-10 of Sec. 16-3. With reference

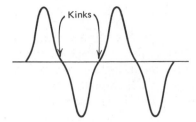

Figure 16-21   Crossover distortion.

to this figure it is clear that a Q-point at zero base current distorts because of the curvature of the characteristic at low current levels. For a sinusoidal input the waveform of the output voltage appears as in Fig. 16-21. The kinks, which result from gradual cutoff, are referred to as *crossover distortion*.

Although not usually practical, crossover distortion can be reduced by using a high effective generator resistance R_g, which drives a base current proportional to the signal voltage on the conducting half-cycle. Multistage signal feedback is nearly always employed, and this reduces all forms of nonlinear distortion, including crossover distortion. For large signals the kinks of Fig. 16-21 have little effect, but when the signal is small, crossover distortion that results from class B operation is intolerable. *The kinks are nearly eliminated by biasing the stage slightly above cutoff*, which gives a small dc current through the battery even when the signal is zero. This is class AB operation, even though such amplifiers are oftentimes referred to as class B.

HIGH-POWER CONSIDERATIONS

A 50 W public-address amplifier can easily be designed using the circuit of Fig. 16-19 as the output power stage, with a supply voltage of 15 and an R_L' of 1.5 ohms (see P16-16). At full power the maximum collector current of each transistor is perhaps 9 A. At such a high current the β_F is likely to be low, probably from 5 to 20. For a β_F of 10 the maximum base current is 0.9 A. Assuming each half of the secondary of the input transformer works into an incremental resistance of 2 ohms, the total power required from the secondary is 0.81 W for a sinusoidal signal. Multistage signal feedback would surely be used. Some form of bias stability is essential. One possible arrangement is to connect a 0.1 ohm 10 W resistor between ground and the common emitters so that each emitter current passes through this resistor. A bypass capacitor could not be used because the dc level varies with the signal, and furthermore, to effectively bypass the small resistance would require a great deal of capacitance. The signal feedback that results from R_E reduces distortion. However, this is a dubious benefit, because multistage feedback is more effective and can be accomplished without sacrificing output power.

TRANSFORMERLESS CIRCUITS

The amplifier of Fig. 16-22 is designed for push-pull operation without an output transformer. Resistors R_1 and R_2 provide suitable bias for active-mode operation, along with signal feedback. Transistor Q_1 and its associated circuitry constitute an emitter follower and the circuit of Q_2 is a CE amplifier. However, with each of the matched transistors replaced with a linear incremental model using only the h_{ie} and h_{fe} parameters, analysis shows that the individual circuits of Q_1 and Q_2 have *identical voltage gains* except for the phase inversion of the CE stage. Furthermore, they also have *identical output resistances* (see P16-14).

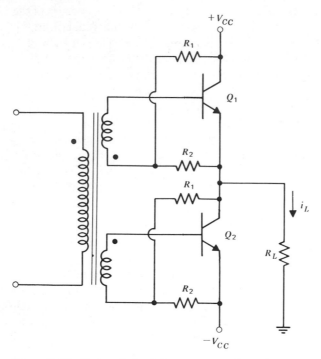

Figure 16-22   Push-pull circuit (no output transformer).

These results show that the signal feedback modifies the circuit in a manner that gives *matched characteristics* to the different configurations. Consequently, class B operation is possible with minimum distortion. Suppose the transistors are biased at cutoff, and a signal is applied. When Q_1 is *on*, Q_2 is *off* and i_L is positive. On the other hand, when Q_1 is *off*, Q_2 is *on* and i_L is negative.

There are various ways to eliminate the input transformer of a push-pull amplifier. When the power requirement of the driver is low, the *phase splitter* of Fig. 16-23 can replace the transformer. Analysis shows that the output voltages

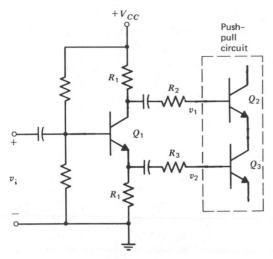

Figure 16-23   Phase splitter.

v_1 and v_2 are almost equal and opposite (see P16-17). Disadvantages are the necessity for maintaining balanced loads and the low voltage gain, which is less than unity because of the large amount of negative feedback. Another phase-inverting circuit (see P16-18) is the emitter-coupled amplifier of Fig. 16-24, which was examined in Sec. 14-1. However, the most common and easiest way to eliminate the input transformer is to remove the need for phase inversion. This is accomplished by using a PNP and an NPN transistor in the push-pull circuit,

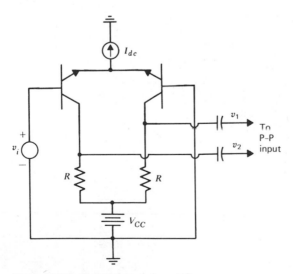

Figure 16-24   Emitter-coupled amplifier.

in place of two identical devices. Fortunately, the arrangement also permits removal of the output transformer. In many applications the size, weight, and cost of high-quality iron-core transformers are prohibitive. The transformer introduces nonlinear distortion, and because it attenuates very low and high frequencies, it contributes to *frequency distortion*. In addition, the design of feedback circuitry is complicated by transformer phase shifts.

COMPLEMENTARY AMPLIFIERS

The output stage of a push-pull amplifier requiring no transformers is shown in Fig. 16-25. The circuit is designed so that the point between the diodes is at ground potential when the input signal is zero, and the diodes provide each base-emitter input with about 0.6 V. The no-signal dc bias is such that the output transistors are approximately at cutoff. By thermally bonding the diodes to the heat sinks of their respective transistors, the proper bias is maintained at all temperatures. When the signal voltage at the input is positive, the NPN transistor Q_1 is *on*, but Q_2 is *off*, and i_L is positive. In the next half-cycle the PNP transistor

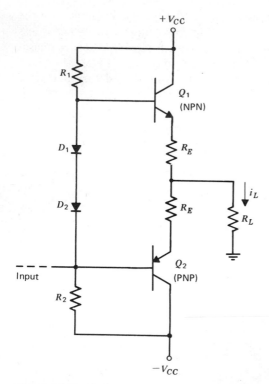

Figure 16-25 Push-pull class B complementary amplifier.

Q_2 is *on*, Q_1 is *off*, and i_L is negative. The use of an NPN and a PNP transistor in the output stage is referred to as *complementary symmetry*, and the circuit is called a *complementary amplifier*.

The emitter-follower configuration of Fig. 16-25 has several important advantages over other arrangements. The bias circuitry is relatively simple, and the low output impedance is desirable because typical loads are only a few ohms. There is considerable tolerance to mismatch in the complementary pair. Because of the small and nearly constant value of v_{BE} in the active mode, the instantaneous output voltage at the emitter differs from the input voltage at the base by only a small constant value. Therefore, the voltage gain is approximately unity, nearly independent of β_F and V_{BE}. In addition to the resultant tolerance to mismatch, this property provides reasonably linear voltage gain at different signal levels. The class A driver can be direct coupled, feeding into the high input impedance typical of emitter followers.

Complementary amplifiers not utilizing Darlington pairs are normally used in amplifiers with outputs in the 3 W to 5 W range. For larger outputs the class A driver with its typical inefficiency will consume appreciable power, and the heat sink becomes more expensive. Also, the increased drain on the power supply results in the introduction into the driver stage of larger ripple voltages, which are amplified by the push-pull stage and produce hum in the output. This hum is kept within tolerable limits only with the use of excessively large capacitors in the power-supply filter circuits. At higher power levels each transistor of Fig. 16-25 is replaced with a "composite" device, or *compound* transistor, referred to as a *Darlington pair*. Two audio amplifiers with Darlington pairs are examined in the following section.

16-6. CLASS B AMPLIFIER DESIGN

Examples of Darlington compound transistors are shown in Fig. 16-26. Each configuration is a two-stage direct-coupled arrangement having characteristics similar to those of a single transistor with a very large current gain β_F. The labeled terminals B, E, and C correspond to those of the equivalent NPN or PNP transistor. The Darlington NPN compound transistor employs two NPN devices, whereas the complementary pair uses an NPN and a PNP transistor. Packaged monolithic Darlington-connected power transistors are available with a variety of current, voltage, and power ratings. Typical values of β_F range from 1000 to 10 000. With the resistors R_1 and R_2 of Fig. 16-26 omitted, the overall β_F is approximately equal to the product of the individual betas. This is true for both the dc and ac betas (see P16-19).

In a complementary amplifier either of the NPN configurations can be used in conjunction with either of the PNP configurations. With diode biasing as in Fig. 16-25 a combination of the NPN Darlington transistor of Fig. 16-26a and the PNP complementary Darlington transistor of Fig. 16-26d requires only

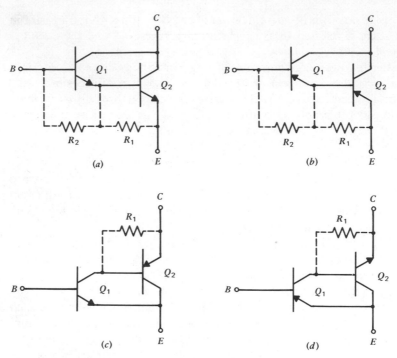

Figure 16-26   Compound transistors $(R_1 \approx 150\,\Omega, R_2 \approx 10\,\mathrm{k\Omega})$. (a) NPN Darlington. (b) PNP Darlington. (c) NPN complementary Darlington. (d) PNP complementary Darlington.

three diodes, because with the emitters connected together there are three base-emitter junctions between the two base terminals. The use of the Darlington pairs of Fig. 16-26a and b requires four diodes.

The resistors of Fig. 16-26 labeled R_1 provide a path for the leakage currents of the driver and output transistors. When each of the junctions is reverse biased, with R_1 omitted the leakage currents of the reverse biased junctions flow through the output transistor, generating higher junction temperatures and possibly starting a thermal-runaway condition. In addition, when the compound transistor is conducting in the forward direction, the resistor R_1 prevents the current of the input transistor from becoming excessively low at small load currents, with consequent low beta. Resistor R_1 is typically about 150 ohms and is normally used in both discrete and integrated circuitry. Resistor R_2 with a value of about 10 kΩ is common in monolithic circuits. By providing an additional path for leakage currents, it further enhances thermal stability.

In monolithic Darlington-pair circuits the driver transistor is included in the power transistor package. The designer does not specify and select driver transistors, and driver heat sinks are eliminated. Advantages are easier design, simpler and smaller circuits, and lower cost. The design of a 15 W audio amplifier will now be considered, with this design followed by a brief discussion of a 50 W amplifier.*

SPECIFICATIONS

The amplifier must be capable of supplying a continuous output of 15 W to an 8 ohm speaker with less than 0.5% harmonic distortion. A bandwidth from 50 Hz to 20 kHz is desired. The full output power must be obtained with an input of approximately 1 V rms supplied to an input impedance of 10 kΩ. The specified maximum ambient temperature is 55°C. A single unregulated power supply will be used, with the voltage dropping by perhaps 10% at full load.

The circuit[8] of Fig. 16-27 will be used because it is simple and adequate. The CE input stage gives the necessary voltage gain, and a V_D multiplier provides the

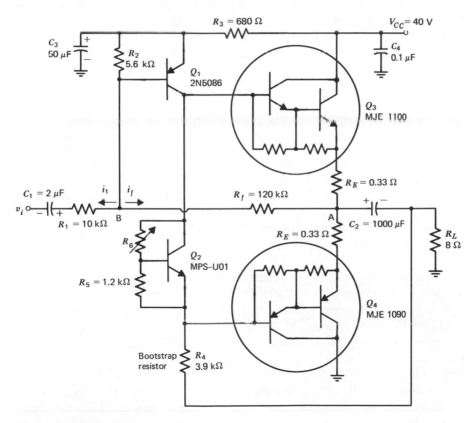

Figure 16-27 A 15-W audio amplifier.

* The circuits examined were taken from an application note prepared by R. G. Ruehs of Motorola Semiconductor Products, Inc., Refer to Reference 8.

dc bias for the complementary emitter followers consisting of IC Darlington-connected power transistors. Although the transistors and the parameter values are given, let us examine the circuit from the viewpoint of a design engineer.

THE SUPPLY VOLTAGE

An rms voltage of 11 V across the 8 ohm load resistance delivers 15 W. For a sine wave the maximum load voltage V_m is 15.5 V and the corresponding maximum load current is 1.94 A. With no signal present, the circuit is designed so that the voltage v_A at the input side of the output coupling capacitor is $0.5V_{CC}$, and when the signal produces an output voltage with a maximum value of 15.5, the voltage v_A is $15.5 + 0.5V_{CC}$. After addition of 1.9 V for v_{CE} and 0.6 V for the full-load voltage drop across R_E, this becomes $18 + 0.5V_{CC}$, which must equal V_{CC}. Therefore, a full-load supply voltage of 36 V is satisfactory. Because the unregulated supply may drop its voltage by 10% at full load, we choose a 40 V supply.

The allowance of 1.9 V for the minimum value of v_{CE} is reasonable. According to the data sheet the saturation voltage at 2 A is 1.4 V, and the additional 0.5 V margin is desirable because of increased nonlinear distortion as saturation is approached.

POWER TRANSISTORS AND HEAT SINKS

In order to estimate the maximum power dissipation in an output transistor of a class B amplifier with a single voltage supply, an approximate analysis will be made. With the emitter resistors removed, the maximum value of the instantaneous output voltage across the load R_L is assumed to be $0.5V_{CC}$. With a sinusoidal input signal the full-drive instantaneous voltage v_{CE} and the current i_C of the NPN transistor become

$$v_{CE} = 0.5V_{CC}(1 - \sin \omega t) \qquad i_C = \frac{0.5V_{CC}}{R_L} \sin \omega t \qquad (16\text{-}28)$$

The instantaneous power dissipated in the output transistor is the product $v_{CE}i_C$, with v_{CE} and i_C given by (16-28) during the conducting half-cycle. The power is zero during the nonconducting half-cycle. The average power is easily determined to be $0.017V_{CC}{}^2/R_L$. This should be increased about 5% to include the dissipation of the driver transistor of the Darlington pair. Thus the total power dissipation P_T of the compound transistor, assuming the instantaneous sinusoidal output voltage has a peak value of $0.5V_{CC}$ and assuming a *resistive* load, is approximately

$$P_T = \frac{0.018V_{CC}{}^2}{R_L} \qquad (16\text{-}29)$$

For $V_{CC} = 40$ V and $R_L = 8$ ohms, each compound transistor dissipates 3.6 W, or less. The selected transistors must be capable of dissipating this power with heat sinks of reasonable size. With one transistor fully *on*, the other transistor has almost the entire supply voltage across its collector-emitter terminals. Therefore, the breakdown voltage $V_{(BR)CEO}$ should be at least 40 V. Also, each transistor should be capable of carrying a peak current of $(2P_o/R_L)^{1/2}$, which is 1.94 A for an output power P_o of 15 W.

The power transistors selected satisfy the requirements. Each has a collector-emitter breakdown voltage of 60 V and can carry a continuous current of 5 A. With an allowable junction temperature of 150°C, a maximum ambient temperature of 55°C, and a dissipation not exceeding 3.6 W, the junction-to-ambient thermal resistance must not exceed $(150 - 55)/3.6$, or 26.4°C/W. Allowance of 1.9°C/W for the specified thermal resistance from the junction to the case and 0.5°C/W for the resistance from the case to the heat sink leaves a thermal resistance not exceeding 24°C/W for the heat sink.

There are advantages in using a heat sink with a lower thermal resistance. As we shall see, the values required for the emitter resistances are proportional to the junction-to-ambient thermal resistance. Another consideration is the partial protection from an accidental short-circuited load. The designer recommends a heat sink of 7°C/W, which makes $R_{\theta JA}$ equal to about 9.4°C/W. This choice allows smaller values for the emitter resistances and provides safety for several minutes from a short-circuit at 25°C ambient temperature. A suitable heat sink based on the curves of Fig. 16-1 has an area of about 170 cm^2.

There is yet another consideration. Operation must be wholly confined to the safe operating region as specified on the data sheet in the form of that of Fig. 16-3. Equations 16-28 with $V_{CC} = 40$ V and $R_L - 8$ ohms should be examined to determine that the safe operating region is not exceeded at any instant of time. Such is the case for the selected transistors.

THE DC BIAS CIRCUITRY

With the input voltage $v_i = 0$ the circuit is quiescent. Transistors Q_1 and Q_2 are biased in the active mode, and transistors Q_3 and Q_4 are biased near cutoff with collector currents of 20 mA. Analysis shows that the terminal voltage V_t of the V_D multiplier is approximately

$$V_t \approx V_{BE2}\left(1 + \frac{R_6}{R_5}\right) \tag{16-30}$$

Note that *the voltage is independent of the current supplying the network*. However, the expression is valid only if the transistor beta is sufficiently large (see P16-22). A Darlington pair is sometimes used for Q_2.

The voltage V_t must equal the sum of the base-emitter voltages of both compound transistors biased near cutoff. Assuming this sum is 2.4 V, with $R_5 = 1.2$ kΩ and $V_{BE2} = 0.7$ V, the resistance R_6 should be 2.9 kΩ. It is adjusted to give quiescent power-transistor collector currents of 20 mA. This is almost at cutoff, and the small currents nearly eliminate crossover distortion. Technically, operation is class AB.

The voltage v_A at the left side of the output coupling capacitor should be $0.5V_{CC}$, or 20 V. Assuming this is the case, the base of the compound transistor Q_4 has a potential of 18.8 V, with $V_{BE4} = -1.2$ V at 20 mA. Therefore, the current through R_4 is 18.8 divided by 3.9 kΩ, or 4.82 mA. This is the collector current of Q_1 and it flows through the speaker coil to ground. Because it is important for Q_1 to have a very small base current, a high-beta transistor was selected. The data sheet specifies a value of β_{F1} between 150 and 600, and V_{BE1} is assumed to be -0.65. The current through R_2 is 0.65/5.6, or 0.116 mA, and the current out of the base of Q_1 is 0.017 mA for an assumed beta of 280. Adding these gives 0.133 mA, which is the current through the 120 kΩ feedback resistor R_f. With 16 V across R_f the voltage v_A is 20 V as assumed. Because β_{F1} is large, the uncertain and temperature-sensitive base current of Q_1 has only a small effect on this voltage.

Resistor R_f provides dc feedback, which tends to maintain v_A at 20 V. To illustrate, suppose v_A drops to 19 V. This increases the current through R_f, resulting in a larger current out of the base of Q_1. The collector current of Q_1 rises in magnitude, increasing the potential at the base of the compound transistor Q_4. This raises v_A back toward its proper value of 20 V. A value of v_A appreciably different from this will reduce the allowed output voltage swing.

BOOTSTRAP CURRENT DRIVE

Let us suppose that the lower terminal of R_4 is disconnected from the speaker and connected directly to ground. This is done in some circuits. There is no problem with current drive for Q_3. When v_i is negative, the signal at the collector of the input transistor Q_1 is positive and Q_3 conducts. Transistor Q_4 is *off*. Each Darlington-connected transistor has a minimum β_F of 1000. For a peak load current of about 2 A, the maximum current that must be supplied by Q_1 to Q_3 is 2 mA. Under quiescent conditions the collector current of Q_1 is 4.8 mA. Thus the available current is ample. As v_A rises, the current i_f through R_f decreases and i_1 through R_1 increases. At node B the voltage v_B remains nearly constant.

However, the current drive for Q_4 is inadequate. When v_i is positive, giving a negative signal voltage at the collector of Q_1, the power transistor Q_4 conducts, and v_A is less than 20 V by the magnitude of the instantaneous output voltage. The voltage at the upper terminal of R_4 is less than v_A by two diode drops. Let us suppose the instantaneous value of the signal is sufficiently large to make this voltage equal to 4 V. With R_4 connected to ground its current is only 1 mA, and this is the sum of the positive current out of the collector of Q_1 and the positive current out of the base of the compound transistor Q_4. Clearly there is insufficient current for adequate bias of Q_1 and for driving Q_4, and the output voltage could not be as large as was assumed. Reducing R_4 to increase the current is unsatisfactory because this reduces the voltage gain of the CE input stage and also increases the power dissipation under quiescent conditions.

The current-drive problem is avoided in the circuit of Fig. 16-27 by connecting R_4 to the output. In this circuit with Q_4 conducting, *the potentials at the terminals of R_4 rise and fall simultaneously, maintaining a constant current through this resistor.* The 4.8 mA current is adequate for both the bias of Q_1 and the base drive of Q_4. The arrangement is called a *bootstrap circuit.* During the half-cycle when Q_3 is *off,* there is no appreciable power drawn from the voltage supply. In this time interval the source of the energy supplied to the load, as well as that dissipated in R_E and Q_4, is the stored energy of the bootstrap capacitor C_2 which also serves as the coupling capacitor. The purpose of the bootstrap circuit is to provide the necessary current drive when the signal is large. With inadequate drive, the output power is reduced and distortion is increased.

A disadvantage of the circuit of Fig. 16-27 is the flow of the quiescent bias current of R_4 through the speaker coil. If desired, this can be avoided by connecting R_4 to ground, dividing it into two equal resistances and connecting the center point through a capacitor to the speaker. Amplifiers with both positive and negative voltage supplies, requiring no output coupling capacitor, often use this type of circuit. The bootstrap capacitor is usually from 10 to 100 μF.

EMITTER RESISTORS

In the discussion of thermal runaway in Sec. 16-1 it was noted that the problem is minimized by means of adequate heat sinks and bias stabilization. The stabilization is usually accomplished by employing multistage dc feedback, which has already been considered, and also by adding emitter resistors. Although the problem of thermal stability is complicated, an approximate analysis[7] based on certain idealizations and simplifications indicates that a suitable value for R_E in class B complementary amplifiers is given by

$$R_E \approx 0.0025 V_{CE} R_{\theta JA} \qquad (16\text{-}31)$$

V_{CE} is the quiescent collector-emitter voltage, and for the amplifier of Fig. 16-27 it is 20 V.

With the specified heat sink the junction-to-ambient thermal resistance is 9.4°C/W. According to (16-31), R_E should be 0.47 ohm or larger. The equation is not precise, and the value of 0.33 ohm specified by the designer is satisfactory. From (16-31) we note that increasing the size of the heat sink lowers the thermal resistance and allows a smaller value of R_E. In addition to improving thermal stability, the emitter resistors are helpful in establishing the proper quiescent current for minimizing crossover distortion. The disadvantage, of course, is the reduction in available output power. Bypass capacitors are not appropriate.

VOLTAGE GAIN AND INPUT RESISTANCE

In addition to providing dc feedback the resistor R_f gives signal feedback. The amplifier is inverting because of the 180° phase shift of the CE input stage. The output signal voltage is sampled at point A and a feedback signal current proportional to the sampled voltage is fed back into the node at the base of Q_1. This voltage-sample current-sum type of negative feedback is the same as that previously encountered in our study of the inverting OP AMP. As in the case of the operational amplifier, the input resistance excluding R_1 is reduced by this feedback to a very low value. Therefore, the input resistance seen by v_i is approximately R_1, or 10 kΩ,

As we might expect from our study of operational amplifiers, the voltage gain is approximately $-R_f/R_1$, or -12. For an rms input of about 1 V the voltage gain is sufficient for developing full output power. Once again we find that the voltage gain is largely determined by precision resistors and is nearly independent of temperature-sensitive transistor parameters. Another very important function of the feedback is to reduce distortion. The specified minimum harmonic distortion would be unobtainable without the use of negative feedback. Feedback concepts are discussed in greater depth in Chapters 20 and 21.

COUPLING AND DECOUPLING NETWORKS

The 2 μF electrolytic capacitor at the input is in series with an input resistance of approximately 10 kΩ. At the lower frequency of 50 Hz the reactance of the coupling capacitor is 1.59 kΩ. Calculations show that the signal voltage at 50 Hz is reduced only 1.2% by this reactance.

At the output node A the signal voltage is developed across the 1000 μF capacitor in series with the 8 ohm speaker. At 50 Hz the reactance of the capacitor is 3.2 ohms, which reduces the load voltage by 7%. The phase shift of 22° is more significant. However, the coupling capacitors have sufficient capacitance and adequate voltage ratings.

The decoupling network consisting of R_3, C_3, and C_4 rejects ripple voltages from the power supply. It is especially important to prevent these voltages from feeding into the CE input stage. Here they would be amplified, producing hum in the speaker. The voltage supplied to the base and emitter circuits of Q_1 is that across the 50 μF capacitor C_3, and this voltage is nearly free of ripple.

OVERLOAD PROTECTION

With the selected heat-sink thermal resistance, at an ambient temperature of 25°C the amplifier can tolerate a short-circuit at the output for several minutes without damage. However, overload-protective circuitry is certainly desirable, and such protection is usually a requirement.

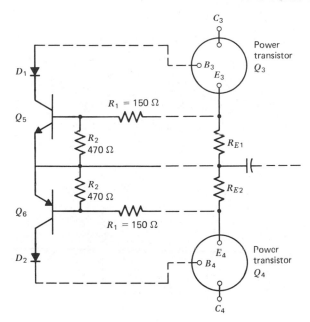

Figure 16-28   Overload protection circuit

A circuit suitable for this purpose is shown in Fig. 16-28. It is connected between the base-emitter terminals of the compound transistors. Normally Q_5 and Q_6 are *off*. In the event of an accidental overload, the current through R_{E1} turns Q_5 *on*. Base current is drawn away from the power transistor Q_3, thereby limiting its current. The bottom half of the circuit operates similarly. The diodes prevent the collector-base voltages of Q_5 and Q_6 from being forward-biased during normal operation.

A 50 WATT AMPLIFIER

The amplifier that has been examined can deliver an output up to 60 W by adding an additional CE amplifier at the input and replacing the bootstrap current-drive network with a current source, along with other modifications. However, a somewhat different configuration will be considered here. The network[8] is shown in Fig. 16-29. The use of two voltage supplies, with one supplying a positive voltage and the other a negative voltage, allows direct coupling to the 4 ohm speaker. Under quiescent conditions the output voltage is zero, and a coupling capacitor having a thousand or more microfarads is not needed.

Bootstrap circuitry requires a capacitor. At low frequencies the capacitor prevents the circuit from developing sufficient current drive at high power levels,

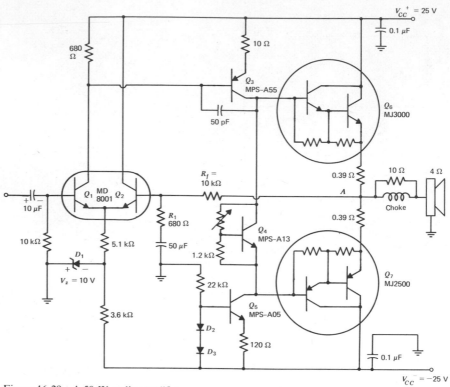

Figure 16-29   A 50-W audio amplifier.

and distortion results. An advantage of this amplifier is the elimination of boot-strap circuitry. Instead, the constant-current source associated with transistor Q_5 provides adequate current for driving the power transistor Q_7. Assuming diode drops of 0.7 V, the collector current of Q_5 is 5.8 mA.

Zener diode D_1 fixes the current through the emitter resistor of the emitter-coupled input stage at 1.8 mA. For betas of 50 the base currents of Q_1 and Q_2 are 18 μA. This gives -0.2 V at the base of Q_1, and the voltage at the base of Q_2 must be the same because of the coupled emitters. Therefore, the quiescent output voltage v_A at node A is zero. *Thus the emitter-coupled differential input stage maintains the quiescent output voltage at zero.*

The signal feedback circuit consists of R_f and R_1. This voltage-sample voltage-sum feedback is the same as that of the noninverting OP AMP. It lowers the output impedance, raises the impedance looking into the base of Q_1, and provides a voltage gain that is approximately $(1 + R_f/R_1)$. Accordingly, the input resistance of the amplifier is that of the bias resistor at the input, or 10 kΩ. The voltage gain is 15.7. An input of 0.9 V rms develops the full output power of about 50 W.

The Zener diode not only sets the bias current of the differential input stage but also provides ripple filtering of the negative supply. The choke at the output prevents high-frequency oscillations, which tend to occur with capacitive loads. Further enhancement of high-frequency stability is furnished by the 50 pF capacitor of Q_3. For a maximum ambient temperature of 55°C, the thermal resistance of the heat sink should not exceed 4°C/W. Although the bandwidth extends from 10 Hz to 50 kHz, operation above 30 kHz at full power must be avoided. Protective circuitry such as that of Fig. 16-28 is strongly recommended.

The design and analysis of this section have been based on the assumption of a constant resistive load. Actual loudspeakers have appreciable reactance, and *both the resistance and the reactance vary with frequency.* The peak current of a resistive load flows through a power transistor when its collector-emitter voltage is a minimum. However, such is not the case when the impedance is complex (see P16-24), and *the dissipation within the device is substantially increased.*[9] This is another good reason for employing a heat sink with considerably less thermal resistance than that needed for a resistive load.

16-7. INTEGRATED CIRCUIT POWER AMPLIFIERS

There are two basic types of integrated-circuit (IC) power amplifiers. One of these is the monolithic amplifier, which has all elements, except perhaps the coupling capacitors, contained within a single silicon crystal. The major problem is the removal of heat from the very small volume. Another difficulty is the cost of making NPN and PNP complementary transistors on the same chip. Monolithic amplifiers are limited to a few watts or less of output power. The other basic type is the *hybrid* power amplifier, which consists of discrete transistors mounted on a ceramic substrate, along with the necessary resistors and capacitors. The resistors are made of thick-film pastes placed on the substrate, and these resistive pastes are composed of metallic powder mixed with ground glass or other material. Chip capacitors are often used, which consist of a number of thin layers of ceramic dielectric sandwiched between even thinner layers of metallic electrodes. Connections are made internally by means of conducting film or wire bonding techniques. A unit perhaps $10 \times 5 \times 1$ cm in size may deliver 80 W or more to an 8 ohm loudspeaker over a bandwidth from 25 Hz to 25 kHz. The circuit employed would most likely be a complementary-symmetry class AB amplifier with composite transistors. Feedback would be used to stabilize the circuit and reduce the distortion.

A power amplifier can be made from an operational amplifier by adding a suitable power output stage. An example is the 500 mW audio amplifier of Fig. 16-30. The 741 OP AMP feeds into a class B complementary output stage consisting of a matched pair of silicon transistors. Biasing by the diode and the

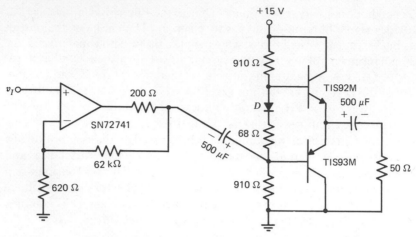

Figure 16-30   500-mW audio amplifier.[5]

68 ohm resistor provides adequate voltage for almost complete elimination of crossover distortion. An output of 500 mW at less than 2% distortion is possible, and the power can be increased to about a watt at higher distortion. With a heat sink the output power can be raised further.

The operational amplifier has a voltage gain of 100, and that of the emitter-follower output stage is unity. The bandwidth extends from 5 Hz to 20 kHz. Capacitive coupling blocks the small dc offset voltage at the output of the OP AMP, and the 200 ohm resistor assures stability. For two-channel stereo applications the SN52558 could be used. It contains two 741 operational amplifiers, each with its own inputs and output.

There are monolithic integrated circuits designed specifically for use as power drivers. An example is the 540 class AB amplifier by Signetics, which has differential inputs and two outputs. These can supply up to ±150 mA to a pair of complementary power transistors. The output power rating from perhaps 5 to 50 W depends on the selection of the output transistors, the applied voltage, and the output impedance. Such integrated circuits lower the cost by reducing the number of components.

REFERENCES

1. J. F. Pierce and T. J. Paulus, *Applied Electronics*, Charles E. Merrill Publishing Co., Columbus, Ohio, 1972.
2. P. M. Chirlian, *Electronic Circuits: Physical Principles, Analysis, and Design*, McGraw-Hill, New York, 1971.
3. D. L. Schilling and C. Belove, *Electronic Circuits: Discrete and Integrated*, McGraw-Hill, New York, 1968.

4. J. A. Walston, J. P. Miller, and others, *Transistor Circuit Design*, McGraw-Hill, New York, 1963.
5. *Linear and Interface Circuits Applications*, Texas Instruments Inc., Dallas, Texas, 1974.
6. J. Lindmayer and C. Y. Wrigley, *Fundamentals of Semiconductor Devices* (Chapter 7), D. Van Nostrand, Princeton, New Jersey, 1965.
7. G. C. Haas, "Design Factors and Considerations in Full Complementary Symmetry Audio Power Amplifiers," *Journal of the Audio Engineering Society*, **16**, No. 3, July 1968.
8. R. G. Ruehs, "15 to 60 Watt Audio Amplifiers Using Complementary Darlington Output Transistors," Application Note 483B, Motorola Semiconductor Products Inc., Phoenix, Arizona, 1972.
9. R. B. Peterson, "Basic Design of Medium Power Audio Amplifiers," Application Note 484A, Motorola Semiconductor Products Inc., Phoenix, Arizona, 1972.

PROBLEMS

Section 16-1

16-1. The 2N5301 NPN silicon transistor of Example 1 of Sec. 16-1 has a maximum allowable junction temperature of 200°C and JC and JA thermal resistances of 0.875°C/W and 35°C/W, respectively. For a free-air temperature of 40°C find the maximum allowable dissipation (a) with no provision for cooling, (b) with the heat sink of Fig. 16-1 with a side length of 30 cm (allow 1°C/W for the mica washer), and (c) with a heat sink having a total CA thermal resistance of 1.5°C/W. [*Ans:* (c) 67.4 W]

16-2. A transistor with a maximum allowable junction temperature of 175°C can dissipate 100 W at a case temperature of 25°C. (a) What is the allowable dissipation at a case temperature of 100°C? (b) What is the allowable dissipation at 25°C free-air temperature using the heat sink of Fig. 16-1 with a side length of 15 cm? Assume 1°C/W for the mica washer. (c) With the heat sink of part (b) calculate T_J, T_C, and T_{HS} when T_A is 35°C and P_D is 10 W.

16-3. A power amplifier, designed for an ambient temperature range from 25°C to 80°C, uses a silicon transistor whose room-temperature Q-point is to be 0.3 A, 10 V, with a corresponding junction temperature of 130°C, and with β_F having its minimum value of 25 at this T_J. At the maximum ambient temperature of 80°C, the maximum power hyperbola and the dc load line intersect at 0.5 A, at which point T_J is at its limit of 200°C. The maximum β_F at this temperature is 75. I_{CBO} is specified to be less than 1 μA at 25°C, and (16-3) applies. Assuming V_{BE} decreases 2.5 mV/°C, if the total base resistance (including r_x) is 100 ohms, find the minimum required emitter resistance R_E, using (15-17). Note that T_1 is 130°C. (*Ans:* 5.6 Ω)

16-4. In (2-44), which gives the multiplication factor M, V_a is 100 and n is 3 for a certain power transistor. Calculate $V_{(BR)CEO}$ if β_F is 100 and also for $\beta_F = 15$, assuming V_{BE} is 1 V.

Section 16-2

16-5. For the amplifier of Fig. 16-4, deduce that R_L is $V_{CC}^2/4P$, with P denoting the power represented by the hyperbola that is tangent to the load line. For a transistor with $V_{(BR)CEO}$ of 80 V and a maximum allowable dissipation of 8 W, calculate the load resistance that gives the most voltage gain along with the maximum possible output power, and determine the Q-point (I_C, V_{CE}). With V_{CC} constant, what effect does increasing R_L have on the maximum output power and also on the voltage gain?

16-6. The amplifier of Fig. 16-4 has a 60 V supply and a 150 ohm load resistance, with the Q-point at the middle of the load line. The junction-to-ambient thermal resistance of the transistor is 20°C/W. For an ambient temperature of 25°C, determine the junction temperature when the sinusoidal input is adjusted for maximum output power and also when the signal is zero.

16-7. The transformer-coupled amplifier of Fig. 16-6, with the load lines of Fig. 16-8, has $V_{CC} = 23$, $R_E = 5\,\Omega$, $R_B = 10\,\Omega$, and $V_{CE}I_C = 4$ W. Assuming a transformer efficiency of 75%, calculate η when the sinusoidal signal is adjusted so that the minimum value of v_{CE} is 1 V. Also, find the power to R_E and the power dissipated at the collector junction. (Ans: $\eta = 29.4\%$)

16-8. A transformer-coupled amplifier in the form of Fig. 16-6 is to operate in an ambient temperature range from 25°C to 60°C with a 90 V supply. The dc resistance of the transformer primary is 100 ohms. The quiescent collector current is to have minimum and maximum values I_{C1} and I_{C2} of 25 and 30 mA. A heat sink is selected so that the maximum junction temperature will not exceed 190°C, which gives a safety margin of 10°C below the allowed value, and at this temperature the maximum β_F is 200. The data sheet specifies a maximum I_{CBO} of 50 μA at 150°C and (16-3) applies. Assume V_{BE} decreases 2.5 mV/°C. (a) For an assumed V_{CE} of 80 V, which allows for an estimated 10 V drop across R_E and R_p, calculate the lowest junction temperature T_1 with no signal. (b) At T_1 assume the minimum β_F is 35 and calculate the minimum required R_E for $R_B = 1$ kΩ.

Section 16-3

16-9. For the CE class A power amplifier described by Figs. 16-10 and 16-12, calculate the second-harmonic percentage distortion D_2 for $R_g = 0$ and also for $R_g = \infty$. For each case use the Q-point of the figures and the maximum possible V_m with i_B maintained greater than 1 and less than 25 mA.

16-10. For $i_C = 100(1 + v_g) + 20(1 + v_g)^2$ mA in a class A power amplifier, determine I_C and I_{dc} in mA and the percent distortion. The input v_g is sinusoidal with a maximum value of 0.5 V.

Section 16-4

16-11. The 500 mW amplifier of Fig. 16-16 with the numerical values shown is in an ambient temperature of 25°C, and the β_F of the silicon power transistor is 100. From the data sheet I_{CBO} is 50 μA at 150°C, V_{BE} is 0.5 V at 129°C, and saturation occurs at 1 V. The total thermal resistance R_θ is 65°C/W. With a sinusoidal signal determine the maximum output power without gross distortion, and find the collector-circuit efficiency at this output. Transformer efficiency is 75%. [To find the Q-point, use (16-3), (15-7), and Fig. 15-11. First, obtain the Q-point approximately and calculate the corresponding T_J. Then repeat the procedure.]

16-12. For the 500 mW amplifier of Fig. 16-16, repeat the design of Sec. 16-4, using the calculated value of 441 ohms for R_{dc} in place of the assumed 300 ohms. Both the Q-point and R_L' are changed. Find the maximum power to R_L, and determine the proper values for R_E, R_4, and R_5. (Ans: $R_5 = 1.60$ kΩ)

16-13. Suppose that each silicon transistor in the 500 mW direct-coupled amplifier of Fig. 16-18 has a β_F of 100. The ambient temperature is 25°C, the junction voltages are $V_{BE}' = 0.7$ and $V_{BE}'' = 0.5$, and I_{CO}'' is 20 μA. Calculate the Q-points (I_C, V_{CE}) for each transistor in mA and V.

Section 16-5

16-14. (a) In the push-pull circuit of Fig. 16-22 with Q_2 off, replace Q_1 with an incremental model using h_{ie} and β, replace the secondary winding with a voltage source V_s having its plus terminal at the dot, and deduce that the voltage gain is

$$\frac{V_o}{V_s} = \frac{-(\beta R_1 + \beta R_2 + R_2)R_L}{h_{ie}(R_1 + R_2 + R_L) + (\beta + 1)R_2 R_L + R_1 R_2}$$

(b) In the model with $V_s = 0$, replace R_L with a voltage source and deduce that the output resistance is

$$R_o = \left[h_{ie}\left(1 + \frac{R_1}{R_2}\right) + R_1 \right] \bigg/ \left(\beta + 1 + \frac{h_{ie}}{R_2}\right)$$

16-15. Repeat P16-14 for the circuitry of Q_2, with Q_1 off, deducing the same relations.

16-16. The push-pull amplifier of Fig. 16-19 has the characteristics and load line of Fig. 16-20 with $V_{CC} = 15$ V and $R_L = 16\ \Omega$. The center-tapped primary of the output transformer has 68 turns of wire with a total resistance of 0.44 ohm. The secondary has 120 turns with a resistance of 1 ohm. Core losses and the loss in R_1 are negligible. To avoid saturation v_{CE} must not be less than 2 V. For a sinusoidal input, determine the maximum power to R_L and calculate the collector-circuit efficiency. (*Ans:* 55.2%)

16-17. In the phase splitter of Fig. 16-23 assume the transistor Q_1 can be modeled with r_x, r_π, and β_o. The measured values of h_{ie} and h_{fe} are 500 Ω and 100. With both R_2 and R_3 zero, and $R_1 = 800$ ohms, find v_1/v_i and v_2/v_i. The load impedances from B_2 to ground and also from B_3 to ground are each 200 ohms.

16-18. In the emitter-coupled amplifier of Fig. 16-24 assume that the matched transistors can be modeled with r_x, r_π, and β_o. The measured values of h_{ie} and h_{fe} are 500 ohms and 100. With $R = 800$ ohms, find v_1/v_i and v_2/v_i. The load impedances from each of the two outputs to ground are each 200 ohms.

Section 16-6

16-19. With R_1 and R_2 omitted, the Darlington-connected transistor of Fig. 16-26a is biased in the active mode. Let parameters β_F, h_{ie}, and h_{fe} apply to the compound transistor. These same symbols with subscripts 1 and 2 apply to the driver transistor Q_1 and the power transistor Q_2, respectively. (a) Find β_F, h_{ie}, and h_{fe} in terms of the parameters of Q_1 and Q_2. Assume the h_{re} and h_{oe} parameters are negligible. (b) If each of the two individual devices has dc and ac betas of 50, base-emitter quiescent voltages of 0.7 V, and short-circuit input resistances of 500 ohms, calculate β_F, V_{BE}, h_{ie}, and h_{fe}. Neglect I_{CO}.

16-20. A class B complementary amplifier with a 4-ohm resistive load has emitter resistors of 0.4 ohm and a single supply of 40 V. The quiescent voltage at the amplifier side of the 2500 μF coupling capacitor is 20 V. With maximum applied signal, the load current is $4 \sin \omega t$ amperes. (a) Determine $v_{CE}(t)$ of a power transistor. Calculate (b) the average power supplied to the load, (c) the average power dissipated in each power transistor (excluding the driver), and (d) the maximum value of the instantaneous dissipation within a transistor (*Ans*: 22.7 W).

16-21. A class B complementary amplifier similar to that of Fig. 16-29 has supply voltages of ± 36 V and a direct-coupled resistive load of 8 ohms. The emitter resistors are each 0.4 ohm. For a load current of $4 \sin \omega t$ amperes, calculate the average power (a) supplied to the load, (b) dissipated in each emitter resistor, and (c) dissipated in each power transistor. (d) Estimating the power dissipation in the driver transistors and front-end circuitry to be 2 W, calculate the collector-circuit efficiency η (*Ans*: 68.3 %).

16-22. Deduce that $V_t = V_{BE}[1 + \beta R_6/R_5(\beta + 1)] + R_6 I/(\beta + 1)$ for the V_D multiplier of Fig. 16-27, with I denoting the current supplied to the network and V_t denoting the supplied voltage. Then deduce (16-30).

16-23. In the power amplifier of Fig. 16-27 change V_{CC} to 46 V, R_f to 130 kΩ, R_2 to 5 kΩ, and R_4 to 4.7 kΩ. With no signal assume V_{BE1} is -0.65 V, β_{F1} is 280, and V_{BE4} is -1.2 V. (a) Calculate I_{C1} and the voltage v_A of node A. (b) With full load the supply voltage drops 10%, and 3 V must be allowed for voltages across the conducting power transistor and its emitter resistor at the peak current. What is the full-load output power?

16-24. A class B complementary amplifier similar to that of Fig. 16-29 has supply voltages of ± 28 V. Emitter resistances are negligible. A current of $3 \sin \omega t$ amperes flows through the load impedance of $8\underline{/\theta}$ ohms. (a) For $\theta = 0$, calculate the average power supplied to the load and also that dissipated in each power transistor. (b) Repeat (a) with $\theta = 60°$. [*Ans*: (b) $P_D = 17.7$ W]

Chapter Seventeen
FET Linear Circuits

The physical mechanisms occurring within field-effect transistors were examined in our study of Chapter 3. Here we consider some of the properties of linear amplifiers utilizing FETs. We have learned that three basic BJT configurations are classified as CE, CC, and CB amplifiers. The corresponding FET circuits are CS, CD, and CG, with S, D, and G denoting the source, drain, and gate terminals, respectively. These are analyzed in this chapter by employing an approximate incremental model, with the analyses serving to illustrate the major characteristics of each.

FETs are utilized in operational amplifiers to provide exceptionally high input impedance, and the OP AMP considered in this chapter is particularly interesting. It has a high-impedance MOSFET input stage, which employs BJTs as active loads. Most of the voltage gain is provided by an internal BJT common-emitter stage, which drives a CMOS output configuration.

The design of a direct-coupled FET-BJT cascade is undertaken, with dc voltage feedback giving the required stabilization. Then we examine power FETs. Their applications are numerous and varied, especially in the areas of power amplification and switching. Problems associated with single-stage bias circuits are discussed, and a new technique quite useful for designing such circuits is presented. In later chapters various networks containing one or more field-effect transistors are investigated.

17-1. SINGLE-STAGE AMPLIFIERS

Three basic configurations are the common-source (CS), common-drain (CD), and common-gate (CG) amplifiers. Analysis of each will utilize a linear circuit based on the short-circuit admittance (y) parameters, which were discussed in Sec. 13-5. As the device is assumed to be biased in the saturation (pinch-off) region, the square law (3-3) applies. Let us consider the low-frequency CS y parameters.

INCREMENTAL MODEL

The CS short-circuit input and reverse transfer admittances y_{is} and y_{rs} are essentially zero, because the gate current is extremely small for both MOSFETs and JFETs. The short-circuit forward transfer admittance y_{fs} at low frequencies is the transconductance g_{fs}, and y_{os} is g_{ds} which is the reciprocal of r_{ds}. Hence, the two-generator model of Fig. 13-24 reduces to that of Fig. 17-1a. At high frequencies the incremental capacitances are added as shown in Fig. 17-1b.

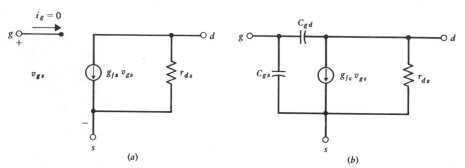

Figure 17-1   FET incremental models. (a) Low-frequency model. (b) High-frequency model.

In Chapter 12 we learned that the linear hybrid-π model is an excellent representation of a BJT provided the incremental base-emitter voltage v_{be} is less than 5 mV. The dynamic range of the FET model of Fig. 17-1 is much greater. This is true because the transconductance curve i_D-versus-v_{GS} is not nearly as nonlinear as the exponential curve i_C-versus-v_{BE} of the BJT. Based on the square law the harmonic distortion of a depletion-type FET is less than 5% as long as the amplitude V_m of a sinusoidal gate-source signal voltage is less than $0.2V_P \times \sqrt{I_D/I_{DSS}}$ (see P17-1). For example, with $V_P = 4$ and $I_D = 0.25I_{DSS}$ an amplitude of 400 mV results in 5% distortion.

THE COMMON-SOURCE AMPLIFIER

A typical CS amplifier, such as that of Fig. 17-29 of Sec. 17-6, has the mid-frequency incremental model of Fig. 17-2. The output voltage v_o is $-g_{fs}v_{gs}$ times

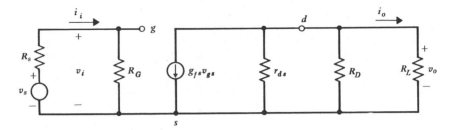

Figure 17-2   CS midfrequency incremental model.

the equivalent load consisting of r_{ds}, R_D, and R_L in parallel. As v_{gs} is v_i, the voltage gain v_o/v_i is easily found. We deduce that A_v, A_i, R_i, and R_o are

$$A_v = -g_{fs}(r_{ds}\|R_D\|R_L) \qquad R_i = R_G$$

$$A_i = \frac{A_v R_G}{R_L} \qquad\qquad R_o = r_{ds}\|R_D \tag{17-1}$$

The transducer gain G_T is the ratio of $v_o{}^2/R_L$ to $v_s{}^2/4R_s$. As v_s equals $v_i(R_s + R_G)/R_G$, we obtain

$$G_T = \left(\frac{A_v R_G}{R_s + R_G}\right)^2\left(\frac{4R_s}{R_L}\right) \tag{17-2}$$

Let us consider a numerical example. Suppose g_{fs} is 2 mmhos, r_{ds} is 40 kΩ, R_s is 100 kΩ, R_G is 2 MΩ, and R_D and R_L are each 20 kΩ. From (17-1) and (17-2) we find that

$$A_v = -16 \qquad A_i = -1600 \qquad R_i = 2 \text{ M}\Omega$$
$$R_o = 13.3 \text{ k}\Omega \qquad G_T = 4640 = 37 \text{ dB}$$

Typically the voltage gain of a CS amplifier is between 5 and 50, and A_i is quite large. The input resistance is simply the 2 MΩ gate resistance. Resistance R_o is fairly large, and the transducer gain is high. The phase inversion corresponds to that of the CE amplifier, with A_v and A_i negative.

THE MILLER EFFECT

At high frequencies the capacitances become important, and the model is that of Fig. 17-3a. For sinusoidal time variations the current through C_{gd} equals the admittance $j\omega C_{gd}$ times the voltage $V_i - V_o$ across C_{gd}. Let us assume that the capacitor current $j\omega C_{gd}(V_i - V_o)$ is very small compared with the current $g_{fs}V_i$ of the controlled source. Then V_o is approximately $-g_{fs}R_L'V_i$, with R_L' denoting

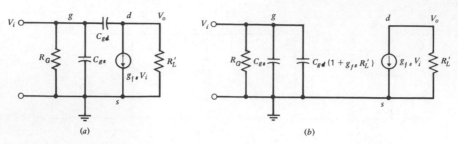

Figure 17-3   CS high-frequency incremental models. (*a*) Actual circuit. (*b*) Approximate circuit.

the *effective* load resistance. Accordingly, the current through C_{gd} can be expressed as $j\omega C_{gd}(1 + g_{fs}R_L')V_i$, which is assumed to be negligible compared with $g_{fs}V_i$. The assumption is clearly valid if and only if ω is restricted as follows:

$$\omega \ll \frac{g_{fs}}{C_{gd}} \qquad \omega \ll \frac{1}{R_L'C_{gd}} \qquad (17\text{-}3)$$

With the frequency limited by (17-3), the current leaving the input node through C_{gd} is $j\omega C_{gd}(1 + g_{fs}R_L')V_i$ and that entering the output node from C_{gd} is negligible. These conditions are satisfied by the circuit of Fig. 17-3*b*, which has the capacitance $C_{gd}(1 + g_{fs}R_L')$ connected in parallel with C_{gs}. Consequently, the circuit of Fig. 17-3*b* is a good approximation.

In most practical cases the inequalities of (17-3) are valid throughout the useful frequency range of the amplifier. From Fig. 17-3*b* we note that C_{gd}, multiplied by $1 + g_{fs}R_L'$, shunts the input. In comparison with C_{gs}, the effect of C_{gd} is amplified by a factor approximately equal to the magnitude of the midband voltage gain $g_{fs}R_L'$. This enhancement of C_{gd} is known as the *Miller effect*.

From examination of Fig. 17-3*b* it is evident that C_{gd} lowers the input impedance at high frequencies. This results in a reduction in transducer gain. The same detrimental effects are present in CE amplifiers because of C_μ. Let us note that the approximate model of Fig. 17-3*b* is valid for FET circuits only *in the CS configuration with transmission in the forward direction at frequencies restricted according to (17-3)*. It must *not* be used when the load impedance is complex. The Miller effect is discussed further in Sec. 18-6.

THE SOURCE FOLLOWER

A common-drain (CD) amplifier, or *source follower*, is shown in Fig. 17-4. This is basically the same bootstrap circuit that was employed in Sec. 13-3 for the CC amplifier. We know that the voltage gain of this type of circuit is approximately $+1$. Hence there is practically no incremental voltage across the large resistor R_3, and the current in R_3 is completely negligible. Therefore, v_o is

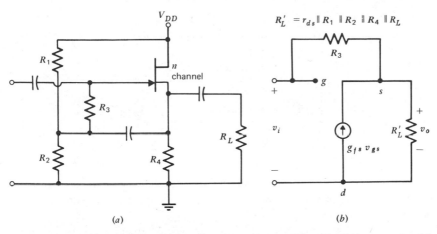

Figure 17-4   CD amplifier and incremental model. (*a*) CD amplifier. (*b*) Midfrequency model.

$g_{fs}v_{gs}R_L'$. The gate-source voltage v_{gs} equals $v_i - v_o$, and the voltage gain A_v is easily determined to be

$$A_v = \frac{1}{1 + (1/g_{fs}R_L')} \qquad (17\text{-}4)$$

Usually the product $g_{fs}R_L'$ is sufficiently large to give a voltage gain only slightly less than unity, and there is no phase inversion. The input current is $(v_i - v_o)/R_3$. From this and (17-4) we find that the input resistance is

$$R_i = (1 + g_{fs}R_L')R_3 \qquad (17\text{-}5)$$

Typically, R_i is tens of megohms. The input impedance consists of R_i in parallel with the incremental capacitance C_{gd}, which directly shunts the input. Capacitance C_{gs} is in the bootstrap circuit and has very little effect. The CD amplifier is especially important because of its very high input resistance and very low input capacitance. The midfrequency current gain can be calculated from A_v and R_i using the expression

$$A_i = \frac{A_v R_i}{R_L} \qquad (17\text{-}6)$$

The output conductance can be found from the circuit of Fig. 17-5, in which R_L has been eliminated. The voltage $v_{gs} = -(i_1 + g_{fs}v_{gs})R_3$. From this relation and a loop equation around the outside loop we can eliminate v_{gs} and find v_1/i_1. The conductance G_o can be expressed as

$$G_o = \left(\frac{1 + g_{fs}R_3}{R_s + R_3}\right) + g_{ds} + G_1 + G_2 + G_4 \qquad (17\text{-}7)$$

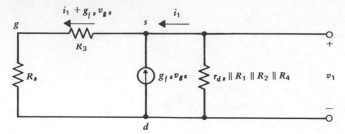

Figure 17-5   Circuit for finding G_o of a CD amplifier.

The first term on the right is approximately g_{fs} in most cases. The reciprocal of G_o is the output resistance R_o, which is usually no more than a few hundred ohms.

The results of the analysis reveal that source followers have incremental characteristics quite similar to those of emitter followers. Each has a large input impedance, a small output impedance, considerable current gain, a voltage gain slightly less than one, and zero midband phase shift. Variations in the output voltage approximately follow those of the input.

Another CD circuit extensively used is that of Fig. 17-6. The resistors labeled R_G, R_S, and R_D are those of the two-battery equivalent circuit of Fig. 13-4 of Sec. 13-1, and V_G is zero. The resistor R_S provides self-bias, with the quiescent voltage V_{GS} equal to $-I_D R_S$. The self-bias procedure described near the end of Sec. 17-6 is quite satisfactory for selecting the proper quiescent current I_{D1} and the bias resistors. In the midfrequency model we note that v_{gs} equals $v_i - v_o$. Precise analysis using three node equations is not difficult, but we shall follow a simpler method that gives excellent results.

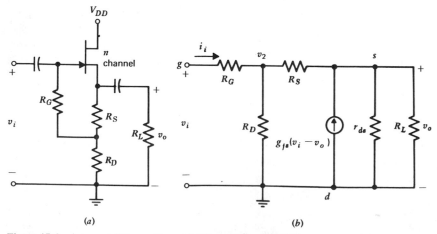

(a)   (b)

Figure 17-6   A second CD amplifier. (a) CD amplifier. (b) Midfrequency model.

With R_S small compared with R_G, the voltage across R_G is approximately $v_i - v_o$. Because A_v is nearly unity, this voltage is almost zero. Furthermore, R_G is typically several megohms. Thus it is apparent that R_G can be regarded as infinite in order to determine the voltages of the circuit. This places $R_S + R_D$ in parallel with R_L. The voltage v_o is the product of the equivalent load resistance R_L' and $g_{fs}(v_i - v_o)$. From this we find A_v to be that given by (17-4) with R_L' denoting the equivalent resistance of $R_S + R_D$, r_{ds}, and R_L in parallel.

The voltage v_2 is $v_i A_v R_D/(R_D + R_S)$, because $v_i A_v$ is v_o and the resistances act as a voltage divider. The input current i_i is $(v_i - v_2)/R_G$. Upon substitution for v_2, the input resistance R_i is determined to be

$$R_i = \frac{R_G}{1 - A_v R_D/(R_D + R_S)} \qquad (17\text{-}8)$$

From A_v and R_i the current gain is calculated from (17-6).

The CD amplifier of Fig. 17-6 has about the same incremental performance as that of Fig. 17-4. Its bias circuitry is not as satisfactory because it has a drain resistance R_D and uses self-bias. However, it eliminates one capacitor, and this is especially important in integrated circuits.

THE COMMON-GATE AMPLIFIER

In the CG amplifier of Fig. 17-7 there is no external gate resistance, and the bias voltage V_{GS} is $-I_D R_S$. The midfrequency circuit is in a form especially suitable for nodal analysis. From two node equations we deduce that

$$A_v = \frac{g_{fs} + g_{ds}}{g_{ds} + G_D + G_L} \approx g_{fs}(R_D \| R_L) \qquad (17\text{-}9)$$

$$G_i = g_{fs} + g_{ds} + G_S - \frac{g_{ds}(g_{fs} + g_{ds})}{g_{ds} + G_D + G_L} \approx g_{fs} \qquad (17\text{-}10)$$

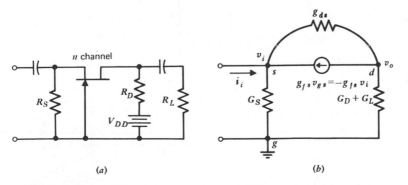

Figure 17-7 CG amplifier. (a) CG amplifier. (b) Midfrequency model.

Using the approximate relations, the current gain is

$$A_i \approx \frac{R_D}{R_D + R_L} \qquad (17\text{-}11)$$

The CG amplifier has properties somewhat similar to those of the CB amplifier. Because of the low input impedance and low power gain, there is little use of this circuit at audio frequencies. Capacitance C_{gs} shunts the input and C_{gd} shunts the output; in tuned RF amplifiers these incremental capacitances are not detrimental, being in parallel with the LC tuned circuits. There is no feedback capacitance between the input source and the output drain except for the extremely small C_{ds}, which is almost always negligible even at high frequencies. Both JFETs and MOSFETs are used in VHF and UHF amplifiers in the CG configuration. With a properly adjusted source impedance the noise figure is very low.

FET-BJT CASCADES

Amplifiers with an FET input stage followed by one or more BJT stages are common. The FET provides high input impedance and large current gain, and the BJT gives both voltage and current gain. An example of such an amplifier is shown in Fig. 17-8. The input comes from a ceramic pickup that delivers about 200 mV of signal. Because of the large capacitive impedance of the source, a high input resistance is essential for obtaining good response at low audio frequencies, and a CD input stage satisfies this requirement. The output is a CE class A high-voltage power stage that provides substantial voltage gain and sufficient power

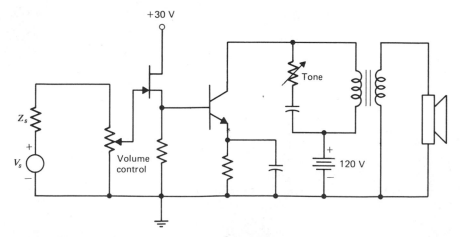

Figure 17-8   Line-operated phono amplifier.

to drive the speaker. Because the dc sources are obtained from rectification of the 60 Hz, 120 V line voltage, the amplifier is said to be line-operated. An FET-BJT cascade is examined in Sec. 17-3.

17-2. FETs IN OPERATIONAL AMPLIFIERS

Certain applications require operational amplifiers with extremely high input resistances and unusually low input bias currents. In such cases it may be advantageous to use an OP AMP with a MOSFET differential input stage, and such a network is examined here. First, however, let us briefly consider a JFET differential amplifier.

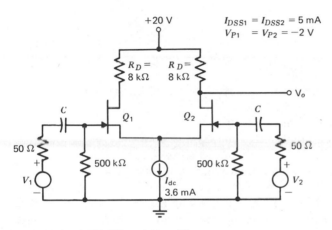

Figure 17-9   JFET differential amplifier.

The network with input capacitive coupling is shown in Fig. 17-9. Because the transistors are assumed to be matched, each drain current is 1.8 mA, and this equals $I_{DSS}(1 - V_{GS}/V_P)^2$. Using 5 mA for I_{DSS} and -2 for V_P, we find that V_{GS} is -0.8 V. For zero gate current the voltage of the coupled JFET source terminals is $+0.8$ V. Subtraction of the $I_D R_D$ drop of 14.4 from 19.2 gives V_{DS}, which is 4.8 V for each transistor. The transconductance g_{fs} is $(-2/V_P)\sqrt{I_D I_{DSS}}$, or 3 mmhos.

The midband incremental model with the transistor parameter r_{ds} neglected is shown in Fig. 17-10. Because the 500 kΩ gate resistors do not appreciably affect the gain, they are omitted. Application of the voltage law to the loop from s to g_1 to ground to g_2 and back to s shows that $V_{gs1} - V_{gs2} = V_1 - V_2$. The currents entering the source node must add to zero, which requires that $V_{gs1} = -V_{gs2}$. From these two relations we find that

$$V_{gs1} = -V_{gs2} = \tfrac{1}{2}(V_1 - V_2) \tag{17-12}$$

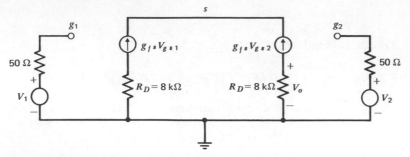

Figure 17-10   Incremental model of JFET amplifier.

As V_o is $-g_{fs}R_D V_{gs2}$, it follows that the differential gain is

$$\frac{V_o}{V_1 - V_2} = \tfrac{1}{2}g_{fs}R_D = 12 \qquad (17\text{-}13)$$

The effect of a finite incremental resistance r_{ds} is to reduce the gain moderately (see P17-5).

Dual monolithic JFETs, which are especially attractive for use in low-noise differential amplifiers with high input impedance, are commercially available. The FETs of a chip normally have exceptionally well-matched parameters and characteristics, and the matching is nearly independent of temperature.

An operational amplifier with FETs is the CA3130B, with the schematic diagram[3] of Fig. 17-11. It is especially interesting because of the use of both MOSFETs and BJTs in the same monolithic circuit. Also, the output stage employs class A complementary symmetry (COS), referred to as *COS/MOS*, or *CMOS*, with matched n-channel and p-channel devices. The monolithic circuit contains one n-channel and eight p-channel enhancement-mode MOSFETs, along with three NPN bipolar transistors, nine diodes including a Zener, and six resistors. Shown are supply voltages of ± 7.5 and a load of 2 kΩ. Because there is no internal ground, the circuit can operate from a single supply, with $V_{CC}{}^+ = 15$ and $V_{CC}{}^- = 0$.

The input stage has a very high input impedance, typically 1.5×10^{12} ohms, which is largely that of diodes D_5 through D_8 used to protect the oxides of the MOSFET gates. Bipolar transistors Q_9 and Q_{10} and associated circuitry provide the proper active loads for converting the differential outputs of the input stage into a single-ended drive for Q_{11}, while maintaining equal bias currents in the differential stage. There are three stages of amplification. The first one is the differential input stage with a typical voltage gain of 5. This is followed by a CE stage consisting of Q_{11} and its load. Because of the large effective load resistance presented by MOSFETs Q_3 and Q_5, the CE stage has an unusually high gain of about 6000. The complementary common-source-connected output stage con-

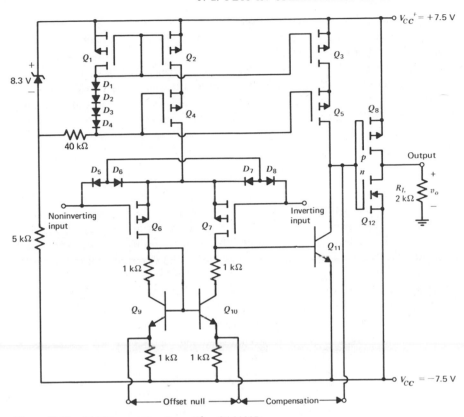

Figure 17-11   CMOS operational amplifier CA3130B.

tributes additionally, providing a voltage swing that can approach the supply voltages, and the overall dB voltage gain (20 log A_V) is typically 110.

Transistors Q_1, Q_2, and Q_3 at the top of the diagram are mirror-connected matched devices with Q_1 designed so that its drain current is about 0.2 mA (see P17-6). Because the three transistors are identical and have the same values of V_{GS}, their drain currents are equal. The gate of Q_1 is connected to its drain, thereby assuring operation in the pinch-off region. The current source Q_2 supplies 0.2 mA to the differential stage through Q_4 which serves to increase the CMRR. Transistor Q_3 is also a current source, supplying the same current to the CE transistor Q_{11} through Q_5. The series connection of Q_3 and Q_5, often referred to as a *cascode* arrangement, gives an effective load resistance for the CE stage that is very large, being equal to the sum of r_{ds5} and $r_{ds3}(1 + g_{fs5}r_{ds5})$ as shown in P17-8. With the inputs adjusted so that the dc output offset voltage is zero, the bias current of the output stage is about 8 mA and its gain is typically from 10 to 30 (see P17-7). Cascode amplifiers are examined in Sec. 19-3.

Inclusion of n-channel and p-channel MOSFETs on the same chip with BJTs enables the circuit designer to utilize the optimum properties of each. The advantage of the high input impedance of the MOSFETs is somewhat offset by the large $1/f$ noise below a few kHz, which is much worse than in either JEFTs or BJTs. Another disadvantage is the complexity of the fabrication process, which requires several additional processing steps. Undoubtedly, the use of analog and digital ICs containing FETs and BJTs will continue to grow. In fact, the circuitry of such a chip is easily designed to perform both analog and digital functions.

17-3. AN FET-BJT AMPLIFIER

It has been noted that FET-BJT cascades have certain unique advantages. The first (FET) stage provides high input impedance, large current gain, and low noise if a JFET is used. The second (BJT) stage, assuming the CE connection, gives both voltage and current gain, along with the capability of operating at values of v_{CE} as low as a few tenths of a volt. A circuit of a CD–CE phono amplifier with a transformer-coupled load was shown in Fig. 17-8, and the schematic diagram of a CS–CE amplifier is presented in Fig. 18-32, accompanied by design problem P18-15. Although a number of interesting circuit configurations are worthy of study, we shall restrict our attention here to a direct-coupled CD–CE amplifier with a resistive load.

CD-CE AMPLIFIER

The schematic diagram is that of Fig. 17-12, where V_i is specified as less than 5 mV rms. The input CD stage employs a p-channel enhancement-mode MOSFET, which is assumed to have static drain characteristics in the saturation region that are described mathematically by the equation

$$I_D = k(V_{GS} - V_T)^2 \quad \text{with} \quad |V_{DS}| > |V_{GS} - V_T| \quad (17\text{-}14)$$

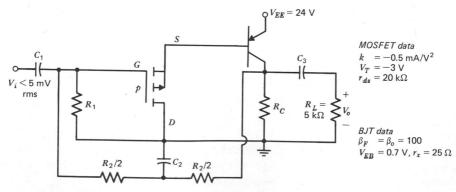

Figure 17-12   CD-CE amplifier.

The threshold voltage V_T is -3 V and k is -0.5 mA/V^2. The incremental output resistance r_{ds} of the MOSFET is assumed to be 20 kΩ and the transconductance g_{fs} is

$$g_{fs} = 2k(V_{GS} - V_T) = 2\sqrt{kI_D} \qquad (17\text{-}15)$$

Bias stabilization is accomplished by means of dc feedback through R_2, with signal feedback eliminated by the capacitor C_2 which acts as a short-circuit at signal frequencies.

Figure 17-13 shows the midband model. Because the 5 kΩ load resistance R_L is much smaller than both r_o and R_2, these resistances are omitted in the output stage. Assuming r_{ds} is large compared with $r_x + r_\pi$, the current I equals $g_{fs}V_{gs}$,

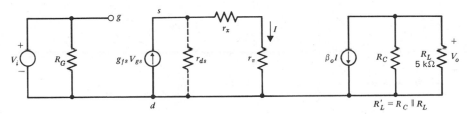

Figure 17-13   Incremental model.

and V_o is $-g_{fs}\beta_o R_L' V_{gs}$ with $R_L' = R_C \| R_L$. The gate-source voltage V_{gs} is the sum of V_i and $-(r_x + r_\pi)g_{fs}V_{gs}$, from which we find that V_i is $[1 + g_{fs}(r_x + r_\pi)]V_{gs}$. From the expressions for V_o and V_i, the voltage gain V_o/V_i is determined to be

$$\frac{V_o}{V_i} = \frac{-g_{fs}\beta_o R_L'}{1 + g_{fs}(r_x + r_\pi)} \qquad \text{with} \qquad R_L' = R_C \| R_L \qquad (17\text{-}16)$$

Our objective is to select the quiescent conditions to give the greatest possible voltage gain. For $\beta_o = 100$ and $q/kT = 40$, the incremental resistance r_π is $2.5/|I_C|$ kΩ with I_C in mA. From (17-15) with $k = -0.5$ mA/V^2, we find that g_{fs} is $1.41\sqrt{|I_D|}$. Because I_D is $0.01I_C$, g_{fs} becomes $0.141\sqrt{|I_C|}$ mmhos and the product $g_{fs}r_\pi$ is $0.353/\sqrt{|I_C|}$. Clearly, the denominator of (17-16) is minimized and g_{fs} of the numerator is made reasonably large by selecting a large value for I_C. Because of the proportionality between I_D and I_C, the transconductance g_{fs} of the FET is determined by the selection of the quiescent collector current. With $V_{EE} = 24$ and with 2 V allowed for V_{EC}, which is adequate for V_i less than 5 mV, the product $I_C R_C$ is limited to 22 V. A large collector current I_C necessitates a small R_C which shunts the load and reduces the gain. For values of I_C between 4 and 8 mA, with β_o assumed to be constant, calculations reveal that the gain is within a few percent of the maximum.

The actual choice of I_C might depend on the variation of beta with this current. Let us select -5 mA. Then R_C is 4.4 kΩ for a 22 V drop, which makes $V_{EC} = 2$. For a dc beta of 100, I_D is -0.05 mA, and from (17-14) V_{GS} is found to be -3.32 V.

With reference to Fig. 17-12, V_{SD} is 23.3 V, and hence V_{GD} must be 20 V. With a 22 V drop across R_C we select R_1 and R_2 so that $22R_1/(R_1 + R_2)$ is 20. Suitable choices are 5 MΩ and 500 kΩ, and the completed design is shown in Fig. 17-14.

For the numerical values of Fig. 17-14 the transconductance g_{fs} is 0.315 mmho, g_m is 200 mmhos, r_π is 0.5 kΩ, and the voltage gain determined from (17-16) is -63. The actual gain is slightly lower because of the effect of r_{ds}. The rather low gain is due in large part to the small value of g_{fs}. Changing the parameters of the circuit to give a larger I_D and hence a larger g_{fs} also increases I_C, requiring a smaller R_C. The advantage of the increased g_{fs} is offset by the greater shunting effect of R_C. The gain that has been calculated is near the maximum that can be obtained from the particular configuration of Fig. 17-12 with a 24 V supply and the specified transistors.

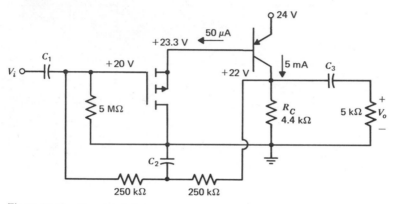

Figure 17-14   Completed design.

The use of RC coupling between the stages would permit an increased drain current with a larger g_{fs} and a greater gain. Another way to increase the gain is to utilize a transformer-coupled load, such as was done in the CD–CE phono amplifier of Fig. 17-8. This allows larger values of I_C and I_D as well as a larger effective load resistance.

In direct-coupled BJT amplifiers the design and analysis of bias stabilization is based on the current that is fed back. This was discussed in Sec. 15-6 and applied to several examples. However, because FETs are voltage actuated, the quantity fed back is a voltage. In the network of Fig. 17-14 suppose the dc beta of the transistor increases because of a rise in temperature, causing I_C to increase. This raises the voltage across R_C, thus increasing the gate potential and reducing V_{GS}. Both I_D and I_B drop, counteracting the increase in I_C. For example, the change in I_C is less than 1% for a change in β_F from 100 to 200. For a change in V_T from -3 to -2 V, resulting from transistor replacement, the change in I_C is only 5% (see P17-9).

CD-CE AMPLIFIER WITH ZENER DIODE

Some improvement in the gain can be obtained by incorporating a Zener diode as shown in the diagram of Fig. 17-15. With this arrangement the MOSFET current can be fixed independently of I_C. The parameters have been chosen to make I_C equal to -5 mA as before, and I_D is -5.85 mA (see P17-10). The transconductances g_{fs} and g_m are 3.42 and 200 mmhos, respectively. Effectively in series with r_π, which is 500 ohms, is the 25 ohm base resistance r_x and the product $(\beta_o + 1)r_d$, with r_d denoting the 5 ohm incremental resistance of the diode. Hence the input impedance of the second stage is about 1 kΩ. In the incremental model R_3 is in parallel with the 20 kΩ resistance r_{ds}, and the combination shunts the input impedance of the BJT. With R_3 equal to 1000 ohms about half of the incremental current $g_{fs}V_{gs}$ of the MOSFET is lost in R_3 and r_{ds}.

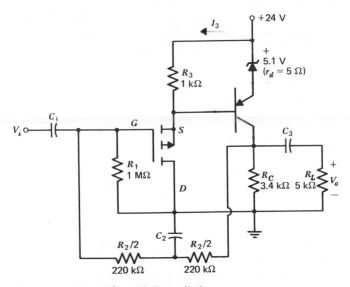

Figure 17-15   Amplifier with Zener diode.

Although we could select a larger R_3, the advantage is offset by the reduced value of g_{fs}. Replacing R_3 with a current source has obvious advantages. However, this requires another transistor and will not be done here.

A small quiescent collector current might seem desirable, because this would allow a larger R_C. The disadvantages, however, are a larger r_π, a greater dynamic resistance r_d of the Zener diode, and probably a lower β_o. Here again, the selection of I_C is a compromise. For the network of Fig. 17-15 the voltage gain is -122 (see P17-11). This is nearly twice that of the amplifier of Fig. 17-14.

Zener diodes are available with breakdowns from about 2 V to several hundred volts. For reference voltages below a few volts *forward-biased* diodes can be used, with several in series if desired, and diodes designed for this purpose are known as *stabistors*. The reference voltage of a Zener diode is somewhat temperature sensitive, with the temperature coefficient defined as the percent increase in the reference voltage per °C rise in temperature. The coefficient is negative for breakdown below about 5 V and positive when the breakdown is above this value. Usually it is between $\pm 0.1 \%/°C$.

In the network of Fig. 17-15 the temperature variation of the reference voltage of the diode is not serious because the dc feedback stabilizes the collector current. Of greater importance is the incremental diode resistance r_d, specified as 5 ohms. This resistance occurs because the breakdown characteristic is not precisely vertical. Minimum values of r_d, commonly from 2 to 50 ohms depending very much on the current, are found in diodes with reference voltages between 5 and 10 V. The resistance is much higher in Zener diodes at lower voltages, and it decreases appreciably with an increase in the quiescent current. For example, the 5 ohm resistance of the diode of the network of Fig. 17-15 is perhaps 2 ohms at 20 mA, whereas at 1 mA it may be around 15 ohms. Because r_d in Fig. 17-15 is in series with the emitter, only a few ohms degrade the voltage gain considerably. Therefore, a substantially smaller quiescent current would be intolerable. The Zener diode is sometimes replaced with a resistor selected to give the desired voltage drop. When this is done, a bypass capacitor is essential to avoid serious loss of voltage gain.

Although the CD–CE amplifier has lower voltage gain than the CS–CE configuration, its frequency response is superior. The gate-drain capacitance C_{gd} of the FET shunts the input, and if the source impedance is low, C_{gd} is not significant. Of course, it has absolutely no effect on the ratio V_o/V_i of Fig. 17-12. Capacitance C_{gs} is located between the input and output terminals of the CD stage. However, the voltage across it is smaller than the input voltage because of the less-than-unity noninverting gain of the stage, and C_{gs} can usually be neglected. Loss of gain at high frequencies is mainly determined by the relatively large equivalent input capacitance of the CE stage. This capacitance, which is enhanced by the Miller effect, is shunted by r_π and the low output impedance typical of CD amplifiers. Thus its effect is less than that of the same CE stage preceded by a common-source input stage.

17-4. POWER FETs

The JFETs and MOSFETs that have been considered thus far have lateral structures that transfer charge through a channel parallel to and adjacent to the surface. Because it is difficult to transmit appreciable heat energy away from a small surface region, the standard FETs have low power-handling capability, with current levels usually restricted to less than several hundred milliamperes. There are several types of vertical-structure FETs designed for larger currents and higher power ratings. In these devices the free electrons leaving the doped

N-type source region at the surface flow downward through a *vertical* channel to the N-type epitaxial layer constituting the drain, and this layer is on a heavily doped N^+ substrate. With the drain terminal connected to the substrate, the device can be thermally bonded to a heat sink in order to lower the case-to-ambient thermal resistance. In the saturation region most of the power dissipated within the device is at the pinched-off part of the channel, located at the drain-end PN junction. The junction-to-case thermal resistance is substantially lower for a vertical FET than for a lateral type. A vertical MOS device is commonly referred to as a *VMOS* transistor, with V denoting *vertical*.

A VMOS STRUCTURE

One technique used to obtain a vertical channel is to form a V groove that penetrates into the surface of the silicon. The cross-section of such a VMOS device is shown in Fig. 17-16. Fabrication is by the planar diffusion process. On the N^+ substrate is initially grown an N^- epitaxial layer. In the layer a rather lightly doped P-type *body* diffusion is made, followed by an N^+ source diffusion.

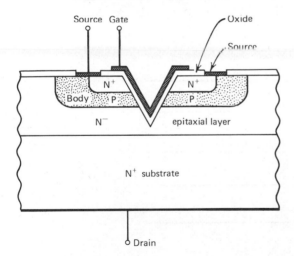

Figure 17-16   VMOS cross-section.

Next, the V groove is etched, with this groove extending into the N^- epitaxial layer as illustrated. An oxide coating is grown, and metallization provides the gates and desired connections, including a connection between the N^+ source and the P body.

The enhancement-mode device requires a positive gate voltage to form the conducting channels, which appear under the oxide on *both* surfaces of the P body. Electrons flow from the source through the two channels to the N^-

epitaxial layer, then passing through the N^+ substrate to the drain terminal at the bottom. The space-charge layer at the reverse-biased junction between the P body and the N^- epitaxial layer is largely within the more lightly doped epitaxial layer, and this results in higher drain-source breakdown voltage. Another advantage of the structure is the reduction in the gate-drain capacitance C_{gd}, which results because the metal gate has a minimum overlap of a drain region that is very lightly doped. Diffusion depths determine the channel lengths, rather than mask spacings as in standard devices, and these depths are better controlled. Consequently, channel lengths of about 1.5 μm can be used, compared with about 4 μm for conventional devices fabricated without benefit of ion implantation. Presently under development are various types of vertical power FETs without V grooves, including both enhancement-mode and depletion-mode types.

OUTPUT CHARACTERISTICS

The drain characteristics of an n-channel enhancement-mode VMOS transistor are shown in Fig. 17-17. Note that the vertical scale is in amperes. These curves are typical for a Siliconix VMP 1. This device[5] has a threshold

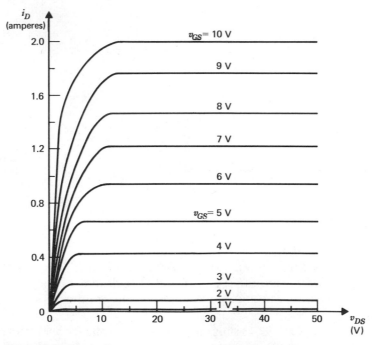

Figure 17-17   Drain characteristics of n-channel enhancement-mode VMOS transistor (Siliconix VMP 1).

voltage greater than 0.5 V and usually about 1 V. The drain-source breakdown voltage is 60 V or more, and the gate-drain capacitance C_{gd} is typically 7 pF.

The incremental resistance r_{ds} of an FET biased in the saturation region *is inversely proportional to the quiescent drain current I_D* provided the temperature in the channel is maintained constant. Consequently, at the high current levels of a power device a low value of r_{ds} is expected. However, because the pinch-off portion of the channel of the VMOS transistor extends mostly into the depletion region of the lightly doped epitaxial layer, an increase in v_{DS} has very little effect on the portion of the channel not pinched off, and accordingly, the effect on the current is slight. Hence r_{ds} is quite large, as indicated by the nearly zero slope of the curves of Fig. 17-17. At a drain current of 100 mA with a drain-source voltage of 25 V, the incremental resistance r_{ds} is typically 5 kΩ, which corresponds to a resistance of about 500 kΩ at 1 mA.

In the vicinity of the origin the drain characteristics have steep slopes, indicating low values of resistance. The higher resistance of lateral FETs is largely attributable to the ohmic resistance of the surface metal required for contact with the drain. For the VMP 1 the drain-source *on* resistance $r_{DS(on)}$ at 500 mA with $v_{GS} = 10$ V is typically 2 ohms.

Below about 400 mA the spacing of the curves of Fig. 17-17 obey the square law (3-3), and the transconductance g_{fs} is proportional to the square root of I_D. However, above this current level the curves are almost evenly spaced, indicating that g_{fs} is reasonably constant, independent of v_{GS} and v_{DS}. The square law of (3-3), which is inapplicable here, is based on the assumption that the current density within the portion of the channel not in pinch-off is proportional to the electric field (Ohm's law). At the high current densities of the power VMOS transistor, an increase in the electric field does not increase the current density nearly as much as predicted by Ohm's law, and the change in i_D is proportional to the change in v_{GS}, which controls the depth of the channel. The transconductance g_{fs} is typically 270 mmhos when the current is above 400 mA, with lower values at the lower currents. The rather constant transconductance provides increased linearity in large-signal amplifiers.

VMOS ADVANTAGES

In comparison with BJTs the VMOS transistors have the advantage of extremely high input impedance. Because the drive current to the gate is very low, high-impedance drivers can be used, and the power drawn from such drivers is negligible.

There is no problem with *current hogging*. If the current density at a particular location of a channel region between the source and drain increases above that of other channel regions, the temperature at this point rises. The negative temperature coefficient of the current, caused by reduced charge-carrier mobility at higher temperatures, tends to lower the current. Thus the current densities throughout the channel width, normal to the cross-section of Fig. 17-16, are automatically equalized. The same process tends to equalize the currents of FETs connected in parallel. In contrast, the positive temperature coefficient of the collector current of bipolar transistors gives serious current hogging problems when such devices are connected in parallel. In addition, secondary breakdown and thermal runaway can occur in power BJTs from localized heating and hot spots. No such problems are encountered with FETs. Class B power amplifiers using bipolar transistors require emitter resistors of about 0.3 ohm. In class B power amplifiers using field-effect transistors, source resistors are not needed.

Transistors have many important applications in power switching circuits. A BJT operating as a switch is normally in cutoff when the switch is *off*, and it is in saturation when the switch is *on*. During turn-off the storage-delay time required to remove the excess charges of the neutral regions of the BJT slows the switching speed considerably. Because the FET has no such storage of minority and majority carriers, it can switch power circuits much faster. In Fig. 17-18 is an

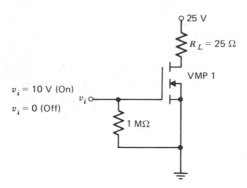

Figure 17-18   Elementary power switching circuit.

elementary switching circuit.[5] With the switch *on*, the channel resistance is only about 2 ohms, and the current through the load is nearly 1 A, providing a power of 25 W. When it is *off*, the current is the drain cutoff current $I_{D(off)}$, which is less than 0.5 μA. Typical values of turn-on and turn-off times are 4 ns each. The corresponding turn-on and turn-off times with the VMOS transistor replaced with a BJT would likely be from 10 to 100 times greater.

There are also disadvantages, of course. Many bipolar power transistors have lower junction-to-case thermal resistances. For the VMOS transistor VMP 1 this resistance is 5°C/W. The minimum value of v_{CE} for a BJT transistor operating in the active mode in a power amplifier is lower than that of a VMOS device operating in the pinch-off region. Consequently, in class B power amplifiers the device dissipation corresponding to a specified maximum output power to a specified load impedance is less for bipolar transistors than for FETs. The bipolar transistors can provide greater voltage gain than can the FETs, because of larger values of transconductances. Available are many different bipolar junction transistors, both NPN and PNP types, with various current, voltage, and power ratings. In 1977 power FETs are n-channel only, and complementary source-follower outputs are not possible. However, p-channel power FETs will un-doubtedly be developed, as will devices with higher current, voltage, and power ratings.

PARALLEL AND SERIES OPERATION

The current capacity of power FETs is easily increased by connecting several devices in parallel. For example, the VMP 1 has a maximum allowable con-tinuous drain current of 2 A. By connecting three of these in parallel, the current capacity is increased to 6 A. The three drain terminals are connected directly together to form a single drain terminal, and the same is done for the three gates and the three source terminals. From the incremental model of the circuit it is easily determined that the transconductance of the composite device is the sum of the individual transconductances, and the incremental drain resistance is the parallel equivalent of the individual drain resistances.

Series connections are also feasible. Shown in Fig. 17-19 is a switching circuit[3] with a supply voltage of 100 V. Each of the two transistors has a drain-source breakdown voltage of at least 60 V. For a VMOS device to enter breakdown,

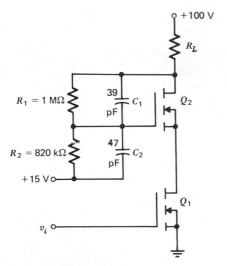

Figure 17-19   Power switching circuit using VMP 1 transistors.

there must be a drain current, and drain current in the circuit of Fig. 17-19 is possible only if both transistors conduct. Therefore, with the switch *off*, the supply voltage is divided so that neither transistor is in breakdown. The resistors R_1 and R_2 fix the gate voltage of Q_2 at 53 V with respect to ground when the switch is *off*. The 15 V at the bottom terminal of R_2 is used, rather than ground, to provide an adequate positive value for v_{GS2} when the switch is *on* (see P17-13). Capacitors C_1 and C_2 allow charge on the gate of Q_2 to be changed very rapidly for fast switching. Their function is similar to that of the speed-up capacitor of Fig. 6-10 of Sec. 6-5.

VMOS POWER AMPLIFIERS

In Fig. 17-20 is the schematic diagram[5] of a class A audio power amplifier using a VMP 1 power transistor. With reference to the drain characteristics of Fig. 17-17, the dc load line is vertical with $v_{DS} = 28$ V, assuming the dc resistance of the transformer primary is negligible. The quiescent gate-source voltage V_{GS1} of the output transistor is 680/4680 times 28, or approximately 4 V. This establishes the Q-point on the dc load line at a drain current of 420 mA. Because the impedance ratio of the transformer is 48/8, corresponding to a turns ratio of

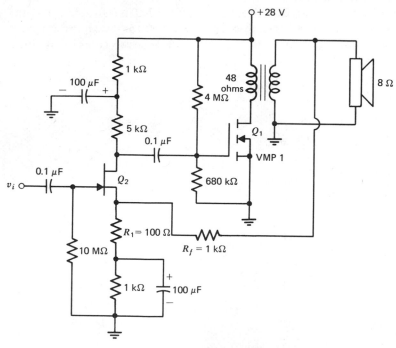

Figure 17-20   Four-watt class A audio amplifier using VMOS power transistor.

2.45, the effective ac load resistance is 48 ohms. The ac load line passes through the Q-point with a slope of $-1/48$, or -0.021 A/V.

Both the JFET driver and the VMOS power stage are inverting amplifiers. Hence the two-stage amplifier is noninverting, and the feedback through R_f is similar to that of the noninverting OP AMP of Fig. 14-11 of Sec. 14-3. The approximate voltage gain V_o/V_i is $1 + R_f/R_1$, or 11. In order to develop 4 W of output power across the 8 ohm speaker, an rms output voltage of 5.7 V is needed, and the required rms input voltage is about 0.5 V. The amplifier has a bandwidth from 100 Hz to 15 kHz with an output distortion around 2% at 3 W. It is suitable for the audio output stage of inexpensive radio, TV, and phonograph amplifiers.

The output stage of a class AB power amplifier is shown in Fig. 17-21. Because the VMOS power transistors are both n-channel devices, *the circuitry is designed so that the CS amplifier associated with Q_4 has characteristics that are quite similar to those of the source follower of Q_2.* This is accomplished by the voltage-sample current-sum feedback through R_F of Q_4. In class A operation an analysis reveals (see P17-15) that the voltage gain of the cascade consisting of Q_3 and Q_4 equals that of the cascade of Q_1 and Q_2, except for a change in sign. Furthermore, both power transistors present the same output impedance to the load (see

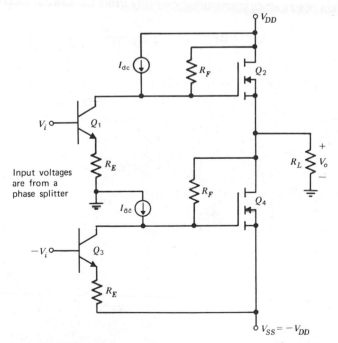

Figure 17-21   Output stage of class AB push-pull amplifier using VMOS power transistors.

P17-16). Consequently, class AB operation with minimum distortion is possible. The arrangement is similar to the push-pull circuit of Fig. 16-22 of Sec. 16-5. A phase-splitter provides the equal-and-opposite input voltages.

A practical high-quality audio amplifier using the output configuration of Fig. 17-21 and designed to deliver 40 W to an 8 ohm speaker has been built and tested.[6] Each output power device is a composite transistor consisting of three VMOS transistors connected in parallel, and the individual transistors are similar to the VMP 1. They are biased for class AB operation and mounted on suitable heat sinks with fins. Voltage supplies are ± 36 V. The bandwidth extends from 1 Hz to 800 kHz, and the slew rate exceeds 100 V/μs. At 40 W and 1 kHz the harmonic distortion is typically 0.04%. Output short-circuit protection is provided.

Power FETs also have important high-frequency applications. For example, a 145 MHz linear amplifier using the VMP 1 can deliver 5 W peak envelope power when used as a transmitter.[5] At both low and high frequencies there are many linear and switching circuits that are ideally suited for VMOS, and the future of these relatively new devices is quite promising. In addition, there are low-power VMOS integrated circuits in development for use in digital electronics. The advantage here is extremely dense circuitry.

17-5. BIAS CIRCUITS

ENHANCEMENT-MODE MOSFET BIAS CIRCUITS

Figure 17-22 shows two possible bias circuits. In circuit (a) the dc feedback through the 2 kΩ source resistance provides stability, and in circuit (b) this is accomplished by the feedback from drain to gate. The two-battery equivalent

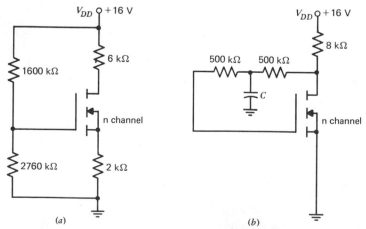

(a)   (b)

Figure 17-22   Enhancement-type MOSFET bias circuits.

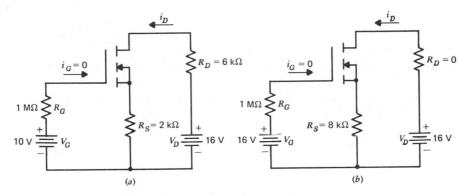

Figure 17-23   Two-battery equivalent circuits.

circuits are those of Fig. 17-23. Note particularly that circuit (b) has $V_G = V_{DD}$, and R_S is the 8 kΩ load resistance. The gate current of a MOSFET at room temperature is typically 1 pA, given on data sheets as I_{GSS}. Although this current increases almost exponentially with temperature, still it is usually negligible provided R_G is not more than a few megohms. We will assume that i_G is zero.

Both circuits (a) and (b) have the same dc load line, which is drawn on the drain characteristics of Fig. 17-24. For circuit (a) the input load-line equation and the saturation-mode transfer-characteristic equation are, using kΩ, mA and V with $k = 0.025$ and $V_T = 2$,

$$v_{GS} = 10 - 2i_D \qquad i_D = 0.025(v_{GS} - 2)^2 \qquad (17\text{-}17)$$

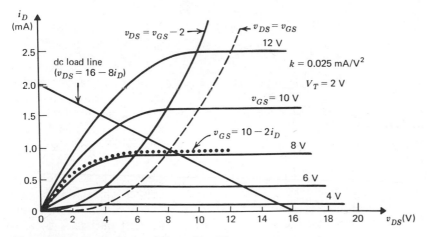

Figure 17-24   Drain characteristics and load lines.

The solution of these gives $I_D = 0.94$ mA and $V_{GS} = 8.1$ V. From I_D and Kirchhoff's voltage law, V_{DS} is calculated to be 8.5 V. A graphical procedure can also be used to find the Q-point. The *dotted* curve of Fig. 17-24 is a point-by-point plot of the input load-line equation, and its intersection with the dc load line locates the Q-point.

For circuit (b), $v_{GS} = 16 - 8i_D$, and similar analytical and graphical techniques apply. The analytical solution gives $I_D = 0.97$ mA, and V_{GS} and V_{DS} are both 8.2 V. Rather than plot the input load line, the graphical solution is more easily accomplished by plotting the curve $v_{DS} = v_{GS}$. This is shown on the figure as a *dashed* line, and because the curve is to the right of the line separating the saturation and nonsaturation regions, operation in the proper region is assured. The Q-point is, of course, at the intersection with the dc load line.

Both bias circuits (a) and (b) produce approximately the same Q-point. To ascertain which one is the better, let us refer to Fig. 17-25 which shows the

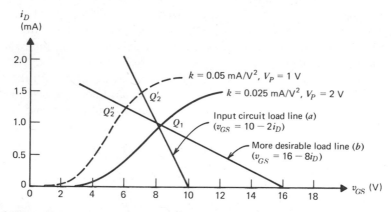

Figure 17-25　Transfer characteristics and load lines.

transfer characteristics and the input load lines. The solid-line dynamic transfer characteristic is a plot of i_D versus v_{GS} with points taken from the dc load line of Fig. 17-24. The straight lines are the input load lines of circuits (a) and (b). Both give nearly the same quiescent point Q_1. Now let us refer to the dashed transfer characteristic, obtained from the same dc load line and a set of drain characteristics similar to those of Fig. 17-24, but the different FET has $k = 0.05$ mA/V² and $V_P = 1$ V. The load line of circuit (a) gives a quiescent point Q_2' in the linear region which is unsatisfactory for a transistor of an amplifier. This is not possible with the configuration of (b), and the point Q_2'' is properly within the pinch-off region. Thus the bias circuit of Fig. 17-22b is more stable than that of (a).

The characteristics of a particular type of MOSFET vary considerably from

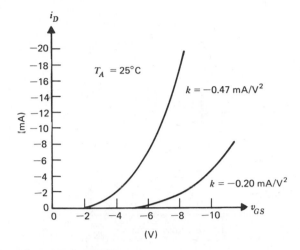

Figure 17-26   Extreme transfer curves for the 3N163.

one device to another. For example, the data sheet of a 3N163 p-channel enhancement-type MOSFET specifies that $I_{D(on)}$ is between -5 and -30 mA at 25°C with $V_{DS} = -15$ and $V_{GS} = -10$ V. The threshold voltage $V_{GS(th)}$ is between -2 and -5 V. With $V_T = -2$ V and $I_D = -30$ mA at $V_{GS} = -10$ V, the square-law equation gives $k = -0.47$ mA/V^2, and this static transfer characteristic is shown in Fig. 17-26. Also shown is the characteristic for $V_T = -5$ V and $I_D = -5$ mA at $V_{GS} = -10$ V, and the k for this curve is -0.20 mA/V^2. *The actual transfer curve at 25°C can be anywhere between the two.* It could start at -2 on the v_{GS} axis and go to -5 mA at $v_{GS} = -10$ V, for a k of -0.078 mA/V^2. For a curve starting at -5 V and going to -30 mA at $v_{GS} = -10$, k is -1.2 mA/V^2. The ratio of the maximum k to the minimum k is over 15, which is typical. The bias circuit must be designed to give a suitable quiescent point for any transfer characteristic between the two extremes. The circuit of Fig. 17-22b is satisfactory in most cases. The purpose of the capacitor C is to eliminate signal feedback.

DEPLETION-MODE FET BIAS CIRCUITRY

A common bias circuit for a depletion-mode device is that of Fig. 17-27. In the two-battery equivalent circuit note that V_G is positive, although both V_{GS} and I_G are negative, and I_G is very small compared with I_D. The gate current is not assumed to be zero in this discussion, because it is often quite important at elevated temperatures in depletion-mode FETs. This is especially true for the JFET, and the bias procedure presented here is equally applicable to this device.

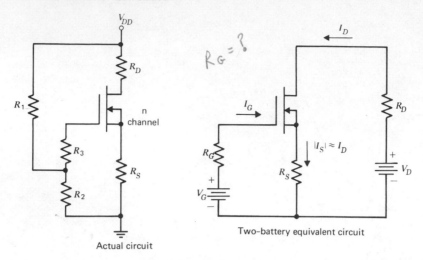

Figure 17-27   Depletion-type MOSFET bias circuit.

First, let us consider the limiting values of the drain current. Equations 15-4 and 15-5 of Sec. 15-1, deduced from the load lines of Fig. 15-5, apply to FETs as well as to BJTs. In FET terminology these are

$$|I_{D1}| \geq I' + \frac{V_m}{R_{ac}} \tag{17-18}$$

$$|I_{D2}| \leq \frac{|V_D| - |V_{P2}| - V_m}{R_D + R_S} \tag{17-19}$$

Here, R_{ac} is the incremental resistance between the drain-source terminals *external* to the transistor. For a CS amplifier with R_S bypassed, R_{ac} is $R_D \| R_L$. An additional restriction on the maximum quiescent drain current I_{D2}, which ensures that operation is confined to the depletion mode, is

$$|I_{D2}| < |I_{DSS1}| - \frac{V_m}{R_{ac}} \tag{17-20}$$

with I_{DSS1} denoting the smallest possible I_{DSS}. This is evident from Fig. 15-5, noting that V_m/R_{ac} is the maximum deviation of i_D from its quiescent value. For MOSFETs the restriction of (17-20) is unnecessary, because there is no objection to instantaneous excursion into the enhancement mode. However, JFET operation must always be confined to the depletion mode. In (17-18) I' is the magnitude of the minimum instantaneous drain current that is allowed. This should be at least an order of magnitude (10) greater than the gate leakage current I_{GSS} at the highest temperature. The maximum I_{GSS} is typically no more than a few nanoamperes for MOS transistors and perhaps a few microamperes for JFETs, and the square law is invalid for such drain currents. Another consideration is the fact that the noise figure is high at very low currents. As a rule, it is good practice to restrict the minimum quiescent drain current I_{D1} to values equal to or greater than 1% of I_{DSS1}. Voltage V_m is the maximum deviation of v_{DS} from its quiescent value V_{DS}. For a CS amplifier with a bypassed source resistor, this is simply the amplitude of the output voltage. In (17-19) the maximum possible pinch-off voltage V_{P2} has replaced V' in (15-5). This inequality, along with that of (17-20), maintains operation in the depletion-mode saturation region. Of course, I_{D2} is the maximum drain current, which is greater than the minimum I_{D1}.

We learned earlier that the threshold voltage of an enhancement-type FET drops slightly as the temperature T rises. However, the opposite is true for depletion-type FETs, with the magnitude of the pinch-off voltage increasing. Although the temperature effect on V_P is usually neglected in MOS circuit design, this is not always justified for the JFET. Temperature also affects the carrier mobility and hence the channel resistance. The result is that the drain current I_{DSS} decreases with a rise in T.

With reference to the extreme static transfer characteristics of Fig. 17-28, V_{P1} is the lowest possible pinch-off voltage at the lowest temperature T_1, whereas V_{P2} is the highest V_P at T_2. At the maximum temperature T_2, I_{DSS1} is the lowest I_{DSS}, and I_{DSS2} is the highest I_{DSS} at T_1. These parameters are obtained from information given on the data sheet. The actual transfer characteristic is somewhere between the two extremes for all temperatures from T_1 to T_2.

From Fig. 17-23 the input-circuit load-line equation is found to be

$$v_{GS} = -i_D R_S + V_G - i_G R_G \tag{17-21}$$

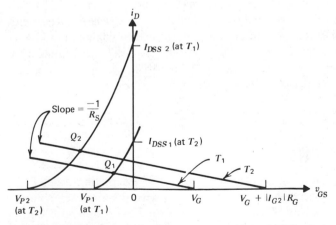

Figure 17-28   Extreme transfer characteristics and input load lines.

The gate current i_G is almost always negligible at the lowest temperature. Hence, we let $I_{G1} = 0$ and use I_{G2} to denote the maximum gate current at T_2. This is found from a curve of I_{GSS} versus temperature, usually given on the data sheet. At the quiescent point Q_1 of Fig. 17-28, which is on the load line at temperature T_1, the current and voltage are I_{D1} and V_{GS1}. The values at Q_2 are I_{D2} and V_{GS2} and from (17-21) we obtain the equations

$$V_G = I_{D1}R_S + V_{GS1} \tag{17-22}$$

$$V_G = I_{D2}R_S + V_{GS2} - |I_{G2}|R_G \tag{17-23}$$

Subtraction eliminates V_G, and the result can be expressed in the form

$$R_S = \frac{|V_{GS2} - V_{GS1}| + |I_{G2}|R_G}{|I_{D2} - I_{D1}|} \tag{17-24}$$

The voltages V_{GS1} and V_{GS2} are found from the square law to be

$$V_{GS1} = V_{P1}\left(1 - \sqrt{\frac{I_{D1}}{I_{DSS1}}}\right) \tag{17-25}$$

$$V_{GS2} = V_{P2}\left(1 - \sqrt{\frac{I_{D2}}{I_{DSS2}}}\right) \tag{17-26}$$

From (17-19) the maximum allowed R_D is

$$R_D = \frac{|V_D| - |V_{P2}| - V_m}{|I_{D2}|} - R_S \tag{17-27}$$

For maximum and minimum drain currents very close together, R_S is large according to (17-24). A large R_S may limit R_D, which shunts the load in the CS configuration, to a value too low for useful gain. On the other hand, if I_{D2} is substantially increased, R_D may again be too small as indicated by (17-27). *The problem confronting the designer is the proper choice of the extreme drain currents.* Once reasonable values for I_{D1} and I_{D2} are determined in some manner, V_{GS1} and V_{GS2} are calculated from (17-25) and (17-26). Then R_S and R_D are found from (17-24) and (17-27) after choosing a suitable value for R_G. A logical procedure for selecting the extreme drain currents is presented in the section that follows.

17-6. DEPLETION-MODE FET BIAS DESIGN

DESIGN EQUATIONS

A technique for determining the drain current I_{D1} of a depletion-mode FET, along with the proper corresponding value of I_{D2}, with these chosen so as to maximize the voltage gain, is developed here. Then a detailed design procedure is presented. It is helpful to express R_D directly in terms of I_{D1} and I_{D2}. This is easily done by eliminating V_{GS1} and V_{GS2} in (17-24) with the aid of (17-25) and (17-26). Then the result is substituted in (17-27), and we obtain the following equations

$$R_S = \frac{B}{|I_{D1}|} - \frac{C}{\sqrt{|I_{D1}|}} \tag{17-28}$$

$$R_D = \frac{A - B}{|I_{D1}|} + \frac{C}{\sqrt{|I_{D1}|}} \tag{17-29}$$

with

$$A = \frac{1}{a}(|V_D| - |V_{P2}| - V_m) = \frac{1}{a}(A_1) \tag{17-30}$$

$$B = \frac{1}{a - 1}(|V_{P2}| - |V_{P1}| + |I_{G2}|R_G) = \frac{1}{a - 1}(B_1) \tag{17-31}$$

$$C = \frac{1}{a - 1}\left(\frac{\sqrt{a}|V_{P2}|}{\sqrt{|I_{DSS2}|}} - \frac{|V_{P1}|}{\sqrt{|I_{DSS1}|}}\right) \tag{17-32}$$

and with a denoting the ratio I_{D2}/I_{D1} For a fixed ratio a, the symbols A, B, and C represent positive constants.

The voltage V_D must be sufficiently large so that, for a selected current ratio a, the constant A is greater than B. Otherwise, R_D of (17-29) is less than $C/\sqrt{|I_{D1}|}$, and this is simply too small for useful gain, regardless of the value of I_{D1}. Hence we shall assume that $A > B$. We note from (17-29) that increasing I_{D1} decreases the maximum value allowed for R_D.

The current ratio a is selected so that R_D is a maximum. A large A and a small B are desired. The minimum ratio of unity makes B infinite by (17-31), but a large ratio a makes A small. Because the second term on the right of (17-29) is usually a rather small part of R_D and has little effect on the selection of a, *the maximum R_D occurs approximately at the value of a that makes $(A - B)$ a maximum, regardless of the actual choice of I_{D1}.* We find this value from

$$\frac{d}{da}(A - B) = \frac{d}{da}\left(\frac{A_1}{a} - \frac{B_1}{a - 1}\right) = 0 \tag{17-33}$$

which gives

$$a = \frac{1 + \sqrt{B_1/A_1}}{1 - B_1/A_1} = \frac{I_{D2}}{I_{D1}} \tag{17-34}$$

In most cases a is between 2 and 3.

The voltage gain of a CS amplifier is shown in Sec. 17-1 to be

$$A_v = -g_{fs}(r_{ds}\|R_D\|R_L) = \frac{-g_{fs}R_L'R_D}{R_L' + R_D} \tag{17-35}$$

572 FET LINEAR CIRCUITS

with R_L' denoting the equivalent resistance of the RC coupled load R_L in shunt with the incremental resistance r_{ds}. Defined as the reciprocal of the slope of the drain characteristic at the Q-point, r_{ds} is inversely proportional to I_D. At current levels of only a few milliamperes, it is typically between 40 kΩ and 500 kΩ. The mutual conductance g_{fs} is defined as $\partial i_D/\partial v_{GS}$, or the slope of the transfer characteristic at the Q-point. As shown by (3-12) of Sec. 3-4 it can be expressed in the form

$$g_{fs} = \frac{2}{|V_P|}\sqrt{I_D I_{DSS}} \tag{17-36}$$

The maximum g_{fs} in the depletion mode occurs at a drain current I_D equal to I_{DSS}, and g_{fs} approaches zero as I_D becomes very small.

DC CIRCUIT DESIGN FOR A CS AMPLIFIER UTILIZING A DEPLETION-MODE FET

The basic design equations have been developed. Let us consider a CS amplifier with a load resistance R_L which is RC coupled. For any specified load we want the maximum voltage gain. Examination of (17-35) shows that both R_D and g_{fs} should be as large as possible. However, increasing I_D to increase g_{fs} lowers the maximum R_D. A major problem is the proper selection of I_D. It is helpful to examine A_v carefully. Both R_D and g_{fs} can be eliminated by means of (17-29) and (17-36). With this done at the quiescent point Q_1 the result can be put in the form

$$A_{v1} = \left[\frac{-2R_L'\sqrt{|I_{DSS1}|}}{|V_{P1}|}\right]\left[\frac{(A-B)\sqrt{|I_{D1}|} + C|I_{D1}|}{A - B + C\sqrt{|I_{D1}|} + R_L'|I_{D1}|}\right] \tag{17-37}$$

From $dA_{v1}/dI_{D1} = 0$, the maximum gain is found to occur at a drain current that is

$$|I_{D1}| = \left[\frac{(A-B)(C + \sqrt{(A-B)R_L'})}{(A-B)R_L' - C^2}\right]^2 \tag{17-38}$$

The design of a CS amplifier can be accomplished as follows:

1. If V_{DD} is not fixed by other considerations, choose it as large as possible without exceeding the minimum breakdown voltages of the device. A high voltage makes the constant A of (17-30) large.
2. Select R_G to provide the required high input impedance, but no larger than necessary. It is desirable that $I_{G2}R_G$ in (17-31) be small compared with the difference between the extreme pinch-off voltages, but of course, this is not always possible.
3. Calculate the current ratio a, using (17-34). The constants A_1 and B_1 in (17-34) are found from (17-30) and (17-31). Then calculate A, B, and C.

4. The minimum r_{ds} should be used to determine R_L', as this is the worst case. Obtain this value from the maximum incremental conductance given on the data sheet, usually referred to as y_{os}. Then find I_{D1} from (17-38). The product aI_{D1} gives I_{D2}. Calculate R_S and R_D from (17-28) and (17-29). Recall that I_{D1} must be greater than $(I' + V_m/R_{ac})$ by (17-18). Also, for operation wholly confined to the depletion mode I_{D2} must be less than $(|I_{DSS1}| - V_m/R_{ac})$ by (17-20). If necessary, adjust I_{D1}. Do not change the current ratio a.

5. Equations 17-22 and 17-25 determine V_G. Convenient values in the kilohm range are selected for R_1 and R_2 of Fig. 17-27 so that $V_G = R_2 V_{DD}/(R_1 + R_2)$. R_3 is the difference between R_G and $R_1 \| R_2$.

6. At each extreme quiescent point Q_1 and Q_2 the voltage gain should be determined from (17-35), with g_{fs1} and g_{fs2} obtained from (17-36).

It is interesting to note from (17-38) that the value of I_{D1} that gives maximum gain decreases as R_L' increases. In fact, for the theoretical case of an infinite R_L' the current I_{D1} is zero; for this case g_{fs} is zero, R_D is infinite, and the gain $g_{fs1}R_D$ is a maximum. Actually, R_L' can be no greater than r_{ds}, and I_{D1} can be made only as small as permitted by inequality (17-18). Usually the load R_L is specified and hence is not a design parameter, but for appreciable voltage gain this resistance should be large, and the larger the better.

Example

In this example, $I_{DSS1} = -15$, $I_{DSS2} = -50$, $V_{P1} = 4$, and $V_{P2} = 9.5$, with all units expressed in mA, kΩ, and V. The p-channel FET is in a CS amplifier, and the minimum r_{ds} is 20 kΩ. A maximum peak-to-peak output voltage of 5 V is required. The value of V_{DD} is -30 V; R_G is 1000 kΩ and the gate current is negligible. Design the circuit with an RC-coupled load of 5 kΩ.

Solution

We determine that V_m is 2.5 V, A_1 is 18, B_1 is 5.5, and the current ratio a is found from (17-34) to be 2.24. The values of A, B, and C are 8.05, 4.45, and 0.79, respectively. For $r_{ds} = 20$ kΩ and $R_L = 5$ kΩ, R_L' is 4 kΩ, and (17-38) gives $I_{D1} = -1.44$ mA. In addition, I_{D2} is aI_{D1}, or -3.21 mA. From (17-28) and (17-29) we find that R_S is 2.44 kΩ and R_D is 3.17 kΩ. We determine V_{GS1} from (17-25) to be 2.76, and V_{GS2} is 7.09 V. From (17-22) V_G is calculated to be -0.74 V. This voltage is obtained by selecting $R_1 = 119$ kΩ and $R_2 = 3$ kΩ, although other choices are possible and acceptable.

At the extreme Q points the transconductances are $g_{fs1} = 2.32$ and $g_{fs2} = 2.67$ mmhos. As the effective load is 1.77 kΩ, consisting of $R_D \| R_L'$, the respective voltage gains at Q_1 and Q_2 are -4.1 and -4.7. These are low but rather typical of FETs operating into loads of only a few kilohms.

Figure 17-29 shows the amplifier of the example with standard resistance values. We note that R_D is less than R_L. In CE amplifiers we usually select R_C

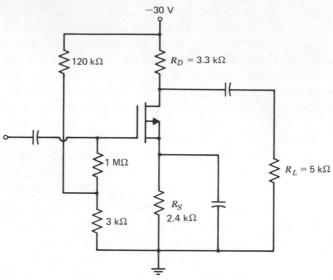

Figure 17-29   CS amplifier.

greater than R_L. The reason for the relatively low R_D is that voltage is nearly always at a premium in the CS amplifier, and this also accounts largely for the basic difference in the bias-circuit design procedure. At I_{D2} the voltage across R_S is 7.7, which is the amount required for supplying the rather large V_{GS2} of about 7 V with a negative V_G. Also, V_{DS2} is -11.7 V, which is the value required to insure operation in the saturation region of the drain characteristics at every instant. The corresponding voltages of the CE circuit are considerably lower. Of course we can increase R_D but only by reducing I_D, g_{fs}, and the gain. Our design is based on the proper compromise between I_D and R_D to obtain the maximum voltage gain. If a very large V_{DD} can be used, we would make R_D much greater than R_L and select the maximum possible I_D. However, this is seldom the case, and furthermore, most FETs break down above about 30 to 40 V.

SELF-BIAS

In the example the calculated value of V_G was small, being -0.74 V. With only a small loss of gain (see P17-23) the circuit of Fig. 17-29 can be simplified by omitting the 120 kΩ resistor, which makes V_G zero. Of course, the values of R_S and R_D must be recalculated. With V_G zero the gate-source voltage V_{GS1} equals $-I_{D1}R_S$ by (17-22), and the circuit is said to be *self-biased*. A design freedom is lost because we cannot use (17-34) to select the current ratio a independent of R_S. To design a self-bias circuit we might assume that (17-34) is approximate and use it to determine I_{D1} from (17-38) as before. This will probably give a quiescent current near the optimum. Then R_S is calculated from (17-22) and (17-25) with $V_G = 0$, I_{D2} from (17-23) and (17-26), and finally R_D from (17-27). If the current ratio I_{D2}/I_{D1} is substantially different from the value of a obtained from (17-34), we can select another value for I_{D1} and repeat the calculations, hoping to obtain a better circuit with more gain.

DISCUSSION

In the design of CE single-stage BJT bias circuits it was noted in Sec. 15-5 that a compromise was required in the selections of R_B and R_C. Making R_B very large so as to avoid appreciable shunting of the input requires a reduced R_C which shunts the load, whereas a smaller R_B permits a larger R_C. A similar relation exists between the gate resistance R_G and the drain resistance R_D of a CS amplifier. Increasing R_G increases the constant B_1 of (17-31), and this in turn reduces the maximum allowable R_D given by (17-29). The advantage of the higher input resistance is somewhat offset by the reduced R_D and lower gain. Again the designer must compromise.

Emphasis has been placed on the CS configuration because it is especially important and the bias-circuit design is rather critical. Except for (17-34), (17-37), and (17-38), the equations that have been presented apply to all bias circuits. It should be understood that (17-24) and (17-28) specify the minimum value of R_S and (17-27) and (17-29) give the maximum allowed R_D. CD and CG amplifiers with practical bias circuits were considered in Sec. 17-1. All equations are expressed in a form suitable for both n- and p-type depletion-mode FETs.

REFERENCES

1. W. Gosling. W. G. Townsend, and J. Watson, *Field-Effect Electronics*, Wiley-Interscience, New York, 1971.
2. L. J. Sevin, Jr., *Field-Effect Transistors*, McGraw-Hill, New York, 1965.
3. "COS/MOS Operational Amplifiers," File No. 817, RCA, 1974.
4. *Linear Integrated Circuits and MOS Devices* (Application Notes), RCA, 1974.
5. L. Shaeffer, "VMOS–A Breakthrough in Power MOSFET Technology," AN76-3, Siliconix Inc., Santa Clara, Calif., May 1976.
6. L. Shaeffer, "The MOSPOWER FET Audio Amplifier," DA76-1, Siliconix Inc., Santa Clara, Calif., May 1976.

PROBLEMS

Section 17-1

17-1. With $v_{GS} = V_{GS} + V_m \cos \omega t$, show from the square law (3-8) that the harmonic distortion D of a depletion-mode FET is $(V_m/4V_p)\sqrt{I_{DSS}/I_D}$. Then deduce that the distortion is 5% for V_m equal to $0.2V_P\sqrt{I_D/I_{DSS}}$.

17-2. The CS high-frequency incremental model of Fig. 17-3a has $g_{fs} = 2$ mmhos, $R_G = 1$ MΩ, $R_L' = 8$ kΩ, $C_{gs} = 4$ pF, and $C_{gd} = 2$ pF. At $\omega = 10^6$ find the complex voltage gain A_v and the input admittance Y_i in mmhos in polar form from the exact model. Repeat using Fig. 17-3b, and compare results.

17-3. Using kΩ, mA, mmhos, and V, the CD amplifier of Fig. 17-4 has R_1, R_2, R_3, R_4, and R_L equal to 40, 60, 2000, 20, and 80, respectively. The supply voltage V_{DD} is 30, and the source resistance R_s is 200. The JFET has $I_{DDS} = 4$, $V_P = -4$, and $r_{ds} = 40$. Calculate g_{fs} and find A_v, A_i, R_i, and R_o. Refer to (3-12).

17-4. Resistance $R_G = 2$ MΩ, $R_S = 400$ Ω, $R_D = 6$ kΩ, $R_L = 4$ kΩ, and $V_{DD} = 30$ in the CD amplifier of Fig. 17-6. The JFET has $I_{DSS} = 4$ mA, $V_P = -4$, and $r_{ds} = 20$ kΩ. Find the Q-point (I_D, V_{DS}), and calculate A_v, R_i, and A_i. Refer to (3-12).

Section 17-2

17-5. For the JFET differential amplifier of Fig. 17-9 assume that each transistor has an incremental resistance r_{ds} of 30 kΩ, and calculate the difference-mode voltage gain. (*Ans*: 9.47)

17-6. In the CMOS operational amplifier of Fig. 17-11 assume that transistor Q_1 has $k = -0.2$ mA/V^2 and $V_T = -1.2$ V. Also assume that diodes D_1 through D_4 leak 116 μA to ground so that the current through the 40 kΩ resistor is less than the drain current of Q_1 by this amount. The voltage across each of these diodes is 0.7 V. Calculate I_{D1} in mA.

17-7. In the CMOS operational amplifier of Fig. 17-11 assume that the matched transistors of the complementary class A output stage have $k = \pm 0.2$ mA/V^2 and threshold voltages of magnitude 1.2 V. With the inputs adjusted so that the output voltage is zero, calculate the quiescent drain current in mA. Then model the transistors using the single parameter g_{fs}, and calculate the voltage gain of the output stage. (*Ans*: 7.94, -10.1)

17-8. For the cascode arrangement of Q_3 and Q_5 in Fig. 17-11, deduce that the incremental resistance r between the collector of Q_5 and $V_{CC}{}^+$ is the sum of r_{ds5} and r_{ds3} $(1 + g_{fs5}r_{ds5})$. Note that the two gates are connected incrementally, and v_{gs3} is zero. Assume r_{ds3} and r_{ds5} are each 40 kΩ, I_D is -0.2 mA, and k is -0.025 mA/V^2. With reference to (17-15), calculate r.

Section 17-3

17-9. For the CD–CE amplifier of Fig. 17-14, with V_{EB} constant at 0.7 V, find the percent increase in the quiescent collector current when a temperature rise causes β_F to change from 100 to 200. Also, calculate the percent increase with β_F constant at 100 when the MOSFET is replaced with one having the same k but with $V_T = -2$ V. Refer to Fig. 17-12 for data.

17-10. From the parameters of the network of Fig. 17-15 calculate I_C and I_D in mA. The MOSFET has $k = -0.5$ mA/V^2 and $V_T = -3$, and the BJT has $\beta_F = 100$ and $V_{EB} = 0.7$ V.

17-11. For the network of Fig. 17-15 calculate the midband voltage gain. The MOSFET has $k = -0.5$ mA/V^2, $V_T = -3$, and $r_{ds} = 20$ kΩ. The BJT has $\beta_o = 100$ and $r_x = 25$ ohms. Assume $q/kT = 40$. I_D is -5.85 mA and I_C is -5 mA.

17-12. The amplifier of Fig. 17-12 is redesigned to operate with $V_{EE} = 12$ V. The resistors are selected so that the collector current is 5 mA and V_{EC} is 2 V. Calculate the voltage gain. Do not neglect r_{ds} and assume q/kT is 40.

Section 17-4

17-13. In the switching circuit of Fig. 17-19, R_L is 20 ohms. Because the load current exceeds the two-ampere rated current of the VMP 1, it is necessary to use six transistors. Sketch the complete circuit. With the switch *on*, determine the load current and also v_{GS2}, assuming each transistor has a channel resistance of 2 ohms. In addition, calculate the gate voltage of Q_2 with respect to ground when the switch is *off*.

17-14. For the class A amplifier of Fig. 17-20 with the drain characteristics of the power transistor given in Fig. 17-17, determine the quiescent drain current I_{D1} and the voltage V_{GS1}. Deduce the equation of the ac load line relating i_{D1} and v_{DS1}, noting that the effective load at the primary side of the transformer is 48 ohms. With a sinusoidal waveform and an output of 4 W, determine the maximum and minimum values of both i_{D1} and v_{DS1}. Assume an ideal transformer and use SI units.

17-15. In the circuit of Fig. 17-21, assume that each BJT has $\beta_o \gg 1$ and $\beta_o R_E \gg h_{ie}$, and each VMOS transistor has $r_{ds} \gg R_L$. Transistor parameters h_{re} and h_{oe} are negligible. (a) With Q_3 and Q_4 treated as *off*, sketch the incremental model of the circuit. From the model and the stated assumptions. show that V_{gd}/V_i equals $-R_F/R_E$, that $V_{gs} = V_{gd} - V_o$, and that the voltage gain V_o/V_i is $-(R_F/R_E)[g_{fs}R_L/(1 - y_{fs}R_L)]$. (b) With Q_1 and Q_2 *off*, sketch the incremental model. From this and the stated assumptions deduce that V_o/V_i is the same as that of (a) provided $g_{fs}R_F \gg 1$.

17-16. In the circuit of Fig. 17-21, assume the transistors can be modeled using h_{ie}, β, g_{fs}, and r_{ds}. (a) With Q_3 and Q_4 *off* and $V_i = 0$, sketch the incremental model of the circuit. Replace R_L with a voltage source V_o and find the output resistance in terms of the network parameters. (Note that $V_{gs} = -V_o$.) (b) Repeat (a) with Q_1 and Q_2 *off*, instead of Q_3 and Q_4. (Note that $V_{gs} = V_o$).

Section 17-5

17-17. A p-channel enhancement-mode MOSFET has $k = -0.04$ mA/V^2 and $V_T = -4$ V. The bias circuit is that of Fig. 17-22b with V_{DD} changed to -16. Find the Q-point (I_D, V_{DS}).

17-18. The MOSFET in the circuit of Fig. 17-30 has the drain characteristics of Fig. 17-24. Find the Q-point (I_D, V_{DS}).

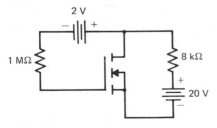

Figure 17-30 For P17-18.

Section 17-6

17-19. The data of the example of Sec. 17-6 apply to this problem, except that R_L is infinite. Redesign the circuit, calculate the voltage gains at Q_1 and Q_2, and compare with the results of the example.

17-20. The current $I_{DSS1} = 5$, $I_{DSS2} = 50$, $V_{P1} = -0.5$, $V_{P2} = -5$, with units of mA, kΩ, and V used in this problem. The n-channel FET is in a CS amplifier with an RC-coupled load of 3.5 kΩ and a bypassed source resistance. $V_{DD} = 20$. The minimum r_{ds} is 21. At the highest temperature the maximum gate current is 0.0001 mA and a gate resistance of 5000 kΩ is desired. The input is only a microvolt. For maximum voltage gain determine proper values for R_S and R_D in the circuit of Fig. 17-27. Also calculate the voltage gain at the extreme quiescent points Q_1 and Q_2 (Ans: -14.3, -6.9).

17-21. The current $I_{DSS1} = 1$, $I_{DSS2} = 3$, $V_{P1} = -0.5$, $V_{P2} = -2.5$, in units of mA, kΩ, and V. The n-channel JFET is in a CS amplifier. V_{DD} is 20 and a maximum peak-to-peak output of 3 V is required. At the highest temperature the maximum gate current is 0.001 mA and R_G is 1000 kΩ. The source resistance is bypassed. For $R_L = 15$ and $r_{ds} = 30$ kΩ find optimum values for R_S and R_D, and calculate the voltage gains at Q_1 and Q_2. Do not use I_{D2} greater than 0.79 (explain this limitation).

17-22. The data of P17-21 apply to this problem, except that V_{DD} is 30 V, R_L is 45 kΩ, and r_{ds} is 90 kΩ. Calculate the maximum voltage gains at Q_1 and Q_2 and compare with the values obtained in P17-21.

17-23. In the example of Sec. 17-6 redesign the circuit for self-bias, specifically determining the proper values of R_S and R_D. Use the same I_{D1} as in the example, and follow the procedure suggested at the end of Sec. 17-6. (Ans: $R_D = 3.05$ kΩ)

Chapter Eighteen
Frequency Response

Thus far our study of amplifiers has concentrated on performance in the mid-frequency region and on the design of the bias circuitry. We are now ready to examine the effects of any capacitances that may be present. The coupling and bypass capacitors often used are in the microfarad range, from a fraction of a microfarad to perhaps several hundred microfarads. Although these capacitors behave like short-circuits in the midfrequency region, at low frequencies their impedances are large enough to reduce the gain appreciably. The device capacitances are in the picofarad range and approximate open circuits in the midfrequency region. However, at high frequencies their impedances become small, causing the gain to drop. Accordingly, there are three distinct frequency regions.

In the low-frequency region the microfarad coupling and bypass capacitors, if any, must be included in the incremental models. In the midfrequency region the model is resistive, with the microfarad capacitances treated as shorts and the picofarad capacitances as opens. Finally, at high frequencies we consider the detrimental effects of the picofarad device capacitances. In a high-quality audio amplifier the midband range may extend from 20 Hz to 20 kHz, whereas in a video amplifier the response might be "flat" from dc to 50 MHz.

We begin with a study of the poles and zeros of an amplifier gain function because they determine the frequency response, the transient response, and the stability. Also, in many cases their use simplifies both analysis and design.

18-1. POLES AND ZEROS

COMPLEX FREQUENCY

An electric-current time function of particular importance to the analysis and design procedures related to poles and zeros is

$$i = |I|e^{\sigma t} \cos(\omega t + \theta) = \text{Re}(|I|e^{j\theta}e^{\sigma t}e^{j\omega t}) \tag{18-1}$$

with Re denoting *the real part of* the expression that follows it. A convenient form of (18-1) is

$$i = \text{Re}(Ie^{st}) \qquad \text{with } s = \sigma + j\omega \tag{18-2}$$

The complex current I is $|I|e^{j\theta}$, and of course, the symbol $|I|$ denotes the magnitude of I. The real part σ and the imaginary part ω of the *complex frequency s* represent real numbers that may be positive, zero, or negative. Although a negative frequency ω has no particular significance, it substitutes into (18-1) as a negative number. This presents no problem, because $\cos -x$ equals $\cos x$.

In Fig. 18-1 are plots of i versus t for positive, zero, and negative values of σ. With σ positive the amplitude of the cosine function increases exponentially with time, and the amplitude decreases for negative σ. The special case of $\sigma = 0$ is the familiar steady-state sinusoidal time function.

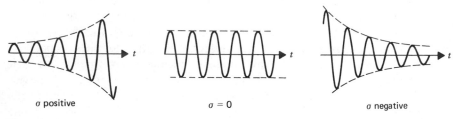

σ positive $\qquad\qquad\qquad\qquad \sigma = 0 \qquad\qquad\qquad\qquad \sigma$ negative

Figure 18-1   Plots of i versus t from (18-1).

In Fig. 18-2 are the three types of waveforms for $\omega = 0$. For both σ and ω equal to zero, the current is steady with time. The other two curves are exponential functions.

From (18-2) we find the voltage $L\, di/dt$ across an inductance to be $\text{Re}(sLIe^{st})$ when the current i has the complex frequency s. The complex voltage V is sLI, and the ratio V/I is the impedance sL. Similarly, the impedance of a capacitance is $1/sC$. Accordingly, to determine the steady-state response of a network excited by a source having a complex frequency, we simply assign to each inductance the impedance sL and to each capacitance the impedance $1/sC$.

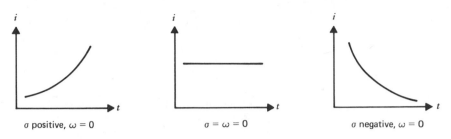

σ positive, ω = 0   σ = ω = 0   σ negative, ω = 0

Figure 18-2   Plots of (18-1) for ω = 0.

The excitation may vary with time in a manner similar to any of the waveforms of Figs. 18-1 and 18-2, but it must have been applied at time minus infinity. The symbol V is used to signify a complex voltage and I denotes a complex current. The analysis follows any convenient procedure, treating the network as though it is a "dc network" in which some impedances depend on s. After the s-domain response is obtained, the time-domain response is found from (18-2) or its voltage equivalent.

When the excitation is applied at a time other than minus infinity, the response has both transient and steady-state components. Fortunately, another interpretation can be given to s, assuming the network has no initial energy storage. Selecting time zero to be the instant at which the signal is applied, we can consider s as the Laplace transform variable with V and I signifying the transforms of v and i. The transform of $L\,di/dt$ is sLI, and that of the voltage across a capacitor is I/sC. These are the same mathematical forms as for the case of excitation at a complex frequency.

Example 1

Analysis of a network in the s-domain gives $I_o/I_i = s/(s + 3)$. (a) Find the steady-state response $i_o(t)$ if the input current $i_i(t)$ is $2e^{-3t} \cos 4t$. (b) Determine $i_o(t)$ when $i_i(t)$ is a unit step function applied at time zero.

Solution

(a) We see that $I_i = 2$ and $s = -3 + j4$. Therefore, I_o is $2(-3 + j4)/j4$, or $2.5\underline{/36.9°}$. In accordance with (18-1), i_o is $2.5e^{-3t} \cos(4t + 36.9°)$. (b) The Laplace transform of $i_i(t)$ is $1/s$. Hence I_o is $1/(s + 3)$. The inverse transform is e^{-3t}, which is $i_o(t)$ for $t > 0$.

THE POLES AND ZEROS OF A GAIN FUNCTION

For a linear network with no independent sources other than that at the input, the gain A is the ratio of an output voltage or current to an input voltage or current. The complex frequency of the signal is s, and each passive element can be replaced with the impedance R, sL, or $1/sC$. Analysis can be made by means of a set of node or loop equations expressed in terms of s. The gain can be obtained from these as a ratio of determinants, and each determinant can be expanded into a polynomial in s. The result has the form

$$A(s) = \frac{a_o + a_1 s + a_2 s^2 + \cdots}{b_o + b_1 s + b_2 s^2 + \cdots} \qquad (18\text{-}3)$$

The real coefficients a and b are functions of the RLC circuit constants and the constants associated with any dependent sources. By factoring the polynomials the gain A becomes

$$A(s) = \frac{K(s - z_1)(s - z_2)(s - z_3)\ldots}{(s - p_1)(s - p_2)(s - p_3)\ldots} \qquad (18\text{-}4)$$

Each *singularity* z_1, z_2, p_1, p_2, etc., can be real or complex. However, because the a and b coefficients are real, a complex root must be one of a pair of complex conjugates. For example, if the polynomial is a cubic, one root certainly is real and the other two are either real or a pair of complex conjugates.

For a source with a complex frequency s equal to z_1, the gain A of (18-4) is obviously zero. Hence z_1, and also z_2, z_3, etc., are called the *zeros* of the gain function. On the other hand, if s equals p_1 or one of the other p-singularities of the denominator, the gain is infinite, and these constants are referred to as the *poles* of A.

Suppose a network containing one or more capacitors has initial energy storage but no input signal. As the capacitors discharge, producing an output, the gain A is infinite. Thus in accordance with (18-4) any complex frequencies present are those of the poles. A negative real pole gives a damped exponential at the output. A pair of complex conjugates yields oscillations, and in the event σ is positive, the oscillations build up until the nonlinearity of an active device prevents further rise. This often happens in high-gain multistage amplifiers with excessive feedback, which takes place through poorly designed networks or perhaps through the stray capacitance or resistance of an improperly wired network. Such unstable amplifiers oscillate even without initial energy storage. There are always noise signals at all radio frequencies present in the source resistance at the input of an amplifier, and if σ is positive, a signal at the radian frequency of the complex pair of poles builds up rapidly. Care must be taken to avoid poles with positive σ.

If A of (18-4) is determined from the low-frequency incremental model of an amplifier, the midfrequency gain A_o is A with s equal to infinity. Because A_o is neither zero nor infinite, *the number of poles must exactly equal the number of zeros*, and A_o equals K. To illustrate, suppose an amplifier has a midband gain of 1000, with zeros at 0 and -10 and with poles at -100 and -500. For this case the low-frequency gain in SI units is

$$A(s) = \frac{A_o(s - z_1)(s - z_2)}{(s - p_1)(s - p_2)} = \frac{1000s(s + 10)}{(s + 100)(s + 500)}$$

The expression on the right gives the midband gain of 1000 when s approaches infinity, which makes short-circuits out of the coupling and bypass capacitors. The picofarad capacitances associated with the active devices are, of course, not included in the low-frequency model.

On the other hand, the high-frequency incremental model does not include the coupling and bypass capacitors. Hence it reduces to the midfrequency model when s is replaced with zero, which makes open-circuits out of the picofarad capacitors. Let us assume the high-frequency gain of a certain amplifier in SI units is

$$A(s) = \frac{K(s - z_1)}{(s - p_1)(s - p_2)} = \frac{10^9(s - 10^{10})}{(s + 10^7)(s + 10^8)} = \frac{-10^4(1 - s/10^{10})}{(1 + s/10^7)(1 + s/10^8)}$$

We note that A is zero when s is 10^{10} and also when s is infinite. Accordingly, the two zeros are at $+10^{10}$ and infinity. The poles are -10^7 and -10^8. When s is replaced with zero, A becomes -10^4, which is A_o. As in this example, the number of poles in a high-frequency gain function is always at least one greater than the number of finite zeros, because all practical amplifiers have gains that approach zero as the frequency approaches infinity. *With the zeros at infinity included, and there may be more than one, the number of zeros equals the number of poles.*

RULES FOR DETERMINING THE NUMBER OF POLES OF A NETWORK GAIN FUNCTION

Before pursuing the design or analysis of an amplifier it is usually helpful to know the number of poles of A, and this is easily determined by following a few elementary rules. Because the gain function is assumed to have the form of (18-4), with A denoting V_o/V_i, I_o/I_i, V_o/I_i, or I_o/V_i, networks with internal independent sources are excluded. Each dependent voltage source must be in series with a resistor. Also, each dependent current source must be in parallel with a resistor. In addition, we require the proportionality constants of the dependent sources to be independent of s.

A *capacitor loop* is defined as a closed path around an electrical network that passes through capacitors only. A *capacitor cutset* is any closed surface in space that cuts the network through capacitors only. For the usual case of a planar circuit, defined as one whose schematic can be drawn in a plane without crossovers, a capacitor cutset becomes any closed path in the plane that cuts the network through capacitors only, as shown in Fig. 18-3. *Inductor loops* and *cutsets* are similarly defined.

Subject to the stated restrictions the rules are as follows:

1. Because ideal transformers are not energy-storage devices and hence do not contribute to the number of poles, connect the input terminals of each ideal transformer directly to the output terminals and remove the coils.

2. If the input quantity of the gain A is V_i, short the input so as to make $V_i = 0$, regardless of whether the circuit is driven with a current or voltage source. If a shorted branch results, remove it from the network. This is justified because a branch shunted across V_i has no effect on A.
3. On the other hand, if the input quantity of A is I_i, open the input so as to make $I_i = 0$, regardless of the actual excitation. If a dangling branch with one end open results, remove it. This is justified because a branch in series with the source current I_i has no effect on A.
4. Replace any one capacitor in a capacitor loop with an open. Replace any one capacitor in a cutset with a short. Repeat these in any order until all capacitor loops and cutsets are eliminated. Normally, cutsets are not encountered in either the low-frequency or high-frequency incremental model of an amplifier, and capacitor loops appear only occasionally in the high-frequency model.
5. Apply rule 4 to inductances that may be present. Ignore mutual inductances of coupled coils. Except for RF tuned amplifiers, oscillators, and power amplifiers with transformer coupling, inductances are seldom found in electronic circuits.
6. *The number of remaining capacitors and inductors represents the number of independent energy-storage elements, which equals the number of poles.* The method fails only in certain rare cases in which a zero exactly cancels a pole (see P18-7).

Example 2

For the network of Fig. 18-3 determine the number of poles of the gain function I_o/I_i.

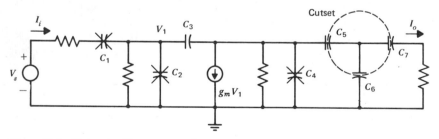

Figure 18-3   Network with seven capacitors.

Solution

Remove C_1 by rule 3. Replace C_2 with an open by rule 4. Also, replace C_4 with an open and C_5 with a short, again by rule 4. Capacitors C_3, C_6, and C_7 remain, and there are three poles. The crosses on C_1, C_2, and C_4 of Fig. 18-3 indicate opens, and C_5 is short-circuited.

18-2. FREQUENCY RESPONSE CURVES

In a great many amplifiers the poles are real numbers, and assuming the amplifier is stable, these must be negative. Also, the zeros are usually real, either positive or negative. For the gain functions considered throughout this section, we shall assume that all poles and zeros are real, with the poles restricted to negative values. Because complex-conjugate singularities are of special interest in feedback amplifiers, the treatment of their effects on the response is deferred to the chapter on the stability of feedback amplifiers, specifically Sec. 21-2. Our objective is to learn how to plot curves of gain-versus-frequency with ease and accuracy for an amplifier whose poles and zeros are known. We begin with the low-frequency gain function A, taken for convenience to be the voltage ratio V_o/V_i.

LOW-FREQUENCY BODE PLOTS

In the real frequency domain $s = j\omega$, and (18-4) becomes

$$A(\omega) = \frac{A_o(j\omega - z_1)(j\omega - z_2)\cdots}{(j\omega - p_1)(j\omega - p_2)\cdots} = \frac{A_o(1 + jz_1/\omega)(1 + jz_2/\omega)\cdots}{(1 + jp_1/\omega)(1 + jp_2/\omega)\cdots} \quad (18\text{-}5)$$

The constant K in (18-4) has been replaced with A_o, which is valid for the low-frequency case. It is evident from (18-5) that $A = A_o$ when ω is infinite. The expression on the right side of (18-5) was found by dividing each factor by $j\omega$, justified because the number of poles and finite zeros are always equal.

The voltage gain can be put in the form

$$A(\omega) = \frac{|A_o|\sqrt{1 + (z_1/\omega)^2}\sqrt{1 + (z_2/\omega)^2}\cdots}{\sqrt{1 + (p_1/\omega)^2}\sqrt{1 + (p_2/\omega)^2}\cdots} \underline{/\theta} \quad (18\text{-}6)$$

with the angle θ given by

$$\theta = \underline{/A_o} + \arctan\frac{z_1}{\omega} + \arctan\frac{z_2}{\omega} + \cdots - \arctan\frac{p_1}{\omega} - \cdots \quad (18\text{-}7)$$

Of course, the angle of A_o is either $0°$ or $180°$. The common logarithm of the magnitude of A is found to be

$$\log|A| = \log|A_o| + 0.5\log\left[1 + \left(\frac{z_1}{\omega}\right)^2\right] + \cdots - 0.5\log\left[1 + \left(\frac{p_1}{\omega}\right)^2\right] - \cdots$$

$$(18\text{-}8)$$

Response curves of amplifiers are often plotted in decibels. A *decibel of voltage gain* is defined as

$$dB = 20\log_{10}\left|\frac{V_o}{V_i}\right| \quad (18\text{-}9)$$

Multiplication of the individual terms of (18-8) by 20 gives the voltage gain in decibels.

Let us consider an amplifier with one pole p_1 and one zero z_1, with $z_1 = 0$. From (18-8) we find the dB voltage gain to be

$$20 \log|A| = 20 \log|A_o| - 10 \log\left[1 + \left(\frac{p_1}{\omega}\right)^2\right] \qquad (18\text{-}10)$$

For $\omega \gg |p_1|$, this becomes

$$20 \log|A| = 20 \log|A_o| \qquad (18\text{-}11)$$

For $\omega \ll |p_1|$, the relation is

$$20 \log|A| = 20 \log|A_o| - 20 \log|p_1| + 20 \log \omega \qquad (18\text{-}12)$$

The dB of voltage gain versus $\log \omega$ is plotted in Fig. 18-4. The straight lines are plots of (18-11) and (18-12) and are called the *asymptotes*. To the right of the intercept the slope of the asymptote is zero. To the left it is 20, often stated as *20 dB per decade*, which signifies that the voltage gain changes by 20 decibels when the frequency changes by a factor of 10. Such a slope can equivalently

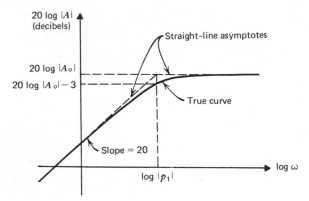

Figure 18-4 Voltage gain versus frequency for a 1-pole amplifier with $z_1 = 0$.

be stated as 6 dB per octave, with an *octave* denoting a change in the frequency by a factor of 2. Because subtraction of (18-11) and (18-12) gives $\omega = |p_1|$, this is the value of ω at the point where the two asymptotes meet. From (18-10) we determine that the true curve at $\omega = |p_1|$ is below the intercept by 10 log 2, or 3 dB. Sketches of asymptotes, such as those of Fig. 18-4, are called *Bode plots*.

A fairly accurate response curve for the one-pole amplifier is obtained in the following manner. We sketch the midfrequency asymptote $20 \log|A_o|$. At the point on this line where ω equals the pole magnitude we draw a straight line with a slope of 20 dB per decade. The true curve is drawn through the point that is 3 dB below the intercept, with the curve approaching the asymptotes as the distance from the point increases. At values of ω equal to $2|p_1|$ or $0.5|p_1|$, each of which is an *octave* away from the break point, the true curve is determined from (18-10) to be below the respective asymptote by 1 dB.

It is easy to show from (18-10) that at $\omega = |p_1|$ the gain $|A|$ is $|A_o|/\sqrt{2}$, or $0.707|A_o|$. For a resistive load the output power is proportional to the square of the voltage. Therefore, with a constant input, the power to the load at this ω is one-half that in the midband range. Thus, the *break frequency*, defined as that at which the output voltage drops to $1/\sqrt{2}$ times its midband value, is also called the 0.707 *frequency*, the 3 dB *frequency*, the *half-power point*, the *corner frequency*, or the *band edge*.

Next let us consider a one-pole gain function having $|z_1|$ greater than zero but less than $|p_1|$. Equation 18-8 multiplied by 20 becomes

$$20 \log|A| = 20 \log|A_o| - 10 \log\left[1 + \left(\frac{p_1}{\omega}\right)^2 \right] + 10 \log\left[1 + \left(\frac{z_1}{\omega}\right)^2 \right]$$

(18-13)

For ω considerably greater than $|p_1|$, (18-11) is still applicable. If ω is much greater than $|z_1|$ and also small compared with $|p_1|$, (18-12) applies. Finally, for ω very much less than $|z_1|$, from (18-13) we obtain

$$20 \log|A| = 20 \log|A_o| - 20 \log\left|\frac{p_1}{z_1}\right|$$

(18-14)

The asymptotes from the midband region down to $\omega = |z_1|$ are the same as before. At $|z_1|$ and below, the asymptote is that of (18-14), which represents a straight line with zero slope. As ω is lowered from its midfrequency values, we note that the term 20 log ω is introduced into the equation of the asymptote (18-12) when the pole is passed. Passing the zero introduces an equal negative term, and consequently (18-14) is independent of frequency.

The response is shown in Fig. 18-5. In both figures the asymptotes are those of (18-11), (18-12), and (18-14). They are drawn in this manner regardless of the separation between the singularities. In Fig. 18-5a the zero is much less than the pole. At $\omega = |p_1|$ the effect of the zero is negligible, and the true curve is *below* the break point by 3 dB. At $\omega = |z_1|$, we find from (18-13) that 20 log$|A|$ is given by (18-14) with 3 dB added to the right side. Thus, at the break point corresponding to the zero the true curve is *above* the asymptote by 3 dB. In Fig. 18-5b the pole and the zero are nearly equal. Although the true curve is no longer off 3 dB at the break points, an approximate response curve is easily sketched as shown. Clearly, a pole and a zero that are close together have effects

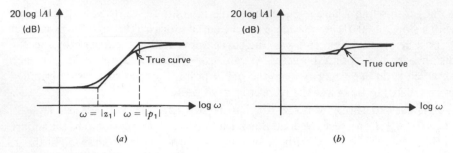

Figure 18-5 Voltage gain vs. frequency for $|p_1| > |z_1| > 0$. $(a)\,|z_1| \ll |p_1|$. $(b)\,|z_1| \approx |p_1|$.

that tend to cancel. Of course, if they are equal, they disappear from the gain function.

For an amplifier with two or more poles in the low-frequency gain function the straight-line asymptotes are determined similarly. *As the frequency is lowered from the flat midband region, the slope of the asymptote increases by 20 each time a pole is passed and by -20 when a zero is passed.* The change is 20 dB per decade. *At each pole the true curve is 3 dB below the break point, and at each zero it is this same amount above the point, provided other singularities are at least a decade away.*

The manner in which the angle θ of A, or the phase of A, varies with frequency is also of considerable importance. Plots of θ versus ω can be made from (18-7). In the midband region the arctangents in the equation are nearly zero, and θ is the angle of A_o, which is $0°$ or $180°$. For a one-pole amplifier with $z_1 = 0$ and zero phase shift in the midband, θ equals the arctangent of $|p_1|/\omega$. As ω is lowered from the midfrequency region, θ increases from $0°$ to a value of $45°$ at the pole and to $90°$ at zero frequency. For ω equal to $2|p_1|$ and $0.5|p_1|$ the respective angles are $26.6°$ and $63.4°$. These must be changed by $\pm 180°$ if the amplifier is either CE or CS.

HIGH-FREQUENCY BODE PLOTS

When the voltage gain A is determined from the high-frequency incremental model, it is convenient to write (18-4) in the form

$$A(s) = \frac{A_o(1 - s/z_1)(1 - s/z_2)\cdots}{(1 - s/p_1)(1 - s/p_2)\cdots} \qquad (18\text{-}15)$$

Because any practical amplifier has zero gain at an infinite frequency, A has at least one zero at infinity. For sinusoidal time variations,

$$A(\omega) = \frac{A_o(1 - j\omega/z_1)(1 - j\omega/z_2)\cdots}{(1 - j\omega/p_1)(1 - j\omega/p_2)\cdots} \qquad (18\text{-}16)$$

Recalling that the poles and zeros are assumed to be negative real numbers, $A(\omega)$ can be expressed as

$$A(\omega) = \frac{|A_o|\sqrt{1 + (\omega/z_1)^2}\,\sqrt{1 + (\omega/z_2)^2}\cdots}{\sqrt{1 + (\omega/p_1)^2}\,\sqrt{1 + (\omega/p_2)^2}\cdots}\,\underline{/\theta} \qquad (18\text{-}17)$$

with the phase θ being

$$\theta = \underline{/A_o} - \arctan\frac{\omega}{z_1} - \cdots + \arctan\frac{\omega}{p_1} + \cdots \qquad (18\text{-}18)$$

Let us consider an amplifier with one pole p_1 and a zero at infinity. From (18-17) we obtain

$$20\log|A| = 20\log|A_o| - 10\log\left[1 + \left(\frac{\omega}{p_1}\right)^2\right] \qquad (18\text{-}19)$$

For $\omega = 0$, the last term on the right side of (18-19) is zero, and the midband asymptote is the same as before. For $\omega \gg |p_1|$, the relation becomes

$$20\log|A| = 20\log|A_o| + 20\log|p_1| - 20\log\omega \qquad (18\text{-}20)$$

A plot of $20\log|A|$ versus $\log\omega$ from this equation is a straight line that intercepts the midband asymptote at $\omega = |p_1|$. The slope is -20 dB per decade. A little reflection on this result and a close examination of (18-17) reveals that the high-frequency asymptotes are determined in a manner similar to the method used at low frequencies. We conclude that, *as the frequency is increased from the flat midband region, the slope of the asymptote changes by -20 dB per decade each time a pole is passed and by $+20$ when a zero is passed. At each pole the true curve is 3 dB below the breakpoint, and at each zero it is this same amount above the point, provided other singularities are at least a decade away.*

The upper 3 dB frequency ω_2 is defined similarly to the lower 3 dB frequency ω_1. The difference $\omega_2 - \omega_1$ is the *bandwidth* BW of the amplifier in radians per second. Usually ω_2 is much greater than ω_1 so that the BW is approximately ω_2. Because the bandwidth is mainly determined by the largest pole in the low-frequency region and the smallest pole in the high-frequency region, these are referred to as the *dominant poles.*

The phase of a high-frequency gain function is determined from (18-18), which is somewhat similar to (18-7) for the low-frequency case. For a one-pole amplifier with the zero at infinity and zero phase shift in the midband, θ is the negative of the arctangent of $\omega/|p_1|$. As ω is increased from the midfrequency region, θ changes from $0°$ to a value of $-45°$ at the pole and to $-90°$ at an infinite frequency. For ω equal to $0.5|p_1|$ and $2|p_1|$ the respective angles are $-26.6°$ and $-63.4°$. These must be changed by $\pm180°$ if the amplifier is either CE or CS.

Example 1

A CE amplifier with a midband voltage gain of -100 has poles at -10^3 and $-10^6 s^{-1}$. One zero is at $s = 0$ and the other is at infinity. (a) Plot the frequency response curves. (b) Determine the response when the input voltage is a step function of magnitude 0.01 V.

Solution

(a) A knowledge of the poles, the zeros, and A_o is sufficient for obtaining the magnitude response curve of Fig. 18-6. The phase response, determined from (18-7) and (18-18), is given in Fig. 18-7. (b) The mathematical expression for the voltage gain in SI units is

$$\frac{V_o}{V_i} = -100\left(\frac{s}{s + 1000}\right)\left(\frac{1}{1 + s/10^6}\right) = \frac{-10^8 s}{(s + 1000)(s + 10^6)}$$

Using Laplace transforms, V_i is $0.01/s$, and V_o becomes

$$V_o = \frac{-10^6}{(s + 1000)(s + 10^6)} = \frac{-1}{s + 1000} + \frac{1}{s + 10^6}$$

In the time domain the inverse transform shows that

$$v_o(t) = -e^{-1000t} + e^{-10^6 t}$$

The curve of v_o versus t is sketched in Fig. 18-8. The rapid initial rise is caused by the charging of the picofarad capacitances of the device, and the much slower exponential decay is the result of the charging of the coupling capacitor.

Example 2

In SI units the voltage gain of an amplifier is determined to be

$$A(s) = 10^{13}\left[\frac{s(s + 100)}{(s + 10)(s + 1000)}\right]\left[\frac{s - 10^8}{(s + 10^6)(s + 10^7)}\right] \qquad (18\text{-}21)$$

Find A_o, plot the asymptotes of the decibel gain, sketch the true curve, and determine the bandwidth.

Solution

The poles in the low-frequency region are -10 and -1000, and the zeros are 0 and $-100 s^{-1}$. At high frequencies the poles are -10^6 and $-10^7 s^{-1}$. One zero is at $+10^8$ and the other is infinite. Because the dominant poles, -1000 and -10^6, are a decade or more removed from other singularities, the 3 dB frequencies ω_1 and ω_2 are 1000 and 10^6. The bandwidth is 10^6 radians per second. The midband gain A_o is -10^8, found from (18-21) by letting s approach infinity in the bracketed low-frequency expression and replacing s with zero in the high-frequency bracketed expression.

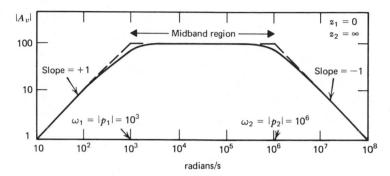

Figure 18-6   Amplitude response of CE amplifier.

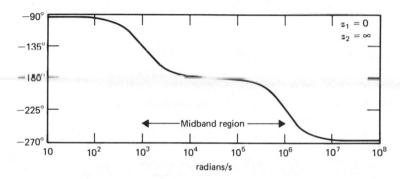

Figure 18-7   Phase response of CE amplifier.

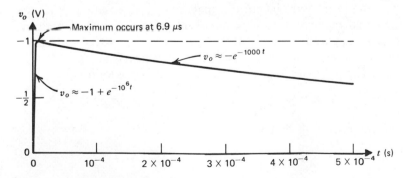

Figure 18-8   Step-function response of CE amplifier.

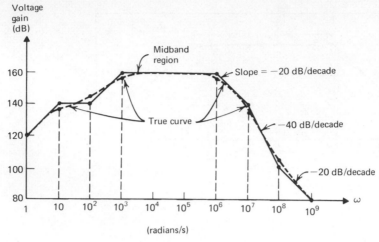

Figure 18-9   Frequency response of an amplifier.

The asymptotes and the true curve are shown in Fig. 18-9. Note that, when the frequency is increased from 0^- to plus infinity, each time a zero is encountered the slope increases by 20 dB per decade and each time a pole is encountered the slope decreases by 20. Also, the true curve is 3 dB above or below each break point.

18-3. POLES AND TIME CONSTANTS

There are important relations between the poles of a gain function and the time constants of the network. Two of these are presented in the first part of this section as (18-22) and (18-23), followed by their derivations. We shall restrict our attention to linear networks containing no inductors.

Suppose the amplifier model has three capacitors C_1, C_2, and C_3. With the source at the input replaced with its internal impedance, let R_{1s} denote the effective resistance in parallel with C_1 when C_2 and C_3 are replaced with short-circuits. Symbols R_{2s} and R_{3s} denote the corresponding resistances associated with C_2 and C_3. The *short-circuit time constants* are defined as follows:

$$\tau_{1s} = R_{1s}C_1 \qquad \tau_{2s} = R_{2s}C_2 \qquad \tau_{3s} = R_{3s}C_3$$

The sum of the reciprocals of these is

$$\sum_{j=1}^{3} \frac{1}{\tau_{js}} = \frac{1}{\tau_{1s}} + \frac{1}{\tau_{2s}} + \frac{1}{\tau_{3s}}$$

A relation quite useful in the design and analysis of low-frequency amplifier circuits is

$$-(p_1 + p_2 + \cdots) = \sum_j \frac{1}{\tau_{js}} \tag{18-22}$$

This mathematical statement says that the negative of the sum of the poles equals the sum of the reciprocals of the short-circuit time constants. It applies to a network with any number of capacitors. Although capacitor loops are not allowed, these are normally not encountered in low-frequency incremental models.

For the network with three capacitors the *open-circuit time constants* are

$$\tau_{1o} = R_{1o}C_1 \qquad \tau_{2o} = R_{2o}C_2 \qquad \tau_{3o} = R_{3o}C_3$$

Here, R_{1o} denotes the effective resistance in parallel with C_1 when the other capacitors are removed, or open-circuited, and R_{2o} and R_{3o} are defined similarly. The following relation is useful in design and analysis of high-frequency amplifier circuits:

$$-\left(\frac{1}{p_1} + \frac{1}{p_2} + \cdots\right) = \sum_j \tau_{jo} \tag{18-23}$$

It states that the negative of the sum of the reciprocals of the poles equals the sum of the open-circuit time constants. Any number of capacitors, and also capacitor loops, are allowed.

PROOF OF (18-22)

To justify (18-22) we examine a linear network with only two capacitors. The independent source at the input is replaced with its internal impedance. Let us connect a voltage source V_1 across C_1 and V_2 across C_2. The sources have the complex frequency s, and they and the capacitors are shown in Fig. 18-10. The remainder of the network is within the block and without energy storage.

An expression for the current I_1 of the source V_1 will be found using superposition. With $V_2 = 0$, which gives a short-circuit across C_2, the admittance

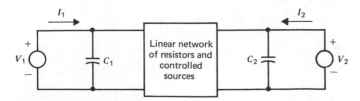

Figure 18-10   Network with two capacitors.

shunting C_1 is the conductance G_{1s}. This is the reciprocal of R_{1s} as defined earlier. Hence, I_1 is $(G_{1s} + sC_1)V_1$. Now with $V_1 = 0$, which shorts C_1, the current I_1 is $G_{12}V_2$. The symbol G_{12} denotes the short-circuit reverse trans-conductance. Addition of the two expressions gives I_1, and the equation for I_2 is similar. The results are

$$\begin{bmatrix} I_1 \\ I_2 \end{bmatrix} = \begin{bmatrix} G_{1s} + sC_1 & G_{12} \\ G_{21} & G_{2s} + sC_2 \end{bmatrix} \begin{bmatrix} V_1 \\ V_2 \end{bmatrix} \qquad (18\text{-}24)$$

Let Δ denote the determinant of the coefficient matrix of (18-24), and let s_1 and s_2 denote the roots of the equation $\Delta = 0$. At either frequency s_1 or s_2 the currents I_1 and I_2 of (18-24) are zero even though V_1 and V_2 are not. For I_1 and I_2 both zero, the sources supply no energy and can be removed without affecting any voltages or currents of the network. Clearly, the roots s_1 and s_2 of the equation $\Delta = 0$ are the natural frequencies of the network, which are the poles p_1 and p_2 of an appropriate gain function $A(s)$. Therefore, with K denoting a constant, $K(s - p_1)(s - p_2)$ equals Δ. Hence,

$$K(s - p_1)(s - p_2) = (G_{1s} + sC_1)(G_{2s} + sC_2) - G_{12}G_{21} \qquad (18\text{-}25)$$

or,

$$s^2 - (p_1 + p_2)s + p_1p_2 = s^2 + \left(\frac{G_{1s}}{C_1} + \frac{G_{2s}}{C_2}\right)s + \frac{G_{1s}G_{2s} - G_{12}G_{21}}{C_1C_2} \qquad (18\text{-}26)$$

We note that

$$-(p_1 + p_2) = \frac{1}{R_{1s}C_1} + \frac{1}{R_{2s}C_2} \qquad (18\text{-}27)$$

This is (18-22) for a network with two capacitors. The proof is similar for a circuit with any number of capacitors, and the result is (18-22). If a capacitor loop is present, we cannot connect independent voltage sources across each capacitor; thus the derivation is invalid, and (18-22) does not apply.

PROOF OF (18-23)

We again consider a two-capacitor network, but independent current sources are inserted in series with the capacitors as shown in Fig. 18-11. The voltage V_1 across the current source I_1 will be determined by superposition. With $I_2 = 0$, which removes C_2, the resistance in series with C_1 is R_{1o}, and V_1 is $(R_{1o} + 1/sC_1)I_1$. With $I_1 = 0$, which open-circuits C_1, the voltage V_1 is $R_{12}I_2$. Here, R_{12} denotes the open-circuit reverse transfer resistance. Addition of the two expressions gives V_1, and V_2 is determined similarly. The results are

$$\begin{bmatrix} V_1 \\ V_2 \end{bmatrix} = \begin{bmatrix} R_{1o} + 1/sC_1 & R_{12} \\ R_{21} & R_{2o} + 1/sC_2 \end{bmatrix} \begin{bmatrix} I_1 \\ I_2 \end{bmatrix} \qquad (18\text{-}28)$$

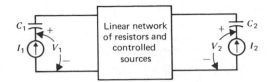

Figure 18-11   A second two-capacitor network.

The roots of the equation obtained by setting the determinant of the co-efficient matrix of (18-28) equal to zero are the poles of an appropriate gain function. At these complex frequencies V_1 and V_2 are zero even though I_1 and I_2 are not, and the current sources of Fig. 18-11 can be replaced with short-circuits without affecting the currents of the network. With K denoting a constant the poles satisfy the relation

$$K\left(1 - \frac{p_1}{s}\right)\left(1 - \frac{p_2}{s}\right) = \left(R_{1o} + \frac{1}{sC_1}\right)\left(R_{2o} + \frac{1}{sC_2}\right) - R_{12}R_{21} \quad (18\text{-}29)$$

After multiplying through by s^2, we obtain

$$K\{s^2 - (p_1 + p_2)s + p_1 p_2\} = (R_{1o}R_{2o} - R_{12}R_{21})s^2 + \left(\frac{R_{1o}}{C_2} + \frac{R_{2o}}{C_1}\right)s + \frac{1}{C_1 C_2}$$
$$(18\text{-}30)$$

We now equate the coefficients of the s terms and also the constants on the two sides of the equation. Division of the two equations that result gives (18-23) for a two-capacitor network. The proof is similar for a circuit with any number of capacitors, and there is no restriction on capacitor loops. The poles of $A(s)$ are often called the *natural frequencies* of the network.

18-4. CE AMPLIFIER AT LOW FREQUENCIES

When coupling and bypass capacitors are present in a CE amplifier, the low-frequency gain function has poles and zeros that cause the gain to drop as the frequency is lowered. The circuit designer must select values for these capacitors to give the desired 3 dB frequency ω_1. One method, of course, is trial and error, in which the designer simply guesses at suitable values and then analyzes the circuit, perhaps with the aid of a computer. If the value of ω_1 is unsatisfactory, the process is repeated, and so forth. This is not only tedious but may lead to a poor choice of values for some capacitors, and a better method is available. We shall examine a one-stage CE amplifier having both a coupling capacitor C_c and an emitter bypass capacitor C_e.

SELECTION OF THE COUPLING CAPACITOR

First, let us consider the amplifier of Fig. 18-12, which has only the single capacitor C_c. In low-frequency circuits it is sometimes convenient to use the following units:

milliamperes	volts	milliseconds	kilohertz	(18-31)
kilohms	millimhos	microfarads	henrys	

This is a consistent set that will be used throughout this and the next section unless specifically stated otherwise. Both s and ω are in reciprocal milliseconds.

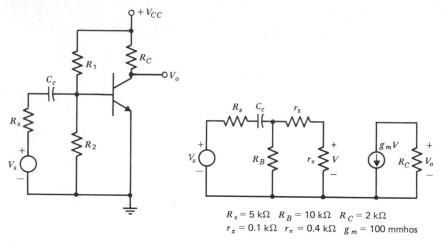

$$R_s = 5 \text{ k}\Omega \quad R_B = 10 \text{ k}\Omega \quad R_C = 2 \text{ k}\Omega$$
$$r_x = 0.1 \text{ k}\Omega \quad r_\pi = 0.4 \text{ k}\Omega \quad g_m = 100 \text{ mmhos}$$

Figure 18-12   CE amplifier and model with C_c.

The voltage gain A_{Vs} is V_o/V_s, which has a midfrequency value A_o found from Fig. 18-12 to be -13.9. At $s = 0$ the impedance $1/sC$ is infinite, which makes the capacitor an open circuit, and V_o is zero. Hence the network zero is at $s = 0$. By (18-22) the pole is $-1/R_{cs}C_c$, with R_{cs} denoting the resistance facing C_c with $V_s = 0$ and all other capacitors shorted. There are no other capacitors, and R_{cs} is the sum of R_s and the parallel equivalent of R_B and $r_x + r_\pi$. This is evident from the incremental model of Fig. 18-12. The expressions for the pole and the zero are

$$p = \frac{-1}{[R_B \| (r_x + r_\pi) + R_s]C_c} \qquad z = 0 \qquad (18\text{-}32)$$

For the numerical values given in Fig. 18-12, R_{cs} is 5.48 kΩ and the pole is $-1/(5.48C_c)$. The magnitude of p is ω_1. Let us assume that a 3 dB frequency of 0.03 kHz, or 30 Hz, is desired. Then ω_1 is 0.1885 ms^{-1}, and the required value of C_c is 0.97 μF. Note that the capacitor is chosen so that the reciprocal of the time constant equals ω_1. In analysis we would first determine the time constant from the given C_c and then calculate ω_1. The magnitude response is that of Fig. 18-4 and the phase response is the low-frequency portion of Fig. 18-7, with $A_o = -13.9$ and $p_1 = -0.1885$ ms^{-1}.

In selecting f_1 we should make it no lower than required. The cost and size of C_c are considerations. Flicker ($1/f$) noise becomes more objectional as f_1 is reduced. Also, with high gain at low frequencies the output capacitance of the power-supply filter in a multistage amplifier must be quite large, or feedback through this impedance may cause oscillations.

SELECTION OF THE EMITTER BYPASS CAPACITOR

Next, let us examine the circuit of Fig. 18-13. This is the amplifier previously analyzed, except that C_c is eliminated and R_e and C_e have been added. The numerical values of Fig. 18-12 apply and R_e is 0.4 kΩ.

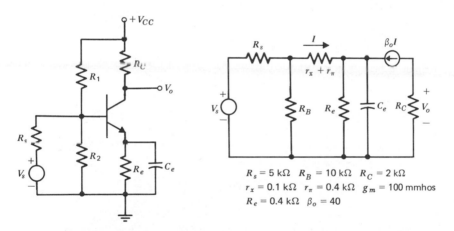

$R_s = 5$ kΩ $R_B = 10$ kΩ $R_C = 2$ kΩ
$r_x = 0.1$ kΩ $r_\pi = 0.4$ kΩ $g_m = 100$ mmhos
$R_e = 0.4$ kΩ $\beta_o = 40$

Figure 18-13 CE amplifier and model with C_e.

The emitter-to-ground impedance is $R_e/(1 + sR_eC_e)$, which is infinite when s equals $-1/R_eC_e$. At this frequency the sum of I and $\beta_o I$ of Fig. 18-13 must be zero, and so are I and V_o. Thus the zero z of the gain function is $-1/R_eC_e$.

The short-circuit time constant τ_s is $R_{es}C_e$, with R_{es} denoting the effective resistance in parallel with C_e when V_s is zero. Of course, in networks with only a single capacitor there is no distinction between short-circuit and open-circuit time constants. The incremental models of Fig. 18-14 apply, and R_{es} is the resistance looking into the terminals labeled V_1. The models are identical, except that the one on the right has been rearranged for convenience with the source representing V_1 at the left side and I_3 denoting $-I$.

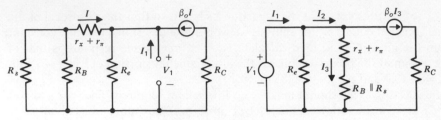

Figure 18-14   Models for determining R_{es}, or V_1/I_1.

Resistance R_{es} is R_e in parallel with V_1/I_2. Current I_2 is $(\beta_o + 1)I_3$, and I_3 is $V_1/(r_x + r_\pi + R_B \| R_s)$. From these we find R_{es} to be

$$R_{es} = R_e \left\| \left[\frac{r_x + r_\pi + R_B \| R_s}{\beta_o + 1} \right] \right. \tag{18-33}$$

Substitution of numerical values gives $R_{es} = 0.0758$ kΩ, and the pole and the zero are

$$p = \frac{-1}{R_{es}C_e} = \frac{-1}{0.0758C_e}$$

$$z = \frac{-1}{R_e C_e} = \frac{-1}{0.4C_e} \tag{18-34}$$

As before, we select ω_1 at 0.1885 ms^{-1} for a 3 dB frequency of 30 Hz. We note from (18-34) that the pole is more than five times greater than the zero. Thus, *as an approximation we select the magnitude of* p *equal to* ω_1. This gives $C_e = 70$ μF, and p is -0.1885 and z is -0.0357 ms^{-1}. The voltage gain in the s-domain is

$$A_{Vs}(s) = \frac{-13.9(s + 0.0357)}{s + 0.1885} \tag{18-35}$$

In the real frequency domain we replace s in (18-35) with $j\omega$. It is convenient to divide both numerator and denominator by $j\omega$ and put the resulting expression for A_{Vs} in the form $|A_{Vs}|\underline{/\theta}$, with

$$|A_{Vs}| = 13.9 \left[\frac{1 + (0.0357/\omega)^2}{1 + (0.1885/\omega)^2} \right]^{1/2} \tag{18-36}$$

$$\theta = -180° + \arctan \frac{0.1885}{\omega} - \arctan \frac{0.0357}{\omega} \tag{18-37}$$

The frequency response curves are shown in Fig. 18-15. At high frequencies C_e is a short-circuit and the gain is flat. At very low frequencies C_e becomes approximately an open-circuit. The gain is again flat but relatively much smaller than the midband gain. Because the value of the zero is about 20% of that of the

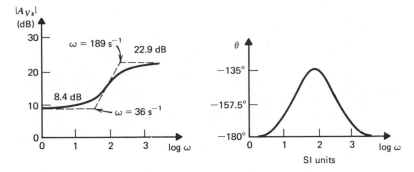

Figure 18-15 Frequency response curves of CE amplifier with C_e.

pole, it makes the actual 3 dB angular frequency ω_1 slightly less than the pole magnitude. This is indicated by the magnitude curve of the figure, noting that the zero tends to reduce the drop in the low-frequency gain caused by the pole.

The exact ω_1 can be found from (18-36). When ω equals ω_1, $|A_{Vs}|$ is $|A_o|/\sqrt{2}$. Therefore, at this frequency the quantity within the brackets must equal 0.5, and its reciprocal is

$$\frac{1 + (0.1885/\omega_1)^2}{1 + (0.0357/\omega_1)^2} = 2 \tag{18-38}$$

This is easily solved, giving $\omega_1 = 0.182 \text{ ms}^{-1}$, and f_1 is 29 Hz. The difference between f_1 and the design objective of 30 Hz is insignificant.

The phase response of Fig. 18-15 was obtained from (18-37). At high frequencies θ is $-180°$ (or $+180°$). At the upper breakpoint the pole contributes a phase shift of $+45°$ and the zero introduces $-11°$. At very low frequencies the network is resistive, the pole and the zero give phase shifts of $+90°$ and $-90°$, respectively, and θ is again $-180°$.

NETWORK DESIGN WITH INTERACTING CAPACITORS C_c AND C_e

Now that we have examined the effects of C_c and C_e individually, we are ready to consider the problem of selecting these capacitors when both are present, as in the circuit of Fig. 18-16. The specified numerical values are those used previously, and again we select 30 Hz as the desired f_1. The network has two poles and two zeros. By inspection of the circuit we find the zeros to be

$$z_1 = 0 \qquad z_2 = \frac{-1}{R_e C_e} \tag{18-39}$$

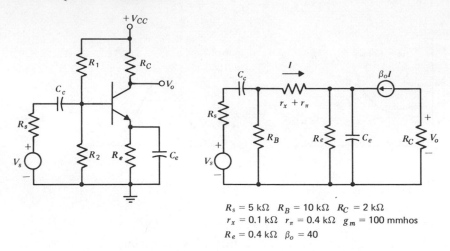

$$R_s = 5 \text{ k}\Omega \quad R_B = 10 \text{ k}\Omega \quad R_C = 2 \text{ k}\Omega$$
$$r_x = 0.1 \text{ k}\Omega \quad r_\pi = 0.4 \text{ k}\Omega \quad g_m = 100 \text{ mmhos}$$
$$R_e = 0.4 \text{ k}\Omega \quad \beta_o = 40$$

Figure 18-16   CE amplifier and model with C_c and C_e.

At $s = z_1$, the coupling capacitor C_c is an open-circuit, and at z_2 the emitter-to-ground impedance is infinite, which makes I zero in Fig. 18-16.

The short-circuit time constants are $R_{cs}C_c$ and $R_{es}C_e$. The expression for R_{cs} is indicated in (18-32), with the same numerical value of 5.48 kΩ as before. R_{es} is given by (18-33) and has a value of 0.0758 kΩ. In accordance with (18-22) the negative of the sum of the poles is

$$-(p_1 + p_2) = \frac{1}{5.48C_c} + \frac{1}{0.0758C_e} \tag{18-40}$$

We wish to select suitable values for C_c and C_e to obtain a 3 dB frequency of 30 Hz, or $\omega_1 = 0.1885 \text{ ms}^{-1}$. For a one-pole amplifier we found that choosing the capacitance so as to make the pole magnitude equal to ω_1 gave excellent results. In the case of a two-pole amplifier that has its dominant pole much greater than the other singularities, we know that ω_1 and the dominant pole have nearly identical magnitudes, and furthermore, this one pole approximately equals the sum of the poles. Hence,

$$\omega_1 \approx -\sum_j p_j = \sum_j \frac{1}{\tau_{js}} \tag{18-41}$$

Actually, the sum of the poles is always somewhat greater in magnitude than ω_1. It is easy to show (see P18-9) that the ratio $(p_1 + p_2)/\omega_1$ is -1.09 for poles a decade apart, provided there are no significant zeros. For p_2 one-fifth of p_1 this ratio becomes -1.16 and it is -1.29 for equal poles. In the specific circuit under consideration there is interaction between the capacitors, which results in "pole splitting," and the two poles cannot possibly be equal. However, equal poles can occur in other circuits. *Unless we know that the response is dominated by a single pole, a good design procedure is to assume that the sum of the poles is* $-1.15\omega_1$, *or*

$$1.15\omega_1 \approx -\sum_j p_j = \sum_j \frac{1}{\tau_{js}} \qquad (18\text{-}42)$$

When the capacitors are selected in accordance with this relation, the actual 3 dB frequency is about 5% too high for poles a decade apart, and 12% too low for identical poles. It is nearly always within 10% of the design objective. In fact, (18-42) usually leads to excellent results when applied to amplifiers with more than two poles.

For a one-pole amplifier we select the capacitor so that the reciprocal of the time constant equals the specified ω_1, but with two or more poles we shall use either (18-41) or (18-42). In certain cases we may wish to choose the capacitors so that only one of them contributes significantly to the sum of the reciprocals of the short-circuit time constants. Although this procedure probably will not give a good choice of capacitors insofar as costs are concerned, it does ensure that the response is dominated by a single capacitor and hence a single pole. *Whenever one time constant is selected so that its reciprocal is at least* 10 *times greater than the sum of the reciprocals of the others,* (18-41) *applies. In all other cases we shall use* (18-42). Because of the approximations involved, it is good procedure to analyze the network after the design is completed in order to ensure that the specifications are satisfied. For complicated circuits computer analyses are desirable. When either the design or analysis is based on (18-41), the true ω_1 is always somewhat lower than the value in the equation, assuming more than one pole.

From (18-40) and (18-42) with $\omega_1 = 0.1885$ ms^{-1} we obtain

$$\frac{1}{5.48C_c} + \frac{1}{0.0758C_e} = 0.217 \text{ ms}^{-1} \qquad (18\text{-}43)$$

Often a good initial trial is to pick values that make the time constants equal, or nearly so. For $C_c = 1.68$ and $C_e = 122$ μF, each time constant is 9.2 ms and (18-43) is satisfied. Perhaps a more suitable value for the emitter capacitance is 100, and this requires that C_c be 2.15 μF. Selecting these values, we find the two time constants $R_{cs}C_c$ and $R_{es}C_e$ to be 11.8 and 7.58 ms, respectively. Because ω_1 is determined from the reciprocal of the time constants, the capacitance C_e has the greater effect in reducing the gain at low frequencies. Now we resort to analysis to verify that the design is satisfactory.

AMPLIFIER ANALYSIS

The incremental model of Fig. 18-16 is redrawn in Fig. 18-17 with numerical values indicated. Two loop equations are

$$V_s = \left(15 + \frac{0.465}{s}\right)I_1 - 10I_2 \tag{18-44}$$

$$0 = -10I_1 + \left(10.5 + \frac{16.4}{1 + 40s}\right)I_2 \tag{18-45}$$

The ratio I_2/V_s is easily found by solving (18-45) for I_1 and substituting the result in (18-44). As V_o equals $-80I_2$, the voltage gain V_o/V_s is determined to be

$$A_{Vs} = \frac{-13.9s(s + 0.025)}{(s + 0.188)(s + 0.029)} \tag{18-46}$$

In reciprocal milliseconds the two zeros are 0 and -0.025, the dominant pole is -0.188, and the smaller pole is -0.029. Because the zero and the pole that are nearly equal have cancelling effects on the frequency response, the 3 dB frequency is determined solely by the dominant pole. Therefore, ω_1 is 0.188

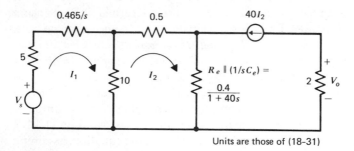

Units are those of (18-31)

Figure 18-17   Amplifier model. ($C_c = 2.15$, $C_e = 100 \ \mu F$.)

ms^{-1}, f_1 is 30 Hz, and this result is indeed excellent. The reciprocals $1/R_{cs}C_c$ and $1/R_{es}C_e$ of the short-circuit time constants are 0.085 and 0.132 ms^{-1}, respectively, whereas the magnitudes of the poles are 0.029 and 0.188 ms^{-1}. Clearly, each capacitor contributes partly to each pole, and thus the capacitors are *interacting*.

A third time constant enters into the calculations if the output of the CE amplifier is RC coupled to a load R_L. However, because of the isolating current source in the incremental model of Fig. 18-16, the coupling capacitor C_{c2} at the output does not interact with C_c and C_e. Equation 18-43 applies provided we add the term $1/(R_C + R_L)C_{c2}$ to the left side (see P18-10). For a CE cascade in which each stage has one coupling capacitor C_c along with an emitter bypass capacitor

C_e the short-circuit time constants of each stage are determined independently of the other stages, and (18-42) is utilized to select suitable capacitance values. In most cases the lowest cost and the smallest weight and size are obtained by selecting the capacitors so that the dominant pole is largely determined by C_e. Consequently, the short-circuit time constant associated with C_e is usually chosen smaller than the other time constants of (18-42).

18-5. CS AMPLIFIER AT LOW FREQUENCIES

LOW FREQUENCY DESIGN

To illustrate further the design procedures at low frequencies the CS amplifier of Fig. 18-18 will be examined. The units employed are those used in the preceding section, given by (18-31). We need the short-circuit time constants. With

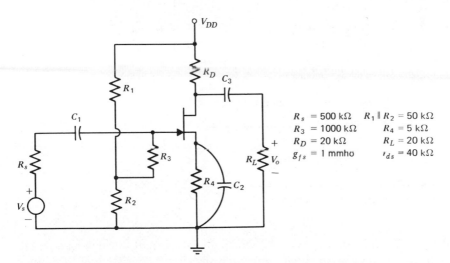

$R_s = 500 \text{ k}\Omega$ $R_1 \| R_2 = 50 \text{ k}\Omega$
$R_3 = 1000 \text{ k}\Omega$ $R_4 = 5 \text{ k}\Omega$
$R_D = 20 \text{ k}\Omega$ $R_L = 20 \text{ k}\Omega$
$g_{fs} = 1 \text{ mmho}$ $r_{ds} = 40 \text{ k}\Omega$

Figure 18-18   FET amplifier

reference to the incremental model of Fig. 18-19 it is clear that the time constant associated with C_1 is independent of the drain-source circuit. Thus C_1 does not interact with the other capacitors, and one of the poles of the voltage gain A_{Vs} is $-1/R_{1s}C_1$, with

$$R_{1s} = R_s + R_3 + R_1 \| R_2 \qquad (18\text{-}47)$$

To find R_{2s} associated with C_2 we use the circuit of Fig. 18-20, in which C_1 and C_3 are replaced with shorts. Resistance R_{2s} is the parallel equivalent of R_4

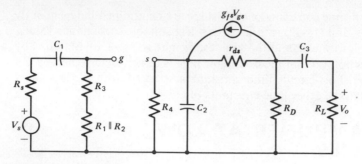

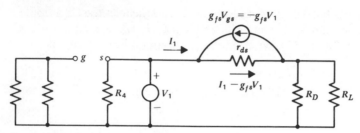

Figure 18-19   Low-frequency incremental model.

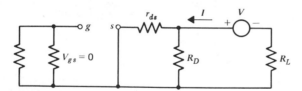

Figure 18-20   Model for determining R_{2s}.

Figure 18-21   Model for determining R_{3s}.

and V_1/I_1. We note that $V_{gs} = -V_1$, which gives a current through r_{ds} equal to $I_1 - g_{fs}V_1$. A single loop equation around the circuit consisting of the source V_1, r_{ds}, and the equivalent load is sufficient for determining V_1/I_1, and R_{2s} is

$$R_{2s} = R_4 \left\| \left(\frac{R_D \| R_L + r_{ds}}{1 + g_{fs}r_{ds}} \right) \right.$$ (18-48)

In the network of Fig. 18-21 capacitors C_1 and C_2 are replaced with shorts, and R_{3s} is V/I. By inspection this is

$$R_{3s} = R_L + r_{ds} \| R_D$$ (18-49)

Let us digress a moment and assume that r_{ds} is so large that it can be neglected. For this case R_{1s} is the same as specified by (18-47), R_{2s} is $R_4 \| (1/g_{fs})$ by (18-48), and R_{3s} is $R_L + R_D$ by (18-49). Furthermore, the three capacitors are non-interacting, and each time constant determines a pole. For the numerical values given in Fig. 18-18, R_{1s}, R_{2s}, and R_{3s} are 1550, 0.833, and 40 kΩ, respectively. For a 3 dB frequency of 30 Hz, ω_1 is 0.1885 ms^{-1}, and from (18-42) we obtain

$$\frac{1}{1550C_1} + \frac{1}{0.833C_2} + \frac{1}{40C_3} = 0.217 \text{ ms}^{-1} \qquad (18\text{-}50)$$

Because the capacitors are noninteracting, *the negative of each of the three terms on the left represents a pole.* If we select the capacitors so that the time constants are equal, the poles are identical. For this choice the capacitances C_1, C_2, and C_3 are 0.0089, 16.6, and 0.35 μF, respectively, and each pole is -0.072 ms^{-1}. Because of the coupling capacitors there are two zeros at $s = 0$, and the third zero z_3 is $-1/R_4C_2$, or -0.012 ms^{-1}. The voltage gain becomes

$$A_{Vs} = \frac{A_o s^2(s + 0.012)}{(s + 0.072)^3} \qquad (18\text{-}51)$$

From this result the lower 3 dB frequency is calculated to be 22 Hz, which is substantially below the design objective of 30 Hz. Furthermore, the effect of r_{ds} reduces f_1 to an even lower value. The substantial discrepancy appears because (18-42) has its greatest error when the poles are identical. Therefore, whenever noninteracting capacitors are encountered, the error introduced by the design equation can be minimized by selecting capacitor values so as to separate the poles. Of course, other considerations may be more important.

Suppose we select $C_1 = 0.1$, $C_2 = 6.47$, and $C_3 = 1$ μF. For these values (18-50) is satisfied, and the poles are -0.0065, -0.186, and -0.025 ms^{-1}. The zero z_3 is -0.031 ms^{-1}. These give a voltage gain that is

$$A_{Vs} = \frac{A_o s^2(s + 0.031)}{(s + 0.0065)(s + 0.186)(s + 0.025)} \qquad (18\text{-}52)$$

The zero nearly cancels a pole and the smallest pole is fairly negligible. Thus ω_1 is 0.186 ms^{-1}, which makes $f_1 = 29$ Hz. The design is satisfactory as long as r_{ds} is negligible. However, with the values selected for the capacitances used in the incremental model of Fig. 18-19 with $r_{ds} = 40$ kΩ, a detailed analysis reveals that f_1 is 26 Hz. The advantage of neglecting r_{ds} is that the expression for A_{Vs}, given by (18-52), is easily obtained without writing loop equations. For a more careful design the effect of r_{ds} should be included, and this will now be done.

With $r_{ds} = 40$ kΩ, the short-circuit resistances R_{1s}, R_{2s}, and R_{3s} are found from (18-47), (18-48), and (18-49) to be 1550, 0.980, and 33.3 kΩ, respectively, and (18-42) becomes

$$\frac{1}{1550C_1} + \frac{1}{0.98C_2} + \frac{1}{33.3C_3} = 0.217 \text{ ms}^{-1} \qquad (18\text{-}53)$$

Although C_1 does not interact with the other capacitors, C_2 and C_3 interact with each other because of r_{ds}. Because this interaction is small, selecting equal time constants gives poles fairly close together. In fact, for this choice, C_1 is 0.0089, C_2 is 14.1, and C_3 is 0.415 μF, and analysis of the circuit shows that the poles are -0.099, -0.072, and -0.046 ms^{-1}. The corresponding 3 dB frequency is 23 Hz. Poles close together tend to split into complex conjugates when there is stray or applied negative feedback, and instability can result. This is discussed in Chapter 21. Perhaps a better choice is to make $C_1 = 0.1, C_2 = 5.65,$ and $C_3 = 1$ μF. These selections satisfy (18-53), give unequal time constants, are suitable values, and are used in the analysis that follows.

AMPLIFIER ANALYSIS

To verify the suitability of the design, we need to find the poles and zeros. Two zeros are at $s = 0$ and the third is $-1/R_4 C_2$, or -0.035 ms^{-1}. The pole associated with C_1 is $-1/R_{1s}C_1$, or -0.0065 ms^{-1}. The other two poles are determined from analysis of the incremental model of Fig. 18-22.

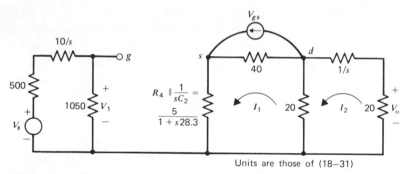

Units are those of (18–31)

Figure 18-22   Model with $C_1 = 0.1, C_2 = 5.65, C_3 = 1$ μF.

Clearly, V_1/V_s is $1050/(1550 + 10/s)$, or $0.677s/(s + 0.0065)$. The ratio V_o/V_1 can be obtained from the drain-source circuit. As g_{fs} is 1 mmho, the current source is V_{gs}, which is

$$V_{gs} = V_1 - \frac{5I_1}{1 + s28.3} \tag{18-54}$$

The mesh currents are $V_{gs}, I_1,$ and I_2. Two loop equations in terms of the mesh currents, with (18-54) used to replace V_{gs}, are

$$0 = \left(60 + \frac{5}{1 + s28.3}\right)I_1 - 20I_2 - 40\left(V_1 - \frac{5I_1}{1 + s28.3}\right) \tag{18-55}$$

$$0 = -20I_1 + \left(40 + \frac{1}{s}\right)I_2 \tag{18-56}$$

Solving for the ratio I_2/V_1 and multiplying this by -20 to obtain V_o/V_1, we find that

$$\frac{V_o}{V_1} = \frac{-8s(s + 0.035)}{s^2 + 0.21s + 0.00468} \tag{18-57}$$

The quadratic is easily factored, and the product of V_o/V_1 with the expression previously found for V_1/V_s gives the voltage gain to be

$$A_{Vs} = \frac{-5.42s^2(s + 0.035)}{(s + 0.185)(s + 0.025)(s + 0.0065)} \tag{18-58}$$

The sum of the poles is -0.217 ms^{-1}, which is the negative of the sum of the time constants of (18-53).

In the real frequency domain we replace s with $j\omega$. After dividing the numerator and denominator of (18-58) by $(j\omega)^3$, it becomes obvious that the magnitude of the gain is

$$|A_{Vs}| = 5.42 \left\{ \frac{1 + (0.035/\omega)^2}{[1 + (0.185/\omega)^2][1 + (0.025/\omega)^2][1 + (0.0065/\omega)^2]} \right\}^{1/2} \tag{18-59}$$

When $\omega = \omega_1$, $|A_{Vs}|$ is $5.42/\sqrt{2}$. It follows from this and (18-59) that

$$\frac{[1 + (0.185/\omega_1)^2][1 + (0.025/\omega_1)^2][1 + (0.0065/\omega_1)^2]}{1 + (0.035/\omega_1)^2} = 2$$

The effects of the zero and the pole at -0.025 tend to cancel, and the smallest pole is almost negligible. Hence, ω_1 approximately equals the magnitude 0.185 ms^{-1} of the dominant pole. This could have been determined directly from (18-58) for this particular case, but (18-59) and (18-60) were given to illustrate the general procedure. Actually, by trial and error ω_1 is found from (18-60) to be 0.182 ms^{-1}, and f_1 is 29 Hz. This result is quite satisfactory.

The time-constant method based on (18-41) or (18-42) is applicable to any low-frequency design. It is also valuable in analysis. With all components of the circuit specified, the short-circuit time constants are calculated, and either (18-41) or (18-42) is used to determine the approximate 3 dB frequency (see P18-11). Usually, the dominant pole is at least twice as great as the closest singularity, in which case the results are fairly accurate.

18-6. HIGH-FREQUENCY AMPLIFIER DESIGN

PROCEDURE

At high frequencies the device capacitances are included in the incremental model of an amplifier, and they give poles and zeros that cause the gain to drop off in this frequency region. The dominant pole p_1 is the smallest one, and assuming it is much smaller in magnitude than the next larger singularity, the upper 3 dB frequency ω_2 is simply the magnitude of p_1. Furthermore, for this case the sum of the reciprocals of the poles is about the same as $1/p_1$. From (18-23) we obtain

$$\frac{1}{\omega_2} \approx -\sum_j \frac{1}{p_j} = \sum_j \tau_{jo} \qquad (18\text{-}61)$$

When the open-circuit time constants are selected in this manner, the true ω_2 is always greater than the values used in the equation, assuming more than one pole.

Now let us suppose that the upper 3 dB frequency is largely determined by two poles p_1 and p_2 that are not widely separated. It is easy to show (see P18-16) that the sum of the reciprocals of the poles is $-1.09/\omega_2$ for poles a decade apart, provided there are no significant zeros. For p_2 five times p_1, this sum is $-1.16/\omega_2$ and it is $-1.29/\omega_2$ for equal poles. Thus *a good design procedure is to assume that*

$$\frac{1.15}{\omega_2} \approx -\sum_j \frac{1}{p_j} = \sum_j \tau_{jo} \qquad (18\text{-}62)$$

When the open-circuit time constants are selected in accordance with this relation, the actual 3 dB frequency is about 5% too low for poles a decade apart, and 12% too high for identical poles. It is nearly always within 10% of the design objective, and in fact, (18-62) usually leads to excellent results when applied to amplifiers with more than two poles.

As we shall see, one-stage CE and CS amplifiers with resistive loads have two poles that are very widely separated. The poles of an amplifier are also widely separated when one open-circuit time constant is much larger than any of the others. Therefore, *for one-stage CE and CS amplifiers with resistive loads, and for any amplifier in which one time constant is at least ten times greater than the sum of the others, we shall use* (18-61). *For all other cases,* (18-62) *applies.* Because the equations are approximate, the design should be checked through analysis.

In numerical work with high-frequency incremental models it is usually convenient to use the following system of units, which constitute a consistent set:

milliamperes	volts	nanoseconds	gigahertz
kilohms	millimhos	picofarads	microhenrys

$$(18\text{-}63)$$

We shall assume these units throughout this section unless specifically stated otherwise. Frequency s is in reciprocal nanoseconds, with $1 \text{ ns} = 10^{-9}$ s. A gigahertz is 1000 MHz, or 10^9 Hz.

In Fig. 18-23 is the circuit of a typical CE amplifier. Because the effective load resistance R_L is less than 10% of r_o, we can neglect r_μ in the incremental

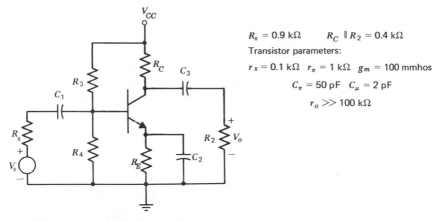

$R_s = 0.9$ kΩ $R_C \parallel R_2 = 0.4$ kΩ

Transistor parameters:

$r_x = 0.1$ kΩ $r_\pi = 1$ kΩ $g_m = 100$ mmhos

$C_\pi = 50$ pF $C_\mu = 2$ pF

$r_o \gg 100$ kΩ

Figure 18-23   CE amplifier with resistive load.

model. The resistance r_o is in parallel with R_L and can also be omitted with negligible error. For the sake of simplicity let us assume that R_B, defined as the parallel equivalent of R_3 and R_4, is large enough to justify its elimination, and the high-frequency incremental model becomes that of Fig. 18-24. Clearly, the gain function has two poles.

THE MILLER EFFECT

In the network of Fig. 18-24 the current through C_μ is usually very small compared with that of the controlled source. For this case the output voltage V_o is approximately $-g_m R_L V_1$. Therefore, the current through C_μ, which equals the product of the admittance $j\omega C_\mu$ and the voltage $V_1 - V_o$, can be expressed as $j\omega C_\mu(1 + g_m R_L)V_1$. In order for this to be negligible compared with the current $g_m V_1$ of the controlled source, the frequency ω must be much smaller than both g_m/C_μ and $1/R_L C_\mu$. However, for such limited values of ω the model of Fig. 18-25 becomes a good approximation. Clearly, the capacitance $C_\mu(1 + g_m R_L)$ gives the correct input node current resulting from C_μ, and the negligible output-node current due to this capacitance is simply zero in the model.

From the results it follows that the *CE approximate model of Fig. 18-25 is valid provided* (1) *transmission is in the forward direction*, (2) *the load is resistive*, and (3) *the radian frequency* $\omega \ll 1/R_L C_\mu$. Not enumerated is the requirement that ω be much less than g_m/C_μ, because this condition is always satisfied. The ratio g_m/C_μ is much greater than ω_T by (12-47) of Sec. 12-6, and practical radian frequencies are substantially less than ω_T. In fact, the hybrid-π model is invalid at frequencies greater than about $\omega_T/3$. All three of the enumerated conditions must be satisfied. When they are, analysis and design are simplified

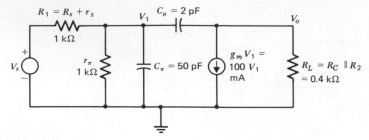

Figure 18-24   High-frequency model with R_B, r_μ, and r_o neglected.

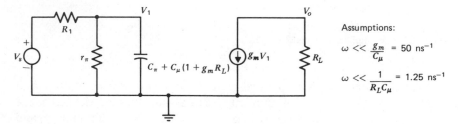

Assumptions:

$$\omega \ll \frac{g_m}{C_\mu} = 50 \text{ ns}^{-1}$$

$$\omega \ll \frac{1}{R_L C_\mu} = 1.25 \text{ ns}^{-1}$$

Figure 18-25   Approximate model based on the Miller effect.

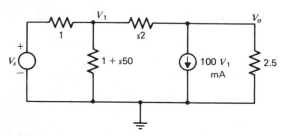

Figure 18-26   Exact s-domain model, using admittances in mmhos.

considerably by use of the approximate model. In particular, the one-capacitor model must *not* be used when the load is appreciably reactive, as in the case of a CE amplifier feeding into a second CE stage.

As indicated by the circuit of Fig. 18-25 the effect of C_μ on the input capacitance of the CE amplifier is enhanced by the magnitude of the midband gain $g_m R_L$. This is known as the *Miller effect*. It is caused by feedback provided by C_μ, with the current fed back approximately equal to the product of the output voltage and the admittance of C_μ. Because V_o is much greater than V_i, the effect of C_μ on the input impedance is amplified.

The load resistance of a CE amplifier is nearly always large enough to ensure that ω is much less than $1/R_L C_\mu$ at frequencies in the vicinity of and below the upper 3 dB breakpoint. This is true in all examples which we shall encounter in this book and in most practical amplifiers. Therefore, the requirement is not usually a serious limitation.

The circuit of Fig. 18-25 has only one capacitor, and of course, only one pole. It follows that the upper break frequency of the CE amplifier is determined from a single dominant pole. Consequently, we conclude that the second pole of the two-capacitor circuit of Fig. 18-24 *must be much greater than the dominant pole*.

In Sec. 17-1 we found that a CS amplifier with a resistive load also has Miller effect and thus can be represented by a similar approximate model having only one capacitor. Therefore, the larger pole p_2 of this configuration is likewise much greater than p_1. Because both CE and CS amplifiers *with resistive loads* have high-frequency responses that are governed by a single dominant pole, the use of (18-61) in design is accurate even though there are two capacitances and two open-circuit time constants associated with each exact model.

The voltage gain V_o/V_s of the approximate model of Fig. 18-25 has the form $A_o/(1 - s/p_1)$. With the capacitor removed, the midfrequency gain A_o is easily found, and according to (18-61), the pole p_1 is the negative of the reciprocal of the time constant. The expressions for A_o and p_1 are

$$A_o = \frac{-\beta_o R_L}{R_1 + r_\pi} \qquad p_1 = \frac{-1}{[R_1 \| r_\pi][C_\pi + C_\mu(1 + g_m R_L)]} \qquad (18\text{-}64)$$

For the numerical values given in Fig. 18-24, A_o is -20 and p_1 is -0.01515 ns^{-1}. The voltage gain becomes

$$A_{Vs} = \frac{-20}{1 + s/0.01515} \qquad (18\text{-}65)$$

The upper 3 dB frequency ω_2 is 0.01515 ns^{-1}, and f_2 is 0.00241 GHz, or 2.41 MHz.

The break frequency can also be found from the open-circuit time constants of the exact model of Fig. 18-26. With $V_s = 0$ and C_μ open, the resistance R_{1o} facing C_π is 0.5 kΩ, and τ_{1o} is 25 ns. With C_π open, the resistance R_{2o} facing C_μ is determined from the network of Fig. 18-27. Resistance R_{2o} is the ratio V/I.

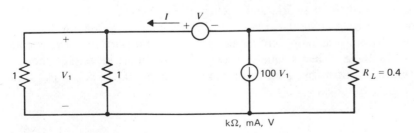

$k\Omega$, mA, V

Figure 18-27   Model for determining R_{2o}.

Voltage V_1 equals $0.5I$, and the current $100V_1$ is $50I$. Application of Kirchhoff's voltage law to the outside loop gives $V = 20.9I$. The open-circuit time constant τ_{2o} is $R_{2o}C_\mu$, or 41.8 ns, and the sum of the two time constants is 66.8 ns. By (18-61) ω_2 is the reciprocal of this, or 0.01497 ns^{-1}, and f_2 is 2.38 MHz.

CE AMPLIFIER ANALYSIS

To satisfy ourselves that the preceding approximate results are reasonable, let us find the exact voltage gain from the network of Fig. 18-26. Two node equations are

$$V_s = (2 + s52)V_1 - s2V_o$$
$$0 = (100 - s2)V_1 + (2.5 + s2)V_o$$

(18-66)

From these we determine the gain V_o/V_s to be

$$A_{Vs} = \frac{-20(1 - s/50)}{(1 + s/0.01504)(1 + s/3.3)} = \frac{-20}{1 + s/0.01504}$$

(18-67)

The right side of (18-67) is correct because the zero at $+50$ ns^{-1} and the pole at -3.3 ns^{-1} are so large that they can be regarded as infinite. Mathematically, their effect on ω_2 is insignificant. The transition frequency ω_T of the transistor is $g_m/(C_\pi + C_\mu)$, or 1.92 ns^{-1}, corresponding to an f_T of 306 MHz. Both the zero and the larger pole are greater than ω_T. The hybrid-π model is invalid at such high frequencies. It follows that, *whenever a network using a hybrid-π model yields singularities greater in magnitude than ω_T, these singularities are meaningless and should be made infinite.* Henceforth this will be done. From (18-67) we find that ω_2 is 0.015 ns^{-1} and f_2 is 2.39 MHz. This exact value agrees closely with those obtained from the approximate model (2.41 MHz) and from the sum of the two open-circuit time constants (2.38 MHz). Stated on Fig. 18-25 are the assumptions on which the approximate model is based. We note that these are reasonably well satisfied at frequencies as high as $10\omega_2$.

The two zeros of the amplifier were found from circuit analysis to be 50 ns^{-1} and infinite. In most high-frequency networks the zeros are easily obtained from inspection of the network, and such is the case here. In Fig. 18-24 the capacitor C_π becomes a short-circuit at an infinite frequency, making both V_1 and V_o zero. To find the finite zero z_1 let us refer to that portion of the circuit shown in Fig. 18-28. When s equals z_1, both V and I must be zero as shown. The current $sC_\mu V_1$ through C_μ equals that of the source $g_m V_1$, or $sC_\mu = g_m$. This value of s is z_1, which is

$$z_1 = \frac{g_m}{C_\mu}$$

(18-68)

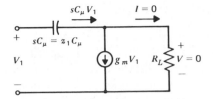

Figure 18-28   Network for determining
the finite zero z_1.

THE GAIN-BANDWIDTH PRODUCT

In the CE amplifier that we have analyzed, the designer has some freedom in selecting R_L. Expressions for A_o and the dominant pole p_1 are presented in (18-64), and f_2 is $|p_1|/2\pi$. Using the numerical values of Fig. 18-24, but with R_L treated as a variable, we find that

$$A_o = -50R_L \qquad f_2 = \frac{1.59}{R_L + 0.26} \text{ MHz} \qquad (18\text{-}69)$$

Increasing R_L raises the midfrequency gain while reducing the bandwidth, and vice versa. Suppose, for example, a bandwidth of 3 MHz is desired. From (18-69) we find that R_L must be selected as 0.27 kΩ, and A_o becomes -13.5. Lowering R_L from 400 ohms to 270 ohms increases the bandwidth from 2.4 to 3.0 MHz, but the midfrequency gain is reduced from 20 to 13.5.

The term $A_o\omega_2$ is $-\beta_o R_L/[(R_1 r_\pi)(C_\pi \mid C_\mu + C_\mu g_m R_L)]$, with this gain-bandwidth product obtained from (18-64). Replacing r_π with β_o/g_m, the expression can be put in the form

$$A_o\omega_2 = \frac{-R_L/R_1}{(1/\omega_T) + C_\mu R_L} \qquad (18\text{-}70)$$

This result shows quite clearly that *a large gain-bandwidth product requires a transistor with a high transition frequency and a small C_μ*. For the amplifier of Fig. 18-23, f_T is 306 MHz, ω_T is 1.92 ns^{-1}, C_μ is 2 pF, and R_L is 0.4 kΩ. In the denominator of (18-70), $1/\omega_T$ is 0.52 and $C_\mu R_L$ is 0.80 ns. Thus the detrimental effect of C_μ, which is enhanced by the Miller effect, is greater than that of $1/\omega_T$, which determines C_π. Because this is usually the case, *a small C_μ is especially desirable for transistors designed for use at high frequencies*. Furthermore, great care in circuit construction is essential to minimize base-collector capacitance due to proximity of wires and components. This stray capacitance is in parallel with C_μ, is also enhanced by the Miller effect, and similarly degrades high-frequency performance.

HIGH-FREQUENCY CS AMPLIFIER DESIGN

Figure 18-29 illustrates a CS amplifier and its high-frequency incremental model. The load consists of a 10 kΩ resistor in parallel with a 150 pF capacitor, which represents the input impedance of a CE stage that is driven by the CS amplifier. Let us suppose that we wish to design the network for a bandwidth of 60 kHz, if possible. We must select the proper value for R_D. Then the supply voltage and the source resistance can be chosen to give the desired Q-point. At this Q-point let us assume that the transistor parameters are those specified on the figure.

The incremental model has three capacitors and one capacitor loop. Therefore, the network has two poles. Because of C_{gs} there is a zero at infinity. The finite zero is g_{fs}/C_{gd}, or $+0.5$ ns^{-1}. The midfrequency voltage gain is $-g_{fs}R_L$, and R_L is the parallel equivalent of r_{ds}, R_D, and R_1.

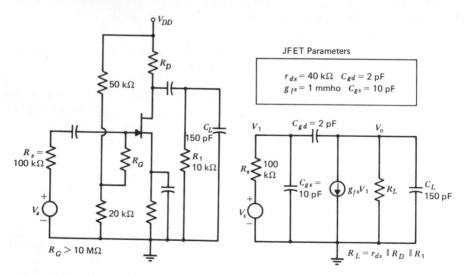

Figure 18-29 CS amplifier and high-frequency model.

The open-circuit time constants are easily found in terms of R_L. The time constant τ_{1o} associated with C_{gs} is $R_s C_{gs}$, or 1000 ns. The resistance R_{2o} facing C_{gd} is the sum of R_s and $(1 + g_{fs}R_s)R_L$, determined from a circuit similar to that of Fig. 18-27. Thus R_{2o} is $100 + 101R_L$, and τ_{2o} is $200 + 202R_L$. The third open-circuit time constant τ_{3o} is the product $R_L C_L$, or $150R_L$. The sum of the time constants is $1200 + 352R_L$, and this must equal $1.15/\omega_2$ by (18-62). As we desire f_2 to be 6×10^{-5} GHz, ω_2 is 37.7×10^{-5} ns^{-1}, and (18-62) becomes

$$3050 = 1200 + 352R_L \qquad (18\text{-}71)$$

From this we find the proper value of R_L to be 5.26 kΩ, and the corresponding value of R_D is 15.3 kΩ.

An exact analysis of the network is readily accomplished by writing node equations at nodes labeled V_1 and V_o, and solving for the ratio V_o/V_s. The result can be put in the form

$$\frac{V_o}{V_s} = \frac{-5.26(1 - s/0.5)}{(1 + s/3.71 \times 10^{-4})(1 + s/2.82 \times 10^{-3})} \tag{18-72}$$

From this relation f_2 is found to be 58.1 kHz, which is reasonably close to our design objective of 60 kHz. *It should be noted that the one-pole model of the form of Fig. 18-25 does not apply here, because the load is not a pure resistance.* Multistage amplifiers at high frequencies are considered in Chapters 19 and 21. (Also, see P18-15.)

REFERENCES

1. P. E. Gray and C. L. Searle, *Electronic Principles*, Wiley, New York, 1969.
2. D. L. Schilling and C. Belove, *Electronic Circuits: Discrete and Integrated*, McGraw-Hill, New York, 1968.
3. P. M. Chirlian, *Electronic Circuits: Physical Principles, Analysis, and Design*, McGraw-Hill, New York, 1971.
4. J. Millman and C. C. Halkias, *Integrated Electronics: Analog and Digital Circuits and Systems*, McGraw-Hill, New York, 1972.

PROBLEMS

Section 18-1

18-1. In the network of Fig. 18-30 determine the number of poles of (a) the voltage gain V_o/V_i, and (b) the current gain I_o/I_i. Repeat (a) and (b) for the case in which the transformer is not ideal (treat each coil as an inductance). [*Ans:* (a) 3]

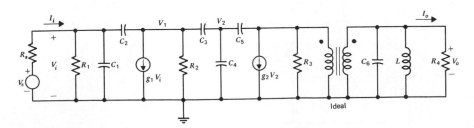

Figure 18-30 Network for P18-1.

18-2. Analysis of the high-frequency model of a poorly designed feedback amplifier reveals that the voltage gain V_o/V_i in SI units is

$$A(s) = \frac{6s - 106}{(s + 5)(s^2 - 2s + 101)}$$

(a) Find the poles and zeros of A. (b) If $v_i(t)$ is a unit impulse, which has a Laplace transform of unity, the output V_o equals A. Deduce that the response $v_o(t)$ is $-e^{-5t} + e^t \cos 10t$. Comment on the stability.

18-3. A voltage v_i is applied to an RL series network and v_o is taken across the resistance. $R = 1$, $L = 1$, and V_o/V_i is $1/(s + 1)$ in SI units. (a) For $v_i = 4e^{-5t}$ for all values of t, find V_i, s, and the steady-state response $v_o(t)$. For $v_i = 2 \cos (t + 30°)$, use the s-domain gain function to find $v_o(t)$. (c) For v_i a unit step function at $t = 0$, find $v_o(t)$ for $t > 0$, using the s-domain gain function. (*Ans:* (a) 4, -5, $-\exp - 5t$]

Section 18-2

18-4. Analysis of the low-frequency s-domain model of an amplifier in SI units yields the voltage gain

$$A(s) = \frac{-10^5 s(s + 10)}{(s + 100)(s + 1000)}$$

Using scales similar to those of Fig. 18-9, sketch the straight-line asymptotes of the dB gain versus ω. Determine the true dB gain at $\omega = 100$ and also at $\omega = 10$. What is the lower 3 dB frequency f_1?

18-5. The high-frequency gain function of an amplifier in SI units is

$$A(s) = \frac{-5 \times 10^5(s - 10^{10})(s - 10^{12})}{(s + 10^7)(s + 10^8)(s + 5 \times 10^8)}$$

Determine the midband dB gain A_o, the poles, the zeros, and the BW in MHz. Also, plot the dB gain versus ω from 10^6 to 10^{10}, using scales similar to those of Fig. 18-9.

18-6. An amplifier has a midband voltage gain of 100, a pole at -10^6, and a zero at infinity. Determine the response $v_o(t)$ to an input step voltage of magnitude 0.01, applied at time zero. Also, calculate the rise time, defined as the time required for the output voltage to rise from 10 to 90% of its final value. Units are SI.

Section 18-3

18-7. (a) By direct analysis of the network of Fig. 18-31 the ratio V_o/V_s is easily determined as a function of s. From this ratio find the poles. (b) Calculate the short-circuit and

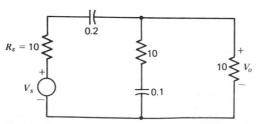

Figure 18-31 For P18-7.

open-circuit time constants and verify that (18-22) and (18-23) are satisfied by the results. (c) Deduce that the network has only a single pole if $R_s = 5$, because of direct cancellation of a pole and a zero. Units are SI.

Section 18-4

18-8. For the CE amplifier of Fig. 18-12 find the 3 dB frequency f_1 if C_c is 0.5 μF. Also, determine f_1 for the amplifier of Fig. 18-13 if C_e is 40 μF. Do not neglect the effect of the zero. (*Ans*: 58.1, 50.6)

18-9. For a low-frequency amplifier model having two poles, and with both zeros at $s = 0$, show that

$$\left(1 + \frac{a}{x}\right)\left(1 + \frac{1}{x}\right) = 2$$

with $x = (\omega_1/p_2)^2$ and $a = (p_1/p_2)^2$. (a) For equal poles show that ω_1 is $-1.55p_2$, and then deduce that $(p_1 + p_2)/\omega_1$ is -1.29. (b) For $p_1 = 0.2p_2$, show that the ratio is -1.16.

18-10. In the CE amplifier of Fig. 18-16, add an RC coupled load to the output with $C_{c2} = 4$ μF and $R_L = 2$ kΩ. The input coupling capacitor C_{c1} is 2.15 μF and C_e is 100 μF. these being the same values that were selected in the design procedure of Sec. 18-4 so as to make $f_1 = 30$ Hz. (a) Using (18-42), determine the approximate value of f_1. (b) Deduce that A_{Vs} is the expression of (18-46) multiplied by $0.5s/(s + 0.0625)$ in the units of (18-31). (c) From A_{Vs} determine the true value of f_1, using a trial-and-error procedure.

Section 18-5

18-11. For the FET amplifier of Fig. 18-18, use (18-42) to find the approximate f_1, with $C_1 = 0.01$, $C_2 = 4$, and $C_3 = 0.4$ μF. (An exact analysis gives $f_1 = 46$ Hz; see P18-12). Do not neglect r_{ds}.

18-12. For the FET amplifier of Fig. 18-18 with $C_1 = 0.01$, $C_2 = 4$, and $C_3 = 0.4$ μF, find the poles, the zeros, and A_o for the voltage gain V_o/V_s. Then determine the lower 3 dB frequency f_1.

18-13. Suppose the FET amplifier of Fig. 18-18 has $R_s = 100$, $R_4 = 4$, $R_D = R_L = r_{ds} = 10$, $g_{fs} = 2$, and a total gate resistance $R_G = 500$, using the units of (18-31). With the aid of (18-42), determine values of C_1, C_2, and C_3 to give an approximate 3 dB frequency of 50 Hz, with the values selected so that $\tau_{1s} = 3\tau_{2s} = \tau_{3s}$. Then employ direct circuit analysis to determine the poles, the zeros, and the true value of f_1 in hertz.

Section 18-6

18-14. Assume that the CE amplifier of Fig. 18-23 has $R_s = 5$, $R_3 = 180$, $R_4 = 60$, $R_E = 4.1$, $R_C = 12$, and $R_2 = 4$, with values in kΩ. V_{CC} is 12 V. At the Q-point at 1 kHz the transistor has $h_{ie} = 5.5$ kΩ, $h_{fe} = 100$, $h_{re} = 10^{-4}$, and $h_{oe} = 0.004$ mmho. The transition frequency f_T is 150 MHz, and C_μ is 2 pF. Assume q/kT is 40, β_F is 100, and $V_{BE} = 0.7$ V. Using the approximate one-pole model based on the Miller effect, find the bandwidth in kHz. (*Ans*: 453)

18-15. The CS–CE cascade of Fig. 18-32 is to be designed for a BW of 60 kHz. Assume R_G is greater than 10 MΩ and can be neglected. Using the units of (18-63), at the selected Q-point the JFET has $g_{fs} = 1$, $r_{ds} = 40$, $C_{gs} = 10$, and $C_{gd} = 2$. The BJT has $g_m = 5$,

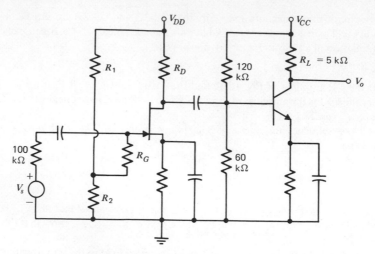

Figure 18-32 CS-CE amplifier for P18-15.

$r_\pi = 20$, $C_\pi = 15$, $C_\mu = 5$, and r_x, r_o, and r_μ can be neglected. (a) Sketch the incremental model using the approximate network of Fig. 18-25 for the CE stage, and select R_D so that the sum of the open-circuit time constants satisfies (18-62). (b) Using the model of (a), determine the poles, the zeros, and the midfrequency gain for the ratio V_o/V_s. (c) From the gain function calculate f_2. [*Ans*: (c) 57.9 kHz]

18-16. For a high-frequency amplifier with two poles, and both zeros at infinity, show that

$$\left(1 + \frac{a}{x}\right)\left(1 + \frac{1}{x}\right) = 2$$

with $x = (p_1/\omega_2)^2$ and $a = (p_1/p_2)^2$. (a) For equal poles show that $1/\omega_2$ is $-1.55/p_1$, and that $1/p_1 + 1/p_2$ equals $-1.29/\omega_2$. (b) For $p_2 = 5p_1$, show that the sum of the reciprocals of the poles is $-1.16/\omega_2$.

18-17. Assume the BJT of the CD–CE amplifier of Figs. 17-12 and 17-14 of Sec. 17-3 has $C_\mu = 5$ pF and a transition frequency f_T of 50 MHz. Using the Miller-effect capacitance, calculate the upper 3 dB frequency in kHz. The capacitances of the MOSFET can be neglected. Assume $q/kT = 40$ and do not neglect r_{ds}.

Chapter Nineteen
Wideband Multistage Amplifiers

In previous chapters we have considered a number of single-stage and multistage amplifiers, and others are examined in later chapters. Here we investigate certain amplifiers intended for use *over a wide band of frequencies*. Specifically, the design concepts of the previous chapter are applied to the important emitter-coupled configuration. Then the circuitry and characteristics of a typical integrated-circuit video amplifier are analyzed. In addition, we shall study the cascode amplifier, which is especially well-suited for use with integrated-circuit supergain transistors.

19-1. EMITTER-COUPLED AMPLIFIER DESIGN

The design techniques of the preceding chapter are, of course, applicable to multistage amplifiers. To illustrate, we shall consider the design of an emitter-coupled amplifier consisting of an input CC stage followed by a CB stage. The schematic diagram is shown in Fig. 19-1, and we note that it is quite similar to the differential amplifier examined in Sec. 14-1.

In the base circuit of transistor Q_2 the bias resistor R_s is bypassed by C_1. The current source I_{dc} is assumed to be ideal, maintaining a current that is steady with time. Such a source, which is described later in this section, is an open-circuit to the incremental currents, and the incremental network is that of Fig. 19-2.

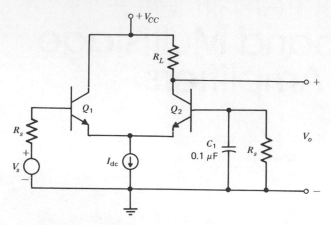

Figure 19-1   An emitter-coupled amplifier.

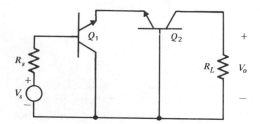

Figure 19-2   Incremental network, CC-CB amplifier.

MIDBAND VOLTAGE GAIN

In Secs. 13-3 and 13-4 we learned that the base-width-modulation resistances r_μ and r_o can be neglected in both CC and CB amplifiers, when transmission is in the forward direction, provided the load resistance of the stage is small compared with r_o. This was shown in Sec. 14-1 to be applicable to the type of amplifier considered here. Accordingly, the midfrequency incremental model of Fig. 19-2 is that of Fig. 19-3.

From the model we find that

$$(\beta_{o1} + 1)I_1 = (\beta_{o2} + 1)I_2 \tag{19-1}$$

$$V_s = (R_s + r_{x1} + r_{\pi1})I_1 + (r_{x2} + r_{\pi2})I_2 \tag{19-2}$$

If the transistors have identical betas, the node equation (19-1) shows that the incremental base currents I_1 and I_2 are equal. The output voltage V_o is $\beta_{o2}I_2 R_L$.

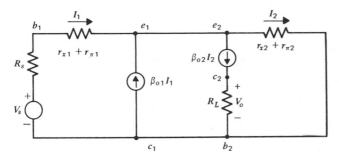

Figure 19-3   Midfrequency model.

From this and (19-1) and (19-2) it is easy to deduce the voltage gain V_o/V_s, which is

$$\frac{V_o}{V_s} = \frac{\beta_{o2} R_L}{(R_s + r_{x1} + r_{\pi 1})(\beta_{o2} + 1)/(\beta_{o1} + 1) + r_{x2} + r_{\pi 2}} \tag{19-3}$$

Assuming each transistor has an h_{ie} of 1 kΩ and a beta of 100, the voltage gain for source and load resistances of 1 kΩ is 33.

THE HIGH-FREQUENCY MODEL

To illustrate the design procedure let us suppose that an emitter-coupled amplifier of the form of Fig. 19-1 is required to have a bandwidth of at least 10 MHz. As this bandwidth is rather wide, we must select transistors with moderately high transition frequencies and small values of C_μ. From the data sheet for the selected transistors we find that, at a quiescent current of 1 mA, the minimum f_T is 500 MHz, C_μ is less than 2 pF, β_o is 80, and h_{ie} is 2.2 kΩ at 1 kHz. The source resistance R_s is specified to be 800 ohms. R_L *will be chosen for maximum gain with the specified* BW *of* 10 MHz.

For convenience we shall use units of kΩ, mA, V, mmhos, nanoseconds, GHz, and pF. With each transistor biased at 1 mA, the transconductance g_m at room temperature is 40, and from g_m and β_o we determine that r_π is 2 kΩ. As h_{ie} is 2.2, the base resistance r_x is 0.2 kΩ. We shall use the maximum value of 2 pF for C_μ, and the maximum C_π is found from the minimum f_T and (12-47) to be 10.7 pF. Summarizing, we have the following parameters:

$$r_{x1} = r_{x2} = 0.2 \text{ k}\Omega \qquad g_{m1} = g_{m2} = 40 \text{ mmhos} \qquad C_{\mu 1} = C_{\mu 2} = 2 \text{ pF}$$
$$r_{\pi 1} = r_{\pi 2} = 2 \text{ k}\Omega \qquad \beta_{o1} = \beta_{o2} = 80 \qquad C_{\pi 1} = C_{\pi 2} = 10.7 \text{ pF}$$

In Fig. 19-4 is shown the complete high-frequency incremental model. This network can be utilized to find the sum of the open-circuit time constants in terms of R_L, evaluating R_L by equating the sum to $1.15/\omega_2$, in accordance with

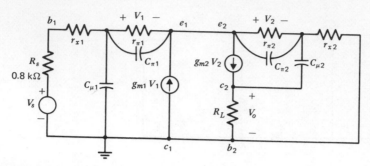

Figure 19-4   High-frequency model.

(18-62). The frequency ω_2 denotes the specified BW in reciprocal nanoseconds, and ω_2 is $2\pi f_2$, or 0.02π ns^{-1}. After determining R_L, we shall calculate the midband voltage gain using (19-3). Then in order to verify that a satisfactory design has been accomplished, the bandwidth will be found from direct analysis of the network.

For transistors with identical parameters a node equation relating the currents at the node of the emitters of Fig. 19-4 reveals that $V_1 = V_2$. Therefore, the net current fed into the node from the two dependent sources is zero. For this case these two sources can be replaced with a single source supplying the same current between ground and node c_2, as shown in Fig. 19-5. The elements $r_{\pi 1}$, $r_{\pi 2}$, $C_{\pi 1}$, and $C_{\pi 2}$ can then be combined into a single parallel combination of $2r_\pi$ and $0.5C_\pi$, with the voltage V_3 across the combination equal to $2V_1$. *Thus the network of Fig. 19-5 is exactly equivalent to that of Fig. 19-4.* It has the advantage of one less energy-storage element and one fewer node, and we shall use it for the design procedure.

From Fig. 19-4 we may be led to believe that there are four poles, because there are four capacitors and no capacitor loops. However, the rules of Sec. 18-1 for determining the number of poles are inapplicable because the dependent current sources are not shunted by resistors as required. Actually, for identical

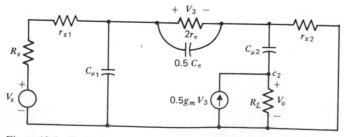

Figure 19-5   Equivalent model for identical transistors.

transistors the voltages across $C_{\pi1}$ and $C_{\pi2}$ are equal, and thus these capacitors are not independent energy-storage elements. As indicated by the network of Fig. 19-5, the combination of $C_{\pi1}$ and $C_{\pi2}$ leads to only one pole. Accordingly, the gain function has three poles, two finite zeros, and one zero at infinity due to $C_{\mu1}$. An analysis of the network of Fig. 19-4 yields four poles, but one of these is cancelled by an identical zero.

OPEN-CIRCUIT TIME CONSTANTS AND SELECTION OF R_L

To find the open-circuit resistance R_{1o} facing $C_{\mu1}$ we use the network of Fig. 19-6, obtained from Fig. 19-5 with the capacitors eliminated. By inspection, R_{1o} is the parallel equivalent of the two resistive branches, or 0.808 kΩ. The open-circuit time constant is

$$\tau_{1o} = R_{1o}C_{\mu1} = 0.808 \times 2 = 1.616 \text{ ns} \qquad (19\text{-}4)$$

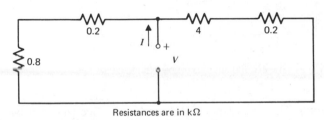

Resistances are in kΩ

Figure 19-6   Network for determining R_{1o} facing $C_{\mu1}$.

The resistance R_{2o} facing $0.5C_{\pi}$ is found from the circuit of Fig. 19-7 to be the parallel equivalent of 4 and 1.2, or 0.923 kΩ. The open-circuit time constant τ_{2o} is

$$\tau_{2o} = R_{2o}(0.5C_{\pi}) = 0.923 \times 5.35 = 4.938 \text{ ns} \qquad (19\text{-}5)$$

To find R_{3o} facing $C_{\mu2}$ we refer to Fig. 19-8. The dependent source $0.5g_m V$ of Fig. 19-5 has been replaced with $\beta_o I_1$, or $80I_1$. Let $I_1 = 1$ mA. Then I_2 is 25, found from Kirchhoff's voltage law applied to the outside loop, and the current through the source V is 26 mA. From the second loop equation V is determined to be $5 + 106R_L$, and the ratio of the source voltage to the source current is $0.192 + 4.08R_L$ kΩ. Thus the open-circuit time constant in ns is

$$\tau_{3o} = R_{3o}C_{\mu2} = (0.192 + 4.08R_L)(2) = 0.384 + 8.16R_L \qquad (19\text{-}6)$$

We determine R_L by equating the sum of the time constants to $1.15/\omega_2$, which is 18.3 for $\omega_2 = 0.02\pi$. Therefore,

$$1.616 + 4.938 + 0.384 + 8.16R_L = 18.3 \text{ ns} \qquad (19\text{-}7)$$

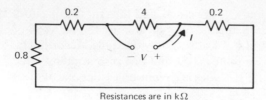

Resistances are in kΩ

Figure 19-7 Network for determining R_{2o} facing $0.5C_\pi$.

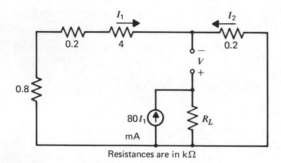

Resistances are in kΩ

Figure 19-8 Network for determining R_{3o} facing $C_{\mu 2}$.

Solving gives $R_L = 1.39$ kΩ. This is the maximum load resistance that can be used for a BW not less than 10 MHz. The corresponding midband voltage gain V_o/V_s is found from (19-3) to be 21.

NETWORK ANALYSIS

To check the design we shall analyze the network of Fig. 19-5. The gain function has the form

$$\frac{V_o}{V_s} = \frac{A_o(1 - s/z_1)(1 - s/z_2)}{(1 - s/p_1)(1 - s/p_2)(1 - s/p_3)} \tag{19-8}$$

The midband gain A_o is 21. We need to find the three poles and the two finite zeros, from which the bandwidth can be calculated.

The network of Fig. 19-5 is redrawn in Fig. 19-9 in a form suitable for nodal analysis. The input voltage source has been converted to a current source by means of Norton's theorem, and admittances are given. The node equations in V, mA, mmhos, and ns are

$$V_s = (1.25 + 7.35s)V_1 - (0.25 + 5.35s)V_2 + 0$$
$$0 = -(0.25 + 5.35s)V_1 + (5.25 + 7.35s)V_2 - 2sV_o$$
$$0 = -20V_1 + (20 - 2s)V_2 + (0.719 + 2s)V_o$$

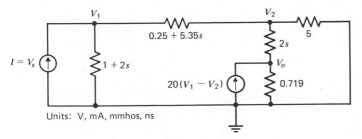

Units: V, mA, mmhos, ns

Figure 19-9   Incremental model with admittances.

From these the poles are found to be -0.0632, -0.427, and -8.1 ns^{-1}, and the zeros are complex conjugates with values of $-1.9 \pm j2.4$ ns^{-1}. The largest pole and both zeros are so large in magnitude compared with the dominant pole that they have no appreciable effect on the bandwidth. Treating them as infinite, the gain function of (19-8), with s in ns^{-1}, becomes

$$\frac{V_o}{V_s} = \frac{21}{(1 + s/0.0632)(1 + s/0.427)} \tag{19-9}$$

At the upper 3 dB radian frequency ω_2 the magnitude of the denominator of (19-9), with s replaced by $j\omega_2$, equals $\sqrt{2}$. Therefore,

$$\left[1 + \left(\frac{\omega_2}{0.0632}\right)^2\right]\left[1 + \left(\frac{\omega_2}{0.427}\right)^2\right] = 2 \tag{19-10}$$

Solving for ω_2 and dividing by 2π, we find that the bandwidth is 9.9 MHz for the hybrid-π values that were used in the calculations. Because we assumed the minimum f_T and the maximum C_μ for each transistor, the actual bandwidth might be substantially greater.

THE DIRECT CURRENT SOURCE

To complete the design we need to provide a suitable source I_{dc} for the network of Fig. 19-1. We could use a large resistor in series with a battery, but a more satisfactory method is to employ a transistor that is biased to give the desired current. In Fig. 19-10a is shown transistor Q_3 in a network nearly equivalent to an ideal current source, and its two-battery equivalent circuit is shown in Fig. 19-10b. The collector goes to the coupled emitters of Fig. 19-1. These emitters have a quiescent voltage with respect to ground of approximately -0.7 V, because each $I_B R_s$ voltage drop in the network of Fig. 19-1 is very small.

Each stage of the CC-CB amplifier is biased at a collector current of 1 mA and β_o is 80. Assuming the dc beta is also 80, the sum of the emitter currents is

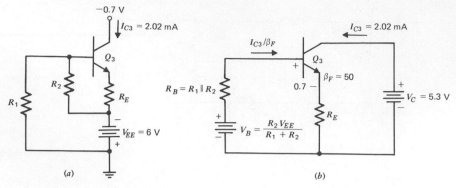

Figure 19-10   Direct-current source. (a) Network. (b) Two-battery equivalent circuit.

2.02 mA, and this is I_{C3}. For good stabilization of I_{C3} the voltage V_{EE} should be as large as possible. Let us assume that a 6 V supply is available for V_{EE}, which makes V_C of Fig. 19-10b equal to 5.3 V. The collector-emitter voltage of Q_3 must be greater than a few tenths of a volt to avoid the saturation region. However, the larger we make V_{CE} the smaller is $C_{\mu 3}$. Selecting 2 V for V_{CE} leaves 3.3 V for R_E. Assuming Q_3 has a dc beta of 50, the current through R_E is 1.02 times I_{C3}, or 2.06 mA, which requires that R_E be 1.60 kΩ.

Application of Kirchhoff's voltage law to the base-emitter circuit gives the equation

$$V_B = 0.04R_B + 0.7 + 3.3 \qquad (19\text{-}11)$$

with R_B in kΩ. For good stability R_B should be small compared with $\beta_F R_E$. An excessively small R_B heavily loads the battery V_{EE}. Let us select 4 kΩ for R_B, which fixes V_B at 4.2 V. Both R_1 and R_2 are found from (13-6), with V_{CC} replaced with V_{EE}, to be 5.7 and 13.3 kΩ, respectively.

The circuit that has been designed is shown in Fig. 19-11. The 800-ohm resistor in the base circuit of Q_2 matches the source resistance and thus tends to maintain dc balance. With each emitter at -0.7 V, the collector-emitter voltages of Q_1 and Q_2 are 6.7 and 5.3 V, respectively, assuming I_{C2} is 1 mA. These unequal voltages cause an imbalance in the currents, but this imbalance is small because the collector current of a BJT in the active region is nearly independent of V_{CE}.

The incremental resistance between the collector of Q_3 and ground is very large. At low frequencies this resistance is 720 kΩ for assumed values of 0.05, 100, and 5000 kΩ for r_x, r_o, and r_μ, respectively. If the current source provided by Q_3 were replaced with a 720 kΩ resistor in series with a battery, the battery voltage would have to be nearly 1500 V. At the upper 3 dB frequency of 10 MHz the incremental impedance of Q_3 is much lower, due primarily to the capacitance

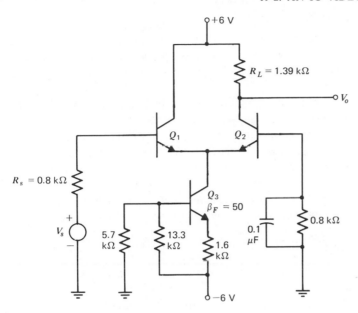

Figure 19-11   The complete network.

$C_{\mu 3}$ that supplies signal current from the collector of Q_3 to the resistors of the base-emitter circuit, and the signal voltage across these resistors is amplified. The effect is substantially reduced by connecting a microfarad bypass capacitor from the base of Q_3 to ground. This passes the small current that flows through $C_{\mu 3}$ directly to ground. With the capacitor present, which places Q_3 in the CB configuration, the resistance is increased from 720 kΩ to over 2 MΩ at moderate frequencies (see P19-4), and the impedance at 10 MHz is approximately that of $C_{\mu 3}$ (see P19-5).

When time constants are used in the design of either the low-frequency or high-frequency circuit of an amplifier containing a direct-current source such as that of Fig. 19-10, it is best to disregard the time constants associated with device and bypass capacitors of the current source. As long as the incremental impedance is sufficiently large to justify the assumption of an ideal current generator, the *net effect* of the poles and zeros contributed by the source to the amplifier gain function is negligible.

19-2. AN IC VIDEO AMPLIFIER

A *video* amplifier is one designed to amplify signals over a wide bandwidth, often from dc to at least 10 MHz, and perhaps to 200 MHz or more, with a voltage gain typically about 100, or 40 dB. The input stage is most likely an emitter-coupled differential amplifier with a direct-current source transistor, and there is usually a differential output. In most amplifiers the output voltage swing is restricted to only a few volts. Because loads are generally capacitive, a low output impedance is desired, and emitter-follower output stages are common.

In addition to general-purpose video and pulse amplification, IC video amplifiers have many specialized applications. To mention a few, they find extensive use in radar data processing, as components in LC tuned feedback oscillators and crystal oscillators at frequencies of 50 MHz or more, in high-frequency variable-phase-shift networks, in fast signal-shaping circuits that utilize the amplifier as a limiter, and in conjunction with JFET inputs to give high-input-impedance wideband amplifiers. We shall examine in this section a representative IC video amplifier—the SN7510.

THE INTEGRATED CIRCUIT

Figure 19-12 is the schematic diagram[1] of the IC. The pins used for external connections are identified in the top view of the package diagram, illustrated in Fig. 19-13. The first stage, consisting of transistors Q_1 and Q_2 and associated circuitry, is an emitter-coupled differential amplifier with each output fed into a single-ended common-collector stage. The signals at the emitters of Q_3 and Q_4 are then amplified by a second differential amplifier with transistors Q_5 and Q_6, and again each output is supplied to a single-ended CC stage. Basically, the IC is a two-stage differential amplifier with the stages separated by emitter followers that prevent interaction. The emitter followers associated with transistors Q_7 and Q_8 provide low output impedances (see P19-9) that enable the devices to drive capacitive loads at high frequencies.

Let us now examine the dc circuit. In Sec. 13-1 it was noted that, because the quiescent conditions are not very dependent on β_F and V_{BE}, a good procedure when these are unknown is to assume typical values. A value of 50 was suggested for β_F and 0.7 V for each V_{BE} and each diode voltage, and these values are assumed here. As the stages are direct-coupled and interacting, several simultaneous equations are required. It is convenient to write the equations in terms of the collector currents. This can be done by noting that each base current is $0.02I_C$ and each emitter current is $-1.02I_C$. With the two inputs grounded, three loop equations relating I_{C1}, I_{C3}, and I_{C5}, in units of kΩ, mA, and V, are easily found to be

$$5.3 = 2.06I_{C1} + 0.8368I_{C3} - 0.0156I_{C5}$$
$$0.7 = 1.055I_{C1} + 0 - 0.7548I_{C5} \qquad (19\text{-}12)$$
$$3.9 = -0.0151I_{C1} - 0.7956I_{C3} + 1.775I_{C5}$$

With reference to Fig. 19-12 the first of Equations 19-12 is around the loop from ground to $+V_{CC}$ and through R_1, R_2, and R_3. The current through R_2

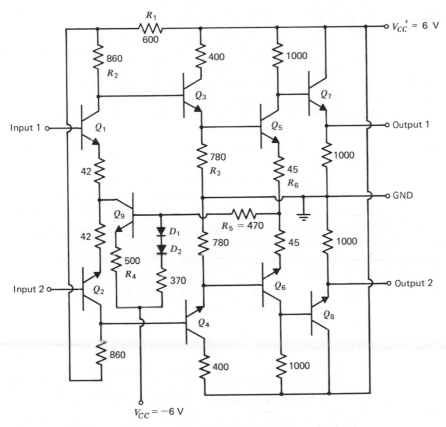

Figure 19-12   Schematic diagram of the SN7510. (Resistances are in ohms)

is $I_{C1} + 0.02I_{C3}$, and twice this current flows through R_1. That through R_3 is $1.02I_{C3} - 0.02I_{C5}$. The second of (19-12) is around the base-emitter circuit of transistor Q_9. The current through R_4 is $1.02 \times 2.04I_{C1}$ and that through the 370-ohm resistor is $2.04(I_{C5} - 0.02I_{C1})$. The third equation is around the loop from ground through R_3, R_6, R_5, and the diodes to $-V_{CC}$. Equations 19-12 are easily solved. From the solution and an additional loop equation appropriately

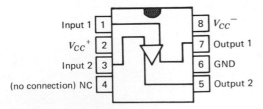

Figure 19-13   Package diagram (top view).

selected, we can determine I_{C7}. The four quiescent currents are

$$I_{C1} = 2.40 \text{ mA} \qquad I_{C3} = 0.47 \text{ mA}$$
$$I_{C5} = 2.43 \text{ mA} \qquad I_{C7} = 2.76 \text{ mA} \tag{19-13}$$

From the known currents and the schematic diagram, the collector-emitter voltages are readily found (see P19-6).

Collector currents of both Q_1 and Q_2 pass through R_1, which serves to drop the voltage supplied to the first differential stage to a value suitable for proper bias conditions. Because the incremental currents from transistors Q_1 and Q_2 flowing through R_1 are equal in magnitude and opposite in sign, a bypass capacitor is unnecessary. Resistor R_1 could be eliminated by adding 1200 ohms to each of the 860-ohm resistors. Although this would not affect the bias conditions and would increase the gain, the bandwidth would suffer. The diodes of the network have voltages that decrease as the temperature rises, partially compensating for changes in the base-emitter voltages. In particular, they tend to maintain the currents of transistors Q_9, Q_1, and Q_2 fairly constant (see P19-10). Note that the buffers associated with Q_3 and Q_4 have low bias currents for high input impedance, whereas the output emitter followers have relatively large bias currents for low output impedance.

VOLTAGE GAIN

The small-signal voltage gain is readily determined from the incremental network. In Fig. 19-14 the IC is represented by the conventional block diagram, the source resistances are 50 ohms, the load resistances are 5 kΩ, and operation is single-ended. We shall assume that each β_o is 50 and that each r_x is 50 ohms. For the first stage, $I_{C1} = I_{C2} = 2.40$ mA by (19-13), and for $q/kT = 40$ the transconductances are 96 mmhos. From g_m and β_o we find r_π to be 520 ohms. At 100 kHz the incremental model is that of Fig. 19-15, with units of kΩ, mA, and V. The loading effects of Q_3 and Q_4 are negligible because the input impedances of the CC stages are very large. From the model we find that V_{b3}/V_s is 7.78.

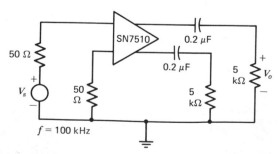

Figure 19-14 Single-ended configuration.

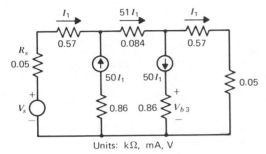

Figure 19-15 First-stage model (transistors Q_1, Q_2).

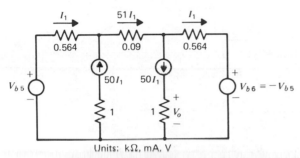

Figure 19-16 Second-stage model (transistors Q_5, Q_6).

The output of the first stage is fed into a CC amplifier with a voltage gain that is nearly unity. The second differential stage, with transistors Q_5 and Q_6, has two input voltages that are equal in magnitude and opposite in phase. As the collector currents are 2.43 mA from (19-13), r_π is 514 ohms, and the sum of r_x and r_π is 564 ohms. From the model of Fig. 19-16 the gain V_o/V_{b5} is found to be 17.5. Because $V_{b3} \approx V_{b5}$, the approximate overall gain V_o/V_s is the product of the gains of the two differential stages, which is 136, or 43 dB. According to the manufacturer's data sheet, the voltage gain is generally between 60 and 120, with a typical value of 90, or 39 dB.

The CC amplifiers between the two differential stages provide sufficient isolation to justify the analysis of each differential stage independently of the other. Furthermore, the CC amplifiers at the output isolate the second differential amplifier from the load. Because the isolation is effective at high frequencies as well, the dominant poles of the entire circuit are simply those of each differential amplifier.

THE EMITTER RESISTORS

It has been mentioned that the emitter resistors of an emitter-coupled amplifier provide both dc and ac feedback. To show how they contribute to equalization of the direct currents, let us consider a numerical example.

Example 1

The partial circuit of Fig. 19-17 has two emitter resistors R, and the base input terminals are connected to ground. Because of a mismatch the active-mode Ebers–Moll equations for the two transistors have slightly different values of I_{ES}, and the direct currents I_1 and I_2 of Fig. 19-17 are

$$I_1 = 40 \times 10^{-14}e^{40V_{BE1}} \qquad I_2 = 35.5 \times 10^{-14}e^{40V_{BE2}} \qquad (19\text{-}14)$$

with $q/kT = 40$ and with the currents in mA. Calculate I_1 and I_2 with $R = 0$. Then determine the value of R required to make $I_1 = 1.02$ and $I_2 = 0.98$ mA.

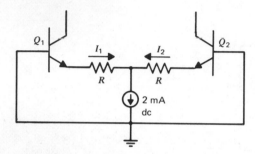

Figure 19-17 Partial circuit with emitter resistors.

Solution

Using kΩ, mA, and V, the node and loop equations of the network of Fig. 19-17 are

$$I_1 + I_2 = 2 \qquad V_{BE1} - V_{BE2} = R(I_2 - I_1) \qquad (19\text{-}15)$$

For $R = 0$ the voltages V_{BE1} and V_{BE2} are equal according to (19-15), and from (19-14) and (19-15) we find these voltages to be 0.715. Both I_1 and I_2 are found from (19-14) to be 1.06 and 0.94 mA, respectively.

For $I_1 = 1.02$ and $I_2 = 0.98$ mA, V_{BE1} is 0.714 and V_{BE2} is 0.716 by (19-14). From these and (19-15) the required value of R is 50 ohms. The results show that the emitter currents differ by 12% with $R = 0$, whereas the difference is only 4% when 50-ohm resistors are present.

The emitter resistors also improve the bandwidth, raise the input impedance, and aid in stabilizing the incremental performance. For example, if the ac betas increase because of a rise in temperature, the incremental voltage across the emitter resistors rises, increasing the input resistance and reducing the input current. This tends to counteract the effect of the increased beta, as demonstrated in the following example.

Example 2

Using the network of Fig. 19-15, calculate the percentage increase in the gain V_{b3}/V_s when the β_o of the transistors increases from 50 to 60 because of a rise in temperature. Then repeat the calculation with the 0.084 kΩ resistance removed.

Solution

With $\beta_o = 50$ the voltage gain is 7.78, and it is 8.11 with $\beta_o = 60$. The increase is 4.2%. With the 0.084 kΩ resistor removed, the voltage gain is 35 for a beta of 50 and 42 for a beta of 60, for an increase of 20%. The stabilization is considerably improved by the emitter resistors, but the loss of gain is substantial.

In Sec. 14-1 it was shown that the emitter resistors enhance the linearity of the dynamic transfer characteristic, thereby extending the useful dynamic range. This was demonstrated in the solution of P14-5. Other advantages of signal feedback are discussed in greater detail in Chapters 20 and 21. In the IC video amplifier SN7510 small emitter resistances are included in both differential stages.

The operating frequency range of the SN7510 is from dc to 100 MHz or more, and the bandwidth is typically 40 MHz. Input and output resistances are about 6000 ohms and 35 ohms, respectively. The maximum common-mode input voltage is ± 1 V, and the CMRR is approximately 80 dB. In single-ended operation with a 5 kΩ load the maximum peak-to-peak output is around 4.5 V, and with the output adjusted to one volt rms the harmonic distortion has a typical value of 2%. When the inputs are connected to ground, a small dc voltage of perhaps several tenths of a volt or more is present between outputs 1 and 2. This is called the differential output offset voltage V_{DO}.

PULSE RESPONSE

Video amplifiers are commonly used to amplify high-frequency pulses, and the rise and fall times are of particular importance. A pulse input can be regarded as the sum of two step functions, one being a positive step at time zero and the other being a negative step of equal magnitude applied at time T, with T denoting the pulse width. Let us consider a video amplifier that has a high-frequency dominant pole which is a negative real number much smaller in magnitude than any other pole or zero. For this case the gain V_o/V_s has the form $A_o/(1 + s/\omega_2)$, with A_o denoting the midfrequency gain and with ω_2 denoting the upper 3-dB frequency, which equals the magnitude of the dominant pole. If $v_s(t)$ is a unit step function, the Laplace transform V_s is $1/s$, and the s-domain output voltage is

$$V_o = \frac{A_o}{s(1 + s/\omega_2)} = A_o\left(\frac{1}{s} - \frac{1}{s + \omega_2}\right) \qquad (19\text{-}16)$$

From this we find the time-domain output $v_o(t)$ to be, for $t > 0$,

$$v_o(t) = A_o(1 - e^{-\omega_2 t}) \tag{19-17}$$

We note that $v_o(t)$ rises exponentially from zero at time zero to A_o at time infinity. Let t_1 and t_2 be the times at which $v_o(t)$ is 10% and 90%, respectively, of its final value. Then

$$1 - e^{-\omega_2 t_1} = 0.1 \qquad 1 - e^{-\omega_2 t_2} = 0.9 \tag{19-18}$$

From these the *rise time* $t_2 - t_1$ is found to be

$$t_2 - t_1 = \frac{\ln 9}{\omega_2} \approx \frac{2.2}{\omega_2} \tag{19-19}$$

Note the close relationship between the rise time and the bandwidth.

For the SN7510 the BW is typically 40 MHz, which corresponds to a frequency ω_2 of 0.25 reciprocal nanosecond. By (19-19) the rise time is determined to be 8.8 ns. The fall time, which is defined similarly, should be about the same. This numerical value agrees closely with the data sheet, which specifies typical rise and fall times of 10 ns. The response of the SN7510 to a rectangular input pulse is shown in Fig. 19-18.

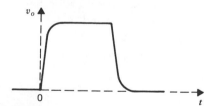

Figure 19-18 Pulse response of video amplifier.

Many applications of video amplifiers require a very flat frequency response and a near-zero phase shift over the entire pass band. Because multistage feedback is likely to produce peaking and phase shifts, such feedback is usually avoided in the integrated circuits. If feedback is desired, it can of course be applied externally.

IC video amplifiers are quite similar to operational amplifiers. The primary differences between the two are the bandwidth and gain. In general, compensated operational amplifiers have substantially lower bandwidths, typically around 10 to 100 Hz, and much higher voltage gains, perhaps 100 dB or more. The input stage often consists of emitter followers or FETs in order to obtain high impedance, and operational amplifiers are normally used in conjunction with multistage feedback, which increases the BW to perhaps 100 kHz more or less.

19-3. THE CASCODE AMPLIFIER

In the network of Fig. 19-19 the two transistors are said to be in *cascode*, with the collector of Q_1 connected in series with the emitter of Q_2. A similar FET arrangement was encountered in Sec. 17-2. The first stage is a common-emitter amplifier, which is followed by a common-base output stage. As we shall see, the CE-CB configuration has certain very desirable properties.

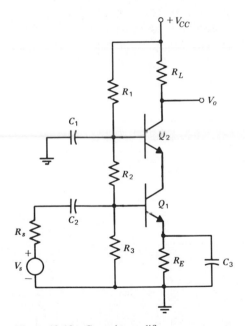

Figure 19-19 Cascode amplifier.

The bias circuit is easily designed, with resistor R_E serving to stabilize the current. Resistances R_1, R_2, and R_3 are selected to give suitable bias voltages (see P19-12) along with a reasonable value of effective base resistance R_B for the input stage. In the incremental circuit R_2 and R_3 shunt the input and R_1 is absent. Coupling and bypass capacitors are chosen to provide the desired low-frequency response (P19-13). The quiescent currents of integrated circuits are normally fixed by a current source.

SMALL-SIGNAL ANALYSIS

With R_2 and R_3 assumed to be large enough to be negligible, the high-frequency incremental network is that of Fig. 19-20. Because there are four capacitors and no capacitor loops, we expect four high-frequency poles and three finite zeros plus one at infinity due to C_π of the input transistor. Using units of mA, kΩ, mmhos, V, pF, ns, and GHz unless otherwise stated, with a quiescent current of 1 mA and identical transistors, let us assume that the following parameters apply:

$$R_s = 1 \qquad \beta = 100 \qquad r_x = 0.1 \qquad C_\mu = 4$$
$$R_L = 2 \qquad g_m = 40 \qquad r_\pi = 2.5 \qquad C_\pi = 100 \tag{19-20}$$

The midband network is easily analyzed. The first stage has a voltage gain V_a/V_s of -0.715 and a current gain I_a/I_1 of -100. The unusually low voltage gain is a consequence of the very small effective load resistance, which is the input resistance (26 ohms) of the CB stage. The second stage has a voltage gain of 76.9 and a current gain of 0.99, values which are reasonable for a CB amplifier.

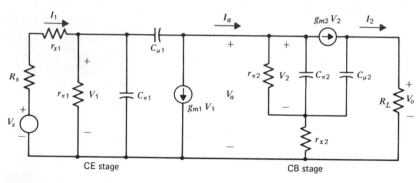

Figure 19-20   Incremental model of cascode amplifier.

Overall, the voltage and current gains are -55 and -99, respectively. It is interesting to note that similar gains would be obtained from the CE stage terminated in a load of 2 kΩ with the CB stage eliminated. Thus, the midband performance of a BJT cascode amplifier is about the same as that of a typical one-stage CE amplifier.

The upper break frequency is easily estimated by assuming that the input impedance of the CB stage at this frequency is approximately the same 26 ohms of the midband region. This gives a value of 1.04 for the product of g_m and the effective load resistance. Accordingly, the input capacitance of the first stage is the sum of C_π and $2.04C_\mu$, or 108 pF. Note that *there is very little Miller-effect multiplication of C_μ because of the small effective load*. The equivalent resistance in shunt with this capacitance is 0.764 kΩ, determined from $(R_s + r_x)$ in parallel with r_π. The time constant is 82.5 ns and its reciprocal is 0.0121. Using this value for ω_2, we find the bandwidth to be 1.93 MHz.

It is easy to justify the assumption that the capacitances of the CB amplifier are reasonably negligible at the break frequency. We note from the incremental model that, because r_{x2} is quite small, $C_{\pi 2}$ is almost in shunt with the very low input resistance of the second stage. Consequently, its effect is negligible. Also, $C_{\mu 2}$ is approximately in shunt with R_L, and the combination gives a time constant of 8 ns and a pole at about -0.125 ns^{-1}. Although this is a rough approximation, the pole is considerably greater in magnitude than ω_2, which indicates that the bandwidth is determined mainly by the CE stage, as assumed.

An exact analysis of the incremental model reveals that the poles are -0.0124, -0.10, -0.40, and -3.26 ns^{-1}. Two zeros are $0.20 \pm j0.98$ ns^{-1}, and the third is 10 ns^{-1}. These singularities give a bandwidth of 1.94 MHz, which agrees closely with our estimate.

ADVANTAGES OF CASCODE AMPLIFIERS

Now suppose that the CB stage is replaced with a 2 kΩ resistor so that the CE stage has the same gain as the cascode amplifier. The Miller-effect multiplication of C_μ is $1 + g_m R_L$, or 81, and the time constant associated with the effective input capacitance is 324 ns, giving a bandwidth of 491 kHz. A major advantage of the cascode circuit is readily apparent. In comparison with a single-stage CE amplifier with the same midband gain, the cascode network has a substantially higher break frequency. For the numerical example considered, the bandwidth is greater by a factor of 4. When used in an integrated circuit, as illustrated in the differential amplifier of Fig. 19-22, the cost is only an additional transistor for each cascode, and IC transistors are cheap.

In addition to the larger gain-bandwidth product, an advantage is that the output transistor can operate with a much larger collector voltage, permitting a greater output voltage swing. The collector voltage of a CB stage is limited by the avalanche voltage of the collector junction, whereas the collector voltage of a CE stage is limited by the smaller sustaining voltage, discussed in Sec. 2-6.

BJT DIFFERENTIAL AMPLIFIERS WITH HIGH INPUT IMPEDANCE

Many applications of differential amplifiers require an extremely small input offset current along with a high input impedance. An obvious solution that is often employed is to add emitter-follower input stages to the differential amplifier and to bias these stages at very low current levels, which unfortunately reduces the gain. Such a network is shown in Fig. 19-21. Transistors Q_2 and Q_5 constitute an emitter-coupled differential pair. Emitter followers are used at the inputs and outputs, and current mirrors supply the bias currents.

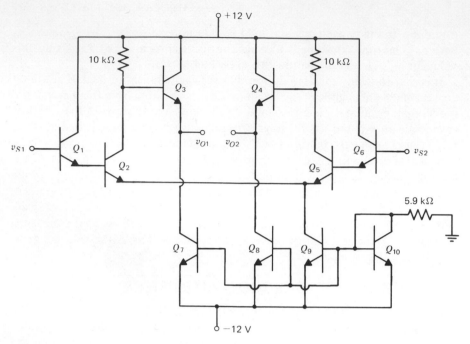

Figure 19-21   Differential amplifier with emitter-follower inputs.

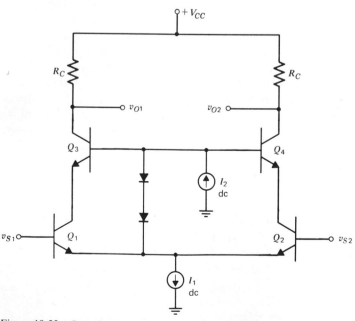

Figure 19-22   Cascode differential amplifier.

638

Another method is to use the cascode differential amplifier of Fig. 19-22 with *supergain*, or *super-beta*, transistors at the two inputs. These are IC devices with extremely thin base regions, having a base width of 0.1 μm or less. The very thin base gives a beta in excess of 2000. With an input current of 1 nA and a beta of 2000 the collector current is 2 μA, which is adequate for obtaining appreciable gain. However, because of the thin base, punch-through occurs at collector-base voltages of the order of a volt, and base-width modulation is quite large. Accordingly, super-gain transistors must be operated with the collector-base bias voltage at about zero and with low voltage gain. The cascode network of Fig. 19-22 is designed to satisfy these requirements. Because R_C must be large, perhaps a megohm or more, an IC *pinch resistor* might be used. This consists of a lightly doped p-channel surrounded by an N-type region with the junction reverse biased. Reasonable gain is possible along with input currents less than a nanoampere (P19-11). Note that the collector-base bias voltages of Q_1 and Q_2 are zero.

REFERENCES

1. *The Linear and Interface Circuits Data Book*, Texas Instruments Inc., Dallas, Texas, 1973.
2. P. E. Gray and C. L. Searle, *Electronic Principles*, Wiley, New York, 1969.
3. V. H. Grinich and H. G. Jackson, *Introduction to Integrated Circuits*, McGraw-Hill, New York, 1975.

PROBLEMS

Section 19-1

19-1. For the CC–CB amplifier designed in Sec. 19-1 we found that a load resistor of 1.39 kΩ gave a BW of 10 MHz and a midband gain V_o/V_s of 21. Using the same network with the same transistor parameters, calculate the midband voltage gain when R_L is increased just enough to reduce the BW to 5 MHz.

19-2. Suppose the emitter-coupled amplifier of Fig. 19-1 has a 50 ohm potentiometer connected between the emitters of Q_1 and Q_2, with the variable tap connected to the direct-current source, for the purpose of controlling the division of current. The effect of the incremental model is to add 50 ohms between the emitters of Figs. 19-1, 19-3, and 19-4. With $R_s = 800$ ohms and $R_L = 1.39$ kΩ, and with the same transistor parameters used in Sec. 19-1, calculate (a) the midband gain V_o/V_s, (b) the four open-circuit time constants in nanoseconds, and (c) the bandwidth in MHz, using (18-62). [*Ans*: (c) 14.2; 14.9 from computer analysis]

19-3. For the emitter-coupled amplifier of Fig. 19-1 with the high-frequency model of Fig. 19-5, each of the identical transistors has $r_x = 200$ ohms, $g_m = 100$ mmhos, $\beta_o = 100$, and $C_\mu = 3$ pF. The source resistance is 800 ohms. A midband gain V_o/V_s of 40 and a BW of 3 MHz are desired. Determine the minimum required transition frequency f_T in MHz. Use (18-62).

19-4. The transistor in the direct-current source of Fig. 19-11 has a collector current of 2 mA. At room temperature g_m is 80 mmhos, and both the dc and ac betas are 50. Assume r_x is 50 ohms, r_o is 100 kΩ, and r_μ is 5 MΩ. (a) Calculate the incremental resistance in kΩ between collector and ground. (b) Repeat (a) but with a bypass capacitor that incrementally connects the base to ground. The nodal method of analysis is suggested.

19-5. The transistor in the direct-current source of Fig. 19-11 has the hybrid-π parameters of P19-4 and C_π and C_μ are 20 pF and 2 pF, respectively. With the base connected to ground through a microfarad bypass capacitor, calculate the incremental impedance in kΩ between collector and ground at 10 MHz. Compare the result with the impedance of C_μ. Neglect r_x, which has very little effect. (Ans: $-j7.9$)

Section 19-2

19-6. Using the currents of (19-13), calculate V_{CE1}, V_{CE3}, V_{CE5}, V_{CE7}, and V_{CE9} for the network of Fig. 19-12 with the two inputs grounded. The dc beta is 50 and each V_{BE} is 0.7 V.

19-7. For the SN7510 having the schematic diagram of Fig. 19-12 and with both inputs connected to ground, calculate the total power dissipation in mW. Use the currents of (19-13). The dc beta is 50.

19-8. Using the network of Fig. 19-16, calculate the percent increase in the gain V_o/V_{b5} when the β_o of the transistors increases from 50 to 60 because of a rise in temperature. Assume resistance values are unchanged. Then repeat with the 0.09 kΩ resistance removed. Also, calculate the ratio of the voltage gain with the emitter resistors removed to that with the resistors present, assuming β_o is 50. (Ans: 3.5, 20, 5.1)

19-9. For the SN7510, with the schematic diagram of Fig. 19-12, the inputs are grounded and the quiescent currents are those of (19-13). Each transistor has an ac beta of 50 and a base resistance r_x of 50 ohms. Assume $q/kT = 40$ and calculate the output impedance in ohms, at 100 kHz, between output 1 and ground.

19-10. For the SN7510 with the inputs connected to ground calculate the quiescent collector currents in mA of transistors Q_1, Q_3, Q_5, and Q_7 at an elevated temperature that makes each diode voltage and each V_{BE} equal to 0.4 V. Use the diagram of Fig. 19-12 with $\beta_F = 50$. Compare the results with those of (19-13).

Section 19-3

19-11. The cascode differential amplifier of Fig. 19-22 has supergain transistors at the inputs, with dc and ac betas of 2000. The collector currents are 2 μA, giving $g_m = 0.08$ mmho for each active device. Transistors Q_3 and Q_4 have betas of 80, and 1 MΩ pinch resistors are used for R_C. Assuming r_x, the incremental resistances of the diodes, and the base-width-modulation parameters are negligible, calculate the gain v_{o1}/v_{S1} with $v_{S2} = 0$.

19-12. The bias circuit of the cascode amplifier of Fig. 19-19 is to be designed for a 5 mA collector current for the input stage. Voltage V_{CC} is 9 V and R_L is 400 ohms. Select R_3 equal to $20R_E$ with R_E chosen for a 1 V drop. Assuming V_{BE} is 0.7 and beta is 50 for each transistor, determine R_1 and R_2 in kΩ so that the transistors have identical collector-emitter voltages.

19-13. The capacitors C_1, C_2, and C_3 of the cascode amplifier of Fig. 19-19 are to be selected to give a lower break frequency of 150 Hz. For a quiescent current of 5 mA, each transistor has $r_\pi = 500$ ohms and $\beta_o = 100$. Assume r_x is negligible. Resistance R_E is 200 ohms, R_s and R_L are each 400 ohms, and R_1, R_2, and R_3 are 8, 6, and 4 kΩ, respectively. Determine C_1, C_2, and C_3 in μF, with these capacitances selected to give identical short-circuit time constants for C_1 and C_2 and a time constant for C_3 that is 10% of each of the others. First, draw the complete low-frequency model without crossovers, placing the base of Q_2 at the extreme right side. (*Ans*: $C_3 = 138$)

19-14. For the cascode amplifier of Fig. 19-19 with the incremental model of Fig. 19-20 and the parameters of (19-20), calculate the open-circuit time constants in ns associated with $C_{\pi 1}$, $C_{\mu 1}$, $C_{\pi 2}$, and $C_{\mu 2}$. From these and (18-62), estimate the upper 3 dB frequency. (*Ans*: 1.96 MHz)

19-15. The amplifier of Fig. 19-23 employs Darlington pairs to obtain low input bias currents and high input resistance. Assume the dc and ac betas of each transistor are 100, each base-emitter voltage is 0.6, q/kT is 40, and r_x is negligible. (a) Calculate V_{CE2} and I_{B1} in nA. (b) Find the voltage gain and the input resistance in MΩ.

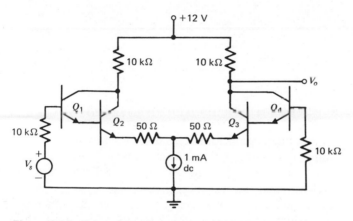

Figure 19-23   Compound differential amplifier for P19-15.

Chapter Twenty
Feedback
Amplifiers

We have learned that bias stabilization is accomplished by means of negative feedback. For example, a one-stage CE amplifier usually has an emitter resistor. This introduces feedback, which tends to keep the dc collector current fairly constant even though moderate changes may occur in β_F, T, V_{CC}, or one or more of the network parameters. On several occasions we have encountered direct-coupled amplifiers in which multistage feedback was used to stabilize the quiescent points of the transistors. In a somewhat similar manner feedback can stabilize the incremental performance, making the gain nearly insensitive to variations in the Q-points of the transistors, to fluctuations in the parameters of the active devices, and to changes in the passive components and voltage supplies of the amplifier. In fact, *the gain can be made to depend almost entirely on one or more temperature-insensitive precision resistors.* We have observed this in our study of operational amplifiers and multistage power amplifiers.

Some other advantages are an increased bandwidth, a reduction in nonlinear distortion, an improvement in the signal-to-noise ratio when noise is introduced internally, and some control over the values of the input and output impedances. Although gain is reduced, the advantages are so great that negative feedback is used to some extent in nearly all electronic amplifiers. In this chapter we examine the properties of feedback amplifiers, along with techniques for design and analysis. Problems of instability associated with multistage feedback amplifiers are treated in Chapter 21.

642

20-1. BASIC PRINCIPLES

The signal-flow block diagram of an *idealized* feedback amplifier is illustrated in Fig. 20-1. The input signal x_i and the output x_o may be phasor voltages, or they may be phasor currents, or one may be a voltage and one a current. As we shall see, their definitions are selected according to the particular type of feedback configuration. The arrows on the block diagram indicate that, in the idealized system, forward transmission from input to output occurs through the basic amplifier, whereas reverse transmission is through the feedback network.

Let us assume that operation is in the midfrequency region, with the networks being resistive. The input signal x_i and the signal Bx_o that is fed back through the B network have the same dimensions; that is, they are both voltages or both currents. For negative feedback they are in time phase, with x_i greater than Bx_o. The output of the summing network is their difference $x_i - Bx_o$, which is smaller than x_i. The amplifier output x_o is A times this difference, with A denoting the gain of the basic amplifier. Because A is also the gain of the system

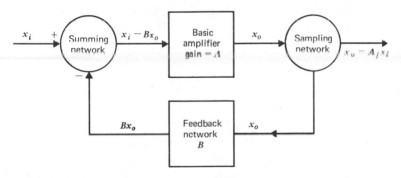

Figure 20-1   Block diagram of idealized feedback amplifier.

when the feedback loop is disconnected, making Bx_o zero, A is referred to as the *open-loop gain*. In contrast, the ratio x_o/x_i of the feedback amplifier is the *closed-loop gain* A_f, with the subscript f signifying the gain with feedback. For this discussion we assume that A depends only on the components within the basic amplifier, including the useful load, and that the *feedback factor* B depends only on the parameters of the feedback network. The sampling network samples the desired output quantity, which is either the output voltage or current, and supplies this signal to the feedback network.

There are four distinct combinations, and each is important. One of these is the case in which both x_i and x_o are voltages. The open-loop gain A, the closed-loop gain A_f, and the feedback factor B are voltage ratios, and the system is often referred to as *voltage-sample voltage-sum*. If both x_i and x_o are currents, the quantities A, A_f, and B are current ratios, and the system is referred to as *current-sample current-sum*. A third possibility is for x_i to denote a current and x_o a voltage. Both A and A_f are transimpedances, with each being a ratio of a voltage to a current, and B is a transadmittance. This is known as *voltage-sample current-sum*. Finally, if x_i is a voltage and x_o a current, A and A_f are transadmittances, B is a transimpedance, and the feedback is *current-sample voltage-sum*.

With reference to Fig. 20-1 it is clear that the open-loop gain A is

$$A = \frac{x_o}{x_i - Bx_o} = \frac{x_o/x_i}{1 - B(x_o/x_i)} = \frac{A_f}{1 - BA_f} \qquad (20\text{-}1)$$

Solving for A_f gives

$$A_f = \frac{A}{1 + AB} \qquad (20\text{-}2)$$

In the midfrequency region the output and both inputs of the summing network of Fig. 20-1 are in time phase when the feedback is negative, or *degenerative*. Thus the ratio of Bx_o to $x_i - Bx_o$ is positive, and this ratio equals the product AB, as indicated by the block diagram. Accordingly, in (20-2) the product AB, which is called the *loop transmission* T, is positive. This makes the closed-loop gain A_f smaller in magnitude than the gain A of the basic amplifier. With the input x_i removed, it is clear from Fig. 20-1 that the loop gain around the closed loop is $-AB$, with the negative sign resulting from inversion of the feedback signal at the summing network. Thus the loop transmission T is defined here as the negative of the loop gain.

Let us suppose that the amplifier is designed so that T is very much greater than unity. For this case (20-2) becomes

$$A_f \approx \frac{1}{B} \qquad (20\text{-}3)$$

The result indicates that the gain of the amplifier depends only on the feedback factor B. If the B network is made of precision resistors with zero temperature coefficients, A_f is a constant, *independent of temperature, transistor parameters, voltage supplies, and amplifier components as long as* T *is large*. By utilizing a lot of feedback, an electronic amplifier becomes a precision network. It is easy to deduce from (20-2) that

$$\frac{dA_f}{A_f} = \left(\frac{1}{1 + AB}\right)\left(\frac{dA}{A}\right) = \left(\frac{1}{1 + T}\right)\left(\frac{dA}{A}\right) \qquad (20\text{-}4)$$

The expression $1 + T$ is called the *desensitivity* of the amplifier.

To illustrate, suppose A is -4950 and B is -0.02, giving $T = 99$, which is large compared with unity. With the aid of (20-3) we note that the gain is reduced by feedback from -4950 to -50. This loss is the cost of a desensitized amplifier.

From (20-4) we find that the fractional change dA_f/A_f in A_f equals 0.01 times the fractional change dA/A in A. Suppose a variation in some quantity of the basic amplifier causes a 10% reduction in the open-loop gain A. The corresponding drop in the closed-loop gain A_f is only 0.1%.

Both A and B are real numbers in the midfrequency range, and if their product is negative, the feedback is said to be *positive*, or *regenerative*. For AB between zero and -1 the closed-loop gain is greater in magnitude than the open-loop gain, as indicated by (20-2). This larger gain is at the expense of increased sensitivity to small changes in the transistor and network parameters, a reduction in bandwidth, greater nonlinear distortion, and degradation in the signal-to-noise ratio. Because of these detrimental effects, positive feedback is seldom introduced intentionally, except in oscillators and certain specialized networks.

When the loop transmission is negative with a magnitude greater than unity, the amplifier becomes unstable and oscillates. To illustrate this effect let us refer to Fig. 20-2 in which A is -100 and B is 0.02, giving $AB = -2$. With $x_i = 0$, a

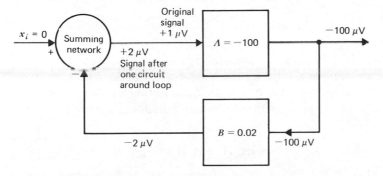

Figure 20-2   An unstable feedback amplifier.

$1\ \mu V$ signal into the basic amplifier produces an output of $-100\ \mu V$. After returning through the B and summing networks the signal into the amplifier is $+2\ \mu V$. This process continues until the signal becomes so large that the nonlinearity of the active devices reduces the gain A to -50, making AB exactly -1. The oscillations are now of constant amplitude, and the closed-loop gain of (20-2) is infinite. The signal initiating the build-up might be the thermal noise of a resistor or the transient disturbance that occurs when the circuit is energized. The frequency of oscillation is that which makes the imaginary part of AB equal to zero. Clearly, too much regenerative feedback results in an unstable amplifier.

Most practical feedback amplifiers can be arranged in the form of the idealized block diagram of Fig. 20-1 only by making suitable approximations. Sometimes either the summing or the sampling network, or perhaps both, are internal to the basic amplifier. The open-loop gain A of the basic amplifier is not completely independent of the feedback network that loads its output. Some forward transmission occurs through the feedback loop, and some reverse transmission takes place through the basic amplifier. However, by utilizing reasonable approximations, one can nearly always organize the actual circuitry into the idealized form. Then by properly defining A_f, A, and B, the feedback equation (20-2) results, and basic feedback concepts can be applied to the design process.

20-2. SOME FEEDBACK EFFECTS

It has been noted that feedback affects the bandwidth, the signal-to-noise ratio, nonlinear distortion, and the input and output impedances. Let us examine each of these briefly.

BANDWIDTH

Suppose the low-frequency incremental model of an amplifier has one pole and one zero, with the zero at $s = 0$. The gain $A(\omega)$ becomes

$$A(\omega) = \frac{A_o}{1 + j\omega_a/\omega} \tag{20-5}$$

with A_o denoting the midfrequency open-loop gain and with ω_a signifying the lower 3 dB radian frequency. If the feedback network is made of resistors, then $B = B_o$, a constant independent of frequency. Substitution for A and B into the feedback equation (20-2) gives, with a little manipulation, the expression

$$A_f = \frac{A_{fo}}{1 + j\omega_a/(1 + A_o B_o)\omega} \tag{20-6}$$

with $A_{fo} = A_o/(1 + A_o B_o)$. The lower 3 dB frequency ω_1 of the feedback amplifier is that value of ω that makes the j-term of (20-6) equal to unity. Hence,

$$\omega_1 = \frac{\omega_a}{1 + A_o B_o} \tag{20-7}$$

If the basic amplifier has $f_a = 30$ Hz, and if the midfrequency loop transmission is 9, the feedback amplifier has a lower 3 dB frequency of 3 Hz.

Assuming the high-frequency model of the basic amplifier has one pole and a zero at infinity, the gain $A(\omega)$ has the form

$$A(\omega) = \frac{A_o}{1 + j\omega/\omega_b} \tag{20-8}$$

With $B = B_o$, a constant independent of frequency, the closed-loop gain is determined from (20-2) to be

$$A_f = \frac{A_{fo}}{1 + j\omega/(1 + A_o B_o)\omega_b} \tag{20-9}$$

The value of ω that makes the j-term unity is ω_2, or

$$\omega_2 = (1 + A_o B_o)\omega_b \tag{20-10}$$

If the basic amplifier has $f_b = 3$ MHz, and if $A_o B_o$ is 9, the feedback amplifier has an upper 3 dB frequency of 30 MHz. The results clearly show that negative feedback increases the bandwidth by reducing the lower 3 dB frequency and raising the upper 3 dB frequency, although these benefits are accompanied with a sacrifice in gain. Because B is independent of frequency, the approximate equation $A_f \approx 1/B$ indicates that A_f is independent of frequency as long as the loop transmission AB is large. This explains the increased bandwidth.

SIGNAL-TO-NOISE RATIO

Shown in Fig. 20-3 is an amplifier with noise present at its input. This noise may be hum introduced by the power supply, pick-up from a nearby circuit, or thermal noise from a resistor. For $A_1 = 10$ the output x_o is

$$x_o = 10x_i + 10x_n \tag{20-11}$$

The first term on the right is the output signal and the second term is the output noise. The signal-to-noise ratio S/N is

$$\frac{S}{N} = \frac{x_i}{x_n} \tag{20-12}$$

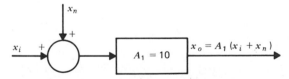

Figure 20-3   Amplifier with noise x_n at its input.

This ratio is the same at both the input and output terminals. Introduction of a feedback loop has no effect on this ratio because both the signal and the noise are reduced the same amount.

Now suppose this amplifier is preceded by a noiseless amplifier having a gain A_o of 100, with a feedback loop added as shown in Fig. 20-4. The output of the first stage is $A_o(x_i - Bx_o)$ and the input to the second stage is the sum of this output and x_n. Therefore,

$$x_o = A_1[x_n + A_o(x_i - Bx_o)] \tag{20-13}$$

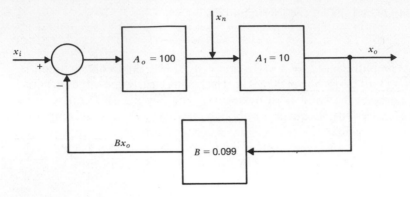

Figure 20-4   Feedback amplifier with internal noise.

Solving for x_o gives

$$x_o = \left(\frac{A_o A_1}{1 + A_o A_1 B}\right) x_i + \left(\frac{A_1}{1 + A_o A_1 B}\right) x_n \qquad (20\text{-}14)$$

For $A_o = 100$, $A_1 = 10$, and $B = 0.099$, this becomes

$$x_o = 10 x_i + 0.1 x_n \qquad (20\text{-}15)$$

Comparison of the result with that of (20-11) reveals the greatly improved signal-to-noise ratio. From (20-14) and (20-15) this ratio is found to be

$$\frac{S}{N} = \frac{A_o x_i}{x_n} = \frac{100 x_i}{x_n} \qquad (20\text{-}16)$$

For the same values of x_i and x_n the output signals of the amplifiers of Figs. 20-3 and 20-4 are identical, but the output noise of the feedback amplifier is only 1% of that of the amplifier of Fig. 20-3. In the feedback amplifier of Fig. 20-4 the noise at the input of the second stage consists of the noise x_n plus the noise out of the first stage which came from the feedback loop. These two noise voltages are out of phase and tend to cancel.

The block diagram of Fig. 20-4 might represent a two-stage power amplifier. The large bias current of the second stage may cause hum to be introduced from an inadequately filtered power supply. This hum is x_n. In the first stage the bias current is lower, and consequently, the power-supply hum is mostly eliminated by a decoupling circuit. Although the addition of the first stage along with the feedback network adds complexity to the circuitry, the reduction in the output noise is considerable, and other benefits of negative feedback are present as well.

NONLINEAR DISTORTION

To illustrate the effect of feedback on nonlinear distortion let us suppose that the output x_o is a nonlinear function of the input x_2 of the basic amplifier in accordance with the relation

$$x_o = 100(x_2 + 0.5x_2{}^2) \qquad \text{for positive } x_2 \qquad (20\text{-}17)$$

The input x_2 is $x_i - Bx_o$ from the block diagram of Fig. 20-1, and the x parameters as used here denote instantaneous values. Replacing x_2 in (20-17) with $x_i - Bx_o$ and solving for x_i, we find that

$$x_i = Bx_o + \sqrt{0.02x_o + 1} - 1 \qquad (20\text{-}18)$$

For any specified feedback factor B this equation can be used to sketch a curve showing the relationship between the input and output quantities.

Two such curves are presented in Fig. 20-5, along with the numerical data obtained from (20-18) and used to sketch the curves. The one on the left side

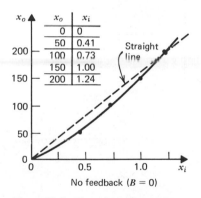

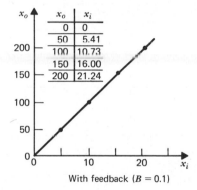

Figure 20-5   Two plots of x_o versus x_i.

corresponds to no feedback ($B = 0$) and the one on the right is for a feedback amplifier with a loop transmission of 10, or $B = 0.10$. Note that the feedback amplifier requires a larger input x_i in order to give the same output. This is the cost of the improved linearity. It is easy to explain the reduction in nonlinear distortion. For a large loop transmission, $A_f \approx 1/B$, and this equation indicates that the gain depends only on the linear B-network.

INPUT AND OUTPUT IMPEDANCES

Earlier it was mentioned that feedback changes the input and output impedances. Each of these is raised or lowered depending on which of the four basic

types of feedback networks is employed. Let us first consider the input impedance. The feedback quantity is either a voltage or a current. If it is a voltage, the
summing network introduces a voltage at the input of the basic amplifier that is
less than the voltage at the input of the feedback amplifier. Because the current
through this network does not change, as indicated in Fig. 20-6a, the input
impedance of the feedback amplifier is greater than that of the basic amplifier.
On the other hand, it the quantity fed back is a current, the current at the input
of the basic amplifier is less than the current at the input of the feedback amplifier.
In this case the voltage is unchanged, as shown in Fig. 20-6b, and the input
impedance of the feedback amplifier is less than that of the basic amplifier.

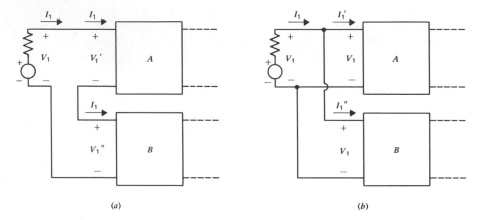

(a)   (b)

Figure 20-6   Voltage-sum and current-sum networks. (a) Voltage sum, Z_i large. (b) Current sum,
Z_i small.

The general effect on the output impedance is easily deduced from the basic
feedback equation. For convenience let us assume that the loop transmission is
sufficiently large to justify use of the approximate equation $A_f \approx 1/B$, or $x_o \approx$
x_i/B. Suppose the input x_i is maintained constant and the load resistance is
varied. The output x_o is constant as long as the loop gain is large. For current
sampling, as illustrated in Fig. 20-7a, the output x_o is a current; thus the amplifier
output simulates an ideal current source with an infinite output impedance.
Because the output current is not absolutely constant, the output impedance is
not infinite, but it is larger than that of the basic amplifier.

For voltage sampling, as shown in Fig. 20-7b, x_o is a voltage. Consequently,
the amplifier simulates an ideal voltage source with zero output impedance.
Actually, Z_o is small, perhaps much smaller than that of the basic amplifier.

Examination of the summing and sampling networks reveals that an input
series connection of the amplifier and feedback two-ports, as in Fig. 20-6a,

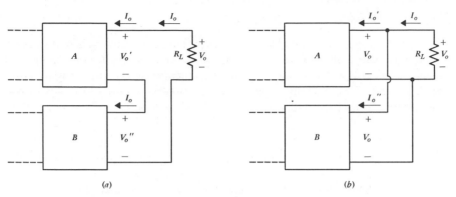

Figure 20-7 Current-sampling and voltage-sampling networks. (a) Current sampling, Z_o large. (b) Voltage sampling, Z_o small.

raises the input impedance, and a series connection at the output, as in Fig. 20-7a, *raises the output impedance*. The *parallel* connections of Figs. 20-6b and 20-7b *lower the impedances*. Equations for calculating the input and output impedances are presented in later sections.

20-3. VOLTAGE-SAMPLE CURRENT-SUM FEEDBACK

When the basic amplifier has a phase shift of 180° in the midfrequency range, as in a one- or three-stage CE amplifier or an inverting OP AMP, the type of feedback most often used is voltage-sample current-sum. This is sometimes called *parallel-parallel* feedback, because the terminals of the amplifier and the feedback network are connected in parallel at both the input and output, as shown in Fig. 20-8.

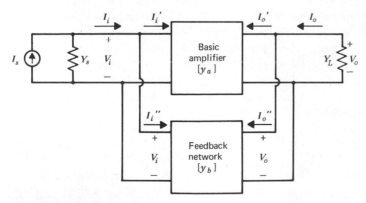

Figure 20-8 Block diagram of parallel-parallel feedback network.

THE y-PARAMETER MODEL

In Fig. 20-8 a Norton equivalent circuit of the source is employed. The symbol $[y_a]$ denotes the matrix of the y-parameters of the basic amplifier, and $[y_b]$ signifies the corresponding matrix of the feedback network. It is easy to show (see P20-4) that the composite y-parameters of two-port networks in a parallel-parallel arrangement are determined by adding the respective y-parameters; that is,

$$[y] = [y_a] + [y_b] \tag{20-19}$$

This states that y_i equals $y_{ia} + y_{ib}$, y_f equals $y_{fa} + y_{fb}$, etc. It follows that the terminal voltages and currents are unchanged if the two-ports of Fig. 20-8 are replaced with the single two-port of Fig. 20-9.

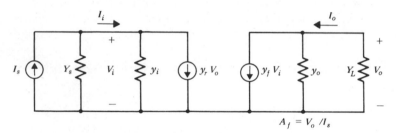

Figure 20-9 y-parameter model of feedback amplifier.

trans impedence $\dfrac{V_o}{I_s}$

With reference to Fig. 20-9 the voltage gain A_v is V_o/V_i, the current gain A_i is $-I_o/I_i$, the input impedance Z_i is V_i/I_i, and Z_o is V_o/I_o with $I_s = 0$. The symbols A_v, A_i, Z_i, and Z_o are used consistently to represent these ratios. We need to define the closed-loop gain A_f suitably. Because the network has voltage sampling and current summing, A_f is the ratio of the output voltage V_o to an input current, and it is convenient to select the short-circuit current I_s of the source. Thus A_f is the transimpedance V_o/I_s. From node equations written for the network of Fig. 20-9 it is straightforward (see P20-5) to show that A_f is

$$A_f = \frac{V_o}{I_s} = \frac{A}{1 + AB} = \left(\frac{T}{1 + T}\right)\frac{1}{B} \tag{20-20}$$

with the loop transmission $T = AB$, and with the open-loop gain A and the feedback factor B given by (20-21). It is also easy to deduce (see P20-6) equations

loop transmission $T = AB$

open loop gain A

feed back factor B

(20-22) and (20-23) for A_v, A_i, Z_i, and Z_o. The relations are

$$A = \frac{-y_f}{(y_i + Y_s)(y_o + Y_L)} \qquad B = y_r \qquad (20\text{-}21)$$

$$A_v = \frac{-y_f}{y_o + Y_L} \qquad A_i = \frac{A_v Z_i}{Z_L} \qquad (20\text{-}22)$$

$$Z_i = \frac{1}{y_i(1 + T) + T Y_s} \qquad Z_o = \frac{1}{y_o(1 + T) + T Y_L} \qquad (20\text{-}23)$$

Let us now briefly examine the results. The y-parameters are the composite parameters of (20-19). We would like the feedback factor B to depend only on the elements of the B network. In addition, it is certainly desirable that the open-loop gain A be reasonably independent of the parameters of the feedback network. Fortunately, in nearly all practical feedback amplifiers,

$$|y_{fa}| \gg |y_{fb}| \qquad \text{and} \qquad |y_{ra}| \ll |y_{rb}| \qquad (20\text{-}24)$$

These inequalities state that forward transmission occurs through the basic amplifier and reverse transmission is through the feedback network. From (20-21) and (20-24) we note that the open loop gain A depends on the feedback network only to the extent that y_{ib} and y_{ob} load the input and output, respectively, and usually these loading effects are slight. Furthermore, the feedback factor B approximately equals y_{rb}. Thus B *depends only on the elements of the feedback network*, as does the closed-loop gain A_f provided $T \gg 1$.

From (20-24) and the expression for the voltage gain it follows that the reduction in A_v results from the loading effect of y_{ob}, which appears in the denominator as part of y_o. Because this loading effect is slight, especially in multistage feedback amplifiers, the feedback has only a second-order effect on A_v. However, A_i, Z_i, and Z_o are substantially lowered, in accordance with the expressions for these quantities given in (20-22) and (20-23). For the special and *unrealistic* case of very large source and load resistances and for negligible values of y_{ib} and y_{ob}, the equations show that A_i, Z_i, and Z_o are each reduced approximately by the factor $1 + T$, and this is a first-order reduction (see P20-7). For a given T the reductions are even greater with smaller values of R_s and R_L.

In (20-23) it may appear that Z_i depends on Y_s, and Z_o on Y_L. Of course this is not the case. The loop transmission T is a function of both Y_s and Y_L, and it is convenient to have the impedances expressed in terms of T, as done in (20-23).

A AND B NETWORKS

Equations 20-21 for A and B, and the inequalities of (20-24), lead to the network of Fig. 20-10 in which the feedback amplifier is represented by A and B

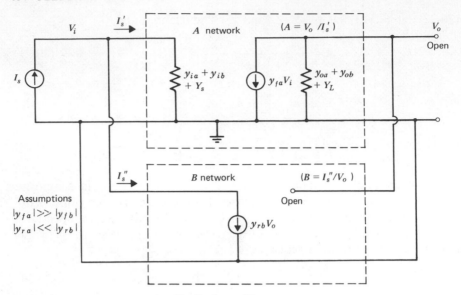

Figure 20-10   *A* and *B* networks of feedback amplifier.

networks. *The A network differs from the basic amplifier of Fig. 20-8 in that it contains the loading effects of the feedback network, as represented by* y_{ib} *and* y_{ob}. *The B network differs from the feedback network in that it has only a single y-parameter* y_{rb}. *However, the configuration of Fig.* 20-10 *is the same as that of Fig. 20-8 with negligible* y_{fb} *and* y_{ra}. *This is evident from the realization that* y_{ib} *in Fig. 20-8 is in parallel with* y_{ia}, *and* y_{ob} *is in parallel with* y_{oa}. Although the equations (20-21) for *A* and *B* are exact, the formation of *A* and *B* networks as shown in Fig. 20-10 is justified only if the inequalities of (20-24) are reasonable. Such is nearly always the case.

Let us compare the variables of Fig. 20-10 with those of the idealized feedback amplifier of Fig. 20-1. The input x_i to the summing network is I_s, the feedback quantity Bx_o is I_s'', and the input to the *A* network is $x_i - Bx_o$, or $I_s - I_s''$, or I_s'. At the output the sampled quantity x_o is V_o, which appears across the output of the *B* network. The open-loop gain *A* is the transimpedance V_o/I_s', and *B* is the transadmittance I_s''/V_o, both of which are given by (20-21). The loop transmission *T* is the product *AB*, or I_s''/I_s', which is dimensionless. Of course, *T* must be positive in the midfrequency region. Although *B* is independent of the amplifier parameters, the open-loop gain *A* depends on the feedback parameters y_{ib} and y_{ob}. However, these usually have little effect on *A*.

In parallel-parallel feedback amplifiers the feedback is normally obtained simply by connecting a resistor from the output node to the input node, as shown in Fig. 20-11a. The configuration can be rearranged in the form of a parallel-

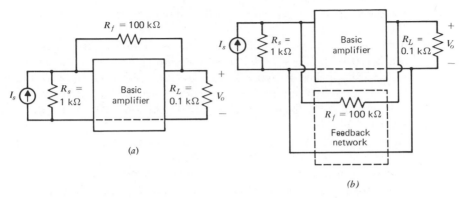

Figure 20-11 Parallel-parallel feedback amplifier.

parallel connection of two-port networks, as shown in Fig. 20-11b. The y-parameters of the feedback network are

$$y_{ib} = y_{ob} - G_f \qquad y_{rb} = y_{fb} = -G_f \qquad (20\text{-}25)$$

The feedback factor B is y_{rb}, or $-G_f$. Because B is negative, the open-loop gain A must also be negative in order to make T positive as required. Because of this, *parallel-parallel feedback is used with inverting amplifiers, which have a phase shift of* $180°$ *in the midfrequency region.* An example is the inverting OP AMP.

ANALYSIS AND DESIGN

The equations that have been presented can be used in various ways to analyze and design parallel-parallel feedback amplifiers. Suppose, for example, the basic amplifier of Fig. 20-11 has the midband model of Fig. 20-12, and we wish to determine the desensitivity $1 + T$ and the closed-loop gain A_f. An excellent procedure, which we shall commonly employ, is to form the A and B networks and to utilize network analysis to find the desired quantities. From (20-25) and Fig. 20-10 we note that the A network consists of the basic amplifier

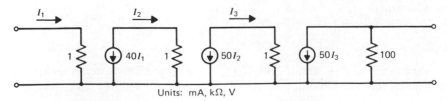

Figure 20-12 Midband model of the basic amplifier of Fig. 20-11.

with the feedback resistor R_f placed across the input and also across the output, as shown in Fig. 20-13.

The open-loop gain is easily found from the network of Fig. 20-13. Let $I_s' = 1$ mA. Then I_1 is 0.4975, I_2 is -19.9, I_3 is 995 mA, and V_o is -4965 V. The ratio V_o/I_s' is -4965 kΩ, which is A. Figure 20-10 shows the B network, where B is y_{rb}, which is $-1/R_f$, or -0.01 mmho; T is the product AB, or 49.65. From (20-20) the closed-loop gain A_f, defined as V_o/I_s, is found to be -98 kΩ, and the desensitivity $1 + T$ is 50.65.

Now suppose that we wish to find A_v, A_i, R_i, and R_o in addition to the closed-loop gain and the desensitivity. When this much information is desired, the simplest procedure is to form the A and B networks as before, and then to calculate the composite y-parameters directly from these networks. The parameter y_i is defined as the sum of y_{iu} and y_{ib}, and y_o is the sum of y_{oa} and y_{ob}. Because y_{fb} is assumed to be negligible, y_f is y_{fa}. With reference to Fig. 20-10 it is clear that y_i, y_o, and y_f are the parameters of that part of the A network *between the Norton*

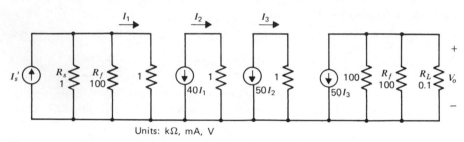

Units: kΩ, mA, V

Figure 20-13   The A network of the amplifier of Figs. 20-11 and 20-12.

equivalent circuit of the source and the load admittance. Specifically, they are not functions of the source and load admittances. The parameter y_r is y_{rb} of the B network, and this is the feedback factor B. From the y-parameters and the source and load admittances the desired gains and impedances can be calculated from (20-20) through (20-23). Alternate equations suitable for finding A_v, A_i, R_i, and R_o are given in Appendix B.

Using the network of Fig. 20-13 with I_s', R_s, and R_L removed, it is easy to deduce that the admittances y_i, y_f, and y_o in mmhos are 1.01, 10^5, and 0.02, respectively, and y_r is -0.01. From these and the source resistance R_s of 1 kΩ and the load R_L of 0.1 kΩ, with the aid of (20-20) through (20-23), we find that

$$A_v = -9980 \qquad A_i = -990 \qquad R_i = 9.9 \text{ ohms} \qquad R_o = 2.0 \text{ ohms}$$

The values of T and A_f are, of course, the same as previously determined. For comparison the corresponding values of the quantities with R_f removed are

$$A_v' = -9990 \qquad A_i' = -99{,}000 \qquad R_i' = 1 \text{ k}\Omega \qquad R_o' = 100 \text{ k}\Omega$$

As expected, the change in A_v is small. However, the other quantities are reduced to about 1 % or less of their values without feedback.

Earlier it was noted that the loop transmission T is the ratio I_s''/I_s' of Fig. 20-10. Suppose the input port of the B network is disconnected from the input of the A network and shorted. Clearly the ratio I_s''/I_s' is unchanged because I_s'' is produced by a current source, and the result suggests the network of Fig. 20-14 as a convenient configuration for the determination of T. For the amplifier of Figs. 20-11 and 20-12 the network of Fig. 20-14 is precisely that of the A network of Fig. 20-13 (see P20-8), and I_s'' is the current in R_f at the output. Although the procedure for determining T directly from the network of Fig. 20-14 has no particular advantage in analytical work, it is sometimes useful in the laboratory.

Another method often used to find the loop transmission of a feedback amplifier consists of removing the input signal and determining either the voltage gain or the current gain around the closed AB loop. As shown in Sec. 20-1 the loop transmission T is the negative of the loop gain. To illustrate the procedure,

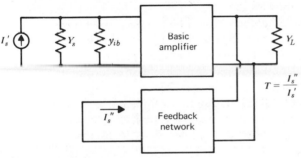

$$T = \frac{I_s''}{I_s'}$$

Figure 20-14   Network for determination of T.

let us refer to Fig. 20-10 with $I_s = 0$. Let $V_i = 1$ V, calculate $y_{rb} V_o$, and multiply this by the input impedance of the A network to obtain the loop gain. Alternatively, we can let $y_{fa} V_i = 1$ A, calculate V_i, and multiply this by y_{fa} to get the gain. In both cases the negative of the loop gain is the product AB of (20-21), provided inequality (20-24) is valid. The method is applicable for each of the four basic feedback configurations.

In many parallel-parallel feedback amplifiers the loop transmission is sufficiently large to justify treating the gain A_f as $1/B$, or $1/y_{rb}$. For a feedback network consisting of R_f connected between the input and output nodes, y_{rb} is $-1/R_f$. Therefore, V_o/I_s is approximately $-R_f$ provided T is large. The short-circuit current I_s of the source equals V_s/R_s, with V_s denoting the open-circuit voltage and R_s denoting the source impedance. Accordingly,

$$\frac{V_o}{V_s} \approx - \frac{R_f}{R_s} \qquad \text{(provided } T \gg 1) \qquad (20\text{-}26)$$

The approximation is fairly good as long as T is greater than 10. For such cases the considerable feedback invariably lowers the input impedance to a value that is small compared with R_s, which justifies the assumption that the input current I_i equals I_s, or $I_i \approx V_s/R_s$. For a resistive load, V_o is $-I_o R_L$, and it follows from these relations and (20-26) that the current gain is

$$A_i \approx \frac{-R_f}{R_L} \quad \text{(provided } T \gg 1) \tag{20-27}$$

Equations 20-26 and 20-27 are valuable in both analysis and design of amplifiers with substantial feedback.

The design of a parallel-parallel feedback network is quite similar to the analysis procedure. First, the A network is formed, as was done in Fig. 20-13. Because R_f is unknown, it is initially either estimated or treated as infinite, and the approximate open-loop gain A is calculated. Then the feedback factor B is selected so as to give the desired desensitivity $1 + T$. The value of R_f is found from B, and analysis is employed to check the design (see P20-9). If necessary, the process is repeated beginning with the value of R_f that has been calculated.

The inverting operational amplifier of Fig. 14-11 of Sec. 14-3 utilizes voltage-sample current-sum feedback. Normally the loop transmission T is so large that the closed-loop gain V_o/I_s equals $1/B$, with $I_s = V_s/Z_1$ and $B = -Y_f$. For this case V_o/V_s is given by (20-26). Equations 20-23 are useful in determining the input and output impedances. It is easy to show that T is $A_v Z_1/(Z_1 + Z_f)$ provided the basic OP AMP without feedback has an input impedance Z_i much greater than $Z_1 \| Z_f$ and an output impedance Z_o much less than $Z_L \| Z_f$; A_v is the open-circuit voltage gain of the operational amplifier. This is discussed further in P20-10 and also in Sec. 21-9.

A discrete circuit with voltage-sample current-sum feedback is the 15-W audio amplifier of Fig. 16-27 of Sec. 16-6. The feedback is provided by R_f which is connected between the input and output nodes as in Fig. 20-11, and the approximate voltage gain is given by (20-26).

20-4. VOLTAGE-SAMPLE VOLTAGE-SUM FEEDBACK

When the basic amplifier has a phase shift of zero degrees in the midfrequency range, as in a two-stage CE amplifier or a noninverting OP AMP, the type of feedback most often used is voltage-sample voltage-sum. Because the amplifier and the feedback network can be connected in the form of two-ports having their inputs in series and their outputs in parallel, as shown in Fig. 20-15, the arrangement is sometimes referred to as *series-parallel* feedback. Clearly the output voltage is sampled, and the feedback network introduces a voltage into the input of the amplifier for comparison. The amplifier is noninverting.

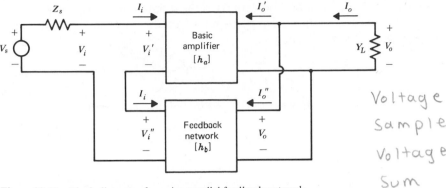

Figure 20-15   Block diagram of a series-parallel feedback network.

THE h-PARAMETER MODEL

In Fig. 20-15 a Thévenin equivalent of the source is employed. It is easy to show (see P20-11) that the composite h-parameters of two-port networks in a series-parallel arrangement are determined by adding the respective parameters; that is,

$$\lfloor h \rfloor = \lfloor h_a \rfloor + \lfloor h_b \rfloor \qquad (20\text{-}28)$$

The hybrid parameters were defined and discussed in Sec. 12-4. Expressions for the current and voltage gains and the input and output impedances of a two-port network are given in Appendix B in terms of the parameters and the source and load impedances.

Examination of Fig. 20-15 shows that the summing network consists of the series arrangement at the input, and the quantity fed back is V_i''. Thus the closed-loop gain A_f is the ratio of the sampled output voltage V_o to an input voltage, and it is convenient to select the open-circuit voltage V_s of the source. Accordingly, A_f is the voltage ratio V_o/V_s. The basic amplifier and the feedback network can be combined into a single two-port that utilizes the composite h-parameters of (20-28), and the configuration of Fig. 20-15 becomes that of Fig. 20-16.

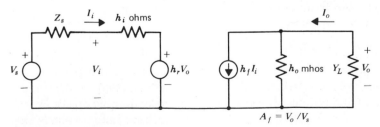

Figure 20-16 h-parameter model of feedback amplifier.

From Kirchhoff's voltage law applied to the input loop of Fig. 20-16 and the current law applied to the output node, we obtain the h-parameter equations of the equivalent two-port. These are

$$V_i = h_i I_i + h_r V_o$$
$$I_o = h_f I_i + h_o V_o \tag{20-29}$$

The parameter h_i is the short-circuit input impedance V_i/I_i of the two-port, and h_f is the forward short-circuit current ratio I_o/I_i. The parameter h_o is the open-circuit output admittance I_o/V_o, and h_r is the reverse open-circuit voltage ratio V_i/V_o.

From the definitions of the parameters we can find h_{ia}, h_{ra}, h_{fa}, and h_{oa} of a specified basic amplifier. For a given feedback network we can find h_{ib}, h_{rb}, h_{fb}, and h_{ob}. Analysis of the feedback amplifier could be made by obtaining the composite parameters from (20-28), and then using either Fig. 20-16 or the equations of Appendix B. However, for amplifier design and also for the determination of the desensitivity of a given network, it is more appropriate to follow the procedure of forming A and B networks as was done in the preceding section.

From the network of Fig. 20-16, it is straightforward (see P20-12) to show that A_f is

$$A_f = \frac{V_o}{V_s} = \frac{A}{1 + AB} = \left(\frac{T}{1 + T}\right)\frac{1}{B} \tag{20-30}$$

with $T = AB$, and with A and B given by (20-31). It is also easy to deduce (20-32) and (20-33) for A_i, A_v, Z_i, and Z_o. The method is similar to that of P20-6.

$$A = \frac{-h_f}{(h_i + Z_s)(h_o + Y_L)} \qquad B = h_r \tag{20-31}$$

$$A_i = \frac{-h_f Y_L}{h_o + Y_L} \qquad A_v = \frac{A_i Z_L}{Z_i} \tag{20-32}$$

$$Z_i = h_i(1 + T) + TZ_s \qquad Z_o = \frac{1}{h_o(1 + T) + TY_L} \tag{20-33}$$

Note that A_v is the ratio V_o/V_i, whereas A_f is V_o/V_s.

In nearly all practical feedback amplifiers, the forward transmission is almost entirely through the basic amplifier and the reverse transmission is largely through the feedback network. Hence,

$$|h_{fa}| \gg |h_{fb}| \qquad \text{and} \qquad |h_{ra}| \ll |h_{rb}| \tag{20-34}$$

From (20-31) and (20-34) we observe that the open-loop gain A depends on the feedback network only to the extent that the impedance h_{ib} and the admittance h_{ob} load the input and output, respectively, and these loading effects are usually slight. Also, B is approximately h_{rb}, and thus the feedback factor depends only on the elements of the feedback network, as desired.

We have learned that a series connection at the input raises the input imped-ance. This is shown mathematically by (20-33). Also shown by (20-33) is the reduction in Z_o as a consequence of the parallel connection at the output. These are first-order effects. From (20-32) we note that the change in A_i results from the small loading effect of h_{ob}, which shunts a portion of the current from the load. However, the reduction in A_v is quite substantial provided there is appreciable feedback.

In series-parallel feedback amplifiers the feedback is normally obtained from a resistive network with the form shown in Fig. 20-17, which also shows the approximate h-parameter model. Because forward transmission through the feedback network is assumed to be negligible, the current source that depends

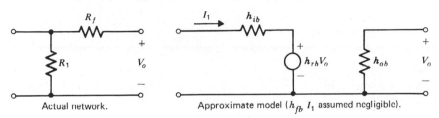

Actual network.   Approximate model ($h_{fb} I_1$ assumed negligible).

Figure 20-17 Feedback network and approximate h-parameter model.

on h_{fb} is omitted in the model. The open-circuit reverse voltage ratio h_{rb} of the feedback network is approximately equal to the feedback factor B, as indicated by (20-31) and (20-34). Hence,

$$B \approx h_{rb} = \frac{R_1}{R_1 + R_f} \qquad (20\text{-}35)$$

Because B is positive, the open-loop gain A of the amplifier must also be positive in the midband range.

A AND B NETWORKS

From (20-31) for A and B and from the inequalities of (20-34) we deduce the A and B networks of Fig. 20-18. *The A network is simply the basic amplifier of Fig. 20-15 with h_{ib} in series with the input, with h_{ob} in shunt with the output, and with reverse transmission eliminated. The B network has only the single h-parameter h_{rb}.* However, the configuration of Fig. 20-18 is the same as that of Fig. 20-15 provided h_{fb} and h_{ra} are negligible. This equivalence results from the fact that the impedances h_{ia} and h_{ib} are in series and the admittances h_{oa} and h_{ob} are in parallel.

Once the A and B networks have been determined, the procedure is similar to that of the preceding section. The open-loop gain A can be calculated directly

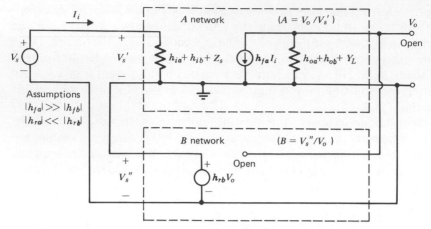

Figure 20-18   *A* and *B* networks.

from the *A* network and *B* is found from (20-35). If A_v, A_i, Z_i, and Z_o are desired in addition to *T*, the composite *h*-parameters are deduced. *From that part of the* A *network which is between the Thévenin equivalent circuit of the source and the load admittance*, the parameters h_i, h_f, and h_o are found, and h_r is determined from either the *B* network or (20-35). Then the desired quantities are calculated from (20-30) through (20-33).

In many cases the loop transmission *T* is sufficiently large to justify treating A_f as $1/B$, or $1/h_{rb}$. From (20-35) this gives

$$\frac{V_o}{V_s} \approx \frac{R_1 + R_f}{R_1} \qquad \text{(provided } T \gg 1) \qquad (20\text{-}36)$$

When (20-36) applies, the feedback nearly always makes the input impedance much greater than the source impedance. Consequently, V_s is approximately equal to V_i, and (20-36) also represents the terminal voltage gain A_v. The results are fairly accurate as long as *T* is greater than 10.

VOLTAGE-SUM CIRCUITS

There are practical problems associated with implementing series-parallel feedback. For example, suppose the basic amplifier of Fig. 20-15 is a CS-CE configuration and the feedback network is that of Fig. 20-17. Because the lower input terminal and lower output terminal of the basic amplifier are common, the resistor R_1 of the feedback network is short-circuited. This problem can be eliminated by using an *isolation transformer* at the left side of the feedback network, as illustrated in Fig. 20-21. Such a device has an iron core, because it must approximate an ideal transformer. The turns ratio could be one-to-one, and there should be no phase shift. The purpose is simply to break the short-circuit across R_1. When an isolation transformer is used, the source and load are not permitted to have a common ground. This is objectional, along with the size, weight, and cost of the transformer. Consequently, isolation transformers are avoided in electronic circuits whenever possible.

The output terminals of the feedback network can be connected across the load, so there is no problem with voltage sampling at the output. The only problem is the voltage comparison at the input. Fortunately, there are two ways to accomplish this that are quite satisfactory and convenient, and these are examined in the remainder of this section.

One widely used method is to employ an emitter-coupled amplifier at the input, as shown in Fig. 20-19. From the incremental model of Fig. 20-20 we note that the output of the emitter-coupled stage is taken from the collector of Q_1, giving a phase shift of 180°. The CE amplifier associated with Q_3 gives the additional 180° phase shift that is required. Resistors R_1 and R_f constitute the feedback network. This network samples the output voltage and introduces the feed-

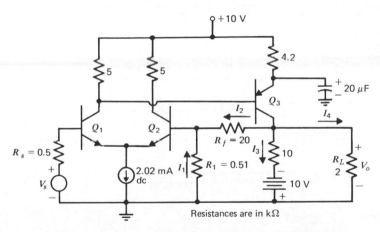

Figure 20-19 Series-parallel feedback using an emitter-coupled amplifier.

back voltage V_f into the base of Q_2. For matched transistors Q_1 and Q_2 biased at the same Q-points, *the networks of Figs. 20-20 and 20-21 are exactly equivalent* insofar as the terminal variables are concerned. The incremental base currents I_i of Q_1 and Q_2 are equal, and furthermore, I_i is the same in the configurations of both Figs. 20-20 and 20-21. This equality arises because I_i is the difference $V_s - V_f$ divided by the sum of R_s and the base-emitter impedances of the two transistors. In the network of Fig. 20-20 we are utilizing the basic property of the differential amplifier, which was discussed in Sec. 14-1. Clearly, the equivalent model of Fig. 20-21 has the series-parallel arrangement. Note the use of an

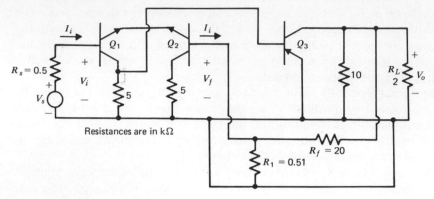

Figure 20-20   Incremental model.

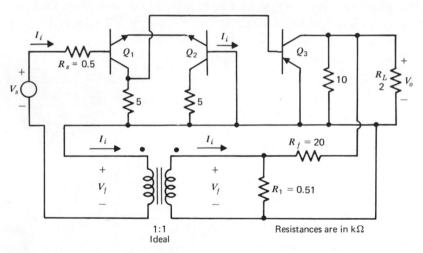

Figure 20-21   Equivalent incremental model.

isolation transformer to avoid a short-circuit across R_1. *The presence of both inverting and noninverting inputs eliminates the need for this device in the actual circuit.*

Let us assume that each transistor of Fig. 20-19 has dc and ac betas of 100, a base-emitter voltage of 0.7, and an h_{ie} of 3 kΩ. It is easy to show (see P20-13) that each collector current is 1 mA and that there is no dc voltage across the load, justifying the omission of a blocking capacitor. The A network at midfrequencies is that of Fig. 20-22. As in Fig. 20-18, the input is loaded with the series resistance h_{ib}, which is the parallel equivalent of R_1 and R_f, and the output is loaded with the shunt resistance $1/h_{ob}$, consisting of the sum of R_1 and R_f. The B network,

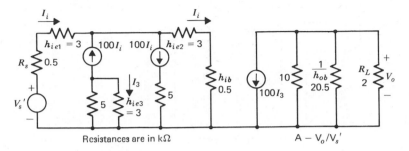

Figure 20-22   The A network of the emitter-coupled feedback amplifier.

which is shown in Fig. 20-18, has the single parameter h_{rb}, and this is the composite parameter h_r. The value of B is found from (20-35), with $R_1 = 0.51$ and $R_1 = 20$ kΩ. The parameters h_i, h_f, and h_o are determined from the A network of Fig. 20-22. Equations 20-31 through 20-33 can then be used to complete the analysis (see P20-14). We observe that, once the A and B networks are formed, a little circuit experience makes the remainder of the analysis a rather routine process.

A second way to accomplish voltage comparison at the input is to use a CE amplifier at the input with the resistor R_1 of the feedback network inserted between the emitter and ground, as shown in Fig. 20-23. For simplicity, the batteries are omitted in the figure. The first stage could just as well be a CS amplifier, and the stage or stages that follow can have any configuration provided the overall phase shift of the multistage amplifier is zero.

The network of Fig. 20-24 is readily deduced from those of Figs. 20-23 and 20-17. The voltage source $h_{rb} V_o$ in series with the emitter can be removed and replaced with the two sources shown as broken circles in Fig. 20-24. In any three-terminal network a voltage source in one leg can be replaced with identical sources in each of the other two legs without affecting the terminal conditions; it is easy to show that the input and output loop equations are unchanged. In the low and midfrequency ranges the source $h_{rb} V_o$ at the collector of Q_1 is in series with the current source of the hybrid-π model. Because a voltage source in series with a current source has no effect on the terminal conditions, it can be omitted from the circuit as done in Fig. 20-25. Let us note, however, that at high frequencies the capacitor C_μ has a connection between the current and voltage sources, and the omission of the voltage source in Fig. 20-25 is not justified. Even so, we shall use the circuit of Fig. 20-25 for amplifier design, recognizing that omission of the generator introduces error at high frequencies. Analysis of the exact circuit can always be employed to ascertain that a design based on approximations is reasonable.

In Fig. 20-25 the generator $h_{rb} V_o$ is in series with the base, but its position has been shifted so as to locate it in the B network. The A and B networks now have the

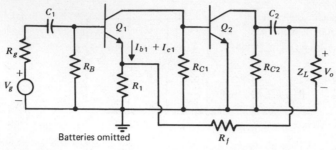

Figure 20-23 Series-parallel feedback using an emitter resistor.

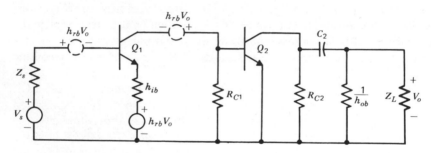

Figure 20-24 Equivalent network.

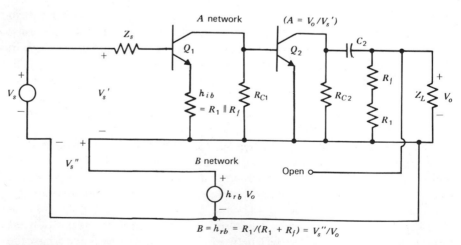

Figure 20-25 The A and B networks.

666

series-parallel form. In the low and midfrequency regions the only approximation involved in the determination of the A and B networks is the neglect of forward transmission in the feedback circuit. Because the input current of this network is $(\beta_o + 1)I_{b1}$, which is evident from Fig. 20-23, we require that

$$|h_{fa}| \gg (\beta_o + 1)|h_{fb}| \qquad (20\text{-}37)$$

with $h_{fb} = -R_1/(R_1 + R_f)$. The inequality is satisfied in nearly all amplifiers with series-parallel feedback. However, at high frequencies the result is not as satisfactory, because the omitted voltage source in series with the collector of Q_1 introduces some error. It should be carefully noted that V_s and Z_s in Fig. 20-25 *are the components of the Thévenin equivalent circuit looking back from the base of Q_1* in Fig. 20-23.

For direct measurement of the loop transmission T in the laboratory the input terminals of the feedback network are disconnected from the basic amplifier and replaced with a resistance equal to the parallel equivalent of R_1 and R_f, which is h_{ib}. The voltage V_s'' across the open-circuited input terminals of the feedback network is measured when a voltage V_s', which is now the same as V_s, is applied to the input. The loop transmission T is V_s''/V_s'. The loading of the amplifier at its output, represented by h_{ob} in Fig. 20-18, is accomplished by the actual feedback network.

The noninverting operational amplifier of Fig. 14-11 of Sec. 14-3 employs voltage-sample voltage-sum feedback. The loop transmission T is usually so large that (20-36) applies. With A_v denoting the open-loop voltage gain of the operational amplifier it is easy to show that T is $A_v Z_1/(Z_1 + Z_f)$ provided the basic OP AMP without feedback has an input impedance Z_i much greater than $Z_1 \| Z_f$ and an output impedance Z_o much less than $Z_L \| Z_f$. This is discussed further in P20-16 and also in Sec. 21-9. Another example encountered earlier is the 50 W audio amplifier of Fig. 16-29 of Sec. 16-6. The feedback circuit is similar to that of Fig. 20-19.

The two basic types of feedback that have been considered utilize voltage sampling that provides a low output impedance. Because this is generally a design objective, these feedback arrangements are of greater practical importance than the two current-sampling types presented in the next two sections. However, there are many useful applications of current-sampling feedback.

20-5. CURRENT-SAMPLE CURRENT-SUM FEEDBACK

We know that amplifiers commonly have loads that can be represented by a resistance in parallel with a capacitance, and such amplifiers should have low output impedances in order to maintain a fairly constant voltage over a wide frequency range. However, there are important applications for amplifiers designed to deliver an output current that is proportional to the input signal. For example, one used to drive deflection coils in a magnetic-deflection oscilloscope definitely should have a high output impedance that tends to maintain a load current nearly independent of frequency. The current-sample current-sum feedback configuration, also referred to as *parallel-series* feedback, is characterized by low input and high output impedances, with a stabilized I_o/I_s ratio. It is particularly suitable for use with a basic noninverting amplifier, having a phase shift of zero degrees in the midband region.

Shown in Fig. 20-26 is a parallel-series configuration. The output current is sampled, and the feedback network introduces a current into the input of the amplifier for comparison. When two-ports are arranged with their inputs in

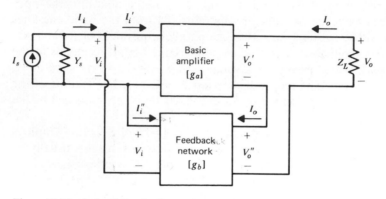

Figure 20-26   A parallel-series feedback network.

parallel and their outputs in series, the composite g-parameters are determined by adding the respective parameters; that is,

$$[g] = [g_a] + [g_b] \tag{20-38}$$

With reference to Fig. 20-26 the composite g-parameters are defined by the relations

$$I_i = g_i V_i + g_r I_o$$
$$V_o = g_f V_i + g_o I_o \tag{20-39}$$

Note the mixed dimensions, as in the case of the h-parameters. Expressions for the current and voltage gains and the input and output impedances of a two-port network are given in Appendix B in terms of the g-parameters.

It is convenient to define the closed-loop gain A_f as the ratio of the sampled output current I_o to the short-circuit current I_s of the source. In Fig. 20-26 the basic amplifier and the feedback network can be replaced with a circuit model based on the composite g-parameters of (20-38) and the equations of (20-39), somewhat as was done with h-parameters in Fig. 20-16. From this network can be found (see P20-17) the design equations that follow:

$$A_f = \frac{I_o}{I_s} = \frac{A}{1 + AB} = \left(\frac{T}{1 + T}\right)\frac{1}{B} \tag{20-40}$$

$$A = \frac{-g_f}{(g_i + Y_s)(g_o + Z_L)} \qquad B = g_r \tag{20-41}$$

$$A_v = \frac{g_f Z_L}{g_o + Z_L} \qquad A_i = \frac{A_v Z_i}{Z_L} \tag{20-42}$$

$$Z_i = \frac{1}{g_i(1 + T) + TY_s} \qquad Z_o = g_o(1 + T) + TZ_L \tag{20-43}$$

As before, T is AB.

AMPLIFIER CHARACTERISTICS AND DESIGN

Practical amplifiers nearly always have

$$|g_{fa}| \gg |g_{fb}| \qquad |g_{ra}| \ll |g_{rb}| \tag{20-44}$$

Thus the feedback factor B is g_{rb}, which depends on the resistances of the feedback network. From (20-43) it is evident that Z_i is substantially lowered and Z_o is substantially raised, as expected, assuming the loop transmission T is reasonably large. The voltage gain A_v is reduced only slightly, and this is simply because of the loading effect of g_{ob}, which is only a small part of g_o in (20-42). However, the reduction in A_i is a first-order effect, as indicated by the equations. Note that the terminal current gain A_i is $-I_o/I_i$, whereas the closed-loop gain A_f is the ratio I_o/I_s.

Formation of the A and B networks follows a procedure similar to that of the preceding two sections. The parameters g_{ib} and g_{ob} of the feedback network are shifted into the A network, leaving the B network with only the one parameter g_{rb}. From the A network can be found the open-loop gain A and the composite parameters $g_i, g_f,$ and g_o. The parameter g_r is g_{rb}, or B, which is determined from the feedback circuit. For the usual case of a network of resistors having a common ground, g_r is negative. This is a consequence of selected reference directions, along with the defining relationship given by the first of equations (20-39) with $V_i = 0$. To give the required positive loop transmission T in the midband range, A must also be negative. By (20-40) A is I_o/I_s with the feedback factor $B = 0$, and when this ratio is negative, the voltage gain A_v must be positive. Parallel-series feedback is normally used only with amplifiers having positive values of midband voltage gain, for otherwise the feedback network would have to provide voltage phase inversion.

AN EXAMPLE

Shown in Fig. 20-27 is a current-sample current-sum feedback circuit[4] suitable for use as the preamplifier of a servo amplifier. Although the source impedance Z_s might be nonlinear, variable, and a function of frequency, I_i is independent of Z_s because of the very low input impedance resulting from the current-sum feedback arrangement. Furthermore, the output current I_o is independent of R_L for values of load resistance small compared with the 20 kΩ collector resistance. The current-sample configuration provides an extremely large output impedance looking into the collector of Q_2.

The current sampled is the emitter current of Q_2. However, because the transistor beta is 50 or more, the incremental emitter and collector currents are

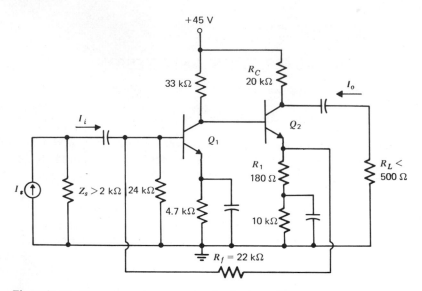

Figure 20-27   Current-sample current-sum feedback amplifier.

almost equal, and because R_L is much smaller than R_C, nearly all of the incremental collector current flows through the load. Thus the arrangement effectively samples I_o. The use of an unbypassed emitter resistor for sampling approximately the output current of a CE stage is convenient. If R_L were not small compared with R_C, then the parallel equivalent of these would be treated as the effective load resistance for use in the equations.

The feedback network consists of R_1 and R_f, and the reverse short-circuit current ratio g_{rb} is $-R_1/(R_1 + R_f)$, which is B. The open-loop gain can be found from the A network, obtained by disconnecting the feedback loop, placing $R_f + R_1$ in shunt with the input, and placing $R_f \| R_1$ between the emitter of Q_2

and ground. The value of T is calculated from A and B. An alternate procedure for finding T is to form a circuit somewhat similar to that of Fig. 20-14 using the proper loading at the input of the amplifier and, of course, the actual feedback connections at the output. The current ratio I_s''/I_s' can be either calculated or measured.

For the network of Fig. 20-27 the loop transmission is sufficiently large (see P20-18) to estimate A_f from $1/B$. This gives a closed-loop gain I_o/I_s of -120. The two direct-coupled CE stages are noninverting, and dc feedback provides excellent bias stability (see P20-19).

20-6. CURRENT-SAMPLE VOLTAGE-SUM FEEDBACK

The fourth basic feedback configuration samples the output current. A voltage proportional to this current is fed back to the input for voltage comparison. The closed-loop gain A_f is the transadmittance I_o/V_s. The A and B networks have series connections at both input and output, and the feedback is often referred to as *series-series*. Both the input and output impedances are substantially raised by series-series feedback, and the terminal voltage gain A_v is appreciably lowered. However, the reduction in A_i is second-order. This type of feedback is normally used only with basic amplifiers having a midband phase shift of $180°$.

THE z-PARAMETER EQUATIONS

A block diagram of a series-series feedback amplifier is shown in Fig. 20-28. It is easy to show that the composite z-parameters are

$$[z] = [z_a] + [z_b] \tag{20-45}$$

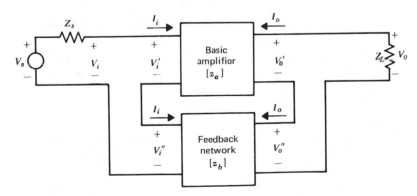

Figure 20-28 A series-series feedback configuration.

With reference to Fig. 20-28 the composite z-parameters are defined by the relations

$$V_i = z_i I_i + z_r I_o$$
$$V_o = z_f I_i + z_o I_o \qquad (20\text{-}46)$$

Clearly, each parameter is an impedance that is found with either the input or output port open-circuited, and they are referred to as the open-circuit impedance parameters. The respective input and output impedances are z_i and z_o, and the respective forward and reverse transfer impedances are z_f and z_r. Expressions for A_v, A_i, Z_i, and Z_o in terms of the z-parameters are given in Appendix B.

Let us replace the basic amplifier and the feedback network of Fig. 20-28 with a single z-parameter circuit using composite parameters satisfying (20-46). From the network are found the design equations that follow:

$$A_f = \frac{I_o}{V_s} = \frac{A}{1 + AB} = \left(\frac{T}{1 + T}\right)\frac{1}{B} \qquad (20\text{-}47)$$

$$A = \frac{-z_f}{(z_i + Z_s)(z_o + Z_L)} \qquad B = z_r \qquad (20\text{-}48)$$

$$A_i = \frac{z_f}{z_o + Z_L} \qquad A_v = \frac{A_i Z_L}{Z_i} \qquad (20\text{-}49)$$

$$Z_i = z_i(1 + T) + TZ_s \qquad Z_o = z_o(1 + T) + TZ_L \qquad (20\text{-}50)$$

Practical amplifiers nearly always have

$$|z_{fa}| \gg |z_{fb}| \qquad \text{and} \qquad |z_{ra}| \ll |z_{rb}| \qquad (20\text{-}51)$$

The formation of the A and B networks is based on the validity of these inequalities.

SAMPLE AND SUMMING CIRCUITS

In the previous section we learned that the output current of a CE amplifier can be sampled approximately by adding a small unbypassed resistor between the emitter and ground, as shown in Fig. 20-27. The symbol Z_L in the equations must then be regarded as representing the equivalent impedance from the collector to ground. For an emitter-follower output stage the resistor is placed in series with the collector, and Z_L becomes the impedance from emitter to ground. Voltage summing at the input is usually accomplished by means of either a differential input stage or an unbypassed emitter resistor of a CE input stage. Such voltage-sum networks were examined in Sec. 20-4 and illustrated in Figs. 20-19 and 20-23.

For a specified desensitivity $1 + T$ *the loss of gain is considerably less when the feedback loop includes several stages than it is when local feedback is applied to individual stages* (see P20-20). However, single-stage feedback is occasionally used. Examples that we have encountered include the unbypassed emitter resistors of the class B power amplifiers of Sec. 16-6 and also those of the emitter-coupled stages of the video amplifier of Sec. 19-2. Figure 20-29 illustrates the circuit of an elementary one-stage CE amplifier with current-sample voltage-sum feedback, and the bias circuitry is omitted for simplicity. Although the two incremental models are clearly identical, the one of Fig. 20-29b is arranged in the form of two-port networks in a series-series arrangement. Each z-parameter of the feedback network equals R_1.

The A circuit is found by removing the feedback, with a resistor of value R_1 placed in series with R_s at the input and a second resistor of the same value connected in series with R_L at the output. The feedback factor B equals R_1. For a sufficiently large loop transmission T the closed-loop gain I_o/V_s is approximately

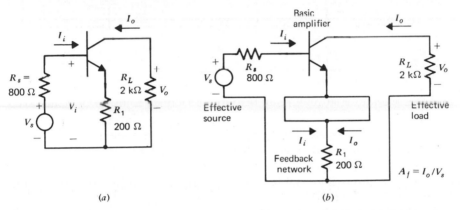

(a)   (b)

Figure 20-29   Identical incremental models of CE amplifier. (a) Circuit. (b) Series-series two-port networks.

$1/B$. This equals $1/R_1$, giving a voltage ratio V_o/V_s of $-R_L/R_1$. For greater accuracy circuit analysis is required. Suppose h_{ie} is 1 kΩ and beta is 100. From the A network the open-loop gain is found to be 50 mmhos, and AB is 10. Using (20-47), we obtain a value of 4.55 mmhos for the closed loop gain. This gives -9.1 for V_o/V_s, compared with -10 calculated from the approximate relation.

A multistage amplifier with current-sample voltage-sum feedback is illustrated in Fig. 20-30. The required phase inversion of the basic amplifier is obtained by taking the output of the emitter-coupled amplifier from the collector of Q_1. The output emitter follower has a 5 kΩ collector resistor used for current sampling. Resistors R_1, R_2, and R_3 constitute the feedback network, and the specified values give a feedback factor B of 0.05. For greater accuracy this should be multiplied by $\beta_o/(\beta_o + 1)$, which properly relates the feedback to the output

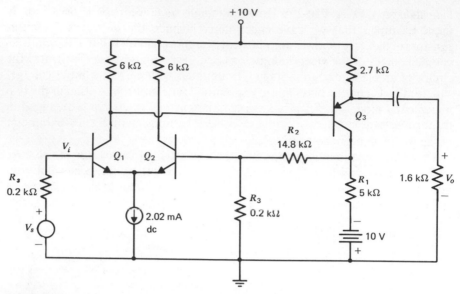

Figure 20-30   Current-sample voltage-sum feedback amplifier.

incremental emitter current rather than to the sampled collector current. This is not usually necessary, however. The feedback raises both the input and output impedances, stabilizes the ratio I_o/V_s, reduces A_v, and has very little effect on A_i. With transistor parameters specified in P20-21, analysis reveals that the amplifier has a voltage gain of -17 and a desensitivity of 6. At the output the impedance of the emitter follower is 5.6 kΩ, which is considerably higher than the typical value of a few tens of ohms.

The loop transmission T can be found directly from the network. To accomplish this, the feedback network is disconnected from the basic amplifier at the input and replaced with z_{ib}. The loop transmission T is the ratio of the open-circuit voltage across the input port of the feedback network to the open-circuit voltage of the source.

20-7. DESIGN TECHNIQUES

Approximate procedures frequently enable us to estimate with reasonable accuracy, simply by inspection of the circuitry, the gains and terminal impedances of a feedback amplifier. For a large T the closed-loop gain A_f is approximately $1/B$, and as we have seen, B *is the appropriate reverse matrix parameter.* Also, with a large T a parallel connection at the input gives an input resistance R_i so small that it can often be regarded as zero, and a series connection yields a value so large that it can often be regarded as infinite. A similar statement applies to the output resistance R_o.

For example, the series-parallel amplifier of Fig. 20-19 of Sec. 20-4 has B, or h_{rb}, equal to $R_1/(R_1 + R_f)$, which is 0.0249. The gain A_f is V_o/V_s, and assuming T is large, this is approximately $1/B$, or 40. With R_i regarded as infinite, V_i equals V_s, and the terminal voltage gain is also 40. These results were deduced without circuit analysis. Actually, as shown in P20-14, the true values of A_f and A_v are 39.1, and R_i is 246 kΩ, with $T = 34$.

With a basic amplifier specified, suppose we wish to design a feedback network to provide suitable desensitivity. Our first decision regards the type of feedback to be employed. For an amplifier with a midband phase shift of zero, we should select either series-parallel or parallel-series feedback, whereas one of the other two types is appropriate when the phase shift is 180°. Otherwise, the feedback network must introduce a phase shift of 180° to make T positive, and this requires reactances or transformers. Having narrowed the choice to two, the selection is based on the desired impedance levels, although practical circuit considerations often influence the decision. Because most loads can be represented by a resistor that is shunted by a small capacitance, voltage sampling is usually preferred, giving a low output impedance.

Having selected the type of feedback, we form the A network. However, we need values of the input and output matrix parameters of the feedback network that is to be designed. One procedure is to assume the loading effects of the parameters are negligible. This is usually reasonable for feedback over at least three stages. Perhaps a better way is to estimate the parameters by means of the approximate procedures. From the desired closed-loop gain, or from the specified desensitivity and the estimated open-loop gain, we find the required feedback factor B, which determines the reverse matrix parameter. In some cases the feedback network has only a single resistor, and its value is fixed by B. However, if B is a ratio of resistances, the values are selected to give the proper ratio, *while minimizing the loading effects on the* A *network*. From these resistances the input and output matrix parameters are determined approximately, and the A network is formed.

The open-loop gain A can be found directly from the A network. An alternate procedure is to calculate A from the composite input, forward, and output matrix parameters, which are also useful in finding the terminal gains and the input and output impedances. These parameters are determined from that part of the A network between the source and load impedances. From A and the required desensitivity the feedback factor is found, and the resistances of the feedback network are then chosen to give the proper B and to minimize loading of the A network. The procedure is repeated only if the input and output matrix parameters of the feedback network differ substantially from those used to find A, but seldom is this necessary.

In nearly all multistage and most single-stage feedback amplifiers the forward matrix parameter of the basic amplifier is large compared with that of the feedback network and the reverse matrix parameter of the amplifier is much smaller than that of the feedback network. For such cases the method is precise with one exception, which occurs when there is current sampling by an emitter resistor in a CE output stage. For this situation the calculated input matrix parameter, the loop transmission, and the output impedance are likely to be significantly in error, but other quantities determined from the A and B networks are reasonably accurate. When a collector resistance in a CC output stage is employed for current sampling, B is the product of $\beta_o/(\beta_o + 1)$ and the reverse matrix parameter of the feedback network, with β_o being that of the output transistor.

For each basic type of feedback the closed-loop gain was defined in a manner that enabled us to express A_f in the form $A/(1 + AB)$, with B depending on the resistors of the feedback network. The output quantity of A_f is either V_o or I_o, depending on which of these is sampled. The input quantity is either V_s or I_s, depending on whether the compared quantity is a voltage (series connection) or a current (parallel connection).

A multistage amplifier with a large amount of feedback is likely to oscillate. The problem has been completely ignored in the networks that have been presented. However, the design of feedback amplifiers must carefully consider the stability problem, and this is the subject of the next chapter. Let us now consider two illustrative examples.

Example 1

In Fig. 20-31 is an emitter-coupled (EC) amplifier followed by CB and CC stages. We wish to design a feedback network to give a voltage ratio V_o/V_s of -40, and a very low output impedance is required. All transistors have dc and ac betas of 100. Assume q/kT is 40 and each base-emitter voltage is 0.7 V. The incremental ohmic base resistances r_{x1}, r_{x2}, and r_{x3} are each 100 ohms, and r_{x4} is negligible. Also negligible are the base-width-modulation parameters r_o and r_μ.

Solution

Units are in kΩ, mmhos, mA, and V. The quiescent collector currents are needed. Current source Q_5 has R_B, R_E, and V_B equal to 5 kΩ, 4.5 kΩ, and 10 V, respectively. We find I_{C5} to be 2.02, and the quiescent collector currents of Q_1 and Q_2 are each 1 mA.

In the partial bias circuits of Fig. 20-32, for Q_3 the values of R_B, R_E, and V_B are 5.39 kΩ, 5.1 kΩ, and 5.91 V, respectively, which yield $I_{C3} = -1$ mA. The corresponding values for Q_4 are 22 kΩ, 0.43 kΩ, and 4.78 V, and I_{C4} is 6.24 mA. From the collector currents we can calculate the transconductances of the transistors, and r_π is determined from g_m and β_o. The sum of r_x and r_π is h_{ie}. We find that h_{ie1}, h_{ie2}, and h_{ie3} are each 2.6, and h_{ie4} is 0.4 kΩ.

The EC amplifier has a midband phase shift of 180°, and the other stages introduce no additional phase shift. Hence it is convenient to select either parallel-parallel or series-series feedback. We choose the former to give the desired low output impedance. The feedback is accomplished by connecting a resistor R_f

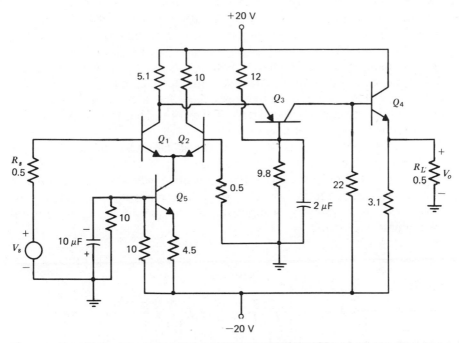

Figure 20-31   RC-CB-CC amplifier with resistances in kilohms.

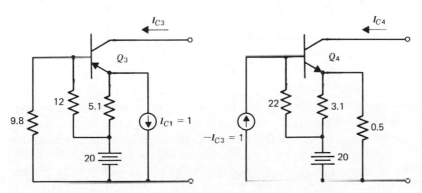

Figure 20-32   Partial bias circuits with units of kΩ, mA, V.

from the output node E_4 to the base B_1 of Q_1. We expect the voltage ratio V_o/V_s to be approximately $-R_f/R_s$. For a gain of -40 the feedback resistor should be approximately 20 kΩ. This is only an estimate for use in the A network. The resistor R_f introduces no appreciable dc feedback because the potential of B_1 is -0.005 and that of E_4 is -0.060 V. If necessary, the 3.1 kΩ resistor should be adjusted to make the dc component of the output voltage equal to zero.

To find A let I_s' in the A network of Fig. 20-33 be 1 mA. Then it is easy to deduce that I_1 is 0.0788 mA. The input resistance of the CB stage is 2.6/101, or 0.0257 kΩ, and the current $101I_2$ can be found from $100I_1$ with the aid of the current-division theorem. Current I_2 is 0.0776 mA. A portion of the source $100I_2$ is shunted through the 22 kΩ resistor and the remainder I_3 is supplied to the CC stage. This stage has an input resistance of 43 kΩ, which is 0.4 plus the product of 101 and the effective load resistance of 0.421. Thus I_3 is $-100I_2(22/65)$, or -2.63 mA, and V_o is -112, which is A.

The closed-loop gain V_o/I_s is $0.5V_o/V_s$. We must choose B to make $A_f = -20$, with $A = -112$. Thus B is -0.0411, and $-1/B$ is 24.3. A resistor of 24.3 kΩ connected from the output node to the input node gives the right amount of feedback for a voltage ratio V_o/V_s of -40. The desensitivity $1 + T$ is 5.6. The feedback substantially lowers the input and output resistances and the terminal current gain A_i (see P20-22).

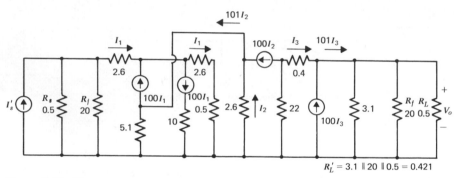

$$R_L' = 3.1 \parallel 20 \parallel 0.5 = 0.421$$

Figure 20-33   The A network with units of kΩ, mA, V.

Although the internal CB stage has no current gain, it presents a low input resistance to the output of the EC amplifier. Consequently, most of the current of the source $100I_1$ goes into the CB stage and thence to the parallel combination of the 22 kΩ resistor and the high input resistance of the CC stage. Clearly the CB configuration acts as an impedance transformer with a near-unity current gain. The output CC stage also acts as an impedance transformer, but with a near-unity voltage gain. Thus the CB-CC cascade is characterized by low input and output resistances and both current and voltage gain. The 22 kΩ bias resistor shunts the high input resistance of the output stage and reduces the gain. It can be replaced with a direct-current source with a high incremental resistance, but this requires an additional transistor.

Example 2

The source-coupled (SC) stage of Fig. 20-34 is followed by a CE-CC cascade. Design a feedback network to give a desensitivity $1 + T$ of 10. The input im-

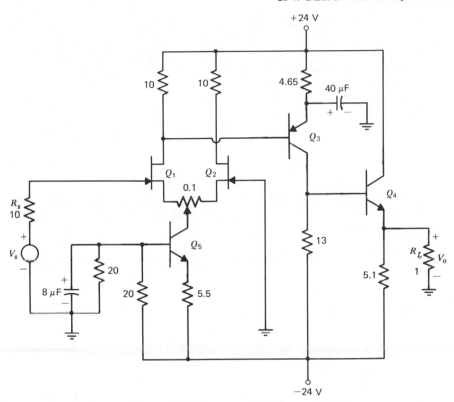

Figure 20-34   SC-CE-CC amplifier with resistances in kilohms.

pedance must remain high and a low output impedance is desired. The JFETs are matched, with $I_{DSS} = 4$ mA, $V_P = -4$, and $r_{ds} = 40$ kΩ. Each BJT has dc and ac betas of 100. The emitter-base voltages are 0.7, q/kT is 40, and r_{x3} and r_{x4} are 50 ohms.

Solution

Unless otherwise stated, units are in kΩ, mmhos, mA, and V. The bias circuit of Q_5 has R_B, R_E, and V_B equal to 10 kΩ, 5.5 kΩ, and 12 V, respectively, and beta is 100. Accordingly, I_{C5} is 2, and I_{D1} and I_{D2} are each 1 mA. The bias circuit of Q_3 has $R_B = 10$ kΩ, $R_E = 4.65$ kΩ, and $V_B = 10I_{D1} = 10$ V. As beta is 100, I_{C3} is -1.94 mA. For Q_4 we find that $R_B = 13$ kΩ, $R_E = 0.836$ kΩ, and $V_B = 13|I_{C3}| - 20.07$, or 5.15 V, giving $I_{C4} = 4.57$ mA. The dc output voltage is -0.074 V.

From I_D, I_{DSS}, and V_P we determine that g_{fs} is 1 mmho. From the quiescent collector currents, the ac betas, and the specified values of r_x, we find that h_{ie3} is 1.34 and h_{ie4} is 0.60 kΩ. The only suitable feedback is that utilizing voltage sampling and voltage summing.

In the A network of Fig. 20-35 the resistance R shunting the input is for the purpose of avoiding infinites in the determination of the composite h-parameters (see P20-23). We shall regard R as infinite. Note that the loading effect of h_{ib} is of no consequence. We shall select the resistors of the feedback network so that $1/h_{ob}$ is about 40 kΩ, as indicated.

To determine A, let V_s' be 1. Current I_1 can be calculated from two loop equations. The equation around the loop from s_1 to ground to s_2 and back to s_1 can be put in the form $V_{gs1} - V_{gs2} = 2.28 I_1$. That around the loop from g_1 to ground to g_2 to s_2 and s_1 and back to g_1 requires that $V_{gs1} - V_{gs2}$ equals $1 - 0.1 I_1$. From these we find that I_1 is 0.42 mA. Using the current-division theorem gives $I_2 = -0.37$ mA. The input resistance of the CC stage is 83 kΩ, determined by adding 0.60 to the product 101×0.819. Again using current division, we find that I_3 is $37 \times (13/96)$, or 5.01 mA. The output voltage V_o is $101 I_3$ times the effective load resistance, and V_o equals A. We obtain $A = 414$.

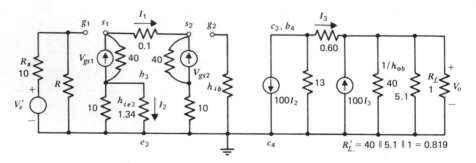

Figure 20-35   The A network with units of kΩ, mA, V.

The specified desensitivity requires that AB be 9. Thus we shall design the feedback network so that B is 0.0217, using a resistor R_1 between the gate of Q_2 and ground and a resistor R_f connected from the gate to the output node. We find that B is h_{rb}, which is $R_1/(R_1 + R_f)$. Suitable values are 865 ohms for R_1 and 39 kΩ for R_f.

REFERENCES

1. P. E. Gray and C. L. Searle, *Electronic Principles*, Wiley, New York, 1969.
2. J. F. Pierce and T. J. Paulus, *Applied Electronics*, Merrill, Columbus, Ohio, 1972.
3. J. Millman and C. C. Halkias. *Integrated Electronics*, McGraw-Hill, New York, 1972.
4. J. F. Cleary, Editor, *Transistor Manual*, General Electric Co., Syracuse, New York, 1964.

PROBLEMS

Section 20-1

20-1. An idealized feedback amplifier, represented by the block diagram of Fig. 20-1, has an open-loop voltage gain A of -1000 and a feedback factor B of -0.049. If the phasor output voltage V_o is 2, calculate the input voltage V_i and the voltage supplied to the summing network by the feedback network. Also, determine the ratio of the voltage at the input of the basic amplifier to V_i.

Section 20-2

20-2. The output stage of a power amplifier has a gain A_1 of -20. At the input is objectional hum introduced from the power supply, giving a signal-to-noise S/N ratio of 5. In order to increase this ratio at the output, a noiseless amplifier stage with a gain of 50 is inserted immediately ahead of the power stage, along with two-stage feedback. The feedback factor B is adjusted so that the net gain is -20 as before. Calculate B and the ratio S/N at the output

20-3. The function $x_i = Bx_o + \sqrt{0.02x_o + 1} - 1$ was given as (20-18), and plots of x_o versus x_i were presented in Fig. 20-5 for $B = 0$ and also for $B = 0.1$. (a) With no feedback ($B = 0$), suppose x_i is $0.5(1 + \cos \omega t)$, causing x_o to vary between 0 and 150. Calculate the second harmonic percent distortion. (b) With $B = 0.1$, deduce that x_o is $10x_i + 11 - \sqrt{20x_i + 121}$. Using the Taylor's expansion at $x_i = 0$ to find the first two terms of the series, deduce that x_o is $9.09(x_i + x_i^2/242 - \cdots)$. Refer to (12-13). For $x_i = 8(1 + \cos \omega t)$, which again causes x_o to vary between 0 and 150, calculate the second-harmonic percent distortion, using only the first two terms of the series expansion. Compare the results of (a) and (b).

Section 20-3

20-4. Write the y-parameter equations for both the basic amplifier and the feedback network of Fig. 20-8. Add the corresponding equations, noting that I_i is the sum of I_i' and I_i'' and I_o is the sum of I_o' and I_o''. From the results, deduce (20-19).

20-5. With reference to the network of Fig. 20-9, write two node equations relating I_s, V_i, and V_o, and solve for the gain A_f, with $A_f = V_o/I_s$. Show that the result can be put in the form $A/(1 + AB)$ with A and B defined by (20-21).

20-6. Using the network of Fig. 20-9, write two node equations that relate I_i, V_i, and V_o. From these and from T as defined by (20-21), derive the expressions for A_v and Z_i given by (20-22) and (20-23). Deduce A_i in the form shown in (20-22). Also, deduce Z_o from Z_i and symmetry.

20-7. Suppose the source resistance R_s of a parallel-parallel feedback amplifier is so large that Y_s in (20-21) is negligible, making T independent of R_s. (a) For this case deduce that Z_i is $(1/y_{ia})$ divided by $1 + T$, provided y_{ib} is small compared with y_{ia}. (b) With the same assumptions, deduce that A_i is also reduced approximately by the same factor $1 + T$. (c) Then show that Z_o is reduced by $1 + T$ provided the load resistance is so large that T does not depend on R_L and provided y_{ob} is negligible.

20-8. For the amplifier of Figs. 20-11 and 20-12, form the network of Fig. 20-14 and calculate the ratio I_s''/I_s', which is T. Compare the network used to find T with the A network of Fig. 20-13.

20-9. For the amplifier of Figs. 20-11 and 20-12, redesign the feedback network, selecting R_f so as to give a desensitivity of 10. Then calculate the closed-loop gain A_f in kilohms. Determine A_v, A_i, R_i, and R_o and compare with the values given in Sec. 20-3 for the amplifier with no feedback. Also, calculate V_o/V_s, with V_s denoting the open-circuit voltage of the source (*Ans*: -500).

20-10. (a) For the inverting OP AMP of Fig. 14-11 of Sec. 14-3, having the incremental model of Fig. 14-13a with $Z_2 = Z_1 \| Z_f$, form the A network and determine A in terms of A_v, Z_i, Z_o, $Z_1 \| Z_f$, and $Z_L \| Z_f$. Let $I_s = V_s/Z_1$. (b) Deduce that T is $A_r Z_1/(Z_1 + Z_f)$ provided $Z_i \gg Z_1 \| Z_f$ and $Z_o \ll Z_L \| Z_f$. (c) Calculate T, V_o/V_s, the input resistance including Z_1, and the output resistance using (20-23), for $A_r = 10^5$, $Z_i = 1$ MΩ, $Z_o = 100$ ohms, $Z_1 = 1$ kΩ, $Z_f = 20$ kΩ, and $Z_L = 10$ kΩ. (*Ans*: 4680, -20, 1 kΩ, 0.021 ohm)

Section 20-4

20-11. Deduce (20-28) for the series-parallel configuration of Fig. 20-15.

20-12. With reference to the network of Fig. 20-16, write one loop and one node equation relating V_s, I_i, and V_o, and solve for the gain A_f, with $A_f = V_o/V_s$. Show that the result can be put in the form of (20-30) with A and B defined by (20-31).

20-13. In the feedback amplifier of Fig. 20-19 assume that each transistor has $\beta_F = 100$ and a base-emitter voltage of 0.7 V. Noting that each I_B is $0.01 I_C$ and each I_E is $-1.01 I_C$, and using units of kΩ, mA, and V, write four loop equations and three node equations relating the currents I_{C1}, I_{C2}, I_{C3}, I_1, I_2, I_3, and I_4. Verify that these equations are satisfied for currents with respective values of 1, 1, -1, 0.0098, 0.0002, 1, and -0.0002 mA.

20-14. For the amplifier of Fig. 20-19 calculate the desensitivity and the closed-loop gain V_o/V_s. Also find the terminal voltage gain A_v, the current gain A_i, the input resistance R_i, and the output resistance R_o. The A network is that of Fig. 20-22.

20-15. For an amplifier of the form of Fig. 20-23 the resistances R_B, R_{C1}, and R_{C2} are so large they can be omitted, and V_g and R_g become V_s and R_s respectively. Both R_g and the load resistance are 500 ohms, and each transistor has $h_{ie} = 1$ kΩ and $h_{fe} = 50$. With $R_1 = 40$ ohms estimate from (20-36) the value of R_f that should be used to give a voltage ratio V_o/V_s of 30. With this R_f calculate the open-loop gain A, and from the result determine more accurately the value of R_f in kilohms for a closed-loop gain of 30. (*Ans*: 1.32)

20-16. (a) For the noninverting OP AMP of Fig. 14-11, sketch the incremental circuit using the model of Fig. 14-12. With $Z_2 = Z_1 \| Z_f$, form the A network and determine A in terms of A_v, Z_i, Z_o, $Z_1 \| Z_f$, and $Z_L \| Z_f$. (b) Deduce that T is $A_r Z_1/(Z_1 + Z_f)$ provided $Z_i \gg Z_1 \| Z_f$ and $Z_o \ll Z_L \| Z_f$. (c) Calculate T, V_o/V_s, and the input and output resistances using (20-33), for $A_v = 10^5$, $Z_i = 1$ MΩ, $Z_o = 100$ ohms, $Z_1 = 1$ kΩ, $Z_f = 20$ kΩ, and $Z_L = 10$ kΩ.

Section 20-5

20-17. In Fig. 20-26 replace the basic amplifier and the feedback network with a g-parameter model that satisfies (20-39). The model should contain a current source $g_r I_o$ and a voltage source $g_f V_i$. From the network deduce (20-40) through (20-43).

20-18. For the two-stage amplifier of Fig. 20-27, assume h_{ie1} is 1300 ohms, h_{ie2} is 1000 ohms, and each transistor has an ac beta of 50. For a source resistance R_s of 5 kΩ and a load of 400 ohms, calculate the open-loop gain A, the feedback factor B, the loop transmission T, the current ratio I_o/I_s, and the input impedance in ohms. (*Ans*: -1390, -0.00812, 11.3, -111, 78)

20-19. For the feedback amplifier of Fig. 20-27, assume each V_{BE} is 0.7 V and each dc beta is 50. Express each branch current in terms of I_{C1}, I_{C2}, and the direct current I fed back through R_f to the input node. From three suitably chosen loop equations determine I_{C1} and I_{C2} in mA, and deduce that I is $12.5 I_{B1}$, which is sufficiently large for excellent stability.

Section 20-6

20-20. (a) Each stage of a two-stage amplifier has internal feedback. Each has an open-loop gain A_1 and a feedback factor B_1. Deduce that the desensitivity is $(1 + \sqrt{AB_1})$, with A denoting the overall open-loop gain A_1^2. (b) Suppose the internal feedback is removed and replaced with multistage feedback with a feedback factor B_2. Deduce that the desensitivity is $(1 + AB_2)$. (c) If the feedback factors B_1 and B_2 of the two cases are such that the desensitivities are equal, calculate the ratio of the closed-loop gains for $A = 10^4$ and $B_2 = 0.0049$. Note the improved gain when feedback is multistage. From (20-4) the desensitivity is $(dA/A)/(dA_f/A_f)$.

20-21. In the feedback amplifier of Fig. 20-30 transistors Q_1 and Q_2 have $h_{ie} = 2.5$ kΩ and Q_3 has $h_{ie} = 1.5$ kΩ. Each β_o is 100. (a) Form the A network and determine z_i, z_f, and z_o from that part of the A network between R_s and the effective load R_L of 1 kΩ. (b) Calculate T and A_f. (c) Find A_v, A_i, R_i, and R_o. Let I_o be the incremental emitter current I_{e3} and give answers in kΩ.

Section 20-7

20-22. For the amplifier of Fig. 20-31 with parallel-parallel feedback using a resistance of 24.3 kΩ for R_f, from the composite y-parameters found from the A and B networks determine the terminal gains A_v and A_i and the input and output resistances R_i and R_o in ohms. The data of Example 1 apply.

20-23. With $R_1 = 865$ ohms and $R_f = 39$ kΩ, add R_1 and R_f to the amplifier of Fig. 20-34 and Example 2, forming series-parallel feedback. From the A and B networks find the composite parameters h_i, h_r, h_f, and h_o in terms of the resistance R shown in Fig. 20-35 at the input. Then calculate T, A_f, and the output resistance R_o, as R approaches infinity.

20-24. From the voltage-sample current-sum operational amplifier of Fig. 14-13b of Sec. 14-3, form the A network with $I_s = V_s/Z_1$. (a) In terms of A_v and the impedances, find A, B, T, and A_f, with $A_f = V_o/I_s$. Then deduce (14-19) for A_{Vs}. (b) With $A_v = 10^4$, $Z_1 = Z_2 = 1$ kΩ, $Z_i = Z_f = Z_L = 50$ kΩ, and $Z_o = 100$ ohms, calculate the loop transmission T and the voltage gain A_{Vs}.

20-25. Assume the noninverting OP AMP of Fig. 14-11 of Sec. 14-3 has $Z_1 = Z_2 = 1$ kΩ, $Z_f = 100$ kΩ, and $Z_L = 10$ kΩ. The amplifier has $A_v = 10^5$, $Z_i = 500$ kΩ, and $Z_o = 100$ ohms. Without making approximations, form the A network. From h-parameters calculate the closed-loop gain A_f, the input resistance R_{in} in MΩ, and the output resistance R_{out} in ohms. Check your results using the approximate equations (14-21) through (14-23).

20-26. For the CE amplifier of Fig. 15-16 of Sec. 15-6, with $R_g = 800\ \Omega$ and $R_L = 300\ \Omega$, assume the ac betas are 100, h_{ie1} is 4 kΩ, and h_{ie2} is 1.7 kΩ. (a) Replace V_g, R_g, and the base bias resistors with an equivalent circuit consisting of V_s and Z_s, and form the A network. Find the composite h-parameters and determine the desensitivity. (b) Calculate V_o/V_g and the input and output resistances (Ans: 10.0, 31.5 kΩ, 41 Ω).

Chapter Twenty One
Frequency and Transient Response of Feedback Amplifiers

Although the design of feedback amplifiers was treated in the preceding chapter, no design is complete without consideration of the frequency and transient response. A large amount of feedback often leads to excessive peaking in the frequency response curve, as well as overshoot accompanied by damped oscillations in the transient response to a step function. In fact, instability as evidenced by sustained oscillations results if the feedback causes any of the poles to move into the right-half plane. As we shall see, it is possible to add circuit components so as to modify the poles and zeros of the loop transmission in a way permitting increased feedback with acceptable response; this is referred to as *compensation*. Let us begin with a discussion and review of the effect of feedback on an amplifier having a single low-frequency pole and a single high-frequency pole.

21-1. ONE-POLE AMPLIFIERS

The open-loop gain $A(s)$ of an amplifier having one pole p_a in the low-frequency region with the finite zero at $s = 0$ is

$$A(s) = \frac{A_o s}{s - p_a} = \frac{A_o}{1 - p_a/s} \qquad (21\text{-}1)$$

The midband gain is A_o, the pole p_a lies on the negative real axis of the complex plane, and the zero is at the origin. Assuming a feedback network of resistors,

684

the feedback factor B equals its midband value B_o at all frequencies. From (21-1) and the relation $A_f = A/(1 + T)$, we determine the closed-loop gain to be

$$A_f(s) = \frac{A_{fo}}{1 - p_1/s} \qquad \text{with } p_1 = \frac{p_a}{1 + T_o} \qquad (21\text{-}2)$$

where T_o is the midband loop transmission $A_o B_o$, and of course, A_{fo} is the midband closed-loop gain. The result shows that the pole p_1 with feedback equals the value of the open-loop pole p_a *divided by the midband desensitivity* $1 + T_o$.

The position of p_1 in the complex plane depends on the midband loop transmission T_o. A plot of the paths of the poles of an amplifier as T_o is increased from zero to infinity is referred to as the *root locus*. For this case there is only a single pole p_1. It coincides with p_a when T_o is zero and it moves toward the origin as T_o approaches infinity. Because the pole is confined to the left-half plane for all values of T_n, the amplifier is unconditionally stable. The root locus shown in Fig. 21-1a consists of that part of the real axis between p_a and the origin. If the zero z_a is not at the origin but is located at some point to the right of p_a, the pole p_1 approaches z_a as T_o approaches infinity (see P21-2).

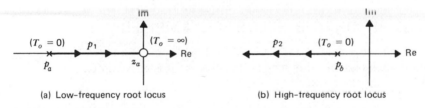

(a) Low-frequency root locus   (b) High-frequency root locus

Figure 21-1   Root locus plots for one-pole amplifiers.

Next let us consider the high-frequency region. For a single pole p_b the open-loop gain is

$$A(s) = \frac{A_o}{1 - s/p_b} \qquad (21\text{-}3)$$

For $B = B_o$ the closed-loop gain is

$$A_f(s) = \frac{A_{fo}}{1 - s/p_2} \qquad \text{with } p_2 = p_b(1 + T_o) \qquad (21\text{-}4)$$

The pole p_2 with feedback equals the value of the open-loop pole p_b *multiplied by the midband desensitivity* $1 + T_o$. The root locus is shown in Fig. 21-1b. Note that p_2 moves from p_b, with $T_o = 0$, to minus infinity as T_o approaches infinity. Again the amplifier is unconditionally stable, because the only pole is confined to the left-half plane.

We observe that the product $A_{fo}p_2$ with feedback equals the product $A_o p_b$ without feedback. Thus feedback can be used to trade gain for bandwidth, *with the gain-bandwidth product maintained constant as the feedback is increased.* Amplifiers with a single low-frequency pole and a single high-frequency pole have no peaking in the frequency response, and there is no overshoot in the transient response to a step function. This was shown in Figs. 18-6 and 18-8 of Sec. 18-2.

21-2. HIGH-FREQUENCY RESPONSE OF A TWO-POLE AMPLIFIER

We shall now examine in some detail the high-frequency response of a feedback amplifier with a loop transmission $T(s)$ that is

$$T(s) = \frac{T_o}{(1 - s/p_a)(1 - s/p_b)} \tag{21-5}$$

Transmission T is AB and the midband value T_o is $A_o B_o$. The two poles p_a and p_b are on the negative real axis, and the zeros of T are infinite.

THE POLES OF THE FEEDBACK AMPLIFIER

The closed-loop gain A_f is $A/(1 + T)$. For a resistive network, B is B_o, a constant independent of frequency. In this case it is easy to deduce with the aid of (21-5) that $A_f(s)$ is

$$A_f(s) = \frac{A_{fo}}{1 - \dfrac{(p_a + p_b)s}{p_a p_b(1 + T_o)} + \dfrac{s^2}{p_a p_b(1 + T_o)}} \tag{21-6}$$

with $A_{fo} = A_o/(1 + T_o)$. By setting the denominator of (21-6) equal to zero and solving for s we find that the poles p_1 and p_2 of the feedback amplifier are

$$p_1, p_2 = \frac{p_a + p_b}{2} \left(1 \pm \sqrt{1 - \frac{4 p_a p_b (1 + T_o)}{(p_a + p_b)^2}} \right) \tag{21-7}$$

Numerical subscripts are used in this chapter to denote poles and zeros of the closed-loop gain, whereas letter subscripts denote those of T.

It is convenient to define the *quality factor* Q by the relation

$$Q = \sqrt{m(1 + T_o)} \tag{21-8}$$

with

$$m = \frac{p_a p_b}{(p_a + p_b)^2} = \frac{p_b/p_a}{(1 + p_b/p_a)^2} \tag{21-9}$$

In terms of Q the poles of (21-7) are

$$p_1, p_2 = 0.5(p_a + p_b)(1 \pm \sqrt{1 - 4Q^2}) \qquad (21\text{-}10)$$

The root locus, shown in Fig. 21-2, can be deduced from (21-10). For $T_o = 0$, Q is $\sqrt{m}$, and it follows from (21-10) that p_1 and p_2 are equal to p_a and p_b. *Thus the root locus begins at the poles of the loop transmission* T. This is true not only for this case but also for amplifiers with any number of poles. In fact, it applies to situations in which one or more of the poles of T are contributed by the B network (see P21-3). Increasing T_o increases Q. From (21-10) we note that the smaller pole becomes larger and the larger pole becomes smaller, so that the poles move toward one another. They meet when Q becomes 0.5.

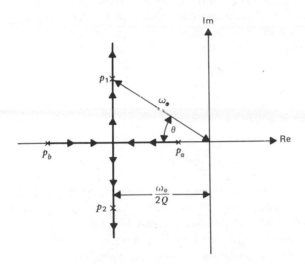

Figure 21-2   Root locus of two-pole amplifier.

For Q greater than 0.5 the poles of (21-10) can be expressed as

$$p_1, p_2 = 0.5(p_a + p_b)(1 \pm j\sqrt{4Q^2 - 1}) \qquad (21\text{-}11)$$

The real part of each pole is a constant independent of T_o, but the imaginary parts increase as T_o increases, approaching plus and minus infinity as indicated in Fig. 21-2. It is convenient to define ω_o as follows:

$$\omega_o = -Q(p_a + p_b) = \sqrt{p_a p_b (1 + T_o)} \qquad (21\text{-}12)$$

From (21-11) we find that the magnitude of each complex-conjugate pole is ω_o and each real part equals $-\omega_o/2Q$. From the geometry of Fig. 21-2 it is evident that Q and the *pole angle* θ are related, with

$$\cos \theta = \frac{1}{2Q} \tag{21-13}$$

The angle θ lies between the negative real axis and the segment drawn from the origin to the complex pole p_1, as illustrated, and θ has values between zero and 90°.

EFFECTS OF COMPLEX-CONJUGATE POLES

In Sec. 18-2 Bode plots were developed to show the effect of real poles and zeros on the frequency response of an amplifier. Our interest here is in the effect of complex poles. First, let us consider the extreme case of poles that have an angle θ near 90°, which gives complex-conjugate poles that are almost pure imaginaries. Assuming p_1 and p_2 are $\pm j\omega_o$, the s-domain closed-loop gain A_f is

$$A_f(s) = \frac{A_{fo}}{(1 - s/p_1)(1 - s/p_2)} = \frac{A_{fo}}{1 + s^2/\omega_o^2} \tag{21-14}$$

In the real frequency domain s is $j\omega$ and A_f becomes

$$A_f(\omega) = \frac{A_{fo}}{1 - \omega^2/\omega_o^2} \tag{21-15}$$

For small values of ω, the amplifier is in the midband region and A_f is A_{fo}. For sufficiently large ω the gain is decreasing 40 dB per decade, just as required by the Bode plot of a two-pole amplifier. However, when ω is near ω_o the dB gain is large with the gain approaching infinity as ω approaches ω_o. Clearly, *complex-conjugate poles result in peaking in the frequency response provided the pole angle is sufficiently large, and the peaking occurs in the vicinity of the magnitude ω_o of a complex pole.*

The mathematical behavior of complex-conjugate zeros with sufficiently large imaginary components is somewhat similar. However, because the zeros are in the numerator of the gain function, *the dB gain dips at frequencies in the vicinity of the magnitude of a complex zero.* We encountered a pair of complex-conjugate zeros in the computer solution of the network of Fig. 19-9 of Sec. 19-1. Seldom do they occur within the useful frequency range of an amplifier.

To investigate in greater depth the peaking effect of complex poles it is convenient to express the gain $A_f(s)$ of (21-6) in the form

$$A_f(s) = \frac{A_{fo}}{1 + s/Q\omega_o + s^2/\omega_o^2} \tag{21-16}$$

With $s = j\omega$ this becomes

$$A_f(\omega) = \frac{A_{fo}}{\sqrt{(1 - \omega^2/\omega_o^2)^2 + \omega^2/Q^2\omega_o^2}} \Big/ -\tan^{-1}\left(\frac{\omega/Q\omega_o}{1 - \omega^2/\omega_o^2}\right) \quad (21\text{-}17)$$

From (21-17) we find that the gain peaks for Q greater than 0.707 (see P21-4), with the peak occurring at a frequency and with a maximum given by

$$\omega = \omega_o\sqrt{1 - 0.5/Q^2} \qquad A_{f(max)} = \frac{Q|A_{fo}|}{\sqrt{1 - 0.25/Q^2}} \quad (21\text{-}18)$$

Normalized frequency responses for different values of Q are shown in Fig. 21-3. The two lower curves have values of Q of 0.5 and 0.707, corresponding to respective pole angles of zero and 45°, and these do not peak. For $Q = 1$ the pole angle is 60°, and the maximum point is marked on the curve with an x. The other two curves peak almost precisely at the magnitude ω_o of a complex pole of the two-pole feedback amplifier.

The response to a step function is of particular importance in many amplifiers. Suppose that $A_f(s)$ represents the voltage ratio V_o/V_s. If $v_s(t)$ is a step function of magnitude $1/A_{fo}$, then the Laplace transform $V_s(s)$ of the input voltage is $1/s$ times this magnitude. From (21-16) the transform $V_o(s)$ of the output voltage is determined to be

$$V_o(s) = \frac{\omega_o^2}{s(s^2 + \omega_o s/Q + \omega_o^2)} \quad (21\text{-}19)$$

With the aid of a suitable table of transforms the output voltage $v_o(t)$ is found. The result is

$$v_o(t) = 1 - \sqrt{\frac{4Q^2}{4Q^2 - 1}}\, e^{-\alpha t} \sin[(\sqrt{4Q^2 - 1})\alpha t + \cos^{-1} 0.5/Q] \quad (21\text{-}20)$$

with $\alpha = 0.5\omega_o/Q$, which is the magnitude of the real part of the complex-pole pair. This can be used to plot $v_o(t)$ as a function of αt for any selected value of Q equal to or greater than 0.5.

Shown in Fig. 21-4 are the step-function responses for four different values of Q. The output voltage for $Q = 0.5$ approaches its final value gradually, with no overshoot and no oscillations. However, for $Q > 0.5$ the curves approach unity with damped oscillations. For the normalized response of Fig. 21-4 the overshoot is the maximum v_o less 1. The percent overshoot is given on the illustration for each of the four curves.

To plot curves such as those of Fig. 21-4 from (21-20) it is helpful to know the maximum and minimum points. These are found by setting the time derivative

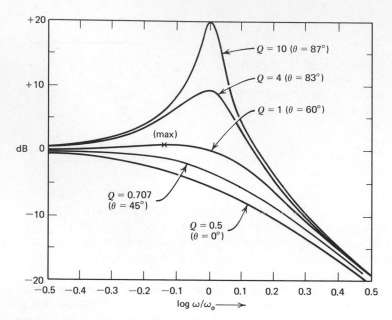

Figure 21-3   Normalized frequency response.

$$[\omega_o = -Q(p_a + p_b)]$$

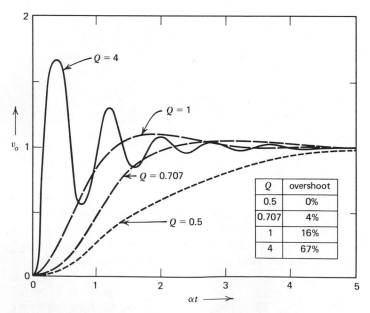

Q	overshoot
0.5	0%
0.707	4%
1	16%
4	67%

Figure 21-4   Normalized step-function response.

690

of v_o equal to zero and solving for αt (see P21-5); then the expression for αt is substituted into (21-20). The results are

$$v_{o(max,min)} = 1 - (-1)^n \exp\left(\frac{-\pi n}{\sqrt{4Q^2 - 1}}\right) \tag{21-21}$$

$$\alpha t_{(max,min)} = \frac{\pi n}{\sqrt{4Q^2 - 1}} \tag{21-22}$$

with $n = 1, 2, 3$, etc. From (21-21) with $n = 1$ the percent overshoot is determined to be

$$\text{overshoot} = \exp\left(\frac{-\pi}{\sqrt{4Q^2 - 1}}\right) \times 100\% \tag{21-23}$$

A satisfactory value of Q depends, of course, on the particular application. As we have seen, for $Q > 0.5$ the poles are complex, and complex poles produce overshoot in the transient response to a step function, accompanied by damped oscillations. In some applications overshoot must be completely avoided. For example, any visible overshoot in the display of an oscilloscope is certainly intolerable. *In addition to overshoot, which is present when $Q > 0.5$, there is peaking in the frequency response when Q is greater than* 0.707. Frequencies in the input signal that are within the band of the peaking are amplified more than others. This produces a ringing effect in the output if the peaking is appreciable. In audio amplifiers an overshoot of perhaps 5% to 10% is often acceptable, corresponding to values of Q from 0.72 to 0.85. Overshoots of about 5% are satisfactory in TV amplifiers. In fact, a small overshoot is desirable because of the decreased rise time. The curves of Fig. 21-4 clearly show the improved rise time for the larger values of Q (also, see P21-6). Because a fast rise time is especially important in many servo amplifiers, such amplifiers are often designed with substantial overshoot as well as with peaking in the frequency response.

The restriction on Q places serious limitations on the amount of feedback that can be utilized in a two-pole amplifier; Q equals $\sqrt{m(1 + T_o)}$ by (21-8), and m is a function of the ratio of the poles of T, with m given by (21-9). These relations were used to obtain the curves of Fig. 21-5, which show T_o versus the pole ratio for different Q. If the basic amplifier has a pole ratio that is moderate, say 5 or less, note that T_o is limited to rather small values, assuming Q is reasonably restricted. For example, suppose a maximum Q of 0.707 is desired for an amplifier with a pole ratio of 2. For this case, T_o must not exceed 1.25. If T_o is increased to 10 for improved desensitivity, Q becomes 1.56, which gives an overshoot of 35% as well as appreciable peaking in the frequency response.

In the event that the maximum allowable Q results in a midband loop

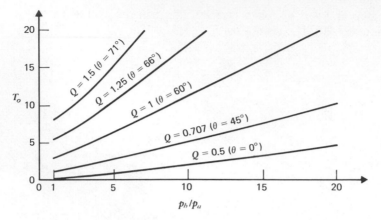

Figure 21-5 T_o versus p_b/p_a for different Q.

transmission that is too small for adequate desensitivity, changes must be made that modify the root locus appropriately. For example, a capacitor might be added to the basic amplifier so as to produce an additional pole of T that is much smaller in magnitude than the original poles. Thus the amplifier resembles one with a single dominant pole, and a larger T_o can be used. Although the bandwidth is substantially reduced, this procedure is easy to implement. It is often used to stabilize operational amplifiers. Other methods of compensation will be considered later.

The upper 3 dB frequency ω_2 of an amplifier having the gain function of (21-14), with complex poles p_1 and p_2 that are $-\alpha \pm j\beta$, can be expressed (see P21-7) as follows:

$$\omega_2 = [\sqrt{(\alpha^2 - \beta^2)^2 + (\alpha^2 + \beta^2)^2} - \alpha^2 + \beta^2]^{1/2} \qquad (21\text{-}24)$$

This can also be used for amplifiers with more than two poles provided the other poles are much larger than the magnitude of the complex-conjugate pair.

The results deduced in this section apply precisely only to the particular case of a two-pole amplifier with both zeros at infinity. A further restriction appears from our tacit assumption that the poles of T remained constant as T_o was increased. *In fact, the resistors of the feedback network load the A network, and consequently, adjusting* T_o *by changing these resistors certainly affects the poles of* T. *In practical cases the effect is often small.* Many amplifiers with more than two poles have high-frequency responses that are governed almost entirely by the two smallest poles, in which case the results of this section apply approximately. We shall now examine a two-pole feedback amplifier in the low-frequency region.

21-3. LOW-FREQUENCY RESPONSE OF A TWO-POLE AMPLIFIER

The feedback amplifier to be considered here is one in which the low-frequency incremental model yields a loop transmission $T(s)$ that is

$$T(s) = \frac{T_o s^2}{(s - p_a)(s - p_b)} = \frac{T_o}{(1 - p_a/s)(1 - p_b/s)} \tag{21-25}$$

The low-frequency poles p_a and p_b are on the negative real axis, and the zeros of $T(s)$ are at the origin. One way of obtaining such a loop gain is to use a resistive feedback network with an amplifier having two coupling capacitors, one at the input and one at the output.

THE POLES OF THE FEEDBACK AMPLIFIER

For $B = B_o$, a constant independent of frequency, it is easy to deduce that the closed-loop gain $A_f(s)$ is

$$A_f(s) = \frac{A_{fo}}{1 - \dfrac{p_a + p_b}{s(1 + T_o)} + \dfrac{p_a p_b}{s^2(1 + T_o)}} \tag{21-26}$$

By setting the denominator of (21-26) equal to zero, the low-frequency poles of the feedback amplifier are found to be

$$p_1, p_2 = -\frac{\omega_o}{2Q}(1 \pm \sqrt{1 - 4Q^2}) \tag{21-27}$$

with

$$\omega_o = \sqrt{\frac{p_a p_b}{(1 + T_o)}} \tag{21-28}$$

Equations 21-8 and 21-9 of the preceding section define Q and m; that is, Q is $\sqrt{m(1 + T_o)}$ and m is $p_a p_b/(p_a + p_b)^2$. However, note that ω_o does not have the same mathematical form of ω_o of the high-frequency amplifier. For $Q > 0.5$ the poles are complex, and it is clear from (21-27) that ω_o is the magnitude of each complex pole. This is the identical interpretation given to ω_o in the high-frequency case.

The root locus is presented in Fig. 21-6. As T_o increases from zero, the poles p_1 and p_2 move from p_a and p_b toward one another. At a value of T_o that makes $Q = 0.5$ the poles coincide. It is easy to show that further increases in T_o cause them to move along circular paths that approach the origin as T_o approaches infinity (see P21-8). In the previous section we learned that the root locus starts at the poles of $T(s)$ in all cases. In Fig. 21-6 we observe that the root locus

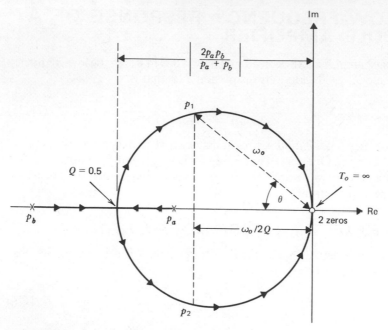

Figure 21-6   Root locus of low-frequency two-pole amplifier.

ends at the zeros of $T(s)$. This was also found to be true in the root loci of Figs. 21-1 and 21-2, and in fact, it is always true. Thus *the poles of a feedback amplifier move from the poles of* $T(s)$ *to the zeros of* $T(s)$ *when the midband loop transmission* T_o *is increased from zero to infinity.*

The pole angle θ is defined as before and illustrated on the figure. The real part of each complex pole is $-0.5\omega_o/Q$ and the pole magnitude is ω_o. Accordingly, $\cos\theta = 0.5/Q$, which is the same relationship obtained in the preceding section and given as (21-13).

FREQUENCY AND STEP-FUNCTION RESPONSES

In the real-frequency domain the closed-loop gain of (21-26) can be put in the form

$$A_f(\omega) = \frac{A_{fo}}{\sqrt{(1 - \omega_o^2/\omega^2)^2 + \omega_o^2/Q^2\omega^2}} \Big/ {-\tan^{-1}\left(\frac{\omega_o/Q\omega}{1 - \omega_o^2/\omega^2}\right)} \quad (21\text{-}29)$$

From this we find that the gain peaks for Q greater than 0.707, with the peak occurring at a frequency and with a maximum given by

$$\omega = \frac{\omega_o}{\sqrt{1 - 0.5/Q^2}} \qquad A_{f(\max)} = \frac{Q|A_{fo}|}{\sqrt{1 - 0.25/Q^2}} \quad (21\text{-}30)$$

Comparison of (21-29) with (21-17) reveals that the expressions for the low- and high-frequency gains are the same except that ω/ω_o in (21-29) corresponds to ω_o/ω in (21-17). Thus the frequency at which the peak occurs in the low-frequency case is greater than ω_o by the factor $1/\sqrt{1 - 0.5/Q^2}$, whereas in the two-pole high-frequency case with both zeros at infinity, the peak occurs at a frequency less than ω_o by the factor $\sqrt{1 - 0.5/Q^2}$. This is verified by comparison of (21-30) with (21-18). For example, if the low-frequency complex poles have a Q of unity, the peak occurs at $1.41\omega_o$, but in the high-frequency case with poles having a unity Q the peak occurs at $0.707\omega_o$; in each case the maximum value of A_f is the same. As the feedback is increased with T_o and Q made larger, the peaks in the low-frequency and high-frequency cases approach the product $Q|A_{fo}|$ at frequencies approaching the respective values of ω_o, and these values move away from the midband region at both ends of the frequency band. The normalized frequency-response curves of Fig. 21-3 apply also to the low-frequency case provided the variable of the abscissa is changed from log ω/ω_o to log ω_o/ω.

The transient response to a step function can be determined from the expression of (21-26) for $A_f(s)$, which can be converted into the form

$$A_f(s) = \frac{A_{fo}s^2}{s^2 + \omega_o s/Q + \omega_o^2} \tag{21-31}$$

Let us assume that $A_f(s)$ is the voltage ratio V_o/V_s and that the instantaneous input $v_s(t)$ is a step function of magnitude $1/A_{fo}$. The Laplace transform $V_s(s)$ of $v_s(t)$ is $1/(A_{fo}s)$. From this and (21-31), we find $V_o(s)$ to be

$$V_o(s) = \frac{s}{s^2 + \omega_o s/Q + \omega_o^2} \tag{21-32}$$

For $Q \geq 0.5$, the time-domain output voltage can be determined with the aid of a suitable table of Laplace transforms. The result is

$$v_o(t) = \sqrt{\frac{4Q^2}{4Q^2 - 1}}\, e^{-\alpha t} \sin[(-\sqrt{4Q^2 - 1})\alpha t + \cos^{-1} 0.5/Q] \tag{21-33}$$

with $\alpha = 0.5\omega_o/Q$, which is the magnitude of the real part of the complex-pole pair.

For $Q = 0.5$, from (21-32) we find that $v_o(t)$ is $(1 - \alpha t)\exp(-\alpha t)$, and this function is plotted in Fig. 21-7a. For larger values of Q the response is oscillatory with damping, as shown in Fig. 21-7b for $Q = 4$. In order to avoid excessive oscillations in the transient response and peaking in the frequency response it is good design practice to limit the feedback sufficiently so as to keep the Q of the low-frequency complex poles less than unity.

In Fig. 21-5 of the preceding section several plots of T_o versus the pole ratio p_b/p_a were given. These also apply to the low-frequency case that has been considered. The curves show rather clearly that the midband loop transmission T_o

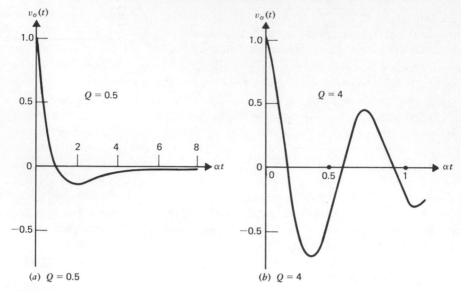

(a) $Q = 0.5$ (b) $Q = 4$

Figure 21-7   Step-function response of low-frequency feedback amplifier for $Q = 0.5$ and $Q = 4$. Note the different scales for αt. The poles of T are p_a and p_b and both zeros are at the origin of the complex plane.

can be made large as long as the pole ratio is large. Fortunately, the circuit designer has control over these low-frequency poles through the selection of the coupling and bypass capacitors. *When a large amount of feedback is to be used, the designer simply selects the capacitors so that the magnitude of the dominant pole is much greater than all the others.* This is accomplished by choosing one capacitor so that its short-circuit time constant is considerably smaller than those of the others, which gives a root locus similar to that of a one-pole amplifier. In the high-frequency region the problem is somewhat more complicated because the poles are not so easily controlled.

Example

Figure 21-8 is a simplified low-frequency model of a three-stage CE amplifier with parallel-parallel feedback. (a) Determine the midband loop transmission T_o that gives complex poles with a pole angle of 45°. (b) If C_2 is increased, a larger value of T_o can be used. With $\theta = 45°$ and $T_o = 15$, what capacitance is required?

Solution

Units used are kΩ, milliseconds, and μF. (a) With R_f infinite, the noninteracting coupling capacitors have time constants of 1 and 10 ms. Hence the poles are -1 and -0.1 ms^{-1}. From (21-9) we find m to be 0.0826. The value of Q is 0.707 for a pole angle of 45°, and (21-8) gives $T_o = 5.05$. This result can also be

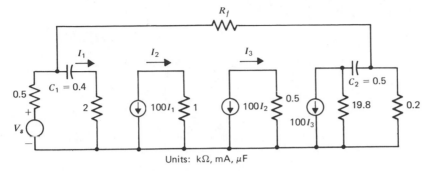

Units: kΩ, mA, μF

Figure 21-8   The low-frequency incremental model of a three-stage CE feedback amplifier.

deduced from the curves of Fig. 21-5. (b) For $T_o = 15$ and the same Q, the value of m is 0.03125 from (21-8), and the required pole ratio is determined from (21-9) to be 30. The dominant pole associated with C_1 is -1. Thus the smaller pole is $-1/30$ ms^{-1}, which is $-1/(20C_2)$. Therefore, C_2 should be 1.5 μF.

The lower 3 dB frequency ω_1 of the gain function of (21-25) in which the poles are complex conjugates $(-\alpha \pm j\beta)$ is

$$\omega_1 = [\sqrt{(\alpha^2 - \beta^2)^2 + (\alpha^2 + \beta^2)^2} + \alpha^2 - \beta^2]^{1/2} \qquad (21\text{-}34)$$

The derivation of (21-34) is similar to the procedure of P21-7.

21-4. MULTIPOLE FEEDBACK AMPLIFIERS

We are now ready to investigate feedback amplifiers with more than two poles. As mentioned in the previous section, the low-frequency response can usually be made to depend primarily on a single pole, or perhaps two poles, by properly selecting the coupling and bypass capacitors. Thus we shall concentrate our attention here on the high-frequency gain, which presents a more difficult problem. Because the high-frequency response of most amplifiers depends mainly on the three lowest poles and because the zeros are quite commonly so large that they can be neglected, the example chosen to illustrate the basic principles is a feedback amplifier having an open-loop gain $A(s)$ given by

$$A(s) = \frac{-A_o p_a p_b p_c}{(s - p_a)(s - p_b)(s - p_c)} \qquad (21\text{-}35)$$

where A_o is the midband gain, the poles p_a, p_b, and p_c are negative and real, and the feedback network is assumed to be resistive. Thus $B = B_o$ and $T_o = A_o B_o$.

THE ROOT LOCUS

The closed-loop gain A_f is $A/(1 + AB)$, and from this and (21-35) we find that

$$A_f(s) = \frac{-A_o p_a p_b p_c}{(s - p_a)(s - p_b)(s - p_c) - T_o p_a p_b p_c} \tag{21-36}$$

The denominator of (21-36) must equal $(s - p_1)(s - p_2)(s - p_3)$ for all values of s, with p_1, p_2, and p_3 denoting the poles of the feedback amplifier. This expression, as well as that of the denominator of (21-36), can be written as the sum of s^3, $b_2 s^2$, $b_1 s$, and b_o, with the b coefficients denoting constants. By equating the respective b coefficients of the two expressions, we obtain the following relations:

$$p_1 + p_2 + p_3 = p_a + p_b + p_c \tag{21-37}$$

$$p_1 p_2 + p_2 p_3 + p_3 p_1 = p_a p_b + p_b p_c + p_c p_a \tag{21-38}$$

$$p_1 p_2 p_3 = p_a p_b p_c (1 + T_o) \tag{21-39}$$

The root locus is shown in Fig. 21-9. As T_o is increased from zero to infinity, the poles move along the indicated paths from the poles p_a, p_b, and p_c of the open-loop gain to infinity. The starting points are evident from examination of (21-37) through (21-39) with $T_o = 0$. From these same equations we can show that the poles must approach infinity as T_o increases without limit. From (21-39) it is clear that at least one pole must be infinite for an infinite T_o. Because the sum of

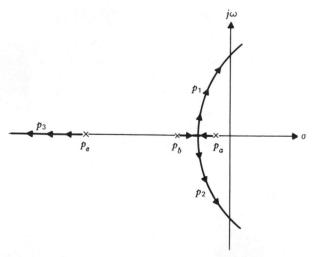

Figure 21-9   Root locus of three-pole amplifier with the zeros at infinity.

the poles must remain constant in accordance with (21-37), it follows that a second pole must be infinite and of opposite sign. The complex-conjugate pair dictates that the third pole also must approach infinity.

The real-axis portions of the root locus are easily deduced from the denominator of (21-36). For s equal to a pole this denominator must be zero, or

$$(s + |p_a|)(s + |p_b|)(s + |p_c|) = -T_o|p_a p_b p_c| \qquad (21\text{-}40)$$

Suppose $|p_c| > |p_b| > |p_a|$. When s is negative and real with a magnitude less than $|p_a|$, the left side of (21-40) is positive, and the equation has no solution for values of T_o between zero and infinity. However, when s is real with any value between p_a and p_b, the left side is negative, and there is a value of T_o that satisfies the equation. For s real and having any value between p_b and p_c, the left side is positive and no solution exists. Finally, for s real, negative, and larger in magnitude than p_c, a value of T_o can be found that satisfies the equation. Thus the real-axis portions of the root locus of Fig. 21-9 are theoretically verified.

As shown in Fig. 21-9, the pole p_3 starts at p_c and moves along the real axis to minus infinity. The poles p_1 and p_2 start at p_a and p_b and move together, then splitting into complex conjugates. The sum of the three poles is constant, independent of T_o, in accordance with (21-37). Therefore, as p_3 moves to the left, the complex conjugates shift to the right. For T_o sufficiently large these poles move into the right-half plane and the amplifier oscillates continuously. The value of T_o giving a pole angle of $90°$ is an absolute maximum. Of course, T_o should be substantially less than this in order to avoid the undesirable effects of complex-conjugate poles with large Q.

Let us summarize, and also generalize without proof, three basic root-locus rules that have been deduced here and in previous sections. They apply to both low-frequency and high-frequency gain functions with poles and zeros allowed in the feedback factor B as well as in the open-loop gain A.

1. The root locus of each pole of $A_f(s)$ starts on a pole of $T(s)$ and ends on a zero of $T(s)$.
2. Portions of the real axis to the left of an odd number of *left-half-plane* poles and zeros of $T(s)$, including zeros at the origin if any, are included in the root locus. Other portions of the real axis in the left half-plane are excluded.
3. The sum of the poles is constant, independent of T_o, provided the number of poles of $T(s)$ is at least two greater than the number of finite zeros of $T(s)$.

FEEDBACK DESIGN

The problem confronting the designer is that of selecting the proper T_o to give complex-conjugate poles with a specified Q. For complex-conjugate poles it is convenient to let p_1 and p_2 be $-\alpha \pm j\beta$ with both α and β positive.

The third pole p_3 is negative and real. In terms of α and β, (21-37) through (21-39) become

$$-2\alpha + p_3 = p_a + p_b + p_c \tag{21-41}$$

$$\alpha^2 + \beta^2 - 2\alpha p_3 = p_a p_b + p_b p_c + p_c p_a \tag{21-42}$$

$$p_3(\alpha^2 + \beta^2) = p_a p_b p_c(1 + T_o) \tag{21-43}$$

From these we deduce the following equations which simplify the design procedure:

$$3\alpha^2 - \beta^2 + 2\alpha(p_a + p_b + p_c) + p_a p_b + p_b p_c + p_c p_a = 0 \tag{21-44}$$

$$p_3 = p_a + p_b + p_c + 2\alpha \tag{21-45}$$

$$T_o = \frac{p_3(\alpha^2 + \beta^2)}{p_a p_b p_c} - 1 \tag{21-46}$$

The first of these equations is that of (21-42) with p_3 eliminated with the aid of (21-41). The other two are those of (21-41) solved for p_3 and (21-43) solved for T_o.

For an amplifier with known poles p_a, p_b, and p_c suppose we wish to design the feedback network to give a value of T_o such that the complex poles have a specified Q. The pole angle is related to Q by the expression $\cos \theta = 0.5/Q$, which is (21-13), and the ratio β/α equals $\tan \theta$. From these it is easy to show that

$$\frac{\beta}{\alpha} = \sqrt{4Q^2 - 1} \tag{21-47}$$

For $Q = 0.5$, both the pole angle and β are zero, and a Q of 0.707 gives an angle of $45°$ with $\beta = \alpha$. With $Q = 1$ corresponding to $\theta = 60°$, β is $\alpha\sqrt{3}$. A pole angle of $90°$ makes Q infinite and $\alpha = 0$. These are the cases of greatest interest. The first step in the design procedure is to solve (21-44) and (21-47) for α and β. Then p_3 is found from (21-45), and finally, T_o is calculated from (21-46). The method is illustrated in the example that follows.

Example

Figure 21-10 is a diagram of an amplifier driven from a 200 ohm source and supplying power to a 1 kΩ load. The type of amplifier is unspecified. It could be an integrated-circuit inverting operational amplifier. Possibly it consists of a discrete noninverting emitter-coupled stage followed by a high-gain CE–CC configuration, or perhaps the arrangement is that of a three-stage CE cascade. In any event, the amplifier is inverting.

From manufacturer's specifications, or from analysis of the incremental circuit, or from experimental data, let us suppose the midband voltage gain V_o/V_s without feedback is determined to be $-50\,000$, with high-frequency poles located at 600 kHz, 5 MHz, and 20 MHz. All zeros and any additional poles are large enough to be regarded as infinite. A suitable feedback network is to be

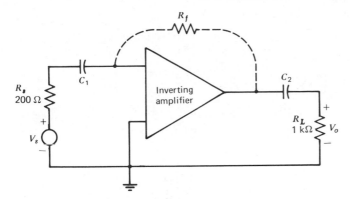

Figure 21-10 Inverting amplifier.

designed to provide a bandwidth of at least 3 MHz, which is about five times that of the basic amplifier.

The voltage-sample current-sum feedback to be used reduces the voltage gain. For a one-pole amplifier we learned in Sec. 21-1 that the gain-bandwidth product remains constant when feedback is added. For our three-pole amplifier we might expect a reduction in the voltage gain by a factor of about 5 to -10^4, more or less. This is the design objective. Investigate conditions for complex-conjugate pole angles of (a) $0°$, (b) $45°$, and (c) $90°$.

Solution

Unless otherwise stated, units employed are kΩ, mA, V, ns, GHz, and pF. The feedback network is simply the resistor R_f shown in Fig. 21-10. Loading effects of this resistor on the A network are assumed to be negligible, and as we shall see, this assumption is well justified. The gain $A(s)$ is V_o/I_s, with the short-circuit current I_s of the source equal to $5V_s$ mA. As the midband ratio V_o/V_s is $-50\,000$, the corresponding open-loop gain A_o is -10^4 kΩ. With s in reciprocal nanoseconds, $A(s)$ can be expressed as

$$A(s) = \frac{-10\,000}{(1 + s/0.00377)(1 + s/0.0314)(1 + s/0.126)} \qquad (21\text{-}48)$$

The poles of (21-48) are -0.00377, -0.0314, and -0.126 ns^{-1}, corresponding to respective frequencies of 600 kHz, 5 MHz, and 20 MHz. We know that T_o is $A_o B_o$, or $10^4/R_f$ with R_f in kΩ, and the root locus is similar to that of Fig. 21-9. With no feedback the bandwidth is found from (21-48) to be 590 kHz. If possible, we want to select R_f so as to increase the bandwidth to at least 3 MHz with a voltage gain of $-10\,000$ or more.

(a) For coincident poles Q is 0.5, the pole angle is zero, and $\beta = 0$. Equation (21-44) becomes

$$3\alpha^2 - 0.322\alpha + 0.00455 = 0 \qquad (21\text{-}49)$$

This gives $\alpha = 0.0167$ ns^{-1}. From (21-45) p_3 is -0.13 ns^{-1}, and T_o is 1.43 from (21-46). As T_o is $10^4/R_f$, R_f is 7.0 MΩ. The midband gain A_{fo} is $A_o/(1 + T_o)$, or -4120 kΩ, and the midband voltage ratio is $5A_{fo}$, or $-20\,600$. Accordingly,

$$\frac{V_o}{V_s} = \frac{-20\,600}{(1 + s/0.0167)^2(1 + s/0.13)} \qquad (21\text{-}50)$$

From (21-24) with $\beta = 0$ the BW is found to be 1.70 MHz, which is somewhat lower than desired. The larger pole has no effect on the first two significant figures of the BW.

(b) For $Q = 0.707$ the pole angle is $45°$, and $\alpha = \beta$. Replacing β in (21-44) with α and solving give $\alpha = 0.0156$ ns^{-1}. From (21-45) we find that p_3 is -0.13 ns^{-1}, and T_o is 3.24 from (21-46). The corresponding value of R_f is 3.1 MΩ. A_{fo} is -2360 kΩ, and the midband voltage ratio is $-11\,800$. It follows that

$$\frac{V_o}{V_s} = \frac{-11\,800}{(1 - s/p_1)(1 - s/p_2)(1 - s/p_3)} \qquad (21\text{-}51)$$

with $p_1, p_2 = -0.0156 \pm j0.0156$ and $p_3 = -0.13$ ns^{-1}. From (21-24) with $\alpha = \beta$, the upper 3 dB radian frequency ω_2 is found to be $\alpha\sqrt{2}$, and f_2 is 3.5 MHz. The larger pole has no effect on the first two significant figures of f_2. Both the midband gain and the BW exceed the specifications.

(c) The amplifier approaches instability when the pole angle approaches $90°$. With $\alpha = 0$ in (21-44), β is found to be 0.0675, and p_3 is -0.161 ns^{-1}, determined from (21-45). Using (21-46), we find T_o to be 49, and the corresponding R_f is 204 kΩ. For R_f less than this the amplifier oscillates.

A practical circuit for the amplifier considered in the example is shown in Fig. 21-11. The feedback is provided by a T-network of resistors rather than by a

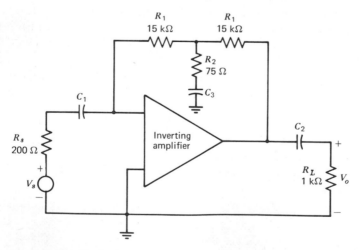

Figure 21-11 The feedback amplifier.

single resistor R_f. The arrangement allows the use of more practical values for precision resistors in place of megohms of resistance as indicated in the solution. Perhaps the two resistors labeled R_1 and capacitor C_3 are a part of the inverting amplifier, designed to give dc feedback for stabilizing the bias circuit of direct-coupled stages. In this event the addition of R_2 permits the same network to be used also for signal feedback. If the basic amplifier is such that dc feedback is undesirable, a blocking capacitor should be added to the feedback loop.

The T-network has a reverse short-circuit transfer admittance y_{rb} equal to $-R_2/(R_1^2 + 2R_1 R_2)$, which is -0.00033 mmho for the values shown on the figure. This feedback factor is equivalent to that of a single resistor R_f of 3.0 MΩ. Therefore, as indicated in part (b) of the preceding solution, it is evident that the complex poles have a pole angle of about $45°$, giving a midband voltage gain of $-11\,800$, a bandwidth of 3.5 MHz, and a midband desensitivity $1 + T_o$ of 4.2. The specifications are adequately satisfied.

At very low frequencies the impedance of C_3 becomes large, resulting in increased feedback. A value of C_3 too small gives complex poles and sharp peaking in the low-frequency region. With C_1 and C_2 selected for a break frequency of about 100 Hz, C_3 should be chosen so that any peaking is below 10 Hz. Several hundred microfarads of capacitance would probably be required.

21-5. GAIN AND PHASE MARGINS

Feedback amplifier design based on root locus requires a knowledge of the poles of the loop transmission, which must be determined from either network analysis or from laboratory data. It is possible and in many cases it is quite advantageous to base the design on the *measured* sinusoidal response of the loop transmission. As we shall see, the experimental data can be related to the steady-state and transient response of the feedback amplifier.

LOOP-TRANSMISSION MAGNITUDE AND PHASE PLOTS

Methods for the direct measurement of T were discussed in the preceding chapter. For each of the four basic configurations the procedure consists of first disconnecting the input terminals of the feedback network from the basic amplifier. For current comparison at the input, an admittance equal to the appropriate input matrix parameter y_{ib} or g_{ib} is then added in parallel with the source, the input at the left side of the feedback network is shorted, and T is the ratio I_s''/I_s'. This is illustrated in Fig. 20-14. For voltage comparison, an impedance equal to the appropriate matrix parameter h_{ib} or z_{ib} is inserted in series with the source, and T is the ratio of the open-circuit voltage at the input of the feedback network to the open-circuit voltage of the source. The modified network is used to obtain data for plotting curves of both the magnitude and phase of $T(\omega)$ versus frequency. An alternate procedure is to determine the magnitude and phase of $A(\omega)$ versus frequency, with $T(\omega)$ calculated from the product AB.

Although the main advantage of the method lies in the use of measured characteristics, it is convenient for illustrative purposes to consider a loop transmission T having a specified form. We shall use the T of the example of the preceding section, with a midband value of 6.10. As shown in P21-11, *this gives a pair of complex-conjugate poles in the high-frequency region having a Q of unity*. With the poles in reciprocal nanoseconds, the function is

$$T(s) = \frac{6.1}{(1 + s/0.00377)(1 + s/0.0314)(1 + s/0.126)} \tag{21-52}$$

In the real-frequency domain $T(\omega) = |T(\omega)| \underline{/\phi}$, with

$$|T(\omega)| = \frac{6.1}{\left[1 + \left(\dfrac{\omega}{0.00377}\right)^2\right]^{1/2}\left[1 + \left(\dfrac{\omega}{0.0314}\right)^2\right]^{1/2}\left[1 + \left(\dfrac{\omega}{0.126}\right)^2\right]^{1/2}} \tag{21-53}$$

and

$$\phi = -\tan^{-1}\left(\frac{\omega}{0.00377}\right) - \tan^{-1}\left(\frac{\omega}{0.0314}\right) - \tan^{-1}\left(\frac{\omega}{0.126}\right) \tag{21-54}$$

In Fig. 21-12 are shown Bode plots of $T(\omega)$, consisting of magnitude and phase plots versus frequency. The data were obtained from (21-53) and (21-54). The magnitude curve is in decibels with the scale on the left and the phase curve is in degrees with the scale on the right. Note that the zero of the decibel scale coincides with $-180°$ of the phase scale.

The closed-loop gain A_f is $A/(1 + T)$, and A_f is infinite for T equal to $1/\underline{-180°}$. In fact, as shown in Sec. 20-1, if the magnitude of T exceeds unity when the phase ϕ of T is $-180°$, the amplifier is unstable and oscillates. This implies complex-conjugate poles in the right-half plane. An amplifier that is almost unstable has high-Q complex-conjugate poles in the left half-plane. We are definitely interested in the degree of stability, and there are two concepts commonly used as indicators. One of these is the *phase margin*, defined *for the high-frequency case* as follows:

$$\text{phase margin} = 180° + \phi_1 \tag{21-55}$$

with ϕ_1 denoting the phase of $T(\omega)$ when the magnitude $|T(\omega)|$ is unity, or zero dB. The other is the *gain margin*, defined as the reciprocal of the magnitude of T, or $1/|T(\omega)|$, when the phase ϕ of $T(\omega)$ is $\pm 180°$. In decibels this is given by the relation

$$\text{gain margin} = -20 \log_{10}|T_1(\omega)| \tag{21-56}$$

with $|T_1(\omega)|$ denoting the magnitude of $T(\omega)$ when ϕ is $\pm 180°$.

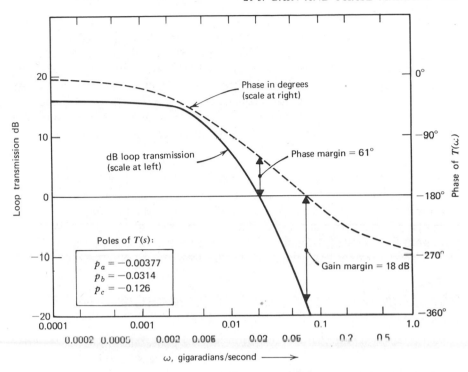

Figure 21-12   Magnitude and phase plots of $T(\omega)$.

In Table 21-1 are shown several calculations made from (21-53) and (21-54). From these data and (21-55) and (21-56) we find that the phase margin is 61° and the gain margin is 7.9, or 18 dB. Both are indicated on the plots of Fig. 21-12. Note that a stable amplifier has a positive phase margin and a gain margin greater than unity, or zero dB. For the amplifier represented by the curves of Fig. 21-12, an increase in the feedback factor B that causes an increase in T of 18 dB or more gives instability. For practical amplifiers a phase margin of at least 60° and a gain margin of at least 4 (or 12 dB) are usually required.

Table 21-1. Calculations from (21-53) and (21-54)

ω	$|T(\omega)|$	dB Loop Transmission	Phase ϕ of $T(\omega)$
0	6.10	15.7	0°
0.0191	1	0	−119°
0.0675	0.126	−18.0	−180°

Example

A feedback network is to be added to an amplifier having a midband open-loop gain A_o of 80 dB and zero phase shift. The maximum midband loop transmission T_o is to be used, consistent with a phase margin of at least 60° and a gain margin of 12 dB or more in the high-frequency region. Accordingly, the gain $A(\omega)$ is measured at high frequencies that make the angle of A equal to $-120°$ and also $-180°$, and the respective gains are 52 dB and 41 dB. Assume the loading of the feedback network has a negligible effect on A. Determine T_o.

Solution

At the frequency giving a phase of $-120°$, A is 52 dB, or 398. For $|AB| = 1$, B is 0.00251, and the phase margin is 60°. At the frequency giving a phase of $-180°$, A is 41 dB, or 112. For $|AB| = 0.25$, corresponding to a gain margin of 12 dB ($T = -12$ dB), B is 0.00223. The value of A_o is 80 dB, or 10^4. With the minimum B selected, T_o becomes 22.3. This value gives a gain margin of 12 dB and a phase margin greater than 60°. The feedback factor B is -53 dB.

RELATIONSHIP BETWEEN PHASE MARGIN AND THE Q OF COMPLEX-CONJUGATE POLES

For a two-pole amplifier it is easy to relate the phase margin to the Q of the poles, as defined by (21-8). First, let us consider a feedback amplifier with a loop transmission having two high-frequency poles p_a and p_b. With both zeros at infinity the loop transmission $T(s)$ is given by (21-5). In the real-frequency domain ($s = j\omega$) this becomes

$$T(\omega) = \frac{T_o}{1 - \omega^2/p_a p_b - j\omega(p_a + p_b)/p_a p_b} \tag{21-57}$$

The imaginary term in the denominator can be replaced with $(\omega^2/p_a p_b m)^{1/2}$, because the sum $p_a + p_b$ is equal to $-(p_a p_b/m)^{1/2}$. Making this change, we find that the magnitude and phase ϕ of $T(\omega)$ are

$$|T(\omega)| = \frac{T_o}{[(1 - \omega^2/p_a p_b)^2 + \omega^2/p_a p_b m]^{1/2}} \tag{21-58}$$

$$\phi(\omega) = -\tan^{-1}\left[\frac{(\omega^2/p_a p_b m)^{1/2}}{1 - \omega^2/p_a p_b}\right] \tag{21-59}$$

Let ω_c denote the radian frequency at which the magnitude of $T(\omega)$ is unity. Equating (21-58) to unity and solving for $\omega_c/(p_a p_b)^{1/2}$, we obtain

$$\frac{\omega_c}{\sqrt{p_a p_b}} = \left[\frac{(m^2 T_o^2 - m + 0.25)^{1/2} + m - 0.5}{m}\right]^{1/2} \tag{21-60}$$

The phase margin is $180° + \phi_1$, with ϕ_1 denoting ϕ at ω_c. From this and (21-59) we find the phase margin P.M. to be

$$\text{P.M.} = 180° - \tan^{-1}\left[\frac{(\omega_c{}^2/p_a p_b m)^{1/2}}{1 - \omega_c{}^2/p_a p_b}\right] \tag{21-61}$$

This can be expressed in the form

$$\text{P.M.} = \tan^{-1}\left[\frac{(\omega_c{}^2/p_a p_b m)^{1/2}}{\omega_c{}^2/p_a p_b - 1}\right] \tag{21-62}$$

Note that both (21-61) and (21-62) give a phase margin between zero and 90° when $\omega_c{}^2 > p_a p_b$, and a phase margin between 90° and 180° when $\omega_c{}^2 < p_a p_b$. Thus they are clearly equivalent. For the two-pole amplifier being considered, the phase margin is restricted to values between zero and 180°.

The desired relationship between the phase margin and Q is obtained by using (21-60) to eliminate ω_c in (21-62). With the product mT_o in the result replaced with $Q^2 - m$, justified by (21-8), we find that

$$\text{P.M.} = \tan^{-1}\frac{(\sqrt{(Q^2 - m)^2 - m + 0.25} + m - 0.5)^{1/2}}{\sqrt{(Q^2 - m)^2 - m + 0.25} - 0.5} \tag{21-63}$$

Recall that the constant $m = (p_b/p_a)/(1 + p_b/p_a)^2$, and m has a value between 0.25 (for equal poles) and zero (for an infinite pole ratio).

In Fig. 21-13 are shown plots of the phase margin versus Q for pole ratios of 1, 5, and 20, with the data used to plot the curves calculated from (21-63). For T_o less than unity the phase margin is undefined. As Q approaches $(2m)^{1/2}$, each of the curves of Fig. 21-13 approaches a phase margin of 180°. For Q greater

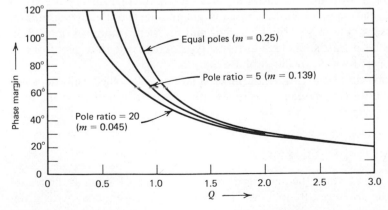

Figure 21-13   Phase margin versus Q for a two-pole feedback amplifier with the poles of $T(s)$ having ratios of 1, 5, and 20.

than unity, examination of the curves reveals that the phase margin is nearly independent of the pole ratio and has values given approximately by the relation

$$\text{P.M.} \approx \frac{60°}{Q} \qquad \text{for } Q > 1 \tag{21-64}$$

Next, let us consider a feedback amplifier with $T(s)$ having two low-frequency poles p_a and p_b and both zeros at the origin. The loop transmission is given by (21-25). With s replaced with $j\omega$, and $p_a + p_b$ replaced with $-(p_a p_b/m)^{1/2}$, it is easy to show that the magnitude and phase of $T(\omega)$ are

$$|T(\omega)| = \frac{T_o}{[(1 - p_a p_b/\omega^2)^2 + p_a p_b/\omega^2 m]^{1/2}} \tag{21-65}$$

$$\phi(\omega) = \tan^{-1}\left[\frac{(p_a p_b/\omega^2 m)^{1/2}}{1 - p_a p_b/\omega^2}\right] \tag{21-66}$$

The procedure follows that of the high-frequency case. We let ω_c be the frequency at which the magnitude of $T(\omega)$ is unity. Then we equate (21-65) to unity and solve for $(p_a p_b)^{1/2}/\omega_c$. The result is substituted in (21-66) to obtain the phase ϕ_1 at frequency ω_c. Because this is between zero and $+180°$, the phase margin *for the low-frequency case* is defined as

$$\text{phase margin} = 180° - \phi_1 \tag{21-67}$$

The analysis reveals (see P21-16) that the phase margin is precisely that of (21-63). Thus the curves of Fig. 21-13 and the approximate expression of (21-64) apply to the two-pole low-frequency feedback amplifier as well as to the high-frequency case.

For amplifiers with more than two poles the curves of Fig. 21-13 give an approximate relation between the phase margin and the Q of the two dominant poles. Because we know how the frequency and transient responses depend on Q, we now have at least some idea as to the type of response to expect from an amplifier designed with a specified phase margin.

The gain margin of a two-pole amplifier is infinite. The phase shift does not become $180°$ until the frequency becomes either zero or infinite, and the loop transmission is zero.

CROSSOVER FREQUENCY

The frequency at which the magnitude of $T(\omega)$ is unity is known as the *phase-margin*, or *gain-crossover*, frequency. The high crossover frequency ω_c is given by (21-60) and the corresponding expression for the low-frequency region is indicated in P21-16.

Regardless of the number of poles and zeros of $T(s)$, a feedback amplifier with a phase margin of $60°$ has a loop transmission of $1\underline{/-120°}$ at the high crossover frequency, and the closed-loop gain is

$$A_f = \left(\frac{1\underline{/-120°}}{1 + 1\underline{/-120°}} \right) \frac{1}{B} = \frac{1\underline{/60°}}{B} \qquad (21\text{-}68)$$

We know that A_{fo} is $T_o/[(1 + T_o)B_o]$. Assuming B is independent of frequency, A_f of (21-68) has a magnitude greater than $0.707A_{fo}$ provided T_o is greater than 2.5. For these usual conditions it is clear that *the upper* 3 dB *frequency exceeds the crossover frequency when the phase margin is* $60°$ (see P21-17). It can be shown similarly that the lower 3 dB frequency of a feedback amplifier with a low-frequency phase margin of $60°$ is below the crossover frequency, provided T_o exceeds 2.5.

21-6. DOMINANT-POLE COMPENSATION

When appreciable feedback is applied to a multistage amplifier, we have learned that high-Q complex poles and perhaps instability are likely to result. For a two-pole amplifier the curves of Fig. 21-5 show the limit on T_o for a specified Q and the curves are at least rough approximations for amplifiers having more than two poles. It is often desirable to add to a basic amplifier a considerable amount of desensitivity, even though the feedback exceeds that permitted by the specified Q, or the specified phase and gain margins. This can be done only by suitable modification of the basic amplifier or its feedback network or both.

A procedure extensively used consists of changing one or more capacitances of the basic amplifier so as to increase the pole ratio p_b/p_a without adding additional poles. As shown by the curves of Fig. 21-5, an increased pole ratio allows a greater T_o. It is important to add or subtract the correct amount of capacitance at the proper places, or otherwise the result might be detrimental.

In the low-frequency region the coupling or bypass capacitance having the most influence on the dominant pole of the basic amplifier can be substantially reduced, thus moving the pole along the real axis away from the origin and increasing the ratio of this pole to the next smaller one. An alternate procedure, which does not appreciably affect the lower 3 dB frequency, is to increase all other capacitances so as to move the smaller poles closer to the origin, with the dominant pole remaining constant or nearly so. In the high-frequency region the dominant pole p_a is determined largely by the device capacitance C_1 having the greatest open-circuit time constant R_1C_1. Accordingly, *the compensating capacitor should be placed in parallel with* C_1. This lowers the magnitude of p_a, thereby increasing the pole ratio p_b/p_a. The use of capacitance to change the pole ratio is referred to as *dominant-pole compensation.*

DESIGN PROCEDURE

To illustrate the procedure we shall consider the amplifier of Fig. 21-11 with the open-loop gain $A(s)$ of (21-48). Let us assume that the specifications call for a desensitivity of 25 with a phase margin of at least 60° and a gain margin of 12 dB or more. Because A_o is -10^4 by (21-48), the desired desensitivity is accomplished by replacing the 75 ohm resistor of the feedback network with a resistance of 580 ohms, thereby making y_{rb} equal to -0.0024 mmho, which is B. In P21-11 it was shown that a midband loop transmission T_o of 6.1 gave complex poles with a Q of unity. Compensation is clearly required.

Because we must increase the capacitance associated with the *largest* time constant, our first problem in the design of dominant-pole compensation is to locate this capacitance. Frequently, it is the device capacitance $C_{\mu 2}$ of transistor Q_2 of an intermediate CE stage of the amplifier. Let us assume this is the case. Accordingly, we shall add a capacitor between the base and collector of Q_2.

The next problem is the determination of the proper amount of capacitance to add. A value too small gives insufficient compensation and one too large excessively reduces the bandwidth of the basic amplifier. Although the feedback that is employed to give the specified desensitivity results in a substantially increased bandwidth, it should be noted that the full amount of midband desensitivity $1 + T_o$ applies only throughout *the passband of the basic amplifier*, because outside this passband T is less than T_o. Therefore, *it is important to use the minimum capacitance required for adequate compensation*.

An experimental trial-and-error procedure is normally used. This consists of selecting a particular value of C, adjusting the feedback network for the required desensitivity, and obtaining data for plotting magnitude and phase curves of $T(\omega)$. The procedure is repeated until a satisfactory value of C is found, as determined from the measured phase and gain margins. Earlier it was mentioned that, in the event it is inconvenient to find $T(\omega)$ directly, the open-loop gain $A(\omega)$ can be measured and $T(\omega)$ can be calculated from AB. For the amplifier of Fig. 21-11, opening the feedback loop may change the bias conditions. However, $A(\omega)$ can be measured simply by replacing the 580 ohm resistor of the feedback network, which replaced the 75 ohm resistor, with a short-circuit.

The experimental work is simplified if an approximate value of C can be estimated. Although a multistage amplifier normally has more than two high-frequency poles, a reasonable estimate of the capacitance can be made by considering only the two lowest poles. Because compensation moves the dominant pole p_a to a much lower value, the magnitude of the complex-conjugate pair formed when feedback is applied is most certainly small in comparison with the other poles. From the specified phase margin an approximate value of Q is found from the lower curve of Fig. 21-13, which has the smallest value of m. This curve is reasonably accurate for pole ratios greater than 20. Then we determine the required value of m from the relation $Q^2 = m(1 + T_o)$, which is

(21-8). The pole p_a is calculated from m by assuming that the next larger pole p_b is unchanged by the addition of the capacitor. Unfortunately, this assumption is poor and can be expected to introduce considerable error. An estimate of C is obtained by equating the open-circuit time constant associated with this capacitance to the reciprocal of the calculated value of the magnitude of p_a.

As shown on Fig. 21-12 the poles of the gain function of the incremental model of Fig. 21-10 are -0.00377, -0.0314, and -0.126 ns^{-1}. A compensating capacitor is to be connected at the proper location, which we earlier assumed to be between the base and collector of the transistor of the second stage. For the specified phase margin of $60°$ we estimate Q to be 0.9, using the lower curve of Fig. 21-13. With $T_o = 24$, which gives the specified desensitivity, m is found from (21-8) to be 0.0324. The corresponding pole ratio p_b/p_a is 28.9. Assuming p_b remains constant at -0.0314 when C is added, we calculate the required value of p_a to be -0.0011 ns^{-1}. Let us suppose that $C_{\mu 2}$ is 6 pF and that it shunts a resistance of 30 kΩ when all other capacitances are treated as open-circuits. Equating the open-circuit time constant $30(C + 6)$ ns to the magnitude of the reciprocal of the calculated p_a yields an estimated value of 24 pF for C.

EXPERIMENTAL DATA

The next step in the design is to connect a 24 pF capacitor between the base and collector of the intermediate-stage transistor and to obtain experimental data relating the magnitude and phase of $A(\omega)$ to ω throughout the high-frequency region. In the circuit of Fig. 21-11 the short-circuit current of the source is $5V_s$ mA. Therefore, the open-loop gain A is $0.2V_o/V_s$ kilohms with the feedback resistor R_2 shorted out. By multiplying the measured A at each frequency by B, which is -0.0024 mmho, we obtain T.

In Fig. 21-14 are shown magnitude and phase plots of $T(\omega)$, assumed to apply to the circuit of Fig. 21-11 with the selected compensating capacitor added. Although not essential to this discussion, the plots were obtained from a compensated amplifier having the circuit of Fig. 21-11, with the basic amplifier consisting of a three-stage CE cascade designed with poles approximately the same as those of (21-48) prior to compensation. As indicated on the curves, the phase margin of the compensated amplifier is $64°$ and the gain margin is 22 dB. The 3 dB frequency is about 0.001 gigaradian/s, or 160 kHz. Compensation has reduced the bandwidth of the basic amplifier from 590 kHz to 160 kHz. The gain-crossover frequency is 0.022 ns^{-1}, or about 3.5 MHz. Because the feedback is substantial and the phase margin exceeds $60°$, the bandwidth of the feedback amplifier exceeds the crossover frequency. This was discussed in the preceding section.

The poles of $T(s)$ can be found from the curves of Fig. 21-14. The 3 dB frequency is estimated at 0.001 ns^{-1}. This is the approximate value of the magnitude of p_a and is close to the estimate of 0.0011 determined in the design

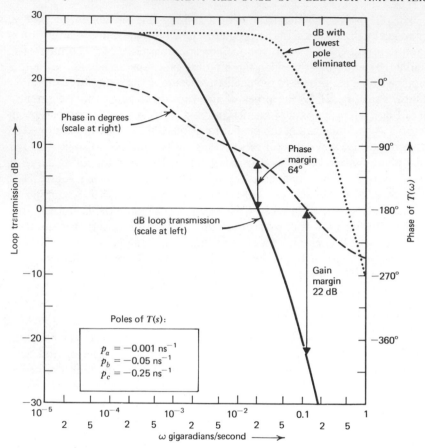

Figure 21-14   Magnitude and phase plots of $T(\omega)$ for the compensated amplifier.

procedure and used to calculate the compensating capacitance. We now simply reconstruct the magnitude plot with p_a eliminated. To do this, at each value of ω we increase the ordinate by $10 \log[1 + (\omega/p_a)^2]$ dB. The curve that results is also shown in Fig. 21-14. It has a corner frequency of about 0.05 ns^{-1}, which is the magnitude of p_b. The third pole is most easily determined from the phase curve by eliminating the effects of both p_a and p_b. This is done by adding to each ordinate the sum of $\tan^{-1}|\omega/p_a|$ and $\tan^{-1}|\omega/p_b|$, and the phase of $-45°$ then occurs at a radian frequency equal to the magnitude of p_c. Using -0.001 for p_a and -0.05 for p_b, this procedure gives -0.25 ns^{-1} for p_c.

The method described is useful for experimental determination of the poles and zeros of a basic amplifier. It is accurate only if adjacent singularities are separated by at least two octaves. An alternate procedure consists of locating approximate straight-line segments of the magnitude curve, with slopes of 20, 40, and 60 dB/decade. These are extended as straight lines, and the intersections locate the singularities. Although easier, this latter method is usually less accurate.

The poles that have been found correspond to frequencies of 160 kHz, 8 MHz, and 40 MHz. Those of the uncompensated amplifier are at frequencies of 600 kHz, 5 MHz, and 20 MHz. Note that *compensation changes the values of all the poles*, not just that of the dominant pole. The pole ratio p_b/p_a is 50, which is considerably larger than the estimated value of 29 based on the assumption of a constant p_b. It was noted earlier that this assumption was rather poor.

The equation for $T(s)$ based on the poles found from the curves of Fig. 21-14 is

$$T(s) = \frac{24}{(1 + s/0.001)(1 + s/0.05)(1 + s/0.25)} \qquad (21\text{-}69)$$

From the feedback factor B, which is -0.0024 mmho, and the loop transmission given by (21-69), we can determine the closed-loop gain $A_f(s)$. Its midband value is -400 kΩ, corresponding to a voltage gain V_o/V_s of -2000. In reciprocal nanoseconds the real pole of A_f is found to be -0.26 and the complex conjugates are $-0.023 \pm j0.027$, which have a pole angle θ of $50°$ and a Q of 0.77. The bandwidth of the feedback amplifier, calculated from (21-24), is 6.1 MHz. This compares with a BW of 160 kHz for the compensated basic amplifier. The full desensitivity $1 + T_o$ of 25 is applicable only at frequencies less than 160 kHz. Above this, $|T|$ drops off with frequency.

In the low-frequency region compensation can be accomplished similarly. This was examined briefly in the example of Sec. 21-3, and the procedure is further illustrated in the problem that follows.

Example

The low-frequency incremental model of a certain amplifier has three non-interacting coupling capacitors C_1, C_2, and C_3 with respective values of 0.2, 0.01, and 0.4 μF. The associated short-circuit resistances R_{1s}, R_{2s}, and R_{3s} are 1, 25, and 1.25 kΩ, respectively, and the corresponding poles are -5, -4, and -2 reciprocal milliseconds. Resistive feedback is to be added to provide a desensitivity of 40. With the dominant pole of -5 kept constant, compensate the amplifier for a phase margin of $45°$.

Solution

Units of kΩ, μF, and ms will be used. The three zeros that result from the coupling capacitors are at the origin, and $T(s)$ has the form

$$T(s) = \frac{T_o s^3}{(s - p_a)(s - p_b)(s - p_c)} \qquad (21\text{-}70)$$

For resistive feedback it follows that the poles of A_f are the roots of the equation

$$T_o s^3 + (s - p_a)(s - p_b)(s - p_c) = 0 \qquad (21\text{-}71)$$

Solving the cubic, using the specified poles and $T_o = 39$, we find the poles of A_f are -0.76 and $0.24 \pm j1.12$ ms^{-1}. Thus the amplifier is unstable, and compensation is required. We need to estimate suitable values for C_2 and C_3.

For a phase margin of 45° the lower curve of Fig. 21-13 gives a pole Q of 1.2, and with $T_o = 39$ we find m from (21-8) to be 0.036. For a dominant pole p_b of -5, the next lower pole p_a is determined from m to be -0.2 ms^{-1}. In the corresponding high-frequency case, when T_o was increased we found that the largest pole moved away from the complex-conjugate pair that formed, thereby reducing its effect. However, in the amplifier considered here all three poles move to the right, converging on the three zeros at the origin. Consequently, we would be wise to select one of the capacitors so that its corresponding pole is as close to zero as practical.

Let us increase C_2 from 0.01 to 4 μF, giving a pole at -0.01 ms^{-1}, which is nearly inconsequential. The desired pole at -0.2 is obtained by making C_3 equal to 4 μF also. With $T_o = 39$ and with poles of -5, -0.2, and -0.01 ms^{-1}, it is easy to show from (21-70) that $T(\omega)$ is $1\underline{/136°}$ at a crossover frequency of 31.5 Hz. Thus the phase margin is 44°. The gain margin is 23.6 dB, because $T(\omega)$ is -0.066 at a frequency of 7 Hz. Two of the poles of A_f are complex, of course, and the lower 3 dB frequency of the feedback amplifier is 18 Hz (see P21-18).

At a given frequency below crossover the phase of $T(\omega)$ is more negative for the amplifier with dominant-pole compensation than for the uncompensated one, which is evident from comparison of the phase curves of Figs. 21-12 and 21-14. For this reason dominant-pole compensation is sometimes called *lag compensation*. However, at and above the crossover frequency the contribution of the dominant pole to the phase is approximately the same for each case, being almost $-90°$.

21-7. FEEDBACK COMPENSATION

A second widely used and effective technique for compensation consists of adding a zero to the open-loop gain A of the basic amplifier or to the feedback factor B, with the zero located on the negative real axis. The choice is made to modify suitably the root locus and the Bode plots of $T(\omega)$. A pole is also added, of course. However, the design is such that the magnitude of the high-frequency pole is so large that it can be neglected.

To illustrate the procedure let us consider an amplifier with a high-frequency gain $A(s)$ having three poles p_a, p_b, and p_c with the zeros at infinity. The gain $A(s)$ is that of (21-35). The feedback factor B is assumed to have a zero z_a on the negative real axis, with $B(s)$ having the form

$$B(s) = B_o\left(1 - \frac{s}{z_a}\right) \tag{21-72}$$

From A and B the loop transmission T is found to be

$$T(s) = \frac{T_o(p_a p_b p_c/z_a)(s - z_a)}{(s - p_a)(s - p_b)(s - p_c)} \tag{21-73}$$

Because z_a is negative, the zero contributes a positive, or *leading*, angle to the phase $\phi(\omega)$ of $T(\omega)$, thereby making the phase less negative. Accordingly, the addition of a zero to $T(s)$, by modification of either the basic amplifier or the feedback network, is commonly referred to as *phase lead* compensation. In the case being considered here the modification is made in the feedback network, and the phase lead compensation that results is often called *feedback* compensation.

The closed-loop gain can be expressed as

$$A_f(s) = \frac{-A_o p_a p_b p_c}{(s - p_a)(s - p_b)(s - p_c) + T_o(p_a p_b p_c/z_a)(s - z_a)} \tag{21-74}$$

From the denominator we note that the coefficient of the s-term is a function of both T_o and z_a, and the constant term depends on T_o. Desired values of T_o and z_a can be obtained by proper design of the feedback network, and these values control the poles p_1, p_2, and p_3 of A_f.

THE 3-POLE BUTTERWORTH DESIGN

A design of particular importance consists of locating the three poles so that

$$p_1, p_2 = -\alpha \pm j\sqrt{3}\,\alpha \qquad p_3 = -2\alpha \tag{21-75}$$

Each of the poles has the same magnitude 2α and the pole angle of the complex pair is $60°$. This is called the *Butterworth* pole configuration.

The denominator of (21-74) equals $(s - p_1)(s - p_2)(s - p_3)$, and for the poles of (21-75) the closed-loop gain of (21-74) can be expressed in the form

$$A_f(s) = \frac{8\alpha^3 A_{fo}}{s^3 + 4\alpha s^2 + 8\alpha^2 s + 8\alpha^3} \tag{21-76}$$

with A_{fo} denoting A_f at $s = 0$. By equating the corresponding s-coefficients of (21-74) and (21-76) and solving for α, T_o, and z_a, we obtain the following design equations:

$$\alpha = -0.25(p_a + p_b + p_c) \tag{21-77}$$

$$T_o = \left(\frac{-8\alpha^3}{p_a p_b p_c}\right) - 1 \tag{21-78}$$

$$z_a = \frac{T_o p_a p_b p_c}{8\alpha^2 - p_a p_b - p_b p_c - p_c p_a} \tag{21-79}$$

The Butterworth configuration has some especially interesting properties. Even though the pole angle of the complex poles is 60°, there is *no peaking* in the frequency response. This is attributable to the effect of the real pole. The upper 3 dB frequency ω_2 is 2α. These statements can be verified with the aid of (21-76) written in the real frequency domain (see P21-20). The pole arrangement provides the maximum bandwidth for the condition of no peaking in the frequency response. The overshoot in the transient response to a step function is moderate at 8.1% (see P21-21). Butterworth poles are discussed further in Sec. 23-5 in conjunction with Butterworth active RC filters.

In Fig. 21-15 is shown the root locus. With $T_o = 0$ the poles p_1, p_2, and p_3 coincide with p_a, p_b, and p_c. As T_o is increased, p_3 moves from p_c toward the zero z_a, and p_1 and p_2 move together and split into complex conjugates. Because the sum of the poles must remain constant, the complex conjugates move to the left, *approaching infinity along a vertical asymptote* as T_o approaches infinity and p_3 approaches z_a. The positions indicated for p_1, p_2, and p_3 are those of the Butterworth configuration.

Let us now redesign the feedback network of the amplifier of Fig. 21-10 having the open-loop gain $A(s)$ of (21-48) of Sec. 21-4, with feedback compensation

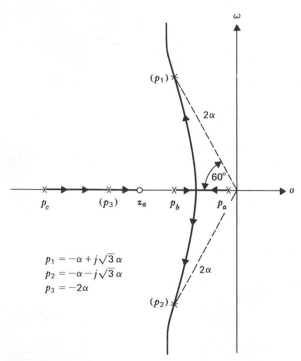

Figure 21-15 Root locus.

used to give the Butterworth pole configuration. Unless otherwise stated, units are understood to be those of (18-63), which we have been using for high-frequency circuits. The value of A_o is -10^4 kΩ, the poles p_a, p_b, and p_c are -0.00377, -0.0314, and -0.126 ns^{-1}, and the bandwidth of A is 590 kHz. From (21-77) through (21-79) we find that α is 0.040 ns^{-1}, T_o is 34.1, and z_a is -0.0603 ns^{-1}. The midband feedback factor B_o is T_o/A_o, or -0.00341 mmho, and the bandwidth of the feedback amplifier is 2α, or about 13 MHz.

The feedback network should give the proper B_o and z_a. A possible choice is to connect a single resistor R_f between the input and output nodes, with a capacitor C_f placed in parallel with R_f. The transadmittance y_{rb} is $-G_f - sC_f$. Therefore,

$$B = -G_f - sC_f = -0.00341\left(1 + \frac{s}{0.0603}\right) \text{mmho} \qquad (21\text{-}80)$$

This gives $R_f = 290$ kΩ and $C_f = 0.057$ pF. The result is somewhat surprising. The required capacitance is extremely small, and yet, without it the amplifier would be unstable. A serious problem arises in implementing this compensating network. Practical carbon resistors have parasitic capacitances that range from about 0.1 to 1 pF. *Thus the amplifier is likely to be overcompensated, with too much feedback capacitance, even with the capacitor* C_f *omitted.* Clearly, we must find a better arrangement. We do not want the compensation to depend on a parasitic capacitance of unknown magnitude.

It is desirable and practical to employ once again a feedback network having a T-configuration, and this is done in the network of Fig. 21-16. The total dc resistance of the feedback loop is 30 kΩ as in Fig. 21-11, and bias conditions

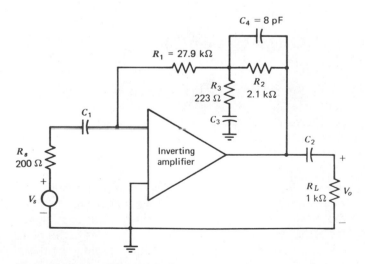

Figure 21-16 Amplifier with feedback compensation.

are unchanged. The feedback factor B, or y_{rb}, has the approximate midband value B_o and zero z_a that are desired, and in addition, it has a pole at 99 MHz.

The feedback factor B can be expressed as $B_o(1 - s/z_a)/(1 - s/p_a)$, with B_o denoting the midband value and with z_a and p_a denoting the respective zero and pole of y_{rb}. Analysis shows (see P21-22) that B_o is $-R_3/(R_1R_2 + R_2R_3 + R_3R_1)$, the zero z_a is $-1/R_2C_4$, and the pole p_a is $-(1/C_4)(1/R_1 + 1/R_2 + 1/R_3)$. These relations are helpful in choosing proper values for the network parameters. The specified parameters of the feedback circuit of Fig. 21-16 satisfy the preceding relations, giving B_o and z_a as required by (21-80). The value of C_4 is small enough so that its loading effect on the A network is negligible but large enough to minimize effects of parasitic capacitances less than a picofarad. The rather small value selected for R_3 gave a very large value for the undesirable pole p_a.

The compensating capacitance C_4 should be adjusted to give satisfactory high-frequency performance based on experimental data. Such an adjustment is always necessary because of simplifying assumptions of the design procedures, uncertain transistor parameters, and stray capacitance present in practical amplifiers.

In the example that has been considered, a desensitivity of 35 is achieved. If a larger value is desired, a combination of dominant-pole compensation and feedback compensation can be used. A capacitor C_1 is added at the appropriate place to lower the dominant pole. Then the feedback network is designed to give the Butterworth pole configuration for the three smallest poles. It may be necessary to try several values of C_1 to obtain the desired desensitivity, and the use of a computer to find the poles eliminates the tedium. *Because* C_1 *lowers the bandwidth of the basic amplifier, and hence the passband of the full amount of desensitivity, its value should be no larger than necessary.*

21-8. POLE-ZERO (LAG-LEAD) COMPENSATION

Modification of the basic amplifier of a feedback system so as to add a pole to the gain function produces a phase lag in the loop transmission $T(\omega)$, causing the angle of T to be more negative. The addition of a zero on the negative real axis has an opposite effect, giving a phase lead. In *pole-zero* compensation, also called *lag-lead* compensation, both a pole and a zero are added, with the pole smaller in magnitude than the zero. Thus the combination contributes *a lagging phase at the lower frequencies and a leading phase at the higher frequencies.*

The additional pole is somewhat lower than the dominant pole without compensation and the zero is usually in the approximate vicinity of the next higher pole, with the phase lead of the zero partially cancelling the phase lag of this pole. The results are similar to the combined effects of dominant pole compensation, discussed in Sec. 21-6, and phase-lead compensation, discussed in Sec. 21-7.

Because of the phase lead of the zero, the dominant pole does not need to be lowered as much as in dominant-pole compensation. Consequently, *the full desensitivity is applicable over a wider bandwidth*, and this leads to superior performance. In particular, *the rise time of the response to a step-function input is smaller*.

ILLUSTRATIVE FEEDBACK AMPLIFIER

To illustrate the procedure we shall use the feedback amplifier of Fig. 21-17, consisting of an emitter-coupled stage followed by a CE stage. The feedback is voltage-sample voltage-sum, or series-parallel. This is the same amplifier shown in Fig. 20-19 of Sec. 20-4 except for the addition of the RC compensating network connected between the base of Q_3 and the incremental ground. Unless otherwise stated all units are those of (18-63), which are convenient for high-frequency circuits. The three transistors are assumed to be identical, and each is biased at 1 mA. We assume that g_m is 40 mmhos, r_π is 2 kΩ, β_o is 80, C_π is 80, and C_μ is 5 pF. The amplifier has been designed to give a midband voltage gain V_o/V_s of 39 with a desensitivity of 36.

Bode plots of $T(\omega)$ can be obtained experimentally by constructing the network of Fig. 21-18. The feedback loop has been disconnected from the base of Q_2, and a resistor of 0.5 kΩ has been connected between this base and ground, representing the parallel equivalent of the two resistors of the feedback network, which is h_{ib}. The quiescent conditions are not appreciably changed. As indicated, $T(\omega)$ is the ratio V_s''/V_s'.

Although Bode plots can be obtained from laboratory measurements, we shall determine them from the gain $A(s)$ found from Fig. 21-19, with the aid of a computer. In the model the capacitance C_5 of 393 pF is the Miller-effect capacitance of Q_3, and the lag-lead RC compensation network is in parallel with C_5. The effective load resistance R_L' is 1.54 kΩ, which is the parallel equivalent of 2, 10, and 20.51 kΩ. Although the network has six capacitors and no capacitor loops, the gain function has only five poles, four finite zeros, and a zero at infinity due to C_5. As noted in Sec. 19-1, capacitors C_1 and C_3 have identical voltages, are not independent energy-storage elements, and can be combined into a single capacitance with the node at V_7 eliminated. With the compensating network removed, the poles and zeros are each reduced by one.

The analysis is easily accomplished by converting the source into a Norton equivalent circuit, changing element values into admittances, and writing node equations. With the source current I of the Norton circuit equal to unity, V_s' is 0.5. From the circuit we note that V_o is $-61.65V_5$. Thus the open-loop gain A, which is V_o/V_s', becomes $-123.3V_5$. From the feedback network we find B to be 0.02487, and it follows from A and B that $T(s)$ is $-3.066V_5$. Voltage V_5 can be found as a function of s from the set of node equations written for the nodes labeled V_1 through V_5, and a computer can be used to determine the midband

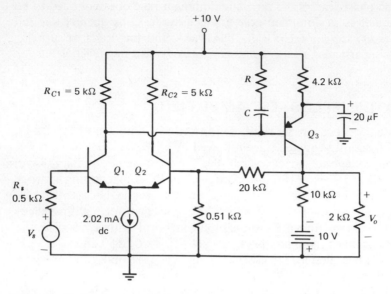

Figure 21-17   Feedback amplifier.

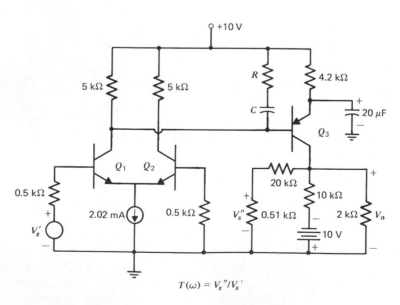

$$T(\omega) = V_s''/V_s'$$

Figure 21-18   Network for measurement of $T(\omega)$.

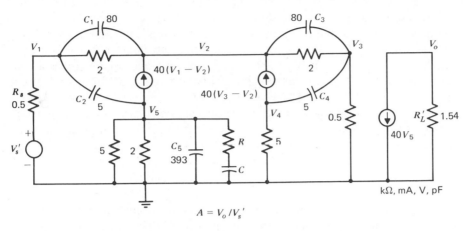

Figure 21-19   Network for calculation of $A(\omega)$.

value and the poles and zeros. With the compensation network removed, the loop transmission $T(s)$ is found to be

$$T(s) = \frac{35(1 + s/0.0338)(1 - s/0.28)(1 + s/0.85)}{(1 + s/0.00152)(1 + s/0.00449)(1 + s/0.212)(1 + s/1.24)} \qquad (21\text{-}81)$$

with the zeros and poles in reciprocal nanoseconds. The computer solution gave an additional pole and an additional zero, each equal to -0.506 ns^{-1} and exactly cancelling one another. One of the three finite zeros is in the right half-plane. This is detrimental to the phase margin because it not only gives a phase lag but also tends to increase the loop gain.

THE COMPENSATING RC CIRCUIT

The Bode plots of $T(\omega)$, which are discussed later, indicate that compensation is certainly desirable, if not necessary. We shall employ pole-zero compensation, which is usually accomplished by connecting a series RC branch at a suitable location. We want the RC network to have considerable influence on the dominant pole. In emitter-coupled amplifiers the branch is often connected directly between the two collectors. As in the case of dominant-pole compensation, a good procedure is to place the branch in parallel with the capacitance having the largest open-circuit time constant. With reference to Fig. 21-19 the time constants of the capacitors labeled C_1, C_2, C_3, C_4, and C_5 are 17, 67, 17, 227, and 561 ns, respectively. These were found from the incremental model using methods similar to those of Sec. 19-1. Thus the logical location of the RC compensating network is that shown in the figure. We now need to find suitable values for R and C.

The addition of the RC network not only contributes a pole and a zero but also changes the poles of the basic amplifier, which complicates the problem. The added zero is $-1/RC$. This value of s makes the impedance of the RC branch equal to zero, thereby eliminating forward transmission. Usually a reasonable choice is to make the zero somewhat larger in magnitude than the second lowest pole, which is -0.0045 ns^{-1}. *Such a value contributes an appreciable phase lead at the crossover frequency, with the undesirable increase in the* dB *gain due to this zero kept at a minimum.* The zero is often chosen to be about one-half the gain-crossover frequency of the uncompensated amplifier. This crossover frequency is 0.016 ns^{-1}, determined from the Bode plots or from T, given by (21-81). Let us try a value of -0.005 ns^{-1} for the zero, with R and C selected so that their product is 200 ns.

We want the compensating network to reduce the dominant pole. It follows that R and C must be chosen so that the associated open-circuit time constant is about the same order of magnitude as, or perhaps substantially greater than, the time constant of 561 ns associated with C_5. The larger we make the time constant, the smaller will be the dominant pole. Let us try a capacitance of 400 pF, which requires that R be 0.5 kΩ in order to give a zero of -0.005 ns^{-1}. These values contribute a time constant of 772, thus increasing the sum of the time constants from 889 to 1661 ns. Accordingly, we expect the dominant pole to be lowered to almost one-half its original value. We hope this will prove to be sufficient.

With $R = 0.5$ kΩ and $C = 400$ pF the loop transmission $T(s)$ is found from the network of Fig. 21-19, with the aid of a computer, to be

$$T(s) = \frac{35(1 + s/0.005)(1 + s/0.0338)(1 - s/0.28)(1 + s/0.85)}{(1 + s/0.000762)(1 + s/0.00397)(1 + s/0.0112)(1 + s/0.211)(1 + s/1.24)}$$

$$(21\text{-}82)$$

with the zeros and poles in reciprocal nanoseconds. Let us compare this with the uncompensated $T(s)$ of (21-81). As expected, the RC network has contributed a zero with a value of -0.005 ns^{-1} without affecting the other zeros. The poles of -0.21 and -1.24 are not appreciably changed. However, the two lower poles of -0.00152 and -0.00449 have been replaced with three poles of -0.000762, -0.00397, and -0.0112 ns^{-1}. The dominant pole has been reduced as expected.

RESULTS AND DISCUSSION

Shown in Fig. 21-20 are magnitude and phase plots of the loop transmission $T(\omega)$ for both the uncompensated and compensated cases. These were obtained from (21-81) and (21-82), but of course, it is often convenient to determine these plots experimentally. The dashed curves refer to the compensated amplifier. The decibel loop transmission is somewhat lower for the compensated amplifier, due mostly to the smaller dominant pole. The result is a reduced gain-crossover

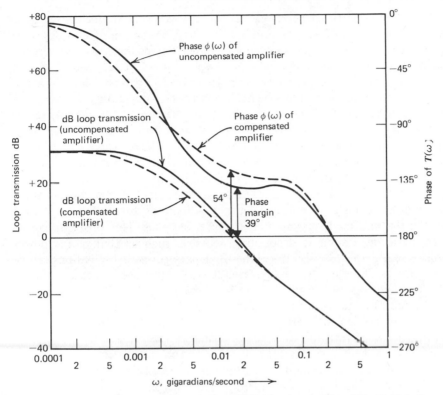

Figure 21-20   Plots of the magnitude and phase of $T(\omega)$ for the amplifier of Fig. 21-17, both with and without compensation.

frequency, which is 0.0146 compared with 0.0159 ns^{-1} for the uncompensated case. Of greater significance are the phase relationships in the vicinity of crossover. The phase of the compensated amplifier is smaller in magnitude than that of the uncompensated amplifier, and this smaller phase means an increased phase margin. As shown on the Bode plots, compensation has increased the phase margin from 39° to 54°. Observe the larger spacing between the gain and phase curves of the compensated amplifier. The improvement is considerable.

A computer analysis of the uncompensated amplifier reveals that the poles are $-0.0057 \pm j0.0147$, -0.21, and -1.24, and the zeros are -0.034, $+0.28$, and -0.85, in reciprocal nanoseconds. The pole angle of the complex-conjugate pair is 68.8°. At 2.2 MHz the frequency response peaks, with a gain that is $1.6A_{fo}$. The peaking is attributable mainly to the complex poles, but the zero at -0.034 has a significant effect. The bandwidth is 3.8 MHz.

For the compensated amplifier the poles are $-0.0082 \pm j0.0133$, -0.0051, -0.21, and -1.24 ns^{-1}, and the pole angle of the complex pair is 58.3°. The zeros are -0.0050, -0.034, $+0.28$, and -0.85 ns^{-1}. The effects of the pole and the zero at about -0.005 nearly cancel. At 1.75 MHz the frequency response peaks, with a gain of $1.2A_{f_o}$. Although this is much better, it is usually desirable to maintain the peak at less than $1.1A_{f_o}$. Either the feedback can be reduced a small amount or the compensating network can be redesigned. The zero at -0.034 contributes appreciably to the peaking. The bandwidth is 3.4 MHz.

Other values of R and C might be more appropriate, depending on the design objectives. If bandwidth is not of particular importance, a considerably larger value of C might be selected so as to lower further the dominant pole of the basic amplifier. We might be wise to reduce R so as to move the zero added to $T(s)$ to a larger value. The final design should be checked in the laboratory.

We have examined the instability problem resulting from the addition of a feedback network. There are other subtle ways feedback can occur and these sometimes lead to instability. Low-frequency oscillations can result from positive feedback through inadequately filtered power supplies. This problem can be solved by using a suitable decoupling network, such as that of the power amplifier of Fig. 16-27 of Sec. 16-6. High-frequency *parasitic* oscillations can be quite serious. These are due to positive feedback through stray capacitance associated with network elements and leads, particularly those of a common power supply. Filters consisting of small resistances and picofarad capacitances are useful in eliminating parasitic oscillations, and sometimes helpful is the addition of a small resistor (10 to 50 ohms) in series with the base of a trouble-some transistor, which increases the effective r_x. Above a megahertz, an electro-lytic capacitor can become inductive, producing undesirable feedback along with parasitic oscillations. The obvious solution is to add a parallel capacitor having a value in the range from 50 to 1000 pF.

The stability of feedback systems is treated extensively in texts on control theory, and the many principles, the stability criteria, the root-locus rules, and the compensation techniques given in these texts are equally applicable to feedback amplifiers. However, some of the techniques depend on the type of feedback, positive or negative, and *the "negative feedback" of an amplifier might well be classified as positive by a control-theory text.* Accordingly, when applying procedures from such a text, it is important to determine the feedback class-ification.

The loop transmission $T(s)$, which is the product $A(s)B(s)$, can be expressed in the form

$$T(s) = \frac{\pm K(s - z_a)(s - z_b) \cdots}{(s - p_a)(s - p_b)(s - p_c) \cdots} \tag{21-83}$$

with K positive. Electronics engineers regard the feedback as negative when the midband loop transmission T_o is positive. For a high-frequency transmission function with poles on the negative real axis, T_o is given by (21-83) with $s = 0$, or

$$T_o = \frac{\pm K(-z_a)(-z_b) \cdots}{|p_a p_b p_c \cdots|} \tag{21-84}$$

However, most control-theory texts define the feedback as negative if the sign ahead of K in (21-83) is positive, and the feedback is positive if the sign is negative. Note that $T(s)$ of (21-81), when arranged in the form of (21-83), has a negative sign ahead of K even though T_o of (21-84) is positive. Thus the feedback amplifier of Fig. 21-17 would be classified by these texts as a positive-feedback system. In fact, any feedback amplifier with a high-frequency loop transmission $T(s)$ that has an odd number of zeros in the right half-plane is so classified. The problem does not arise with low-frequency gain functions or with high-frequency amplifiers having a $T(s)$ with an even number of zeros in the right half-plane. For such cases electronic and control engineers are in agreement.

21-9. OP AMP COMPENSATION

Some operational amplifiers are internally compensated, such as the 741, the 307, and the 771. In each of these a 30 pF MOS feedback capacitor is internally connected across an intermediate amplifier stage to give a dominant pole located at a very low frequency, typically about 8 Hz. From this frequency the voltage gain drops 20 dB per decade (6 dB per octave) to a value of unity at approximately 1 MHz. Because the frequency corresponding to the second pole is substantially greater than this, for closed-loop gains appreciably greater than unity the stability is typical of that of one-pole amplifiers. Accordingly, these OP AMPs are stable even when connected as voltage followers.

OP AMP 748 is similar to the 741 except that the 30 pF capacitor across the Darlington pair is omitted and pins are provided on the package for connection of an external compensation network. By using a 30 pF capacitor, performance is the same as that of the 741. However, for applications requiring wider bandwidths or higher slew rates, the required compensation capacitance is

$$C \geq \frac{30R_1}{R_1 + R_f} \, \text{pF} \qquad (21\text{-}85)$$

with R_1 and R_f defined on the diagram of Fig. 21-22. Normally, C should not be less than 3 pF. The closed-loop gain A_V of the inverting amplifier is $-R_f/R_1$. As this gain is reduced in magnitude, the compensation capacitance must be increased. For a voltage-follower circuit, which has the noninverting configuration, R_1 is infinite, R_f is zero, and C must be at least 30 pF.

DOMINANT-POLE (PHASE-LAG) COMPENSATION

This is quite common. In many OP AMPs it is accomplished simply by connecting between the compensation terminals a capacitor from about 3 to 50 pF. The proper value should be determined from the data sheet supplied by the manufacturer. For operational amplifiers 301, 308, 748, 770, and 777 the specified value to use is that calculated from (21-85), and the dominant pole of the open-loop gain A_v of a typical IC is at the approximate frequency of $300/C$ hertz with C in pF.

Example 1

Suppose a 748 operational amplifier in the noninverting configuration has an open-loop gain A_v of 200 000 and a closed-loop gain of 50. Using the minimum required compensation capacitance, determine the closed-loop bandwidth and compare this bandwidth with that of an identical 741. Assume $300/C$ represents the dominant pole of the open-loop gain and the only other significant pole is at 5 MHz.

Solution

We find that $(R_1 + R_f)/R_1 = 50$, and by (21-85) the capacitance C is 0.6 pF, but a minimum capacitance of 3 pF should be used. This gives a pole at 100 Hz. Because the amplifier has voltage sampling and voltage summing the desensitivity $1 + T_o$ is 200 000/50, or 4000. With poles p_a and p_b located at 100 Hz and 5 MHz, the poles p_1 and p_2 of the feedback amplifier are found from (21-8) through (21-10) to be at 438 kHz and 4.6 MHz. Note that the dominant pole has shifted about 9% more than the value given by (20-10) for a one-pole amplifier. The bandwidth is 438 kHz.

For the 741 OP AMP the compensating capacitance of 30 pF gives a dominant pole at 10 Hz. Thus the BW is $10(1 + T_o)$, or 40 kHz, with the larger pole having negligible effect. This BW is only 9% of that of the externally compensated 748.

POLE-ZERO COMPENSATION

Another compensation network of considerable importance consists of a capacitor C_c in series with a resistor R_c, with the combination connected from a suitable point of the transmission path to ground. The network is shown in Fig. 21-21, and the current source I and the resistance R represent the Norton

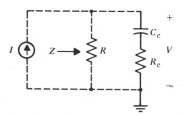

Figure 21-21 $C_c R_c$ compensation network.

equivalent circuit looking back from the compensation network. The impedance Z across the terminals of the current source can be expressed in the form

$$Z = \frac{K(s - z_a)}{s - p_a} \qquad (21\text{-}86)$$

with $K = R_c R/(R_c + R)$, $z_a = -1/C_c R_c$, and $p_a = -1/[C_c(R_c + R)]$. Clearly, the voltage V of Fig. 21-21 is zero at the zero of the impedance, and V is infinite at the pole. Thus the compensation adds a zero and a pole to the open-loop gain A_v of the amplifier. The zero is at a frequency ω of $1/C_c R_c$ and the pole is at a lower frequency of $1/[C_c(R_c + R)]$.

The OP AMP terminal labeled COMP 1, or perhaps LEAD COMP, is connected internally to a point having a low or moderate impedance to ground. This point may be located at the output of a common-collector stage or at the input of a common-emitter amplifier. The COMP 2 or LAG COMP terminal is a high-impedance point. In the 748, for example, it is the junction at the output of a CE stage and the input of an emitter follower, and both present high impedances. Now suppose the $C_c R_c$ network is connected from the LEAD terminal to ground, with the LAG terminal open. Because of the moderate value of R, the compensation network introduces a pole and a zero not widely separated, and the result is pole-zero (lag-lead) compensation. On the other hand, if the $C_c R_c$ network is connected between the LAG terminal and ground, the large value of R gives a pole very much smaller than the zero. This is essentially dominant-pole (phase-lag) compensation. However, the zero contributes a phase-lead effect that is significant.

Some OP AMPs, such as the 709, have enough gain and phase shift around internal feedback loops to cause oscillations even with no external feedback, and compensation is always essential. Others have stable open-loop operation, requiring compensation only when the closed-loop gain is less than a certain amount. Data sheets often provide the necessary information for designing simple and satisfactory networks for phase-lag (dominant-pole), phase-lead (zero-addition), and lag-lead (pole-zero) compensation. Improved performance is often obtained from other networks based on the design techniques of the preceding sections. The input and output terminals, as well as the compensation terminals, can be and often are used for connection to compensating networks.

LOOP TRANSMISSION

In Fig. 21-22 are shown typical inverting and noninverting connections. The network of Fig. 21-22a has voltage sampling and current comparison, and the open-loop gain V_o/I_s is found by removing the feedback and shunting both the input and output with R_f, as indicated by Fig. 20-13 of Sec. 20-3. At the input

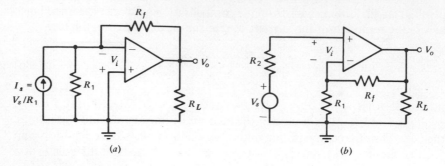

Figure 21-22   Basic OP AMP feedback configurations. (a) Inverting amplifier. (b) Noninverting amplifier.

the voltage V_i is the product of $-I_s$ and the parallel equivalent of R_1, R_f, and R_i. Because R_i is always much larger than R_1, we deduce that

$$V_i \approx \frac{-I_s R_1 R_f}{R_1 + R_f} \tag{21-87}$$

At the output, R_o is in series with the parallel equivalent of R_f and R_L. Assuming R_o is negligible, the output voltage V_o is $A_v V_i$. From this and (21-87) we find that the open-loop gain A, which is the ratio V_o/I_s, is

$$A = \frac{-A_v R_1 R_f}{R_1 + R_f} \tag{21-88}$$

The loop transmission T is BA, with the feedback factor $B = -1/R_f$. Therefore,

$$T = \frac{A_v R_1}{R_1 + R_f} \tag{21-89}$$

The closed-loop gain V_o/I_s is $A/(1 + T)$, with A and T given by (21-88) and (21-89). As I_s is V_s/R_1, the closed-loop voltage gain A_V is

$$A_V = \frac{-[R_f/(R_1 + R_f)]A_v}{1 + T} \tag{21-90}$$

From this and (21-89), it follows that $A_V = -R_f/R_1$ for A_v sufficiently large.

The noninverting network of Fig. 21-22b has voltage sampling and voltage comparison. The open-loop gain is found from the network with the feedback resistor R_f disconnected and placed in parallel with R_1 and a second resistor of value $R_f + R_1$ shunted across R_L. Voltage V_i approximately equals V_s because of the large R_i of the OP AMP. The output voltage V_o is $A_v V_i$, again assuming that R_o is negligible compared with the parallel equivalent of R_f and R_L. Therefore, the open-loop gain is simply A_v. The feedback factor B is $R_1/(R_1 + R_f)$, which gives a loop transmission T identical to that of (21-89). The closed-loop gain for the noninverting configuration is

$$A_V = \frac{A_v}{1 + T} \qquad (21\text{-}91)$$

From this and (21-89), it follows that $A_V = (R_1 + R_f)/R_1$ for A_v sufficiently large.

To design compensating networks using the equations developed here we need to know A_v as a function of frequency. This is found either experimentally or from specifications of the appropriate data sheet. *The loop gain of (21-89) applies to both inverting and noninverting amplifiers.* The following examples illustrate design techniques, which are basically no different for OP AMPs than for other types of feedback amplifiers.

Example 2

(a) Suppose an internally compensated operational amplifier having the magnitude Bode plot of Fig. 21-23 is connected as a unity-gain inverting amplifier. Find the low-frequency loop transmission in dB and determine the phase margin. (b) Repeat the calculations for a unity-gain noninverting amplifier.

Solution

(a) Equation 21-89 gives T_o, with $A_v = 10^5$ and $R_1 = R_f$. Therefore, T_o is 5×10^4, or 94 dB. With $R_1/(R_1 + R_f)$ equal to 0.5, the loop transmission is unity

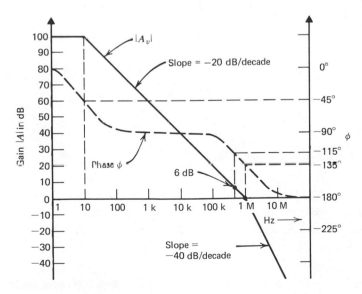

Figure 21-23   Bode plots of compensated OP AMP.

at the frequency at which A_v is 2, or 6 dB. From the dashed curve of Fig. 21-23, the corresponding phase of A_v is $-115°$, providing a phase margin of 65°.

The phase curve was sketched from the known poles. At the first pole, located at 10 Hz, the phase is $-45°$. At 5 and 20 Hz the respective values are $-26.6°$ and $-63.4°$. This was discussed in Sec. 18-2. The phase at the second breakpoint is $-135°$, with 90° of this contributed by the first pole.

(b) For the unity-gain noninverting amplifier, $R_1/(R_1 + R_f)$ is unity, and T_o becomes 10^5, or 100 dB. From (21-89) it is evident that T has a magnitude of unity at the frequency at which A_v is also unity. This frequency is 1 MHz. The corresponding phase of A_v is $-135°$, for a phase margin of 45°. The amplifier is stable with a phase margin of at least 45° for up to 100 dB of low-frequency loop transmission, giving voltage gains of unity or more.

Example 3

In the noninverting amplifier of Fig. 21-24 the proper values of R_c and C_c for the 702 OP AMP are specified in SI units by the manufacturer as follows:

$$R_c = \frac{20(R_1 + R_f)}{R_1} \qquad C_c = \frac{2 \times 10^{-7}}{R_c} \qquad (21\text{-}92)$$

The effective internal resistance R between the lag terminal and ground is 3.4 kΩ. Assume the uncompensated OP AMP has an open-loop voltage gain of

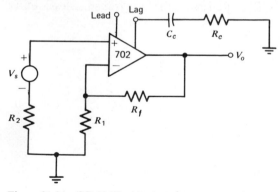

Figure 21-24 OP AMP with phase-lag compensation.

68 dB and poles at 1 and 8 MHz, with the other poles so large that they can be neglected, and assume these poles are not appreciably affected by R_c and C_c. Using the specified values of R_c and C_c, determine the phase margin when the voltage gain is 20 dB.

Solution

For a gain of 20 dB, A_{Vo} is 10, and from (21-92) we find that R_c is 200 ohms and C_c is 1000 pF. With $R = 3.4$ kΩ, the pole p_a of (21-86) is -0.0002778 ns^{-1}, corresponding to a frequency of 44.2 kHz, and the zero is -0.005 ns^{-1}, corresponding to 796 kHz. As A_o is 68 dB, or 2510, the open-loop voltage gain is

$$A(s) = \frac{2510(1 - s/z_a)}{(1 - s/p_a)(1 - s/p_b)(1 - s/p_c)} \qquad (21\text{-}93)$$

with $z_a = -0.005$, $p_a = -0.0002778$, $p_b = -0.006283$, and $p_c = -0.05027$ ns^{-1}. For a voltage gain of 20 dB, or 10, the feedback factor $B = 0.1$.

Multiplication of (21-93) by B gives the loop transmission T. In the frequency domain T can be expressed as

$$T = \frac{251(1 + jf/0.796)}{(1 + jf/0.0442)(1 + jf)(1 + jf/8)} \qquad (21\text{-}94)$$

with f in MHz. It is easy to deduce by trial-and-error that the magnitude of T is unity at a frequency of 9.16 MHz. At this crossover frequency the phase of T is $-137°$, giving a phase margin of 43°. The result could have been determined by Bode plots of the magnitude and phase of T. Without compensation the poles at 1 and 8 MHz give a phase margin of only 11°.

The poles of the closed-loop gain can be determined from (21-93) and B. With the aid of a programmable calculator the poles are found to be -0.00494 and -0.0669 $\underline{/+67°}$ in reciprocal nanoseconds. As the feedback is increased from zero, the smallest pole moves toward the zero at -0.005 ns^{-1}, *effectively cancel-ing it when the feedback becomes appreciable.* The other two poles move together, split into complex conjugates, and move vertically away from the real axis. Although the added pole and zero cancel one another, their presence increases stability. Because of the importance of the zero, compensation is lag-lead. The complex poles produce a small amount of peaking as indicated by Fig. 21-3 of Sec. 21-2. The bandwidth of the feedback amplifier, determined from the poles and zeros, is about 15 MHz. Higher-order poles not considered here will reduce this to about 10 MHz.

Example 4

Suppose the inverting amplifier of Fig. 21-22a uses an OP AMP with poles at 1, 4, and 40 MHz and with $A_{vo} = 10^4$. Using no compensation, what is the minimum voltage gain possible without oscillations?

Solution

With f in MHz, A_v of the OP AMP is

$$A_v = \frac{10^4}{(1 + jf)(1 + jf/4)(1 + jf/40)} \qquad (21\text{-}95)$$

By (21-89) the loop transmission T is

$$T = \frac{10^4 R_1/(R_1 + R_f)}{(1 + jf)(1 + j0.25f)(1 + j0.025f)} \qquad (21\text{-}96)$$

From either a phase plot or from a calculator the phase of T is found to be $-180°$ at a frequency of 14.3 MHz. At this frequency the magnitude of T is determined from (21-96) to be unity when $(R_1 + R_f)/R_1$ is 177, giving a voltage gain $-R_f/R_1$ of -176, or 44.9 dB. For gains less than 44.9 dB compensation is essential.

Example 5

For the inverting amplifier of Example 4 design a phase-lead feedback network to give a voltage gain of 44.9 dB with a 45° phase margin.

Solution

A compensating capacitor will be placed in shunt with R_f. In (21-89) for T the resistance R_f must be replaced with Z_f. With the ratio $R_f/R_1 = 176$, which gives the specified 44.9 dB gain, it is easy to show that (21-89) becomes

$$T = 0.00565 A_v \left(\frac{1 + jf/f_1}{1 + jf/f_2} \right) \qquad (21\text{-}97)$$

with $f_1 = 1/(2\pi R_f C_f)$ and $f_2 = 177 f_1$. Because the pole at f_2 is so much larger than the zero at f_1, we can treat f_2 as infinite. In effect, the compensating capacitor adds only a zero to the loop gain. As we learned in Sec. 21-7, the zero of the feedback network shifts the poles of the closed-loop gain without adding any additional poles or zeros other than those that can be regarded as infinite.

Using (21-95) to replace A_v in (21-97), with f_2 treated as infinite and with frequencies in MHz, we obtain

$$T = \frac{56.5(1 + jf/f_1)}{(1 + jf)(1 + j0.25f)(1 + j0.025f)} \qquad (21\text{-}98)$$

The problem now is to select f_1 so that T becomes $1 \; \underline{/-135°}$ at the unknown crossover frequency, thereby giving the specified phase margin. A computer was employed to determine that T has the desired value at a crossover frequency of 18.6 MHz with $f_1 = 13.1$ MHz. The zero at f_1 is obtained with a product $R_f C_f = 12.1$ with R_f in kΩ and C_f in pF. Let us select 2 kΩ for R_f and 6 pF for C_f. The specified voltage gain is obtained by choosing 11.4 ohms for R_1.

The bandwidth of the feedback amplifier is 19.9 MHz (see P21-28). This is readily found from (21-90) by replacing T and A_v, using (21-98) and (21-95). It is not necessary to replace R_f in (21-90) with Z_f, because the loading effect of C_f on the open-loop gain is negligible. Analysis reveals that the poles in reciprocal nanoseconds are -0.149 and $-0.124 \; \underline{/\pm 57°}$. The arrangement is nearly that of a Butterworth configuration (see P21-29).

Increasing C_f moderately above the calculated value improves the phase margin but adversely affects the bandwidth. For example, if C_f is doubled, the phase margin is increased to 52° with a reduced bandwidth of 7.2 MHz. It is interesting to note that, for gains less than 40 dB, feedback compensation is unable to provide the required phase margin, as indicated by a computer analysis. Another type of compensation would be required.

REFERENCES

1. P. E. Gray and C. L. Searle, *Electronic Principles*, Wiley, New York, 1969.
2. J. F. Pierce and T. J. Paulus, *Applied Electronics*, Merrill, Columbus, Ohio, 1972.
3. J. Millman and C. C. Halkias, *Integrated Electronics*, McGraw-Hill, New York, 1972.
4. *Linear and Interface Circuits Applications*, Texas Instruments, Inc., Dallas, Texas, 1974.
5. E. R. Hnatek, *Applications of Linear Integrated Circuits*, Wiley, New York, 1975.

PROBLEMS

Section 21-1

21-1. Assume the low-frequency closed-loop gain of a feedback amplifier is that of (21-2), with A_f denoting the voltage ratio V_o/V_s. If $v_s(t)$ is a unit step function applied at time zero, deduce that the *sag*, or *tilt*, at time t_1 is $(1 - \exp p_1 t_1) \times 100\%$, with sag defined by the relation

$$\text{sag at time } t_1 = \frac{v_o(0) - v_o(t_1)}{v_o(0)} \times 100\% \qquad (21\text{-}99)$$

If the basic amplifier has a lower 3 dB frequency of 100 Hz, calculate the midband loop transmission T_o required for a sag of only 10% at time $t_1 = 0.01$ second.

21-2. The open-loop gain of a low-frequency amplifier has a midband gain A_o, a zero z_a, and a pole p_a. Both the zero and the pole are negative, with p_a greater in magnitude then z_a. With $B = B_o$, deduce that the closed-loop gain is

$$A_f(s) = \frac{A_{fo}}{1 - (p_a - z_a)/[(s - z_a)(1 + T_o)]}$$

For $T_o = 0$ and also for T_o infinite, find the values of the pole p_1 of $A_f(s)$ recalling that this pole is the value of s that makes A_f infinite.

Section 21-2

21-3. For $A = A_o/(1 - s/p_a)$ and $B = B_o/(1 - s/p_b)$, deduce that A_f is the expression of (21-6) with the numerator multiplied by $(1 - s/p_b)$. For $p_a = -0.05$ and $p_b = -0.95$ reciprocal nanosecond, deduce from calculations that the poles p_1 and p_2 of A_f are p_a and p_b when $T_o = 0$. Also, calculate the values of T_o that give coincident poles at -0.5 ns^{-1} and complex-conjugate poles with a pole angle at $60°$. (*Ans*: 4.26, 20)

21-4. Deduce (21-18) from (21-17). Also, for p_a and p_b equal to -0.005 and -0.015 ns^{-1}, $T_o = 47$, and $A_{fo} = 1000$, calculate the peak gain and the frequency at the peak in MHz.

21-5. From the step-function response of (21-20), deduce that the values of αt at the maximum and minimum points are those of (21-22), and show that the maxima and minima of the response v_o are those of (21-21). Also, calculate the percent overshoot for a pole ratio p_b/p_a of 2 and a midband loop transmission of 17.

21-6. The loop transmission of a feedback amplifier has poles of -0.02 and -0.10 reciprocal nanosecond, and the input is a step function. (a) With T_o such that the poles of the feedback amplifier are coincident, determine in nanoseconds the rise time, defined in Sec. 19-2. (b) Repeat (a) with T_o increased so that the poles are complex conjugates with a pole angle of $45°$, and note the improvement. Either use the curves of Fig. 21-4 or a trial-and-error procedure. (Ans: 56.0, 25.3)

21-7. Let the poles p_1 and p_2 of the gain function of (21-14) be $-\alpha \pm j\beta$. Deduce that $A_f(s)$ can be expressed as the ratio of $A_{fo}(\alpha^2 + \beta^2)$ to $s^2 + 2\alpha s + \alpha^2 + \beta^2$. Then find $A_f(\omega)$ and deduce that the frequency ω_2 that makes the magnitude of A_f equal to $0.707A_{fo}$ is that given by (21-24).

Section 21-3

21-8. For $Q > 0.5$ the poles of (21-27) are complex, with the real part given by $x = -\omega_o/2Q$ and the imaginary part given by $y = \pm(\omega_o/2Q)(4Q^2 - 1)^{1/2}$. Both x and y are functions of T_o. Deduce that

$$(x + Q\omega_o)^2 + y^2 = Q^2\omega_o^2$$

Then show that $Q\omega_o = -p_a p_b/(p_a + p_b)$, a constant independent of T_o. The result justifies the root locus of Fig. 21-6.

21-9. In SI units the loop transmission of an amplifier in the low-frequency region has poles at -300 and -100 and zeros at the origin. In rectangular form find the complex poles of the feedback amplifier when T_o is 5/3 and also when T_o is 13/3.

21-10. In SI units assume the loop transmission of a feedback amplifier has poles at -200 and -50 and zeros at the origin. The closed-loop gain $A_f(s)$ is the ratio V_o/I_s. If I_s is a step function applied at time zero, calculate the percent sag at $t_1 = 0.001$ for $Q = 0.5$ and also for $Q = 2$. Refer to (21-99) of P21-1. (Ans: 15.1, 1.015)

Section 21-4

21-11. For the amplifier of Fig. 21-10 with $A(s)$ given by (21-48), determine the value of R_f in $M\Omega$ that gives complex poles with a Q of unity, and calculate the midband voltage ratio V_o/V_s. Also, find the BW in MHz, using (21-24), which gives a value about 2% too high.

21-12. For the amplifier of Fig. 21-10 with $A(s)$ given by (21-48), determine the value of T_o that gives complex poles with a pole angle of $70°$. (Ans: 10.4)

21-13. An amplifier has poles at -0.005, -0.05, and -0.2 ns^{-1} with the zeros at infinity. Calculate the midband desensitivity $1 + T_o$ and the BW in MHz, using (21-24), that result from the addition of a resistive feedback network producing complex poles with a Q of 1.2. Assume the feedback network has negligible loading effect on the A network.

Section 21-5

21-14. The amplifier having the loop transmission $T(\omega)$ given by the curves of Fig. 21-12 utilizes resistive feedback, and the loading effect of the feedback network on the open-loop gain is negligible. If the feedback factor is doubled, determine the phase and dB gain margins.

21-15. A voltage-sample current-sum feedback network consisting of a single resistor R_f is to be added to an amplifier with a midband gain V_o/I_s of -10^7 ohms. The gain A at 1 MHz is $-10^6 \angle -120°$ and that at 3 MHz is $+140{,}000$. Determine the value of R_f in $M\Omega$ required to give a phase margin of $60°$, and with this R_f calculate the midband loop transmission T_o and the dB gain margin.

21-16. The loop transmission of a feedback amplifier has two poles p_a and p_b in the low-frequency region and both zeros are at the origin. Let $x^2 = p_a p_b / \omega_c^2$, with ω_c denoting the gain-crossover frequency. Deduce from (21-65) that x is given by the right side of (21-60). Then deduce from (21-66) and (21-67) that the phase margin is the arctangent of $x / [(x^2 - 1)\sqrt{m}]$. Finally, substitute for x and show that the phase margin is that of (21-63).

21-17. The loop transmission $T(s)$ of a feedback amplifier has a midband value of 16, poles at -0.002 and -0.040 ns^{-1}, and zeros at infinity. Calculate the gain-crossover frequency in MHz, the phase margin, and the bandwidth in MHz, using (21-24). Compare the BW with the crossover frequency. (*Ans*: 4.23, 60.7°, 6.97)

Section 21-6

21-18. In the amplifier of the example of Sec. 21-6 the coupling capacitors were selected to give poles of $T(s)$ at -5, -0.2, and -0.01 reciprocal milliseconds. From the complex-conjugate poles of A_f to be obtained from the cubic of (21-71) with $T_o = 39$, determine the pole Q and the lower 3 dB frequency in hertz. (The real pole of A_f is -0.00996.)

21-19. For the amplifier of Fig. 21-11 with compensation added so as to give $T(s)$ of (21-69), deduce the midband loop transmission T_o when the feedback is increased so as to reduce the phase margin to 45°. Then calculate the feedback resistance in kΩ that should replace the 75 ohm resistor of Fig. 21-11 to give this T_o. First, verify from (21-69) that T is -6.26 dB when the phase of T is $-135°$, at $\omega = 0.0386$ ns^{-1}. (*Ans*: 1.30 kΩ)

Section 21-7

21-20. (a) Deduce that the Butterworth configuration has no peaking in the frequency response. Do this by finding the magnitude of $A_f(\omega)$ from (21-76) and showing that the maximum occurs in the midband region ($\omega = 0$). (b) Also, from the expression for the magnitude of $A_f(\omega)$ show that the upper 3 dB frequency ω_2 is the magnitude 2α of each pole.

21-21. The gain $A_f(s)$ of the Butterworth three-pole configuration is given by (21-76). Assume A_f is V_o/I_s, and let $i_s(t)$ be a step function of magnitude $1/A_{fo}$ applied at time zero. Deduce that the Laplace transform $V_o(s)$ of $v_o(t)$ is

$$V_o(s) = \frac{1}{s} - \frac{1}{s + 2\alpha} - \frac{2\alpha}{s^2 + 2\alpha s + 4\alpha^2}$$

The inverse transform of $\beta / [(s + \alpha)^2 + \beta^2]$ is $\exp(-\alpha t)\sin \beta t$. Find $v_o(t)$. The first maximum of $v_o(t)$ occurs at $\alpha t = 2.46$. Calculate the percent overshoot.

21-22. For the feedback network of the amplifier of Fig. 21-16 derive the expressions given in Sec. 21-7 for B_o, the zero z_a, and the pole p_a of the feedback factor $B(s)$. Then redesign the network, determining R_1, R_2, and R_3, so as to give a feedback factor $B(s)$ having a midband value of -0.002 mmho and a zero at 8 MHz. Use 10 pF for C_4 and 50 kΩ for the sum of R_1 and R_2. Also, calculate the pole in MHz.

Section 21-8

21-23. From the incremental model of Fig. 21-19 deduce that the open-circuit time constants of capacitors C_1 and C_2 are those given in Sec. 21-8.

21-24. In the feedback amplifier of Fig. 21-17, change R_{C2} from 5 kΩ to zero and remove the RC compensating network. Then rearrange the incremental model, which is similar to that of Fig. 21-19, so that C_1 and C_3 are combined into a single capacitance (refer to Sec. 19-1). Apply Norton's theorem to the source and find $A(s)$ from suitable node equations, with $A(s) = V_o/V_s'$. (a) Determine T_o. (b) In reciprocal nanoseconds find the zeros and poles of $T(s)$. (c) From $T(\omega)$ determine the phase margin by a trial-and-error procedure. [Ans: (b) −0.268, 0.663; −0.00159, −0.0326, −0.403; (c) 43.2°]

21-25. For the amplifier of P21-24 the open-loop gain $A(s)$ can be expressed in the form

$$A(s) = \frac{-0.165(s^2 - 0.394s - 0.178)}{s^3 + 0.437s^2 + 0.0138s + 2.08 \times 10^{-5}}$$

The feedback factor B is that of the network of Fig. 21-17. Determine A_{fo}. Also, in reciprocal nanoseconds find the zeros and poles of $A_f(s)$.

Section 21-9

21-26. In Fig. 21-22 the OP AMP with $A_{vo} = 10^5$ has a pole at 5 Hz with the next pole greater than 5 MHz. Resistance R_f is 4 kΩ and R_1 is 1 kΩ. Calculate the voltage gain V_o/V_s and the BW in kHz if the amplifier is (a) noninverting and (b) inverting.

21-27. If the inverting OP AMP of Fig. 21-22a has poles at 1, 10, and 50 MHz with $A_{vo} = 100$ dB, what is the minimum voltage gain V_o/V_s in dB that is possible for a phase margin of at least 45°?

21-28. For the inverting OP AMP of Example 5 of Sec. 21-9, deduce the BW in MHz from (21-90), (21-98), and (21-95). Resistance R_f is 2 kΩ, R_1 is 11.4 ohms, C_f is 6 pF, A_{vo} is 10^4, and the poles of A_v are at 1, 4, and 40 MHz.

21-29. The inverting amplifier of Fig. 21-22a has an OP AMP with $A_{vo} = 10^4$ and poles at 1, 4, and 40 MHz. With $R_f = 2$ kΩ, determine the values of R_1 in ohms and C_f in pF to give the maximum bandwidth without peaking. What is the BW in MHz? The capacitor is in parallel with R_f.

Chapter Twenty Two
Oscillators and Tuned Amplifiers

Various oscillators generating a number of waveforms are used in instrumentation and test equipment. A common application of a sinusoidal oscillator is the local oscillator of a superheterodyne radio receiver, and oscillators are present in transmitters that generate high-frequency power. Digital systems employ pulse generators. A familiar application is the light flasher. One type consists of a multivibrator that drives two lamps alternately, and such a flasher is often used on aircraft lights. A construction barricade flasher can run 60 days on a single battery.

In this chapter we examine both *harmonic* and *relaxation* oscillators. In a harmonic oscillator the output is sinusoidal. Usually the transistor operates in the active region and continually supplies power to the associated passive elements. Exceptions are the transistors of high-power class C oscillators, which use tuned LC circuits to convert current pulses to sinusoidal voltages. On the other hand, the output of a relaxation oscillator is nonsinusoidal, depending on the transient rise and decay of voltages in RC or RL circuits. In most cases the transistor alternately switches *on* and *off*. An example is the free-running multivibrator of Sec. 22-4.

We have learned that feedback in an amplifier may become positive at sufficiently high frequencies. If the magnitude of the loop transmission $T(\omega)$ is greater than unity at the frequency at which the angle of T is 180°, the amplifier has poles in the right half-plane and is unstable. An unstable amplifier is, of course, an oscillator. Although building an oscillator is rather simple, it is not so easy to design one that supplies the required power to a specified load with precisely the desired waveform at a stable frequency and constant amplitude.

Radio-frequency tuned amplifiers have circuits similar to those of many LC oscillators. Such amplifiers, along with heterodyning and mixing, are discussed in the last two sections.

22-1. THE WIEN BRIDGE OSCILLATOR

The basic elements of a sinusoidal oscillator are an amplifier and external circuitry that includes a positive feedback network. This network returns a portion of the output voltage or current to the input with the phase needed for oscillation. Because we do not want parasitic capacitances and inductances of the amplifier to affect appreciably the frequency, *the external circuitry must contain reactances that control the phase shifts of the system*, and the parameters are selected to give the proper phase shifts at the desired frequency. Steady-state sinusoidal oscillations occur when the natural frequencies of the network, or the poles of a suitably chosen gain function, contain a complex-conjugate pair on the imaginary axis. All other frequencies should be in the left half-plane.

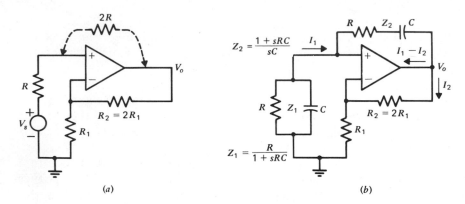

Figure 22-1   Amplifier and oscillator. (*a*) Noninverting amplifier. (*b*) Wien bridge oscillator.

In Fig. 22-1*a* is the diagram of a noninverting amplifier and in Fig. 22-1*b* is the same amplifier with elements added so as to convert it to a Wien bridge oscillator. The noninverting amplifier has a gain of 3, with the phasor voltage V_o equal to $3V_s$. Suppose a resistor of value $2R$ is connected as shown by the dashed lines. One-third of the output voltage is now fed back to the input. The feedback is positive, with the voltage that is fed back being in phase with the input voltage. In fact, for the specified resistors it is precisely equal to V_s at the instant the feedback resistor is connected. Therefore, if V_s is reduced to zero at this instant, in theory the voltages and currents do not change, and the input is supplied from the output through the feedback network.

A serious problem with the circuit that has been described results from parasitic reactances present, which introduce phase shifts. External reactances must be added for phase control. This is done in the Wien bridge oscillator of Fig. 22-1b. It is easy to show that Z_2/Z_1 is $2 + j0$ at a frequency ω equal to $1/RC$, which is the same as the resistance ratio of the noninverting amplifier with the resistor $2R$ present. Because this ratio gives a feedback voltage of precisely the proper magnitude and phase, the network oscillates, and the frequency is determined by the parameters R and C.

A technique useful for analyzing oscillator circuits and for developing design equations is based on a suitable set of loop or node equations. To illustrate, the Wien bridge oscillator of Fig. 22-1b with the resistor $2R_1$ designated as R_2 will be analyzed. Utilizing the virtual short-circuit of the OP AMP, we find from the circuit the following loop equations:

$$\left(\frac{R}{1 + sRC} + \frac{1 + sRC}{sC}\right)I_1 + (R_1 + R_2)I_2 = 0 \qquad (22\text{-}1)$$

$$\left(\frac{R}{1 + sRC}\right)I_1 + R_1 I_2 = 0 \qquad (22\text{-}2)$$

If the circuit oscillates, the currents I_1 and I_2 exist. For nonzero currents, the equations require that the determinant of the coefficients be zero. Setting it to zero and simplifying the result, we obtain

$$s^2 + \frac{s}{RC}\left(2 - \frac{R_2}{R_1}\right) + \frac{1}{R^2C^2} = 0 \qquad (22\text{-}3)$$

For sinusoidal time variations s is $j\omega$. Making this substitution and equating the reals and also the imaginaries to zero, we deduce that

$$\omega = \frac{1}{RC} \qquad R_2 = 2R_1 \qquad (22\text{-}4)$$

Equations 22-4 inform us that R_2 should be made equal to $2R_1$ and that, for this condition, oscillations occur at a radian frequency of $1/RC$. For $R_2 = 2R_1$ the roots of (22-3) lie on the imaginary axis of the complex-frequency plane. If R_2 is less than this, the reduced gain places the two natural frequencies in the left half-plane and there are no oscillations. On the other hand, for $R_2 > 2R_1$ the frequencies are in the right-half plane. In this event oscillations build up until restrained by the nonlinearities of the amplifier.

Let us note that the oscillator has the form of a bridge circuit. Between the output node and ground are two parallel branches. One branch consists of Z_1 and Z_2 and the other consists of R_1 and R_2. The circuit oscillates at a frequency giving a balanced bridge, with Z_2/Z_1 equal to R_2/R_1. The null voltage is that at the input of the OP AMP.

The Wien bridge oscillator is termed a *phase-shift* oscillator, because resistors and capacitors are used to provide the proper phase shift for positive feedback at the desired frequency. *Phase-shift oscillators are normally used for frequencies from zero to about* 100 kHz. *At higher frequencies LC and crystal oscillators are employed.* The preference for the RC phase-shift oscillator in the audio and low radio-frequency range occurs because values of L and C required for LC oscillators in this range are large. Another popular phase-shift oscillator is examined in the following example.

Example

For the RC oscillator of Fig. 22-2, assume that the bias resistors R_2 and R_3 are very large and that R_1 is adjusted so that $R_1 + h_{ie}$ equals R. In terms of R_L, R, and C find how large the transistor β must be to ensure oscillations and find the oscillation frequency.

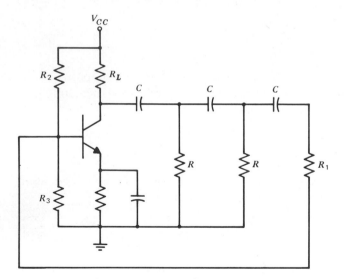

Figure 22-2   RC phase shift oscillator ($R_1 + h_{ie} = R$).

Solution

Because this amplifier is used at low frequencies the transistor in the incremental model of Fig. 22-3 is represented by h_{ie} and the current source βI_3. For convenience the model is arranged with h_{ie} at the extreme right. We first must write three loop equations in terms of I_1, I_2, and I_3. The determinant of the coefficients is then equated to zero, s is replaced with $j\omega$, and two separate

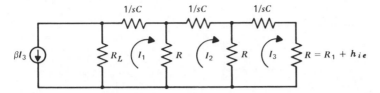

Figure 22-3 Incremental model (R_2, R_3 neglected).

equations are deduced from the reals and the imaginaries. The details are left as an exercise (see P22-1).

The equations resulting from this procedure are

$$R^3 + (\beta + 3)R^2 R_L - X^2(5R + R_L) = 0 \tag{22-5}$$

$$X^2 - 4RR_L - 6R^2 = 0 \tag{22-6}$$

with $X = 1/\omega C$. Using (22-6) to eliminate X in (22-5) and solving for β, we find that

$$\beta = 23 + \frac{29R}{R_L} + \frac{4R_L}{R} \tag{22-7}$$

This is the minimum beta. The oscillation frequency is found from (22-6) to be

$$\omega = \frac{1}{C\sqrt{6R^2 + 4RR_L}} \tag{22-8}$$

AMPLITUDE STABILIZATION

A necessary and sufficient condition for sinusoidal oscillations is the presence of a pair of poles of a feedback system on the imaginary axis with no poles in the right half-plane. In the practical case the poles must initially be to the right of the axis if oscillations are to build up to the desired level. There must be some way for the oscillator to stabilize the amplitude. If no provision is made in the external circuitry, the nonlinearity of the amplifier that results when the signal becomes large will reduce the amplifier gain sufficiently to shift the poles to the left until they reside on the axis. Limiting is accomplished by a nonlinearity that unfortunately produces some distortion.

An example of external limiting is the Wien bridge oscillator of Fig. 22-4. The gain of the operational amplifier is greater than 3 by 1%, which is sufficient to place the poles slightly to the right of the imaginary axis of the s-plane (see P22-3). This provides adequate allowance for component tolerances. Each of the matched diodes is *off* as long as v_o is between ± 10 V. When v_o reaches $+10$, the node voltage v_1 is 3.75 and v_3 is 10/3.03, or 3.30 V. The diode D_1 is forward biased 0.45 V and begins to conduct, reducing the amplifier gain. Diode D_2

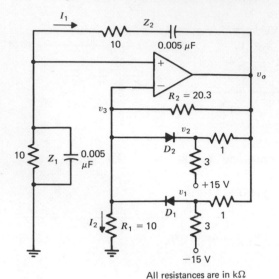

Figure 22-4   Wien bridge oscillator with amplitude limiting.

conducts when v_o becomes -10 V. Thus the output is 20 V peak-to-peak. A limiter is especially important for variable-frequency oscillators, because the amplitude ordinarily varies with the tuning.

FREQUENCY STABILITY

Variations in the supply voltage, temperature, stray capacitance, and the coupled load, along with small shifts in transistor parameters and component values with time, produce changes in the frequency of an oscillator. A good measure of frequency stability is the value of $d\theta/d\omega$ with θ denoting the phase shift of the closed-loop gain. We want the phase to shift appreciably whenever ω changes a very small amount because the change that occurs in θ tends to return the frequency to the proper value. Clearly an amplifier with a stable frequency has a large value of $d\theta/d\omega$, and this value should depend mainly on precision circuit elements.

For the oscillator of Fig. 22-1b the ratio of the output voltage V_o to the voltage across Z_1 at the input is $1 + Z_2/Z_1$. Oscillations occur at the frequency that makes the phase angle θ of Z_2/Z_1 equal to zero. With $s = j\omega$, this angle θ is $-90°$ plus $2 \tan^{-1} \omega RC$, and $d\theta/d\omega$ is found to be

$$\frac{d\theta}{d\omega} = \frac{2RC}{1 + \omega^2 R^2 C^2} \tag{22-9}$$

At the oscillation frequency, $d\theta/d\omega$ equals RC. For high stability R and C should be stable elements of large value. Because ω is $1/RC$, *the stability is best at low oscillation frequencies.*

22-2. LC OSCILLATORS

In most cases LC oscillators are designed so that the frequency of oscillation is approximately the resonant frequency of the LC network. This is often referred to as a "tank" circuit because of its ability to store energy. Thus

$$\omega^2 \approx \frac{1}{LC} \tag{22-10}$$

with L and C denoting the respective values of the effective inductance and the effective capacitance around the LC loop.

To illustrate, let us refer to the circuits of Fig. 22-5, which represent three common oscillators with bias circuitry omitted for simplicity. In the Colpitts configuration of Fig. 22-5a the effective loop capacitance is that of C_1 and C_2 in series, and C in (22-10) is $C_1 C_2/(C_1 + C_2)$. The tank circuit of the Hartley oscillator of Fig. 22-5b has an effective loop inductance equal to that of the inductivity coupled coils. This is $L_1 + L_2 + 2M$, which is the value to be used in (22-10). A single tapped coil on a ceramic core is common. Finally, for the

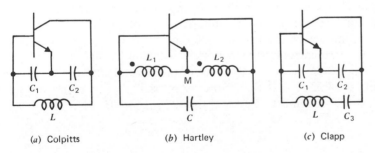

| (a) Colpitts | (b) Hartley | (c) Clapp |

Figure 22-5 LC oscillators with bias circuitry omitted.

Clapp oscillator the effective loop capacitance is that of the three capacitors in series and $1/C$ in (22-10) equals the sum of $1/C_1$, $1/C_2$, and $1/C_3$. These are approximations. In practical oscillators the transistor and bias-circuit parameters and also the small ohmic resistance of the inductance L have some effect on the oscillation frequency, although these effects are usually second-order.

Determination of the oscillation frequency and the requirement on the amplifier gain is easily accomplished at low frequencies by the method of the preceding section. For the Colpitts circuit the low-frequency model is that of

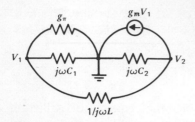

Figure 22-6 Low-frequency incremental model of Colpitts oscillator.

Fig. 22-6, with the transistor represented by the two parameters g_π and g_m. The two ω-domain node equations are

$$0 = \left(g_\pi + j\omega C_1 + \frac{1}{j\omega L}\right)V_1 - \left(\frac{1}{j\omega L}\right)V_2$$

$$0 = \left(g_m - \frac{1}{j\omega L}\right)V_1 + \left(j\omega C_2 + \frac{1}{j\omega L}\right)V_2$$

(22-11)

We equate the determinant of the coefficients to zero. The real part of the equation that results gives ω to be that which has been stated. Substitution of the expression for ω into the imaginary part of the equation reveals that g_m/g_π, which is β, equals C_2/C_1. This is, of course, the minimum beta (see P22-4).

The other circuits can be analyzed similarly. For the tapped-coil Hartley the low-frequency model of the network of Fig. 22-5b is that of Fig. 22-7. The transistor is represented by the parameters r_π and β, and the current-controlled voltage sources represent the emfs of mutual induction. We deduce from the circuit (see P22-5) that the frequency of oscillation is as previously stated and the transistor β must be greater than $(L_1 + M)/(L_2 + M)$.

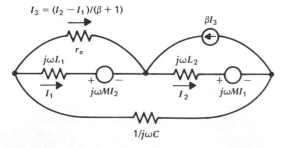

Figure 22-7 Low-frequency incremental model of Hartley oscillator.

The Clapp oscillator is a modified Colpitts circuit. The transistor β must be greater than C_2/C_1 (see P22-6), independent of C_3. If C_3 is much smaller than the other two capacitors, the resonant frequency is nearly independent of C_1 and C_2. An obvious advantage of the Clapp configuration is that the transistor beta can be used to determine suitable values of C_1 and C_2, with C_3 chosen to control the frequency. The result is improved frequency stability.

HIGH-FREQUENCY DESIGN AND ANALYSIS

Although LC oscillators can be used at low frequencies, applications are usually at frequencies above 100 kHz. At high frequencies the inductors and capacitors have more reasonable values, and the frequency stability is greater that that typical of RC phase-shift oscillators. Determination of the necessary conditions for oscillation and the oscillation frequency is more complicated because the transistor parameters are complex functions of frequency. If the frequency is less than about 30% of the transition frequency f_T of the transistor, the wide-band hybrid-π model is adequate. However, LC oscillators often operate in the VHF range (30–300 MHz) or higher. At such frequencies a set of complex matrix parameters are used to represent the active device. These must be measured at the proper frequency, or determined from suitable data sheets if possible.

Most oscillator circuits can be arranged in the form shown in Fig. 22-8. The amplifier and the feedback network are connected, with the dashed lines ignored, in a series-series arrangement of two-ports. This is similar to the feedback configuration of Fig. 20-28 of Sec. 20-6, and the z parameter matrix equations of (20-46) apply. Because V_i and V_o in these equations are zero, the determinant of the coefficients, which is $z_i z_o - z_r z_f$, must be zero for oscillations to occur. The composite z parameters are found by adding the corresponding parameters of the amplifier and the feedback network.

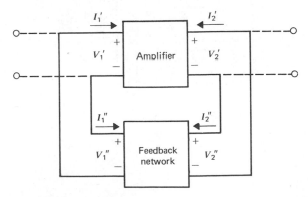

Figure 22-8   Feedback oscillator configuration.

An alternate viewpoint is to consider the configuration of Fig. 22-8 as a parallel-parallel connection of two-ports, as indicated by the dashed lines at the input and output terminals. This arrangement is similar to that of Fig. 20-8 of Sec. 20-3 with I_s, Y_s, and Y_L all equal to zero. However, it is necessary to reverse the reference directions of the voltages and currents of the feedback network in order to provide identical input voltages, and also output voltages, as required. Thus we can use y parameters instead of z parameters if we choose to do so. Equating the determinant of the coefficients of the y parameter equations to zero, we find that

$$(y_{ia} + y_{ib})(y_{oa} + y_{ob}) - (y_{ra} + y_{rb})(y_{fa} + y_{fb}) = 0 \qquad (22\text{-}12)$$

with the subscripts a and b referring to the amplifier and the feedback network, respectively. Equation 22-12 can be written in terms of z parameters if desired, but these are seldom used.

Often h parameters are employed. The equation corresponding to (22-12) is

$$(h_{ia} + h_{ib})(h_{oa} + h_{ob}) - (h_{ra} - h_{rb})(h_{fa} - h_{fb}) = 0 \qquad (22\text{-}13)$$

The negative signs ahead of h_{rb} and h_{fb} are a consequence of the conventional reference directions. If the reference directions at the output of the feedback network are reversed, we have a series-parallel arrangement except that the reference directions of the feedback network no longer conform to convention. If any doubt exists it is suggested that (22-13) be deduced directly from (22-12) using the matrix interrelations of Appendix B (see P22-7).

To illustrate the use of the design equations a Colpitts oscillator will be used. First, we must arrange the network of Fig. 22-5a in the feedback form of Fig. 22-8, and two acceptable configurations are shown in Fig. 22-9. If we wish to work with CE matrix parameters, the arrangement of Fig. 22-9a is used. In

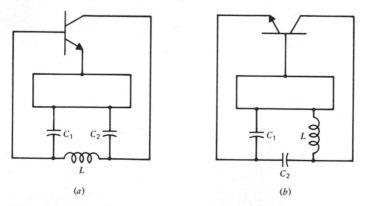

(a)   (b)

Figure 22-9   Alternate arrangements of Colpitts oscillator in the feedback form of Fig. 22-8. (a) CE configuration. (b) CB configuration.

many cases it is more convenient to work with CB parameters, and the circuit of Fig. 22-9b is used. The networks are exactly equivalent. They apply to the same oscillator, and it makes no difference whether the oscillator is actually connected as a CE circuit or a CB circuit, or perhaps neither. As has been done here, the network of a feedback oscillator can usually be sketched so that the amplifier is CE or CB, whichever is more convenient.

Let us select the circuit of Fig. 22-9a, along with CE y parameters. The parameters of the feedback network are

$$y_{ib} = j\omega C_1 + \frac{1}{j\omega L} \qquad y_{rb} = \frac{-1}{j\omega L}$$

$$y_{fb} = \frac{-1}{j\omega L} \qquad y_{ob} = j\omega C_2 + \frac{1}{j\omega L} \tag{22-14}$$

The requirements for oscillation and the oscillation frequency are determined by substituting these expressions and the CE y parameters of the transistor into (22-12). The equation that results is complicated because the transistor parameters are complex.

Example

A Colpitts oscillator at 45 MHz is to be designed using a Texas Instruments chip-type N20 NPN silicon transistor. At this frequency at a bias point of 10 mA and 15 V, the manufacturer gives y_{ie}, y_{fe}, y_{re}, and y_{oe} to be $11 + j9$, $180 - j160$, $j0.28$, and $0.16 + j0.68$ mmhos, respectively. For convenience we shall use units of kΩ, mA, V, pF, μH, ns, and GHz.

Solution

Even at high frequencies we expect the product LC to be approximately $1/\omega^2$, and for $\omega = 0.283$ the product is 12.5 ns^2. Choosing 0.2 μH for L will give an effective capacitance C of about 62 pF, which is a suitable value. The amplifier parameters in (22-12) are the CE y parameters of the transistor, and we substitute the numerical values of these into the equation. The y parameters of the feedback network are those of (22-14) with $\omega = 0.283$ ns^{-1}, $1/\omega L = 17.7$ mmhos, and C_1 and C_2 unknown. These parameters are also substituted into (22-12). The equation that results contains the unknowns C_1 and C_2. Equating both the real and imaginary parts to zero and solving for C_1 and C_2, we find that the respective values are 71 500 and 59.7 pF (see P22-8).

The value of C_1 is a maximum. Feedback is increased by reducing C_1, which raises its impedance. A substantially lower value should be used to ensure that the circuit will oscillate. The effective capacitance C of the tank circuit is $C_1 C_2/(C_1 + C_2)$, which is 59.6 pF. Using this value in the approximate relation $1/LC$ for ω^2 gives a frequency of 46.1 MHz, whereas the frequency specified by the exact relation (22-12) is 45.0 MHz. The approximate relation is reasonable for most practical LC oscillators.

Only a few of the many different types of LC oscillators have been considered here. There are a variety of LC configurations with numerous different bias arrangements. For example, the Hartley oscillator has CE, CB, and CC circuits, and some of these use two separate inductively-coupled coils rather than a tapped coil. Tuned-base tuned-emitter oscillators have a series LC network between base and ground and another one between emitter and ground, with coupling provided by the active device.

Push-pull circuits are common, and there are class B and class C oscillators in addition to the class A types examined here. Above 50 MHz the *common-base oscillator* is popular. It has a tank circuit at the output with feedback to the input provided by a capacitor. Sometimes this feedback capacitor is omitted, in which case feedback takes place through the capacitances of the active device. An example with a crystal is shown in Fig. 22-11a.

A coil has ohmic resistance, and when power is drawn from an oscillator, the effective value of this resistance is increased. The *quality factor* Q of the tank circuit is the ratio of the coil reactance ωL to the effective series resistance r. The value of r is always somewhat uncertain, for it is a function of temperature, frequency, and loading, including core losses and radiation. Because frequency deviation from that of (22-10) is inversely proportional to Q, it is important to use high-Q tank circuits for good stability. The analyses made here have ignored coil resistances for simplicity, but their effects are significant. When extreme frequency stability is required, a crystal is usually substituted for the tank circuit.

22-3. CRYSTAL OSCILLATORS

The *piezoelectric effect* is exhibited by many crystals, such as quartz, Rochelle salt, and certain ceramics. Mechanical compression or expansion of the crystal along one axis causes a voltage to appear between opposite faces of a second axis. Conversely, application of a voltage across the faces of the second axis produces compression or expansion, depending on the polarity, along the first axis. Ceramic crystal microphones, headphones, and phonograph pickups often are employed to convert mechanical vibrations into equivalent electrical oscillations. Because the quartz crystal has unusually low mechanical losses, it is extensively used to control the frequency of an oscillator.

A slice of a crystal is mounted in a suitable holder and silver plated on each electrical face for connections. The holder of a high-frequency crystal is about $1 \times 1 \times 0.5$ cm, with two pins for electrical connections. The natural mechanical vibration frequency, which corresponds to the electrical frequency of oscillation, depends on the thickness of the slice and its mode of oscillation. Commercial crystals have a maximum fundamental frequency limited to about 20 MHz, because at higher frequencies the crystal must be cut so thin that it is likely to break.

In Fig. 22-10 is the symbol of a crystal and its approximate electrical equivalent circuit. The inductance L is analogous to the mass of the crystal, the capacitance C_1 is analogous to the crystal elasticity, and the resistance R represents mechanical friction, accounting for power loss. Shunt capacitance C_2 is that of the leads and terminals of the mounting structure, including that of the silver-plated faces. Low-frequency crystals (70 kHz to 1 MHz) are thicker with larger mass and greater frictional losses, and consequently they have relatively large values of L and R. Medium-frequency crystals (1 MHz to 20 MHz) are thinner with lower L and R. High-frequency crystals (20 MHz to 160 MHz) generally operate on *overtones*, which are mechanical vibrations at frequencies above the fundamental resonance.

Some crystals are piezoelectric at overtone frequencies. These are approximately odd harmonic multiples of the fundamental. The third, fifth, seventh, and ninth mechanical overtones are often used. A 160 MHz crystal is probably designed to operate on its ninth overtone. The use of mechanical overtones can

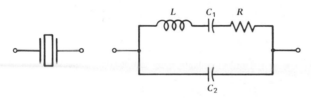

Figure 22-10 Symbol and electrical equivalent circuit of a piezoelectric crystal.

eliminate the need for a frequency multiplier stage. Normally, overtone oscillators use an LC tuned circuit in addition to the crystal so as to provide selectivity that prevents oscillation at the fundamental or at an undesired overtone.

The analogous electrical parameters L and R of a crystal depend a great deal on its physical size, and hence its frequency. For example, a 90 kHz crystal may have an inductance L of 140 H and a resistance R of 3 kΩ, for a Q of 26 400. Capacitances C_1 and C_2 are approximately 0.022 pF and 5 pF, respectively. For a 10 MHz crystal L might be 0.025 H with $R - 20$ ohms, giving a Q of almost 80 000. Capacitance C_1 is 0.01 pF and C_2 is typically 9 pF. Series capacitance C_1 is usually between 0.01 and 0.05 pF and C_2 is commonly between 4 and 10 pF. The very large quality factors, with values generally above twenty thousand, account for the exceptional frequency stability of crystal oscillators. For comparison, quality factors of LC tank circuits are normally from 50 to 200.

A crystal has two resonant frequencies, which are given by the equations

$$\omega_s{}^2 = \frac{1}{LC_1} \qquad \omega_p{}^2 = \frac{1}{LC} \qquad (22\text{-}15)$$

where C denotes the effective capacitance around the loop of the circuit of Fig. 22-10. It follows that $1/C$ is the sum of $1/C_1$ and $1/C_2$. At the series resonant frequency ω_s the impedance of the tank circuit is approximately equal to the small resistance R, whereas at the parallel resonant frequency ω_p the impedance is $1/(\omega_p{}^2 C_2{}^2 R)$, which is very large. For example, the 430 kHz crystal of P22-10 has an impedance of 200 ohms at the series-resonant frequency of 427.5 kHz, but at the parallel-resonant frequency of 429.0 kHz, the impedance is 20 MΩ. As in this example, the series-resonant and parallel-resonant frequencies are nearly equal. In fact, *they seldom differ by more than 1%.* In (22-15) the capacitance C_1 is approximately equal to C, which is $C_1 C_2/(C_1 + C_2)$, because C_2 is always large compared with C_1.

Crystals operate in either the series mode or the parallel mode, depending on the circuit design. In Fig. 22-11 are the diagrams of two CB crystal oscillators. In Fig. 22-11a the crystal operates in the parallel mode, replacing an LC tank circuit, with feedback provided by the capacitor. In the Hartley oscillator of Fig. 22-11b the crystal operates in the series mode. The tank circuit is tuned to the crystal frequency. Only at the series-resonant frequency is there sufficient

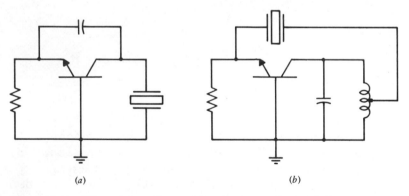

(a)   (b)

Figure 22-11   CB crystal oscillators, bias circuitry omitted. (a) Common-base oscillator, parallel mode. (b) Hartley oscillator, series mode.

feedback to produce sustained oscillations. The crystal is used here as a frequency-selective network. When this circuit is operated at the seventh and higher overtones, the capacitance C_2 of the equivalent circuit of the crystal is usually neutralized by connecting a small inductance in parallel with it and adjusting this inductance to give parallel resonance with C_2. Otherwise, the reactance of C_2 may be sufficiently small to give appreciable feedback at other frequencies, with the result that the LC tank circuit becomes the frequency-controlling network.

Another example of series-mode operation is the *Pierce* oscillator of Fig. 22-12, which is actually a Clapp circuit. Capacitors C_3 and C_4 are much larger than both C_1 and C_2 of the crystal equivalent circuit. Therefore, the resonant fre-

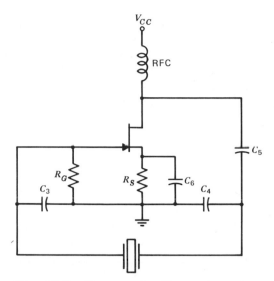

Figure 22-12   Pierce crystal oscillator, series mode.

quency is determined by L and C_1, and capacitors C_3 and C_4 provide the feedback (see P22-11). The radio-frequency choke (RFC) has sufficient inductance to be essentially an open-circuit for ac, while providing a low resistance path for the bias current of the junction FET. Capacitor C_5 connects one side of the crystal to the drain while blocking dc.

A crystal oscillator is often followed by a buffer stage that minimizes the effect of the load on the frequency. An example is shown in Fig. 22-13. The base of Q_1

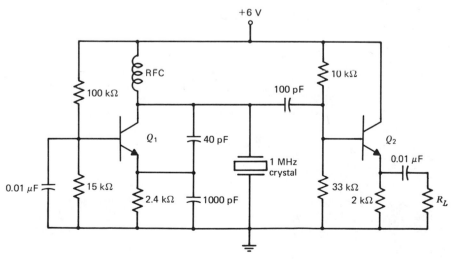

Figure 22-13   Crystal oscillator and CC buffer stage.

is at ac ground, and feedback from the oscillatory circuit to the emitter is accomplished by the 1000 pF capacitor. The emitter-follower output stage has a high input impedance that is insensitive to changes in R_L provided the load is greater than about 10 kΩ. The oscillator has the Clapp configuration, and the frequency is approximately the series resonant frequency of the crystal. Moderate changes in load resistance, temperature, and supply voltage have negligible effect on the oscillation frequency.

22-4. MONOSTABLE AND ASTABLE MULTIVIBRATORS

Both flip-flops and Schmitt trigger circuits are examples of bistable networks. By suppressing one of the stable states a bistable circuit becomes a *monostable multivibrator*, also called a *one-shot*, defined as a network with a single stable state and a *quasi-stable* state. The quasi-stable state can exist only temporarily, with the time determined by certain network parameters. These are usually the values of a resistor and a capacitor. A trigger signal is employed to switch the circuit from its stable state to the quasi-stable state. After a short time, the circuit reverts to its stable state. Because regenerative feedback activates the switching, the transitions are very rapid.

MONOSTABLE MULTIVIBRATOR CIRCUITS

Figure 22-14 is a monostable multivibrator configuration with the input triggering pulse and the output pulse shown. Before time zero the n-channel transistor is *off*, the capacitor is charged, both inputs to the NAND gate are logical 1's, and the output is zero. The input trigger pulse almost instantly discharges C, converting the output to a logical 1 at V_{DD} volts. At time t_1 the trigger pulse returns to zero and the capacitor gradually charges through the

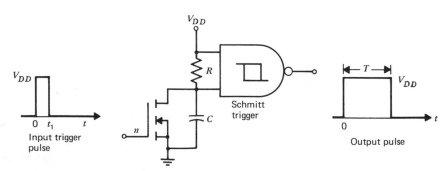

Figure 22-14   Monostable multivibrator.

resistor R. When the voltage v_C rises to the threshold voltage V_P of the Schmitt trigger, the output switches back to zero.

The expression $V_{DD}(1 - e^{-t/RC})$ denotes the voltage across a capacitor when V_{DD} is applied at time zero to a series RC circuit. This equals V_P provided t is replaced with $T - t_1$, noting that t_1 is the instant at which charging begins and T is the output pulse width. Solving for T, we obtain

$$T = t_1 + RC \ln \frac{V_{DD}}{V_{DD} - V_P} \tag{22-16}$$

Usually, t_1 in (22-16) is negligible, and in CMOS circuits the trigger voltage V_P is about $0.5 V_{DD}$.

One of the most important of the many applications of a monostable multivibrator is the generation of a pulse of prescribed amplitude and duration in response to a trigger signal. For the configuration of Fig. 22-14 using CD4093B for the Schmitt trigger, with values of R between 50 kΩ and 1 MΩ and C between 100 pF and 1 μF, the pulse width T may be as low as 4.6 μs and as high as 0.9 s. The results assume V_{DD} is 10 V, V_p is 6 V, and t_1 is negligible. Thus we can easily obtain a wide range of pulse widths by selection of R and C, and the pulse amplitude is $V_{DD} - V_{SS}$.

In Fig. 22-15 is shown a monostable multivibrator configuration made of two NOR gates and an RC network. The second NOR gate is connected as an inverter. At time zero v_3 is logical 1, v_4 is 0, and v_1 is 0. Therefore, v_2 is 1. At time t_1 the trigger pulse is applied, which makes v_2 switch to 0. Because the voltage across C cannot change instantaneously, v_3 also drops to 0, giving a logical-1 output v_4. With v_2 zero, the capacitor charges gradually, causing v_3 to rise exponentially. When v_3 reaches the threshold voltage of the inverter, the output drops to zero. The pulse width is expressed by (22-16) with V_P replaced with the threshold voltage at which the inverter switches states. A monostable multivibrator made from operational amplifiers is examined in Sec. 23-6.

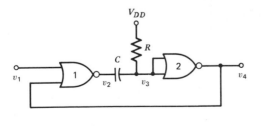
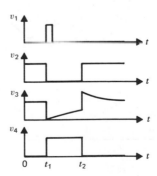

Figure 22-15   A one-shot multivibrator configuration.

ASTABLE MULTIVIBRATORS

An *astable* multivibrator has no stable state. However, there are two quasi-stable states, and the circuit switches back and forth between them, remaining in each one for a period of time determined by network parameters. No trigger pulses are employed. Astable multivibrators are oscillators, and among their many applications are the generation of periodic waveforms and pulse trains.

In Fig. 22-16 is shown an astable multivibrator, consisting of two inverters and an RC network, along with the voltage waveforms. Let us assume the inverters are CMOS with low and high levels of 0 and 10 V. Each inverter output is precisely either 0 or 10 V, whereas the input v_1 can vary slowly between certain limits because it is the voltage of the insulated gate. Furthermore, no current can flow into the input. The only possible current path is between nodes v_2 and v_O. With these assumptions the operation is easily analyzed.

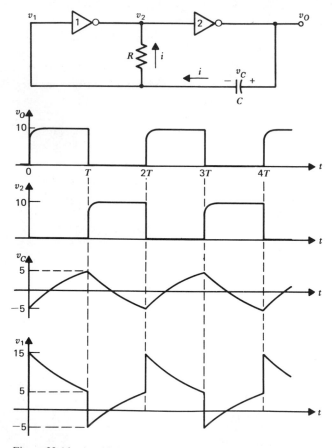

Figure 22-16 Astable multivibrator and waveforms.

At a certain instant, suppose the logic states of v_1, v_2, and v_O are 1, 0, and 1, respectively. It follows that v_1 is greater than the inverter switching voltage of $0.5V_{DD}$, or 5 V. The voltage across R produces a current i, which charges the capacitor causing v_C to rise. Thus v_1 drops, and when it reaches 5 V, the inverters switch states. The respective logic levels of v_2 and v_O are now 1 and 0. The current i reverses, and v_C drops until v_1 rises through 5 V. At this instant the inverters again switch states.

Let us examine the waveforms of Fig. 22-16. Those for v_O and v_2 are the rectangular pulses resulting from the periodic switching of the inverters. The capacitor voltage rises exponentially when v_O is 10 V and falls when it is zero. It varies between plus and minus 5 V, because inverter switching occurs at 5 V, and its average value is zero. The input voltage v_1 is a plot of $v_O - v_C$. When v_O changes abruptly by 10 V, v_1 must do so, too.

In most practical CMOS circuits protective diodes normally included at the input of inverter 1 prevent v_1 from dropping below V_{SS} or rising above V_{DD} by more than a diode drop. In such cases the waveform of v_1 is modified. The abrupt rise from 5 V terminates at 10.7 and the drop from 5 terminates at -0.7. With the RC time constant unchanged, the reduced range of v_1 gives a decreased pulse width T. Also, the voltage across the capacitor changes rather abruptly at these switching instants, because v_C equals $v_O - v_1$ (see P22-14). It is usually necessary to connect a resistor in series with the input of inverter 1 to limit the current through the protective diodes.

Flexibility is added to the astable multivibrator by replacing the input inverter with a two-input NOR or NAND gate. The second input can then be used to turn the oscillator *on* and *off*. Also, it can be used for generation of pulse-modulated waves. The use of a variable capacitor allows the frequency of the output pulses to be easily controlled. In addition, by placing an n-channel transistor in parallel with R, the effective resistance of the combination is determined by the voltage applied to the gate of this transistor. The result is a *voltage controlled oscillator* (see P22-12). The n-channel device and necessary inverters are available in a single package.

In Fig. 22-17*a* is a circuit of a free-running (astable) multivibrator using BJTs. In Fig. 22-17*b* are waveforms for v_{CE2} and v_{BE2}. Those for v_{CE1} and v_{BE1} are similar, only shifted in time by half a period. With the capacitors removed, both transistors are biased in the saturation mode. However, with the capacitors present and assuming the transistors are not both locked-up in saturation, oscillations occur because of positive feedback. Although the locked-up condition is unlikely when the full supply voltage is applied suddenly, additional circuitry is sometimes added to provide sure starting. In oscillation, the state of each transistor alternates between saturation and cutoff at a frequency determined by the network parameters (see P22-16). Suitably modified, the basic circuit shown can give 10 MHz repetition rates with pulse rise times less than 10 ns.

To explain the operation let t_o denote the instant when Q_1 switches *off* to *on*

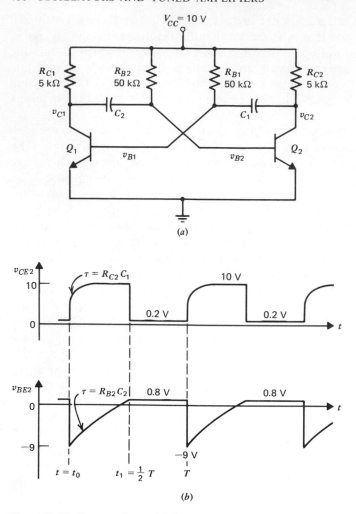

Figure 22-17 Free-running multivibrator, (a) Circuit. (b) Waveforms.

and Q_2 switches *on* to *off*. At time t_o^- just before switching occurs, the collector voltage v_{C1} and the base voltage v_{B1} of Q_1 are 10 and 0.8 V, respectively. For the saturated transistor Q_2 the collector voltage v_{C2} is 0.2 and its base voltage v_{B2} is 0.8 V. There are 9.2 V across C_2 and 0.6 V across C_1.

At t_o^+ immediately after switching, Q_1 is saturated with $v_{C1} = 0.2$ and $v_{B1} = 0.8$ V. The collector voltage v_{C2} of Q_2 is 0.2 V. It rises rapidly toward 10 V as capacitor C_1 charges, with a time constant $R_{C2}C_1$. The base voltage v_{B2} drops from 0.8 at t_o^- to -9 at t_o^+. This change equals that of v_{C1}, with the voltage across C_2 remaining at 9.2 V. Current through R_{B2} now flows into capacitor C_2, causing the base voltage v_{B2} to rise from -9 toward $+10$ with a time constant $R_{B2}C_2$ which is considerably larger than $R_{C2}C_1$. At time t_1 the voltage v_{B2} reaches 0.8 V, switching Q_2 *on* and thereby turning Q_1 *off*. The cycle continues.

An astable multivibrator can be made by cascading two monostable circuits with the output of one serving to trigger the input of the other. This is illustrated in Fig. 23-18 of Sec. 23-4, using two integrated circuits. Then there are various ways to modify one-shot circuits to give free-running oscillators with pulses of desired width and spacing. In addition, both astable and monostable circuits are frequently formed from operational amplifiers.

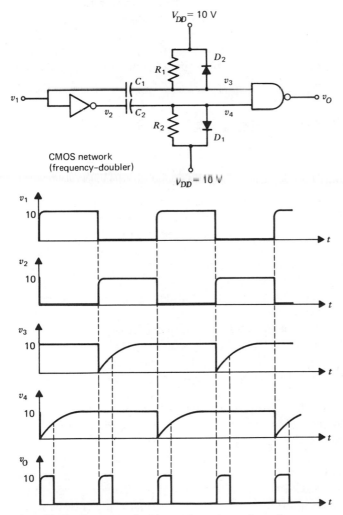

Figure 22-18   Frequency-doubler configuration and waveforms.

FREQUENCY MULTIPLIERS

It is sometimes necessary to divide or multiply the frequency of an oscillator such as an astable multivibrator. We have already considered frequency division. A CMOS frequency-doubler configuration is shown in Fig. 22-18, along with sketches of the waveforms. A 2^n-*multiplier* is obtained by cascading n stages.

The operation is easily explained with reference to the circuit and waveforms of the figure. For simplicity let us assume that the protective diodes D_1 and D_2, which are built into the NAND package, have negligible forward voltage drop. The input v_1 is a pulse train and v_2 is its complement. When v_1 rises from 0 to 10 V, v_3 remains clamped by diode D_2 at 10 V. However, when v_1 drops, v_3 drops the same amount, because capacitor C_1 is uncharged. As C_1 is charged by the current through R_1, the voltage v_3 rises exponentially. For the time interval during which v_3 is less than 5 V, the NAND output v_O is 10 V. An identical relation exists between v_2, v_4, and v_O. Consequently, the output frequency is doubled.

22-5. BLOCKING OSCILLATORS

There are various types of relaxation oscillators. One is the multivibrator of the previous section. Another important type is the *blocking oscillator*. It utilizes an iron-core pulse transformer to obtain the required regenerative feedback. As in the case of the multivibrator, operation can be in either a monostable (triggered) mode or an astable (free-running) mode. The blocking oscillator is often used as a pulse generator in computer circuitry. The astable circuit can supply trigger pulses for synchronizing different kinds of periodic waveforms, such as square waves and sawtooth waves. The monostable circuit can provide abrupt pulses from a slowly varying input trigger signal. Although many forms of blocking oscillators exist, we shall examine one basic configuration for the purpose of learning the principles involved.

In Fig. 22-19a is the circuit of an astable blocking oscillator. For convenience the transformer is assumed to have a turns ratio n of unity. An equivalent circuit suitable for an approximate analysis is that of Fig. 22-19b. The pulse transformer is replaced with an ideal transformer and a magnetizing inductance L as shown. Winding resistances, capacitances, and leakage inductances are neglected, although these significantly affect the rise and fall times of the pulses. With reference directions as indicated, the coil currents labeled i_B are equal and the coil voltages v are also equal, because of the assumption that n is unity.

Initially, suppose transistor Q and diode D are *off* and the magnetizing current i_m through L is zero. The 1 V at terminal V_{BB} produces a base current i_B. According to the diagram of Fig. 22-19b, the collector current i_C equals i_B, because i_m through L is zero and cannot change instantly. Therefore, the transistor immediately becomes saturated. Actually, the change from cutoff to saturation is

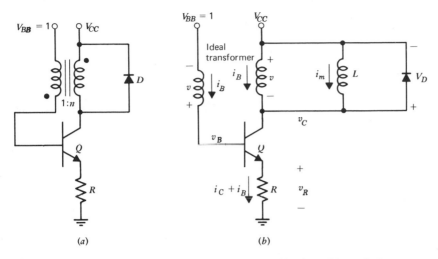

Figure 22-19 Astable blocking oscillator. (*a*) Circuit. (*b*) Circuit used in analysis.

gradual, with the transistor passing through the active mode. However, the positive feedback through the transformer causes the transition to occur very rapidly. Our simplifying assumptions led to the instantaneous change in the transistor state.

With the transistor saturated, both v_{CE} and v_{BE} are reasonably constant. Loop equations around the base-collector and base-emitter circuits reveal that *the coil voltage* v *and the voltage* v_R *across* R *are independent of time as long as the device remains saturated.* The constant voltage v causes i_m to rise from zero, increasing linearly with time. Consequently, the collector current i_C increases, and i_B must decrease to maintain a constant voltage across R. Eventually, i_C becomes βi_B, which is the borderline of active-mode operation.

At this instant i_B is very small and the collector current is approximately i_m. The magnetizing current begins to drop, v changes from positive to negative, and v_B and v_R drop. The regenerative feedback quickly reduces the transistor currents to zero. The coil voltage v becomes $-V_D$ as i_m switches to the diode, v_C is now $V_{CC} + V_D$, and v_B is $1 - V_D$. Because v_R is zero, v_{BE} equals v_B which is about 0.3 V. The diode allows i_m to decay gradually to zero. This holds the transistor in cutoff and also prevents the collector voltage from rising to an excessive value. After a period of time, the current i_m becomes zero, the diode turns *off*, the coil voltage v is no longer clamped at $-V_D$, and the cycle repeats itself.

Suppose the transistor Q is saturated from time 0 to t_1, with $v_{BE} = 0.8$ and $v_{CE} = 0.2$ V. From a loop equation around the base-collector circuit we find that

$$v = 0.5V_{CC} - 0.2 \qquad (22\text{-}17)$$

The coil voltage v equals $L\, di_m/dt$. Because v is constant and the initial current is zero, the current i_m, or $i_C - i_B$, is

$$i_m = i_C - i_B = \frac{vt}{L} \tag{22-18}$$

The sum of i_C and i_B is determined from a loop equation around the base-emitter circuit to be

$$i_C + i_B = \frac{v + 0.2}{R} = \frac{0.5V_{CC}}{R} \tag{22-19}$$

The right side of (22-19) is a consequence of (22-17).

From (22-17) through (22-19) the currents i_C and i_B are found to be

$$i_C = \frac{0.25V_{CC}}{R} + \frac{(0.25V_{CC} - 0.1)t}{L} \tag{22-20}$$

$$i_B = \frac{0.25V_{CC}}{R} - \frac{(0.25V_{CC} - 0.1)t}{L} \tag{22-21}$$

At time t_1 the collector current i_C is βi_B. From this we find that t_1 is

$$t_1 = \frac{L\, V_{CC}(\beta - 1)}{R(V_{CC} - 0.4)(\beta + 1)} \approx \frac{L}{R} \tag{22-22}$$

The maximum value I_o of the magnetizing current i_m occurs at t_1. Substituting for v and t_1 in (22-18), we find that

$$I_o = \frac{V_{CC}(\beta - 1)}{2R(\beta + 1)} \tag{22-23}$$

When the transistor turns *off*, the diode conducts and the coil voltage v is

$$v = -V_D \tag{22-24}$$

This also equals $L\, di_m/dt$. With time zero now denoting the time at which turn-off occurs, the magnetizing current is determined to be

$$i_m = -\frac{V_D t}{L} + I_o \tag{22-25}$$

with I_o given by (22-23). At time t_2 this current is zero. It follows that the *off* time is

$$t_2 = \frac{LV_{CC}(\beta - 1)}{2RV_D(\beta + 1)} \approx \frac{LV_{CC}}{2RV_D} \tag{22-26}$$

The period T is the sum of t_1 and t_2. Using the approximate relations of (22-22) and (22-26), we find that T is

$$T \approx \left(\frac{L}{R}\right)\left(1 + \frac{0.5V_{CC}}{V_D}\right) \tag{22-27}$$

The duty cycle t_1/T is approximately

$$\frac{t_1}{T} \approx \frac{2V_D}{V_{CC} + 2V_D} \tag{22-28}$$

Sketched in Fig. 22-20 are waveforms of i_B, i_C, and v_R for a blocking oscillator having $V_{CC} = 12.6$ V, $V_D = 0.7$ V, $L = 1$ mH, and $R = 1$ kΩ. The pulse width of v_R depends only on L and R and hence is very stable, with a value of 1 μs. The period T is 10 μs, giving a frequency of 100 kHz and a duty cycle of 10%. The duty cycle can easily be increased by increasing V_D. This can be accomplished by adding one or more additional diodes, or perhaps a Zener diode, in series with the diode D. For example, if a 5.6 V Zener diode is in series with D, the voltage V_D is 6.3 V, giving a duty cycle according to (22-28) of 50%. This is a square wave.

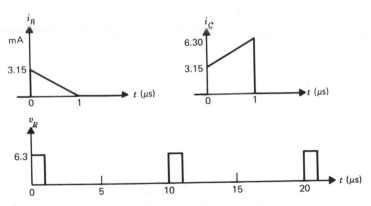

Figure 22-20   Waveforms of blocking oscillator.

The base voltage supply V_{BB} can be obtained from V_{CC} and a voltage divider consisting of two resistors. If V_{BB} *is connected to ground, the blocking oscillator is monostable and must be triggered.* The trigger voltage can be applied to the collector, forcing the transistor into saturation. It can also be applied to the base, feeding a current pulse into the base sufficient to produce saturation. When the active device returns to the cutoff mode, it remains there until it is again triggered.

Pulse widths are normally from a few nanoseconds to perhaps 25 μs, with a rise time about 10% of the pulse width. Duty cycles can be as low as 1% but higher duty cycles are more stable. The pulse transformer has a turns ratio usually between 1 and 5. When used with the high-speed pulses for which it is designed, it approximates an ideal transformer. Output power to a load is frequently obtained from a third winding. The core is made of either high-permeability laminated iron or high-resistivity ferrite.

In addition to the saturated CE blocking oscillator considered here, there are nonsaturated types, both CE and CB. These are used when fast response is especially important.

22-6. TUNED AMPLIFIERS

A tuned amplifier amplifies signals in a selected narrow band of frequencies and suppresses signals at frequencies outside the desired band. The bandwidth is a small fraction of the center frequency. LC tank circuits are commonly used to provide the required gain and selectivity. Although the resistance associated with a high-quality capacitor can be neglected without serious error, that of a coil is quite significant.

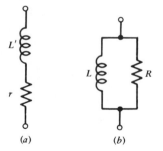

(a)   (b)

Figure 22-21   Equivalent circuits of an inductor. (a) Series form. (b) Parallel form.

Suppose an inductor is represented by a self-inductance L' in series with an ohmic resistance r. This is shown in Fig. 22-21a. The circuit of Fig. 22-21b, which consists of an inductance L in parallel with a resistance R, is an exact equivalent circuit (see P22-19) provided the parameters are related by the expressions

$$L = L'\left(1 + \frac{1}{Q^2}\right) \qquad R = r(Q^2 + 1) \qquad (22\text{-}29)$$

with the quality factor Q defined by

$$Q = \frac{\omega L'}{r} = \frac{R}{\omega L} \qquad (22\text{-}30)$$

The two expressions for Q are identical.

Skin effect causes the inductance L' of a coil to vary slightly with frequency. This can be ignored in most inductors used in tuned amplifiers. Because Q is large at frequencies of interest, $L \approx L'$, and L can also be regarded as a constant. In the passband the ohmic resistance r is approximately proportional to the frequency, which is a consequence of skin and proximity effects and core losses. Therefore, Q is reasonably constant, and R and r vary with frequency by about the same percentage. It is usually satisfactory to treat these as constants, and we shall do so here. Insofar as variation of the parameters with frequency is concerned, it makes little difference whether we use the series or parallel form of the equivalent circuit. Because the parallel form is convenient, it will be employed. The resistance R can be interpreted as representing Q^2r combined with any additional resistance that may shunt the coil, such as a parallel-connected load.

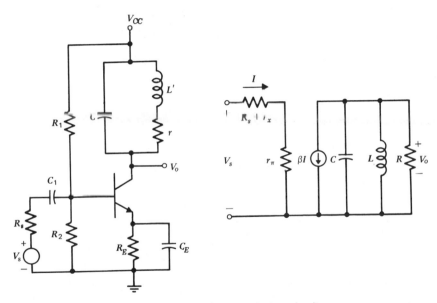

Figure 22-22   Tuned amplifier and approximate equivalent circuit.

In Fig. 22-22 are shown a tuned-amplifier schematic diagram and an approximate equivalent circuit. The hybrid-π capacitances C_π and C_μ have been omitted for simplicity. Because C_μ is assumed to be much smaller than the capacitance C of the load, the device capacitances do not appreciably affect the tuning of the tank circuit. Hence their main effect is to shunt r_π with an impedance, which at resonance is that attributable to the sum of C_π and $C_\mu(1 + g_m R)$. We assume here that the parameters are such so as to make the effect second-order.

At the resonant frequency the tank-circuit impedance is R, and the gain is found from the equivalent circuit of Fig. 22-22 to be

$$\frac{V_o}{V_s} = \frac{-\beta R}{R_s + r_x + r_\pi} \tag{22-31}$$

The load R can be chosen to give the desired gain. At other frequencies the gain is found by replacing R in (22-31) with the impedance Z of the tank circuit. The value of Z is a maximum at resonance. As the frequency is moved above or below ω_o, the gain drops. Also, phase shifts occur. The amplifier must be designed so that the bandwidth is the specified small fraction of ω_o. We shall now investigate this.

The tank-circuit admittance $Y = G + j\omega C + 1/j\omega L$. For convenience the *difference frequency* ω' is defined as

$$\omega' = \omega - \omega_o \qquad \text{with } \omega_o = \frac{1}{\sqrt{LC}} \tag{22-32}$$

Above resonance ω' is positive, and it is negative below resonance. Replacing ω in the expression for Y with $\omega_o + \omega'$, we obtain

$$Y = G + j\omega_o C + j\omega' C + \left[j\omega_o L \left(1 + \frac{\omega'}{\omega_o} \right) \right]^{-1} \tag{22-33}$$

For values of ω' small compared with ω_o, which is a resonable assumption throughout the passband, we have

$$\left(1 + \frac{\omega'}{\omega_o} \right)^{-1} \approx 1 - \frac{\omega'}{\omega_o} \tag{22-34}$$

Using this and replacing the reciprocal of $j\omega_o L$ in (22-33) with $-j\omega_o C$, we obtain

$$Y = G + j2\omega' C = \frac{1 + j2\omega' RC}{R} \tag{22-35}$$

The voltage gain in terms of ω' is found by replacing R in (22-31) with the reciprocal of Y, given by (22-35). The result is

$$\frac{V_o}{V_s} = \frac{-\beta R}{(R_s + r_x + r_\pi)(1 + j2\omega' RC)} \tag{22-36}$$

Half-power points occur when $\omega' = \pm(2RC)^{-1}$. Therefore, the respective lower and upper 3 dB radian frequencies ω_1 and ω_2 are

$$\omega_1 = \omega_o - \frac{1}{2RC} \qquad \omega_2 = \omega_o + \frac{1}{2RC} \tag{22-37}$$

The bandwidth $\Delta\omega$ is the difference $\omega_2 - \omega_1$. From (22-37) this is

$$\text{bandwidth } \Delta\omega = \frac{1}{RC} \tag{22-38}$$

At resonance the quality factor Q_o is $R/\omega_o L$, or $\omega_o RC$. From this and (22-38) we observe that

$$\Delta\omega = \frac{\omega_o}{Q_o} \tag{22-39}$$

The relation clearly shows that a narrow bandwidth is obtained by using a high-Q coil. The use of the equations in design is illustrated by the following example.

Example 1

An intermediate-frequency (IF) amplifier of the form shown in Fig. 22-22 is to be designed for an AM receiver. The amplifier must be tuned to 455 kHz and must have a bandwidth of 10 kHz. The source resistance R_s is 170 ohms, and a voltage gain of -60 is specified. A 2N4996 NPN transistor biased at 5 mA and 10 V will be used. It has a transition frequency f_T of 600 MHz, a capacitance C_μ of 0.5 pF, and a beta of 60. Assume r_x is 10% of r_π.

Solution

Units of kΩ, mmhos, mA, V, pF, μH, ns, and GHz will be used. Assuming q/kT is 40, g_m is 200 and r_π and r_x are 0.3 and 0.03 kΩ, respectively. For a voltage gain of -60, we find from (22-31) with $\beta = 60$ and $R_s = 0.17$ kΩ that the load resistance R is 0.50 kΩ.

We know that C_π is g_m/ω_T less C_μ, and ω_T is 1.2π ns^{-1}. Thus, C_π is 52.6 and C_μ is 0.5 pF. At resonance the equivalent capacitance C_T in parallel with r_π is the sum of C_π and $C_\mu(1 + g_m R)$, or 103 pF. The resonant frequency ω_o is 2.86×10^{-3} ns^{-1}, and $1/\omega_o C_T$ is 3.4 kΩ. Because r_π is only 0.3 kΩ, it follows that C_T can be neglected.

The tank-circuit capacitance C is $1/(R \Delta\omega)$ by (22-38), and the bandwidth $\Delta\omega$ is $2\pi \times 10^{-5}$ ns^{-1}. Therefore, C is 31 800 pF. As this is much larger than C_μ, it is evident that the device capacitances do not affect the tuning. The inductance L required to give the proper resonant frequency in conjunction with C is 3.85 μH, found from (22-32).

The value of Q_o equals the resonant frequency f_o divided by the bandwidth Δf, or 45.5. This equals $R/\omega_o L$ and also $\omega_o RC$. As in this example, the quality factor of the load of a tuned amplifier is usually somewhat lower than that of the tank circuit of an LC oscillator. The reactance $\omega_o L$ is 11 ohms and the effective series resistance r of the coil is 0.24 ohm.

IMPEDANCE TRANSFORMATION

It is frequently desirable to transform impedance levels in radio-frequency circuits in order to allow the use of inductors and capacitors with reasonable values. Transformers, tapped coils (autotransformers), and tapped capacitors are common. Shown in Fig. 22-23 is a resonant circuit with a tapped coil, along with an equivalent circuit based on the assumptions of unity coupling and negligible coil resistance (see P22-20). The equivalent inductance is the coil inductance L divided by $(1 + n)^2$ and the equivalent capacitance is C multiplied by $(1 + n)^2$, with n denoting the turns ratio n_2/n_1, or $\sqrt{L_2/L_1}$. Accordingly, tapped coils are convenient for reducing L and increasing C. They permit the use of coils with larger values of L than specified by the design and the use of capacitors with lower values of C.

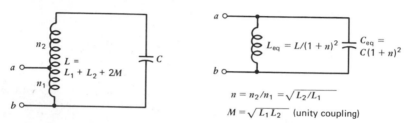

Figure 22-23 Tapped-coil resonant circuit and equivalent network, assuming unity coupling and negligible resistances.

Example 2

In the tuned-amplifier design of Example 1, a coil with 3.85 μH and a capacitor of 31 800 pF are required. Suppose a tapped coil having a total inductance of 360 μH is to be used. Determine the required turns ratio and also calculate the capacitance C.

Solution

The inductance of 360 μH divided by $(1 + n)^2$ must equal 3.85, which gives a turns ratio of 8.67. The value of C is 31 800 divided by $(1 + n)^2$, or 340 pF. For both L and C the values are more suitable than those of Example 1. At 455 kHz it is easier to obtain a high-Q coil with an inductance of 360 μH than it is with one having only 3.8 μH.

22-7. AMPLIFIERS WITH INPUT AND OUTPUT TUNING

Many amplifiers have a tuned circuit at the input and a second one at the output. An example is the input MOSFET stage of the FM receiver front end shown in Fig. 22-24. This input stage provides both station selectivity and voltage

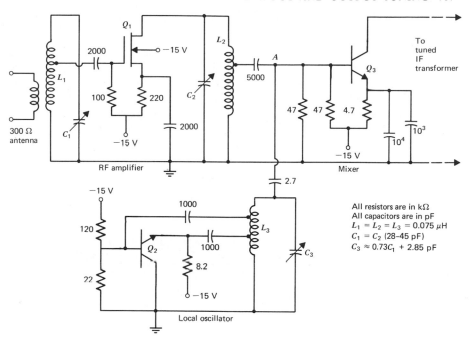

Figure 22-24   An FM receiver front end.[5]

gain. Bipolar transistor Q_2 and its associated circuitry constitute a Hartley CC oscillator. Tank-circuit capacitors C_1, C_2, and C_3 are ganged capacitors that tune the L_1C_1 and L_2C_2 networks to the incoming signal in the band between 88 and 108 MHz and tune the *local-oscillator* L_3C_3 network to a frequency greater than that of the signal by 10.7 MHz. For example, if the receiver is tuned to 100 MHz, the oscillator frequency is 110.7 MHz. Actually, each of the capacitors C_1, C_2, and C_3 consists of a fixed capacitor of perhaps 16 pF, a trimmer capacitor adjustable with a screwdriver from about 2 to 12 pF, and a ganged capacitor which is variable from 5 to 23 pF by rotation of the tuning knob.

The output voltage of the RF amplifier appears between point A and ground. Nearly all of this voltage is across the 2.7 pF capacitor, and the oscillator voltage is across the tank circuit. Therefore, the sum of these two voltages is fed into the *mixer* stage Q_3. The collector current of Q_3 is an exponential function of the input base-emitter voltage in accordance with an Ebers–Moll equation, and i_C can be expanded into a Taylor's series about the operating point. The incremental base-emitter voltage v can be expressed as

$$v = V_s \sin \omega_s t + V_o \sin \omega_o t \qquad (22\text{-}40)$$

with ω_s denoting the signal frequency and ω_o denoting the oscillator frequency. The Taylor's series for i_C has the form

$$i_C = a_o + a_1 v + a_2 v^2 + a_3 v^3 + \cdots \qquad (22\text{-}41)$$

When v of (22-40) is substituted into the series of (22-41), the term $a_2 v^2$ gives sum and difference frequencies (see P22-22). The difference frequency is 10.7 MHz. The load network of the mixer is a tuned transformer consisting of an LC tuned primary and an LC tuned secondary, with the transformer tuned to 10.7 MHz. The other frequencies of i_C, which come from the series of (22-41), are sufficiently far removed so that the load impedance is very small for these. Therefore, the only appreciable voltage fed from the mixer into the intermediate-frequency (IF) amplifier is at 10.7 MHz. The band of frequencies constituting the signal, which has a carrier frequency in the FM band, has been transformed to the intermediate frequency of 10.7 MHz.

The process of shifting the frequency of a carrier to a lower fixed frequency is referred to as *heterodyning*. The advantage is that the IF amplifier cascade, which has a fixed input frequency regardless of that of the received signal, can be designed more easily for improved selectivity and for a flatter passband gain than can a tuned amplifier required to amplify signals having different carrier frequencies.

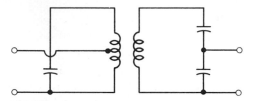

Figure 22-25   Double-tuned coupling network.

In Fig. 22-25 is shown a typical coupling circuit used between the mixer and the first IF stage and also between IF stages. It consists of double-tuned circuits with one having a tapped coil and the other having tapped capacitors. Both techniques provide impedance transformation resulting in higher-Q networks. A flatter passband with sharper cutoff at the band edges and with greater attenuation of unwanted frequencies is obtained by tuning the networks to slightly different frequencies. This is called *stagger tuning*.

INSTABILITY AND ALIGNABILITY

A double-tuned amplifier with input and output tuned circuits, such as the RF amplifier of Fig. 22-24, may have stability and tuning-interaction problems resulting from internal feedback paths. In a CE amplifier feedback occurs through lead and stray capacitance between the collector and base leads and through the device capacitance C_μ, which is typically from 0.5 to 2 pF in high-frequency transistors. There is always some frequency at which the loop gain T is a negative real number. If the magnitude of T at this frequency is greater than unity, the circuit oscillates.

From another viewpoint, suppose the resistive part of the impedance at the input of the transistor is negative at the resonant frequency of the input LC network. A net negative resistance, including the effective shunt resistance of the tank circuit, produces instability (see P22-23). Such a negative resistance results from the inductive load presented by the output LC circuit at frequencies slightly below resonance. An alternate possibility is for oscillations to occur as a consequence of a negative output resistance shunting the output tuned circuit. The "dissipated" power I^2R is negative for R negative, indicating a power source.

Often, an even more serious problem than that of stability is tuning interaction, which makes it difficult for proper tuning to signals of a frequency band. For example, when the tuning of the output LC circuit is changed by varying either L or C, the feedback causes the input impedance to change. As this shunts the input LC circuit, it is detuned. The interaction makes alignment difficult and leads to asymmetric response curves which introduce distortion.

Neutralization is a technique often employed to minimize the difficulties in IF amplifiers, which are designed for a specified carrier frequency. This consists of feeding a signal back through an external network, usually consisting of a single neutralizing capacitor C_N, with the network connected so that the signal fed back cancels that fed back through the active device. The voltage applied to the feedback network must be out of phase with the collector voltage.

Two typical neutralizing networks are shown in Fig. 22-26. In (a) the 180° phase inversion is obtained by means of the transformer. In part (b) the voltage across C_p is 180° out of phase with the collector voltage provided the reactance of C_p is less than that of the inductance L. The resistor R_N is optional. It complicates the network but allows better neutralization. The elements of the feedback circuit must be carefully adjusted at the operating frequency, and they must be readjusted whenever the frequency is changed. The current I_N of the neutralizing network should be equal in magnitude and opposite in phase to that introduced into the base circuit by stray and device capacitances. Typical values of C_N are from 1 to 10 pF.

The need for neutralizing circuits can be avoided by using an active device with a very small capacitance between the input and output terminals and by designing the circuit so that the gain is not too large. Many high-frequency junction transistors have feedback capacitances of about 0.5 pF and field-effect devices with even smaller capacitances are available. For example, the RF amplifier of the FM circuit of Fig. 22-24 was designed for use with RCA MOS

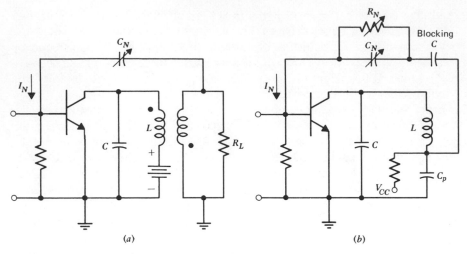

Figure 22-26   Two typical neutralizing circuits.

transistor 40468 having a typical feedback capacitance of only 0.12 pF. This low value allows a voltage gain of 13 dB without neutralization.

In addition to low capacitance there are other reasons for using a MOS transistor in the RF input stage. The high input impedance presents negligible load to the input tuned circuit. An even more important advantage is the larger dynamic range of an FET in comparison with a BJT. Input signals approximately five times greater can be handled without appreciable distortion. This is a consequence of the square-law characteristic relating i_D to v_{GS}, which is not nearly as nonlinear as the exponential characteristic relating i_C to v_{BE}.

The increased linearity minimizes spurious responses. These occur when a harmonic of an unwanted incoming signal mixes with a harmonic of the local oscillator, giving a difference frequency within the IF passband. For example, suppose the receiver is tuned to 100 MHz with an unwanted signal present at 105.35 MHz. The second harmonic (210.7 MHz) of the unwanted signal beating with the second harmonic (221.4 MHz) of the local oscillator tuned to 110.7 MHz gives a difference frequency of 10.7 MHz, which is amplified by the IF section of the tuner. Clearly, the output of the RF amplifier should have very low levels of harmonics of undesired signals.

THE DUAL-GATE MOSFET

Field-effect transistors with two independent insulated gates have wide application, especially in the high-frequency communications area. Shown in Fig. 22-27 are a cross-sectional view, the symbol, and a circuit representation of a dual-gate MOSFET. In normal operation the input signal is applied to gate 1

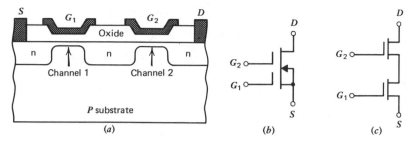

Figure 22-27   Dual-gate MOSFET (*a*) Cross-section. (*b*) Symbol. (*c*) Circuit representation.

with the source grounded. Gate 2 is at ac ground and the output is taken from the drain. With these connections and proper biasing, the circuit representation of Fig. 22-27*c* clearly shows that the configuration is an FET cascode amplifier, having a CS-CG arrangement and similar to the BJT cascode of Fig. 19-19 of Sec. 19-3. We learned there that the effective capacitance between the output and input terminals is unusually small. The same is true of the dual-gate MOSFET and for the same reasons. The low capacitance between the drain and gate 1 is the outstanding feature, with typical values from 0.005 to 0.02 pF.

The low-frequency incremental properties of a CS dual-gate amplifier are similar to those of a single-gate circuit. Input impedance is high and the dynamic

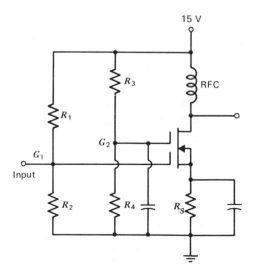

Figure 22-28   Typical dual-gate MOSFET bias circuit.

range is wide. The second gate is often used for automatic gain control. This is possible because the transconductance g_{fs} from gate 1 to drain is a function of the bias voltage of gate 2. Some of the many applications of dual-gate MOSFETs are automatic-gain-control RF amplifiers, frequency converters, product detectors, and color demodulators. When used in tuned RF amplifiers, neutralization is unnecessary. Figure 22-28 is the diagram of a typical bias circuit for a dual-gate MOSFET used in RF amplifier and mixer circuits that do not use automatic gain control (see P22-24).

IC TUNED AMPLIFIERS

An emitter-coupled amplifier of the form of that shown in Fig. 19-1 of Sec. 19-1 can be made into a tuned amplifier by employing an LC circuit at the input and replacing the load resistance with a second LC circuit. Many integrated circuits are available for tuned-amplifier applications, and these normally use emitter-coupled networks and other differential configurations. The LC resonant circuits are connected externally. Frequently, the ICs are designed to perform a variety of functions. For example, the CA3088E 16-lead integrated circuit is designed for use in high-quality AM superheterodyne radio receivers as well as for many other applications up to 30 MHz. It provides RF amplification, mixing, IF amplification, detection, and audio preamplification.

REFERENCES

1. F. K. Manasse, J. A. Ekiss, and C. R. Gray, *Modern Transistor Electronics, Analysis, and Design*, Prentice-Hall, Englewood Cliffs, New Jersey, 1967.
2. R. F. Shea (Editor), *Transistor Circuit Engineering*, Wiley, New York, 1957.
3. D. L. Schilling and C. Belove, *Electronic Circuits: Discrete and Integrated*, McGraw-Hill, New York, 1968.
4. P. E. Gray and C. L. Searle, *Electronic Principles* (Chapter 17), Wiley, New York, 1969.
5. *Linear Integrated Circuits and MOS Devices* (Application Notes), pages 353–418, 74 Databook series SSD-202B, RCA Solid State Division, Somerville, New Jersey, 1973.

PROBLEMS

Section 22-1

22-1. From the incremental model of the phase-shift oscillator of Fig. 22-3 of the example of Sec. 22-1, deduce (22-5) and (22-6) and from these deduce (22-7) and (22-8). Then calculate the minimum β and the oscillation frequency in kHz for $R_L = 2000 \, \Omega$, $R = 1000 \, \Omega$, and $C = 0.01 \, \mu F$.

22-2. For the FET phase-shift oscillator of Fig. 22-29, in terms of r_{ds}, R_L, R, and C determine the oscillation frequency and the minimum value of g_{fs} required. Calculate the

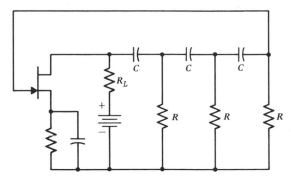

Figure 22-29 For P22-2.

minimum g_{fs} in mmhos and the frequency in Hz for r_{ds}, R_L, and R equal to 60, 15, and 100 kΩ, respectively, with $C = 0.001$ μF.

22-3. For the Wien bridge oscillator of Fig. 22-4, add a voltage source V_s in series with the RC parallel combination. From a loop equation around the outside loop and a second equation relating the voltages across R_2 and Z_2, find the gain function I_2/V_s in terms of s. In rectangular form calculate the complex poles and zeros of the gain in reciprocal ms. Also, determine the oscillation frequency. Assume the diodes are open-circuits. (Ans: 3.18 kHz)

Section 22-2

22-4. From the node equations of (22-11), obtained from the Colpitts oscillator of Figs. 22-5a and 22-6, deduce that ω^2 is $1/LC$, with $C = C_1 C_2/(C_1 + C_2)$, and show that the minimum beta is C_2/C_1.

22-5. From the Hartley oscillator model of Fig. 22-7 write two loop equations in terms of I_1 and I_2. After collecting terms, equate the determinant of the coefficients to zero. From the imaginary part of the result deduce that ω^2 is $1/LC$ with L equal to $L_1 + L_2 + 2M$. Also, from the real part with ω eliminated, deduce that the minimum beta is $(L_1 + M)/(L_2 + M)$. Make no approximations.

22-6. For the Clapp oscillator of Fig. 22-5c, model the transistor using the parameters g_π and g_m, and draw the low-frequency incremental model in a form similar to that of Fig. 22-6. Use the complex admittance for the LC_3 branch. From two node equations deduce that ω^2 is $1/LC$ with $1/C$ equal to the sum of $1/C_1$, $1/C_2$, and $1/C_3$. Then show that the minimum beta is C_2/C_1.

22-7. Deduce (22-13) directly from (22-12) by replacing each y-parameter of (22-12) with equivalent h parameters obtained from Appendix B.

22-8. A Colpitts oscillator of the form of Fig. 22-5a is to have a frequency of 45 MHz. Using 0.2 μH for L, deduce the required values of C_1 and C_2. The CE y parameters of the transistor are those given in the example of Sec. 22-2.

22-9. Sketch the Hartley oscillator of Fig. 22-5b in the form of the CB feedback network of Fig. 22-9b. At the frequency determined from (22-10) calculate in millimhos the y parameters of the feedback network for L_1, L_2, and M equal to 1, 2, and 0.5 μH, respectively, and $C = 100$ pF.

Section 22-3

22-10. For the tank circuit of Fig. 22-10, deduce that the impedances Z_s at ω_s and Z_p at ω_p, between the terminals, are

$$Z_s = \frac{R}{1 + j\omega_s C_2 R} \qquad Z_p = \frac{1}{\omega_p^2 C_2^2 R} - \frac{j}{\omega_p C_2}$$

with the series-resonant and parallel-resonant frequencies ω_s and ω_p given by (22-15). Calculate Z_s in ohms and Z_p in MΩ for $L = 3.3H$, $R = 200\ \Omega$, $C_1 = 0.042$ pF, and $C_2 = 5.8$ pF.

22-11. In the Pierce oscillator of Fig. 22-12, the JFET capacitances C_{gs} and C_{gd} are effectively in parallel with C_3 and the crystal, respectively. Assume the effects of these capacitances are negligible. Both C_5 and C_6 are incremental short-circuits, and R_G and the RFC are incremental open-circuits. Neglecting r_{ds} and capacitance C_2 of the crystal, find ω^2 and the minimum required g_{fs} in terms of L, R, C_1, C_3, and C_4. From the results show that ω^2 and g_{fs} are $1/LC_1$ and $RC_3 C_4 \omega^2$, respectively, provided C_1 is much smaller than both C_3 and C_4. Calculate the frequency in kHz and the required g_{fs} in mmhos for $L = 100$ H and $R = 2$ kΩ, assuming respective values of 0.026, 1000, and 500 pF for C_1, C_3, and C_4. The crystal circuit is that of Fig. 22-10.

Section 22-4

22-12. In the astable multivibrator of Fig. 22-16 with $R = 10$ kΩ and $C = 0.001\ \mu$F, suppose a MOSFET is connected in parallel with R with a variable gate voltage giving a channel resistance between 1 kΩ and infinity. Determine the minimum and maximum frequencies in kHz. Repeat if C is $0.004\ \mu$F. Note that from time zero to T, with T denoting half a period, v_1 decreases exponentially from 15 toward zero with the time constant RC.

22-13. Repeat P22-12, except assume that protective diodes prevent v_1 from exceeding 10.7 V and from dropping below -0.7 V. (*Ans*: 65.7, 723; 16.4, 181)

22-14. Suppose the astable multivibrator of Fig. 22-16 has diodes at the input that prevent v_1 from rising above 10.7 and falling below -0.7 V. Resistance R is 5 kΩ and C is $0.002\ \mu$F. Determine the frequency in kHz and the capacitor voltage v_C at time T^+ and also at time $2T^+$, with T denoting half the period. Sketch and dimension the waveforms for v_0, v_2, v_C, and v_1. Assume *ideal* inverters and 0.7 V diodes.

22-15. The frequency-doubler circuit of Fig. 22-18 has resistors of 1 kΩ each, equal capacitors, and an input pulse train with a period of 0.1 ms. Conducting diode drops are 0.7 V, and the gates switch states at 5 V. What capacitances in μF should be used to give output pulses of width 0.04 ms, and what is the output frequency in kHz? (*Ans*: 0.0577, 20)

22-16. Deduce that the transistors of the free-running multivibrator of Fig. 22-17 are saturated if the capacitors are removed. Assume the betas are 50. Also, with capacitors of $0.01\ \mu$F present, calculate the oscillation frequency in kHz. Use the saturation voltages shown on the waveforms of Fig. 22-17b. (*Ans*: 1.38)

Section 22-5

22-17. The approximate effects of the dominant circuit parameters of the blocking oscillator of Fig. 22-19 can be easily determined by neglecting the small voltage V_{BB}, treating the transistor as a short-circuit when in saturation, and assuming β is very large. With these assumptions find the pulse width t_1 and the period T in terms of V_{DD}, V_D, R, L, and the turns ratio n of the pulse transformer. Calculate t_1 and T in μs for $V_{CC} = 10$ V, $V_D = 4$ V, $R = 500\ \Omega$, $L = 500\ \mu H$, and $n = 4$. (*Ans*: 0.25, 0.75)

22-18. Suppose the blocking oscillator of Fig. 22-19 has respective saturation values of v_{BE} and v_{CE} equal to 0.8 and 0.2 V, and V_{CC} is 12.6 V, V_D is 0.7 V, R is 1 kΩ, and L is 1 mH, which are the values used in the analysis of Sec. 22-5. Following a similar procedure, calculate the pulse width t_1 and the period T in μs for a pulse transformer with a turns ratio $n = 4$. Assume $\beta \gg 1$.

Section 22-6

22-19. (a) With Q defined as $\omega L'/r$, deduce that Q is $R/\omega L$ provided R and L are related to r and L' by (22-29). (b) Show that the networks of Fig. 22-21 have precisely the same complex impedances provided (22-29) and (22-30) apply.

22-20. The model of Fig. 22-30 represents the tapped-coil resonant circuit of Fig. 22-23. Assuming unity coupling, deduce from loop equations that the input admittance is

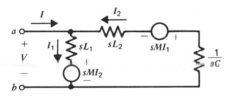

Figure 22-30   For P22-20.

the sum of $sC(1 + n)^2$ and $1/sL_1$, with n defined in Fig. 22-23. Then show that L_1 is $L/(1 + n)^2$. The results justify the equivalent network of Fig. 22-23.

22-21. Design a tuned amplifier similar to that of Fig. 22-22 for a frequency of 1 MHz, a bandwidth of 25 kHz, and a voltage gain of -50. The transistor is biased at 5 mA and 10 V, and $\beta = 40$, $f_T = 800$ MHz, $C_\mu = 0.5$ pF, and $r_x = 50\ \Omega$. The source resistance is 150 ohms and q/kT is 40. (a) Determine Q_o. (b) Find R in ohms. (c) Calculate the reactance in kΩ of the equivalent capacitance C_T in parallel with r_π. (d) Determine C in pF and L in μH. (e) If the LC circuit is replaced with a tapped-coil resonant network of the form shown in Fig. 22-23, determine C in pF and L in μH for a turns ratio $n = 9$. [*Ans*: (e) 127, 199]

Section 22-7

22-22. Suppose the voltage v at the input of an FM mixer stage consists of the sum of the amplified signal at 94.5 MHz and the output of a local oscillator at 105.2 MHz. If the collector current can be expressed as the sum $a_o + a_1 v + a_2 v^2$, determine in MHz all frequencies present in the current.

22-23. A CE amplifier with base and collector tuned circuits has $g_m = 50$ mmhos, $r_\pi = 1$ kΩ, and $C_\mu = 1$ pF. The input LC network is tuned to 100 MHz, C is 40 pF (including effect of C_π), and Q_o is 50. At 100 MHz assume the load LC circuit has an effective impedance of $j10$ ohms. Calculate the total resistance in ohms between base and ground (including r_π), the effective shunt resistance R of the tank circuit, and the resistance reflected into the input by the inductive load. Is the circuit stable?

22-24. Characteristic curves for the dual-gate MOSFET of Fig. 22-28 show that I_D is 10 mA when V_{DS} is 12.5, V_{G1S} is -0.5, and V_{G2S} is 4 V. With $R_3 = 210$ kΩ, determine R_1, R_2, R_4, and R_S in kΩ which give this Q-point and give an incremental input resistance of 25 kΩ at G_1.

Chapter Twenty Three
Applications of Integrated Circuits

A few of the many linear and nonlinear applications of analog integrated circuits are presented in this chapter. The networks have been selected to illustrate some of the basic functions that can be accomplished electronically and to show configurations suitable for the purpose. Usually the circuits are monolithic, but also used are hybrid and multichip ICs as well as networks made of discrete components. Included are analog-computer networks, limiters and comparators, voltage-current converters, analog-to-digital converters, active RC filters, and waveform generation. Most applications involve one or more operational amplifiers.

23-1. ANALOG COMPUTER NETWORKS

Operational amplifiers are extensively used in analog computers, which are networks designed to solve differential equations and to perform certain types of calculations. The calculations often include addition, multiplication, division, differentiation, and integration, as well as operations involving logarithms and exponentiation. We examine here some of the simpler networks, realizing that many practical circuits are considerably more complex, with the increased complexity providing improved performance. In addition to their use in analog computers, the circuits presented here have numerous other applications.

INTEGRATION

An integrating circuit is shown in Fig. 23-1. Treating the input of the operational amplifier as a virtual short-circuit, we find that the current through the source is V_s/R_1, and this same current is $-V_o/Z_f$ with Z_f denoting the impedance of the parallel combination of R_f and C. It follows that

$$V_o = \frac{-V_s R_f}{R_1(1 + sR_f C)} \tag{23-1}$$

By selecting the time constant $R_f C$ considerably greater than the periods of the frequencies of interest, the right side of (23-1) becomes approximately $-V_s/(sR_1 C)$, and the corresponding time-domain relation is

$$v_O(t) = -\frac{1}{R_1 C} \int v_S(t)dt \tag{23-2}$$

The purpose of the resistor R_f is to give dc feedback. Without it the dc input offset voltage and any dc component present in the source are integrated, causing the feedback capacitor to charge continuously until the dc output rises

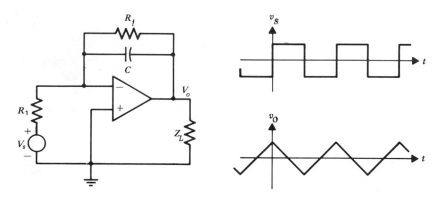

Figure 23-1   Integrator, along with an example of input and output waveforms.

sufficiently to limit the gain (see P23-1). The value of R_f must be small enough to provide adequate dc feedback, but it must be much larger than $1/\omega C$ at the lowest significant frequency of the input waveform. For example, suppose C is 0.03 μF, R_1 is 39 kΩ, and R_f is 390 kΩ. For these values the magnitude of the pole of (23-1) is located at about 14 Hz, and at frequencies considerably higher than this the circuit is an excellent integrator. Furthermore, the dc gain is only 10, or 20 dB. For a square-wave input having a fundamental frequency of 5 kHz, the capacitive reactance at this frequency is about 1 kΩ, which is much less than R_f, and the waveforms of Fig. 23-1 apply. Compensation is essential because of the large amount of feedback at high frequencies. We can omit R_f in integrators of networks having dc feedback applied over several stages in a manner such that input offset voltages are not integrated. Examples are considered in Sec. 23-6.

Associated with the indefinite integral of (23-2) is a constant of integration that depends on the initial condition. With the aid of a battery, a potentiometer, and a switch, a charge can be placed on the capacitor while a second switch simultaneously replaces v_S with a short-circuit. The selected value of the initial voltage across the capacitor is obtained from the desired initial condition. Both summing and integrating can be performed by a single *adder-integrator*, which consists of the summing amplifier of Fig. 14-19 of Sec. 14-4 with R_f replaced by the R_fC feedback network of Fig. 23-1.

DIFFERENTIATION

A differentiating amplifier is shown in Fig. 23-2. Because of the virtual short-circuit, the source current is V_s/Z_1 with Z_1 denoting the series impedance of R_1 and C, and this same current also equals $-V_o/R_f$. Therefore,

$$V_o = \frac{-sR_fCV_s}{1 + sR_1C} \tag{23-3}$$

We select the time constant R_1C much smaller than the periods of the frequencies of interest so that the right side of (23-3) becomes approximately $-sR_fCV_s$. The corresponding time-domain relation is

$$v_o(t) = -R_fC\frac{dv_S}{dt} \tag{23-4}$$

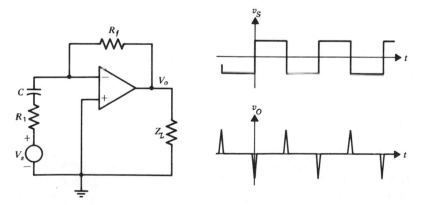

Figure 23-2 Differentiator, along with an example of input and output waveforms.

For example, suppose C is 0.03 μF, R_1 is 510 ohms, and R_f is 39 kΩ. The magnitude of the pole of (23-3) is located at 10.4 kHz. Thus the circuit is a reasonable differentiator of input signals having all significant frequencies well below this value. For the input square wave of Fig. 23-2, at a fundamental frequency of perhaps 1 kHz, the output consists of a series of spikes as shown. The maximum gain of 76.5, or 38 dB, occurs at high frequencies, and consequently, high-frequency noise is a common problem. Because of this, parameter values should be selected so that the pole is no larger than necessary. Omission of R_1 would give excessive gain at high frequencies, along with stability problems, increased noise, and an undesirable input impedance consisting of a pure capacitance.

Although the circuit of Fig. 23-2 is used as a differentiator at frequencies well below the pole of (23-3), it has important applications in the midband region above the pole. Here it is a very stable ac amplifier with a gain $-R_f/R_1$ and with dc blocking at the input. The lower 3 dB frequency is determined by R_1 and C.

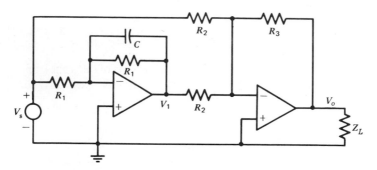

Figure 23-3   Differentiator with constant input impedance.

The differentiator of Fig. 23-2 has an input impedance that varies with frequency. By employing two OP AMPs as shown in Fig. 23-3, a differentiator with a constant input impedance is obtained (see P23-2).

LOGARITHMIC AMPLIFIER

The network of Fig. 23-4 is a logarithmic amplifier, which has an output that is proportional to the logarithm of the input voltage. Because of the virtual short-circuit, the voltage v_{CB} of the transistor is zero, and the transistor is, in effect, a diode. In fact, a diode can be used in place of Q_1.

With v_S positive, v_O is negative and Q_1 is biased in the active mode with

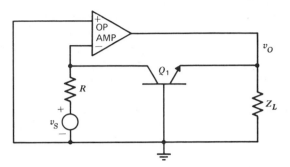

Figure 23-4   Logarithmic amplifier ($v_S > 0$). The feedback element of the inverting amplifier is Q_1.

$v_{BE} = -v_O$. We know from the Ebers–Moll equations that i_C is $\alpha_F I_{ES} \exp(-qv_O/kT)$, and this must equal v_S/R. It follows that

$$v_O = -\frac{kT}{q} \ln \frac{v_S}{\alpha_F I_{ES} R} \tag{23-5}$$

When the signal is small the incremental feedback resistance of the effective diode is fairly large, with reduced feedback and a smaller bandwidth. A large signal gives a low feedback resistance, which may present stability problems Furthermore, (23-5) is very temperature sensitive, mainly because of I_{ES}. These problems are largely eliminated with more complicated circuits employing matched transistors and temperature compensation, but the basic principle is the same. The log amplifier is a nonlinear application of the operational amplifier, as is the antilog amplifier which is examined next.

ANTILOG AMPLIFIER

Shown in Fig. 23-5 is an *antilog amplifier*, also called an *exponentation amplifier*. It has an output that is proportional to the exponential of the input voltage. The

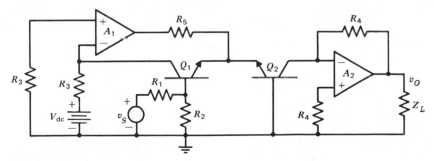

Figure 23-5   Antilog amplifier ($v_S > 0$).

matched transistors Q_1 and Q_2 are biased in the active mode at the same temperature T and with the same values of α_F and I_{ES}. Each collector current has the form $\alpha_F I_{ES} \exp q v_{BE}/kT$. The inputs of the two OP AMPs are assumed to be virtual short-circuits.

The two node equations at the nodes of the inverting inputs are

$$\frac{V_{dc}}{R_3} = \alpha_F I_{ES} \exp \frac{q v_{BE1}}{kT} \tag{23-6}$$

$$\frac{v_O}{R_4} = \alpha_F I_{ES} \exp \frac{q v_{BE2}}{kT} \tag{23-7}$$

Dividing the latter equation by the first gives

$$\frac{v_O R_3}{V_{dc} R_4} = \exp \frac{q(v_{BE2} - v_{BE1})}{kT} \tag{23-8}$$

With the base current of Q_1 assumed negligible, $v_{BE2} - v_{BE1}$ equals $-R_2 v_S/(R_1 + R_2)$. Therefore, (23-8) can be expressed in the form

$$v_O = \frac{R_4 V_{dc}}{R_3} \exp \left[\frac{-q R_2 v_S}{kT(R_1 + R_2)} \right] \tag{23-9}$$

This is the desired result. A cascade of log and antilog amplifiers can be used to raise an input voltage to a power (see P23-3).

MULTIPLICATION AND DIVISION

The diagram of Fig. 23-6 shows how the log and antilog operators can be used for multiplication and division. Multiplication occurs when the inputs to the summing amplifier are added, and division occurs when they are subtracted.

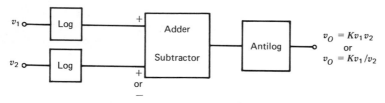

Figure 23-6   Use of operators to achieve multiplication and division.

ANALOG COMPUTATION

The block diagram of Fig. 23-7 is a representation of an analog computer that has been programmed to solve the following differential equation:

$$\frac{d^2 v}{dt^2} + 4 \frac{dv}{dt} + v = 4 \cos t \tag{23-10}$$

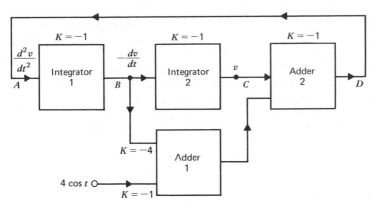

Figure 23-7   Analog-computer block diagram.

The computer consists of two integrators and two adders. Adder 1 has a scale factor of -4 for the input from B, and all other scale factors are -1. The voltage v at C represents the solution. It is supplied to an oscilloscope or a recorder.

Because v represents the output of integrator 2, the voltage v_B at B is $-dv/dt$ and v_A at A is d^2v/dt^2. The inputs to adder 1 are $-dv/dt$ and 4 cos t. Therefore, the output voltage v_D of adder 2 is $-v - 4\,dv/dt + 4$ cos t. Equating the expressions for v_D and v_A gives the differential equation of (23-10).

The capacitors of the integrators must be charged so as to give the proper initial conditions. For example, suppose v is 2 and dv/dt is -3 at time $t = 0$. The output v of integrator 2 equals the voltage across the integrating capacitor, and this voltage must be set at 2 V at time zero. Also, the output v_B of integrator 1, which is $-dv/dt$, equals the voltage across its capacitor, and thus it must initially be 3 V. These voltages are established by means of dc supplies and switching circuits that provide the correct values at precisely the right time, and the oscilloscope or recorder is triggered to commence operation at this instant. Integrators are preferred to differentiators because they are easier to stabilize, are less sensitive to noise, and are convenient for introducing initial conditions.

The types of networks described in this section are available in integrated-circuit form. There are many other functions also available. For example, Analog Devices Model 433 Programmable Multifunction Module gives an output voltage v_O equal to $Kv_1(v_2/v_3)^m$, with inputs v_1, v_2, and v_3. The inputs can be variables or constants, and m can be adjusted to any value between 0.2 and 5. Some ICs contain several types of operators, with external terminals that can be connected so as to provide different configurations with different transfer characteristics, thereby providing considerable flexibility. Digital circuits are often used to perform most of the functions of the analog circuits that have been considered here.

23-2. LIMITERS AND COMPARATORS

A limiter is a network with a transfer characteristic of the general form of that of Fig. 23-8. The circuit shown is an example of a limiter.

Assuming the diodes are ideal with zero resistance and voltage, the output voltage cannot exceed the battery voltage V_B, and this is the limiting value V_{L1}. Also, v_O cannot be less than $-V_A$, which is $-V_{L2}$. For values of v_I that give an output between the two limits, the network provides a linear gain K. For the diode circuit shown, K is unity. Limiters are often referred to as *clipping circuits*, *clamps*, *amplitude selectors*, and *slicers*. With an input having sufficiently large positive and negative amplitudes, the output consists of a slice of the input.

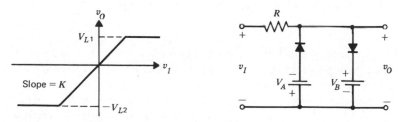

Figure 23-8 Limiter characteristic and network.

Shown in Fig. 23-9 is a *feedback limiter* using an operational amplifier. The inverting-input voltage is zero because of the virtual short-circuit. When v_O is between V_{z1} and $-V_{z2}$ there is insufficient voltage to produce breakdown, the diode circuit is open, and the amplifier has a linear gain $K = -R_f/R_1$. However, when v_O attempts to exceed V_{z1}, the diodes conduct with D_1 in breakdown, and v_O is maintained at V_{z1} assuming the drop across D_2 is negligible. A similar

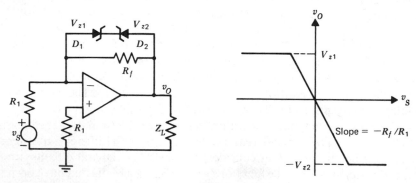

Figure 23-9 Limiter amplifier and characteristics.

process occurs when v_O becomes approximately equal to $-V_{z2}$. The Zener diodes limit the output excursions. Available are double-anode Zener diodes, which have both junctions in a single package. More complicated configurations provide adjustable limits, sharper and faster transition between states, temperature compensation, and reduced effects of diode leakage currents. Multiple inputs are sometimes used, giving a *summer-limiter*.

An obvious application of the limiter of Fig. 23-9 is one that eliminates the danger of overloading the amplifier. The amplifier is protected by using diodes with breakdown voltages at the allowed output-voltage extremes, and the diodes have no effect on normal operation. There are many other specialized uses.

By eliminating R_f in the feedback limiter (see P23-6) we obtain a *comparator*, which is a network that compares two inputs for equality. The characteristic is that of Fig. 23-9 with a slope $-R_f/R_1$ that is infinite. Shown in Fig. 23-10 is a

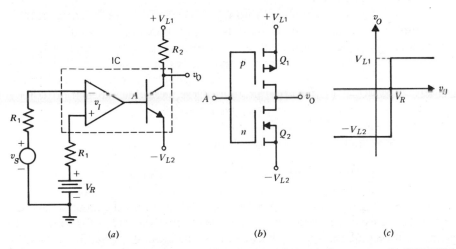

(a) (b) (c)

Figure 23-10 IC comparator diagram and ideal characteristic. (a) Comparator. (b) COS/MOS output. (c) Characteristic.

diagram of an IC comparator and its ideal characteristic. Voltage V_R is the reference voltage at which transition between states occurs, and the differential input voltage v_I is $V_R - v_S$. When $v_S > V_R$, v_I and V_A are negative, and the base-to-ground voltage V_A is larger in magnitude than V_{L2}, which places the transistor in the cutoff mode. With no current through R_2, the output voltage v_O is V_{L1}. For $v_S < V_R$, the voltage V_A is sufficient to saturate the transistor, and $v_O \approx -V_{L2}$. The actual circuitry is such that the transistor is protected from excessive voltage. With $V_R = 0$ the circuit is called a *zero-crossing detector*.

The BJT and its resistor can be replaced with the COS/MOS arrangement of Fig. 23-10b. When V_A is negative, Q_1 is *on* and Q_2 is *off*. There is no current, however, and v_O is V_{L1}. Note that the OP AMP operates open-loop and hence is not ideal. The input offset voltage simply adds to V_R, but this offset is normally only about a millivolt. Thus switching occurs within a millivolt of V_R. Actually, *an operational amplifier with no feedback is a comparator*. A differential input voltage v_I of a few millivolts is sufficient to saturate the output. In most cases the high-level output is at or near $+V_{CC}$, and the low-level output is $-V_{CC}$, or near ground in many amplifiers designed for use as comparators. Limiting networks are usually used to fix output levels at desired values and to prevent saturation of transistors within the OP AMP, which reduces the switching speed.

Unlike the limiter, the comparator reproduces no part of the input signal. There are only two output states, and the signal determines which one exists. The circuit can be used to mark the instant when the input waveform reaches the reference level. The output can activate a lamp or a relay, drive a logic circuit, or perform some other function.

Accurate time measurements are among the many applications of comparators. For example, suppose the voltage used as the time base of a CRO is fed into the comparator, giving a ramp input $v_S = at$ beginning at time zero. When at becomes equal to V_R, the output suddenly changes states. Differentiation of the output produces a sharp pulse at time V_R/a. This pulse can be used as a timing marker. It can control the time interval of the sweep, which is adjusted by changing the dc voltage V_R. Thus we have a *voltage-to-time converter*. Radar systems use comparators for this purpose.

Comparators can produce a series of pulses from a sine wave, accomplished by making v_S of Fig. 23-10 equal to $V_m \sin \omega t$ with $V_m > V_R$. With $V_R = 0$ a symmetrical square wave appears at the output. A simple RC network with a sufficiently small time constant can be employed to differentiate the output signal, producing a train of sharp pulses synchronized with the input sine wave. These may be used as timing markers for the sweep of a CRO or for some other purpose. If desired, the spacing between the pulses can be varied by an audio-frequency signal placed in series with V_R. This gives a relative spacing dependent on the audio signal, and the process is called *pulse-time modulation* with applications in communications systems. A comparator differs from a Schmitt trigger (Sec. 9-6) in that it has no hysteresis.

IC differential comparators are often supplied with two terminals for connection to *strobe* voltages. An input at either strobe less than the specified low-level strobe voltage causes the output to be in the high state regardless of the differential input. With no connections or with both terminals connected to voltages greater than the specified high-level strobe voltage, the output is controlled by the differential input. This feature is especially desirable for comparators used in *analog-to-digital converters* (ADC), which is the subject of Sec. 23-4. Other applications are investigated in Sec. 23-6.

The comparator that has been considered here compares two input voltages, one of which is referred to as the reference voltage. There are two possible outputs, and the one selected by the network depends on whether one input is less than or greater than the other. Another type of comparator compares binary numbers and has three outputs—less than, equal to, or greater than. Still another compares phase or frequency and gives a variable output voltage with a value depending on the relationship between the inputs.

23-3. VOLTAGE-CURRENT CONVERTERS

It is sometimes necessary to generate a current proportional to a voltage. An example of such a requirement is the current-driving of a coil to produce magnetic deflection in a cathode-ray tube. Such a circuit should have a high input impedance so as not to load the voltage source, and a high output impedance for satisfactory current driving. Also desired is a large transconductance g_m, defined as the ratio $\Delta i_L/\Delta v_S$.

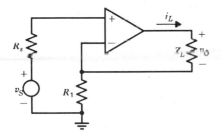

Figure 23-11 Voltage-to-current converter.

Shown in Fig. 23-11 is a *voltage-current converter* using an operational amplifier. The load impedance Z_L serves as the feedback impedance. A very high input impedance, given by (14-22) of Sec. 14-3, is provided by the noninverting configuration, and the output resistance looking back from the terminals of Z_L is $A_v R_1$ (see P23-8), which is also very high. Because of the virtual short-circuit, the voltage and current of R_1 are v_S and i_L. Therefore,

$$i_L = \frac{v_S}{R_1} \qquad g_m = \frac{1}{R_1} \qquad (23\text{-}11)$$

An objection to the circuit is the floating load, having neither terminal at ground.

In Fig. 23-12 is a diagram[1] of the RCA CA3080 *operational transconductance amplifier*, which makes an excellent voltage-current converter. The IC is made

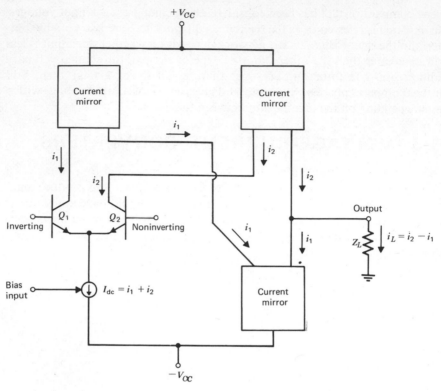

Figure 23-12   Diagram of IC transconductance amplifier.

entirely of transistors and diodes. Together, Q_1 and Q_2 form the differential input stage, with collector currents i_1 and i_2. The arrangement of the three current mirrors provides a load current i_L equal to $i_2 - i_1$. With both inputs grounded, i_1 and i_2 are equal, and i_L is zero. Suppose a positive voltage is applied to the noninverting terminal with the inverting input grounded. This causes i_2 to increase. Because the sum of i_1 and i_2 is constant, equal to I_{dc}, the current i_1 decreases, and i_L is positive and proportional to the differential input voltage. The open-loop transconductance g_m, which is typically 10 mmhos, is proportional to I_{dc}, and I_{dc} is proportional to the current supplied at the *bias input* terminal. The bias input can be used for linear gain control and for gating purposes.

A typical input impedance is 26 kΩ, shunted by 3.6 pF. The output is taken from a class A push-pull circuit with an output resistance R_o of about 15 MΩ, which is shunted by 5.6 pF. The voltage gain depends on the load impedance and can have values from 0 to 150 000 (104 dB). This maximum gain is the product $g_m R_o$, obtained when the load is that of a MOSFET gate. The gain can be increased by connecting a single or multistage COS/MOS amplifier at the output. Applications include many of those of conventional operational amplifiers and also a variety of nonlinear functions such as mixers, multipliers, and modulators (see P23-9).

Photocells and photomultiplier tubes supply a current from a high impedance source. To transform this current source into a low-impedance voltage source we can use an amplifier with low input and output impedances. The operational amplifier in the inverting configuration is suitable for this purpose. Such a *current-voltage converter* is shown in Fig. 23-13. Because of the virtual short circuit, the current through R_s is zero, and v_O is $-i_S R_f$.

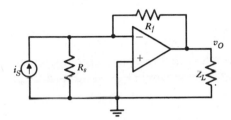

Figure 23-13   Current-to-voltage converter.

Shunting R_f with a capacitor lowers the gain at high frequencies, thereby reducing high-frequency noise. This technique is often used in feedback amplifiers when the signals are restricted to low frequencies. The dc input bias current places a lower limit on the current i_S that can be amplified. By connecting emitter followers in series with each input, the lower limit can be reduced approximately by the beta of the transistors, from perhaps 5 μA to 0.1 μA.

23-4. ANALOG-TO-DIGITAL (A/D) CONVERSION

Various techniques are employed for converting an analog signal into a digital word. Usually an internal or external *reference voltage* V_R is supplied, and the digital output is proportional to the ratio of the analog input voltage V_I to V_R. High accuracy requires that V_R be supplied by a precision voltage source. Each conversion is normally started by an *initiate-conversion* pulse or by a change in a voltage produced by a sensing device or switch. The analog input is sampled and held by a *sample-hold circuit* such as that of Fig. 23-14. With the electronic switch S of this circuit closed, the input voltage V_I charges the capacitor C. When S is opened, the extremely high input impedance of the voltage follower causes the charge to be held on the capacitor without appreciable decay for a least the time

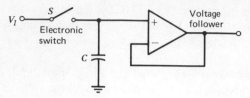

Figure 23-14 Sample-hold circuit.

interval required for conversion. After conversion, the digital word is transferred to a register, which delivers the word to a computer or other network. A basic element of nearly all A/D converters is the analog voltage comparator.

PARALLEL-COMPARATOR ADCs

A simple and very fast A/D converter employs multiple parallel comparators. Such a network with 32 stacked comparators is shown in Fig. 23-15. The small circle on the comparator symbol signifies the inverting input.

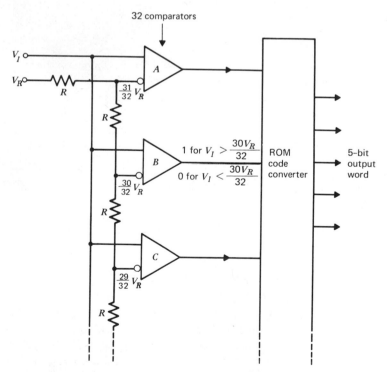

Figure 23-15 Parallel-comparator A/D converter.

Suppose the sampled input V_I is between $(30/32)V_R$ and $(31/32)V_R$. For this case all except comparator A at the top give high-level (logical 1) outputs, and the output of A is a logical 0. The input to the code converter, which is typically a read-only memory as shown, is a binary word consisting of a 0 followed by 31 logical 1's. Because the input voltage is placed in one of 32 discrete levels, the output word of the ROM has only five bits (see P23-11). A disadvantage of the circuit is that 2^n comparators are required for n-bit conversion. This limits the resolution to only four to six bits, but conversion rates of more than 10^8 words per second are possible. Parallel-comparator A/D converters are especially suitable for pulse-code-modulation communication, where course resolution is acceptable and high speed is essential. The number of comparators is reduced in *cascade* arrangements. These can have resolution up to about eight bits, with speeds of 10^6 words per second.

INTEGRATING CONVERTERS

To illustrate the basic principle of an integrating A/D converter, we shall consider a simple digital voltmeter. It is designed to convert an analog dc voltage into a digital signal that activates a light-emitting diode (LED) display. The input V_I controls a gated clock, or pulse generator, so that the number of pulses fed from the clock into a counter is proportional to the voltage. The output of the counter activates a visual display which reads the voltage.

Figure 23-16 shows the block diagram. At the input on the left is a three-position electronic switch, with the setting of the switch determined by the control circuits. In the illustration the switch is connected to ground. The next position is the terminal connected to the sampled input voltage V_I, assumed to be between 0 and 1.99 V dc. For higher values, say from 1.99 to 19.9 V, a 10-to-1 attenuator could be used at the input, changing the output scale by a factor of 10. For negative inputs the voltage could be converted to a positive value by adding a known positive voltage to it. Additional circuitry is required. The third switch position is that of the reference voltage V_R, which has the precise value of -2 V.

The output voltage v_1 of the electronic switch is fed into the integrator and from there to the comparator, used here as a zero-crossing detector. Operation and timing is governed by the control circuitry. With the READ switch open, the electronic switch S_1 is at ground, and the capacitor of the integrator is shorted by the closed switch S_2, making $v_2 = 0$. Also, the output of the gated clock is disconnected from the counter.

At time t_0 suppose the READ switch is closed. In an integrated-circuit A/D converter this is accomplished electronically by an initiate-conversion pulse. Control circuits supply logical 1's to electronic switches S_1 and S_2, to the counter, and to the gated clock. Switch S_1 is set at input V_I and S_2 is opened.

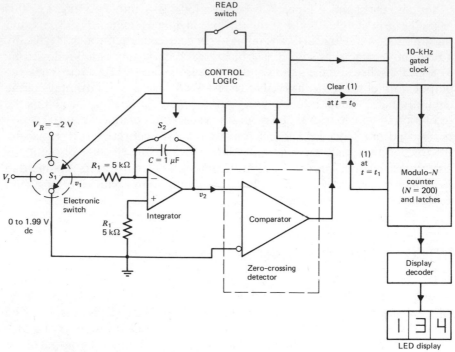

Figure 23-16   Block diagram of digital voltmeter.

The counter is cleared with the display set at 000. Also, the clock is gated *on*, and a pulse train of 10^4 pulses per second enters the counter.

From t_0 to t_1 the dc voltage at the input of the inverting integrator is integrated, and the output $v_2(t)$ is the negative ramp shown in Fig. 23-17. As indicated here, an integrator with a dc input becomes a *ramp generator*. The negative slope of the ramp is proportional to V_I, being $-200V_I$ V/s for the specified values of R_1 and C. For V_I between 0 and 1.99 V, the ramp slope is between 0 and -398 V/s.

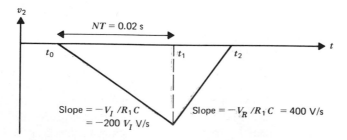

Figure 23-17   Output voltage $v_2(t)$ of integrator.

The negative ramp continues until the modulo-N counter has counted 200 pulses. With the 200th pulse at time t_1, the count changes from 199 to 000, and an "overflow" logical 1 signal is delivered from the counter to CONTROL. The logic circuits cause switch S_1 to shift to position V_R, delivering -2 V to the integrator. As shown in Fig. 23-17, the output $v_2(t)$ is now a ramp with a positive slope of 400 V/s. The counter continues to count the pulses from the clock, commencing at 000 at time t_1.

At t_2 the positive ramp crosses the axis, and the output of the comparator changes from low-level to high-level, delivering a logical 1 to CONTROL. Switch S_1 is returned to ground, S_2 is closed, the clock is gated *off*, the READ switch is opened, and the count stored in the register activates the LED display. Numericals 134 of the display indicate that V_I is 1.34 V.

The waveform of the output $v_2(t)$ of the integrator is drawn in Fig. 23-17 as two straight-line segments with different slopes. Accordingly, the converter is called a *dual-slope integrating ADC*. From the preceding discussion it is evident that the negative ramp is present for a fixed time interval t_1 t_0, which equals the product of the modulo N of the counter and the period T of the pulse train. As N is 200 and T is 10^{-4}, this interval is 0.02 s, independent of V_I. Multiplication by the slope gives $-4V_I$ for the peak value of $v_2(t)$.

The positive ramp has the value $-4V_I$ at time t_1 and a slope of 400 V/s, independent of V_I. Consequently, the interval $t_2 - t_1$ equals $4V_I/400$, or $0.01V_I$ s. Division by the pulse period T of 10^{-4} gives $100V_I$ for the number of pulses occurring in this time interval. For $V_I = 1.34$, the count is 134.

Switch S_2 across the integrating capacitor is necessary to maintain the output voltage of the ramp generator at zero when the READ switch is open. Without this shorting switch the input dc offset voltage and also the voltage across R_1 produced by the dc offset current would be integrated, causing the OP AMP to saturate. When S_2 is opened at the beginning of the conversion cycle, the voltage from these offset quantities is integrated along with V_I. If the undesirable voltage is 1 mV and V_I is 100 mV, an error of 1 % is introduced. Balanced input bias currents have no effect because resistor R_1 is included in series with the noninverting input, as shown in Fig. 23-16. By replacing this resistor with a zero-adjust potentiometer, errors caused by offset currents and voltages can be reduced to a very low value, but adjustment of the potentiometer is critical.

The digital voltmeter can be designed around available MSI integrated circuits. For the integrator a 741 OP AMP is suitable. Although the comparator can be made from a general-purpose operational amplifier, for optimum performance a monolithic voltage comparator would be used. Such integrated circuits provide logic level outputs, are designed for fast operation, and have low values of input offset voltage and current.

The 10 kHz clock can be an astable multivibrator. An example is the CMOS integrated circuit CD4098, consisting of dual monostable multivibrators. By connecting an output of each multivibrator to the proper trigger

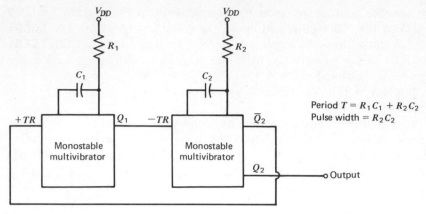

Figure 23-18 Astable multivibrator.

terminal of the other, as shown in Fig. 23-18, astable operation is obtained (see P23-12). The external resistors and capacitors are chosen to give the desired pulse width and repetition rate.

The control circuitry must be designed to set properly switches S_1 and S_2, to gate the clock *on*, and to clear the display when the READ switch is closed. Upon receiving an overflow signal from the counter it must switch S_1 to V_R. Finally, when the comparator gives the signal that the output voltage of the ramp generator rises above zero, the control circuits must gate the clock *off*, activate the display, return S_1 to ground, close S_2, and open the READ switch. The required functions can be accomplished with suitably designed combinational gates and latches available on monolithic chips.

A major advantage of the converter considered here is noise elimination. The integration procedure gives the time average of the input voltage. Consequently, high-frequency noise in low-level dc measurements is rejected. Also, if the observation interval is approximately a multiple of the period of the power-line frequency, noise from this source may be insignificant.

A/D CONVERTERS WITH D/A FEEDBACK

A simplified block diagram of an analog-to-digital converter with digital-to-analog feedback is shown in Fig. 23-19. The input signal $v_I(t)$ is sampled at time t_0 and supplied to the comparator. Assuming v_I is greater than the output v_A of the D/A converter, the comparator output is a logical 1, which instructs the counter to count *up*. As clock pulses enter the counter, the binary output word increases by one bit with each pulse. The D/A converter, which is similar to that of Fig. 14-20 of Sec. 14-4, delivers a staircase output voltage v_A which

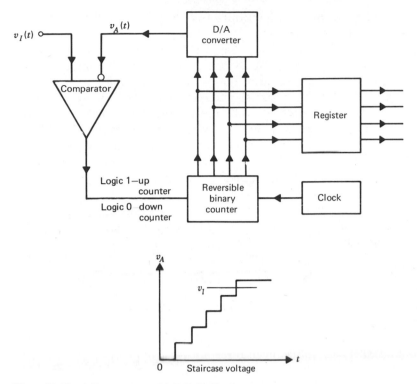

Figure 23-19 A/D converter with D/A feedback.

increases with time in increments as shown. When v_A rises above v_I, the comparator output changes to 0, the count stops, and the binary output word is fed into the output register. This is the digital representation of $v_I(t_0)$.

At time t_1 the analog voltage v_I is again sampled. Let us suppose that v_I has decreased during the interval between t_0 and t_1, so that v_I is less than v_A. The logical 0 present at the output of the comparator instructs the counter to count *down*. The output v_A of the D/A converter is now a downward staircase. When v_A falls below v_I, the comparator output changes to logical 1, the count stops, and the binary word represents $v_I(t_1)$. The process continues. Reversible binary counters that function as up-counters and down-counters depending on the control signal are available in IC form.

There are other types of A/D converters. A *successive-approximation* converter resembles that of Fig. 23-19 except that the counter is replaced with a programmer. This sets the binary number supplied to the D/A circuit, and the output v_A is compared with v_I. The binary word is increased or decreased by the programmer, depending on the comparator output. The process continues, with the binary word successively approaching the correct value. This technique gives high accuracy, with resolution from 8 to 12 bits and speed of the order of 10^5 words per second.

Various different types of monolithic A/D converters are available. An example is the Teledyne CMOS 8702 12-bit converter in a 24-pin dual in-line package. Its operation is somewhat like that of the integrating ADC which has been examined, although there are significant differences. On the chip are a ramp generator, a comparator, a clock, a counter, control logic, and output latches. A reference voltage of about -6.4 V is connected externally. When a logical 1 is applied to the initiate-conversion pin, the cycle commences and cannot be interrupted until completed. Clocked pulses can be supplied to this pin, or if desired, the pin can be connected to a permanent logical 1 voltage, such as the V_{DD} supply. In this latter event operation is free-running, with consecutive conversion cycles occurring without delay between cycles. In conjunction with a decoding network and an LED display, the converter can be used as a digital voltmeter.

23-5. ACTIVE RC FILTERS

A filter allows one or more frequency bands to be transmitted while rejecting signals outside these bands. In a *low-pass* filter the transmitted band extends from zero to some maximum frequency. A *high-pass* filter passes only frequencies greater than a specified value. Then there are *bandpass* and *band-elimination* filters. A *notch* filter is one that removes a very narrow band, and it is sometimes placed in the feedback network of an amplifier to obtain a narrow bandpass filter. Passive filters are composed of resistors, capacitors, and inductors whereas *active RC filters* consist of resistors, capacitors, and active elements. Elimination of inductance is the main advantage of RC filters, which are widely used and readily available. They are often employed to reject noise at frequencies outside the range of interest. Because of frequency limitations of the operational amplifiers, applications are generally at audio frequencies below about 10 kHz. Internally compensated OP AMPs, such as the 741, are restricted to use with filters having pass bands below about 1 kHz.

Figure 23-20 illustrates a low-pass filter which is first-order because it has a single pole. We assume here and throughout this section that the operational amplifier is ideal. Let us define ω_o and the gain K of the basic feedback amplifier and associated RC network as follows:

$$\omega_o = \frac{1}{RC} \qquad K = \frac{R_1 + R_f}{R_1} \tag{23-12}$$

From Fig. 23-20 it is easy to deduce that the voltage gain V_o/V_s is

$$A_V = \frac{K\omega_o}{s + \omega_o} \tag{23-13}$$

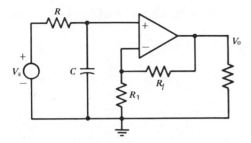

Figure 23-20   First-order low-pass filter.

The pole is $-\omega_o$ and the zero is infinite. The gain is K at low frequencies, it is down 3 dB at ω_o, and it approaches zero at very high frequencies. The pass band extends from zero to ω_o. Buffering between the RC filter and the next stage, or load, is provided by the operational amplifier, which also gives voltage gain. In some applications the OP AMP of Fig. 23-20 is omitted.

The second-order low-pass filter of Fig. 23-21 gives a sharper cutoff. The gain is determined (see P23-13) to be

$$A_V = \frac{K\omega_o^2}{s^2 + (3 - K)\omega_o s + \omega_o^2} \tag{23-14}$$

with ω_o and K defined by (23-12). We need to choose K to give suitable poles. It is easy to show (see P23-14) that the poles of (23-14) are complex conjugates of magnitude ω_o with a specified pole angle θ, provided K is selected in accordance with the relation

$$K = 3 - 2 \cos \theta \tag{23-15}$$

In Sec. 21-2 we found that the sharpest cutoff without peaking occurred for a pole angle of 45°. Choosing this angle, it follows from (23-15) that the amplifier

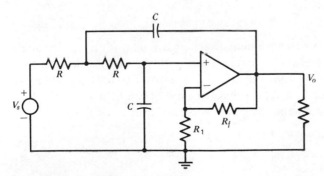

Figure 23-21   Second-order low-pass filter.

gain K should be adjusted to $3 - \sqrt{2}$. We determine from (23-14) that the corresponding 3 dB frequency is ω_o.

A further improvement in performance is obtained from a third-order filter. By connecting the configurations of Figs. 23-20 and 23-21 in cascade with each having the same ω_o although different values of K, the gain becomes the product of those of (23-13) and (23-14), or

$$A_V = \frac{K_1 K_2 \omega_o{}^3}{(s + \omega_o)[s^2 + (3 - K_2)\omega_o s + \omega_o{}^2]} \qquad (23\text{-}16)$$

We found in Sec. 21-7 that maximal flatness without peaking in the response of a three-pole amplifier is obtained when the poles are of equal magnitude and include a complex-conjugate pair with a pole angle of 60°. For this choice, K_2 in (23-16) is determined from (23-15) to be 2.

The pole configurations that have been considered are *Butterworth filters*. In each case the poles are equally spaced around the left-side semicircle of radius ω_o and centered at the origin of the complex plane, as illustrated in Fig. 23-22

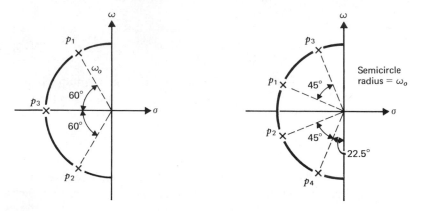

Figure 23-22   Three-pole and four-pole Butterworth arrangements.

for three-pole and four-pole arrangements. Each pole magnitude is ω_o. The angle between two adjacent poles on the circle is π/n, with the order n denoting the number of poles. For odd n, one pole is on the negative real axis. Although we shall not do so here, it can be shown that the ratio of the gain A_V to the low-frequency gain A_{VO} has a magnitude given by the relation

$$\left| \frac{A_V}{A_{VO}} \right| = \frac{1}{\sqrt{1 + (\omega/\omega_o)^{2n}}} \qquad (23\text{-}17)$$

This applies to all orders of Butterworth filters.

It is evident from (23-17) that *the 3 dB frequency of Butterworth filters is the radius ω_o, regardless of the number of poles.* Furthermore, for any ω_o the pass band is quite flat without peaking, and the greater the number of poles, the sharper is the cutoff. Presented in Fig. 23-23 are normalized responses for low-pass Butterworth filters of orders 1, 3, 5, and 7. The plots were obtained from (23-17).

A fourth-order filter is simply a cascade of two stages of the form of Fig. 23-21. For the Butterworth configuration the pole angles are 22.5° and 67.5°, and K_1 and K_2 are found from (23-15). By cascading the one-pole network with the fourth-order filter, we obtain a fifth-order filter. For this case the pole angles of the complex-pairs are 36° and 72°, and these angles and (23-15) determine the required amplifier gains. The design of higher order filters proceeds similarly. *Interaction between the stages is negligible because of the low output impedance of the operational amplifier.*

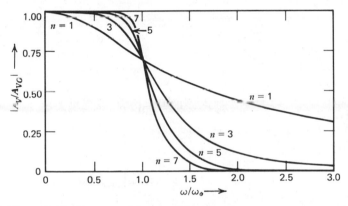

Figure 23-23 Butterworth low-pass filter responses.

Simply by interchanging the resistor R and the capacitor C in the network of Fig. 23-20, we obtain a first-order high-pass filter, and the gain is easily determined to be

$$A_V = \frac{Ks}{s + \omega_o} \qquad (23\text{-}18)$$

The second-order high-pass filter is the network of Fig. 23-21 with each resistor R replaced with a capacitor C and with each capacitor replaced with R. For this case the gain is

$$A_V = \frac{Ks^2}{s^2 + (3 - K)\omega_o s + \omega_o^2} \qquad (23\text{-}19)$$

The high-pass gain functions are quite similar in form to the low-pass functions. In particular, the denominators are the same. Accordingly, the design procedure is unchanged. For each pole angle of a complex-conjugate pair the amplifier gain is calculated from (23-15). Also applicable is (23-17), provided ω and ω_o are interchanged and A_{VO} is replaced with the gain A_{VH} of the high-pass region.

Example

Design a fifth-order high-pass filter with a cutoff frequency of 1 kHz and a voltage gain of 100 in the pass region.

Solution

The filter is shown in Fig. 23-24. It consists of a first-order high-pass stage followed by a cascade of two second-order stages. For each stage R and C are 160 ohms and 1 μF, giving an ω_o of 6250, or approximately 1 kHz. The pole angles are 36° and 72°, and the corresponding amplifier gains are calculated from

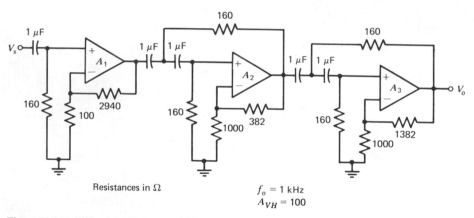

Resistances in Ω

$f_o = 1$ kHz
$A_{VH} = 100$

Figure 23-24   Fifth-order Butterworth high-pass filter.

(23-15) to be $K_2 = 1.38$ and $K_3 = 2.38$. The resistors R_1 and R_f of the second and third stages have been selected, using (23-12), to give these gains. For a total gain $K_1 K_2 K_3$ of 100, K_1 must be 30.4. The feedback resistors of the first stage were selected accordingly.

A bandpass filter is formed by cascading a low-pass filter having a cutoff at f_2 with a high-pass filter having a cutoff at f_1, with $f_1 < f_2$. Frequencies below f_1 and above f_2 are rejected. A band-elimination filter is obtained by connecting the inputs of a low-pass and a high-pass filter, with the outputs fed into a summing amplifier. Assuming the cutoff frequency of the low-pass filter is less than that of the high-pass filter, frequencies between these cutoffs are eliminated.

Our attention has been focused on the Butterworth filter, which is also referred to as a *maximally flat* filter because of the extremely flat response in the pass region. The poles are located on a semicircle in the complex plane, giving a gain characterized by (23-17). There are other arrangements with relative advantages and disadvantages. For example, the poles of the *Chebyshev*, or *equal-ripple*, filter are selected to produce a response containing ripples of equal amplitude in the pass band, with the poles situated on an ellipse in the s-plane. The advantage is a sharper cutoff. Then there are *linear-phase* filters, designed so that the phase of the gain is a linear function of ω, thus eliminating phase distortion. These are sometimes called *maximally flat delay* filters. In addition to the various pole arrangements, there are a number of different types of networks. Only some of the basic concepts and circuits have been examined here.

23-6. WAVEFORM GENERATION

Operational amplifiers are often used to generate periodic waveforms having a specified frequency, waveshape, and amplitude. Here we briefly consider a few such applications.

QUADRATURE OSCILLATORS

A *quadrature oscillator* employs two integrators to give two sinusoidal voltages $90°$ out of time phase. There are several different configurations, one of which is shown in Fig. 23-25. Frequencies are usually between 10 Hz and 100 kHz.

We assume ideal operational amplifiers. The unity-gain inverter makes V_1 equal to $-V_o$ as indicated on the figure. At each inverting input terminal of the OP AMPs of the integrators the sum of the currents is zero. In terms of V_2 and V_o in the s-domain, we find that

$$sC_1 V_2 - \left(\frac{1}{R_1}\right)V_o = 0$$

$$\left(\frac{1}{R_2}\right)V_2 + \left(sC_2 - \frac{1}{R_4}\right)V_o = 0 \tag{23-20}$$

For nonzero voltages the determinant of the coefficients must be zero. Equating it to zero gives

$$s^2 - \frac{s}{R_4 C_2} + \frac{1}{R_1 R_2 C_1 C_2} = 0 \tag{23-21}$$

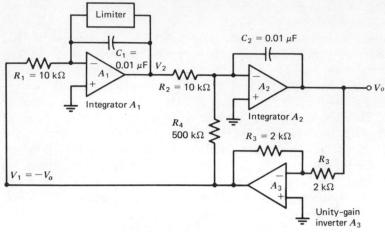

Figure 23-25 Sinusoidal quadrature oscillator.

Using the numerical values of the figure, the complex frequency s is found from (23-21) to be $100 \pm j10^4$ s^{-1}. With the real part σ positive, the natural frequencies of the network lie in the right-half side of the complex-frequency plane. The circuit oscillates at 10^4 radians/s, which is 1.6 kHz.

The purpose of R_4 is to provide the positive σ. With R_4 removed, s would be $\pm j10^4$ and oscillations might not commence. Although σ should be small, it must be large enough to give poles in the right half-plane for nonideal amplifiers

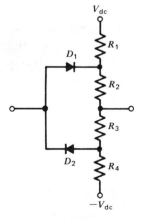

Figure 23-26 Limiting circuit for quadrature oscillator.

with variable parameters. A large value of R_4 makes the pole angle approximately 90° as desired. It is sometimes necessary to adjust the resistance. The positive σ causes the oscillations to build up until the nonlinearity of the limiter of integrator A_1 prevents further increase. Shown in Fig. 23-26 is a suitable limiting circuit. Resistors are selected to provide the desired amplitude of oscillation (see P23-19), and this amplitude is the same for both V_2 and V_o.

The voltages $v_2(t)$ and $v_o(t)$ are 90° out of phase, or in quadrature, because $v_o(t)$ is the integral of $v_2(t)$. The waveforms should be sinusoidal. However, some distortion is introduced by the nonlinearity of the limiting network. At the output of integrator A_1 the voltage $v_2(t)$ can be regarded as consisting of a desired fundamental plus undesired higher frequency harmonics. Assuming limiting is symmetrical on the positive and negative half cycles, only odd harmonics exist. Because the integral of cos ωt is $1/\omega$ sin ωt, the second integrator substantially reduces the magnitudes of the higher-order harmonics relative to the fundamental. Consequently, the output waveform is nearly undistorted.

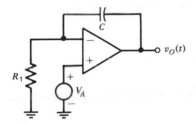

Figure 23-27 Integrator with offset voltage.

The integrating capacitors of the quadrature oscillator are not shunted with resistors. This is unacceptable in a single integrator, except for short time intervals. To illustrate, consider the circuit of Fig. 23-27 in which V_A denotes the input dc offset voltage of the operational amplifier. At the inverting node the current equation is

$$C\frac{d}{dt}(v_O - V_A) = \frac{V_A}{R_1} \tag{23-22}$$

Integration of (23-22) gives v_O to be $V_A t/R_1 C$ plus a constant of integration. Assuming v_O is zero at time zero makes the constant zero. Suppose V_A is 1 mV, R_1 is 1 kΩ, and C is 0.5 μF. Then v_O becomes $2t$, increasing without limit until the amplifier saturates. With C removed, v_O would rise to the saturation value immediately after supply voltages were applied. An input dc offset current through R_1 has an effect similar to that of V_A. However, in the quadrature oscillator of Fig. 23-25 as in many other operational amplifier circuits, there is feedback around the outer loop, dc voltage gains are small, and offset voltages and currents have very little effect.

To show this clearly, let us examine the circuit of Fig. 23-28 with offset voltages and currents included. Analysis of the *dc circuit* shows that the steady components of v_1, v_2, and v_O which exist after a long period of time are indeed small (see P23-20), with the dc component of the output voltage found to be

$$V_O = -V_A + 2V_C + R_1 I_A - R_3 I_C \qquad (23\text{-}23)$$

This is only a few millivolts for values typically encountered. The total voltage $v_O(t)$ consists of a sinusoidal component plus the small dc component given by (23-23). In contrast, in the single integrator of Fig. 23-27 after a long period of time, the voltage V_A appears directly across the input terminals of the operational amplifier, and the output dc voltage is large.

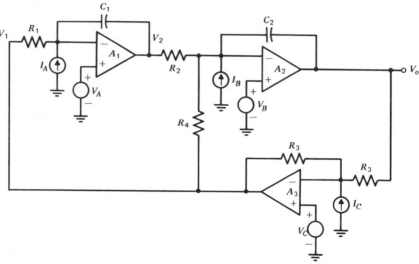

Figure 23-28   Oscillator with dc offset voltages and currents.

SCHMITT TRIGGER

Basic Schmitt trigger circuits were examined in Sec. 9-6. However, many waveform generators use these circuits made from comparator networks, such as the Schmitt trigger of Fig. 23-29. Numerical values have been selected to give a square hysteresis loop crossing the axes at ± 10 V as shown. Regenerative feedback is provided by R_f.

Suppose v_I is positive and sufficiently large to give a positive current i_1 to the parallel combination of diodes D_1 and D_2. Diode D_2 conducts and D_1 is *off*. At the output of A_1 the voltage v_2 is one diode drop below ground. Thus the current

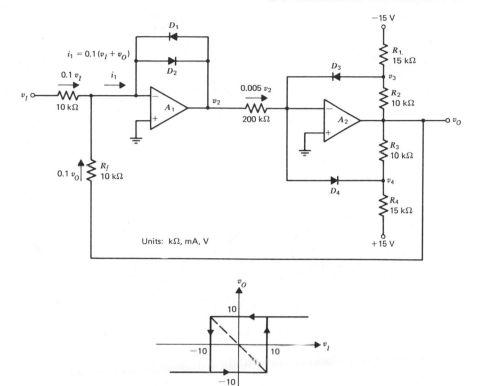

Figure 23-29 Schmitt trigger and transfer characteristic.

$0.005v_2$ mA is negative, although very small. However, it is sufficient to make diode D_3 conduct, with D_4 *off.* Because the current through D_3 is only a few microamperes, v_3 is only slightly above zero. Therefore, the current through R_1 is approximately 1 mA. This same current flows through R_2, making $v_O = +10$ V, and a current of 1 mA is fed back through R_f.

When v_I is decreased, the current i_1 at the input node of A_1 drops. The output voltage stays constant at $+10$ until v_I is lowered to -10 V. At this point the current i_1 is zero. For more negative values of v_I the current i_1 is negative. Hence D_1 conducts with D_2 *off.* This gives a positive output v_2, and D_4 conducts with D_3 *off.* The voltage v_4 is now approximately zero. The 1 mA current through R_3 and R_4 causes v_O to switch to -10 V, and the 1 mA current through R_f is reversed.

For increasing values of v_I the transfer characteristic is explained similarly. It should be noted that v_O is $+10$ *when* i_1 *is positive and it is* -10 *when* i_1 *is negative.* Operating points on the dashed line of the characteristic are possible, with $v_O = -v_I$, but they are unstable. This was discussed in Sec. 9-6.

SQUARE AND TRIANGULAR WAVEFORMS

The Schmitt trigger of Fig. 23-29 can be used in conjunction with an integrator to generate square waves and triangular waveforms. A suitable circuit is shown in Fig. 23-30. For a Schmitt trigger with the characteristic sketched in Fig. 23-29 the output voltage v_O is either $+10$ V or -10 V. Assuming it is positive, the inverting integrator generates a negative ramp. When the ramp output v_I goes through -10 V, the output v_O becomes -10 V and a positive-slope ramp is generated. Output waveforms are indicated on the figure. The oscillation frequency f is $1/4RC$ (see P23-21).

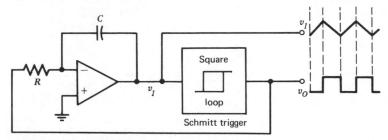

Figure 23-30   Square- and triangular-waveform generator.

At the input of the integrator the dc voltage resulting from offset voltage and current is small compared with the ± 10 V present, and these offsets present no serious problem. Many monolithic circuits use the technique that has been described for generating square and triangular waves. Various modifications are often incorporated to provide additional outputs with other waveforms, including amplitude-modulated and frequency-modulated sine waves.

PRECISION PULSE GENERATOR

A monostable multivibrator made with operational amplifiers is shown in Fig. 23-31. Let us assume that the Schmitt trigger is that of Fig. 23-29, except that a trigger input has been added to the inverting terminal of the input operational amplifier. With the trigger voltage v_2 equal to zero the circuit is stable with v_O at -10 V. For this case the current v_O/R_1 is negative, diode D_1 is *on*, the first OP AMP output is at about -0.7 V, and D_2 is *off*. Resistor R_2 keeps capacitor C from acquiring a charge, and v_I is zero. The negative output v_O gives a negative i_1, as explained in the discussion of Fig. 23-29.

Now suppose a positive trigger pulse in excess of 5 V is applied to the trigger input. During the brief instant when the pulse is present, the trigger input current exceeds the feedback current through R_f, and i_1 becomes positive. Let us recall that a positive i_1 gives a positive output voltage. Therefore, during the

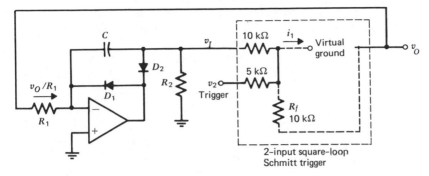

Figure 23-31   Monostable precision-pulse generator.

brief trigger pulse the output voltage v_O switches from -10 to $+10$ V, and i_1 is held at its positive value by v_O when the trigger voltage returns to zero.

A positive current v_O/R_1 now flows into C, generating a ramp voltage v_I with a negative slope $-10/R_1C$. Diode D_1 is *off*, and D_2 turns *on*, giving a current path through the output impedance of the OP AMP to ground. When v_I becomes -10, the Schmitt trigger switches v_O back to its original value of -10 V. Capacitor C discharges, and the circuit returns to its initial stable condition. The rectangular output pulse held the positive value of $+10$ V for the time interval R_1C required for v_I to drop from 0 to -10 V. By adjustment of R_1 or C, the pulse duration R_1C can be fixed precisely at the desired value.

It is evident that applications of operational amplifiers are almost unlimited. Only a few of the more important have been examined in this chapter. New integrated-circuit designs appear constantly, containing one or more operational amplifiers and other basic network configurations. These expand the capabilities and lower the cost of electronic systems.

REFERENCES

1. *Linear Integrated Circuits and MOS Devices* (Application Notes), RCA, 1974.
2. *The Linear and Interface Circuits Data Book*, Texas Instruments, Inc., 1973.
3. J. Millman and C. C. Halkias, *Integrated Electronics*, McGraw-Hill, New York, 1972.
4. J. V. Wait, L. P. Huelsman, and G. A. Korn, *Introduction to Operational Amplifier Theory and Applications*, McGraw-Hill, New York, 1975.
5. V. H. Grinich and H. G. Jackson, *Introduction to Integrated Circuits*, McGraw-Hill, New York, 1975.

PROBLEMS

Section 23-1

23-1. To illustrate the effect of a very large resistance R_f in the integrator of Fig. 23-1, suppose a dc input offset voltage v of 1 mV appears as a step function at time zero, when the power is turned on, with v being effectively in series with the noninverting terminal. With $V_s = 0$, $R_1 = 1$ kΩ, $C = 0.1$ μF, and $R_f = 50$ MΩ, find v_o at times 0^+, 5 s, and infinity. Assume the OP AMP remains ideal even though v_o becomes excessive.

23-2. For the two-amplifier differentiator of Fig. 23-3 find the input impedance in terms of R_1 and R_2, and from two node equations applicable to the OP AMP inverting inputs deduce that

$$\frac{V_o}{V_s} = \frac{-R_1 R_3 Cs}{R_2(1 + R_1 Cs)} \tag{23-24}$$

For $C = 0.1$ μF, $R_1 = R_2 = 300$ ohms, and $R_3 = 10$ kΩ, calculate the input impedance in ohms and determine $v_o(t)$ when $v_s(t) = 5 \sin 100t$.

23-3. Suppose the output v_{O1} of the log amplifier of Fig. 23-4 supplies the input v_{S2} of the antilog amplifier of Fig. 23-5. Assume q/kT is 40 and $\alpha_F I_{ES}$ is 0.1 pA for each transistor. For a design providing an output v_{O2} equal to K times the cube root of v_S for inputs from 0.01 to 10 V, the respective values in kΩ of R, R_1, R_2, R_3, R_4, and R_5 are 100, 2, 1, 500, 0.5, and 0.5, and $V_{dc} = 10$ V. Determine the constant K, and for $v_S = 1$ V calculate the three collector currents in mA.

23-4. For the analog-computer block diagram of Fig. 23-7 assume the output v_B of integrator 1 is initially set at $+3$ V and the output v of integrator 2 is initially at $+2$ V. For $t > 0$ find $v(t)$ and determine v at $t = 1$ s. (Ans: 1.63 V)

Section 23-2

23-5. Using an operational amplifier in cascade with a COS/MOS output as shown in Fig. 23-10b, with supply voltages of ± 10 V, sketch the circuit and the transfer characteristic of the comparator having a reference voltage V_R of -2 V and a source resistance R_1 of 1 kΩ. What is v_o if v_S is precisely -2 V, assuming an input offset voltage such that v_I is $+2$ mV?

23-6. In the network of Fig. 23-9 with $R_1 = 1$ kΩ, $V_{z1} = 8$, and $V_{z2} = 2$ V, add a battery so that the noninverting terminal is $+2$ V and remove R_f. Sketch the circuit and the transfer characteristic. Also, sketch and dimension $v_o(t)$ for $v_S(t) = 2.2 \cos \omega t$. Assume a forward diode drop of 0.7 V.

23-7. For the comparator of Fig. 23-32 assume 0.6 V across a conducting diode and from node equations find v_o in terms of v_I and sketch the transfer characteristic. Also,

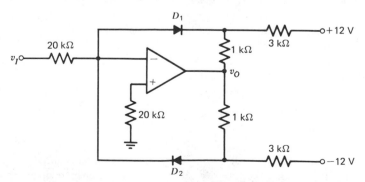

Figure 23-32 Comparator for P23-7.

calculate v_O for $v_I = \pm 2$ V. First deduce that the diodes cannot be both *on* or both *off*. Then show that D_1 is *on* for $v_I > 0$ and *off* for $v_I < 0$. The OP AMP is ideal.

Section 23-3

23-8. Replace the operational amplifier of Fig. 23-11 with the model of Fig. 14-12 of Sec. 14-3, and find the output impedance Z_{out} of the circuit that supplies current to the floating load in terms of A_v, Z_i, Z_o, R_s, and R_1. From the result deduce that Z_{out} is approximately $A_v R_1$ provided Z_i and A_v are reasonably large and Z_o is small. Calculate Z_{out} in MΩ for $A_v = 5000$, $R_1 = R_s = 1$ kΩ, $Z_i = 50$ kΩ, and $Z_o = 200$ ohms, and compare the result with $A_v R_1$. Also, with $Z_L = 800$ ohms, calculate the input impedance in MΩ.

23-9. The *four-quadrant multiplier* of Fig. 23-33 consists of a tranconductance amplifier in cascade with a current-to-voltage converter. It is useful for waveform generation,

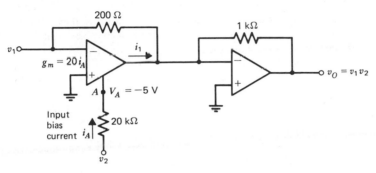

Figure 23-33   Four-quadrant multiplier for P23-9.

modulation, and so forth. The input voltages v_1 and v_2 may have any values between ± 4 V. Deduce that $v_O = v_1 v_2$. Note that g_m, or $-i_1/v_1$, is $20i_A$ with i_A denoting the input bias current. Point A is connected internally to the base of a transistor, and it is fixed at -5 V.

Section 23-4

23-10. A digital voltmeter of the form of Fig. 23-16 is designed to measure input voltage V_I from 0 to 19.99 V. The display has four digits, the modulo of the counter is 2000, and a 1 μF integrating capacitor is used. For a maximum cycle time of 20 ms and a 6 V peak for $v_2(t)$, calculate the required values of V_R, the clock repetition rate f, and R_1. Also, determine the cycle time if V_I is only half its maximum allowed value.

23-11. In the parallel-comparator A/D converter of Fig. 23-15, if V_I is 5.30 and V_R is 6.00 V, what is the 32-bit binary word ABC $\cdots$ fed into the ROM and what is the 5-bit binary output word?

23-12. For the astable multivibrator of Fig. 23-18, sketch outputs Q_2, $\bar{Q}_2$, and Q_1 on three identical time scales assuming $R_1 C_1$ is 80 μs and $R_2 C_2$ is 20 μs. The symbol $+TR$ signifies that triggering occurs on the leading edge of the input pulse and $-TR$ indicates trailing-edge triggering.

Section 23-5

23-13. From Fig. 23-21 deduce (23-14). To do this, first show that the node voltage V_1 at the junction of the resistors R is $V_o(s + \omega_o)/K\omega_o$, utilizing the virtual short-circuit. With this expression used for V_1, a single node equation relating V_o and V_s is sufficient.

23-14. With $K = 3 - 2 \cos \theta$, deduce that the poles of (23-14) are complex conjugates of magnitude ω_o and pole angle θ.

23-15. Design a three-stage sixth-order Butterworth low-pass filter with a cutoff at 1000 Hz, using 0.16 μF capacitors and 1 kΩ resistors for R_f. Calculate the voltage gain in the pass band, and sketch the network.

23-16. From (23-17) for the Butterworth low-pass filter determine the value of ω at which the ratio $|A_V/A_{VO}|$ is a maximum. Also, calculate the decibel drop in gain at $\omega = 2\omega_o$ for both second-order and fifth-order filters.

23-17. Design a Butterworth bandpass filter with a voltage gain of 50 and a pass band from 400 to 500 Hz, consisting of a two-stage fourth order low-pass filter followed by a two-stage third-order high-pass filter. Use 0.1 μF capacitors and 2 kΩ resistors for each R_f. For each stage determine R and R_1 in kΩ, and sketch the complete network.

23-18. Two identical third-order Butterworth filters, each with a pass band from 0 to ω_o, are cascaded. What percent of ω_o is the pass band? For a change in A_V from 96 % to 50 % of A_{VO}, what is the ratio of the change in ω to that of a sixth-order Butterworth with a pass band from 0 to ω_o? Which filter has the sharper cutoff? (*Ans*: 86.3, 1.46)

Section 23-6

23-19. For the sinusoidal quadrature oscillator of Fig. 23-25 deduce that $v_o(t)$ and $v_2(t)$ have approximately the same amplitude. Neglect R_4. Then design the limiting circuit of Fig. 23-26 to provide symmetrical limiting at a peak value of 10 V. Use ± 15 V supplies and select 1 kΩ for R_2. Assume limiting occurs when a diode turns *on* with a forward drop of 0.6 V. Determine R_1, R_3, and R_4. (*Ans*: 1.66, 1, 1.66 kΩ)

23-20. For the quadrature oscillator of Fig. 23-28 with $R_1 = R_2 = 10$ kΩ, $R_3 = 2$ kΩ, and $R_4 = 500$ kΩ, calculate the dc components of v_1, v_2, and v_o that exist after the steady state is reached. Assume the offset voltages V_A, V_B, and V_C are 1, -2, and 3 mV, respectively, and the respective offset currents I_A, I_B, and I_C are $-20, 40$, and 10 nA.

23-21. For the square- and triangular-waveform generator of Fig. 23-30, sketch and dimension v_I and v_O, assuming the Schmitt trigger has the hysteresis loop of Fig. 23-29. Also, deduce that the oscillation frequency f is $1/4RC$.

23-22. Suppose the Schmitt trigger of the monostable multivibrator of Fig. 23-31 has a square hysteresis loop crossing the axes at ± 5 V. For $R_1 = 10$ kΩ and $C = 1$ μF, sketch and dimension $v_O(t)$ and $v_I(t)$ for a 6 V trigger pulse applied at time zero. When v_O drops to -5 V, the Schmitt trigger switches v_O back to its original value. Calculate the reset time, defined as the time required for v_I to return from -5 V to zero. Also, calculate the reset time if a diode in series with a 100 ohm resistor is connected across R_1 so as to reduce this time. Assume diodes are ideal with zero voltage when conducting.

Chapter Twenty Four
Voltage Regulators

Most digital and analog circuits require a regulated power supply. In spite of rather large variations that may occur in the line voltage, the load current, and the temperature, the output dc voltage must be maintained within about 1% of the desired value, more or less depending on the specific application. We begin with a brief discussion of basic rectifier and filter circuits. After this, both linear and switching voltage regulators are examined, with emphasis on applications of various types of monolithic regulators.

24-1. DC POWER SUPPLIES

Quite frequently the power supplies for electronic equipment represent an appreciable portion of the total cost, size, weight, and engineering effort. For an alternating input voltage there are, in general, three basic components of a dc supply. These are shown in the block diagram of Fig 24-1.

The rectifier converts the ac input into a pulsating waveform with both dc and ac components. In certain applications, such as electroplating and battery charging, this output may be adequate, but most applications require filtering to remove the ac components. The output of the filter circuit may provide a suitable supply for phonograph amplifiers and many radio receivers. However, for proper operation of both digital and analog integrated circuits it is usually essential to have well-regulated supplies. In this section we examine briefly the rectification and filtering processes. Regulation is the subject of the rest of the chapter.

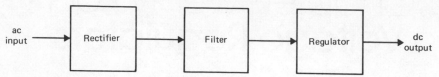

Figure 24-1   Basic components of a power supply.

RECTIFICATION

Shown in Fig. 24-2 is the circuit of a half-wave rectifier consisting of a diode in series with the load resistance. The line-frequency input voltage delivers through the diode a unidirectional current that produces a pulsating voltage across the load. This has a nonzero average value V_{dc}. Thus the rectifier converts an input ac voltage into a pulsating direct voltage. Harmonics of the frequency are present in the output. With an input at 60 Hz the frequencies in the output waveform of Fig. 24-2 are 0, 60, 120, 180, and so forth.

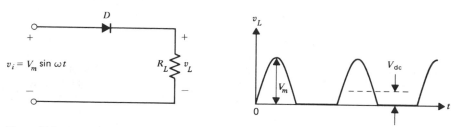

Figure 24-2   Half-wave rectifier and output voltage.

A criterion for specifying the amount of ac present in the output of a power supply is the ripple factor r, defined as

$$r = \frac{V_{ac}}{V_{dc}} = \frac{\sqrt{V_{rms}^2 - V_{dc}^2}}{V_{dc}} \qquad (24\text{-}1)$$

where V_{ac} denotes the rms value of the *alternating components* of v_L, not including V_{dc}. It is easy to deduce that the waveform of Fig. 24-2 has a total rms value of $0.5V_m$ and a dc value of V_m/π. From (24-1) the ripple factor is found to be 1.21. This is much too large for most purposes.

In Fig. 24-3 is a full-wave bridge rectifier. When v_i is positive, diodes D_1 and D_2 conduct. The other two diodes conduct on the next half-cycle. The output voltage has an rms value of $V_m/\sqrt{2}$ and a dc component of $2V_m/\pi$, giving a ripple factor of 0.48 (see P24-1). With identical diodes, only even harmonics of the line frequency appear in the output. For a 60 Hz input the lowest frequency to be

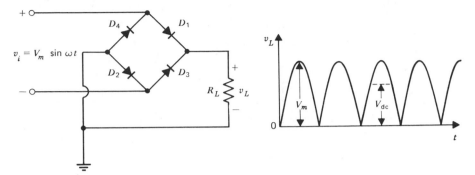

Figure 24-3   Full-wave bridge rectifier and output voltage.

filtered out is 120 Hz, which is twice that of the half-wave circuit. Filtering is easier.

Frequently, the line voltage is too high and must be reduced. In such cases an iron-core transformer can be added at the input. With a center-tapped transformer the full-wave rectifier can be made with two diodes as shown in Fig. 24-4. However, each diode must be capable of withstanding a *peak inverse voltage* equal to the maximum instantaneous value of the voltage across the entire secondary coil.

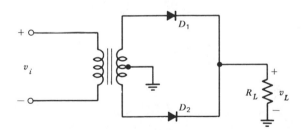

Figure 24-4   Full-wave rectifier circuit.

Finally, let us consider the common three-phase full-wave rectifier circuit of Fig. 24-5. Input voltages v_{ab}, v_{bc}, and v_{ca} from a transformer secondary are $120°$ out of time phase. At any instant diode D_1, D_2, or D_3 conducts, depending on which *terminal* voltage is highest; also, one lower diode conducts determined by the lowest voltage. The three-phase input may come from the coils of an alternator. An example is the typical 12 V system of an automobile. Mounted within the alternator are six diodes, which rectify the voltage, and a capacitor is included for filtering.

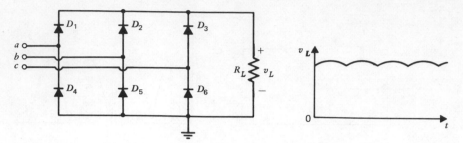

Figure 24-5   Three-phase full-wave rectifier circuit and output voltage.

FILTERS

The ac ripple in the output of a rectifier is smoothed out by a filter circuit. An elementary, widely used type is the *capacitor-input* filter shown in the bridge circuit of Fig. 24-6. The capacitor makes the output waveform quite different from that of the full-wave rectifier of Fig. 24-3.

With reference to Fig. 24-6 diodes D_1 and D_2 conduct from time t_1 to t_2, charging the capacitor C to the peak value of the voltage across the transformer secondary. At t_2 the voltage at point A of the circuit has dropped below v_L and these diodes turn *off*. From t_2 to t_3 all diodes are *off*. Current to the load is

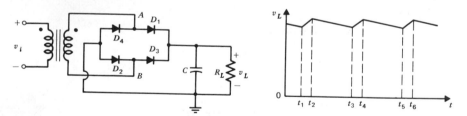

Figure 24-6   Full-wave rectifier with capacitor-input filter and output voltage.

supplied by the energy stored in C, which slowly discharges. At t_3 the voltage at B has risen to a sufficiently high value to turn *on* diodes D_3 and D_4. A current pulse charges C back to the peak value. From t_4 to t_5 the diodes are *off*, and the cycle starts again at t_5.

The ripple is greatly reduced. However, current from the source is supplied in pulses during the conduction intervals of the diodes. These current pulses have peak values considerably greater than the steady load current. The capacitor and the diodes must be chosen to avoid peak currents that might damage the semi-conductor devices.

Filtering is improved by adding one or more LC sections between the input capacitor and the load, as shown in the circuit of Fig. 24-7. Across the input

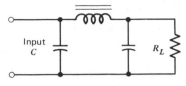 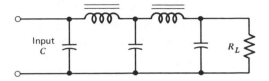

Figure 24-7   Capacitor-input filters.

capacitor is a voltage having a waveform similar to that of Fig. 24-6, but the ripple at the output is much less than that at the input of the filter.

A second basic type is the inductor-input filter, in which the current from the rectifier initially passes through an inductor, or choke. Examples are shown in Fig. 24-8. For full-wave rectifiers with inductor-input filters the voltage waveform at the filter input is the rectified sine wave of Fig. 24-3. At least one diode is conducting at each instant provided L is not too small. Recall that the voltage induced in an inductance is such as to oppose a change in its current.

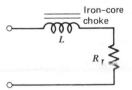 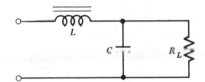

Figure 24-8   Inductor-input filters.

The impedance presented by the inductor to the ac components at the input is large, effectively blocking alternating current. Consequently, the load current is mainly dc, and the output voltage has very little ripple. An advantage over the capacitor-input filter is the reduction in diode peak currents, but this is offset by the requirement for an iron-core inductor. In addition to the size, weight, and cost it has series resistance giving an undesirable dc voltage drop.

24-2. VOLTAGE REGULATION

Line voltages at the input of a power supply often fluctuate by as much as 10 to 20%, causing the output voltage of the filter to vary. Current drawn by the power-supply load may have a wide range of values. In addition, the temperature may change. These effects tend to change the output voltage. A regulator is normally connected between the filter and the load, designed to maintain a nearly constant output voltage for anticipated variations in the input voltage, the load current, and the temperature.

An important figure of merit is the *voltage regulation*. It is defined as the percentage change in the output voltage resulting from a specified change in either the input voltage or the load current. In general,

$$\text{regulation} = \frac{\Delta V_L}{V_L} \times 100\% \qquad (24\text{-}2)$$

with the smallest value of V_L used in the denominator. If the change ΔV_L in V_L is the result of a specified change in the direct voltage at the input of the regulator, the term *input regulation* is used. This is sometimes called *line regulation*. With the input held constant and the load current changed from a minimum to a maximum value, the term *load regulation* is used. Both input and load regulation should be small. Digital and analog integrated circuits typically require voltage supplies with a regulation of 1% or less.

Example

Figure 24-9a shows the circuit of an elementary Zener-diode regulator. Resistance R_1 is 100 ohms. Assume the Zener diode, when in breakdown, can be modeled with a 5 V battery in series with a 5 ohm resistor. (a) With $R_L = 250$ ohms, calculate the input regulation for a change in V_I from 8 to 10 V. (b) With

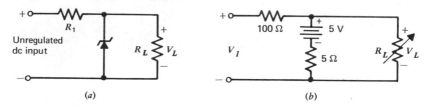

(a)   (b)

Figure 24-9   Elementary Zener-diode regulator and circuit model. (*a*) Regulator. (*b*) Model, assuming $V_L > 5$ V.

$V_I = 10$ V, calculate the load regulation corresponding to a change in the load current from 5 to 40 mA. (c) With $V_I = 10$ V and a load current of 80 mA, determine V_L.

Solution

(a) From the model of Fig. 24-9b with $R_L = 250$ ohms, circuit analysis reveals that V_L changes from 5.047 to 5.140 V when V_I changes from 8 to 10 V. The difference divided by 5.047 gives an input regulation of 1.8%.

(b) With $V_I = 10$, for load currents of 5 and 40 mA the respective values of V_L are found from analysis to be 5.214 and 5.048 V. The difference divided by 5.048 gives a load regulation of 3.3%.

(c) With $V_I = 10$ and a load current of 80 mA, the load voltage is less than 5 V and the diode is an open-circuit. From the series circuit the output voltage is found to be 2 V.

The diode regulator of the preceding example is inadequate for most applications. Regulation is not good. Also, a higher output voltage can be obtained only by replacing the diode with one having a larger breakdown voltage, and such a diode has a higher temperature coefficient. As we shall see, practical configurations normally utilize negative feedback techniques.

SERIES VOLTAGE REGULATOR

A widely used type of regulator is the *series*, or *linear*, regulator. The basic block diagram is shown in Fig. 24-10. Because currents drawn by the current-limit and sample networks are normally very small compared with that of the load, *the pass transistor is effectively in series with the load*. This transistor is biased in the active mode, and V_L equals V_I less V_{CE}. If the load voltage changes for any reason other than adjustments to the control circuitry by the operator,

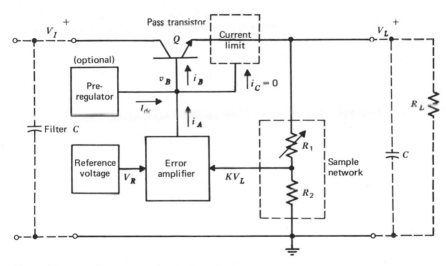

Figure 24-10   Series regulator block diagram.

the feedback causes V_{CE} to change so as to bring V_L back to its proper value. In effect, the pass transistor is a nonlinear resistance in series with the load, with the value of this resistance changed a large amount by a very small change in V_I.

The *sample network* is a resistive voltage divider. It samples V_L and feeds a fraction K of V_L into one of the differential inputs of the *error amplifier*. To keep the current drawn by the sample network small compared with the rated load current, the sum $R_1 + R_2$ should be considerably larger than the rated-load value of R_L. On the other hand, R_1 and R_2 should be small compared with the input resistance of the error amplifier, because we desire K to be independent of parameters of this amplifier. The sum of the two resistors is usually between 100 and 1000 ohms per output volt. It is common practice to connect a potentiometer between the resistors, with the voltage KV_L taken from the sliding tap. This permits the operator to fix the precise value of V_L by adjustment of the feedback.

The other input of the error amplifier is a reference voltage V_R, usually established by a Zener diode and sometimes by a battery cell. Under static conditions the voltages V_R and KV_L are nearly equal, typically differing by no more than a millivolt. Suppose a change occurs in either V_I or the load, causing V_L to rise above its regulated value. The error amplifier is designed so that the increased KV_L causes the output current i_A of the amplifier to decrease. If i_A is negative, it becomes more negative. In turn, the base current i_B of the pass transistor decreases, because it is the sum of i_A and the current I_{dc} from the preregulator. The resultant drop in the collector current lowers the load current, reducing V_L. The action continues until equilibrium is reached with V_L at its proper value.

The base voltage v_B of the pass transistor also changes during the transients. An abrupt increase in V_L produces a momentary rise in v_B that keeps v_{BE} constant. With KV_L greater than normal, the error amplifier begins to reduce v_B. This also reduces v_{BE}, the transistor current, and the load voltage, which gradually but quickly returns to its regulated value. When V_L drops below its normal value, the process is reversed.

OP AMP REGULATOR

The error amplifier of Fig. 24-10 is a high-gain differential amplifier similar to an OP AMP. In fact, it can be replaced by an OP AMP, as shown in Fig. 24-11. Power to the 741 is supplied by the unregulated dc input, with the high side connected to pin $V_{CC}{}^+$ and ground connected to $V_{CC}{}^-$. The direct voltage range of 24 to 30 V is suitable for the operational amplifier.

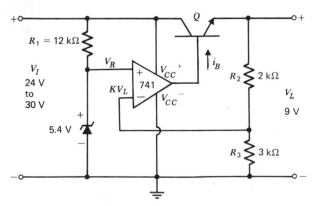

Figure 24-11   OP AMP regulator.

Resistors R_2 and R_3 are specified so as to give an output voltage of 9 V. The load current can vary from 0 to 100 mA with better than 0.1 % load regulation. Input regulation is not as good, mainly because V_R depends to a small extent on V_I. When V_I is increased from 24 to 30 V, the current through the Zener diode increases. Although the incremental resistance R of the diode decreases, the IR product is not constant, and V_R changes slightly. Input regulation is about 0.3 %. Because the preregulator has been omitted, the current out of the operational amplifier is positive, feeding into the base of the pass transistor.

In equilibrium, KV_L differs from the reference voltage V_R just enough to give the required 9.7 V at the output of the OP AMP. Even when the input voltage or the load current is changing, these OP AMP inputs usually differ by no more than a few millivolts. For both the Zener diode and the sample network, loading by the amplifier is negligible. Feedback is voltage-sample voltage-sum, and the output impedance is very low, less than 0.1 ohm.

The pass transistor dissipates up to 2.1 W at the full load of 100 mA. An adequate heat sink is required. The collector-emitter breakdown voltage must be greater than 21 V. With an input voltage V_I of 30, suppose the output of the regulator is short-circuited by accident. The output of the OP AMP is clamped at v_{BE}, which may be about 1 V. Inputs to the amplifier are 5.4 V and zero. Because it has short-circuit protection the OP AMP will not be damaged, but its output current may be as high as 40 mA. For a beta of 30, this gives a collector current of 1.2 A, and v_{CE} is 30 V. The 36 W dissipation will most likely destroy the pass transistor. A current-limit circuit is desirable, along with a pass transistor having a collector-emitter breakdown greater than 30 V.

THE ERROR AMPLIFIER

Many different configurations are used for error amplifiers, two of which are shown in Fig. 24-12. Figure 24-12a illustrates a differential amplifier. The voltage across R_2 is the fixed reference voltage V_R less a diode drop. This voltage establishes the constant current I, which is shared equally by Q_1 and Q_2 when KV_L equals V_R.

Across R_1 is the input voltage V_I less the sum of V_L and a diode drop. Although V_L will not deviate much from its regulated value, the input is unregulated, and i_A changes when V_I changes. This is undesirable, because input regulation suffers and ripple present in V_I is introduced into i_B and amplified by the pass transistor. The circuit would be improved if a preregulator were used in place of R_1, designed to deliver a constant current i_A (see P24-9).

Suppose a change in V_I or in the load causes V_L to drop in value. The current I through R_2 divides unequally, with the smaller portion passing through Q_1. The reduction in i_{C1} causes i_B to increase, and this increases the load current and voltage. Voltage V_L rises until it reaches its regulated value. Sensitive control results from circuit design that makes both I and i_A large compared with i_B.

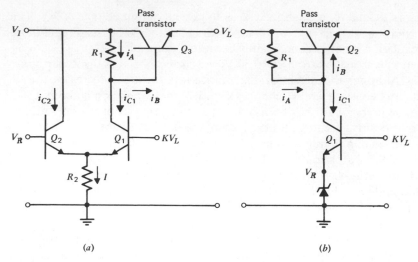

(a)   (b)

Figure 24-12 Elementary error amplifiers. (a) Differential type. (b) Single-transistor amplifier.

A pass transistor with a high beta is desirable, and a Darlington compound device is often used (see P24-7).

In the single-transistor error amplifier of Fig. 24-12b the voltage KV_L is greater than V_R by a diode drop. The current i_A through R_1 supplies both the base current i_B of the pass transistor and the collector current i_{C1} of the amplifier. The circuit is investigated in P24-8.

THE PREREGULATOR

In many regulator circuits it is necessary, or at least desirable, to supply current to the junction between the output of the error amplifier and the base of the pass transistor. This technique allows making the base current equal to the difference between two larger currents, giving better control. This was accomplished in the circuits of Fig. 24-12 by the resistor R_1 connected between the base-collector terminals of the pass transistor. Objections to R_1 are a consequence of the fact that the current through this resistor changes when V_I changes. Also, of course, ripple in the unregulated input voltage is present in the current, part of which feeds the base of the pass transistor. A better procedure usually employed is to replace the resistor with a direct-current source referred to a *preregulator*.

Shown in Fig. 24-13 is a circuit diagram of a practical preregulator connected to a Darlington pass transistor. The diode can be selected so that its Zener voltage changes with temperature at the same rate as V_{EB} of the PNP device, thereby giving a *temperature-compensated* steady current I_A. The current

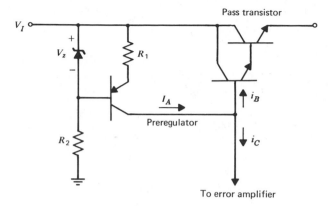

Figure 24-13   Preregulator and pass transistor.

through R_1 is fixed by the constant voltage $V_z - V_{EB}$ appearing across the resistor, and this current should be at least twice the maximum value of i_B. Resistor R_2 establishes the bias current of the diode.

CURRENT LIMIT

To protect the series pass transistor from excessive dissipation resulting from current overloads and short circuits, some form of current limiting is needed. One method often used is shown in Fig. 24-14a. Between the pass transistor and the load is inserted a small resistance R. Its value is selected to give a voltage drop of about 0.7 V when the current approaches the maximum to be tolerated. For

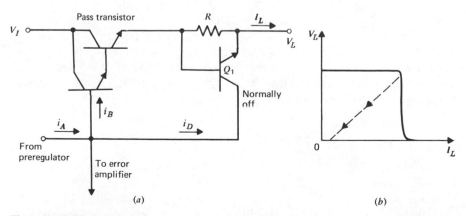

(a)   (b)

Figure 24-14   Current-limit circuit and characteristic. (a) Current-limit circuit. (b) Load characteristic. (Dashed line represents foldback limiting).

currents up to this level, transistor Q_1 is *off*, and R has very little effect. The voltage across it simply adds to the drop across the pass transistor with no appreciable effect on the output voltage.

When the load current exceeds $0.7/R$, transistor Q_1 turns *on*, drawing current away from the base of the pass transistor. Regulation is no longer effective and the output voltage drops. The characteristic is given by the solid line of Fig. 24-14b.

There are other methods. One of these, referred to as *foldback current limiting*, has circuitry designed to give the characteristic of the dashed line of Fig. 24-14b. When the current limit is reached, a reduction in the load resistance causes both V_L and I_L to decrease, and the characteristic moves to the left as R_L is reduced to zero. Increasing R_L while in the limiting mode reverses the foldback process. Foldback-limiting sharpens the limiting process, giving a rather precise maximum current, and it reduces the power dissipation of the pass transistor in event of a short-circuit. An example is considered in the next section.

24-3. MONOLITHIC REGULATORS

An integrated-circuit regulator is basically an operational amplifier combined on a chip with a voltage reference V_R and a series-pass transistor. Internal compensation is normally used to prevent oscillations. The compensated circuit must be stable even with large reactive loads, and overshoot caused by load and line transients must be minimized.

There are many different types of monolithic regulators. Some have adjustable output voltages, whereas others have fixed outputs. A *positive* regulator is designed to operate with a positive input voltage, and the output is also positive. There are *negative* regulators designed specifically for negative voltages. *Dual* types supply both positive and negative fixed outputs, and these are especially well-suited for use with operational amplifiers. To illustrate features typical of general-purpose adjustable regulators, we examine an especially popular one referred to as the 723. Although specifications given here apply to the Texas Instruments SN52723, they are somewhat similar to those for the LAS-723, the μA-723, the LM-723H, and the MC-1723G made by Lambda, Fairchild, National, and Motorola, respectively.

THE 723 REGULATOR

The integrated circuit is available in a dual-in-line package, a flat package, and a circular-type TO package. There are 10 or 11 pins available for external connections, depending on the type package. Input voltages should be between 9.5 and 40 V, with an output adjustable from 2 to 37 V. However, V_I must be greater than V_L by at least 3 V. The load current is restricted to 150 mA, but the regulator can be used to drive an external power transistor to boost the current to several amperes.

Provision is made for adding an external resistor for limiting the output current to the desired value. With V_I increased from 12 to 40 V the input regulation is typically 0.02%, and the load regulation for an output current range from 1 to 150 mA is about 0.03%. For input ripple frequencies from 50 Hz to 10 kHz, ripple attenuation is approximately 74 dB.

Figure 24-15 shows the basic internal configuration of the 723 voltage regulator. There are 11 external pins indicated and labeled. On 10-pin packages the V_z-pin is omitted, as is the illustrated 6.2 V Zener diode. Transistor Q_1 is the pass transistor. Its collector and emitter terminals are connected to pins labeled V_C and V_O for external connections. It can be used to drive the base of an external power transistor, which then becomes the pass transistor.

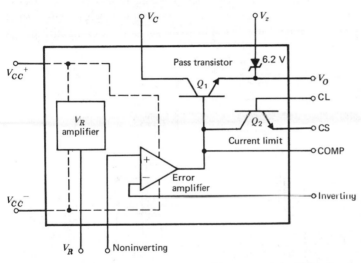

Figure 24-15   Partial circuitry and functional blocks of the 723 voltage regulator.

The base of the current-limit transistor Q_2 is connected to the current-limit (CL) pin, and its emitter is joined to the current-sense (CS) pin. A resistor R can be connected between CL and CS with the external circuitry arranged so that the load current passes through R in a direction making V_{BE2} positive. When the product $I_L R$ reaches about 0.65 V, the current-limit transistor turns *on*, draws current from the base of the pass transistor, and causes the output voltage to drop. The maximum current, occurring with the output short-circuited, is approximately 0.65/R. It is not necessary to make any connections to pins CL and CS if internal current limiting is not desired.

The inverting and noninverting terminals of the error amplifier are brought to pins that can be connected to external voltage-sample and voltage-reference networks. Within the error amplifier is a constant current source that serves as a preregulator. The voltage-reference amplifier contains two Zener diodes, a JFET current-regulator diode, a current-mirror, and amplifier circuitry. It supplies to pin V_R a low-noise temperature-stable reference voltage between 6.95 and 7.35 V. The reference voltage is usually applied to one of the differential inputs of the error amplifier. This may be done either directly, or through a resistor selected to minimize the effect of the input bias current, or through a voltage-divider network designed to reduce V_R.

A POSITIVE-VOLTAGE REGULATOR

Figure 24-16 shows the circuit of a 12 V positive-voltage regulator with current limiting at 100 mA. The reference voltage V_R is externally connected to the noninverting input. Between the pins is a balancing resistor R_3 with a value equal to $R_1 \| R_2$, and it is sometimes omitted. The voltage at the inverting input is

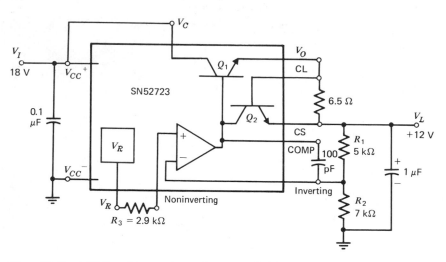

Figure 24-16   A 12-V positive regulator with current limiting at 100 mA.

approximately equal to V_R. For an assumed reference voltage of 7 V, the values selected for R_1 and R_2 of the sample network give an output of 12 V.

In series with the pass transistor is a 6.5 ohm resistor which limits the output current to 100 mA. Any positive input voltage between 15 and 40 V can be used. There are three capacitors; those at the input and output reduce ripple, and the 100 pF capacitor provides stability. The output voltage can be adjusted by changing R_1 and R_2. For $R_1 = 0$ and R_2 infinite, the output is 7 V. Load voltages less than 7 are obtained by connecting the R_1R_2 voltage-divider circuit between V_R and the noninverting input so that only a fraction of V_R is fed to this input (see P24-10). The output is connected directly to the inverting input, usually through a balancing resistor having a value equal to $R_1 \| R_2$. Resistors R_1 and R_2 should be selected so as to draw about 1 mA from pin V_R.

A NEGATIVE-VOLTAGE REGULATOR

In Fig. 24-17 the SN52723 is employed as a negative-voltage regulator, with values chosen to give a load voltage of -24 V. The reference voltage V_R is 7.1 V above V_{CC}^-, which is the load voltage V_L. Therefore, at pin V_R the voltage V_A with respect to ground is the sum of V_R and V_L. Currents at the inputs of the error

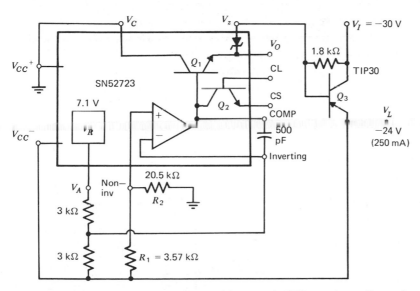

Figure 24-17   Negative-voltage regulator with external PNP transistor. Output is -24 V with a current up to 250 mA.

amplifier are negligible, and the two input voltages are approximately equal. It can be shown (see P24-11) that

$$V_L = \frac{-0.5V_R(R_1 + R_2)}{R_1} \qquad (24\text{-}3)$$

Outputs between -9 and -40 V are possible provided V_I is sufficiently large in magnitude.

The purpose of the 6.2 V internal Zener diode is to give the necessary level shifting. A 723 IC in a 10-pin package does not have this diode and it must be connected externally. The TIP30 pass transistor has an h_{FE} of 40 or more. For the required 250 mA output, the base drive is approximately 6 mA. About half of this is supplied by the 1.8 kΩ base-collector resistor, with the remainder from the internal Zener diode. This arrangement gives good balance and response. If current limiting is desired, an external limit circuit can be added similar to the conventional type used internally in the 723. An adequate heat sink must be provided for the pass transistor.

A 50 W REGULATOR

For power supplies with higher output wattages, a circuit such as that of Fig. 24-18 can be used. The Lambda LAS-723 is designed with a reference voltage V_R of 2.5 V. Because the resistors of the sample network are each 510 ohms, the regulated output voltage is 5 V. The regulator is intended for use with a 10 A, 50 W load. Less than 10 mA of base drive is required by the power Darlington pass transistor, which can dissipate 225 W at a case temperature of 50°C. The 2.2 kΩ resistor prevents the output current of the integrated circuit from falling

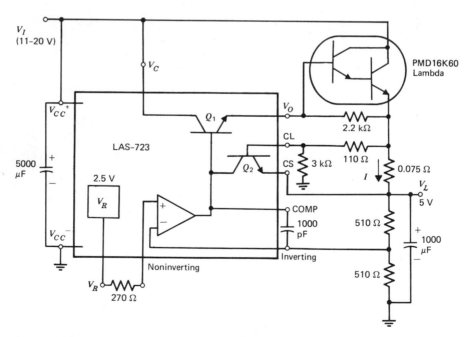

Figure 24-18 A 5-V, 10-A positive-voltage regulator with external boost and foldback current limiting.

below 0.5 mA at low output current levels. This provides increased stability and faster response to transients.

A foldback current-limit circuit is formed by the 0.075 ohm resistor in series with the load, along with the voltage divider between pin CL and the top terminal of this resistor. The base current of the internal current-limit transistor is negligible compared with that through the 3 kΩ resistor. Consequently, we find from application of Kirchhoff's laws that the current I through the 0.075 ohm resistor can be expressed in the form

$$I = \frac{V_{BE2} + (1 - a)V_L}{0.075a} \qquad (24\text{-}4)$$

with a denoting the ratio 3/3.11 of the voltage divider. Current I is approximately equal to the load current.

The current-limit transistor conducts when its base-emitter voltage is about 0.72 V, and when conducting, this value does not change appreciably for values of V_L from 5 to 0. For $V_{BE2} = 0.72$ and $V_L = 5$ V, the current I is found from (24-4) to be 12.5 A, which is the current at the knee of the foldback characteristic. With the same value of V_{BE2} and a short-circuit at the output, I is calculated to be 10 A. In normal operation transistor Q_2 is *off.*

Suppose the load resistance R_L is reduced continuously from an initial high value. When the load current reaches 12.5 A, the base-emitter voltage of Q_2 has risen to 0.72 V, and Q_2 enters the active mode, drawing current from the base of internal transistor Q_1. This causes V_L to drop below its regulated value of 5 V. With V_{BE2} now fixed at about 0.72 V, the load current also drops as required by (24-4). When R_L becomes zero, I is 10 A. If R_L is now increased continuously from zero, the current rises until it becomes 12.5 A with $V_L = 5$. At this operating point transistor Q_2 turns *off,* and operation returns to the normal condition. Without the current-limit voltage divider the constant a of (24-4) would be unity, and I would stay almost constant as V_L drops from 5 to 0. The foldback characteristic gives a rather precise maximum current. Also, it reduces the power dissipation of the pass transistor in event of a short-circuit at the output.

Typically, the regulator of Fig. 24-18 has a line regulation of 0.01% and a load regulation of 0.02%. The output voltage changes about 0.015% for each 1°C change in temperature. With an input of 20 V and a short-circuit at the output, the Darlington pass transistor dissipates almost 200 W. It must have an adequate heat sink. At 12.5 A the 0.075 ohm series resistor dissipates 12 W.

A POSITIVE 105 V REGULATOR

Although limited to a maximum of 40 V at the input, the 723 IC can be used to regulate higher voltages. This is done by "floating" the IC so that it operates between the high output voltage and a still higher level within the safe range. A possible circuit is shown in Fig. 24-19. The circuit should be compensated with a 500 pF capacitor as in Fig. 24-17. Current limiting is accomplished by placing a resistor between pins CL and CS, with this resistor connected in series with the pass transistor.

Resistors R_1 and R_2 have been selected to give an output of 105 V. The Zener diode fixes the voltage between the input terminals V_{CC}^+ and V_{CC}^- at 20 V, well within the safe range. At pin V_R the voltage V_A with respect to ground is $V_R + V_L$, because V_{CC}^- equals V_L. For a reference voltage of 7.1 V, V_A is 112.1 V. An external pass transistor is essential. Without it the input would be connected to pin V_C and a short-circuit at the output would destroy the IC. In the actual configuration, V_C is clamped at 20 V above V_L.

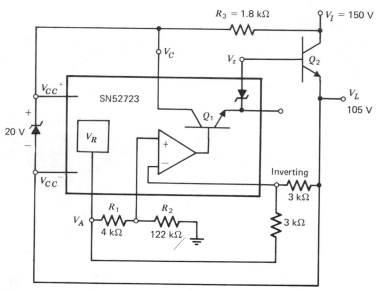

Figure 24-19   Positive 105-V regulator.

The output could be connected directly to the inverting input with the two 3 kΩ resistors omitted. However, it is desirable to draw about 1 mA from the reference-voltage circuit when the internal Zener diode is used. Reducing R_1 and R_2 for this purpose is not practical because of the high voltage V_A. The current through the 3 kΩ resistors provides a satisfactory solution; at the inverting input the voltage is higher than V_L by $0.5V_R$. In normal operation the drop across R_3 is 25 V, and the value of this resistor was chosen to give the desired bias currents to both the IC and the external Zener diode. With a short-circuit at the output the voltage across R_3 rises to 130 V, giving a power dissipation in this resistor of 9.4 W.

FIXED-VOLTAGE REGULATORS

Monolithic fixed-voltage regulators contain an internal sample network in addition to the pass transistor, the error amplifier, and the reference-voltage network. There are only three pins on the package. One pin accommodates the unregulated input, one provides the regulated output, and the third is the common terminal. Each IC provides a regulated output voltage with a fixed value.

To illustrate basic features let us consider the Lambda LAS-1500 and 1800 series. The many devices available have fixed voltages with values from 5 to 28 V and also from -2 to -28 V, with an accuracy of $\pm 5\%$. Each one can deliver about 1.5 A to a load. For the positive regulators the unregulated input from the filter must be at least 2.4 V greater than the output and must not exceed 35 V. Similar restrictions apply to the negative regulators. Line and load regulations are less than 2 and 0.6%, respectively, and the temperature coefficient is less than 0.03%/°C. Ripple attenuation exceeds 58 dB. There is a bias current of about 10 mA out of the common terminal of a positive regulator and into the corresponding terminal of a negative regulator, as shown in Figs. 24-22 and 24-23.

As with all regulators, including the adjustable-voltage types as well as the fixed-voltage types considered here, the heat sink must be adequate to keep the junction temperature of the pass transistor within the specified range. Data sheets specify the maximum operating junction temperature and the required thermal resistances. For the regulators considered here the maximum junction temperature is 150°C and the junction-to-case thermal resistance is 3°C/W. An internal overload control limits the internal power dissipation to 15 W or less.

There are numerous applications. The circuit of Fig. 24-20 is a basic 1.5 A negative-voltage supply, and the circuitry of a positive supply is similar. The dual regulator of Fig. 24-21 provides equal-and-opposite output voltages such as those required by most operational amplifiers. The optional diodes provide current paths from the output terminals to ground in case one of the ICs is *off* for any reason. Conversion of a fixed output into an adjustable one is possible with the circuit of Fig. 24-22 (see P24-12). When current regulation is desired, the circuit of Fig. 24-23 is convenient (P24-13). It is quite similar to that of Fig. 24-22.

HYBRID REGULATORS

Adjustable-voltage and fixed-voltage monolithic regulators can supply load currents of 3 A and more, and some packages can dissipate in excess of 30 W, although rather large heat sinks are required. For substantially greater load currents and power dissipation there are hybrid regulators made by thick-film technology. For example, the adjustable-voltage Lambda LAS-7000 series of hybrid regulators can supply load currents up to 30 A at output voltages from

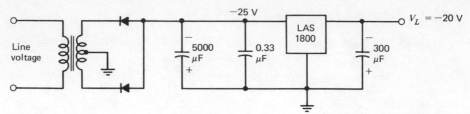

Figure 24-20   Basic 1.5-A negative-voltage supply.

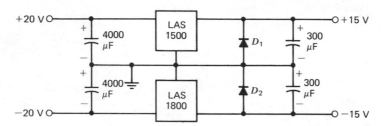

Figure 24-21   Dual-output regulator.

about 5 to 30 V. The pass transistor has a maximum allowed junction temperature of 200°C and a junction-to-case thermal resistance of only 0.44°C/W. In accordance with these specifications, with forced-air cooling used to maintain the case temperature at 50°C the device can safely dissipate 341 W. Package dimensions in centimeters are about 10 × 6 × 3.

Also sold by power-supply manufacturers are submodules that include not only the regulator but also the rectifier and filter. Only a transformer has to be added. They are compact and low in cost, and their popularity with equipment manufacturers is increasing.

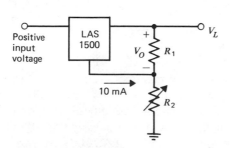

Figure 24-22   Adjustable-voltage regulator.

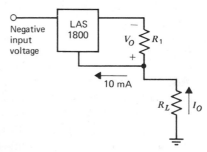

Figure 24-23   Current regulator.

24-4. SWITCHING REGULATORS

In the linear regulator the power dissipated by the pass transistor, which is biased in the active mode, equals the product of the load current and the difference between the input and output voltages. A consequence is very inefficient operation, and the wasted energy adds to the operating cost. This problem is especially serious in the electronic supplies of space vehicles and in all battery-operated equipment. In a switching regulator the pass transistor operates either in cutoff with no power dissipation or in saturation with low dissipation. The transistor rapidly switches back and forth between these states, at frequencies usually from 10 kHz to 100 kHz. Efficiencies are typically from 80 to 90%, and even higher efficiencies are possible. Line-frequency transformers are often omitted from the rectifier circuits and heat sinks are considerably smaller, giving power supplies that are reduced in size and weight. Among the disadvantages are the noise generated, the ripple present at the output, and the increased complexity of the design.

A BASIC SWITCHING REGULATOR

Figure 24-24 shows the diagram of an elementary switching regulator with a dc output of 5 V. The output voltage $v_O(t)$ is fed back to the inverting input of the operational amplifier. At the noninverting input is $v_R(t)$, which is approximately equal to 5 V. However, the small amount of positive feedback through R_f causes v_R to change a small amount when the voltage v_A at the filter input changes.

Suppose v_R is greater than v_O by at least 1 mV. Positive current driven into the base of the NPN pass transistor Q saturates the device. Neglecting the small saturation value of v_{CE}, the voltage v_A at the filter input is 30 V, and diode D is *off*. Because v_O is nearly constant at 5 V, there are 25 V across the 1.2 mH inductor, causing the current i_L through L to increase linearly with time. When i_L becomes

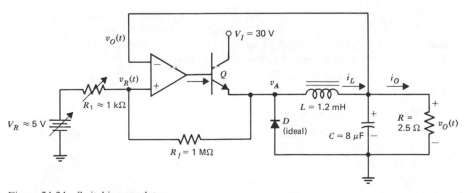

Figure 24-24   Switching regulator.

greater than the load current i_O, capacitor C begins to charge and $v_O(t)$ rises. After a short time the output voltage rises above v_R, which is 5.025 V for the values shown on Fig. 24-24 with $v_A = 30$ V. At this instant the pass transistor turns *off*.

At the instant the pass transistor switches *off*, the current in the inductor is a maximum. Because this current cannot change abruptly, diode D switches *on*. Neglecting the small drop across D, the voltage v_A becomes zero, and v_R is found to be 4.995 V. The drop in v_R from 5.025 to 4.995 is 30 mV, which is the *hysteresis* of the regulator. Across the inductor is the nearly constant output voltage, and i_L decreases almost linearly with time. When i_L becomes less than the load current i_O, the capacitor C begins to discharge, causing v_O to drop. After a short time interval v_O becomes less than v_R, which is 4.995 V. The pass transistor Q turns *on*, diode D turns *off*, v_A rises quickly to 30 V, and the cycle is repeated. Both transistor Q and diode D must be high-speed switching devices.

An approximate, reasonably accurate, and informative analysis will now be made. For simplicity we shall assume negligible the saturation value of v_{CE}, the diode drop, resistance associated with the inductor, and effects of switching transients which occur within the active devices. The filter and associated waveforms are shown in Fig. 24-25, with the load represented by the resistor R.

At the filter input the voltage v_A is a periodic pulse train of period T equal to the reciprocal of the switching frequency f. Each pulse amplitude is the value V_I of the unregulated dc input voltage. The pulse duration t_1 is the *on*-time of the pass transistor, and the *off*-time is t_2. We shall find it convenient to work also with time τ defined as the time in excess of t_1. With V_{dc} denoting the average value of $v_A(t)$, which is also the average value of $v_O(t)$, the times t_1, t_2, and τ can be expressed as

$$t_1 = \left(\frac{V_{dc}}{V_I}\right)T = \textit{on}\text{-time}$$

$$t_2 = T - t_1 = \textit{off}\text{-time} \tag{24-5}$$

$$\tau = t - t_1 = \text{time in excess of } t_1$$

The ratio t_1/T is the *duty cycle*. For a 5 V regulated output with an unregulated input of 25 V, the duty cycle is 5/25, or 0.2. This signifies that the pass transistor conducts during 20% of each period T. The first equation (24-5) follows from the geometry of Fig. 24-25b, and the other two relations are definitions.

Some of the symbols defined on the sketches of Fig. 24-25 and used in the analysis are as follows:

V_{dc} = average value of both $v_A(t)$ and $v_O(t)$
Δ = difference between the maximum and minimum values of $i_L(t)$
$I_1 = V_{dc}/R - 0.5\Delta$ = minimum value of $i_L(t)$
V_1 = output voltage v_O at $t = 0$ and also at $\tau = t_2$
V_2 = output voltage v_O at $t = t_1$ and also at $\tau = 0$
V_{min}, V_{max} = respective minimum and maximum values of $v_O(t)$

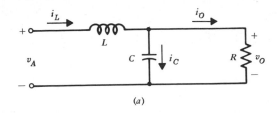

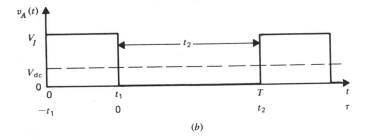

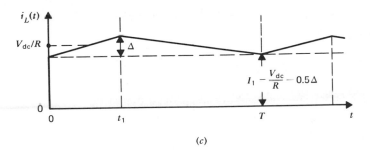

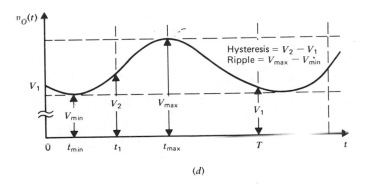

Figure 24-25   Filter and waveforms. (*a*) Filter and load resistor. (*b*) Filter input voltage. (*c*) Inductor current. (*d*) Output voltage.

PASS TRANSISTOR ON $(0 \leq t \leq t_1)$

During the time interval from 0 to t_1 the pass transistor is conducting, v_A equals V_I, and the voltage across L is $V_I - v_O(t)$. Because the regulated output v_O is normally maintained within 1 % of its average value V_{dc}, the inductive voltage drop $L \, di_L/dt$ approximately equals the positive constant $V_I - V_{dc}$. It follows that i_L increases linearly with time. In the interval from t_1 to T the voltage v_A is zero, and $L \, di_L/dt$ is approximately equal to $-V_{dc}$. Accordingly, i_L decreases linearly with time.

The complete waveform is that of Fig. 24-25c. We note that the average value of i_L must be exactly the same as the average load current V_{dc}/R, because the dc component of the periodic current i_C of the capacitor is precisely zero. Therefore, from 0 to t_1 the current i_L can be expressed as

$$i_L(t) = \left(\frac{\Delta}{t_1}\right)t + I_1 \qquad (24\text{-}6)$$

where I_1 is $i_L(0)$, and the slope di_L/dt is Δ/t_1 which equals $(V_I - V_{dc})/L$. Accordingly, the maximum change Δ in i_L is

$$\Delta = \frac{(V_I - V_{dc})t_1}{L} \qquad (24\text{-}7)$$

For a specified switching frequency and given values of V_I and V_{dc}, the on-time t_1 is found from (24-5), and Δ is inversely proportional to the inductance L.

Figure 24-26 shows the time-domain and Laplace-transformed circuits of the filter capacitor C and load R. The Laplace transform $I_L(s)$ of the inductor

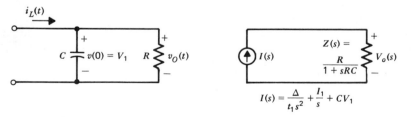

Figure 24-26 Time-domain and Laplace-transformed circuits $(0 \leq t \leq t_1)$.

current $i_L(t)$ of (24-6) is $\Delta/(t_1 s^2) + I_1/s$. Across the capacitor is an initial unknown voltage V_1 to be determined. In the s-domain the charged capacitor can be represented by a current source CV_1 in shunt with the impedance $1/sC$, and the current CV_1 can be added to $I_L(s)$ to give $I(s)$ as shown on the figure. The parallel equivalent of the impedances of the capacitor and the load resistor is denoted $Z(s)$. The product $I(s)Z(s)$ yields $V_o(s)$. With a little manipulation (see P24-14), the inverse transform can be found, and we find $v_O(t)$ to be

$$v_O(t) = V_1 e^{-t/RC} + R\left(I_1 - \frac{RC\Delta}{t_1}\right)(1 - e^{-t/RC}) + \left(\frac{R\Delta}{t_1}\right)t \qquad (24\text{-}8)$$

The constants t_1 and Δ are found from (24-5) and (24-7), and the minimum current I_1 is less than V_{dc}/R by 0.5Δ. However, $v_O(t)$ is not completely known until the constant V_1 in (24-8) is determined, with V_1 denoting $v_O(0)$.

In order to find the ripple we need to know the minimum value V_{min} of $v_O(t)$. This is indicated on Fig. 24-25d. At the minimum the current i_L of the inductor is exactly equal to the minimum load current v_O/R. Of course, the current of the capacitor is zero, being equal to $C\,dv_O/dt$. From (24-8) the instant t_{min} at which the minimum occurs is found (P24-14) to be

$$t_{min} = RC \ln\left[1 - \left(I_1 - \frac{V_1}{R}\right)\left(\frac{t_1}{RC\Delta}\right)\right] \qquad (24\text{-}9)$$

In order to find V_1 we must first investigate conditions with the pass transistor in cutoff.

PASS TRANSISTOR OFF $(0 \le \tau \le t_2)$

A similar analysis procedure can be applied to the circuit to determine its behavior during the *off*-time of the pass transistor. Because it is convenient to refer to the initial time t_1 of this interval as time zero, the symbol τ is used here for the variable time, with τ and t related by $\tau = t - t_1$. Accordingly, the *off*-time extends from $\tau = 0$ to $\tau = t_2$, with t_2 denoting the *off*-time $T - t_1$. In terms of time τ the current $i_L(\tau)$ is

$$i_L(\tau) = -\left(\frac{\Delta}{t_2}\right)\tau + I_1 + \Delta \qquad (24\text{-}10)$$

The slope di_L/dt is negative, equal to $-\Delta/t_2$, and i_L is $I_1 + \Delta$ at $\tau = 0$.

Figure 24-27 illustrates the time-domain and Laplace-transformed circuits. The current $I(s)$ is the sum of CV_2 and the Laplace transform of $i_L(\tau)$, given by

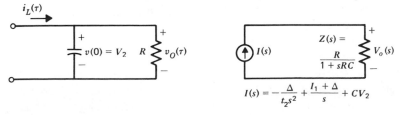

Figure 24-27 Time-domain and Laplace-transformed circuits $(0 \le \tau \le t_2)$.

(24-10). The symbol V_2 denotes the unknown voltage of the capacitor at $\tau = 0$. From the s-domain circuit and transform theory (see P24-15) we find that

$$v_O(\tau) = V_2 e^{-\tau/RC} + R\left(I_1 + \Delta + \frac{RC\Delta}{t_2}\right)(1 - e^{-\tau/RC}) - \left(\frac{R\Delta}{t_2}\right)\tau$$

$$(24-11)$$

The maximum value V_{max} of $v_O(\tau)$ occurs at time $\tau = \tau_{max}$, with τ_{max} given by

$$\tau_{max} = RC \ln\left[1 + \left(I_1 + \Delta - \frac{V_2}{R}\right)\left(\frac{t_2}{RC\Delta}\right)\right] \qquad (24-12)$$

Because we are dealing with periodic waveforms, the voltage $v_O(\tau)$ at $\tau = t_2$, corresponding to $t = T$, must equal V_1 which is $v_O(t)$ at $t = 0$. From this and (24-11) we obtain

$$V_1 = V_2 e^{-t_2/RC} + R\left(I_1 + \Delta + \frac{RC\Delta}{t_2}\right)(1 - e^{-t_2/RC}) - R\Delta \qquad (24-13)$$

Also, the voltage $v_O(t)$ at $t = t_1$ must equal V_2, and from this and (24-8) we find that

$$V_2 = V_1 e^{-t_1/RC} + R\left(I_1 - \frac{RC\Delta}{t_1}\right)(1 - e^{-t_1/RC}) + R\Delta \qquad (24-14)$$

Equations 24-13 and 24-14 can be solved for V_1 and V_2. The results can be expressed as

$$V_1 = RI_1 + \frac{R^2 C\Delta}{t_2}\left[1 - \frac{T}{t_1}\left(\frac{1 - \exp t_1/RC}{1 - \exp T/RC}\right)\right] \qquad (24-15)$$

$$V_2 = R(I_1 + \Delta) + \frac{R^2 C\Delta}{t_2}\left[1 - \frac{T}{t_1}\left(\frac{\exp t_2/RC - \exp T/RC}{1 - \exp T/RC}\right)\right] \qquad (24-16)$$

From these the hysteresis is found to be

$$V_2 - V_1 = R\Delta + \left(\frac{R^2 C\Delta}{t_2}\right)\left(\frac{T}{t_1}\right)\left[\frac{(1 - \exp t_1/RC)(1 - \exp t_2/RC)}{1 - \exp T/RC}\right] \qquad (24-17)$$

The following example illustrates the analysis procedure using the equations that have been developed.

Example

For the switching regulator of Fig. 24-24, suppose V_R and R_1 are adjusted to give a dc output of 5 V and a switching frequency of 20 kHz. Assuming the diode is ideal and neglecting the saturation value of v_{CE}, determine V_R and R_1, the hysteresis, the output voltage $v_O(t)$, and the output ripple. As shown on the figure,

V_I is 30 V and R_f is 1 MΩ. The values of L, C, and R are 1.2 mH, 8 μF, and 2.5 Ω, respectively.

Solution

The period T is $1/f$, or 50 μs, and V_{dc} is specified to be 5 V. With $V_I = 30$ the on-time t_1 is $T/6$, or 25/3 μs, from (24-5). The difference Δ between the maximum and minimum values of i_L is 0.17361 A, by (24-7). The average value of i_L is V_{dc}/R, or 2 A. Subtracting 0.5 Δ from this gives 1.9132 for I_1. The constants V_1 and V_2 are found from (24-15) and (24-16) to be 4.933 and 4.961, respectively. The difference of 28 mV is the hysteresis.

With V_1 and V_2 known, $v_O(t)$ is determined from (24-8) and (24-11). Substitution of numerical values yields the equations

$$v_O(t) = 1.192e^{-50000t} + 3.741 + 52080t$$
$$v_O(\tau) = -0.464e^{-50000\tau} + 5.425 - 10420\tau$$

$$(24\text{-}18)$$

Either from these or from (24-9) and (24-12) we find that t_{min} is 2.70 μs and τ_{max} is 16.0 μs, and V_{min} and V_{max} are 4.924 and 5.050 V, respectively. The difference is 126 mV, which is the peak-to-peak ripple. This is 2.5 % of V_{dc}.

We now need to determine V_R and R_1. When the pass transistor Q is *off*, v_A is zero and v_R is at its minimum value V_1. With Q *on*, v_A equals V_I and v_R has its maximum value V_2. From Fig. 24-24 we find (see P24-16) that these minimum and maximum values of v_R are

$$V_1 = \frac{V_R R_f}{R_f + R_1} \qquad V_2 = \frac{V_R R_f + V_I R_1}{R_1 + R_f}$$

$$(24\text{-}19)$$

With $V_1 = 4.933$, $V_2 = 4.961$, $V_I = 30$, and $R_f = 1$ MΩ, the reference voltage V_R and the resistance R_1 are calculated from (24-19) to be 4.94 V and 930 ohms, respectively.

In the preceding example the hysteresis of 28 mV is a typical value and is reasonable, although a somewhat lower value could be used with reduced ripple. Unfortunately, the peak-to-peak ripple of 126 mV is high. This is a consequence of overshoot, which is one of the problems associated with switching regulators. Each time the pass transistor switches *off*, the current in the inductor is greater than that of the load, and the difference current continues to add charge to the capacitor, briefly increasing $v_O(t)$. The overshoot can be reduced by increasing L and decreasing C, with their product kept constant. For example, if L is increased to 10 mH with C reduced to 1 μF, calculations show that the regulator of the preceding example has a hysteresis of 34 mV for the same 20 kHz switching frequency, and the ripple is reduced to 44 mV (see P24-17).

DESIGN TECHNIQUES

It is convenient to select a monolithic regulator, such as the 723 examined in the preceding section, to provide the reference voltage V_R and the differential amplifier required by a switching regulator. An external pass transistor is normally employed. The circuits are similar to those of linear regulators, but positive feedback is essential to provide the hysteresis that assures switching, along with the high-speed switching diode and the LC filter circuit. Such a circuit is shown in Fig. 24-28.

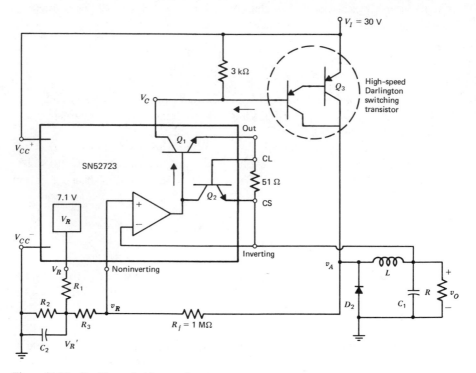

Figure 24-28 Positive switching regulator.

Let us consider the reference-voltage circuit. Capacitor C_2 is about 0.1 μF, which is large enough in value to filter out the ripple present at the switching frequency. Consequently, the effective reference voltage V_R' is constant. For a 5 V output the voltage divider consisting of R_1 and R_2 is selected to reduce the 7.1 V at pin V_R to 5 V at terminal V_R'. Resistors R_f and R_3 provide the positive feedback which gives the necessary hysteresis associated with v_R. The resistance R_3 is typically about 1 kΩ.

For a load resistor of 2.5 ohms drawing an output current of 2 A, inductance L should be selected so that the difference Δ between the maximum and minimum values of the inductor current i_L is less than 20% of the load current. Therefore, Δ should be less than 0.4 A. For an input voltage V_I of 30 V and a specified switching frequency of 20 kHz, the inductance L corresponding to 0.4 A for Δ is found from (24-7) to be 0.52 mH. Because a larger value is better, we choose 1.2 mH for L, assuming an inductance of this value is available.

Selection of the capacitor C_1 is rather critical. A value too large gives a very small hysteresis for the desired 20 kHz switching frequency, perhaps even less than the input offset voltage of the differential amplifier (see P24-20). Without appreciable hysteresis the regulator is linear with the pass transistor operating in the active mode. On the other hand, with too little capacitance the hysteresis is excessive, as is the ripple. It is important that the resonant frequency of the filter circuit be small compared with the switching frequency. In many cases a good initial choice is to make the resonant frequency $1/(2\pi\sqrt{LC})$ approximately $\frac{1}{12}$ of the switching frequency at full load, which gives a capacitance C that is

$$C \approx \frac{4}{Lf^2} \tag{24-20}$$

The value of C found from (24-20) should be substituted into (24-17) to determine that the hysteresis is suitable for the range of values of load resistance R to be encountered. A small hysteresis is desirable in order to minimize the output ripple. However, the minimum hysteresis should be greater than 10 mV in most circuits. If the hysteresis is not reasonable, a new value should be chosen for C. An optimum value can quickly be determined with the aid of a computer.

There is an alternate procedure convenient for computer analysis. It consists of expanding the pulse train at the filter input into a Fourier series and determining the infinite series of the output voltage $v_O(t)$. A plot of v_O versus time can be made using different values of C (see P24-18). The results are used to select the capacitance. However, because the microfarads of electrolytic capacitors may differ substantially from their nominal values, perhaps 30% or more, the final adjustment should be experimental.

In the circuit of Fig. 24-28 the hysteresis associated with v_R forces the pass transistor to be either saturated or cut off. Suppose the pass transistor is saturated. Then v_A is 30 V and the inductor current i_L is increasing. When v_O rises above v_R, which is at the maximum value V_2, the pass transistor abruptly turns *off*, and v_R drops by several tens of mV to its lower value V_1. The difference between v_O and V_1 is now sufficient to assure that Q_1 and Q_3 are *off*. When v_O drops below V_1, both Q_1 and Q_3 turn *on*, and v_R rises abruptly to V_2. Once again the inputs of the high-gain differential amplifier differ by tens of mV, and the drive is sufficient to saturate the pass transistor. The filter parameters L and C

prevent the output voltage from changing abruptly when v_R jumps to a new value.

In many cases values of output voltage greater than the reference voltage V_R of the monolithic circuit are desired. This is accomplished by the same technique considered in the preceding section. The voltage divider formed by resistors R_1 and R_2 is moved to the output and connected so as to feed only a fraction of v_O into the inverting input.

Specially designed fast-switching diodes and pass transistors are essential. Low-frequency power transistors and diodes will quickly overheat if used in high-speed switching regulators. The higher cost of a switching power transistor is an important consideration.

DRIVEN REGULATORS

It has been noted that the switching frequency of a self-oscillating regulator depends on the hysteresis and on the filter parameters L and C. In addition, the frequency is a function of both the input voltage V_I and the load current I_O. Many regulators are designed to operate at a fixed frequency which is synchronized with a square-wave output of a multivibrator.

One common technique is to employ an RC network to integrate the square wave, converting it to a triangular wave with a peak-to-peak value several times greater than the output ripple. A value of 50 mV is usually adequate. With the triangular wave added to the fixed reference voltage V_R, the total reference voltage $v_R(t)$ increases linearly during the interval from 0 to $0.5T$ and decreases linearly from $0.5T$ to T. Each instant its value crosses that of the output voltage $v_O(t)$, the circuit switches. The conventional positive feedback circuit is omitted. The dc level of the output voltage automatically adjusts itself to a few millivolts above or below V_R as necessary to provide the proper duty cycle. Whenever the load is changed, the dc output voltage readjusts itself slightly, but the change is no more than several millivolts.

Although driven switching regulators require additional circuitry, they have some significant advantages. When more than one are used in a system, their operation can be synchronized so as to distribute the input-line switching waveforms uniformly with respect to time. There is greater flexibility in the selection of the filter parameters L and C, and consequently, a lower output ripple can be realized. Furthermore, the stable switching frequency leads to more efficient operation.

REFERENCES

1. E. R. Hnatek, *Applications of Linear Integrated Circuits*, Wiley, New York, 1975.
2. RCA, *Solid-State Power Circuits*, RCA Solid State Division, Somerville, New Jersey, 1971.

3. D. E. Pippenger and C. L. McCollum, *Linear and Interface Circuits Applications*, Texas Instruments, Dallas, Texas, 1974.
4. *The Linear and Interface Circuits Data Book*, Texas Instruments, Dallas, Texas, 1973.
5. Lambda Catalog Vols. 1 and 2, Lambda Electronics, Melville, L.I., New York, 1976.

PROBLEMS

Section 24-1

24-1. For the full-wave bridge rectifier of Fig. 24-3 with ideal diodes and $V_m = 10$ V, deduce from the waveform for v_L the values of v_{rms}, V_{dc}, and V_{ac} and calculate the ripple factor. The definition of V_{ac} follows (24-1). Also, list all frequencies in v_L for a 60 Hz supply.

24-2. Repeat P24-1 for the half-wave rectifier of Fig. 24-2.

24-3. Determine the Fourier cosine series for the rectified sine wave of Fig. 24-3 with $V_m = 3\pi$ V and a line-frequency ω of 1000 radians/s. Using only the first five terms of this series, calculate V_{rms}, V_{ac}, and the ripple factor. Then find these quantities from exact expressions and compare.

24-4. (a) For the half-wave rectifier of Fig. 24-2, with $V_m = 10$ V and $R_L - 100$ ohms calculate the average power to the load and also the power to the load due only to the dc component of the current. (b) Repeat for the full-wave rectifier of Fig. 24-3. [*Ans*: (a) 0.25, 0.10 W]

Section 24-2

24-5. In the OP AMP regulator circuit of Fig. 24-11, assume V_{BE} is 0.7 V, β_F is 25, V_t is 30 V, and the load resistance R_{tt} is 90 ohm. The OP AMP has an open-loop gain A_v of 10^3, an output resistance of 100 ohms, and an input resistance of 2 MΩ. (a) Calculate the power dissipated by the pass transistor. (b) Calculate the difference $9 - V_L$. (c) Determine the current out of the output terminal of the OP AMP in mA. (d) Add a 4.8 kΩ resistor between the collector-base terminals of the pass transistor and repeat (c). State an advantage of this resistor. [*Ans*: (b) 168 μV; (d) -0.314]

24-6. In the diode regulator circuit of Fig. 24-9a having the model of Fig. 24-9b, with $R_L = 125$ Ω calculate the input regulation for a change in V_I from 10 to 12 V. Then calculate the load regulation corresponding to a change in the load current from 0 to 40 mA with $V_I = 12$ V. Also, determine V_L with $V_I = 12$ V and a load current of 100 mA.

24-7. (a) Design a regulator and draw the circuit, which gives an output of 10 V with an input of 20 V. Full load is 10 W. The sample network should have a total resistance of 2 kΩ. V_R is precisely 5 V. Use the differential error amplifier of Fig. 24-12a, assuming identical transistors Q_1 and Q_2 with infinite betas and base-emitter voltages of 0.7 V. Employ a Darlington-pair pass transistor for Q_3, having a beta of 1000 and $v_{BE} = 1.4$ V. Select R_1 and R_2 so that $i_{C1} = i_{C2}$ and $i_A = 10i_B$ at full load. (b) Calculate the full-load pass-transistor dissipation. (c) Calculate i_{C1} and i_{C2} if V_I is reduced to 15 V.

24-8. A regulator uses the error amplifier of Fig. 24-12b with $R_1 = 200$ ohms. Assume the Zener diode can be represented by a 5 V battery in series with 10 ohms. Transistors Q_1 and Q_2 each have betas of 40 with base-emitter voltages of 0.7 V. The two resistors of the sample network are each 2 kΩ, and the input voltage V_I is 16 V. Calculate the load regulation when the load current changes from 0 to 100 mA. To do this, first express all branch currents in terms of I_{C1} and I_{C2} and solve for these currents for each load condition. (*Ans*: 1.034%)

24-9. (a) For the regulator of P24-8 with a load resistor of 200 ohms, calculate the *dB ripple rejection* $20 \log V_i/V_o$, with V_i and V_o denoting the respective input and output ripple voltages. Assume h_{ie} is 50 ohms for the pass transistor and 200 ohms for the error-amplifier transistor, with each h_{fe} being 40. (b) Repeat if R_1 is replaced by a preregulator having an effective R_1 of 10 kΩ. The incremental resistance of the Zener diode is 10 ohms. [*Ans*: (b) 42.05]

Section 24-3

24-10. In the positive regulator of Fig. 24-16, remove R_1, R_2, and R_3. Between pins V_R and noninverting, connect resistor R_4 and connect R_5 between the noninverting pin and ground. Then connect V_L to the inverting pin through R_6 selected to balance R_1 and R_2. Determine R_4, R_5, and R_6 so that V_L is 4 V and the current drawn from pin V_R is 1 mA. Assume V_R is 7.1 V. Sketch the complete circuit.

24-11. For the negative-voltage regulator of Fig. 24-17 deduce (24-3) for V_L in terms of V_R, R_1, and R_2. Then calculate values of R_1 and R_2 for an output of -9 V, with the parallel equivalent of R_1 and R_2 equal to 3 kΩ. V_R is 7.1 V. Sketch the complete circuit, including an external current-limit circuit using an additional transistor and a resistor.

24-12. Suppose the adjustable regulator of Fig. 24-22 has an input of 15 V and uses a fixed-voltage regulator with an output of 5 V. With $R_1 = 290$ ohms, determine R_2 for an output of 12 V.

24-13. The IC of the current regulator of Fig. 24-23 is a fixed-voltage regulator with an output of -20 V. The input is -30 V. Calculate R_1 for an output current of 1 A through a load resistance R_L from 1 to 5 ohms.

Section 24-4

24-14. From the time-domain circuit of Fig. 24-26 with $i_L(t)$ given by (24-6), deduce (24-8) which gives $v_O(t)$ for values of t from 0 to t_1. Also, from the result deduce that the minimum value of $v_O(t)$ occurs at the value of t given by (24-9). Show all steps of the derivations in detail.

24-15. From the time-domain circuit of Fig. 24-27 with $i_L(\tau)$ given by (24-10), deduce (24-11) which gives $v_O(\tau)$ for values of τ from 0 to t_2, with $\tau = t - t_1$. Also, from the result deduce that the maximum value of $v_O(\tau)$ occurs at the value of τ given by (24-12). Show all steps of the derivation in detail.

24-16. From Fig. 24-24 with the diode assumed to be ideal and with the saturation value of v_{CE} neglected, deduce (24-19) for the minimum and maximum values of v_R. Also, determine the hysteresis in terms of V_I, R_f, and R_1. With $R_f = 1$ MΩ and $R_1 = 1$ kΩ, calculate the hysteresis in mV when V_I is 30 V and again when V_I is 10 V.

24-17. For the switching regulator of Fig. 24-24 with L changed to 10 mH and C reduced to 1 μF, suppose V_R and R_1 are adjusted to give a dc output of 5 V and a switching frequency of 20 kHz. Assuming the diode is ideal and neglecting the saturation value v_{CE}, determine V_R and R_1, the hysteresis, the output voltage $v_O(t)$ and $v_O(\tau)$, and the output peak-to-peak ripple.

24-18. For the regulator of the example of Sec. 24-4, select time zero at the midpoint of a 30 V pulse of $v_A(t)$ and determine the Fourier cosine series of this voltage in terms of

$\omega_o t$, with $\omega_o = 2\pi/T$. From the filter circuit of Fig. 24-24 find the ratio V_o/V_a of the phasor voltages in terms of R, L, C, and ω, with $\omega = n\omega_o$. Then write the Fourier series for $v_O(t)$ in terms of R, L, C, and $\omega_o t$. Using the specified numerical values and a suitable computer, obtain a plot of $v_O(t)$ over a complete period. Compare the hysteresis and the ripple with those found in the example.

24-19. Repeat P24-18 with C changed from 8 to 20 μF.

24-20. From (24-17) deduce that the hysteresis approaches zero as C approaches infinity. Calculate the hysteresis for the data of the example of Sec. 24-4 with C changed to 100 μF.

Appendix A
Computer Analysis of Inverter of Section 6-3

In Sec. 6-3 it was shown that the voltages v_{BE} and v_{BC} of a simple transistor inverter can be related by equations of the form of (6-39). This was discussed further in P6-8, P6-9, and P6-14. The equations to be solved have the form

$$\dot{x} = \frac{a_1 + a_2 x + a_3 y + a_4 \exp 40x + a_5 \exp 40y}{a_6 \exp 40x + a_7(a_8 - x)^{-1/3}} \tag{A-1}$$

$$\dot{y} = \frac{b_1 + b_2 x + b_3 y + b_4 \exp 40x + b_5 \exp 40y}{b_6 \exp 40y + b_7(b_8 - y)^{-1/3}} \tag{A-2}$$

The symbols x and y denote v_{BE} and v_{BC}, respectively, and the dot over a symbol denotes a time derivative. The a and b coefficients are constants.

From known initial conditions, (A-1) and (A-2) can be used to calculate initial slopes of $x(t)$ and $y(t)$. A straight segment drawn from the starting point $x(t_o)$ of $x(t)$ with initial slope $\dot{x}(t_o)$ gives an approximate value for the next point x_1 of the curve at time $t_o + h$ with h small. The approximation is excellent provided the time increment h between the points is sufficiently small. A similar procedure is used for $y(t)$. After determination of $x_1(t_1)$ and $y_1(t_1)$, the process is repeated with these calculated values now treated as initial points having new initial slopes $\dot{x}_1$ and $\dot{y}_1$. Because any error in determining $x(t + h)$ from $x(t)$ is cumulative, it is important to make the time increment h very small. A procedure that is efficient with regard to high accuracy and minimum calculations is the fourth-order Runge–Kutta method. Basically, it calculates four values of the derivative between t and $t + h$, with a weighted average of these utilized. The procedure is as follows:

1. Select a suitably small time interval h. For the inverter problem of Sec. 6-3, a study of the time intervals of Table 6-2 of Sec. 6-4 suggests that an interval h of 0.01 ns is probably adequate. Calculations made with $h = 0.001$ ns indicate that 0.01 is sufficiently small, although a larger increment will not suffice.

2. Let t_o denote the initial time zero. Initial values $x(t_o)$ and $y(t_o)$ must be known. Define $t_m = t_o + 0.5h$ and $t_1 = t_o + h$. Calculate $\dot{x}(t_o)$ and $\dot{y}(t_o)$ from (A-1) and (A-2).

3. Calculate $x_a(t_m)$ and $y_a(t_m)$ from the relations

$$x_a(t_m) = x(t_o) + 0.5h\dot{x}(t_o)$$
$$y_a(t_m) = y(t_o) + 0.5h\dot{y}(t_o)$$

4. Calculate $\dot{x}_a(t_m)$ and $\dot{y}_a(t_m)$ from (A-1) and (A-2).

5. Calculate $x_b(t_m)$ and $y_b(t_m)$ from the relations

$$x_b(t_m) = x(t_o) + 0.5h\dot{x}_a(t_m)$$
$$y_b(t_m) = y(t_o) + 0.5h\dot{y}_b(t_m)$$

6. Find $\dot{x}_b(t_m)$ and $\dot{y}_b(t_m)$ from (A-1) and (A-2).

7. Calculate $x_a(t_1)$ and $y_a(t_1)$ using the relations

$$x_a(t_1) = x(t_o) + h\dot{x}_b(t_m)$$
$$y_a(t_1) = y(t_o) + h\dot{y}_b(t_m)$$

8. Find $\dot{x}_a(t_1)$ and $\dot{y}_a(t_1)$ from (A-1) and (A-2).

9. Calculate $x(t_1)$ and $y(t_1)$ using the relations

$$x(t_1) = x(t_o) + \left(\frac{h}{6}\right)[\dot{x}(t_o) + 2\dot{x}_a(t_m) + 2\dot{x}_b(t_m) + \dot{x}_a(t_1)]$$

$$y(t_1) = y(t_o) + \left(\frac{h}{6}\right)[\dot{y}(t_o) + 2\dot{y}_a(t_m) + 2\dot{y}_b(t_m) + \dot{y}_a(t_1)]$$

10. The procedure is now repeated, beginning with $x(t_1)$ and $y(t_1)$, and the process is continued for as long as desired.

11. Data should be printed and plotted at time increments of about $25h$. Also, data and plots for $i_B, i_C, q_F, q_R, q_{SE}$, and q_{SC} can be obtained with the aid of (6-31), (6-32), and (6-35) through (6-38).

For turn-on the initial conditions are those of row one of Table 6-2, and v_I is 3 V. For turn-off the initial conditions are given in row four, and v_I is zero.

Appendix B
Two-Port Matrix Parameters

Matrix Interrelations

To	From [z]	From [y]	From [h]	From [g]
$[z]$	$\begin{bmatrix} z_i & z_r \\ z_f & z_o \end{bmatrix}$	$\dfrac{1}{\Delta^y}\begin{bmatrix} y_o & -y_r \\ -y_f & y_i \end{bmatrix}$	$\dfrac{1}{h_o}\begin{bmatrix} \Delta^h & h_r \\ -h_f & 1 \end{bmatrix}$	$\dfrac{1}{g_i}\begin{bmatrix} 1 & -g_r \\ g_f & \Delta^g \end{bmatrix}$
$[y]$	$\dfrac{1}{\Delta^z}\begin{bmatrix} z_o & -z_r \\ -z_f & z_i \end{bmatrix}$	$\begin{bmatrix} y_i & y_r \\ y_f & y_o \end{bmatrix}$	$\dfrac{1}{h_i}\begin{bmatrix} 1 & -h_r \\ h_f & \Delta^h \end{bmatrix}$	$\dfrac{1}{g_o}\begin{bmatrix} \Delta^g & g_r \\ -g_f & 1 \end{bmatrix}$
$[h]$	$\dfrac{1}{z_o}\begin{bmatrix} \Delta^z & z_r \\ -z_f & 1 \end{bmatrix}$	$\dfrac{1}{y_i}\begin{bmatrix} 1 & -y_r \\ y_f & \Delta^y \end{bmatrix}$	$\begin{bmatrix} h_i & h_r \\ h_f & h_o \end{bmatrix}$	$\dfrac{1}{\Delta^g}\begin{bmatrix} g_o & -g_r \\ -g_f & g_i \end{bmatrix}$
$[g]$	$\dfrac{1}{z_i}\begin{bmatrix} 1 & -z_r \\ z_f & \Delta^z \end{bmatrix}$	$\dfrac{1}{y_o}\begin{bmatrix} \Delta^y & y_r \\ -y_f & 1 \end{bmatrix}$	$\dfrac{1}{\Delta^h}\begin{bmatrix} h_o & -h_r \\ -h_f & h_i \end{bmatrix}$	$\begin{bmatrix} g_i & g_r \\ g_f & g_o \end{bmatrix}$

$\Delta = z_i z_o - z_r z_f$, etc.

The Terminal Properties in Terms of the Matrix Parameters

	Z_i	Z_o	A_i	A_v
z	$\dfrac{\Delta + z_i Z_L}{z_o + Z_L}$	$\dfrac{\Delta + z_o Z_s}{z_i + Z_s}$	$\dfrac{z_f}{z_o + Z_L}$	$\dfrac{z_f Z_L}{\Delta + z_i Z_L}$
y	$\dfrac{y_o + Y_L}{\Delta + y_i Y_L}$	$\dfrac{y_i + Y_s}{\Delta + y_o Y_s}$	$\dfrac{-y_f Y_L}{\Delta + y_i Y_L}$	$\dfrac{-y_f}{y_o + Y_L}$
h	$\dfrac{\Delta + h_i Y_L}{h_o + Y_L}$	$\dfrac{h_i + Z_s}{\Delta + h_o Z_s}$	$\dfrac{-h_f Y_L}{h_o + Y_L}$	$\dfrac{-h_f}{\Delta + h_i Y_L}$
g	$\dfrac{g_o + Z_L}{\Delta + g_i Z_L}$	$\dfrac{\Delta + g_o Y_s}{g_i + Y_s}$	$\dfrac{g_f}{\Delta + g_i Z_L}$	$\dfrac{g_f Z_L}{g_o + Z_L}$

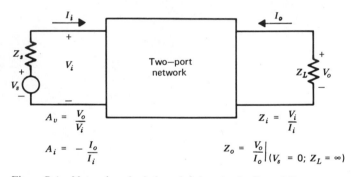

$$A_v = \frac{V_o}{V_i}$$

$$A_i = -\frac{I_o}{I_i}$$

$$Z_i = \frac{V_i}{I_i}$$

$$Z_o = \frac{V_o}{I_o}\bigg|\,(V_s = 0;\ Z_L = \infty)$$

Figure B-1   Network and relations defining A_v, A_i, Z_i, and Z_o.

Index